Milestones in
ANALYTICAL CHEMISTRY

ANALYTICAL CHEMISTRY

EDITOR:
ROYCE W. MURRAY
University of North Carolina

ASSOCIATE EDITORS:
Catherine C. Fenselau, University of Maryland Baltimore County, William S. Hancock, Hewlett Packard, James W. Jorgenson, University of North Carolina, Robert A. Osteryoung, North Carolina State University, Edward S. Yeung, Iowa State University/Ames Laboratory

Editorial Headquarters
Research section:
Department of Chemistry
Venable and Kenan Laboratories
University of North Carolina
Chapel Hill, NC 27599-3290
Phone: 919-962-2541
Fax: 919-962-2542
E-mail: Murray @ uncvx1.oit.unc.edu

Editorial Headquarters
A-page section:
1155 Sixteenth St., N.W.
Washington, DC 20036
Phone: 202-872-4570
Fax: 202-872-4574
E-mail: acx96 @ acs.org

Managing Editor: Mary Warner
Senior Editor: Louise Voress
Associate Editor: Grace K. Lee
Assistant Editor: Felicia Wach
Editorial Assistant: Deborah Noble
Contributing Editor: Marcia Vogel
Head, Graphics and Production: Leroy L. Corcoran
Art Director: Peggy Corrigan
Designer: Sarah Chung
Production Editors: Kathleen Savory, Elizabeth Wood
Electronic Composition: Wanda R. Gordon
Circulation Manager: Leslie Wilson

Publications Division
Director: Robert H. Marks
Head, Special Publications: Anthony Durniak
Head, Journals: Charles R. Bertsch

Journals Dept., Columbus, Ohio
Editorial Office Manager: Mary E. Scanlan
Journals Editing Manager: Kathleen E. Duffy
Associate Editor: Lorraine Gibb
Assistant Editor: Brenda S. Wooten

Milestones in ANALYTICAL CHEMISTRY

EDITED BY
The Washington staff of *Analytical Chemistry*

Mary Warner, Managing Editor
Louise Voress, Senior Editor
Grace K. Lee, Associate Editor
Felicia Wach, Assistant Editor
Deborah Noble, Editorial Assistant

AMERICAN CHEMICAL SOCIETY
WASHINGTON, DC 1994

Library of Congress Cataloging-in-Publication Data

Milestones in Analytical Chemistry/edited by the Washington staff of Analytical Chemistry.

    p.   cm.

Collection of articles published In Analytical Chemistry from 1935 to the present.

    Includes bibliographical references and indexes.

    ISBN 0-8412-2855-8 : $74.95

    1. Chemistry, Analytic. I. American Chemical Society.

II. Analytical Chemistry.

QD75.25.M55 1994

543—dc20

93-51082
CIP

The paper used in this publication meets the minimum requirements of American National Standard for Information Sciences—Permanence of Paper for Printed Library Materials, ANSI Z39.4–1984.

PRINTED IN THE UNITED STATES OF AMERICA

Contents

Preface

OVER THE PAST 65 YEARS *Analytical Chemistry* has published seminal papers on almost every development in the discipline. Our goal in this volume is to present a history of the field of analytical chemistry through original research papers published in the journal from 1935 to the present.

The volume begins with an introduction designed to give an overview of the past six decades of measurement science. The research papers are divided by decade, and each group is preceded by an introduction describing the important events that took place during the decade in terms of science, popular culture, and politics. Each paper is prefaced by a commentary from a prominent scientist in that particular area of analytical chemistry. We hope that these commentaries will provide a historical context for the work described.

This volume has benefited greatly from the involvement of the Journal's Editor, Royce W. Murray, and Associate Editors Catherine Fenselau, William Hancock, James W. Jorgenson, Robert A. Osteryoung, and Edward S. Yeung, who guided us in selecting the articles. We also thank the dedicated group of designers and production staff who worked to make this volume accurate and attractive within the constraints of an unusually short production schedule. Finally, we thank the authors of the research papers included and the countless editors who prepared these papers for publication during the past 65 years.

The Washington staff of *Analytical Chemistry*
American Chemical Society
Washington, DC 20036

December 1993

Introduction

T HE ARTICLES REPRINTED IN THIS COLLECTION PAINT A PICTURE of the truly amazing progress in chemical measurements that analytical chemists have made over the past six decades. The collection encompasses the beginnings of nearly every instrumental method important in chemical analysis today. It is our modern history. It is also a deservedly proud picture of analytical chemistry.

The two decades of the 1930s and 1940s saw the emergence of a broadening range of instrumental approaches to analysis. The power of IR vibration spectroscopy in dealing with quantitative and qualitative problems in industrial organic chemistry became evident. This technique, with spectroscopy as its complement, was introduced in the form of a direct-reading instrument incorporating a photomultiplier tube instead of photographic plates. Early examples of determining trace elements by colorimetry and by flame photometry and of determining specific ions by polarography appeared. As a result of needs during the war years, the mass spectrometer entered as an important analytical tool for petroleum products. Separations methodology, based on liquid adsorption chromatographic columns and on electrolysis, was introduced. The power of the digital computer for data analysis and retrieval was demonstrated in the old punched-card format for storing powder diffraction data.

A 1948 paper on introducing students to the evaluation of analytical data begins with the statement, "The very idea of introducing more subjects of statistics is so important...." That first sentence has proved to have enduring value and can still be said today!

If the 1930s and 1940s represented the stirring of the giant that has become instrumental analysis, the 1950s and 1960s heard the full reverberation of his thundering footsteps. Atomic spectroscopy saw the introduction of a well-designed commercialized flame spectrophotometer as well as the invention and application of the inductively coupled plasma torch, and the principles of atomic absorption spectroscopy and of controlling interferences therein became better understood. In separations science, new modes of separation, including thin-layer chromatography, open-tubular GC, and supercritical fluid chromatography, were invented. Theoretical principles of chromatography saw significant advances, including work by Giddings that helped stimulate the later development of HPLC. In electrochemistry, classic papers by Nicholson and Shain ushered in the cyclic voltammetric technique, which remains the most used of all amperometric electrochemical experiments. Thin-layer electrochemistry was invented and elaborated upon by different modes of potential and current control, and ion-selective potentiometric electrodes showed progress as the fluoride ion and other specific electrodes were introduced into analytical chemistry.

Important papers on the applications of NMR and phosphorimetry analytical chemistry appeared in *Analytical Chemistry* during the 1950s and 1960s, and one of the earliest chemometrics papers—on data smoothing by a least-squares calculation, which is

still used extensively as the Savitsky–Golay filter—appeared then. Analytical chemists began to discern the advantages of combining the talents of different instrumental methods and, as a consequence, hyphenated methods started to appear. Some of the earliest hyphenated methods were the combinations of GC with IR spectroscopy and with MS. Given the flood of ensuing data, the marriage of digital computers to instruments such as the mass spectrometer was an additional and spectacular example of hyphenation. Chemists today are still learning what a wonderful recorder the digital computer is, and mass spectrometrists are still pushing its boundaries of capacity and speed.

Also appearing in 1966 in *Analytical Chemistry* was a description of an instrument for the automated synthesis of peptides, which ultimately revolutionized many aspects of biochemistry and for which the senior author received a Nobel Prize.

From the 1970s to the present, the pace of invention of new instruments and measurement principles, as well as the combining of instrumental approaches, has been especially terrific. To the classical mode of electron impact ionization in MS have been added numerous and important forms of sample introduction and ionization, including chemical ionization, inductively coupled plasmas, thermospray ionization, ion microprobes, electrospray ionization, fast atom bombardment, and matrix-assisted laser desorption/ionization. Large, delicate, nonvolatile biomolecules are no longer safe from the analytical chemist's inquisition of their molecular masses and structure! This has been a truly major revolution in MS. Another important trend has been the use of multiple mass analysis stages; tandem mass spectrometers and collision-induced dissociation techniques have added significantly to the structure-probing capacity of the field.

Separations science has experienced an equally profound revolution since the 1970s. Plasma chromatography incorporates ideas from MS and electrophoresis. Field-flow fractionation extended separation science into the realm of really large molecules and microparticles. Ion chromatography grew out of older ion-exchange principles to provide great power for anion separations. High-performance chromatography saw the successful design of microcolumns and really efficient HPLC separations. The classic paper by Jorgenson and Lukacs on capillary zone electrophoresis started a flood of separation breakthroughs in scope, speed, and efficiency that continues today. Gel-filled electrophoresis capillaries were applied to large biomolecules, and by the use of micelles even neutral samples were shown to be important target analytes. The applicable samples in capillary electrophoresis are so small that the contents of individual, single cells can be introduced and separated. Capillary electrophoresis appears to be a tool that will open numerous new avenues in analytical chemistry and in molecular biology for years to come.

The application of mathematical tools and theory to data acquisition and analysis has been particularly important in continuing developments in spectroscopy since the 1970s. The utility of X-ray spectrometry for trace metal analysis of solids profited greatly from the adoption of semiconductor detectors and the formulation of theoretical models that accounted for X-ray absorption of all the matrix components, allowing true, nondestructive multielement analysis. The Fourier transform has proved immensely powerful in chemistry, and its power was felt early on in IR spectroscopy instrumentation.

Many of the above advances were built on laser radiation, and the importance of this device to analytical chemistry cannot be overemphasized. Lasers, for example, make possible the sensitive detection needed for capillary electrophoresis and provide the energy burst for desorption/ionization in MS sample introduction. Lasers are regularly used to

excite fluorescence in fiber-optic sensors. They have been applied to the study of electrode activation, particularly as regards the chemistry of carbon surfaces. These are mostly fixed-wavelength applications, but wavelength scanning—as in dye lasers applied to atomic fluorescence—has also been exploited.

Electrochemistry has also seen its share of new doors being opened. The idea of chemically modified electrodes ushered in a more molecular approach to and control of the electrode–solution interface that has continued through two decades in the current very active self-assembled monolayer field. Rapid-scan square-wave voltammetry was introduced as an HPLC detector. The introduction of microelectrodes was a powerful addition to the electrochemist's tool bag, giving access to many profound electrochemical problems and sometimes previously forbidding experimental situations. The microelectrode, among its numerous impacts, has even made possible a new form of microscopy: scanning electrochemical microscopy.

Although much of the action of analytical chemistry advances has been in instrumental methods, there remain numerous avenues for important contributions through the analytical chemist simply being a good reaction chemist. This is perhaps best manifested in papers that deal with kinetic analysis, where the nature and pace of the chemical process is the intrinsic source of the analytical signal.

Many, many more important papers have been published in *Analytical Chemistry* than those collected in this volume. It is simply for reasons of space, and for seeking some balance in attention to different periods of time and to different areas, that other equally significant works have not been included. The selection of the articles to be included in the collection was done by *Analytical Chemistry's* capable Washington staff under the leadership of Mary Warner and Louise Voress. They and I apologize to authors whose work has not been included and compliment those whose works are represented.

Royce W. Murray
Department of Chemistry
The University of North Carolina
 at Chapel Hill
Chapel Hill, NC 27599-3290

December 1993

The 1930s & 1940s

The Beginning of a Long Tradition

MANY LONGTIME READERS OF *Analytical Chemistry* do not realize that its long and colorful history has its roots in the 19th century. The *Journal of Analytical Chemistry* made its debut in January 1887. A quarterly publication, it was edited by Edward Hart, head of the chemistry department at Lafayette College. Hart's desire to initiate an analytical chemistry publication stemmed from industries' needs—primarily the steel and iron industries—to find accurate and rapid methods to improve production procedures. The Journal was an ideal forum for disseminating these new advances.

To ensure rapid publication of material, Hart established his own printing company at a barn site. Published articles emphasized applications, including toxicology and forensics and the atomic weight scale. In 1891 the publication's name was changed to the *Journal of Analytical and Applied Chemistry*. The first articles on electrochemical analysis methods appeared under this title.

Year	Event
1885	George Eastman manufactures coated photographic paper Cleveland becomes President
1887	Sir Arthur Conan Doyle writes "A Study in Scarlet"
1893	Henry Ford builds his first car
1894	Lumière invents the cinematograph
1895	Marconi invents radiotelegraphy Chinese–Japanese War ends
1896	A. H. Becquerel discovers radioactivity
1897	McKinley becomes President
1898	The Curies discover radium and polonium
1907	Rasputin gains influence in the court of Czar Nicholas II
1909	Taft is inaugurated as President Freud lectures in the U.S. on psychoanalysis Robert Peary reaches the North Pole
1914	World War I begins Robert Goddard begins rocketry experiments Panama Canal opens
1918	Armistice is signed between the Allies and Germany Daylight savings time is introduced in America
1920	Babe Ruth traded by the Boston Red Sox to the New York Yankees for $125,000 League of Nations forms in Paris; U.S. Senate votes against joining 19th Amendment gives American women the right to vote
1923	Douglas Fairbanks stars in *Robin Hood* Theory of acids and bases is postulated by J. N. Brönsted
1925	J. L. Baird transmits recognizable human features by television Schoolteacher J.T. Scopes goes on trial for violating Tennessee law that prohibits teaching the theory of evolution; defended by Clarence Darrow Heisenberg, Bohr, and Jordan develop quantum mechanics for atoms
1929	Ernest Hemingway's *A Farewell to Arms* is published "Talkies" replace silent films Kodak introduces 16-mm color movie film
1935	President Roosevelt signs Social Security Act
1936	Roosevelt is reelected by a landslide Baseball Hall of Fame is founded in Cooperstown, NY Jesse Owens wins four gold medals in Olympic games in Berlin

In 1893 H. W. Wiley, newly elected president of the ACS, asked Hart to become editor of the *Journal of the American Chemical Society* (*JACS*). Hart accepted and consolidated the two publications. Subscription income doubled because of price increases (from $44.36 in 1892 to $80.75 in 1893) and, although Hart adopted the style of his previous Journal, *JACS* was published monthly instead of quarterly. In 1895 Hart's printing operation moved to a larger location and eventually was renamed the Chemical Publishing Co. Today this company, known as Mack Printing Co., prints all of the ACS journals.

The *Analytical Edition*

The rebirth of the Journal was initiated in 1909 with the establishment of the monthly publication, *Industrial and Engineering Chemistry* (*I&EC*). In 1923, under the direction of Editor Harrison Howe, *I&EC* was cleaved into two publications: the monthly *Industrial Edition*, which published applied chemistry articles, and the *News Edition*, which was designed as a vehicle for news reporting and was published twice a month. From the *Industrial Edition*, a separate *Analytical Edition* was inaugurated in 1929. Published quarterly, this edition provided a new home for analytical chemistry papers that focused primarily on modifications of laboratory equipment and minor improvements of known procedures—many of which were designed to make manual operations less labor-intensive.

The first paper, published in Volume 1, was "Quantitative Analysis with the Spectrograph" by C. C. Nitchie of the N. J. Zinc Co. Volume 2 contained noteworthy reviews from the ACS symposium on analytical chemistry, including "Modern Trends" by H. H. Willard, "Potentiometric Titrations" by N. H. Furman, "Conductometric Titrations" by I. M. Kolthoff, and "Applications of the Photo-Electric Cell" by H. M. Partridge.

The *Analytical Edition* appealed to applied rather than fundamental chemists. As indicated in the policy statement of January 1935, "In it [the *Analytical Edition*] will appear articles dealing with applications of methods of chemical analysis of problems primarily in industrial and engineering chemistry.... Articles dealing with fundamental or theoretical considerations of no immediate application to chemical analysis belong more properly in the *Journal of the American Chemical Society*."

Annual subscription prices were kept low ($1.50 per year), a reflection of the hardships brought on by the stock market crash in 1929 followed by the Great Depression from 1930 to 1934.

In 1936 a separate advisory board for the *Analytical Edition* was established. Initially, one board served all three editions. Although the purpose of the board was not clearly defined, the

founding members included some of chemistry's "giants": B. L. Clarke of Bell Labs and later with Merck, T. R. Cunningham of Union Carbide, N. H. Furman of Princeton, I. M. Kolthoff of the University of Minnesota, G.E.F. Lundell of the National Bureau of Standards, and H. H. Willard of the University of Michigan. Two of the board members were among the leading contributors to the *Analytical Edition* from 1929 to 1939.

Among the many Nobel Prize winners who published in the *Analytical Edition* was The Svedberg of Sweden. After receiving the 1926 Nobel Prize for his ultracentrifuge studies, he authored a review on the ultracentrifuge in 1938.

As noted by Vernon A. Stenger, an Advisory Board member during the early 1950s, the field of analytical chemistry in the 1930s focused on titrimetry, classical gravimetry, colorimetry (which was being replaced by filter photometry and simple spectrophotometry), chromatography (primarily liquid–solid), and emission spectrography. Many of these techniques were used to learn as much as possible about a broad spectrum of materials such as alloys, cement, rubber, natural substances, and compounds in food and water. Unfortunately, many vital developments discovered in physicists' labs, such as mass spectroscopy, were not promoted in the *Analytical Edition*; likewise, significant topics such as IR radiation were covered in optical journals.

The 1940s

The advent of World War II propelled the field of analytical chemistry into a new era. By the 1940s, Europe was already involved in the war; the U.S., the USSR, and Japan would soon be drawn into the conflict.

Although many noteworthy scientific developments were catalyzed by the war effort, the classified nature of the material prevented much of the information from being disseminated in the published literature until the war concluded. In addition, international scientific communication was significantly hampered during the conflict.

Only a handful of U.S. universities, including Caltech, Harvard, the University of Illinois, Iowa State, the University of Michigan, the University of Minnesota, Ohio State University, Princeton University, and Purdue University offered Ph.D.

Editor Harrison Howe promoted his journals and the fields of analytical and industrial chemistry by speaking with members of local ACS sections.

programs in analytical chemistry.

One of the turning points in the Journal's history was Howe's decision to commission Ralph Müller to prepare a series of articles. The first article, which appeared in January 1939, was entitled "Photoelectric Methods in Analytical Chemistry." The second and third series, which filled the entire October 1940 and October 1941 issues, were entitled " American Apparatus, Instruments, and Instrumentation" and "Instrumental Methods of Chemical Analysis," respectively.

Howe died in 1942, and Walter Murphy was appointed Editor. Murphy's editorial responsibilities were numerous; in addition to the *Industrial Edition* and the *Analytical Edition* of *I&EC*, he also edited *Chemical & Engineering News*. There was considerable unrest among the analytical chemistry community. Murphy was approached by a group of scientists who wanted the *Analytical Edition* to become a separate publication of the ACS Division of Analytical Chemistry with a Division-appointed Editor. Murphy rejected this request but promised to withdraw as Editor if significant improvements were not made within a few years.

One of the first things Murphy did was to appoint M. G. Mellon and Müller to the Advisory Board, replacing Furman and Kolthoff. In addition, Lawrence T. Hallett was appointed Associate Editor in 1944. The first analytical chemist hired in an editorial capacity, Hallett had a strong microchemical and instrumentation backgound. Both Murphy and Hallett realized that analytical developments resulting from the war effort would transform the instrumentation industry.

Until then the Advisory Board did not have clearly defined responsibilities. However, Murphy decided to activate the Advisory Board, and often he sought its advice before establishing editorial policies. He enlarged the board to 12 members and instituted the rotation of their terms. The three-year terms are still in effect today.

The Journal gained further momentum with the installation of a monthly feature, "Instrumentation in Analysis," contributed by Müller. The Instrumentation articles first appeared in January 1946 (and continued through 1968) and kept scientists abreast of recent developments in the field of measurement science and in commercial instruments. Among the topics discussed by Müller were automatic analysis, microwave absorption,

1939
World War II begins: Roosevelt declares U.S. neutral
"God Bless America," "Over the Rainbow," and "Beer
    Barrel Polka" are popular songs
Gone with the Wind and *The Wizard of Oz* premiere

1940
Germany invades Norway and Denmark
Churchill becomes Prime Minister
Roosevelt is reelected for a third term
Native Son (Richard Wright), *For Whom the Bell
    Tolls* (Ernest Hemingway), and *Abraham Lincoln:
    The War Years* (Carl Sandburg) are published
Duke Ellington gains fame
Grapes of Wrath, *Fantasia*, *Gaslight*, and *Rebecca*
    premiere
"You Are My Sunshine," "When You Wish Upon a
    Star," and "Blueberry Hill" are popular songs
Giant cyclotron is built at the University of California
New combustion chamber for jet engines is designed
First electron microscope is demonstrated by Radio
    Corp. in New Jersey

1941
Japanese bomb Pearl Harbor
U.S. and Britain declare war on Japan
Germany and Italy declare war on the U.S.; the U.S.
    declares war on Germany and Italy
U.S. savings bonds and stamps go on sale
Citizen Kane and *How Green Was My Valley* premiere

1944
D-Day: Landings are made in Normandy
Roosevelt is elected for a fourth term

1945
Carousel (Rodgers and Hammerstein) premieres
"Bebop" comes into fashion
U.S. drops atomic bombs on Hiroshima and Nagasaki
World War II ends

1946
U.N. General Assembly holds its first session in London
Truman creates the Atomic Energy Commission
All the King's Men (Robert Penn Warren) is published
Xerography process invented by Chester Carlson
Isotope carbon-13 discovered
Joe Louis successfully defends his world heavyweight
    boxing title for the 23rd time

1949
Truman is inaugurated as President
Apartheid is established in South Africa
North Atlantic Treaty is signed in Washington
Israel is admitted into the United Nations
1984 (George Orwell) and *Death of a Salesman*
    (Arthur Miller) are published
South Pacific (Rodgers and Hammerstein) debuts
"Bali Ha'i," "Some Enchanted Evening,"
    "Diamonds are a Girl's Best Friend," and
    "Rudolph, the Red-Nosed Reindeer" are popular
    songs

computers, the transistor, flame ionization detectors, and
cathode ray tubes, to name a few.

By the late 1940s the concept of adding a magazine section
was a logical progression because the editors recognized that
the Journal's readers not only were responsible for developing
instrumentation but also influenced decisions about purchasing
equipment for their laboratories. Furthermore, Hallett realized
that a Washington-based staff of scientists would be needed to
oversee Journal operations. A staff (preferably with industrial
experience) would be responsible for obtaining and editing A-
page material and for assisting with the peer-reviewed research
section.

The Journal's advertisements were considered an ideal forum
for featuring commercialized developments from World War II.
Moreover, these ads were instrumental in keeping subscription
prices low, which in turn stimulated an increase in readership.

The new format of the Journal also prompted the expansion
of the definition of the field. By 1946 papers considered suitable
for publication covered a wide array of topics and disciplines
related to the analytical sciences, including fundamental and
theoretical papers. The first *Annual Review of Analytical
Chemistry*, which included 29 articles on fundamental analysis,
was published in January 1949. The following month, 11 articles
on applications of analytical developments were published.

During this decade, several other postwar analytical
journals, including *Analytica Chimica Acta, Applied
Spectroscopy*, and *Spectrochemica Acta*, evolved. Even
textbooks were produced in response to postwar changes,
although at a much slower rate than the journals. The 1948 text
written by Willard, Merritt, and Dean, entitled *Instrumental
Methods of Analysis*, had a tremendous impact on the
modernization of undergraduate curricula.

In the late 1940s the Journal became an important part of
the professional activities of the ACS Division of Analytical
Chemistry, as outlined in a March 1947 editorial, "The
Profession of Analytical Chemistry." The beginnings of many
current traditions in the analytical community—for example, the
Fisher Award and the Merck Fellowship, which formed the basis
of the Division fellowship program—can be traced back to this
editorial. However, one of the most significant changes of this
decade was the Journal's adoption of its official name, *Analytical
Chemistry*, in January 1948.

From its humble infancy in the 19th century to its
establishment as an autonomous publication in the late 1940s,
the Journal reflected the changes in analytical science and in the
world.
GRACE K. LEE

Milestones in Analytical Chemistry

To avoid co-deposition of arsenic and antimony as interferences in the constant current electrodeposition of copper, it is necessary to oxidize them quantitatively to the metal(V) state, which Furman accomplished with persulfate as the oxidant. An excellent recovery of copper was obtained in the presence of substantial amounts of As and Sb. This paper is an early example of the detailed chemical design of the electrolytic reaction medium in seeking to avoid electrodeposition interferences. Electrogravimetric methods were important instrumental tools in this period; spectrophotometric procedures were in an elementary state, and chromatographic separations and polarography were as yet unknown.

Royce Murray
University of North Carolina–Chapel Hill

Electroanalytical Separations in Ammoniacal Fluoride Solutions

I—Separation of Copper from Arsenic and Antimony[1]

N. HOWELL FURMAN

Frick Chemical Laboratory, Princeton University, Princeton, N.J.

A new procedure has been described for the electrolytic separation of copper from arsenic and antimony in ammoniacal fluoride solution. Some qualitative observations have been made upon the rapidity of the oxidation of trivalent antimony by persulfate under various conditions.

This study originated from a consideration of the problem of electrolytic deposits of copper contaminated with arsenic and antimony. It was shown in preliminary experiments that a known weight of copper could be coated electrolytically with arsenic and antimony, and the deposit could then be dissolved in a mixture of nitric and hydrofluoric acids, after which the procedure described in this paper gave correct results for the copper.

Ammoniacal solutions of quinquevalent arsenic or antimony may, as is well known, be electrolyzed with currents as great as 5 to 10 amperes per square decimeter of cathode surface, without deposition of arsenic or antimony at the cathode. This fact has been made the basis of the separation of a number of metals from arsenic and antimony, for example, cadmium (13), copper (7), nickel (5), or silver (14) from arsenic, or cadmium (15), copper (11, 16) or silver (3) from antimony.

Methods which have been developed previously offer no simple scheme for the solution and complete oxidation of arsenic and antimony when admixed with copper. Bosek (1) has shown conclusively that the complete oxidation of antimony with nitric acid is a matter of very considerable difficulty.

A mixture of copper, arsenic, and antimony, whether derived by electrolytic deposition or otherwise, may be dissolved in a mixture of nitric and hydrofluoric acids (9). The oxidation of the last traces of trivalent arsenic and antimony in such a solution may be effected very readily with potassium persulfate. This method has been studied by McCay in connection with an investigation of the separation of arsenic from antimony (9). He found that some 20 to 30 minutes of boiling of the acid solution with an excess of persulfate would complete the oxidation of arsenic and the antimony. The author has found that the oxidation goes very rapidly in faintly acid solution; 2 to 3 minutes' boiling then suffices to complete the oxidation.

EXPERIMENTAL PROCEDURE

The materials used in the separations were of known purity, being part of a large stock that had been tested in previous investigations (4, 10). In connection with these previous studies it was shown that antimony is not reduced during the electrolysis of an ammoniacal fluoride solution of potassium antimoniate.

Nitric acid solutions of pure electrolytic copper were prepared and the metal was determined by electrodeposition. Table I gives the results.

Table I. Determination of Copper in Nitric Acid Solutions

Nitric acid soln.	Copper found	
Cc.	Soln. I Gram	Soln. II Gram
25	0.2047	0.1991
25	0.2044	0.1994
50	0.4086	0.3983
25	0.2044	....
25	Av. 0.2044	0.1992

[1] Received March 4, 1931.

Weighed amounts of pure antimony and of arsenious oxide were dissolved in 3 to 5 cc. of 48 per cent hydrofluoric acid and 25 cc. of nitric acid (1 volume of acid of sp. gr. 1.42 and 4 volumes of water). A measured volume of standard copper solution was added. Oxidation of arsenic and antimony was completed by adding a moderate excess of potassium persulfate (1 to 2 grams). A few of the oxidations were completed by boiling the strongly acid solution for 30 minutes. Subsequent oxidations were effected by boiling the acid solution 2 to 3 minutes, followed by immediate neutralizations with ammonia. The efficiency of this procedure was established by numerous qualitative tests of which the following is typical: A solution containing 0.12 gram of trivalent antimony, 1.25 cc. of 48 per cent hydrofluoric acid, 1.25 cc. of concentrated sulfuric acid, and 1 gram of potassium persulfate in 35 cc., gave a distinct qualitative test for trivalent antimony after several minutes' boiling. In one instance a positive test was obtained after boiling for 20 minutes, and a further test after 25 minutes showed the oxidation to be complete. A similar solution, previously neutralized with ammonia, then cleared by adding the minimum amount of 6 N nitric acid, was boiled 3 minutes after adding 1 gram of the persulfate; the oxidation was then found to be complete. The qualitative tests were made by the Rose method (2), and by treating the solution with hydrogen sulfide, which under these conditions gives no immediate coloration or precipitate unless trivalent antimony is present.

Copper was separated from the cold, strongly ammoniacal solution by electrolysis after the completion of the oxidation. The total volume was 100 cc., containing 5 to 10 cc. of ammonia of sp. gr. 0.90. The series of separations recorded in Table II was made with stationary electrodes; current density 0.1 to 0.3 amperes per square decimeter and 2 to 4 volts applied at the electrodes. The solution was in a paraffined beaker, and

covered with paraffined split cover glasses. Complete deposition occurred in 5 to 8 hours. A number of electrolyses were allowed to continue overnight (15 to 18 hours) as a matter of convenience.

The copper deposits were carefully tested qualitatively for presence of antimony and arsenic. The tests were generally negative. In one case the residue obtained upon acidifying the sulfosalt filtrate with acetic acid, after removal of the copper sulfide, gave a coloration with hydrogen sulfide like that due to antimony; the color was not as intense as that produced by 0.01 mg. of antimony under similar conditions.

A series of rapid separations was made using current densities of 4 to 8 amperes per square decimeter of cathode at a voltage of 8 to 12. The anode (a platinum-blade stirrer) was rotated at 500 to 700 r.p.m. The copper was deposited completely in 35 to 45 minutes. The results are shown in Table III.

In a subsequent series of determinations, shown in Table IV, both copper and antimony were determined. After the copper had been deposited, the nitrates and fluorides were expelled by evaporation with an excess of sulfuric acid. McCay (9) has shown that this operation may be done in a quartz dish when the antimony is to be determined volumetrically. The reduction was effected by heating with about 1 gram of roll sulfur at about the temperature of boiling sulfuric acid (8), for 30 minutes. The antimony was then determined by titration with 0.05 N potassium permanganate solution which had been standardized against pure dry sodium oxalate (6).

Table II. Separation of Copper from Arsenic and Antimony

Detn.	Antimony present Gram	Arsenic present Gram	Copper present Gram	Copper found[a] Gram	Error Mg.
1	0.1805	....	0.2044	0.2042	−0.2
2	0.4070	....	0.2044	0.2043	−0.1
3	0.3530	....	0.2044	0.2047	+0.3
4	0.2970	....	0.1992	0.1997	+0.5
5	0.1170	0.0758	0.1992	0.1985	−0.7
6	0.2110	0.0987	0.1992	0.1993	+0.1
7	0.1972	0.1485	0.3984	0.3977	−0.7
8	0.1058	0.1582	0.3984	0.3986	+0.2

[a] Values corrected for small amounts of platinum dissolved from anode and deposited at cathode with the copper. In determinations 5 to 8 a hard anode (platinum–iridium) was used and no weighable amount of platinum appeared in copper deposit. In the other cases from 0.2 to 0.7 mg. was found when copper was dissolved in nitric acid. Proof of presence of platinum was obtained by formation of potassium chlorplatinate, or by stannous chloride reduction test.

Table III. Rapid Separation of Copper from Arsenic and Antimony

Detn.	Antimony present Gram	Arsenic present Gram	Copper present Gram	Copper found[a] Gram	Error Mg.
1	0.2342	....	0.1992	0.1997	+0.5
2	0.2021	....	0.3984	0.3977	−0.7
3	0.1733	0.2094	0.1992	0.1987	−0.5
4	0.1028	0.1493	0.1992	0.1998	+0.6
5	0.1082	0.1511	0.3984	0.3981	−0.3

[a] Corrected for platinum (cf. note, Table II); amounts found 0.2 to 0.7 mg.

Table IV. Determination of Both Copper and Antimony

Copper		Antimony	
Present Gram	Found Gram	Present Gram	Found Gram
0.1992	0.1989	0.1211	0.1209
0.3984	0.3986	0.0926	0.0931
0.3984	0.3989	0.1536	0.1531
0.1992	0.1991	0.0880	0.0874

LITERATURE CITED

(1) Bosek, *J. Chem. Soc.*, **67**, 515 (1895).
(2) Fresenius, "Qualitative Analysis," translated by Mitchell, p. 362, Wiley, 1921.
(3) Freudenberg, *Z. physik. Chem.*, **12**, 109 (1893).
(4) Furman, *J. Am. Chem. Soc.*, **40**, 895 (1918).
(5) Furman, *Ibid.*, **42**, 1789 (1920).
(6) McBride, *Ibid.*, **34**, 393 (1912).
(7) McCay, *Chem.-Ztg.*, **14**, 509 (1890).
(8) McCay, *J. Am. Chem. Soc.*, **32**, 1241 (1910); **36**, 2380 (1914).
(9) McCay, *Ibid.*, **36**, 2375 (1914).
(10) McCay and Furman, *Ibid.*, **38**, 640 (1916).
(11) Puschin and Trechinsky, *Chem.-Ztg.*, **28**, 482 (1904).
(12) Rose, *Pogg. Ann.*, **3**, 441 (1924).
(13) Schmucker, *J. Am. Chem. Soc.*, **15**, 195 (1893); *Z. anorg. Chem.*, **5**, 199 (1894).
(14) Smith and Frankel, *Am. Chem. J.*, **12**, 428 (1890).
(15) Smith and Wallace, *Z. anorg. Chem.*, **4**, 273 (1893).
(16) Smith and Wallace, *J. Am. Chem Soc.*, **15**, 32 (1893).

Reprinted from *Ind. Eng. Chem., Anal. Ed.* **1931**, *3*, 217–18.

A method for determining silica in solution is based on the formation of a yellow heteropoly compound, molybdisilicic acid, by the reaction with ammonium molybdate in the presence of mineral acid. Silica solutions and standards were compared by using visual and photoelectric colorimeters, Nessler tubes, and a spectrophotometer, and a better standard was prepared by using buffering K_2CrO_4 solutions with borax. This paper illustrates Mellon's pioneering efforts in which colorimetry and spectrophotometry were used to solve important trace analytical problems, ushering in the era of instrumental analysis.

George Morrison
Cornell University

Colorimetric Standards for Silica

H. W. SWANK AND M. G. MELLON
Purdue University, Lafayette, Ind.

One of the methods recommended for the colorimetric determination of small amounts of dissolved silica depends upon the formation of a yellow heteropoly compound, molybdisilicic acid, by the reaction of ammonium molybdate and the silica in the presence of a mineral acid. Since the usefulness of this procedure for routine work is limited by the instability of solutions of silica and of the yellow color, permanent standards have been proposed by various individuals.

The purpose of the present work was to study, by means of a spectrophotometer, these or other possible standards to determine the system most nearly meeting the requirements of an ideal colorimetric standard. The latter, in addition to being permanent and reproducible, should transmit the same amount of light at each wave length as the substance which it is to match, if the hues are to be identical under all conditions.

In the original recommendation of the method in 1898, Jolles and Neurath (5) used potassium molybdate and nitric acid as reagents and known solutions of silica for comparison. Winkler (14) modified the method by substituting hydrochloric for nitric acid and by introducing aqueous solutions of potassium chromate for standards. A solution containing 0.53 gram per liter of this salt was stated to be equivalent in color to 100 p.p.m. of silica. Dienert and Wandenbulcke (3) used ammonium molybdate and sulfuric acid and, as a standard, an aqueous solution of picric acid containing 35.9 mg. per liter, equivalent to 50 p.p.m. of silica. Atkins (2) employed the same concentration of this standard, but Thresh and Beale (12) recommended 40 mg. Believing the picric acid used by previous workers contained moisture, King and Lucas (7) proposed 25.6 mg. of dried material as equivalent to 50 p.p.m. of silica. Thompson and Houlton (11) recently reported using picric acid standards. Steffens (9) thought the reddish yellow hue of the potassium chromate solution did not match the greenish yellow obtained with silica and ammonium molybdate, an observation confirmed by, the work of Liebknecht, Gerb, and Bauer (8). Winkler's potassium chromate standards were adopted, however, by the American Public Health Association in 1933 (1).

EXPERIMENTAL WORK

Preparation of Materials. Several types of silica solutions were used, two being made from different samples of C.P. sodium silicate and a considerable number from definite amounts of C.P. silica or acid-washed sea sand fused with sodium carbonate. The former were standardized at intervals gravimetrically and kept in Pyrex glass bottles. The latter were discarded if more than 48 hours old. These solutions contained 500 to 5000 p.p.m. of silica and were diluted for use.

Picric acid recrystallized from benzene was used to prepare solutions of the concentrations recommended by King and Lucas and by Dienert and Wandenbulcke. Borax and potassium chromate were recrystallized. All the water used came from a special still built entirely of tin and Pyrex glass. Both C.P. ammonium molybdate and that prepared from molybdic acid and redistilled ammonium hydroxide were used.

Apparatus. A Keuffel and Esser spectrophotometer served for obtaining transmission data, using 20-cm. cells. In contrast to the usual practice with this instrument, the transmissions of the solution and solvent were determined separately and transmittancy calculated from these values. Ordinary visual comparisons were made both in tall-form Nessler tubes with plane glass ends and in a Duboscq-type colorimeter by the substitution method (15).

A photoelectric colorimeter similar to that described by Story and Kalichevsky (10) proved to be both sensitive and

2855-8/94/0008$08.00/0   ©1994 American Chemical Society   *Milestones in Analytical Chemistry*

rapid in determining the effect of different factors on the intensity and stability of the yellow color. Blue glass filters isolated the part of the spectrum in which yellow solutions absorb light. Galvanometer deflections were noted for a constant thickness of solution.

Color Reaction. The selection of satisfactory permanent standards presupposes a knowledge and control of the conditions likely to affect the stability and reproducibility of the color produced with the silica in the unknown. After surveying the work of others upon this point, different variations in reagents, concentrations, and other conditions were tried. On the basis of the results obtained, two reagents were selected for use in the study of permanent standards. For the first, now recommended by the American Public Health Association (A.P.H.A. Reagent), one adds to 100 ml. of unknown 10 ml. of a solution containing 200 ml. of hydrochloric acid (1 to 1), 30 grams of ammonium molybdate, and 400 ml. of water. For the second (Reagent B), adapted from Dienert and Wandenbulcke's recommendation, one adds to 100 ml. of unknown 4 ml. of 10 per cent ammonium molybdate and 1 ml. of 20 per cent sulfuric acid.

There seems to be an optimum amount of acid required for the full development of the yellow color. Either too much or too little gives decreased intensities, presumably from an effect on the heteropoly complex. The acidity of the A.P.H.A. Reagent is near the maximum.

Use of each of the two reagents mentioned gave a reproducible series of results, complete development of color occurring within 5 minutes and no fading within 30 minutes. However, the A.P.H.A. Reagent gave approximately 10 per cent greater intensity of color than Reagent B. The color developed with these reagents is proportional to the silica, at least as high as 75 p.p.m. Transmittancy data were determined for each reagent for concentrations of silica of 10, 20, and 30 p.p.m., thus providing curves with which to compare those of possible permanent standards.

Picric Acid Standards. In Figure 1 are shown spectral transmission curves for solutions of picric acid equivalent to 10 p.p.m., the concentrations being those recommended by King and Lucas and by Dienert and Wandenbulcke. A comparison of these curves with those for silica of the same concentration with Reagent B shows that the standards do not transmit the same amount of light at most wave lengths as the silica solution. The curve for Dienert and Wandenbulcke's solution of acid is fairly close to that for the corresponding silica solution, but no change of concentration would bring together the curves for the standard and the unknown. Similar data were obtained for solutions of picric acid designed to match other concentrations of silica, but only the curves for the latter solutions are shown. Although picric acid solutions show reasonable stability and conformity to Beer's law, the curves show that the system cannot be as satisfactory as one meeting more nearly the criteria of an ideal standard. In Nessler tubes differences are apparent between either of the two concentrations of standard and the corresponding silica solution.

Potassium Chromate Standards. In agreement with the report of Steffens, differences in hue could be noted easily in

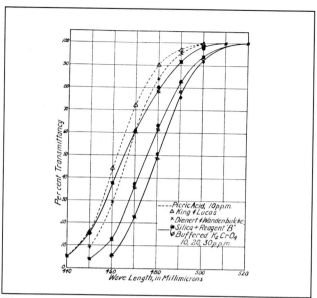

Figure 1. Spectral transmission curves for picric acid standards, buffered potassium chromate standards, and silica solutions with reagent B

Nessler tubes between silica and potassium chromate solutions. The observations of Viterbi and Krausz (*13*), Hantzsch and Clark (*4*), and Jorgensen (*6*) concerning the variability of the chromate-dichromate equilibrium in solutions of potassium chromate gave a possible clue to the objection to this system as a standard. To determine the effect of hydrogen-ion concentration on the hue, a series of solutions equivalent to 10 p.p.m. of silica was prepared in buffer solutions to give pH values of 5, 6, 7, 8, and 9. From the curves in Figure 2 showing their transmittancy data, it is evident that a decrease in hydrogen-ion concentration caused a decided change in hue to a pH value of about 8, beyond which little change occurred. A fresh, unbuffered solution gave a curve indicating a pH value

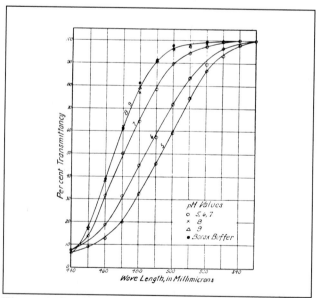

Figure 2. Spectral transmission curves for solution of potassium chromate at different pH values

between 6 and 7. It is evident that such systems may be affected by absorption of carbon dioxide or alkali from the glass.

In view of these facts, an unbuffered solution equivalent to 15 p.p.m. of silica was kept in a flint glass bottle for 3 months. The transmittancy data, shown in Figure 3, show a gradual shift of the curves with time toward the position for an alkaline solution. No appreciable change was noted after the seventy-third day.

The similarity of the curves in the alkaline region to those for silica suggested the use of buffered solutions of potassium chromate as standards. Accordingly, enough of a 1 per cent solution of sodium tetraborate was used in diluting the standards to make the final solution 50 per cent by volume of the buffering reagent. The stability of such a buffered system is indicated by the broken line curve in Figure 3. Measurements with a colorimeter disclosed close conformity to Beer's law for the buffered solutions at least as far as 100 p.p.m. This was checked for the range 10 to 50 p.p.m. with data from a recording photoelectric spectrophotometer. An alkaline buffer makes practical the use of equivalent amounts of potassium dichromate, which, as a primary standard, may be more readily available than the chromate.

As a buffered solution of potassium chromate transmits more light than one unbuffered, it is necessary to use a concentration higher than that recommended by the A.P.H.A. method to give color equal to a given amount of silica. In order to determine the amount to use, measurements were made with three different colorimeters, Nessler tubes, and the spectrophotometer, covering a period of several months and using various solutions of silica and ammonium molybdate. To be equivalent to a silica solution of 100 p.p.m., the amounts recommended are 0.63 and 0.58 gram per liter for the A.P.H.A. Reagent and Reagent B, respectively. One milliliter of this stock solution, diluted with 50 ml. of a 1 per cent solution of sodium tetraborate, $Na_2B_4O_7 \cdot 10H_2O$, and enough water to

make a total of 100 ml., is equivalent to 1 p.p.m. of silica. The details of diluting the standards depend upon the total volume of the molybdate reagent, the acid, and the sample, but in any case the volume of the standard and unknown should be the same.

The curves in Figures 1 and 4 show for three concentrations the good agreement of the data for these standards and for silica solutions treated with one of the two different reagents. In view of these results, the buffered solution of potassium chromate was considered so much better than picric acid that no further work was done upon it.

For use in determining silica in water according to the A.P.H.A. method, the values given in Table I are recommended as a substitute for those on page 67 of the 1933

Table I. Colorimetric Standards for Determination of Silica

Potassium chromate[a] Ml.	Silica[b] P.p.m	Potassium chromate Ml.	Silica P.p.m
0.0	0	8.0	16
1.0	2	9.0	18
2.0	4	10.0	20
3.0	6	11.0	22
4.0	8	12.0	24
5.0	10	13.0	26
6.0	12	14.0	28
7.0	14	15.0	30

[a] 0.63 gram of potassium chromate per liter. The volumes specified are to be diluted with 25 ml. of a 1 per cent solution of borax and enough water to make a total of 55 ml.
[b] When 50 ml. of sample are used, together with 5 ml. of reagent, as recommended in the A.P.H.A. method.

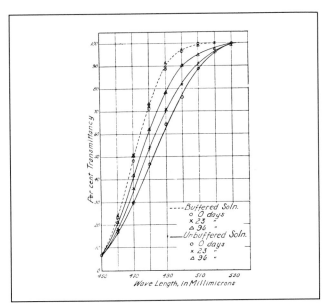

Figure 3. Spectral transmission curves for buffered and unbuffered solution of potassium chromate at different periods of time

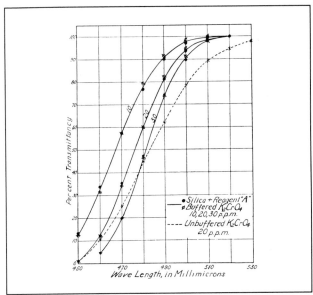

Figure 4. Spectral transmission curves for unbuffered potassium chromate solution and for buffered potassium chromate standards and silica with A.P.H.A. reagent

edition of "Standard Methods of Water Analysis." A careful gravimetric determination of the silica in a well water checked the value obtained colorimetrically with this table.

SUMMARY

A comparison of silica solutions and standards by means of visual and photoelectric colorimeters, Nessler tubes, and a spectrophotometer would seem to justify the following conclusions:

1. Either the A.P.H.A. reagent or an adaptation of that of Dienert and Wandenbulcke is generally satisfactory for producing the yellow color with silica.

2. Solutions of picric acid or unbuffered potassium chromate are not considered as meeting satisfactorily the requirements of a permanent standard.

3. A solution of potassium chromate suitably buffered, as with borax, provides a standard possessing none of the disadvantages of those now recommended.

4. To use the latter solution a new table is recommended.

LITERATURE CITED

(1) Am. Public Health Assoc., "Standard Methods of Water Analysis," p. 66 (1933).
(2) Atkins, *J. Marine Biol. Assoc. United Kingdom*, **13**, 154 (1923); **14**, 89 (1926); **15**, 191 (1928); **16**, 822 (1930).
(3) Dienert and Wandenbulcke, *Compt. rend.*, **176**, 1478 (1923).
(4) Hantzsch and Clark, *Z. physik. Chem.*, **63**, 367 (1908).
(5) Jolles and Neurath, *Z. angew. Chem.*, **11**, 315 (1898).
(6) Jorgensen, *Cereal Chem.*, **4**, 468 (1927).
(7) King and Lucas, *J. Am. Chem. Soc.*, **50**, 2395 (1928).
(8) Liebknecht, Gerb, and Bauer, *Z. angew. Chem.*, **44**, 860 (1931).
(9) Steffens, *Chem.-Ztg.*, **54**, 996 (1930).
(10) Story and Kalichevsky, Ind. Eng. Chem., Anal. Ed., **5**, 214 (1933).
(11) Thompson and Houlton, *Ibid.*, **5**, 417 (1933).
(12) Thresh and Beale, "Examination of Waters and Water Supplies," 3rd ed., p. 347, Churchill, London, 1925.
(13) Viterbi and Krausz, *Gazz. chim. ital.*, **57**, 690 (1927).
(14) Winkler, *Z. angew. Chem.*, **27**, 511 (1914).
(15) Yoe, "Photometric Chemical Analysis," Vol. I, p. 103, John Wiley & Sons, N.Y., 1928.

Received May 3, 1934

Reprinted from *Ind. Eng. Chem., Anal. Ed.* **1934**, *6*, 348–50.

This paper was the primary initiating force for the development of industrial applications of IR spectroscopy. Norman not only introduced many of the concepts and techniques that blossomed into major characterization tools in the 1950s and 1960s, but he also anticipated that the field of IR spectroscopy would grow through the 1980s and 1990s. Many of the techniques Norman discussed would become viable with the introduction of FT-IR. The discussion of qualitative and quantitative uses of IR forms the basis of this article. Norman described the split solvent approach to obtaining full spectral data and realized the value of spectral collections for the identification of materials. Norman also predicted the use of IR measurements in the area of process control, an application that is just being realized today.

Bruce Chase
DuPont

Application of Infrared Spectroscopy to Industrial Research

NORMAN WRIGHT
The Dow Chemical Company, Midland, Mich.

The range of the infrared spectrum discussed in this paper extends from about 2.5 μ (25,000 Å) to approximately 15 μ. Since photographic plates are sensitive to wave lengths no longer than 1.3 μ, the spectrum here mentioned is detected and measured through the heating effect of the radiation, in the present case by a thermopile and galvanometer.

It has long been recognized, through such early work as that of Coblentz (2) in 1905 on the infrared spectra of a large number of compounds, that the selective absorption (or emission) of infrared radiation arises in the mutual vibrations of the atoms constituting the molecules. A molecule does not absorb radiation of all wave lengths but selects only a few narrow wave-length intervals which are known as absorption bands. The resulting absorption pattern is characteristic of the molecule.

Theoretical treatment of the vibrations of molecules and the correlation with infrared spectra has been summarized for the simpler molecules by Dennison (3). He pointed out that the vibration frequencies within a molecule are determined by the masses of the atoms, the strength of the forces which bind them, and the geometrical structure of the molecule. In the case of organic compounds there is only slight dependence on the state of aggregation of the molecules, and the factors mentioned lead to vibration frequencies corresponding to wave lengths lying for the greater part in the spectral range 2.5 to 15 μ.

Inorganic compounds do not present as favorable a field for infrared spectroscopy as do the organic. Chief of the disadvantages encountered with inorganic compounds is the fact that water, the commonest solvent, is nearly opaque to infrared waves longer than 1.5 μ. A second disadvantage is the great width of absorption (or reflection) bands of inorganic compounds. The field of organic chemistry, on the other hand, lends itself particularly well to the methods of infrared spectroscopy, and applications in this field only are discussed in the present paper.

In addition to the general fact that the infrared spectrum of an organic molecule is characteristic of that molecule, it is well known that certain groupings or subgroups of atoms within molecules behave more or less independently of the rest of the molecule and give rise to characteristic absorption bands. For example, the O–H group gives rise to a band in the vicinity of 2.75 μ (in unassociated molecules), irrespective of the type of molecule containing this hydroxyl group. Compounds with the C≡N group possess a band at about 4.45 μ, and those with the C=O group have a band in the interval 5.6 to 5.9 μ. These cases and similar ones have long been known; considerable literature of such subgroup bands or "linkage" bands has been built up in recent years (1, 8, 12).

A great many applications of infrared spectroscopy can be made in the field of industrial organic chemistry, both as a tool for research and for actual control of plant processes. A list of such applications would include: (1) identification of organic compounds, (2) detection and identification of small amounts of impurities in organic compounds, (3) accurate quantitative determination of such impurities, (4) study of reaction mechanisms and speeds, and detection of intermediates, (5) study of isomerism and tautomerism, (6) study of association and compound formation, (7) study of polymerization and copolymerization in the field of plastics, (8) determination of geometrical structures, moments of inertia, and bond lengths, (9) determination of force constants and dissociation constants, (10) calculation of specific heats and other thermodynamic constants, and (11) study of crystal structure through use of polarized radiation. There are no doubt other applications.

2855-8/94/0012$08.00/0 ©1994 American Chemical Society *Milestones in Analytical Chemistry*

In view of the work of Coblentz (2) and others before 1905, pointing directly to many of the above applications, the question arises why these have not long ago been put into industrial practice. The answer lies almost entirely in the severe experimental difficulties encountered by early workers. From the standpoint of optics there was no great obstacle, but the extremely high sensitivity of the radiation-detecting devices to all sorts of external disturbances has been overcome only in recent years. This explains the scarcity of reports on efforts to utilize infrared spectroscopy in industry, particularly in the analytical field.

Two instances of analysis by means of infrared which are no doubt adaptable to industrial use include a method developed by McAlister (9) for the accurate and rapid analysis of carbon dioxide in air, and an ingenious method devised by Pfund (11) for the determination of simple gases such as carbon dioxide, carbon monoxide, and methane without the use of a spectrometer. There are no reports, however, of the industrial use of infrared spectroscopy for quantitative analysis of more complex organic compounds. It is therefore the purpose of this paper to describe in some detail the methods of qualitative and quantitative analysis by means of infrared spectroscopy which are being employed on a routine basis in an industrial laboratory.

APPARATUS

The spectrograph is an automatically recording instrument employing a 60° prism of rock salt in a Littrow mounting. It was designed to meet the requirements of producing and rapidly recording an infrared spectrum of the proper quality and yet retaining the simplicity and dependability vital to industrial application. In some respects it resembles the spectrograph of Strong and Randall (13) which, however, uses the more complicated Wadsworth-Littrow mounting. The instrument was built to order by the shop of the Physics Department of the University of Michigan. A diagram of the optical system is shown in Figure 1.

The source of the infrared radiation is a globar element, A (12.5 cm., 5 inches long), mounted in a water-cooled jacket. The radiation is focused by the concave mirror, B, onto the slit

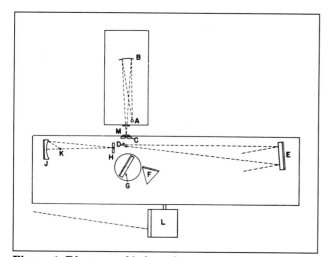

Figure 1. Diagram of infrared spectrograph

of the spectrograph, C. All mirrors of the instrument are aluminized on the front surface. Space is left at M for inserting into the radiation path the cell containing the sample whose absorption spectrum is to be recorded. After passing through the sample and through the slit of the spectrograph the diverging radiation falls on the collimating mirror, E. This mirror is an on-axis paraboloid of 15-cm. (6-inch) diameter and a focal length of 90 cm. (36 inches). The collimated radiation passes through the prism, F, which has faces 10 × 8 cm. and refracting angle about 60°. After passing through the prism the radiation is returned by the plane (Littrow) mirror, G, again passing through the prism and again falling on mirror E. Mirror E then focuses the radiation into a spectrum, a narrow wave-length interval of which is selected by the second slit, H. From the slit the radiation is strongly condensed on the receiver of the thermopile, K, by the ellipsoidal mirror, J, which has focal distances of 7.5 and 37.5 cm. (3 and 15 inches). The thermopile is of the Pfund (10) design and has two junctions; only one of these receives the radiation, the other serving as a compensator.

Connected to this is a galvanometer of moderately high sensitivity (L. & N. Type HS, 1.5-second period), the deflections of which are amplified 10 to 20 times by means of a very simple dry-plate photocell relay. The final deflection is produced in a galvanometer of rugged construction (L. & N. Type R). Each galvanometer is mounted on a type of Julius vibrationless support. It is, of course, possible to achieve deflections of suitable magnitude with a single, very sensitive galvanometer, but such galvanometers usually have periods of from 7 to 10 seconds which would lead to excessive recording times. Use of the amplifier yields a high sensitivity without sacrifice of tune, and permits employment of more stable galvanometers.

Automatic scanning of the spectrum is accomplished by slowly rotating G by a motor drive, causing the spectrum to move slowly over the selector slit, H. Simultaneously the deflection produced in the final galvanometer is photographically recorded on the synchronized drum camera, L.

A photograph of the instrument and room is shown in Figure 2. The radiation source and focusing mirror are enclosed in an airtight case, as is also the spectrometer proper. This allows removal of water vapor from the radiation path by drying agents placed within the cases. The source-box can be readily moved toward or away from the spectrometer to permit insertion of sample cells of different length. The room is lined with sheet metal to shield against the electromagnetic effects of the high-potential spark sources used in the spectrographic laboratory nearby. This instrument has been in constant service for nearly 3 years.

Several records of infrared spectra made with this instrument are shown in Figures 3, 5, and 6. These records (15 cm., 6 inches, wide by 50 cm., 20 inches, long) are graphs of galvanometer deflection produced by the transmitted radiation vs. wave length as indicated in microns along the bottom edge. Zero deflection of the galvanometer is at the bottom edge of the records and is marked by the dots at the beginning and end of each section of the recording. The vertical fiducial lines are photographed on the records at intervals of 20 units

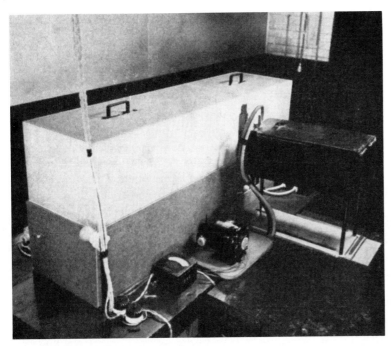

Figure 2. Infrared spectrograph

of a revolution counter geared to the Littrow mirror drive. Infrared wave lengths are easily read from a calibration curve obtained with the use of known standard wave lengths.

The manner in which a record is made is illustrated in Figure 3 by the spectrum obtained with no material other than air in the radiation path.

A record is made by starting at the long wave-length end and recording toward shorter wave-lengths. The radiation, of course, grows more intense following approximately the black body radiation law, approaching a maximum at about 1.5 μ. When the deflection grows too large for the camera, the instrument is stopped and the two slits are narrowed; this decreases the amount of radiation falling on the thermopile and at the same time increases the resolution of the spectrometer. This process is repeated five times during the recording of a spectrum from 14 to 2 μ. The time required for the complete recording is 25 minutes.

The air spectrum (Figure 3) shows the absorption bands of carbon dioxide at 4.26 and at 13.9 μ, as well as the saw-toothed, rotation-vibration band of water vapor extending from 5.5 to 7 μ. A typical spectrum of an organic compound is that of phenylacetylene (Figure 3) in a liquid layer 0.1 mm. thick. The downward dips of the galvanometer indicate the absorption bands or regions of selective absorption characteristic of the phenylacetylene molecule.

SAMPLE PREPARATION

Gases, liquids, and solids can be investigated by means of infrared spec-

troscopy. Examples of the various types of cells and sample preparations are shown in Figure 4.

Gases are placed in evacuable cells of glass having rock-salt windows; a convenient cell length for organic vapors has been found to be about 5 cm. with pressures of the vapor ranging from atmospheric down to 0.1 atmosphere. Organic liquids are observed in the form of thin layers in cells consisting of two carefully ground and polished rock-salt plates separated the proper distance by shims of lead or platinum. Cell thicknesses are usually 0.1 to 0.01 mm., but in special cases may be as great as 1 mm. or even larger. These cells, described in detail elsewhere (5), are constructed without the use of waxes and are sealed tightly enough to hold volatile liquids. Having both an entrance tube and orifice, they may be readily cleaned by forcing cleaning solvents through the cell, avoiding the troublesome necessity of dismantling the cell after each use.

Solids are usually observed in solution if suitable solvents can be obtained. Carbon disulfide and carbon tetrachloride are the most favored solvents because of their high infrared transmission. In some cases—e. g., the plastics—the spectra can be easily obtained by putting the sample into the form of a film having a thickness equivalent to those of the organic liquids mentioned. Solids for which no suitable solvents can be obtained and which cannot be put into film form may be studied in the form of thin layers of finely ground powder wet with carbon tetrachloride or carbon disulfide.

IDENTIFICATION OF UNKNOWNS

The first application of infrared spectroscopy in the analytical field to be considered is the identification of unknown organic compounds which are fairly free of impurities. Dis-

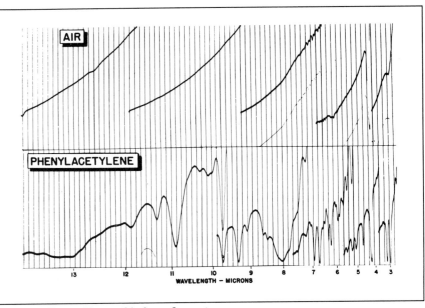

Figure 3. Recordings of infrared spectra

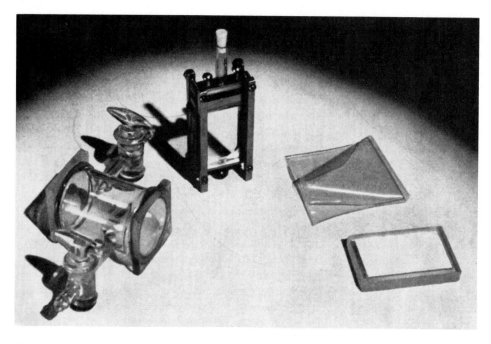

Figure 4. Vapor and liquid cells, plastic film, and powder layer on rock-salt plate

The spectra can be classified according to the presence of the particular subgroups or linkages which produce characteristic infrared bands—for example, compounds containing O–H form one section of the catalog and those containing ≡C–H another, etc. Some repetition is, of course, inevitable. Another typical section of the catalog contains the spectra of the various chlorinated hydrocarbons arranged according to increasing chain length and increasing numbers of substituted chlorine atoms.

Another method of tabulating the spectra which appears to be of great practical value is to classify according to boiling points of the compounds.

tinction is made here between identification and characterization. In the case of compounds not previously prepared or isolated, the infrared method offers very valuable aid in the identification of those groupings and linkages which produce characteristic infrared bands. In such cases it is not usually possible to derive the exact structure of the unknown compound.

A commoner problem, however, is the identification of compounds which are available in the pure state for comparison. Often the circumstances of a particular case provide clues as to the nature of the unknown, and a number of possible compounds may be suggested. It is then necessary merely to obtain and compare the spectrum of the unknown with the spectra of the suggested compounds. When a match is obtained the identification is complete and the problem is solved.

Too often, however, the suggested compounds turn out not to include the unknown. A general technique of identification then must be put into operation which involves the use of a catalog of infrared spectra. This catalog must contain the spectra of a large number of compounds and its usefulness improves as the number increases. The question of systematizing the spectra in such a catalog may depend somewhat upon the type of problem most often encountered in a particular laboratory, but the following general methods have proved satisfactory.

Employing the method of Coblentz (2) for comparison of spectra, the absorption bands of a spectrum are represented as vertical lines plotted on a wave-length (or frequency) scale. Intensities of the absorption bands are indicated by the heights of the lines; for practical purposes these can be conventionalized to three different heights corresponding to strong, medium, or weak. The widths of very strong bands can be represented by correspondingly broad lines. In this manner the spectra of a large number of compounds can be conveniently plotted on an ordinary sheet of graph paper.

In tracing down an unknown the procedure is to plot the spectrum on a strip of graph paper which can be compared directly with the spectra in the catalog. In this way a large number of compounds can be checked in a very brief time. After a match is found the original spectrum records are compared for final verification.

A third method of cataloging employs an index of wave lengths compiled in the manner adopted by Hanawalt and coworkers (6, 7) for x-ray spectra. The bands characteristic of subgroups and linkages are omitted in this index and use is made of the other bands which can be regarded as characteristic of the molecules as a whole. This obviates the difficulty of recording large numbers of compounds having strong bands at almost exactly the same wave length—for example, the band at 3.4 μ of compounds with C–H groups.

DETECTION AND IDENTIFICATION OF IMPURITIES

Detection and identification of impurities constitute a more difficult but much more important use of infrared spectroscopy. Fortunately the spectra of mixtures of organic compounds in gas, liquid, or solid state do not usually differ appreciably from a mere superposition of the spectra of the compounds taken individually. Cases where differences occur are those involving association, polymerization, or compound formation; in fact, these spectral differences are good criteria for such chemical behavior. Extremely small wavelength shifts are usually noticed when a compound is mixed with a highly polar solvent, but this offers no serious difficulty.

The usual method of testing for impurities is simply to compare the spectrum of the sample with that of a purified sample of the same compound. This is demonstrated in Figure 5 by the spectra of two samples of chloroform, one pure and the other showing plainly the superimposed absorption bands of an impurity of ethylidene chloride. When a pure sample of the compound under investigation is not available for comparison,

the presence of impurity can often be detected through the use of partially purified samples which reveal the impurity by the weakening of its absorption bands. Fractional distillation is an enormous aid in this direction. Furthermore, impurity can sometimes be detected by use of the characteristic subgroup bands—e. g., the detection of an aldehyde impurity through the presence of a C=O band in a material not having this group.

Identification of impurities, although complicated by the background spectrum of the main constituent of the sample, is accomplished in much the same manner as already detailed for unknowns in a more refined state. Often many of the bands of the impurity are completely obscured by those of the main constituent and the comparatively few remaining bands must be relied upon for the identification. If only two or three of the absorption bands of the impurity can be picked out and these coincide exactly with bands of a suspected compound, and if all the strong bands of the suspected compound are accounted for as occurring in the spectrum or obscured by the bands of the main constituent, there can be little doubt as to the correctness of the identification. Further corroboration is given if the intensities of the matched absorption bands are correct.

Owing to the fact that fractional distillation is the commonest means of purification, the boiling points of the impurity and the main constituent are usually not far different. Reference to the catalog of spectra classified as to boiling points of the compounds is therefore an efficient means of search.

The view has often been expressed that successful analysis by means of infrared requires that the impurity compound possess characteristic groupings or linkages not present in the primary constituent. An example to the contrary is that of an impurity of 1,2-dibromopropane in 1,3-dibromopropane, the spectra of which are shown in Figure 6. Although these compounds are isomers differing only in geometrical structure, the infrared spectra are seen to be very different. Furthermore the 1,2 isomer can be detected in concentrations as low as 0.3 per cent in mixtures with 1,3-dibromopropane.

QUANTITATIVE ANALYSIS

In addition to detection and identification of impurities, it has been possible in a large number of cases to develop accurate quantitative determinations of organic impurities by means of infrared spectra. The method consists first of selecting an absorption band of the impurity which does not fall too close to bands of the primary constituent of the mixture. This is illustrated in Figure 7, which shows a short wave-length interval of the spectra of samples of chloroform containing methylene chloride in amounts varying from 0 to 4 per cent (by volume). The cell used is 0.1 mm. thick. The absorption band of methylene chloride at 7.93 μ is seen to increase in depth as the concentration increases.

Recording of the spectra as here described in the form of transmitted radiation curves has led to very simple and accurate methods of measuring impurity bands. One method, used in the example of methylene chloride in chloroform, consists of measuring the intensities I_0 and I of the incident and transmitted radiation, respectively, at the wave length of the band. The value of I_0 is obtained by drawing in a straight line tangent to the spectrum curve at the position of the impurity band as shown in Figure 8. The values of $\log_{10} I_0/I$ (actually measured from the spectra of Figure 7) are then plotted against the concentration of methylene chloride in per cent by volume (Figure 8). The result is a straight line through the origin, attesting to the validity of Beer's law for this case. As a matter of fact, no deviation from Beer's law has been found in any of the many cases of organic liquid mixtures which have been encountered where no association occurs.

An analysis of a sample of chloroform for methylene chloride is therefore made in the following manner:

The sample, amounting to less than 0.1 cc., is placed in the cell and inserted in the beam of the radiation entering the spectrograph. The instrument then records the short interval of the spectrum shown in Figure 7. A measurement of $\log_{10} I_0/I$ is made from the record and the percentage of methylene chloride is read directly from the standard curve of Figure 8. The error of this determination amounts to about ±0.05 per cent (of total sample) and the sensitivity is such that measurements can be made in concentrations of methylene chloride as low as 0.05 per cent.

In many cases of quantitative determination, a band of the impurity must be selected which is very close in wave length to bands of the primary constituent. Such a case is that of 1,2-dibromopropane in 1,3-dibromopropane which has already been mentioned (Figure 6). Figure 9 shows short intervals of the spectra of samples containing various percentages of the 1,2 isomer. Here again the cell thickness is 0.1 mm. The absorption band of the 1,2-dibromopropane at 7.29 μ can be seen to deepen as the concentration increases from 0 to 5 per cent. The measurement made in this case is the entirely empirical deflection ratio indicated in Figure 10. A graph of this ratio plotted against per-

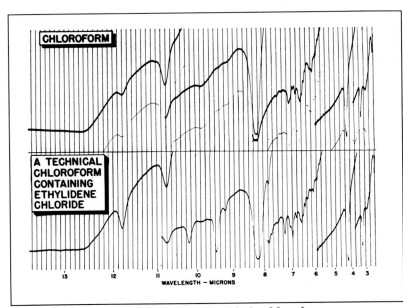

Figure 5. Detection of ethylidene chloride in chloroform

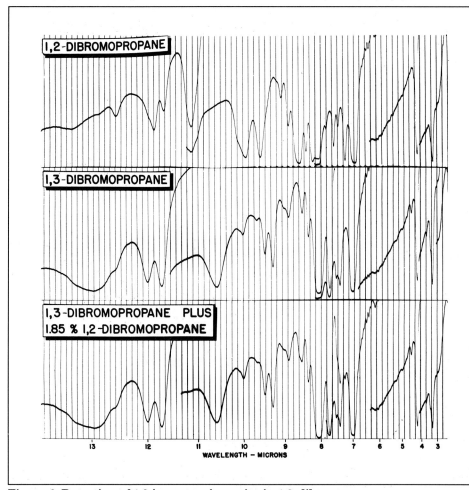

Figure 6. Detection of 1,2 isomer as impurity in 1,3-dibromopropane

be detected. The amount of sample required is less than 0.1 cc.

A similar type of deflection ratio is employed in the case of impurity of propylene dichloride in ethylene dichloride (Figure 11). In this case two absorption bands of propylene dichloride at 7.27 and 8.44 μ can be seen. The deflection ratio used in the measurement of the 7.27 μ band and the resulting curve against concentration of propylene dichloride are shown in Figure 12. The error of the determination is about ±0.05 per cent (of total sample), and the sensitivity is 0.1 per cent. Time required for a complete analysis of a single sample is less than 5 minutes.

As in most methods of quantitative analysis, the infrared method may be complicated by the presence of several impurities; this happens if there is overlapping of the absorption bands. In a considerable number of instances, however, it has been possible to analyze quantitatively two or three different impurities in the same sample. An example is the determination of both methylene chloride and carbon tetrachloride in chloroform.

centage of the 1,2-dibromopropane yields the smooth curve shown in Figure 10.

For this type of measurement the spectrum need not be recorded but instead the galvanometer deflections are read visually on a scale while the instrument automatically scans the short wave-length interval covering the impurity band (Figure 9).

This analysis requires less than 5 minutes for completion from the receipt of the sample to the final reading of the percentage of 1,2-dibromopropane from the graph of Figure 10. The error of the determination is again about ±0.05 per cent (of total sample) and a concentration as low as 0.3 per cent can

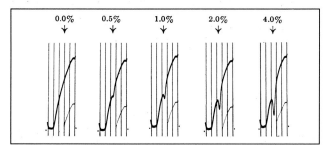

Figure 7. Addition of methylene chloride to chloroform

Spectral interval from 8.35 to 7.50 μ and methylene chloride band at 7.93 μ

The examples of quantitative analysis mentioned and many others have been successfully handled by the instrument and methods described. The materials analyzed include both aliphatic and aromatic hydrocarbons, alkyl and aryl halides, alcohols, phenols, ethers, and others. The amount of sample required places the method in the microanalytical field; furthermore, the sample can be returned unchanged after an infrared analysis. Sensitivities of infrared determinations average from 0.1 to 0.05 per cent (by volume) as the minimum detectable concentration of impurity, and in some cases measurements can be made down to 0.001 per cent.

The described method of obtaining and measuring infrared spectra for quantitative analysis eliminates a number of possible errors. In the first place, no change in cell or cell position is made during a determination. All measurements are made at points of the spectrum sharply defined by the absorption pattern itself; this removes calibration errors which might arise should the method depend upon setting the instrument at definite wave lengths. Furthermore, the use of deflection ratios removes any effect of changes in instrument sensitivity, source intensity, or difference in adjustment of the external optical system. This also corrects for slight differences in cell transmission and for changes in general transmission of the samples caused by presence of free carbon, etc.

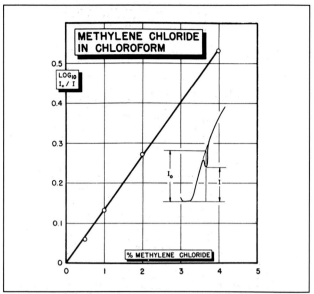

Figure 8. Infrared measurement of methylene chloride in chloroform

Abscissa in per cent by volume

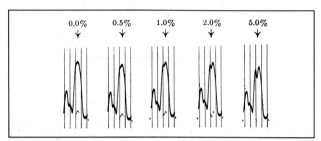

Figure 9. Addition of 1,2-dibromopropane to 1,3-dibromopropane

Spectral interval from 7.70 to 7.00 μ and 1,2-dibromopropane band at 7.29 μ

Further removal of the last-mentioned effects is accomplished by making all measurements at points of the spectrum as close together as possible; incidentally, this allows only a few seconds of time to elapse between readings. Correction for a small amount of stray, short wave-length radiation in the instrument is made, as shown in Figures 8, 10, and 12, by using galvanometer zeros defined by nearby bands in which the absorption is total. This small correction is needed only at wave lengths greater than 6 μ and can be furnished also by using glass in place of the metal shutter.

It may be argued that other more complicated methods of recording infrared spectra which yield percentage transmission or absorption curves directly (4) would be more suitable for quantitative determinations. The spectra shown in Figures 7, 9, and 11 demonstrate, however, that what is to be measured in determining an impurity is the superimposed absorption of a single sharp band of the impurity compound. For this purpose the recording of transmission or absorption percentages would at best give spectrum curves very

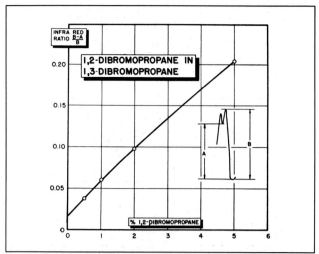

Figure 10. Infrared measurement of 1,2-dibromopropane in 1,3-dibromopropane

little different from those of the present instrument, and identical measurements of the impurity band would have to be made.

DETECTION OF INTERMEDIATES

A study of the reaction mechanism involved in the bromination of chloroform to bromoform affords a simple example of another use of infrared spectroscopy.

The question was the following: Does a molecule of chloroform brominate completely and at once to bromoform, or is the process one of successive replacement of the chlorine atoms with bromine atoms, forming first the intermediate compounds $CHCl_2Br$ and $CHClBr_2$? According to the first process if the bromination of a sample of chloroform is halted at an incomplete stage the reaction products should be chloroform and bromoform only; if the second mechanism takes place the intermediates should be present as well.

The answer to the question is clearly shown by the infrared spectra (Figure 13). Chloroform has a band at 8.24 μ; the corresponding band of bromoform lies at the longer wave length. 8.74 μ. The two intermediates should each have a band falling between those of chloroform and bromoform. The spectrum of the incompletely brominated sample of chloroform shows that the two intermediates are indeed present, two additional bands appearing between those of chloroform and bromoform. The process of bromination is therefore the second mentioned, one of successive replacement of chlorine atoms with bromine atoms.

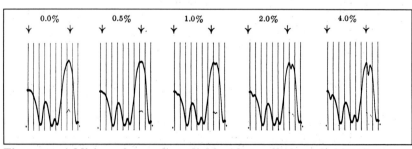

Figure 11. Addition of propylene dichloride to ethylene dichloride

Spectral interval from 8.50 to 7.00 μ, and propylene dichloride bands at 8.44 and 7.27 μ

Any discussion of applications of infrared spectroscopy would be incomplete without at least a brief comparison with the Raman effect. The two methods are similar in that the Raman spectra, although obtained by ordinary photographic spectrographs, correspond to molecular vibration frequencies as do the infrared spectra.

From the standpoint of success in practical applications of the kind discussed in the present paper, however, the two methods differ considerably. The Raman effect has a definite advantage over infrared spectroscopy in being able to study materials in water solutions; but in the general organic field the infrared method is by far the more practicable. Important disadvantages of the Raman method are: (1) the necessity of refining each sample to remove ever-present traces of impurities which fluoresce, masking the spectra; (2) the long exposure times and general difficulty of producing spectra suitable to be micro-photometered; and (3) the insensitivity of the method to compounds in small concentrations.

Advantages of analysis by means of infrared spectroscopy include the following: The amount of sample required is very small; no preliminary refinement of the sample is needed; and the method is sensitive to small concentrations, is accurately quantitative, and requires a very short time for a complete analysis.

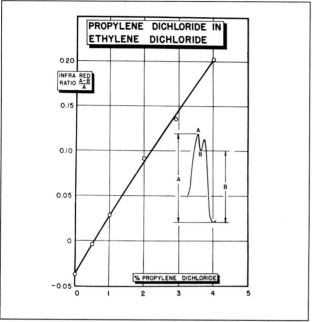

Figure 12. Infrared measurement of propylene dichloride in ethylene dichloride

ACKNOWLEDGMENT

The writer is indebted to J. D. Hanawalt for many valuable suggestions and to L. W. Gildart for very able assistance in the laboratory.

LITERATURE CITED

(1) Barnes, R. B., Bonner, L. G., and Condon, E. U., *J. Chem. Phys.*, **4**, 772 (1936).
(2) Coblentz, W. W., "Investigations of Infrared Spectra", Pub. **35**, Carnegie Institution of Washington, 1905.
(3) Dennison, D. M., *Rev. Modern Phys.*, **3**, 280 (1931).
(4) Gershinowitz, H., and Wilson, E. B., *J. Chem. Phys.*, **6**, 197 (1938).
(5) Gildart, L. W., and Wright, Norman, to be published in *Rev. Sci. Instruments*.
(6) Hanawalt, J. D., and Rinn, H. W., *Ind. Eng. Chem., Anal. Ed.*, **8**, 244 (1936).
(7) Hanawalt, J. D., Rinn, H. W., and Frevel, L. K., *Ibid.*, **10**, 457 (1938).
(8) Lecomte, Jean, "Le spectre infrarouge", Paris, Presses Universitaires de France, 1928.
(9) McAlister, E. D., *Phys. Rev.*, **49**, 704 (1936).
(10) Pfund, A. H., *Rev. Sci. Instruments*, **8**, 417 (1937).
(11) Pfund, A. H., *Science*, **90**, 326 (1939).
(12) Schaefer, C., and Matossi, F., "Das Ultrarote Spektrum", Berlin, Julius Springer, 1930.
(13) Strong, J., and Randall, H. M., *Rev. Sci. Instruments*, **2**, 585 (1931).

Presented before the Division of Industrial and Engineering Chemistry at the 100th Meeting of the American Chemical Society, Detroit, Mich.

Reprinted from *Ind. Eng. Chem., Anal. Ed.* **1941**, *13*, 1–8.

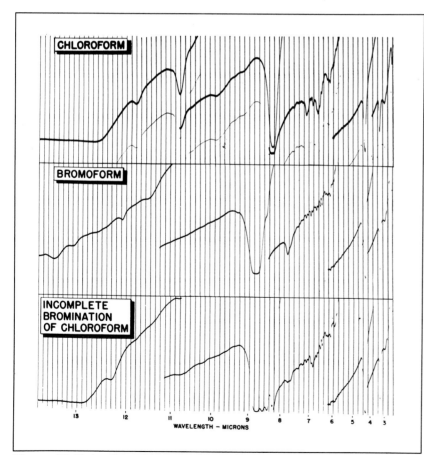

Figure 13. Detection of intermediates in bromination of chloroform to bromoform

This paper is a concise review of the state of LC in 1942. Tracing the early roots of Tswett's work and explaining why the method lay fallow for more than 20 years, Strain succinctly "sells" the capability and usefulness of LC and offers a framework for general operation for the novice in this new area. In appealing to a wide audience, Strain points out the many benefits of applying LC. Indeed, these benefits are the same ones noted by today's HPLC authors: quantitation, determination of purity, concentration by solid-phase extraction, determination of structure, and evaluation of adulteration—to name a few. Strain's only lament was that the use of this technique "remains an art rather than a science." Interestingly, the understanding of the selectivity in LC, which enabled rapid method development, is precisely why the modern technique grew so explosively in the 1970s.

Brian Bidlingmeyer
Sterling Winthrop

Chromatographic Adsorption Analysis

HAROLD H. STRAIN
Carnegie Institution of Washington, Stanford University, Calif.

The separation of mixtures by passage of their solutions through towers or columns of powdered adsorbents (Tswett's chromatographic adsorption analysis) is applicable to the detection, preparation, and estimation of both inorganic and organic substances. This method, in macro and micro modifications, is useful for concentration of solutes from dilute solution, for comparison of substances suspected of being identical, and for estimation of the structure of organic molecules. The new technique is beginning to find application in industrial operations.

In 1906 at the then Russian city of Warsaw, a botanist named Tswett invented a unique adsorption method of chemical analysis that was destined to find wide application in many branches of the natural sciences (52). Results obtained through improvement and application of the method have broken the hardpan in barren fields and have made virgin regions tillable by the chemist.

SEPARATION OF COLORED COMPOUNDS

In the course of his experiments on the pigments of green leaves, Tswett performed a simple experiment that has formed the basis for chromatographic adsorption technique.

In the narrow end of a constricted glass tube he placed a wad of cotton, and above this he tamped successive portions of finely powdered, adsorptive material such as precipitated chalk. This packed tube or adsorption column (Figure 1) was attached to a suction flask, and then a green, petroleum ether extract of dried leaves was passed through it. Under these conditions, some of the pigments, particularly the yellow carotene hydrocarbons, were weakly adsorbed and passed rapidly through the column. Other more strongly adsorbed pigments,

the green chlorophylls and yellow xanthophylls, were held near the top of the adsorbent. In this way, the migrating pigments gradually separated from one another, forming a series of colored bands or zones, called a "chromatogram". The bands always occurred in the same sequence, analogous to the colors in the spectrum.

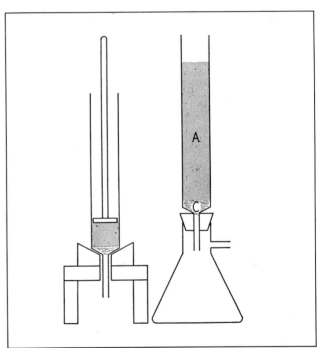

Figure 1. (*Left*) Cork ring support and metal plunger used for packing adsorption column. (*Right*) Commonest arrangement for chromatographic adsorption analysis (*40*)

A, adsorbent; C, cotton

2855-8/94/0020$08.00/0 ©1994 American Chemical Society *Milestones in Analytical Chemistry*

Tswett observed that the separation of the several bands was increased if only a small quantity of the solution was passed into the column and if this was then followed with fresh solvent. The completed separation of the leaf pigments is shown by Figure 2.

Utilization of fresh solvent to complete the separation of the bands is now recognized as an essential step in the resolution of mixtures. It is known as the "development" of the chromatogram. A diagrammatic representation of the development of a chromatogram is provided by Figure 3.

In order to recover the leaf pigments separated on the adsorption column the development with fresh solvent was continued until the pigments in each band were carried successively into the percolate. This provided solutions of the pure pigments. Another procedure was to push the moist adsorbent from the tube. The bands in the cylinder of cohesive adsorbent were then separated from one another with a knife. Each colored band or zone of adsorbent was agitated with a little alcohol or other polar solvent (now called the "eluant"). This liberated the adsorbed pigment. Solutions of the eluted pigment were separated from the adsorbent by filtration. Still another procedure was to dig out the bands separately from the column with a spatula and to elute the pigments from the respective portions with a polar solvent.

As indicated by Figure 2, Tswett separated one yellow carotene, three or four yellow xanthophylls, and two green chlorophylls from leaf extracts. However, only minute quantities of these pigments were obtained. Isolation of crystalline leaf xanthophyll at this time by use of other methods provided a single pigment and gave credence to the assumption that Tswett's adsorption method had produced alteration of a single leaf xanthophyll (54, 55). As a consequence, the adsorption method fell into disrepute, and not until 1931 was its great usefulness rediscovered.

Separation of crystalline, presumably homogeneous, carrot root carotene into several isomeric hydrocarbons by adsorption on columns of specially prepared adsorbents marked the

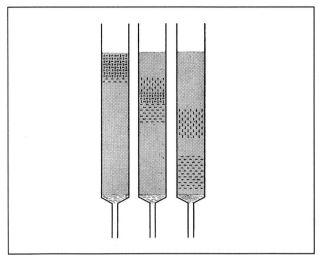

Figure 3. Successive stages in development of a chromatogram

A solution of substances | and – is passed into the column. When the column is washed with fresh solvent, | and – move through the tube at different rates, finally separating completely from each other (40).

beginning of a new era in chromatographic investigations (19, 22, 26, 43). It demonstrated the great selectivity of the method because the carotenes were found to differ only in the position of one double bond in molecules containing 40 atoms of carbon. It proved that sufficient quantities of material for subsequent chemical analysis could be prepared quickly and in good yield. It indicated that the method of preparation and activation of the adsorbent exerted a profound effect upon the applicability of the method.

Because of the importance of the carotenes as provitamins A (14), their resolution by adsorption attracted the attention of workers in many different fields. One result was the substantiation of Tswett's observations pertaining to the resolution of the leaf pigments. With the most selective adsorbents, about a dozen xanthophylls were finally separated from the leaf extracts (45). These advances stimulated a great deal of work on the preparation of various adsorbents for the separation of mixtures of many different types of compounds ranging from the elements themselves to the most complex materials found in organic nature. This in turn led to many new applications of the method, to the perfection of macro- and micro-columns, to improvement in the design of the columns, and to the separation of colorless as well as of colored compounds. Many of the applications of the adsorption method have been reviewed (4, 9, 28, 57, 62, 63), and thorough discussions of the method and extensive bibliographies are to be found in two books (40, 61).

For student practice, the separation of dyestuffs by adsorption upon columns of alumina has been recommended (36). Separation of other materials, including the leaf pigments and the carotenes of carrot roots (40, 43), may be made the basis of lecture demonstrations. As a rule, the bands in a chromatogram gradually become diffuse and indistinct when the flow of solvent through the column is discontinued; hence, for permanent exhibits, columns may be packed with mixtures of

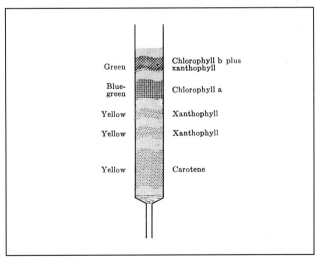

Green	Chlorophyll b plus xanthophyll
Blue-green	Chlorophyll a
Yellow	Xanthophyll
Yellow	Xanthophyll
Yellow	Carotene

Figure 2. Chromatogram obtained by adsorption of leaf pigments under conditions similar to those described by Tswett

stable, dry pigments and siliceous earth, thus yielding permanent replicas of the original chromatograms.

SEPARATION OF COLORLESS SUBSTANCES

Colorless compounds are separable on the adsorption columns, but special methods must be employed to locate these invisible materials. According to one procedure, successive portions of the percolate are collected and analyzed by chemical, optical, or biological methods. In order to use the Toepler *Schlieren* optical method for determination of colorless substances in the percolate, the solution may be passed upward through the adsorbent (49). By another procedure, the column containing the colorless compounds is sectioned empirically and the compounds eluted from the respective sections are examined as just described. Colorless, fluorescent compounds may be observed on the adsorption columns in ultraviolet light (20, 58).

Previous to adsorption some colorless compounds may be converted into colored derivatives that can then be separated on the columns (27, 41). Other colorless compounds may be located on the columns by means of reagents with which they produce color changes. These reagents may be passed through the column (38) or they may be applied to the cylinder of moist adsorbent after it has been pressed from the tube (63). A clever innovation is the adsorption of cations on columns of 8-hydroxyquinoline, with which they form colored compounds and thus become visible (13). Adsorption of a colorless compound with a colored indicator serves to locate the former with respect to the latter (5, 6). The closer together the two are adsorbed, the more precisely one can locate the colorless compound.

APPLICATIONS OF THE METHOD

From the analyst's point of view it is important to know what can be accomplished by use of the adsorption method. To this end it has seemed desirable to classify the various procedures and to illustrate them with selected experiments.

One of the principal uses of adsorption columns is the resolution of mixtures into their constituents. Here all the components are recovered after qualitative or quantitative separation of the mixture. Some examples are summarized in Table I.

Very often it is necessary to use the adsorption method for the preparation of only one or two substances from complex mixtures. Many examples of this application of the method are to be found among the procedures utilized for preparation of specific substances from extracts of biological materials. Minute quantities of enzymes, hormones, vitamins, pigments, and similar materials have been prepared from extracts containing large quantities of other materials. Unwanted materials in these extracts frequently complicate the adsorption procedure, necessitating the use of relatively large quantities of adsorbent. Examples of important biological products prepared or investigated by adsorption methods are reported in Table II.

If quantitative separations of one or two components from complex mixtures are to be made, the adsorption method is often modified so that the materials to be estimated pass

Table I. Mixtures and the Adsorbents and Solvents Used for Their Separation

Mixture	Adsorbent	Solvent
Cations	8-Hydroxyquinoline	Water (13)
	Alumina	Water (38)
Lithium isotopes	Zeolites	Water (47)
Halogens	Magnesium silicate	Carbon tetrachloride (40)
Carotenes	Magnesia	Petroleum ether (43)
	Lime	Petroleum ether (22)
	Alumina	Petroleum ether (26)
Vitamins A$_1$ and A$_2$	Lime	Petroleum ether (19)
Chlorophylls	Sucrose	Petroleum ether (59)
	Inulin	Petroleum ether (30)
	Magnesium citrate	Ether and petroleum ether (30)
Xanthophylls	Magnesia	1,2-Dichloroethane (45)
	Calcium carbonate	Carbon disulfide (60)
Fatty acids	Charcoal	Petroleum ether (8)
Fats	Alumina	Petroleum ether (51)
Polycyclic hydrocarbons	Alumina	Petroleum ether (58)
Thiamin and riboflavin	Decalso and Supersorb	Water (11)

through the column without being adsorbed while the contaminants remain on the column. This reduces losses that often result from incomplete elution of adsorbed compounds. One of the best examples of this application of the adsorption method is the separation of the carotenes from other pigments in leaf extracts. A number of adsorbents and solvents have been utilized for this purpose, as is illustrated in Table III. The method is not restricted to use in this way, because it has been shown that riboflavin and thiamin (vitamins B$_2$ and B$_1$) in food products can be adsorbed and eluted quantitatively (11).

Table II. Substances Prepared by Adsorption of Extracts of Plant and Animal Products

Substance	Adsorbent	Solvent	Source
Vitamin A	Lime	Petroleum ether	Fish liver oil (18)
Vitamin B$_1$	Decalso	Water	Rice hulls (10)
Vitamin B$_2$ (acetylated)	Alumina	Ethyl acetate	Alfalfa (25)
Vitamin D	Alumina	Petroleum ether	Fish liver oil (6)
Vitamin E	Alumina	Petroleum ether	Wheat germ oil (15, 50)
Vitamin K$_1$	Decalso	Petroleum ether	Fish meal (3)
Vitamin K$_2$	Decalso	Petroleum ether	Purified fish meal (29)
Sex hormones	Alumina	Carbon tetrachloride	Human urine (7)
Glucosidase and chitinase	Bauxite	Water	Snails (64)
Anthocyanins	Alumina	Water (21)	. . . .
Pterins	Frankonit	Water	Insects (2)
Carotenes	Magnesia	Petroleum ether	Leaves (30, 44)

Table III. Adsorbents and Solvents for Separation of Carotenes from Other Leaf Pigments

Adsorbent	Solvent
Inulin	Petroleum ether (35, 53)
Sucrose	Petroleum ether (53, 59)
Calcium carbonate	Petroleum ether (53)
Starch	Petroleum ether (46)
Calcium phosphate	Petroleum ether (33)
Soda ash	Petroleum ether (23)
Magnesia (partly activated)	Petroleum ether (17)
Magnesia (activated)	1,2-Dichloroethane (45)

Some materials containing traces of impurities may be purified by passage of the solutions through adsorption columns that retain the impurities. This application of the method is similar to the well-known decolorization of solutions by percolation through towers of charcoal or fuller's earth, as in the industrial clarification of sugar solutions and lubrication oils.

Chromatographic adsorption is often used to determine the purity of chemical compounds. Substances are said to be chromatographically homogeneous if they cannot be purified further with adsorption columns.

Still another use of the adsorption technique is the concentration of solutes from dilute solution, thus eliminating or supplementing the tedious and sometimes destructive process of evaporation. Materials adsorbed by passage of the dilute solution through the column are recovered in greatly concentrated form upon elution from the adsorbent. This procedure has been utilized for the concentration of certain alkaloids and urinary pigments (16, 24).

Two substances suspected of being identical may be compared quickly by use of the chromatographic method. A solution of each compound is adsorbed on separate columns to make certain that the materials yield only one band and are therefore homogeneous. Solutions of the two compounds are then mixed and adsorbed on a fresh column of adsorbent. Formation of a single band indicates that the two substances are very similar or identical. Formation of two bands proves that the compounds are different (28, 37, 40, 61).

The great selectivity of the chromatographic adsorption method makes it extremely useful for the detection of adulterants in technical products. It has been used to prove the presence of artificial coloring in wine and in butter (32, 48).

Determination of the molecular structure of certain types of organic molecules has been facilitated by the use of adsorption methods. Experience has shown that there is a relation between the adsorbability of organic molecules and the architecture of their molecules. The more polar the molecule, the greater the number of polar groups, and the greater the number of double bonds the more strongly is the compound adsorbed. Chemical reactions that alter the molecule also change its relative position upon the adsorption column. The

greatest use of these relationships has been made in studies of the carotenoid pigments (40).

Chromatographic adsorption has found some use for the separation of hydrocarbons and sterols from their addition compounds with trinitrobenzene or picric acid. The nitro compounds are held by the adsorbent and the liberated materials pass into the percolate (34, 35).

For the separation of water-soluble, ionized, or charged compounds the adsorption method may be combined with the electrophoretic method. When electrical potential is applied to the ends of a column upon which ionized substances have been adsorbed, the materials migrate over the surface of the adsorbent although no liquid is permitted to flow through the tube. Thus far, the method has been applied only to the separation of a few dyes (42).

Recent developments indicate that the chromatographic adsorption method may find considerable use in industrial processes (1). Alumina has been utilized for the dehydration of organic solvents (12), and magnesia has been recommended for continuous purification of dry cleaning solvents (39).

APPARATUS AND PROCEDURE

In spite of the hundreds of applications of the adsorption method in the past ten years, utilization of this technique remains an art rather than a science. This is due to the fact that a great many different conditions affect the procedure, often in several unrelated ways. Moreover, there is no adequate theoretical basis for the method to serve as a guide for its use (8, 56).

Some two dozen different types of adsorption columns have been described, a few of which are illustrated in Figure 4. The diameters of the adsorption columns range from 1 or 2 mm. to many centimeters. The lengths are usually 6 to 12 times the diameter, but this proportion is by no means fixed.

The amount of material adsorbed on the columns varies with the activity of the adsorbent, with the solvent, and with the adsorbability of the compound itself. It ranges from a few

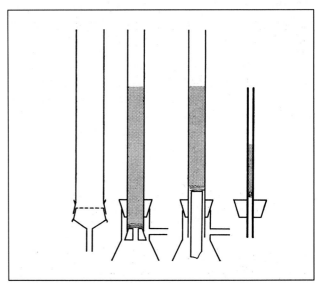

Figure 4. Various types of adsorption tubes (40, 61)

microgms with the smallest columns to a few tenths of a gram with columns several centimeters in diameter. With still larger columns, several grams of material may be prepared.

ADSORBENTS, SOLVENTS, AND ELUANTS

No universal adsorbent or solvent has yet been discovered. Each application of the adsorption method depends upon the careful selection of adsorbents, solvents, and eluants, upon the basis of experience and trial and error. Adsorbents and solvents useful for adsorption of a given substance are usually suitable for adsorption of other materials of the same class.

Particular attention must be paid to the properties of the adsorbent. The adsorptive capacity of most solids varies enormously with the method of preparation and with the presence of impurities and polar solvents such as water. To a considerable extent it is characteristic of the solid itself, as is indicated by the following list of adsorbents arranged in approximate order of their adsorption capacity, the most active recorded first (40):

Activated alumina, charcoal, and magnesia (Micron brand)	Calcium carbonate
Lime	Talc
Magnesia (Merck)	Magnesium citrate
Magnesium carbonate	Inulin
Calcium phosphate	Sucrose, starch

The adsorption capacity of all solids varies with the solvents that are employed. Adsorption is strongest from aliphatic hydrocarbons such as petroleum ether, and is progressively weaker from more and more polar solvents, as illustrated by the following list of solvents arranged in approximate order of their effect on adsorption (adsorption is greatest from those listed first):

Petroleum ether, b. p. 30–50 °	1,2-Dichloroethane
Petroleum ether, b. p. 70–100 °	Alcohols
Carbon tetrachloride	Water
Cyclohexane	Pyridine
Benzene	Acids

Substances adsorbed from a nonpolar solvent may be caused to move through the column at a faster rate by the addition of a more polar liquid to the solvent. This accelerates the development of the chromatogram (51).

For each application of the adsorption technique, the adsorbent must be of just the right activity. If it binds the adsorbed compounds too firmly, a chromatogram cannot be developed. If it is too weakly adsorptive, sufficient quantities of material will not be adsorbed.

The particle size of the adsorbent is of importance because this determines the rate of percolation of the solvent through the column, the definition and evenness of the bands, and the method to be used in packing the adsorption column. Granular adsorbents (about 200-mesh) can be poured into the glass tube and tamped gently. More finely divided materials such as precipitated chalk must be tamped firmly into the column in small portions. Very fine powders should be mixed with a filter aid such as the heat-treated siliceous earth known as Hyflo Super-Cel. These mixtures are pressed rather than tamped into the tubes. Occasionally the adsorbent is made into a slurry with the solvent and this is poured into the glass tube.

Selection of adsorbents and solvents is determined, to a large extent, by the nature of the compounds to be resolved. Some idea of the combinations of adsorbent and solvent used for preparation of various compounds may be gained by perusal of Tables I and II.

Once substances have been separated by adsorption upon Tswett columns, it is essential to be able to release or elute them from the adsorbent. The polar solvents usually used for this are alcohols, or pyridine with alcohols or acetic acid. Materials adsorbed from water can frequently be eluted by changing the pH of the solutions.

POSSIBLE FUTURE DEVELOPMENTS

In view of the spectacular advances made in the preparation of active and selective adsorbents in the past ten years, it seems highly probable that similar developments may continue. The new organic polymers that function as exchange adsorbents have yet to be tested in adsorption columns. Future investigations may be expected to reveal methods whereby adsorbed compounds can be eluted in good yield, thus extending the usefulness of the chromatographic adsorption technique in quantitative analysis. Much remains to be done in order to perfect the methods for location of colorless compounds on the adsorption columns. With the clarification of concepts regarding the process of adsorption itself, further useful applications of the chromatographic method will undoubtedly be made.

LITERATURE CITED

(1) Aluminum Ore Co., East St. Louis, Ill., "Activated Alumina, Its Properties and Uses", 1938.
(2) Becker, E., and Schöpf, C., *Ann.*, **524**, 124 (1936).
(3) Binkley, S. B., MacCorquodale, D. W., Thayer, S. A., and Doisy. E. A., *J. Biol. Chem.*, **130**, 219 (1939).
(4) Brockmann. H., *Angew. Chem.*, **53**, 384 (1940).
(5) Brockmann. H., *Z. physiol. Chem.*, **245**, 96 (1937).
(6) Brockmann, H., and Busse, A. *Ibid.*, **249**, 176 (1937).
(7) Callow, N. H., and Callow, R. K., *Biochem. J.*, **34**, 276 (1940).
(8) Cassidy, H. G., *J. Am. Chem. Soc.*, **62**, 3073, 3076 (1940); **63**, 2735 (1941).
(9) Cassidy, H. G., *J. Chem. Education*, **16**, 88 (1939).
(10) Cerecedo, L. R., and Hennessy, D. J., *J. Am. Chem. Soc.*, **59**, 1617 (1937).
(11) Conner, R. T., and Straub, G. J., *Ind. Eng. Chem., Anal. Ed.*, **13**, 385 (1941).
(12) Derr, R. B., and Willmore, C. B., *Ind. Eng. Chem.*, **31**, 866 (1939).
(13) Erlenmeyer, H., and Dahn, H., *Helv. Chim. Acta*, **22**, 1369 (1939); **24**, 878 (1941).
(14) Euler, B. V., Euler, H. v., and Hellström, H., *Biochem. Z.*, **203**, 370 (1928).
(15) Evans, H. M., Emerson, O. H., and Emerson, G. A., *J. Biol. Chem.*, **113**, 319 (1936).
(16) Fink, H., U.S. Patent 2,072,089 (Mar. 2, 1937).
(17) Fraps, G. S., Kemmerer, A. R., and Greenberg, S. M., *Ind. Eng. Chem., Anal. Ed.*, **12**, 16 (1940); *J. Assoc. Official Agr. Chem.*, **23**, 659 (1940).
(18) Karrer, P., *Helv. Chim. Acta*, **22**, 1149 (1939).

(19) Karrer, P., and Morf, R., *Ibid.*, **16**, 625 (1933).
(20) Karrer, P., and Schöpp, K., *Ibid.*, **17**, 693 (1934).
(21) Karrer, P., and Strong, F. M., *Ibid.*, **19**, 25 (1936).
(22) Karrer, P., and Walker, O., *Ibid.*, **16**, 641 (1933).
(23) Kernohan, G., *Science*, **90**, 623 (1939).
(24) Koschara, W., *Z. physiol. Chem.*, **232**, 101 (1935).
(25) Kuhn, R., and Kaltschmitt, H., *Ber.*, **68**, 128 (1935).
(26) Kuhn, R., and Lederer, E., *Ibid.*, **64**, 1349 (1931).
(27) Ladenburg, K., Fernholz, E., and Wallis, E. S., *J. Org. Chem.*, **3**, 294 (1938).
(28) Lederer, E., *Bull. soc. chim.*, **6**, 897 (1939).
(29) McKee, R. W., Binkley, S. B., MacCorquodale, D. W., Thayer, S. A., and Doisy, E. A., *J. Biol. Chem.*, **131**, 327 (1939).
(30) Mackinney, G., *Ibid.*, **111**, 75 (1935).
(31) *Ibid.*, **132**, 91 (1940).
(32) Mohler, H., and Hämmerle, W., *Z. Untersuch. Lebensm.*, **70**, 193 (1933).
(33) Moore, L. A., *Ind. Eng. Chem., Anal. Ed.*, **12**, 726 (1940).
(34) Pfau, A. S., and Plattner, P. A., *Helv. Chim. Acta*, **19**, 858 (1936).
(35) Plattner, P. A., and Pfau, A. S., *Ibid.*, **20**, 224 (1937).
(36) Rieman, W., *J. Chem. Education*, **18**, 131 (1941).
(37) Scheer, B. T., *J. Biol. Chem.*, **136**, 275 (1940).
(38) Schwab, G. M., and Jockers, K., *Angew. Chem.*, **50**, 546 (1937).
(39) Seaton, M. Y., U.S. Patent 2,077,857 (1937).
(40) Strain, H. H., "Chromatographic Adsorption Analysis", New York, Interscience Publishers (1942).
(41) Strain, H. H., *J. Am. Chem. Soc.*, **57**, 758 (1935).
(42) *Ibid.*, **61**, 1293 (1939).
(43) Strain, H. H., *J. Biol. Chem.*, **105**, 523 (1934).
(44) *Ibid.*, **111**, 85 (1935).
(45) Strain, H. H., "Leaf Xanthophylls", Carnegie Inst. Washington, *Pub.* **490** (1938).
(46) Strott, A., *Jahrb. wiss. Botan.*, **86**, 1 (1938).
(47) Taylor, T. I., and Urey, H. C., *J. Chem. Physics*, **5**, 597 (1937); **6**, 429 (1938).
(48) Thaler, H., *Z. Untersuch. Lebensm.*, **75**, 130 (1938).
(49) Tiselius, A., *Science*, **94**, 145 (1941).
(50) Todd, A. R., Bergel, F., and Work, T. S., *Biochem. J.*, **31**, 2257 (1937).
(51) Trappe, W., *Biochem. Z.*, **305**, 150 (1940); **306**, 316 (1940); **307**, 97 (1941).
(52) Tswett, M., *Ber. deut. botan. Ges.*, **24**, 384 (1906).
(53) *Ibid.*, **29**, 630 (1911).
(54) Willstätter, R., and Mieg, W., *Ann.*, **355**, 1 (1907).
(55) Willstätter, R., and Stoll, A., "Untersuchungen über Chlorophyll, Methoden und Ergebnisse", Berlin, Julius Springer, 1913.
(56) Wilson, J. N., *J. Am. Chem. Soc.*, **62**, 1583 (1940).
(57) Winterstein, A., in G. Klein's, "Handbuch der Pflanzenanalyse", Vol. IV, p. 1403, Wein, J. Springer, 1933.
(58) Winterstein, A., and Schön, K., *Z. physiol. Chem.*, **230**, 146, 158 (1934).
(59) Winterstein, A., and Stein, G., *Ibid.*, **220**, 263 (1934).
(60) Zechmeister, L., and Cholnoky, L. v., *Ann.*, **516**, 30 (1935).
(61) Zechmeister. L., and Cholnoky, L. v., "Die chromatographische Adsorptionsmethode, Grundlagen, Methodik, Anwendungen", 2nd ed., Berlin, J. Springer, 1938; English translation: "Principles and Practice of Chromatography", London, Chapman and Hall, 1941 (N.Y., John Wiley & Sons).
(62) Zechmeister, L., and Cholnoky, L. v., *Monatsh.*, **68**, 68 (1936).
(63) Zechmeister, L., Cholnoky, L. v., and Ujhelyi, E., *Bull. soc. chim. biol.*, **18**, 1885 (1936).
(64) Zechmeister, L., Tóth, G., and Vajda, É., *Enzymologia*, **7**, 170 (1939).

Presented before the Division of Agricultural and Food Chemistry, Symposium on New Analytical Tools for Biological and Food Research at the 102nd Meeting of the American Chemical Society, Atlantic City, N.J.

Reprinted from *Ind. Eng. Chem., Anal. Ed.* **1942**, *14*, 245–49.

Polarographic Determination of Manganese as Tri-dihydrogen Pyrophosphatomanganiate

I. M. KOLTHOFF AND J. I. WATTERS[1]
University of Minnesota, Minneapolis, Minn.

A procedure is presented by which manganese is quantitatively oxidized with lead dioxide to the trivalent state in a medium which contains a large excess of pyrophosphate at a pH smaller than 4. The concentration of the manganic manganese in the complex is determined polarographically. Evidence is given that the violet complex ion is tri-dihydrogen pyrophosphatomanganiate.

The present paper describes a simple procedure in which manganese is oxidized with the aid of lead dioxide to the manganic complex in the presence of a large excess of potassium pyrophosphate at a pH between 2.0 and 4.0. The concentration of manganic complex is determined with the aid of the dropping mercury electrode. Evidence that the complex has the composition tri-dihydrogen pyrophosphatomanganiate is given.

Manganous ion is reduced at the dropping mercury electrode, producing a well-defined diffusion current in neutral alkali chloride as supporting electrolyte (14). The half-wave potential is -1.51 volts vs. the saturated calomel electrode, so the presence of an excess of several metal ions such as ferric, ferrous, cobaltous, nickel, zinc, and copper interferes with the polarographic determination of divalent manganese. Furthermore, the hydrogen wave interferes in acidic solution. The permanganate ion produces a poorly defined wave starting at zero applied e.m.f. Stackelberg et al. (12) concluded that this wave was not suitable for analytical purposes. Verdier (14) found that in an alkaline tartrate solution, a well-defined anodic wave was obtained at -0.4 volt vs. the saturated calomel

[1] Present address, Baker Laboratory, Cornell University, Ithaca, N.Y.

electrode. The manganous ion is apparently oxidized to the trivalent state. A large excess of ferrous iron interferes in this determination. Tri-dihydrogen pyrophosphatomanganiate has the advantage that the manganese is reduced from the tri- to the divalent state at positive potentials vs. the saturated calomel electrode. As the diffusion current can be measured at +0.1 volt to +0.15 volt vs. saturated calomel electrode, it is not necessary to remove dissolved oxygen. A large excess of ions such as ferric, zinc, copper, cobalt, and nickel may be present. Chromium, vanadium, and cerium interfere, since these metals are oxidized to chromate, vanadate, and ceric, respectively, and are also reduced at positive potentials.

Trivalent manganese has been employed in several analytical methods for determining manganese.

Heczko (7) oxidized manganese to the trivalent state with freshly prepared perphosphoric acid in a solution of phosphoric acid and sulfuric acid and titrated the trivalent manganese iodometrically. Lang (10) oxidized manganese with potassium dichromate in a solution of metaphosphoric acid. After reducing the excess dichromate with sodium arsenite solution, he titrated the trivalent manganese with standard ferrous sulfate solution using diphenylamine as the indicator. Hirano (8) estimated manganese colorimetrically by forming the trivalent complex in a solution of sulfuric acid and phosphoric acid. Tomula and Aho (13) estimated manganese colorimetrically by forming a pyrophosphatomangani acid complex in a mixture of pyrophosphoric and sulfuric acids. They employed potassium bromate as the oxidizing agent and removed the liberated bromine through the formation of the slightly dissociated compound, bromine cyanide. This procedure has the disadvantage of the danger associated with the use of acidified cyanide solutions.

2855-8/94/0026$08.00/0   ©1994 American Chemical Society   *Milestones in Analytical Chemistry*

EXPERIMENTAL

The manual apparatus described in previous publications was used. Since the complex slowly oxidizes mercury, a saturated calomel anode was employed as outside electrode. A few polarograms were made using the automatic Heyrovský polarograph. The pH measurements were made with the aid of a Beckman glass electrode pH meter. A thermostatically controlled water bath was used to maintain a constant temperature of 25.0 ± 0.02 °C.

Analytical reagents were employed. A stock solution of 0.1 M manganous sulfate was prepared by dissolving 22.306 grams of manganous sulfate tetrahydrate in water containing 10 ml. of approximately 1 N sulfuric acid and diluting to exactly 1 liter. The molarity was checked by a gravimetric determination in which the manganese was precipitated as manganous ammonium phosphate and weighed as manganous pyrophosphate. The molarity was found to be correct.

SOURCE OF PYROPHOSPHATES

Although orthophosphoric acid may be converted to pyrophosphoric acid if heated at 215 °C., large amounts of silica are dissolved if the conversion is performed in a Pyrex container. Platinum is also attacked. Furthermore, concentrated pyrophosphoric acid is variable in composition (5) and difficult to handle because of its viscosity. Its aqueous solution is unstable, being completely hydrolyzed in a few days.

The authors investigated the use of a mixture of alkali pyrophosphates and strong mineral acids as the source of the dihydrogen pyrophosphate ion. Sodium pyrophosphate is only moderately soluble, 3.16 grams of the anhydrous salt dissolving in 100 ml. of water at 0 °C. and 40.26 grams at 100 °C. Potassium pyrophosphate is extremely soluble and is therefore more suitable for the preparation of a concentrated stock solution. Such a solution is stable for a long time if not acidified (6). The potassium pyrophosphate used in the following experiments was prepared by heating anhydrous dipotassium hydrogen phosphate in a furnace at from 500 ° to 700 °C. for 3 hours.

Sodium pyrophosphate might be used instead of potassium pyrophosphate. In this case it is necessary to weigh out the salt for each determination.

Sulfuric and nitric acids were found satisfactory for the regulation of the acidity. Nitric acid is preferable if an appreciable amount of sulfate is present in the sample, since potassium sulfate, which is only moderately soluble, may precipitate.

SELECTION OF AN OXIDIZING AGENT

Lead dioxide was found to be a satisfactory oxidizing agent. It rapidly oxidized manganous ion to a violet-colored complex in a solution 0.4 M in pyrophosphates having a pH of 4 or less. The excess lead dioxide was removed by filtering through a retentive filter paper, such as Whatman No. 42. By means of titrations with standard ferrous sulfate solution it was shown that the oxidation was quantitative. It was experimentally found that 1 gram of lead dioxide could oxidize 388 mg. of manganous ion to the trivalent complex ion. This is close to the stoichiometric amount—namely, 460 mg. Practically, a large excess of lead dioxide, approximately 1 gram for each 50 mg. of manganese in the solution was used to hasten and ensure quantitative oxidation. It was found that the oxidation of a 0.01 M solution of manganous ion was complete after 3 minutes of shaking with this excess of lead dioxide if the pH was smaller than about 4.0 and a large excess of pyrophosphate was present. Intermittent shaking for 10 minutes resulted in quantitative oxidation of manganese in solutions as concentrated as 0.02 M in manganous ion. Variable concentrations of plumbous lead remained in solution, presumably as a dihydrogen pyrophosphato complex, even when appreciable concentrations of sulfates were present.

SUPPRESSION OF THE MAXIMUM

In the absence of capillary-active substances a large maximum in the current-voltage curve of the manganic complex was obtained. Furthermore, the galvanometer oscillations were irregular at potentials more negative than the region (+0.1 to +0.3 volt $vs.$ S. C. E.) in which the maximum occurred. These disturbances persisted in the presence of various agents generally used for the elimination of maxima. Gelatin, starch, soap, tylose, camphor, thymol, methyl orange, tropeolin 00, methyl red, and methylene blue were unsatisfactory in the suppression of the maximum.

It was found that if the solution was 0.02 per cent in agar the galvanometer oscillations became regular and diffusion current measurements could be made at potentials more negative than +0.16 volt (S. C. E.) using the manual polarograph. Many of the quantitative measurements were made with the manual apparatus, using agar to suppress the maximum. However, when polarographic determinations were attempted using the automatic Heyrovský polarograph, agar was found unsatisfactory. Temporary irregularities, which did not interfere when the manual instrument was used, were frequently recorded in the polarogram.

It was then found that if the final solution was 0.04 per cent in peptone (Merck's, dried, from meat) excellent waves were obtained with either the manual apparatus or the polarograph. It was necessary to add the peptone prior to the oxidation by lead dioxide. If added after the oxidation, a small amount of the complex was reduced.

After the experimental work was concluded the authors found that 0.1 per cent gum arabic is just as effective as peptone. It is stable for a long time and does not decompose the complex if added after the oxidation.

Within experimental error, the same diffusion current per millimolar concentration of the manganic complex was obtained whether 0.05 to 0.2 per cent agar, 0.02 to 0.1 per cent peptone, or 0.1 per cent gum arabic was used to eliminate the maximum.

The procedure followed in the preliminary experiments consisted of preparing solutions containing manganous ion and various concentrations of pyrophosphate at various pH values and in the presence of various maximum suppressors. The manganese was oxidized by shaking intermittently with lead dioxide for 5 to 10 minutes and the solution was filtered. Current voltage measurements were made using a portion of the filtrate. The removal of air is not necessary, since the diffusion current of the manganic complex is obtained before the oxygen wave starts.

Figure 1 illustrates the characteristics of the wave.

The solution used to obtain curve 1 was 0.005 M in manganic manganese, 0.4 M in sodium pyrophosphate, 0.8 to 0.9 M in nitric acid, and contained 0.1 per cent gum arabic. The pH was 2.00. The solution used to obtain curve 2 contained the same constituents, except that no manganese was added. The solutions were treated with lead dioxide as explained above. Tank hydrogen was passed through the solution in the cell for 15 minutes to remove most of the oxygen. In curve 2 the current from +0.2 to −0.4 volt (S. C. E.) was the residual or condenser current and was practically zero at −0.1 volt (S. C. E.). The anodic wave just beyond +0.25 volt is due to the dissolution of mercury. This wave begins at a lower positive potential than in the corresponding polarograms in Figures 2 and 3, owing to the presence of chloride ion introduced by prolonged contact with the saturated calomel anode. The wave beginning at −0.4 volt (S. C. E.) is due to the reduction of the plumbous lead introduced as a result of shaking with lead dioxide. The wave starting at −1.3 volt (S. C. E.) is due to the reduction of hydrogen ions. In curve 1, the wave beginning at about +0.3 volt is due to the reduction of the trivalent manganic complex to the divalent state. The second wave for the reduction of plumbous lead is larger than in curve 2 as a consequence of the reduction of more lead dioxide in the oxidation process. The true diffusion current is the difference between the magnitude of the apparent diffusion current measured and the residual current measured at the same potential, using a blank solution containing no manganese or a portion of the original solution before treatment with lead dioxide.

The half-wave potential of the wave due to the reduction of the manganic complex is obviously not a function of the standard potential of the manganous complex-manganic complex couple. The potential of a platinum electrode in a similar solution in which one-half of the complex had been reduced with ferrous sulfate solution was about +0.78 volt (S. C. E.) when the pH was 2.06.

EFFECT OF pH

The pH may be of influence for several reasons. From the Nernst equation it is readily seen that the oxidation potential of lead dioxide is a function of the pH. Theoretically the oxidation potential is decreased by approximately 0.12 volt for each unit increase in pH. The kind of acid pyrophosphate which preponderates is determined by the pH, as is the stability of complex ions containing ionizable hydrogen atoms. It will be shown in a subsequent paper that the wave due to the reduction of ferric iron in pyrophosphate solutions is shifted to more negative potentials if the acidity is decreased. This wave interferes if the pH is much smaller than 2.

Experiments 1 to 8, Table I, were performed to determine the pH range in which the violet complex could be quantitatively obtained with a large excess of pyrophosphate present.

Samples containing 5 ml. of 0.1 M manganous sulfate were transferred to a 100-ml. volumetric flask, and 20 ml. of 2 M potassium pyrophosphate were added to each flask to obtain a 0.4 M concentration of pyrophosphate. Various volumes of sulfuric acid (1 to 1) were added to regulate the pH. After addition of 2 ml. of 2 per cent agar and dilution to 100 ml., the oxidation and diffusion current measurements were carried out as in the general procedure.

All the solutions having a pH of 3.78 or smaller had, within experimental error, the same diffusion current and the rich violet color. However, in experiments 2 and 3, the solutions having a pH close to 5 appeared orange red to orange brown

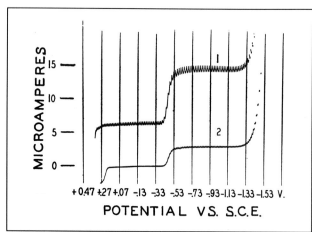

Figure 1. Polarogram

1. 5 millimolar tri-dihydrogen pyrophosphatomanganiate, 0.4 M sodium pyrophosphate, 0.8–0.9 M nitric acid, 0.1 per cent in gum arabic, pH = 2.0; oxidized with lead dioxide; hydrogen passed through solution 15 minutes
2. Like 1 except no manganese present

Table I. Effect of pH

Sample, 5 ml. 0.1 M MnSO$_4$ to produce a final concentration of 5 millimolar in manganese.

Procedure followed, except pH varied. Final pyrophosphate concentration was 0.4 M, agar 0.04 per cent.

Capillary I. $m^{2-3}t^{1/6}$ = 1.314 mg.$^{2/3}$ sec.$^{-1/2}$ at $E_{d.\ e.}$ = +0.1 volt. Av. i_d/millimolar concn. Mn^{III} = 1.54 microamperes.

Experiment no.	H$_2$So$_4$ (1:1) used Ml.	pH of final solution	Color of final solution	i_d at +0.1 volt (S. C. E.) corrected for residual current Microamperes
1	2	....	Brown	7.34
2	4	5.05	Orange brown	7.64
3	4	5.00	Orange brown	7.69
4	4.5	3.78	Violet	7.78
5	5	1.90	Violet	7.82
6	5	1.90	Violet	7.78
7	10	0.90	Violet	7.79
8	20	0.10	Violet	7.84

in color and the diffusion current was somewhat smaller than at lower pH. After completing the experiments, the pH of the solution used in experiment 4 was varied by adding nitric acid or ammonium hydroxide dropwise. When the pH was increased to 4.52, an orange cast was observed. When the pH was decreased below 4, the complex again appeared violet. A brown precipitate of manganese dioxide always formed in the brown solutions upon standing a few hours. The violet solutions were unchanged after 48 hours.

EFFECT OF PYROPHOSPHATE CONCENTRATION

A series of experiments was performed in which the pH was maintained close to 4.0 while the concentration of pyrophosphate was varied. The same final concentration of manganese, 0.005 M, was employed; 0.02 per cent agar was used to suppress the maximum. Before making the polarographic measurements in experiments 1 through 13 in Table II, sufficient potassium nitrate was added to 40 ml. of the oxidized solutions to make the concentration of potassium ions in all solutions 1.6 M. The solutions were then diluted to 50 ml. The oxidation to the trivalent complex was not complete until the pyrophosphate concentration was 0.2 M or 40 times larger than that of manganese.

Experiments 14 through 17 were then performed to determine the maximum concentration of manganese which could

be oxidized to the trivalent state, using a final concentration of 0.4 M potassium pyrophosphate. Since a constant large excess of potassium pyrophosphate was present, no addition of potassium nitrate or dilution was necessary before making diffusion current measurements. Experiment 14 showed that 10 millimolar manganous ion was quantitatively converted to the trivalent complex at pH 3.9. However, in experiment 15, when the concentration of manganese was increased to 20 millimolar, only 90 per cent of the manganese was converted to the trivalent complex at pH 3.3.

In experiments 16 and 17, 20 millimolar manganous ion and 0.4 M potassium pyrophosphate were again employed but the acidity was increased. In experiment 16, in which the pH was 2.2, a quantitative oxidation to the manganic complex was obtained. In experiment 17, in which a 7 N excess of free sulfuric acid was present, the oxidation to the trivalent complex was also quantitative. However, the complex was not stable in strong sulfuric acid. After 5 days the violet color of the complex prepared in 7 N sulfuric acid had disappeared, while the complex prepared at pH 2.2 showed no appreciable decrease in color.

It may be concluded that manganous ion, up to a concentration of 20 millimolar, can be quantitatively oxidized to the violet manganic complex in a solution 0.4 M in potassium pyrophosphate at a pH of about 2.2. At appreciably lower pH

Table II. Effect of Pyrophosphate Concentration

Sample, 5 to 20 ml. 0.1 M MnSO$_4$ to produce a final concentration of 5 to 20 millimolar manganese.

Procedure followed except final pyrophosphate concentration varied. Agar, 0.04 per cent, used to suppress maximum. In experiments 1 through 13, 40 ml. of sample diluted to 50 ml. after oxidation and KNO$_3$ added to obtain 1.6 M potassium ion. Experiments 14 through 17 were not diluted.

Capillary II. $m^{2/3}t^{1/6} = 1.967$ mg.$^{2/3}$ sec.$^{-1/2}$ at $E_{d.\ e.} = +0.1$ volt. Av., i_d/millimolar concn. Mn^{III} = 2.30 microamperes.

Experiment no.	2 M potassium pyrophosphate added Ml.	Final millimolar pyrophosphate concn.	Final millimolar concn. of Mn	Final pH	Color of final solution	True diffusion current at +0.1 volt (S. C. E.) (i_d Apparent—i_r) Microamperes	Estimation of Mn oxidized to trivalent state %
1	0.5	10	5	3.30	Violet brown	3.55	39
2	0.75	15	5	3.30	Violet brown	3.85	42
3	1.00	20	5	3.30	Violet brown	4.10	45
4	1.25	25	5	3.90	Brown violet	4.50	49
5	1.50	30	5	4.09	Brown violet	5.67	62
6	1.75	35	5	4.11	Rose violet	6.37	69
7	2.0	40	5	4.00	Violet	7.19	78
8	3.0	60	5	3.98	Violet	7.50	82
9	4.0	80	5	4.03	Violet	7.85	85
10	5.0	100	5	3.78	Violet	8.35	91
11	7.5	150	5	3.98	Violet	8.82	96
12	10.0	200	5	3.90	Violet	9.18	100
13	20.0	400	5	3.90	Violet	9.17	100
14	20.0	400	10	3.90	Violet	23.0	100
15	20.0	400	20	3.30	Violet	41.4	90
16	20.0	400	20	2.30	Violet	46.2	100
17	20.0	400	20	7 N H$_2$SO$_4$	Violet	46.3	100

the complex is not very stable and the wave due to the reduction of the ferric complex interferes.

EFFECT OF AIR ON THE WAVE

The following experiments were performed to determine the effect of oxygen of the air on the value obtained for the diffusion current.

All the solutions were finally 0.4 M in potassium pyrophosphate. The pH was adjusted to about 2.3 with sulfuric acid (1 to 1), using thymol blue as the indicator. Peptone (0.04 per cent) was used to suppress the maximum. Each solution was treated with 1 gram of lead dioxide.

In experiment 1, Table III, a blank determination was carried out using only the reagents. The measurements were made without removing air from the solution. The measurements were repeated after nitrogen had been passed through the solution rapidly for 2 minutes (columns 2 and 3, Table III). There was no further decrease in the current at these positive potentials after nitrogen had been passed through the solution for a total of 10 minutes.

In experiment 2, the procedure used in the previous experiment was followed but the solution was also 0.004 M in manganous ion before the oxidation. No reduction of oxygen was found at +0.15 volt (S. C. E.). The small value of 0.03 microampere at +0.1 volt (S. C. E.), due to reduction of oxygen, was measured in both the blank and the solution containing manganese. At 0.0 volt (S. C. E.) the effect due to oxygen was no longer the same in the blank and the solution containing manganese. It follows that it is not necessary to remove oxygen if the diffusion current is measured at +0.1 to +0.15 volt (S. C. E.).

These measurements using the manual instrument were substantiated with polarograms obtained with the Heyrovský polarograph. The polarogram (curve 2, Figure 2), was made

without oxidizing the manganese and without removing the air. Another portion of the same solution was oxidized with lead dioxide to obtain curve 1. Air was not removed.

The wave starting at +0.05 volt (S. C. E.) in both curves is due to the first step of the reduction of oxygen. The wave starting at −0.4 volt (S. C. E.) in curve 1 is due to the reduction of plumbous ion.

REAGENTS

Potassium Pyrophosphate, 2 M. Slowly pour 132 grams of anhydrous potassium pyrophosphate into about 150 ml. of water while stirring. Dilute to exactly 200 ml.

Nitric Acid (1 to 1). Dilute 100 ml. of freshly boiled concentrated nitric acid with 100 ml. of water.

Sulfuric Acid (1 to 1). Dilute 100 ml. of concentrated sulfuric acid with 100 ml. of water.

Gum Arabic, 2 per cent. Add 1 gram of gum arabic to 49 grams of water and heat to boiling while stirring. This solution is stable for several days.

Peptone, 2 per cent. Dissolve 1 gram of Merck's peptone, "dried", from meat, in 49 grams of boiling water. This solution may be kept in a stoppered bottle containing a few milliliters of carbon tetrachloride, kept cool, or prepared fresh after 48 hours.

Thymol Blue Solution, 0.1 per cent. Mix 0.1 gram of dry thymol blue with 21.5 ml. of 0.01 N sodium hydroxide in a mortar and dilute to 100 ml. with water.

PROCEDURE

Polarographic Determination of Manganese as Tridihydrogen Pyrophosphatomanganiate. The sample to be determined may contain from 1 to 100 mg. of manganese as the nitrate, sulfate, or perchlorate in approximately 50 ml. of aqueous solution. Iron may be present in quantities up to 0.2 gram, but chromium, vanadium, and cerium interfere.

Transfer the sample quantitatively to a 100-ml. volumetric flask, and add enough nitric acid (1 to 1) or sulfuric acid (1 to 1) to make the total amount of mineral acid present equivalent to 12 ml. of nitric acid (1 to 1) or 4.5 ml. of sulfuric acid (1 to 1). Slowly add 20 ml. of 2 M potassium pyrophosphate to the solution while swirling. Add 2 or 3 drops of thymol blue. Add more acid or ammonium hydroxide dropwise until the color becomes orange, corresponding to a pH of 2.0 to 2.4. Add 5 ml. of 2 per cent gum arabic or 5 ml. of 2 per cent peptone. Dilute to the mark and mix thoroughly.

Table III. Effect of Air on the Wave

Solution 0.004 M Mn, 0.4 M $K_4P_2O_7$, 0.08 percent peptone, pH = 2.3 (approx.) using H_2SO_4 (1.1) and thymol blue.
Capillary III. $m^{2/3}t^{1/6}$ = 2.150 mg.$^{2/3}$sec.$^{-1/2}$ at $E_{d.\ e.}$ = +0.1 volt.

E_c (S. C. E.)	Experiment 1, no Mn present			Experiment 2, 4 millimolar Mn^{III}		
	Air not removed	N$_2$ passed through solution 2 minutes	Current due to air in blank	Air not removed	N$_2$ passed through solution 2 minutes	Current due to air in Mn solution
			Microamperes			
+0.25	−4.35	−4.35	0	4.30	4.30	0
+0.24	−2.86	−2.86	0	5.55	5.55	0
+0.22	−0.56	−0.56	0	....	....	....
+0.20	−0.38	−0.38	0	9.43	9.43	0
+0.15	−0.20	−0.20	0	9.65	9.65	0
+0.1	−0.13	−0.16	+0.03	9.84	9.81	+0.03
+0.075	0.00	....	....	9.89	9.84	+0.05
+0.05	+0.13	−0.10	+0.23	9.98	9.86	+0.12
+0.025	+0.24	....	....	10.06	9.88	+0.18
0.00	+0.41	−0.06	+0.47	10.16	9.90	+0.26

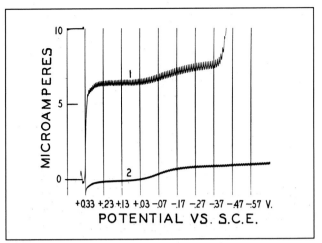

Figure 2. Polarogram
1. 5 millimolar tri-dihydrogen pyrophosphatomanganiate, 0.4 *M* sodium pyrophosphate, 0.8–0.9 *M* nitric acid, 0.1 per cent gum arabic, pH = 2.00; oxidized with lead dioxide; air not removed
2. Solution used in 1 before oxidation with lead dioxide

Transfer a few milliliters to a polarographic cell and measure the residual current as described below for the oxidized solution. Add about 1 gram of manganese-free lead dioxide for each 25 to 50 mg. of manganese present and shake intermittently for 5 to 10 minutes. Filter through a dry retentive filter paper such as Whatman No. 42 into a clean polarographic cell. Rinse out the cell with the first portion of the filtrate and collect a suitable volume of the filtrate.

Use an external anode. Five milliliters of C.P. carbon tetrachloride may be added to cover the mercury drops collecting at the bottom of the cell. Just before starting the measurements insert the dropping mercury electrode. Measure the current at +0.1 volt and +0.15 volt (S. C. E.) if the manual instrument is used.

If the automatic recording instrument is available, use the 2-volt accumulator. Turn the switch to "anodic and cathodic polarization". The applied potential is zero when the sliding contact is at the exact center of the resistance wire. Regulate the voltage so that each abscissa line corresponds to 0.1 or 0.05 volt. Choose a galvanometer sensitivity such that the wave has a suitable height from +0.1 to +0.15 volt (S. C. E.). Start the polarogram at +0.35 volt (S. C. E.). If iron is present, stop the motor as soon as the light beam moves past the camera, then with the sliding contact disengaged record the zero current line for a distance of 1 or 2 mm. on the sensitized paper. It is desirable to run the residual current on the same polarogram.

PROPORTIONALITY BETWEEN DIFFUSION CURRENT AND CONCENTRATION OF MANGANESE

The results reported in Table IV show that proportionality exists between diffusion current

and the concentration over a wide range of concentrations.

The deviation from the average was more than 1 per cent only when very low concentrations of manganese were used.

The determinations were made according to the general procedure. A few polarograms are given in Figure 3. Even at very low manganese concentrations, oxygen from the air does not interfere.

STRUCTURE OF MANGANI PYROPHOSPHORIC ACID

If manganic orthophosphate is dissolved in meta, pyro, or sirupy orthophosphoric acid, violet-colored complex compounds are formed. These compounds have been named mangani metaphosphoric acid, mangani pyrophosphoric acid, and mangani orthophosphoric acid, depending on the acid in which they were prepared.

A study of the literature reveals that the following types of trivalent manganese compounds have been crystallized from solutions of the various acids:

1. From orthophosphoric acid
   $MnPO_4.H_2O$, green gray in color (*4*)
2. From pyrophosphoric acid
   $Mn_4(P_2O_7)_3.14H_2O$, pale violet (*1*)
   $NaMn(P_2O_7).5H_2O$, pale red (*11*)
3. From metaphosphoric acid $Mn(PO_3)_3. H_2O$, red (*1*)

Rosenheim (*11*) observed that many compounds of the type $M^I, MnP_2O_7.3H_2O$, crystallized from solutions containing the mangani pyrophosphoric acid. He suggested the formula $[Mn_{(H_2O)_3}^{P_2O_7}]^-$ for the anion of the violet complex acid in aqueous

---

Table IV. Proportionality between Diffusion Current and Concentration of Manganese

Capillary I $m^{2/3}t^{1/6}$ = 1.314 mg.$^{2/3}$ sec.$^{-1/2}$ at $E_{d. e.}$ = +0.1 volt.
Procedure used except 0.02 per cent agar used and nitrogen bubbled through solution 2 minutes.

Experiment no.	Final millimolar concn. of complex manganic ion	Mn taken Mg.	Apparent diffusion current measured at +0.1 volt Microamperes	True diffusion current (corrected for residual) Microamperes	Diffusion current per millimolar concentration of complex Microamperes
1	0	0	−0.07	....	....
2	0	0	−0.07	....	....
3	0.2	1.1	+0.25	0.32	1.60
4	0.5	2.75	0.69	0.76	1.52
5	1	5.49	1.46	1.53	1.53
6	1	5.49	1.47	1.54	1.54
7	2	10.99	3.02	3.01	1.54$_5$
8	3	16.48	4.57	4.64	1.54$_7$
9	5	27.47	7.68	7.75	1.55$_0$
10	10	54.93	15.40	15.47	1.54$_7$
11	10	54.93	15.32	15.39	1.53$_9$
12	20	109.86	30.70	30.77	1.53$_9$

Av. 1.54

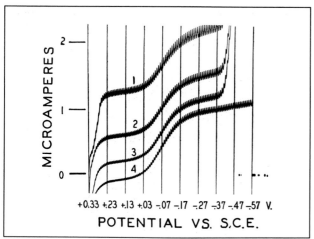

Figure 3. Polarogram

1. 1 millimolar tri-dihydrogen pyrophosphatomanganiate, 0.4 M sodium pyrophosphate, 0.8–0.9 M nitric acid, 0.1 per cent gum arabic, pH = 2.00; air not removed
2. Like 1 but 0.5 millimolar Mn^{III}
3. Like 1 but 0.2 millimolar Mn^{III}
4. Residual current. Solution used in 3 before oxidation with lead dioxide

solution. This formula has been adopted in the literature. The assumption of similarity between the complex in solution and the crystalline form obtained is not justified.

The violet manganic complex with pyrophosphate described in this paper has properties which are very similar to that of the trioxalatomanganiate complex. The latter has the following structure:

$$\left[Mn \left(\begin{smallmatrix} O-C \diagup O \\ O-C \diagdown O \end{smallmatrix} \right)_3 \right]^{-3}$$

which is stabilized by the five-membered chelate ring. A similar structure with a six-membered chelate ring is obtained when the dihydrogen pyrophosphate ion combines with manganic:

$$\left[Mn \left(\begin{smallmatrix} O-H \\ | \\ O \\ O-P \diagup O \\ O-P \diagdown O \\ | \\ O-H \end{smallmatrix} \right)_3 \right]^{-3}$$

Various observations substantiate the conclusion that the dissolved violet complex dealt with in this paper is the tri-dihydrogen pyrophosphatomanganiate of the above structure.

The pH of an alkali dihydrogen pyrophosphate solution is about 4.25. As is evident from Table I, the violet complex is stable at this and lower pH. In order to obtain a stable solution of the violet complex it is necessary to have present a large excess of dihydrogen pyrophosphate. This indicates that the violet complex has ionic bonding and that it has a relatively small stability constant. Cartledge and Nichols (3) in their study of trimalonatomanganiate arrived at a similar conclusion.

When the pH is of the order of 5 or larger dihydrogen pyrophosphate ions in the complex may be replaced by water mol-

ecules, producing the brown-colored complex. This behavior has been observed by Cartledge et al. (2, 3) in the study of trimalonatomanganiate and the trioxalatomanganiate complexes.

Tomula and Aho (13) found in transport measurements that the violet manganic pyrophosphate complex migrates toward the anode. This observation eliminates the possibility that pyrophosphoric acid occupies the coordination places in the complex. The possibility that trihydrogen pyrophosphate ions occupy the coordination positions is not considered probable, because an alkali trihydrogen pyrophosphate solution has a pH of about 1.40 and the complex was found stable at a pH of 4. Evidence that the complex is tri-dihydrogen pyrophosphatomanganiate was obtained from the calculation of the diffusion coefficient of the complex ion. The diffusion coefficient was calculated from the polarographic results with the aid of the Ilkovic equation:

$$i_d = 605 \, n \, C \, D^{1/2} m^{2/3} t^{1/6}$$

By titration with ferrous sulfate it had been found that the manganese in the complex is present in the trivalent form. Hence n in the above equation is equal to one, assuming that the complex ion contains one manganic ion.

The diffusion coefficient was calculated from three sets of measurements with three different capillaries. The results are summarized in Table V.

For the sake of comparison the authors have also calculated the diffusion coefficient of the trioxalatocobaltiate ion from its diffusion current using capillary I. The medium was 1 M in potassium oxalate, 0.2 M in ammonium acetate, and 0.5 M in acetic acid, and contained 0.018 per cent gelatin. The solution had a pH of 5.0. The average value of the diffusion coefficient calculated was 0.523×10^{-5} sq. cm. per second.

Using Jander's (9) expression for the relation between the molecular weights and diffusion coefficients of two closely related species x and k:

$$M_x = \left(\frac{Z_k D_k}{Z_x D_x} \right)^2 M_k$$

in which Z is the specific viscosity of the solution, D, the diffusion coefficient, and M molecular weight, it is possible to cal-

Table V. Diffusion Coefficient of Violet Complex

Capillary no.	m	t at +0.1 volt vs. S. C. E. in 0.4 M pyrophosphate at pH 2.3	$m^{2/3}t^{1/6}$ at +0.1 volt	i_d per millimole per liter	$D \times 10^5$
	Mg./sec.				Sq. cm./sec.
I	1.051	4.10	1.31	1.54	0.375
II	2.111	2.92	1.97	2.30	0.373
III	2.625	2.08	2.15	2.50	0.369
				Av.	0.373 ± 0.003

culate the molecular weight of the manganic pyrophosphate complex. The viscosities of the media in which the diffusion currents of the manganiate and cobaltiate complexes were determined were nearly identical.

The formula of the oxalatocobaltiate has been proved (2) to be $Co(C_2O_4)_3^{---}$, corresponding to a molecular weight of 323. Substituting the various values in the above equation we find:

$$M_x = \frac{(5.23 \times 10^{-6})^2}{(3.75 \times 10^{-6})^2} \times 323 = 628$$

Assigning to the manganic dihydrogen pyrophosphate complex the formula $[Mn(H_2P_2O_7)_n]^{3-2n}$ we calculate from the molecular weight of 628 a value of $n = 3.25$. Since Jander's equation is only approximate we may consider the agreement between the expected value of $n = 3$ and the calculated one of 3.25 as satisfactory. The result of the molecular weight calculation indicates that the complex has the formula $[Mn(H_2P_2O_7)_3]^-$.

LITERATURE CITED

(1) Auger, V., *Compt. rend.*, **133**, 96 (1901).
(2) Cartledge, G. H., and Ericks, W. P., *J. Am. Chem. Soc.*, **58**, 2065 (1936).
(3) Cartledge, G. H., and Nichols, P. M., *Ibid.*, **62**, 3057 (1940).
(4) Christensen, O. T., *J. prakt. Chem.*, **28**, 1 (1883).
(5) Gerber, A. B., and Miles, F. T., *Ind. Eng. Chem., Anal. Ed.*, **10**, 519 (1938).
(6) Gladstone, J. H., *J. Chem. Soc.*, **20**, 435 (1867).
(7) Heczko, T., *Z. anal. Chem.*, **68**, 433 (1926).
(8) Hirano, S., *J. Soc. Chem. Ind.* (Japan), **412B**, 40 (1937).
(9) Jander, G., and Spandau, H., *Z. physik. Chem.*, **A185**, 325 (1939).
(10) Lang, R., *Z. anal. Chem.*, **102**, 8 (1935).
(11) Rosenheim, A., and Triantaphyllides, T., *Ber.*, **48**, 582 (1915); *Z. anorg. Chem.*, **153**, 126 (1926).
(12) Stackelberg, M. V., Klinger, P., Koch, W., and Krath, E., *Tech. Mitt. Krupp, Forschber.* **2**, 59 (1939).
(13) Tomula, E. S., and Aho, V., *Ann. Acad. Sci. Fennicae*, **A LII**, 4 (1939); **A LV**, 1 (1940).
(14) Verdier, E. T., *Collection Czechoslov. Chem. Commun.*, **11**, 216 (1939); **11**, 233 (1939).

Reprinted from *Ind. Eng. Chem., Anal. Ed.* **1943**, *15*, 8–13.

The development of this methodology was prompted by the great demand for aviation gasoline and other petroleum products during World War II. Because of the similarity of certain spectra, such as those of the isomeric butenes, ion beam intensities had to be measured with high accuracy (<1%). Most important, a commercial version of the instrument was built (Figure 3), and this was the beginning of the famous CEC 21-100 series. Not mentioned in the article is the fact that the ion source temperature had to be controlled to ± 0.1° (at 250°) to attain the necessary fragment ion current reproducibility, and the often tedious "computations" were carried out with an electromechanical calculator the size of a typewriter. The availability of a large number of calibration spectra (the API collection of mass spectra) measured on the CEC 21-102 and 103 models was an important selling point and kept the British (Metropolitan Vickers) and Germans (MAT) out of the U.S. market for a long time.

Klaus Biemann
Massachusetts Institute of Technology

The Mass Spectrometer as an Analytical Tool

H. W. WASHBURN, H. F. WILEY, AND S. M. ROCK
Consolidated Engineering Corporation, Pasadena, Calif.

Discussing the mass spectrometer as an analytical tool, the paper describes the method of analysis, presents results obtained with this method, and gives information regarding the commercial instrument that is now in use.

The demands upon the petroleum and chemical industries for large quantities of aviation gasoline, synthetic rubber, and other war materials have resulted in employing many new processes on a large scale. These new processes in general are more complex than those formerly employed in industry. To develop such new processes and to put them into operation necessitates a large amount of preliminary and coordinated analytical work.

These technological advances have therefore required a parallel advance in the art of analysis both as an aid to the development of new processes and for the analytical control of the plants after they have been put into operation.

One of the new methods of analysis which are now being employed is the mass spectrometer method. It is the purpose of this paper to discuss the mass spectrometer as an analytical tool: the method of analysis with the mass spectrometer, a few results obtained with this method, and the commercial instrument now in use.

PRINCIPLE OF OPERATION OF MASS SPECTROMETER

In general, a mass spectrometer is a device for sorting molecules. Figure 1 is a diagrammatic sketch of the Dempster or 180° type of mass spectrometer. The gas mixture to be analyzed enters through the gas inlet to chamber a.

Before the molecules are sorted, they are given an electric charge, so that they can be forced to move by the combined action of electric and magnetic fields. The production of an electric charge on a molecule, or ionization, is accomplished by bombarding the molecules with a stream of electrons in an ionization chamber. In practice the energy in the bombarding electrons is usually made sufficiently high to break the molecules into charged fragments. This feature enables substances of the same molecular weight, but of different molecular structure, to be distinguished from each other (2).

The ionization chamber and associated electron source or filament are shown schematically in Figure 1 at a. The ions are pulled out of chamber a by means of the electric field existing between electrodes d, e, and f. Ions enter the analyzer tube with a high velocity and are sorted out according to their mass by the action of a magnetic field. The magnetic field causes heavy ions to follow a circular path which is of greater

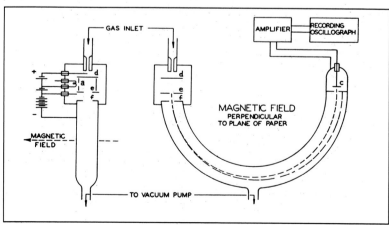

Figure 1. Diagram of mass spectrometer

radius than the path followed by lighter ions. The radius of the path for an ion of given mass can be made smaller or larger by varying the electric field produced between electrodes d, e, and f. Therefore, by gradually varying the electric field, ions of each mass can be caused to fall successively on the collector, c, where their quantity is measured by amplifying and recording equipment. The resulting record shows the relative numbers of ions of each mass which appear at c.

MASS SPECTRA

Figure 2 shows the automatically recorded mass spectra of n-butane and of a paraffin-olefin mixture, C_2 through C_4. It takes about 10 minutes to make a single record. When operating in a routine manner a sample can be run every 20 minutes.

In order to make the records more easily understood it is necessary to explain that there are four traces, all recorded simultaneously and having different degrees of sensitivity. The top trace has unit sensitivity, the second has a sensitivity of one-third, the third a sensitivity of one-tenth, and the fourth or bottom trace a sensitivity of one-thirtieth. This device enables the height of any peak to be recorded within better than 1 per cent accuracy over a range in magnitude of 250 to 1.

The abscissas, or horizontal scale values, on the mass spectra at which peaks occur, represent molecular weights corresponding to the different charged fragments obtained when n-butane or a C_2 to C_4 gas mixture is bombarded with an electron stream. The ordinate or vertical scale at each of these masses is a measure of abundance of the particular fragment which is formed.

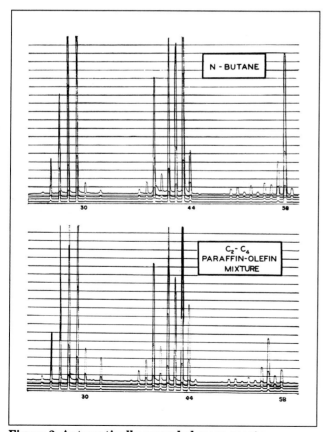

Figure 2. Automatically recorded mass spectra

For any given gas the particular fragments formed depend for the most part upon the number and type of atoms in the molecule. The relative abundance of the different fragments, or pattern of the mass spectrum, depends upon the structure of the molecule.

The mass spectrum, therefore, depends on the structure of the molecule, as well as upon the type of atoms of which the molecule is composed—for example, if this mass spectrum were for isobutane it would have peaks at the same masses but their relative heights would be different. It is this dependence of the mass spectrum upon the structure of the molecule that enables the isomers of a compound to be separately determined.

PRINCIPLES OF ANALYSIS

If now the mass spectrum of an unknown mixture is obtained, employing a suitable technique, it will be a composite mass spectrum and can be considered as being a summation or superposition of the mass spectra of all the components of the mixture. The mass spectrum of the C_2 through C_4 mixture shown is such a composite spectrum. The analysis of the mixture consists of the unraveling of the mixture spectrum. Through methods developed in the Consolidated Engineering Corporation laboratory, this unraveling has been made much simpler than was originally thought possible. One other point of interest here is that all peaks recorded are not generally needed for an analysis. The peaks which are not for calculating the mixture are used as a check on the accuracy of the analysis. They are employed to indicate calculation errors, instrumental errors, or failure to take account of a gas in a mixture. This final check on an analysis has been found to be very valuable.

In order to complete the analysis of a mixture from its composite spectrum, it is necessary to know the mass spectra of all its components. These spectra are obtained by running pure gases through the instrument and are called calibrations. The record of the n-butane in Figure 2 is an example.

An explanation of the general methods of analysis would involve more discussion than space permits. However, the extremely simple examples shown in Table I will serve to clarify some of the previous remarks.

In the first column are listed the masses or molecular weights at which peaks occur on the mass spectrum of the unknown mixture. In the second column are recorded the peak heights read from the automatic record of the mixture.

The analysis of the mixture may be obtained in the following manner: The only components contributing to peaks at mass 57 and 58 are n- and isobutane. The percentages of n- and isobutane are computed from simultaneous Equations 1 and 2 shown at the bottom of the table. In these equations the underlined numbers are taken from calibration records obtained by running pure n- and isobutane. From the determination of the percentages of n- and isobutane and from their calibration spectra, the contributions of n- and isobutane to each of the masses listed can be readily computed. These values are shown in the third and fourth columns.

If now the mass 44 contributions of n- and isobutane are

Table I. Analysis of Depropanizer Overhead

Mass	M, mixture peaks	n-Butane	Isobutane	Propane	Ethane	Methane	Σ, sum of component spectra	M−Σ, residuals
15	167.3	1.0	1.6	30.4	23.0	110.3	166.3	+1.0
16	134.4	0.0	0.0	0.9	0.8	132.7	134.4	0.0
26	146.1	1.6	0.9	37.6	106.4	....	146.5	−0.4
27	361.6	10.8	12.1	183.0	158.0	....	363.9	−2.3
28	777.8	9.3	1.2	275.3	494.4	....	780.2	−2.4
29	593.1	12.5	2.6	467.8	105.6	....	588.5	+4.6
30	130.0	0.3	0.0	10.0	119.7	....	130	0.0
31	2.4				2.5	....	2.5	−0.1
38	24.0	0.5	1.2	21.9	....	....	23.6	+0.4
39	98.0	4.3	8.7	84.6	....	....	97.6	+0.4
40	14.7	0.6	1.3	13.0	....	....	14.9	−0.2
41	94.1	9.0	18.9	65.8	....	....	93.7	+0.4
42	49.9	3.9	15.6	30.2	....	....	49.7	+0.2
43	207.4	32.2	49.8	124.5	....	....	206.5	+0.9
44	145.3	1.0	1.6	142.7	....	....	145.3	0.0
45	4.6	0.0		4.6	....	....	4.6	0.0
50	0.6	0.3	0.3	....	....	....	0.6	0.0
51	0.5	0.3	0.3	....	....	....	0.6	−0.1
52	0.1	0.1	0.1	....	....	....	0.2	−0.1
53	0.6	0.3	0.3	....	....	....	0.6	0.0
54	0.1	0.1	0.0	....	....	....	0.1	0.0
55	0.7	0.3	0.2	....	....	....	0.5	+0.2
56	0.5	0.3	0.2	....	....	....	0.5	0.0
57	2.5	0.9	1.6	....	....	....	2.5	0.0
58	5.0	3.8	1.2	....	....	....	5.0	0.0
59	0.2	0.2	0.0	....	....	....	0.2	0.0

Computation of % of n- and isobutane — Mole %

From peak 57: $\underline{0.503}p_n + \underline{0.654}p_i = 2.5$ (1)

From peak 58: $\underline{2.10}p_n + \underline{0.498}p_i = 5.0$ (2)

		Mole %	
n-Butane	where p_n = % n-butane	= 1.8	
Isobutane	and p_i = % isobutane	= 2.4	
Propane	$(145.3 - 1.0 - 1.60)\underline{0.269}$	= 38.4	(3)
Ethane	$(130.0 - 0.3 - 0 - 10.0)\underline{0.3175}$	= 38.0	(4)
Methane	$(134.4 - 0 - 0 - 0.9 - 0.8)\underline{0.146}$	= 19.4	(5)

Underlined coefficients are obtained from calibrations.

The same procedure is used in calculating the amount of methane as indicated in Equation 5.

If the apparatus were perfect, if there were no errors in the computation, and if all constituents in the mixture were taken into account, the sum of these component spectra for any given mass would be equal to the mixture peak. The comparison of the sums of the component spectra for each mass with the mixture peak of the corresponding mass therefore offers an excellent check on the reliability of the analysis.

The sums of the component spectra are shown in column 8. In order more clearly to show the agreement of this summation column with the mixture column, the difference between these two columns is shown in column 9. Because of the method of computation, the residuals will be zero on the masses used for computing the mixture—viz., 16, 30, 44, 57, 58. The residuals on the peaks not used for determining the analysis of the mixture are a measure of the accuracy of the analysis. In this case all residuals are less than 1 per cent of the respective mixture peaks. This shows that the analysis is of good accuracy and that all substances present in detectable amounts have been accounted for.

The ability of this method to detect unexpected components is illustrated in Tables II and III. The unexpected naphthenes in one case and acetone in the other were detected by the relatively large residuals obtained when the analysis was first made, assuming that these constituents were not in the mixture.

The method of computation illustrated in Table I was described to show how the mass spectra of the components of a mixture superimpose to give the spectrum of the mixture and how the unused peaks may be used to check accuracy of the

subtracted from the mixture 44 peak, the remaining peak is due entirely to propane. The per cent propane in the mixture is computed from this remainder peak and from the sensitivity of the 44 peak to propane. This simple calculation is shown in Equation 3 where the underlined value, 0.269, is the sensitivity in percent per division obtained from the propane calibration.

Similarly, the amount of ethane can be computed by subtracting the 30 peak contributions of the heavier components from the mixture 30 peak. This determination is shown in Equation 4.

Table II. Wet Gas

	Mass spectrometer Mole %	Fractionating column Mole %
N_2	1.1	....
Methane	84.1	87.44
Ethane	4.9	4.60
Propane	4.7	4.30
Isobutane	1.1	0.72
n-Butane	1.7	1.47
Isopentane	0.7	0.55
n-Pentane	0.5	0.47
C_6^+	0.2	0.45
C_4 naphthenes	0.6	....
C_5 naphthenes	0.3	....
C_6 naphthenes	0.1	....

	Time required for analysis by complete direct method (Tables II and III) Min.
Mass spectrometer (instrument time)	
Mixture	25
Prorated calibration	4
Taking data from record (technician's time)	
Mixture	20
Prorated calibration	4
Computing	80
Total man-hours	2.25

Table III. Wet Gas

(Two analyses of this mixture on fractionating column A and two on B are compared with mass spectrometer analysis.)

	Fractionating column A (Mole per cent)		Fractionating column B (Mole per cent)		Mass spectrometer Mole %
H_2	....	....	....	....	0.2
N_2	....	....	....	....	0.4
O_2	....	....	....	....	0.2
CH_4	91.500	91.600	91.517	91.557	89.9
C_2H_6	4.395	4.210	4.372	4.326	4.1
C_3H_8	1.942	1.942	1.968	1.969	2.0
i-C_4H_{10}	0.612	0.583			1.0
n-C_4H_{10}	0.619 / 1.231	0.699 / 1.282	1.255	1.262	0.5
i-C_5H_{12}	0.290	0.272			0.1
n-C_5H_{12}	0.227 / 0.517	0.222 / 0.494	0.491	0.488	0.2
C_6	0.232	0.230	0.236	0.230	....
C_7^+	0.149	0.162	0.161	0.168	....
C_3H_6O	....	....	....	....	1.4

analysis. The particular method used for determining the separate components is applicable to only the simplest of analyses. A large amount of work has been directed to the development of short-cut methods for analyzing more complicated mixtures. The computing manual which explains these methods is over 100 pages in length.

TIMES REQUIRED FOR ANALYSES

The length of time required to make analyses with the mass spectrometer is very short, as is shown in the examples which follow.

Times are shown for the complete direct method of analysis. These times in most cases are somewhat longer than would normally be encountered in production analyses, as will be seen from the explanation below. In three examples (Tables IV, V, VI) times are also shown for the comparison method of analysis, which is faster and may be used for control purposes.

In the complete direct method calibrations of critical components of a mixture are run on the same day on which the mixture is run, and computations are carried through at a large number of masses in order to obtain a check on the accuracy of the analysis as well as to discover unexpected constituents. In general, calibrations will consume about 2 or 3 hours per day of mass spectrometer time when the direct method is used. In the examples which follow, the value given for man-hours includes a prorated time for the calibrations.

If it is not desired to check for unexpected compounds and general accuracy of analyses, it is unnecessary to carry through computations for a large number of masses. Considerable time can therefore be saved in computing and taking data from the records in the partial direct method, since only the masses used directly in the solution need be considered. In some cases both the computing time and record reading time will be cut approximately in half.

The values given for total man-hours in the complete direct method should therefore be considered as upper limits. These times would normally be encountered only when using the mass spectrometer for experimental rather than production analyses.

If it is desired to use the spectrometer for control purposes, the time for analysis can be reduced below that attainable by either direct method, since in general there will be a large number of samples of similar composition. This permits the use of the comparison method of analysis, which eliminates a large portion of both calibration and computation and is therefore faster than the direct method.

The comparison method is essentially a comparison between an unknown sample and a standard sample. If the mass spectrometer is to be used for control purposes—for example, to indicate a change in the composition of a stream—very rapid results can be obtained with the comparison method. Such control can be accomplished merely by making periodic runs on the stream and comparing the mass spectra

Table IV. Synthetic C_1 to C_4 Paraffin–Olefin Mixture

(Composition computed from the manometer synthesis is compared with mass spectrometer analysis.)

	Manometer synthesis	Mass spectrometer	Difference
		Mole per cent	
Methane	11.3	10.9	−0.4
Ethylene	1.5	1.7	+0.2
Ethane	22.1	21.4	−0.7
Propene	11.9	11.8	−0.1
Propane	31.1	31.0	−0.1
Isobutane	5.0	5.0	0.0
Isobutene	4.4	4.7	+0.3
Butene-1	5.0	6.2	+1.2
Butene-2	0.7	0	−0.7
n-Butane	7.0	7.3	+0.3

	Time required for analysis	
	Complete direct method	Comparison (control) method
	Min.	*Min.*
Mass spectrometer (instrument time)		
Mixture	20	20
Prorated calibration	12	4
Taking data from record		
(technician's time)		
Mixture	20	15
Prorated calibration	12	1
Computing	130	80
Total man-hours	3.25	2
For *n*-, isobutane, and total		
butenes only, hour	0.75	....

thus obtained with the mass spectrum of a standard mixture which is run once each day. Any change in the difference between the spectrum of the stream and the spectrum of the standard mixture will indicate a change in the composition of the stream. In addition, the change in composition can be computed from this difference spectrum, or a complete analysis can be made from this difference spectrum and a knowledge of the composition of the standard mixture.

The comparison method of analysis just described may be used to advantage whenever a large number of samples with similar composition are to be run in a single day. There are three main advantages in using this method: (1) Complete calibrations need be made only about once a month, thus reducing to practically zero the prorated calibration time. (2) All data necessary for complete analysis are obtained, but it is unnecessary to carry out computations unless the composition of the sample has changed a sufficient amount to be of interest. This reduces to zero the computing time for some samples. (3) When it is desired to obtain a complete analysis,

the computing time is materially less than required by the direct method. The times given for the comparison method in Tables IV, V, and VI demonstrate the rapidity with which complete analyses can be obtained by this method.

With either the direct or comparison method a large number of samples can be run in a day. For example, with a C_4 feed stream as illustrated in Table V, fourteen samples can be run by the direct method and seventeen samples by the comparison method in an 8-hour day. The corresponding numbers for the C_4 cut free of C_5 shown in Table VI are 16 and 24 samples per 8-hour day.

In some cases when only the abundance of one or two components of a mixture is required, the analysis can be made rapidly, whether or not there are a large number of samples of similar composition. For example, in a C_1 through C_4 paraffin-olefin mixture, if only the mole per cents of *n*- and isobutane and total butenes are desired, the total time is about 40 minutes. In the determination of butadiene in a C_1 to C_4 mixture or in a C_4 cut, the time is only 30 minutes.

ANALYSES MADE WITH THE MASS SPECTROMETER

Some important mixtures which have been analyzed with the mass spectrometer to date are

1. Dry gas
2. Wet gas
3. *n*- and isobutane mixtures
4. C_1 through C_4 paraffin and olefin
5. C_3 through C_5 paraffin and olefin
6. C_4 paraffins, olefins
7. C_4 paraffins, olefins, and diolefin
8. C_5 paraffins, olefins, diolefins, and cyclic
9. C_5 through C_6 paraffins, cyclics, and aromatics
10. Determination of benzene, toluene, and xylenes in gasoline
11. Determination of small amounts of diethylbenzene in ethylbenzene

Examples of depropanizer overhead and *n*- and isobutane mixtures have been previously published (*1*).

Wet Gas Mixture. Table II shows the comparison between the results obtained on a wet gas mixture by a fractionating column and the mass spectrometer. The main point of interest in this analysis is that the mass spectrometer was able to detect small quantities of naphthenes.

Table III shows the comparison between two fractionating columns and the mass spectrometer. Perhaps the most interesting point in this example is the fact that the mass spectrometer was able to determine the presence of a small amount of acetone. The presence of acetone was probably caused by a contaminated sample bottle. If the mass spectrometer analysis is corrected for the presence of acetone, the maximum discrepancy between these corrected results and the fractionating column results is only 0.4 mole per cent.

These two examples illustrate the power of the mass spectrometer method to detect unexpected substances in a sample, even when these substances have the same molecular weight as other components in the mixture—the molecular weight of acetone is the same as that of *n*- and isobutanes.

Table V. Typical Feed Stocks

	Catalytic polymerization			Acid alkylation		
	Manometer	Mass spectrometer *Mole per cent*	Difference	Manometer	Mass spectrometer *Mole per cent*	Difference
Propene	1.0	1.1	+0.1	1.0	1.2	+0.2
Propane	2.1	2.0	−0.1	2.0	2.0	0
Isobutane	9.8	9.2	−0.6	14.8	14.9	+0.1
Isobutene	13.7	12.7	−1.0	0	0	0
Butene-1	15.0	16.5	+1.5	10.0	10.6	+0.6
Butene-2	14.0	14.1	+0.1	5.0	4.6	−0.4
n-Butane	42.4	42.4	0	64.2	63.8	−0.4
Isopentane	0.5	0.5	0	1.0	0.9	−0.1
n-Pentane	0.5	0.6	+0.1	0.5	0.5	0
Pentenes	1.0	0.9	−0.1	1.5	1.5	0

	Time required for analysis	
	Complete direct method *Min.*	Comparison (control) method *Min.*
Mass spectrometer (instrument time)		
Mixture	25	25
Prorated calibration	10	3.5
Taking data from record (technician's time)		
Mixture	25	15
Prorated calibration	10	1.5
Computing	140	75
Total man-hours	3.5	2

C₁ to C₄ Paraffin-Olefin Mixture. Table IV shows the results of the analysis of a synthetic mixture containing C_1 to C_4 paraffins and olefins. In the column labeled "Manometer" is shown the composition of the mixture in mole per cent as synthesized with the aid of a mercury manometer. The next column shows the results of the analysis obtained with the mass spectrometer, while the last column shows the discrepancy between the manometer and mass spectrometer determinations.

The example in Table IV was selected to illustrate what is probably the authors' average accuracy on this type of mixture. In general, the total butenes and separate paraffins are determined to within the same accuracy. However, the separation of the butenes, and particularly the separation of butene-1 and butene-2, is less accurate because of the similarity of their mass spectra. Average errors in determining the separate butenes are about three times the average errors in determining the separate paraffins.

If small amounts of C_5 (say 1 per cent) are present in a mixture such as shown, the times are increased by about 15 minutes. The probable error in separating the butenes is increased about 0.5 per cent.

If larger amounts of C_5 paraffins and olefins are present, the accuracy of the determination of the total butenes is not affected. The determination of the accuracy with which the butenes can be separated in the presence of large amounts of C_5 is at present awaiting additional data.

C₃ to C₅ Paraffin-Olefin Mixture. Table V shows the results obtained on two synthetic mixtures with compositions similar, respectively, to polymerization plant and alkylation plant feed streams. These mixtures contain a small amount of C_5 paraffins and olefins. The results show that good accuracy can be obtained on the separate determination of the C_4's, when as much as 3 per cent of the C_5's is present. The effect on accuracy and analysis time of the presence of larger amounts of pentenes is at present awaiting additional data, as in the previous example.

C₄ Paraffin-Olefin-Diolefin Mixture. Table VI gives analyses of three synthetic C_4 mixtures containing butadiene in addition to paraffins and olefins. In each case the composition as obtained in the manometer synthesis is compared with the mass spectrometer analysis. The presence of small or large amounts of butadiene in a mixture of this type can be detected with an accuracy of about ±0.3 per cent. When the partial concentration of butadiene is about 50 per cent, the average accuracy of its determination is about ±0.7 per cent. The accuracy of determining the other components of these mixtures is only slightly affected by the presence of butadiene.

The third analysis shows a relatively large error in the separation of butene-1 and butene-2. In spite of the fact that large amounts of butenes are present, this error is larger than that normally encountered. It is included here, so that the poor as well as the typically good results will be illustrated.

C₅ Paraffin-Olefin-Diolefin and Cyclic Mixture. Table VII compares three mass spectrometer analyses obtained on three consecutive days on a synthetic C_5 mixture containing paraffins, olefins, isoprene, and cyclopentane, with the composition reported from the synthesis. This mixture did not contain all the C_5 olefins or diolefins, simply because the pure components were not available. However, the mixture could still be analyzed if all components normally encountered in such a mixture were present, although the accuracy in separating the individual olefins would probably not be so good as is shown in this table. "Synthetic" rather than "Manometer" is used in the tables whenever the sample was prepared by a laboratory other than the Consolidated Engineering Corporation.

C₆ Paraffin Mixture. Table VIII shows two analyses of synthetic mixtures containing C_6 paraffins. In each case, the analysis was made, on two different days to see how well the mass spectrometer results would reproduce on this type of mixture.

C₅–C₆ Paraffin-Cyclic and Aromatics. Table IX shows

Table VI. Synthetic C_4 Paraffin-Olefin-Diolefin Mixtures

| | Mixture 1 | | | Mixture 2 | | | Mixture 3 | | |
| | Manometer | Mass spectrometer | Difference | Manometer | Mass spectrometer | Difference | Manometer | Mass spectrometer | Difference |
		Mole per cent			*Mole per cent*			*Mole per cent*	
Propene	0.9	0.4	−0.5	0	0	0	0	0	0
Butadiene	66.6	67.2	+0.6	92.6	92.5	−0.1	3.8	3.8	0
Isobutane	0	0.6	+0.6	0	0	0	0	0	0
Isobutene	0.9	1.0	+0.1	0	0.2	+0.2	30.4	28.9	−1.5
Butene-1	7.1	8.1	+1.0	1.9	2.0	+0.1	31.4	29.2	−2.2
Butene-2	6.9	5.5	−1.4	5.5	5.3	−0.2	34.4	38.1	+3.7
n-Butane	17.6	17.2	−0.4	0	0	0	0	0	....

| | Time required for analysis | |
	Complete direct method *Min.*	Comparison (control) method *Min.*
Mass spectrometer (instrument time)		
Mixture	20	15
Prorated calibration	7	5
Taking data from record (technician's time)		
Mixture	20	15
Prorated calibration	7	5
Computing	95	50
Total man-hours	2.5	1.5
Butadiene only, min.	35	....

Table VII. C_5 Paraffin-Olefin-Isoprene-Cyclopentane Mixture

| | | Mass spectrometer | | |
| | Synthetic | 5-6-42 | 5-7-42 | 5-8-42 |
		Mole per cent		
Isopentane	11.3	10.2	10.4	10.5
Pentene-1	28.8	31.4	25.7	31.6
Isoprene	19.0	19.7	20.3	20.2
n-Pentane	11.9	10.5	11.0	10.4
2-Methylbutene-2	19.3	19.0	20.6	18.7
Cyclopentane	9.7	9.2	12.1	8.6

	Time required for analysis by complete direct method *Min.*
Mass spectrometer (instrument time)	
Mixture	25
Prorated calibration	28
Taking data from record (technician's time)	
Mixture	25
Prorated calibration	28
Computing	195
Total man-hours	5

Table VIII. C_6 Paraffin Synthetics

(Two C_6 mixtures were analyzed by the mass spectrometer method on two successive days. Composition from synthesis is given for comparison.)

| | | Mass spectrometer | | | Mass spectrometer | |
| | Synthetic | 6-25 | 6-26 | Synthetic | 6-25 | 6-26 |
		Mole per cent			*Mole per cent*	
2,2-Dimethylbutane	3.82	3.5	3.5	26.40	26.1	25.9
Cyclopentane	0	0	0	0	0	0
2,3-Dimethylbutane	0	0	0.3	0	0.5	0.2
2-Methylpentane	12.03	12.5	12.3	36.24	35.7	36.4
3-Methylpentane	6.16	6.8	7.3	20.26	20.8	20.6
n-Hexane	77.99	77.2	76.6	17.10	16.9	16.9

	Time required for analysis by complete direct method *Min.*
Mass spectrometer (instrument time)	
Mixture	25
Prorated calibration	12
Taking data from record (technician's time)	
Mixture	25
Prorated calibration	12
Computing	120
Total man-hours	3.25

the analysis of a mixture containing C_5 and C_6 paraffins, cyclics, and aromatics. Since this mixture was not a synthetic, the table shows the comparison between the results obtained with a mass spectrometer and by a combination of refractive index readings and temperature readings taken on a large number of small fractions obtained from a 100-plate fractionating column. The two analyses are in fairly good agreement. Results of the determination of benzene by ultraviolet absorption are also shown.

The time required by the mass spectrometer in this case is extremely short as compared with the fractionating time which in this case was 10 days, 24 hours a day. Even in the case where the unknown mixture contains C_7's in addition to C_5 and C_6, a preliminary relatively rapid fractionation can be made to cut out the C_7's, and the total time for preliminary fractionation and the mass spectrometer analysis is still materially shorter than the time required by fractionation alone.

Gasoline and Aromatics. Table X shows the analysis for the benzene, toluene, and xylene in a synthetic gasoline. If ethylbenzene had also been present in this mixture, its amount could have been separately determined. The major portion of the analysis time given was consumed in preparation of the sample preliminary to its introduction into the mass spectrometer. This preparation is done on an auxiliary apparatus and so does not consume any spectrometer time. The time for the actual spectrometer run and prorated calibration is only 28 minutes. Combined computing and instrument time is thus only 45 minutes. Sample preparation time for this type of mixture can be materially reduced as the bottling method is improved.

Determination of Diethylbenzene in Ethylbenzene. Table XI shows the results of eight tests on the determination of small quantities of diethylbenzene in ethylbenzene. This example is representative of a number of analyses which can be made with the mass spectrometer to check the purity of a particular compound. In this case the minimum amount of impurity that could be detected without making special adjustments is about 0.02 percent. If a small amount (down to 0.02 per cent) of isopropylbenzene were also present in these mixtures, its quantity could be determined.

MASS SPECTROMETER FOR COMMERCIAL USE

In order to make the mass spectrometer useful for routine analysis it was necessary to

Table IX. Mixture of C_5 and C_6 Paraffins, Cyclics, and Aromatics

Hydrocarbon	Fractionation (100-plate) column	Ultraviolet absorption	Mass spectrometer	Difference	
			Mole per cent		
Pentanes	3.8	....	3.7	−1.0	
2,2-Dimethylbutane	14.0	....	15.0	+1.0	
Cyclopentane	1.6	....	2.0	+0.4	
2,3-Dimethylbutane	9.4	....	7.7	−1.7	
2-Methylpentane	33.0	....	32.2	−0.8	
3-Methylpentane	13.9	....	14.6	+0.7	
n-Hexane	12.8	....	13.5	+0.7	
Methylcyclopentane	8.7	....	9.3	+0.6	
Benzene	2.9	2.0	2.1	−0.8	+0.1

	Time required for analysis by complete direct method
	Min.
Mass spectrometer (instrument time)	
Mixture	25
Prorated calibration	23
Taking data from record (technician's time)	
Mixture	25
Prorated calibration	23
Computing	160
Total man-hours	4.25

Table X. Synthetic Gasoline and Aromatics

(Composition obtained from the synthesis is compared with the mass spectrometer analysis for two mixtures containing aromatics and a large number of other hydrocarbons.)

	Synthetic	Mass spectrometer	Difference	Synthetic	Mass spectrometer	Difference
		Mole per cent			*Mole per cent*	
Nonaromatics[a]	49.1	49.1	0	52.9	55.0	+2.1
Benzene	12.7	12.5	−0.2	5.7	5.7	0
Toluene	37.9	38.4	+0.5	21.0	21.8	+0.8
Xylene	....	....	....	20.3	17.5	−2.8

	Time required for analysis by partial direct method
	Min.
Mass spectrometer (instrument time)	
Mixture	25
Prorated calibration	3
Taking data from record (technician's time)	
Mixture	10
Prorated calibration	1
Computing	10
Sample preparation	120
Total man-hours	2 hours 50 minutes

[a] Nonaromatics: n-hexane, 2-methylpentane, 2,3-dimethylbutane, methylcyclopentane, cyclohexane, n-heptane, methylcyclohexane, "isooctane," and "isooctene."

Table XI. Small Amounts of Diethylbenzene in Nearly Pure Ethylbenzene

(Determination of diethylbenzene in ethylbenzene in eight different concentrations. Impurity added in synthesis is compared to mass spectrometer analysis.)

Mixture	Synthetic	Mass spectrometer	Difference
	Mole per cent		
1	0.036	0.033	−0.003
2	0.071	0.066	−0.005
3	0.235	0.22	−0.02
4	0.479	0.44	−0.04
5	0.789	0.76	−0.03
6	2.41	2.2	−0.2
7	4.02	3.6	−0.4
8	7.97	8.0	0

	Time required for analysis by partial direct method
	Min.
Mass spectrometer (instrument time)	
Mixture	25
Prorated calibration	3
Taking data from record (technician's time)	
Mixture	10
Prorated calibration	1
Computing	10
Sample preparation	120
Total man-hours	2 hours 50 minutes

develop an instrument which was reliable, reasonably simple to operate, and at the same time capable of obtaining accurate mass spectra of gas and liquid mixtures.

The accuracy of the Consolidated Engineering Corporation mass spectrometer is demonstrated by the results of the analyses given above.

The commercial apparatus has been designed conservatively, so that it will require a minimum of service. In order to test the reliability of operation, the mass spectrometer in this laboratory was operated in a routine manner over a period of 4 months, during which time from 15 to 20 runs a day were made. This number of runs was obtained in an 8-hour day. However, in order to simulate 24-hour-a-day operation, all voltages were left on at all times during the 4 months with the exception of Sundays. During this period, in addition to many experimental runs, over 200 analyses were carried out on synthetic mixtures made in the laboratory and on mixtures sent in by oil companies which were interested in the possibilities of this new method of analysis.

Figure 3 shows a commercial model of the Consolidated Engineering Corporation mass spectrometer. In the center background is the electromagnet which produces the uniform magnetic field in which the mass spectrometer proper is placed. Immediately to the left of this magnet is the sample inlet system and evacuation apparatus. In the right foreground are the operator's table and the cabinet which houses the automatic recording apparatus and power packs. These power packs supply closely regulated voltages to the spectrometer and electric power to the magnet.

The routine operation of the apparatus for analyzing gas and liquid mixtures is relatively simple. The main operations are (1) the manipulating of the proper stopcocks to introduce the sample into the apparatus and (2) the throwing of the proper switches to initiate the automatic recording of the data.

The data are recorded on photographic paper by a recording oscillograph.

In order to increase the reliability of the instrument and further to simplify its operation, several automatic protective devices have been installed to avoid damage to the instrument in an emergency. To a large extent, these protective features prevent the logs of operating time and allow relatively unskilled operators to take data.

The handling of sampling and sample containers in a refinery may be greatly simplified by using the mass spectrometer method, since the sample required is so minute. Only 0.1 cc. of gas at atmospheric pressure need be introduced into the inlet sample bottle. This extremely small size may be of decided advantage in some laboratory experiments.

In the refinery the size of sample should be picked which is best adapted to easy sampling and handling. Cylinders of about 100-cc. capacity should be very satisfactory.

SUMMARY

The mass spectrometer method has several distinctive features:

1. A large number of samples, in some cases as many as 20, can be analyzed in an 8-hour day with one mass spectrometer.
2. Mixtures containing up to 15 or more components can be analyzed.
3. Results can be computed in such a manner as to indicate whether the analysis is sufficiently accurate, and whether any unexpected components are present.
4. Results are practically independent of the skill and judgment of operators and computers.
5. Only a small sample is required for analysis, 0.1 cc. usually being sufficient.

CONCLUSION

As indicated by the examples given, advances in techniques of analysis using the mass spectrometer have been extremely rapid. As in the case of any new development, the field is constantly expanding. New problems extending the use of the instrument are continually being proposed and solved. Evidence is accumulating which indicates that analyses which are prohibitively time-consuming by other methods will be relatively readily analyzed by means of the mass spectrometer. The examples given in this paper should therefore not be considered as representing the limits of the method.

Figure 3. Commercial model of the consolidated mass spectrometer

ACKNOWLEDGMENT

Cyclopentane used to calibrate the mass spectrometer for examples in this paper was furnished by the A. P. I. Research Group in Pure Hydrocarbons at Ohio State University.

The authors wish to thank the Consolidated Engineering Corporation for permission to publish this material.

LITERATURE CITED

(1) Hoover, Herbert, Jr., and Washburn, Harold, *Calif. Oil World*, **34**, No. 22, 21–2 (Nov., 1941).
(2) Washburn, Harold, Taylor, D. D., Hoskins, E. E., and Langmuir, R. V., *Proc. Calif. Natl. Gas Assoc.*, **15**, No. 11, 7–9 (1940).

Reprinted from *Ind. Eng. Chem., Anal. Ed.* **1943**, *15*, 541–47.

The first papers that described instruments for "flame photometry," which is really a misnomer for flame excitation–atomic emission spectroscopy, appeared in the 1930s—approximately seven decades after Kirchhoff and Bunsen clearly demonstrated the scientific principles involved. These first instruments were striking in their simplicity. Of the early investigators, Barnes et al. departed from the European approaches in one main respect: the use of lower flame temperatures. Barnes et al. reasoned correctly that if fewer sample constituents were excited, the task of locating interference-free lines would be simpler. Although these simple instruments were remarkably successful for the determination of alkali elements, many other investigators turned to higher flame temperatures in order to broaden the elemental coverage of this approach. Eventually, these efforts led to the development of the very high temperature, flamelike, inductively coupled plasmas.

Velmer Fassel
Iowa State University

Flame Photometry
A Rapid Analytical Procedure

R. Bowling Barnes, David Richardson, John W. Berry, and Robert L. Hood
Stamford Research Laboratories, American Cyanamid Company, Stamford, Conn.

A new instrument has been developed to make possible the rapid quantitative determination of the alkali metals (primarily sodium and potassium) in aqueous solution. The principle of operation of the instrument is based upon the quantitative measurement of the characteristic light emitted when a solution of the metal is atomized as a mist into a gas flame. Details of construction and operation are given. An average solution may be analyzed for both the sodium and potassium contents in a few minutes' time with an average accuracy of ± 3% of the amounts of these elements which are present. Several applications of the method are discussed, together with the analytical procedures employed.

The urgent need for extremely rapid and accurate sodium determinations of large numbers of samples led to the development of the flame photometer. With this instrument as many as 150 sodium determinations on water samples low in sodium have been completed in a single hour's time with an average accuracy of about ± 3% of the amount of the sodium present. The flame photometer is simple in construction, its operation requires no excessive training on the part of the analyst, and the samples used require a minimum of preparation. Although originally constructed for performing sodium determinations, its usefulness has been extended to include other alkali metals and various alkaline earths.

A survey of previously used chemical methods, including the zinc uranyl acetate method of Barber and Kolthoff (1), revealed that satisfactory quantitative sodium determinations require a considerable amount of sample preparation and are rather time-consuming. This is particularly true in the cases of samples in which only traces of sodium are present, or which contain other cations in appreciable quantities. Accordingly, a physical approach to this analytical problem was undertaken in the hope that a suitable method could be found which would meet the requirements of analytical speed and accuracy, would necessitate a minimum of sample handling and preparation, and at the same time would not require the use of expensive or complicated equipment. Although many well-known physical methods were tried and tested, only flame photometry was found to satisfy all the above requirements.

Possibly the most widely known spectroscopic phenomenon is the fact that sodium when introduced into a flame emits a characteristic yellow light, its intensity being a function of the amount of sodium present in the flame. This emission by atoms of characteristic radiation when thus excited by high temperatures or by electrical means forms the basis of the analytically important subject of emission spectroscopy. A review of the spectrochemical procedures employed for the determination of sodium indicated that the Lundegårdh method (3, 6–10) has been successfully applied to the analysis of many metallic elements including the alkali metals. In this method, which has recently been modified and called the air-acetylene flame method (2), the sample to be analyzed is put into aqueous solution and sprayed under controlled conditions into an acetylene flame. This is usually accomplished by means of a specially constructed atomizer and burner in which the air and acetylene are carefully regulated to produce constant burning conditions. The light from the Lundegårdh flame enters the slit of a spectrograph and spectrograms are prepared, which in turn may be photometered in order to obtain the intensities of the characteristic spectral lines recorded. After carefully calibrating with solutions of known composition and concentration, it is readily possible to correlate the intensity of a given spectral line, such as the sodium line, of the un-

2855-8/94/0044$08.00/0

known sample with the amount of that particular element present. The use of the high-temperature acetylene flame causes a great many of the metals contained in the sample to emit characteristic radiations. This fact has led most workers to the use of a spectrograph to isolate and measure the intensities of these spectral lines. Although accurate and versatile, this method requires a very considerable amount of manipulation and equipment.

Jansen, Heyes, and Richter (5), by making use of a monochromator to isolate the desired region of the spectrum and a photocell and electrometer to measure the intensity of the lines, successfully determined the alkali metals and alkaline earths in solutions sprayed into a flame, while Heyes (4) later succeeded in applying the method to other metals.

A somewhat simpler technique for the determination of potassium was developed by Schuhknecht (12). The apparatus consists of a Lundegårdh acetylene flame source placed before a box containing a phototube and suitable filters for isolating the red potassium wave lengths. The method of operation is similar to that described below. Apparently the Schuhknecht method has been applied with considerable success in the analysis of samples of soils, fertilizers, and plant materials in Germany. Two instruments based on this principle have been manufactured by German companies (Siemens and Zeiss) and descriptions of both can be found in an article by Schmitt and Breitweiser (11).

Consideration of the flame as a source led to the development of a lower temperature method described below, which is called flame photometry.

FLAME PHOTOMETRY

A very simple instrument has been developed wherein a solution to be analyzed is atomized into the air intake of a flame under controlled conditions and the emitted light characteristic of the element in question is isolated and accurately photometered. If the temperature of the flame is not too high, only a few of the metals which may be present in the sample will be caused by thermal excitation to emit characteristic radiation. In particular, if one uses as a fuel for the flame ordinary illuminating gas or cylinder gas (Pyrofax), only the atoms of the alkali metal and the alkaline earth constituents of the sample will emit light in appreciable amounts. This is particularly important, since it is obvious that the fewer the components of the sample which are excited, the simpler the task of isolating for photometric purposes the light characteristic of any one component.

Figure 1 shows the most prominent wave lengths emitted by various metals when excited by a flame of the type described above. It may be readily demonstrated that any one of these characteristic radiations, if isolated by passage through proper filters, will, upon striking a photosensitive element, give rise to an electrical impulse which is a function of the quantity of the respective metal present in the sample. The above is true provided the burning conditions of the flame and the rate of atomization can be kept constant. The flame photometer is an instrument wherein the above requirements for analysis may be met.

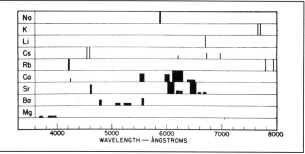

Figure 1. Characteristic flame spectra produced in illuminating gas flame

LABORATORY MODEL FLAME PHOTOMETER

A recent model flame photometer for routine laboratory use is shown in Figure 2. The schematic drawing of this instrument, Figure 3, indicates the arrangement of its various parts.

In successful flame photometry the chief prerequisite is the establishment of sufficiently stable flame conditions, so that solutions of various concentrations may be successively brought to excitation in the flame without such conditions changing sufficiently to influence the quantitative light output being measured.

R and R^2 are pressure regulators controlled by knobs K and K^2, for the atomizer air and the burner gas, respectively, the pressures used being indicated by G and G^2. The operator may readily establish identical conditions each time the photometer is placed in operation. The flow of air through capillary E past the tip of the sample tube, F, atomizes the sample solution which is introduced through the funnel, W. The larger droplets of the solution condense on the inner walls of the glass atomizer bulb, A, and run off to the drain D, while a fine mist of the sample is carried by the air stream to the mixing chamber, C, at the base of the Fisher burner where it is thoroughly mixed with the gas entering the burner. Natural draft created by the burner prevents the escape of any of the mist through the loosely fitting collar of C; indeed, very considerable air in excess of that supplied through the atomizer is drawn in around this collar to support the combustion of the flame.

Whereas the atomizer tips may be made of glass as indicated in Figure 3, a preferred atomizer may readily be made using stainless steel hypodermic needles (see Figure 4). This type of atomizing unit is introduced into a chamber similar to the one in Figure 3, but with a large opening in the top of the bulb into which the unit fits by means of a rubber stopper.

Burner B, Figure 3, is supplied with regulated gas, its flow being controlled by the needle valve, N. In burning certain types of bottled gases, this needle valve has been replaced by a fixed orifice.

Light from the flame passes to the optical system of the photometer through a window, H, in the chimney, Q, so dimensioned as to exclude all light from the grid of the burner and from the unsteady edges and top of the flame. The light then passes horizontally through lens L^2, is reflected downwards by the 45° mirror, M, through lens L and through the filter system housed in container V to the coolest part of the instrument, where it falls upon the surface of the barrier layer photocell, P. Wires, not shown, connect the photocell through the toggle switch, T, to coarse and fine controlling rheostats and finally to the galvanometer jacks, J. The rheostats, S and S^2, enable the operator to control the sensitivity of the instrument by shunting the galvanometer, while retaining the desired damping resistance.

While any galvanometer may be used which has sufficient sensitivity, and a critical damping resistance compatible with the resistance of the photocell, the authors prefer one of the indicating box type as shown in Figure 2.

ANALYSIS FOR SODIUM OR POTASSIUM

In using the flame photometer the procedure is extremely simple. A set of concentrated stock solutions containing, say 1000 p.p.m. of sodium and potassium, respectively, are prepared and stored in glass-stoppered No-Sol-Vit bottles. These stock solutions may be preserved for several months, and from them dilutions to any desired concentration may be made very readily. Once satisfactory conditions have been established for the gas and the air pressures, the diluted standards may be used to calibrate the instrument for a variety of ranges for each element. As long as the flame and atomizer conditions are kept constant, it is usually only necessary to check the upper and lower ends of the proper range calibration curve when placing the instrument in operation.

The procedure for calibrating the instrument for a given metal is as follows:

With the flame burning properly and the correct filters in place, distilled metal-free water is introduced into the atomizer, and the galvanometer is set to read zero.

After the zero reading is set, the sensitivity can be adjusted by the introduction of the standard solution selected to be the

Figure 2. Laboratory model flame photometer

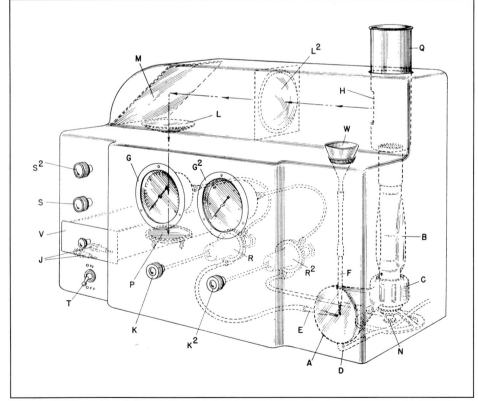

Figure 3. Diagram of flame photometer

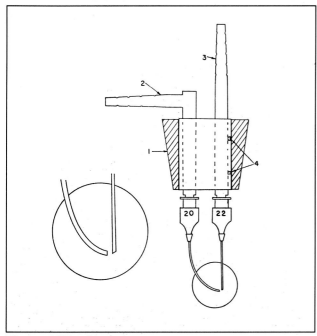

Figure 4. Metal atomizer

Stainless steel hypodermic needles, 20- and 22-gage, used as tips
1. Rubber stopper
2. Air feed
3. Liquid feed
4. Adjusting screws

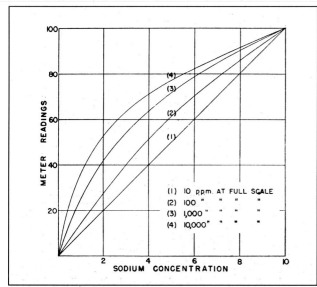

Figure 5. Typical calibration curves for flame photometer

upper end of the desired concentration range. To do this the control rheostats are adjusted so that the galvanometer reads 100. Several standards of lower concentration are introduced in turn and the respective readings noted. These readings, plotted as ordinates versus the concentrations as abscissas, result in a satisfactory calibration curve. Once the proper calibration curve has been established, it is unnecessary to repeat this entire procedure. It is only necessary, at the beginning of each set of readings, to adjust the instrument so

that demineralized water reads 0 and the upper standard chosen reads full scale or 100. For exact work these two readings are checked after the analysis of each three or four unknowns.

For practical reasons it is usually desirable to dilute most unknowns to 100 p.p.m., or less, of the metal; the reason for this is evident from a study of Figure 5, which shows four different calibration curves for a flame photometer of the type shown in Figure 2.

Curve 1, in which 10 p.p.m. are set to read 100 on the galvanometer, is so nearly linear that results may be interpolated directly from the meter readings. Curve 2, 100 p.p.m. full scale, is not so nearly linear, and accordingly the use of a calibration curve is recommended. Curves 3 and 4, with 1000 and 10,000 p.p.m. at full scale, are very appreciably curved and become very low in slope toward the upper end of the range. Probably self-absorption—i.e., absorption of the sodium light by the cooler atoms around the periphery of the flame—plays a role in this dropping off of sensitivity. In general, it will be most convenient to dilute the average sample to a range below 100 p.p.m., so curve 2 may be used. By such dilutions, the dangers of excessive contamination and corrosion of the burner are avoided and smaller samples of the original solutions will be required.

When using such highly diluted samples, care must be exercised to avoid contamination of the sample. Such solutions for example, cannot be kept safely for any length of time in ordinary glass containers, because of the likelihood of contamination from the glass. In transferring these highly diluted solutions, care must be taken not to touch the inside of any of the glassware, stoppers, etc. An interesting demonstration may be made by filling the atomizer funnel with metal-free water and then dipping a finger tip into this water for about one second. A reading of 10 to 20 p.p.m. of sodium is usually obtained.

The sensitivity of the instrument shown in Figure 2 is such that 10 p.p.m. of sodium (or 50 p.p.m. of potassium) may easily be set to read full scale on the galvanometer (100 divisions). The normal rate of atomization of a sample into the instrument is in the order of 5 to 10 ml. per minute. It is thus possible to determine the metal content of a 2- or 3-ml. sample. In general, however, it is preferable to have on hand a sample of approximately 10 ml. for assurance in determining the true rest point of the galvanometer.

ACCURACY OF THE METHOD

In order to determine the average accuracy of the flame photometer, two series of fifty solutions each of sodium and of potassium, respectively, were prepared and submitted to a second staff member for analysis as "unknowns". In preparing these standard unknowns, pure sodium and potassium carbonates were obtained, weighed accurately into volumetric flasks, and made slightly acid by the addition of various acids, thus forming samples of nitrates, phosphates, sulfates, acetates, and carbonates in which the amounts of the various cations were accurately known. Since the purpose of this experiment was to determine the average accuracy of the method under conditions duplicating those of actual daily analyses, the same procedure was followed in each case. Each sample was atomized once in the instrument and one galvanometer reading taken, the operator then proceeding to the next sample. After completing each series of unknowns, the above measurements were repeated. As standard solutions for calibration in these experiments, pure sodium and potassium chloride solutions were used, made up from analytical grade reagents.

Table I is typical of the manner in which the results were recorded and averaged. Ten solutions of sodium carbonate of known concentrations are listed in column I. Column II represents the analysis obtained the first time the solutions were atomized in the instrument, and column IV gives similar data for the repeat analysis. Columns III and V show the errors for the single determinations of analyses 1 and 2, in terms of per cent of the element present. Column VI shows the average error obtained when analyses 1 and 2 were averaged. The average errors (disregarding sign) for the first and second set of ten single determinations, as well as the average error of the duplicate determinations, are shown at the bottom of the table.

Table II represents the summary of ten tables such as Table I. The ten salts used to make the analytical solutions are shown in column I. Average errors for the first and second set of ten single analyses are shown in columns II and III, and the average errors for each group of ten solutions analyzed in duplicate are shown in column IV.

As may be seen, the average error made in 200 single determinations is ±3.0% of the amount of metal present, whereas that for samples determined in duplicate is ±2.8%. Such results compare favorably with those obtained by standard spectrographic procedures and also by chemical procedures where dilute solutions of sodium or potassium are being analyzed.

In certain test cases individual samples have been analyzed repeatedly and the results averaged to obtain even more accurate results. Since the method is so rapid, this procedure may readily be accomplished in a few minutes. Thus, by making a series of ten separate readings on a given sample, checking the instrument calibration between each determination and averaging these results, the average error may be reduced to approximately 1%. In twenty such cases the average error was ±1.4%. All measurements and manipulations for a given sample were completed within 15 minutes.

The precision of the method as expressed by the average deviation of a single measurement from the arithmetical mean of ten single determinations is 2.8% of the mean. This value was also obtained from the analysis of twenty different solutions, ten of sodium and ten of potassium, each solution being analyzed ten times.

ANALYSIS FOR CALCIUM

It has long been established that the flame spectra of a number of elements are sufficiently unique to enable one to use them for analytical purposes.

The extension, however, of the above-described method to the analysis for metals other than sodium and potassium, while theoretically possible, is fraught with certain experimental difficulties. In general, analyses by flame spectra have

Table I. Sodium Carbonate

Solution	I Sodium added P.p.m	II Analysis 1 P.p.m.	III Error %	IV Analysis 2 P.p.m.	V Error %	VI Average error %
1	8.73	8.3	−4.93	8.7	−0.34	−2.63
2	13.1	12.9	−1.53	13.3	+1.53	−0.00
3	21.8	21.2	−2.75	21.5	−1.38	−2.07
4	26.2	26.8	+2.29	26.8	+2.29	+2.29
5	34.9	34.2	−2.00	35.1	+0.57	−0.72
6	39.3	40.4	+2.80	40.4	+2.80	+2.80
7	48.0	47.7	−0.63	48.8	+1.67	+0.52
8	56.8	57.8	+1.76	57.5	+1.41	+1.58
9	61.0	60.2	−1.31	63.0	+3.28	+0.94
10	78.6	80.0	+1.78	80.0	+1.78	+1.78
Average error		2.18			2.01	1.53

Table II. Summary

I Salt	II Average error Analysis 1 %	III Average error Analysis 2 %	IV Average error Analyses 1 and 2 %
Sodium acetate	4.35	3.76	4.07
Sodium carbonate	2.18	2.01	1.53
Sodium nitrate	4.38	5.83	5.11
Sodium phosphate	2.34	1.89	1.92
Sodium sulfate	1.61	1.77	1.47
Potassium acetate	3.85	4.01	3.65
Potassium carbonate	2.42	2.15	1.44
Potassium nitrate	3.41	2.69	3.05
Potassium phosphate	3.17	3.95	3.55
Potassium sulfate	2.52	1.74	2.01
Av.	3.02	2.98	Av. duplicate 2.78

been performed either as qualitative flame tests or as quantitative analyses in which some type of spectrometer was employed. In the case of the former, the human eye alone is used as the detector and since the eye does not resolve mixed colors into their component wave lengths, a wide range of flame shades must be recognized. Many of these shades can be correlated with the simultaneous presence of two or more elements in the sample introduced into the flame. Through the use of a spectrometer of sufficiently high dispersion, the light from flames can readily be resolved into characteristic lines and/or band systems which, when photometered, yield quantitative information regarding the composition of the sample.

In the design of the present flame photometer only a simple filter system has been provided; accordingly, the use of this instrument is limited to those samples containing elements whose flame spectra can be isolated by available filters. In many cases, however, examination of the flame with a small hand spectroscope provides additional information of a qualitative nature.

A considerable amount of work has been done on the application of this method to the determination of calcium. Already the flame photometer has performed well in the analysis of this element in many cases where sodium did not appear in high concentrations. For example, in samples in which only calcium and magnesium are present, the calcium content may be determined readily, as shown in Table III. Unfortunately, in the case of one of the most important clinical analyses—namely, the determination of blood and urine calciums—the situation is not entirely favorable. In such samples sodium is approximately 25 times as prevalent as calcium, and with the best filter system so far found a correction of some 30 to 40% of the calcium reading has been necessary to account for the fraction of the sodium light passing the filters. It will probably be impossible to eliminate the necessity for such a correction by means of simple filters alone, for the sodium flame does actually emit some light in the spectral region of the calcium bands.

Another difficulty encountered in performing calcium analyses arises from the very origin of the calcium spectrum. In contrast to the alkali metals, which emit line spectra that are relatively easy to isolate by filters, calcium emission is largely a band system. Since this band system originates from the thermal excitation of calcium-containing molecules (probably the oxide or chloride), it might be expected that the anion associated with the calcium would exert some influence upon the intensity and character of the emission. Experimentally, this is the case, as may be seen from the calcium calibration curves of Figure 6. From these curves it is clear that the standards used in the flame photometry of calcium must be prepared from the same compound or compounds in the same ratio as that contained in the unknowns being analyzed. Otherwise, chemical pretreatment of all unknowns is necessary in order to convert all of the calcium to the same form.

CONSTRUCTION OF THE FLAME PHOTOMETER

The previous discussion has been limited to the particular flame photometer shown in Figure 2. Obviously the principle of the instrument lends itself to a wide variety of designs, depending very largely upon the specific use to which the photometer is to be put. As a matter of fact, several radically different instruments have been in use for some time. The simplest of these, designed for field work, is portable and makes use of bottled gas and a small motor-driven air compressor.

The photometer itself consists essentially of six parts—the pressure regulators for gas and air, the atomizer, the burner, the optical system, the photosensitive detector, and the instrument for indicating or recording the output of the detector. In the following sections the principal requirements for each of these parts are discussed.

Pressure Regulators. In most laboratories the gas pressure is found to vary to such an extent that adequate regulation is a necessity for accurate operation of the flame photometer. A few companies build pressure regulators which suitably regulate gas pressures at the low pressures generally prevalent in municipal supplies. In certain localities, however, the gas pressure will be found to be inadequate to provide a flame of proper burning characteristics. In such cases, auxiliary booster pumps or other equipment may be necessary.

Always possible is the use of the so-called "bottled gases" typified by one sold under the trade name of Pyrofax. Such a gas supply has been found ideal for the flame photometer since the gas may be supplied to the burner at any desired

Table III. Determination of Calcium

Solution	Calcium added P.p.m.	Magnesium added P.p.m	Calcium determined P.p.m
1	100 (used as standard)	0	(100)
2	100	100	100
3	100	500	100
4	100	1,000	95[a]
5	100	10,000	76[a]
6	50	0	50
7	50	500	50
8	50	1,000	47[a]
9	50	10,000	38[a]

[a] In contrast to what might be expected, that the presence of foreign ions might give rise to positive errors, the presence of such ions actually reduces the overall amount of emitted light. This inhibitory effect seems to be general and must certainly be considered in flame photometry. By using for calibration standard solutions containing approximately the same components as the unknowns, this effect may be minimized.

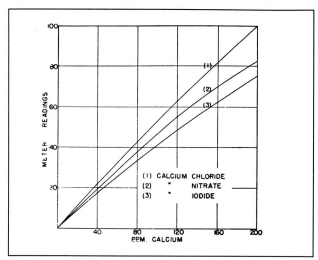

Figure 6. Calibration curves for three calcium compounds

Calcium chloride (200 p.p.m.) used to standardize instrument at full scale

constant pressure. Suitable gages must be provided to indicate the pressures actually prevailing when the instrument is in operation, so that proper adjustment may be made each time the photometer is used. A 15-inch water pressure gage for city gases, a 5-pound gage for "bottled gas", and a 25-pound gage for the air supply are satisfactory.

Burners. The main requirement of the burner is that when supplied with gas and air at constant pressures it shall produce a steady flame.

The temperature of the flame must be high enough to excite the desired metals, primarily the alkali metals, to emit light, but insufficient to excite those which would produce interferences which could not be eliminated by the optical filter system. The Meker-type Fisher burner has been found very satisfactory.

In certain models of the flame photometer, the Fisher blast lamp, which is provided with a pipe connection for introducing air, has proved most advantageous. In this type of flame photometer, all the air supplied to the burner passes through the atomizer. The main advantages of this arrangement are twofold. First, since all the air for the burner may be filtered, the uncontrolled introduction of sodium or potassium in the form of laboratory dust may be avoided. Second, the steadiness of the flame is less influenced by air drafts in the laboratory. One serious drawback, however, does exist in the fact that the pressure through the atomizer must be controlled to a greater extent than in the case of the arrangement wherein the burner draws practically all its air in the normal manner. So far, the use of this type of burner has been reserved for industrial applications where the air has been found to be severely contaminated by dust.

A chimney surrounding the flame serves the purposes of protecting the flame from air drafts, providing a simple means for aperturing the flame so that the unsteady light from the grid of the burner and the edges and tail of the flame will not reach the photocell, and eliminating stray light of the laboratory from the optical system. A glass window over the chimney aperture has, on occasion, been found very useful. Sufficient ventilation in the neighborhood of the flame, in the form of grid openings or louvres, is necessary in order to avoid overheating.

Atomizers. The function of the atomizer is to introduce exceedingly fine droplets of the sample in aqueous solution into the air supply for the burner. Since the sample solution may be either acid or alkaline or contain strong oxidizing or reducing agents, the problem of corrosion is important. Atomizers of glass or stainless steel have been found to give very satisfactory service, and it is probable that atomizers made of hard rubber or a suitable resin would also perform well. Two general types of atomizers have been used in the instruments described in this paper.

The first type is made with two capillary tubes sealed into the walls of a glass flask in such a way that their bores are perpendicular to each other. The blast of air from the tip of one capillary causes a suction in the other sufficient to draw the sample through it. The sample thus entering the air stream is broken into fine droplets, the larger of which collect on the walls of the flask and flow into the drain. The smaller droplets, comprising a virtual fog, are carried by the air stream into the burner where they are thoroughly mixed with the normal burner gases. A second type atomizer may be made as shown in Figure 4, using two stainless steel hypodermic needles. This atomizer has the important advantage of allowing for removal of tips for cleaning or changing to different bore sizes, depending upon the viscosity of the sample to be atomized.

A third type, shown in Figure 7, is constructed of two concentric glass tubes. The inner tube, through which the air passes, is constricted to form an orifice about 1 mm. in diameter, while that of the orifice in the outer tube is about 2 mm. The sample is introduced into the annular space between the two tubes through a side arm. These two tubes may be sealed together to form a single unit or joined by means of an interchangeable ground-glass seal. The latter method allows easy cleaning and adjusting of the tips.

As a result of the suction at the orifice of the sample tube, it is possible to introduce the sample in two ways. Instead of pouring the sample into a funnel as shown in Figure 3, it may be sucked upwards from a beaker through a bent capillary sample tube. The suction is sufficient to raise the sample several inches with ease.

Optical System. The obvious function of the optical system is to collect the light from the steadiest part of the flame, render it monochromatic, and then to focus it onto the photosensitive surface. Although any type of lens may be used, such as a flask filled with water, Fresnel lenses of heat-resisting

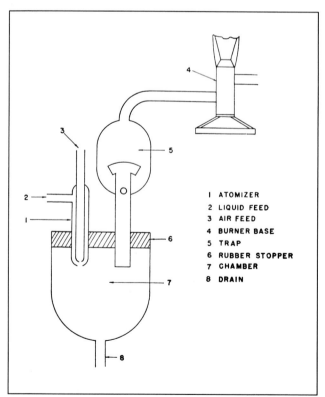

Figure 7. Concentric tube type of atomizing unit
In using this type of atomizer, all the air for the flame passes through the atomizer

glass having high numerical apertures have been found most satisfactory. High optical quality is not essential.

To isolate the desired light, combinations of simple filters have been found satisfactory, the particular filters used depending upon the metal being measured, and the spectral response curve of the detecting device. Table IV lists some of the filter combinations used so far. As the use of the flame photometer is extended, other more refined devices for producing monochromatic light may be required.

Photosensitive Detectors. In the flame photometer, any photoelectric device may be used as a detector, provided it has a response in the part of the spectrum to be used, and a sensitivity high enough for the particular task at hand. Photosensitive surfaces are commercially available which possess spectral sensitivities which vary over wide limits. In certain specific applications a choice of surface may be made, so that the detector is highly sensitive to the characteristic radiation emitted by only one of the elements which may be contained in the sample.

A barrier layer photocell was chosen for the photometer shown in Figure 3 because of its simplicity and its favorable performance characteristics. The sensitivity of this type of photocell is low, as compared with phototubes, and its output must be measured with a galvanometer. It has, however, many advantages for this work: it requires no external electrical power supply; its broad spectral response covers adequately the various wave lengths encountered in the usual sample; it is relatively free from annoying drift and fatigue; and its output for a given light input is steady and reproducible. Its chief disadvantage, its high temperature coefficient, may be minimized by placing it at a cool part of the photometer.

Phototubes of many types may be used if provided with suitable power supplies and amplifiers. Electron multiplier tubes, such as the R.C.A. Type 931-A have been used very successfully in flame photometers. These tubes are extremely sensitive to sodium light and very much less sensitive to that from lithium and potassium, and have the advantage that their output is sufficient to allow the use of rugged types of microammeters or recorders. Their high sensitivity makes possible accurate analytical measurements on solutions containing as little as 0.1 p.p.m. of sodium. One of the main disadvantages of this tube is the fact that a well-regulated power supply is required. Very satisfactory results have been ob-tained using twenty 45-volt B batteries in series, so wired as to provide 90 volts for each stage of the multiplier. The output of this type of tube at constant light has been found to fall off for a period of several hours after the power has been applied; accordingly it is not satisfactory for photometers designed for laboratory analyses, which must be ready at all times for immediate use. Once these tubes have reached equilibrium, their output, is very steady, and their performance entirely satisfactory. A photometer so equipped may be left turned on for very long periods of time, and may thus be used to advantage in installations where analyses are required constantly, three shifts per day.

Electrical Instruments. The particular instrument selected to indicate or record the output of the photoelectric detector will depend in a large measure on the type of detector used and the sensitivity and precision desired.

A multirange meter is convenient if samples of widely different concentrations are encountered, as is also a shunt for varying the individual ranges of the instrument scale. An appreciable period or sluggish response in the meter is advantageous, since it tends to smooth out any slight flicker of the flame. The galvanometer used in Figure 2 is a G.E. Model 32C-245-G9 having a sensitivity of 0.001 microampere per scale division and a period of 4 seconds.

The output of an electron multiplier for solutions containing up to 25 p.p.m. of sodium has been found to be between 10 to 100 microamperes in the various types of flame photometers so far used. This high output permits the use of a rugged pointer-type microammeter. If a potentiometric recorder is used, it is important that it be designed for high damping resistance, so that a certain degree of overdamping is obtained when measurements are made across a fairly high resistance. A micromax with a full-scale range of 100 millivolts, designed for an external circuit resistance of 50,000 ohms, has been used very satisfactorily. With this recorder connected to measure a potential across a 10,000-ohm resistor through which the electron multiplier current is flowing, it is practical to record sodium levels in flowing samples in the range from 5 p.p.m. full scale up to 100 p.p.m. If the 10,000-ohm resistor is variable, the range of the recorder may be varied at will. Another satisfactory method of adjusting the sensitivity range of the photometer when using an electron multiplier is, of course, to vary the voltage applied per stage of the tube until the desired range is obtained.

APPLICATIONS OF FLAME PHOTOMETRY

Generally speaking, the flame photometer may be used wherever the analysis of aqueous samples for their sodium, lithium, potassium, or calcium contents is required. It should be emphasized that this method, as described, is in reality a method for rapidly determining the amount of metal ions in solutions. Before applying it to the analysis of a wide variety of materials a certain amount of sample preparation, which on occasion may be tedious and time-consuming, is still essential. In favor of the method, however, is the fact that, since the number of sources of possible interferences is small, procedures for preparing samples may be employed which in gen-

---

Table IV. Combinations of Corning Glass Filters for Flame Photometry

Element	Combination
Sodium	3482 and 9780
	(H. R. lantern shade yellow + colorimeter blue green)
Potassium	5850 and 2404
	(Blue-purple ultra + H. R. dark red)
Calcium	5120 (2 thicknesses)
	(Didymium) + liquid filter of concd. HCl + $CuCl_2$
Lithium	2404 + liquid filter of concd. HCl + $CuCl_2$

eral are simpler than those normally used. Specifically, it must be recognized that before applying the method to an entirely new material, a careful investigation should be made to determine the possibly large influence of the foreign ions and molecules present upon the light being used for analysis. It has been noted, for example, that the presence of most foreign salts and acids in a solution being analyzed for sodium, potassium, or calcium tends to lower the amount of light emitted by them. Certain organic molecules tend to lower the amount of light emitted, while others increase the light output. In other cases—for example, the determination of calcium in a solution containing considerable sodium—the readings obtained are high because of inability to filter calcium radiation entirely free of sodium. In most cases, however, where solutions of low sodium and potassium concentrations are analyzed, the accuracy of the results obtained is equal to or better than that given by chemical means, and the speed of the method makes it extremely attractive. The number of applications which have already been made is large and is constantly growing, and at this time only a few will be cited.

Perhaps the simplest and most obvious application is to the rapid or continuous analysis of water samples for sodium, potassium, or calcium since in general the concentrations of these metals fall within the most useful range of the photometer without dilution. Both recording and indicating instruments have been used satisfactorily. For many years designers of water-softening installations have been faced with the need for an analytical instrument which would indicate within an extremely short time the exhaustion of the zeolite bed, and if possible initiate the regeneration cycle. Since these water softeners operate by introducing sodium ions into the water in proportion to the calcium and/or magnesium removed, the first indication of the exhaustion of the zeolite bed is a drop in the sodium level of the effluent water. Rapid analyses for sodium allow this exhaustion point to be detected with ease. If a recording flame, photometer is being used, a sudden drop in the recorded sodium level may be used to actuate the necessary valve for transferring the influent to a previously regenerated bed and to start the regeneration cycle for the exhausted bed.

The same type of installation is suitable for use in determining the sodium levels in waters "demineralized" by synthetic ion-exchange resins.

In many industrial processes where it is necessary to remove sodium salts from precipitated products by washing with water, rapid determination of the sodium content of the influent and effluent waters, or a continuous record thereof, makes for accurate control of the washing cycle. Concentrated solutions such as brines or sea water must usually be diluted prior to analysis.

The flame photometer has a definite place in the general analytical laboratory, for it simplifies many of the usually laborious determinations. Among these might be cited the determination of either sodium or potassium in the presence of the other, and the determination of calcium in the presence of magnesium.

Of frequent and important use to the agronomist is a knowledge of the potassium and sodium content of soil samples. Suitable extraction or other processes for bringing these salts into solution have long been used in connection with standard procedures of soil chemistry. Where large numbers of soil samples must be analyzed routinely the flame photometer should prove of great value. In connection with studies of plant metabolism and nutrition this same instrument could be used.

Some of the most interesting uses to which these instruments have already been put concern the analysis of biological materials. In the hands of several clinical research workers, the flame photometer has been used successfully to analyze such materials as whole blood, blood serum, urine, body fluids, and tissue residues. These analyses, which are usually both difficult and time-consuming as a result of the very nature of the samples and the small size of the average samples which are available, may readily be performed. For the determination of sodium in blood serum, a sample of as little as 0.1 ml. can be satisfactorily analyzed after diluting by a factor of 1 to 100. The potassium content of serum could conceivably be determined on a 0.1-ml. sample after diluting to 2.5 ml., but a 0.5-ml. sample is much to be preferred. This fact, plus the speed with which the analyses may be completed, makes possible certain studies which have hereto been impossible. Thus, in the clinical laboratory, the flame photometer has proved itself to be a most valuable instrument in connection with research on metabolic studies, the rapid diagnosis, and the treatment of certain diseases.

ACKNOWLEDGMENT

The authors would like to express their appreciation to the various members of the staffs of the Analytical Groups and of the Physics Division of these laboratories who have assisted in this work. They wish also to acknowledge with thanks the assistance of the members of the several medical research laboratories who have evaluated this method of analysis and whose comments have been most helpful.

LITERATURE CITED

(1) Barber, H. H., and Kolthoff, I. M., *J. Am. Chem. Soc.*, **50**, 1625 (1928).
(2) Cholak, J., and Hubbard, D. M., *Ind. Eng. Chem., Anal. Ed.*, **16**, 728 (1944).
(3) Griggs, M. A., Johnstin, R., and Elledge, B. E., *Ibid.*, **13**, 99 (1941).
(4) Heyes, J., *Angew. Chem.*, **50**, 871 (1937).
(5) Jansen, W. H., Heyes, J., and Richter, C., *Z. physik. Chem.*, **A174**, 291 (1935).
(6) Lundegårdh, H., "Die quantitative Spektralanalyse der Elemente", Jena, Gustav Fischer, Part I, 1929; Part II, 1934.
(7) Lundegårdh, H., *Lantbruks-Högskol. Ann.* (Sweden), **3**, 49 (1936).
(8) Lundegårdh, H., and Philipson, T., *Ibid.*, **5**, 249 (1938).
(9) McClelland, J.A.C., and Whalley, H. K., *J. Soc. Chem. Ind.*, **60**, 288 (1941).
(10) Mitchell, R. L., *Ibid.*, **55**, 269 (1936).
(11) Schmitt, L., and Breitweiser, W., *Bodenkunde u. Pflanzenernähr*, **9–10**, 750–7 (1938).
(12) Schuhknecht, W., *Angew. Chem.*, **50**, 299 (1937)

Reprinted from *Ind. Eng. Chem., Anal. Ed.* **1945**, *17*, 605–11.

This paper is significant for several reasons. First, the authors provided new information that could be used for quantitative and qualitative analysis and included a catalog of spectra and various physical constants (e.g., boiling points). Second, they used new instrumentation that they had previously developed for recording Raman spectra more quickly and reproducibly. Because their spectrograph was equipped with a photomultiplier as the detector, accurate intensity and polarization measurements could be made. Previously researchers had used photographic procedures to record the Raman spectra of hydrocarbons. Photomultiplier detectors became extremely popular in subsequent years and are still used in modern instruments, although recently developed diode arrays and charge-coupled devices have some advantages over photomultipliers. Third, the authors successfully tested their analytical method on mixtures of hydrocarbons containing nine compounds.

Therese Cotton
Iowa State University

Raman Spectra of Hydrocarbons

M. R. FENSKE, W. G. BRAUN, R. V. WIEGAND[1], D. QUIGGLE, R. H. McCORMICK, AND D. H. RANK
School of Chemistry and Physics, The Pennsylvania State College, State College, Pa.

The Raman spectra of 172 pure hydrocarbons are presented both as reproductions of the original records obtained from a recording spectrograph and in tabular form as scattering coefficients and depolarization factors. Data are presented for 76 paraffins, 32 naphthenes, 29 olefins, 3 diolefins, 30 aromatics, and 2 other hydrocarbons. Direct comparisons of spectra are possible because a uniform intensity scale has been used. The spectrograph employed records the Raman spectrum as a graph with coordinates linear in both wave length and intensity. Raman spectra can be used for the qualitative and quantitative analysis of hydrocarbon mixtures. A few nine-component mixtures have been analyzed successfully. In general, Raman spectroscopy in hydrocarbon analyses is best used as a complement to rather than a substitute for the infrared and ultraviolet techniques.

Recent advances in the applications of the newer physical methods of analyses have contributed greatly to the manufacture and quality control of various chemicals. The petroleum industry in particular has used these methods in the conversion and synthesis of various hydrocarbons and in the control of distillation, extraction, and other separational processes in both the laboratory and the refinery. Some of these applications have been described in review articles (9, 17, 18, 24, 27).

Most of the research on spectroscopic methods of analysis has been concerned with the application of infrared and ultraviolet absorption methods (3, 4, 5, 8), and today these are used rather extensively. Many improvements in instrumentation

were made during the war and several good infrared and ultraviolet spectrophotometers are now available commercially. Most of these instruments are of either the direct indicating or recording type and have been designed for maximum economy in work and time.

The application of Raman spectroscopy to the analysis of hydrocarbon mixtures, on the other hand, has been rather limited even though, in many cases, it offers advantages over both the infrared and ultraviolet techniques. Recording spectrometers for this work have not been available and the photographic procedure generally used has not been sufficiently fast or accurate for general analytical purposes. The earlier features of this method of analysis have been described by Goubeau (11) and Glockler (10), and the more recent advances by Stamm (25). All this work was done by recording the Raman spectra photographically. These photographic methods, however, are time-consuming and are always attended by the inherent difficulties of development and photometry.

The work described in this paper was started in 1943 and was directed toward the development of a Raman spectrograph having as the principal objectives: (1) that the instrument be direct recording; (2) that it be semiautomatic; (3) that the actual time for running the instrument and processing the record be as short as possible; and (4) that the spectra be reproducible.

Preliminary investigations (20) showed that the Raman spectra of hydrocarbons could be recorded directly by using a multiplier phototube to scan the spectra. Additional work was carried out at these laboratories to make a direct-recording Raman spectrograph and develop techniques to satisfy the conditions given above. The instrument which was constructed for this purpose has recently been described (23). The analytical results and the spectra of the pure hydrocarbons presented here were obtained with this apparatus.

---

[1] Present address, Montana State College, Bozeman, Mont.

The 1930s and 1940s   2855-8/94/0053$08.00/0   ©1994 American Chemical Society

The basis for applying Raman spectroscopy to hydrocarbon analysis is dependent upon the fact that when a beam of a monochromatic exciting light passes through a transparent medium some of the light is absorbed and may be re-emitted. If this re-emitted light is examined by means of a spectrograph, very weak spectral lines or bands will appear on either side of the line of the exciting light. These weak lines, which are called Raman lines, are characteristic of the substance illuminated and are therefore a "fingerprint" of that substance. The frequency differences between the exciting light and the Raman lines are independent of the frequency of the exciting light—i.e., the frequency differences are the same for exciting lights of different wave lengths. In order to have a convenient system of units and to conform with past usage these frequency differences are expressed as wave number shifts and written $\Delta v \text{cm.}^{-1}$.

It is beyond the scope of this paper to discuss in any detail the theoretical concepts of the Raman effect. The comprehensive monographs of Kohlrausch (14), Hibben (13), Glockler (10), Sutherland (26), and Herzberg (12) have treated rather exhaustively the fundamental principles and reviewed the experimental data. The paper by Stamm (25) is brief but rather complete. To these the reader is referred for more details.

Since the Raman spectra are characteristic of the scattering substances, in both the wave number shift and the intensities of the various lines, they can be used like any other physical property as a means of identification. Experimentally it has been found that for many mixtures of hydrocarbons the intensities of the Raman lines of a constituent are directly proportional to the volume fraction of that constituent present. A qualitative analysis of a mixture may therefore be made by determining the frequencies (wave number shifts) of the various lines of its Raman spectrum and comparing these data with those obtained for pure compounds. The quantitative analysis is made by determining the ratios of the intensities of the Raman lines of a substance in the spectrum of the mixture with those of the same lines in the spectrum of the pure compound.

FIELD OF USEFULNESS

Although, like the ultraviolet and infrared absorption methods, the analysis by Raman spectroscopy involves the measurement of the characteristics of a spectrum, different principles and techniques are used. The types of analyses which can be made, the concentrations of the components which are most suitable for analyses, and the sensitivities of the methods are different. Consequently, the choice of the spectroscopic method to be used for an analysis depends on the operator's knowledge of the type of sample, the number of components present, and the past history and treatment of the sample. No one spectroscopic method is universal in hydrocarbon analytical work. The ultraviolet and infrared absorption methods and Raman spectroscopy will probably be used as complements rather than as substitutes for each other in analytical work.

Ultraviolet absorption spectrophotometry is well suited for the determination of conjugated diolefins and the aromatic hydrocarbons of lower molecular weight. The absorption spectra of the higher molecular weight aromatics—namely, those having nine or more carbon atoms—become similar and accurate analyses are often difficult, if not impossible. The ultraviolet method is normally not sensitive to olefins, paraffins, or naphthenes; it is particularly useful when only a few aromatics must be determined in a mixture of other hydrocarbons.

The limitations of infrared absorption spectrophotometry, on the other hand, are not so specific and analyses can be made on mixtures containing paraffins, olefins, naphthenes, and aromatics. Some difficulties are usually experienced with mixtures of paraffins and naphthenes which contain aromatics; with mixtures of naphthenes and paraffins which contain small amounts of alcohols, ethers, ketones, and esters; with materials containing water; and with certain other systems where a small amount of strongly absorbing compound can mask the absorption of the other materials present.

Both the ultraviolet and infrared methods are sensitive to small amounts of an absorbing material and are often useful for the detection and estimation of small concentrations of certain impurities in an otherwise relatively pure material. Both methods are more accurate for determining small concentrations than they are for large concentrations of an absorbing constituent.

Although in practice most of the hydrocarbon samples subjected to analysis by the infrared absorption methods are relatively simple—that is, they contain from two to four components—as many as nine individual hydrocarbons may be determined in one sample. For ultraviolet absorption analyses the samples should be less complex, and a maximum of perhaps four hydrocarbons which absorb should be determined in a single sample.

Since the relationship between the volume fraction concentration of a compound and the intensity of the Raman lines is a linear one for most hydrocarbon mixtures, the concentrations of compounds as high as 100% can be determined without difficulty. For the infrared and ultraviolet absorption methods the relationship between transmission and concentration is logarithmic. On the other hand, the Raman spectra are usually not sensitive to small concentrations such as 1 to 5% and in a mixture these minor components may even be missed qualitatively. This is no disadvantage when only the concentration of the major constituent is desired.

Mixtures having as many as nine components have been analyzed by their Raman spectra at these laboratories. However, such complex analyses are rather unusual and the study of the spectra and the computations involved become rather tedious. The time required for an analysis depends on the number of components and the similarity of their spectra. Usually about one hour is required for preparing the sample, scanning the spectrum, and processing the record. This is required for all samples. In addition, examination of the spectrograms and the qualitative and quantitative analysis of 2- or 3-component mixtures take about 0.5 to 1 hour except for routine work where the time will be somewhat less. For 5-component mixtures the examination of the spectrograms and the analytical computations may take as long as 3 hours.

The sample size required for the regular Raman tubes is 35 ml., although some analyses have been made, using specially designed tubes, on as little as 12 ml. with the present apparatus. It is believed, however, that recording instruments could be built which would use considerably smaller sample tubes.

DEFINITIONS AND TERMS

The many papers which have been published on the Raman effect have been consistent in most of their designations of the various units. The ones most frequently used to define the Raman effect refer to spectral position, intensity, and degree of polarization.

Spectral Position. The actual position where the Raman frequencies occur in the spectrum is of little importance, since it is an effect which can be produced by an exciting light of any frequency. The important fact is that the frequency difference, preferably measured in the number of vibrations per centimeter, between the exciting radiation and the Raman line is the same no matter where in the spectrum the effect occurs. This difference is usually expressed in wave numbers or vibrations per centimeter and is designated as the Raman shift or wave number shift, $\Delta\nu$ in cm.$^{-1}$.

To express the frequency, ν, in the usual unit, cycles or vibrations per second, would lead to awkward numbers. Hence another unit, obtained by dividing vibrations per second by the velocity of light, c, in centimeters per second, is used. This unit has the dimensions of vibrations per centimeter and is equal to the reciprocal of the wave length, λ:

$$\lambda\nu = c; \therefore \nu/c = 1/\lambda$$

Intensity Measurements. For quantitative analytical work the intensities of each line must be known in addition to the Raman shift. Unfortunately, intensity measurements have not been made on any absolute or comparative basis and each investigator has chosen a system to suit his own work. The most frequently used system is to correlate the intensities of the various lines on a basis of 0 to 10 where 0 is the intensity of the weakest and 10 the intensity of the strongest line in each spectrum. For the correlation of molecular structure and Raman spectra this method has been satisfactory; however, for analytical work there is a serious disadvantage: it does not allow the intensity comparison of a line in one spectrum with a line in another, since each spectrum is usually on a different basis.

In all the work done at these laboratories, intensities have been measured relative to the $\Delta\nu = 459$ cm.$^{-1}$ line of carbon tetrachloride. The unit of intensity is the "scattering coefficient" and is defined as the ratio of the intensity of the hydrocarbon Raman line to that of the $\Delta\nu = 459$ cm.$^{-1}$ line of carbon tetrachloride. Since all the intensities are on this same basis, the spectra of known pure compounds may be compared directly with unknown mixtures and the analysis is straightforward.

Degree of Polarization. The polarization of the Raman lines is defined by the depolarization factor, ρ_n (12, 14), which is the ratio of the intensities of the perpendicular component (the electric vector vibrating in the vertical plane) to the parallel component (the electric vector vibrating in the horizontal plane) of the Raman line. The parallel component is always preponderant and according to theory the value of ρ_n approaches 0 for symmetrical types of vibrations and 6/7 for asymmetrical types.

For most hydrocarbon analytical work this value has little application; however, for the delineation of molecular structure and for the assignment of molecular vibrations it is important.

APPARATUS

A schematic drawing of the instrument and optical path is shown in Figure 1.

The exciting lights, L_1 and L_2, are mercury vapor lamps supplied from a voltage-regulated line. The light from these lamps is focused by means of the cylindrical filter tubes, F_1 and F_2, and concentrated in the sample tube, ST. Scattered light arising from this illumination passes out the end of the sample tube and is directed to the first condensing lens, C_1, by means of diagonal mirror M_2. This condensing lens focuses the light from the sample on the entrance slit, S_1, of the spectrograph. Light entering the spectrograph falls on collimating mirror M_3 and is directed as a parallel beam to the concave diffraction grating. The spectrum diffracted from the grating comes to focus on a parabola passing through the exit slit, S_2, the collimating mirror, M_2, and the grating, G. The grating can be rotated by means of a motor drive so that the spectrum passes the exit slit at a rate of about 11 Å. per minute.

Individual lines are focused on the phototube, P, by means of lens C_2. The phototube (RCA-1P21) used here as a detector is a cascade type which greatly amplifies the initial photocurrent before it leaves the tube. The signal is further magnified by an amplifier and the fluctuations produced in the plate current are passed through the galvanometer, producing rotation of the galvanometer coil. The movement of the coil is recorded by the movement of the light images, I_3 to I_7, produced by the rotation of galvanometer mirror M_1. The movement of the light images together with the movement of the paper past the slit of the recorder produces a continuous curve showing the Raman spectrum of the sample.

To aid in the analysis of the spectrum a mechanism for putting wave length calibration lines on the trace is connected directly to the driving mechanism.

Light Source. The divergent light from the mercury vapor lamps (Type H-1, 400-watt) is focused in the horizontal plane by means of the cylindrical filter tubes, F_1 and F_2, so that an intense beam passes through the center of the Raman tube. Since there is an overlapping of the Raman spectra produced by the 4078 and 4047 Å. lines with that of the 4358 Å. mercury line, these first two must be filtered out. For this reason both tubes are filled with a solution of sodium nitrite which serves as a filter to remove the 4078 and 4047Å. radiation, as well as a small amount of the ultraviolet light which may not have been removed by the glass envelopes of the lamps. For the present unit a tube 6.25 cm. (2.5 inches) in diameter and containing a solution of about 35 grams of sodium nitrite per 100 grams of water has been found to be most satisfactory.

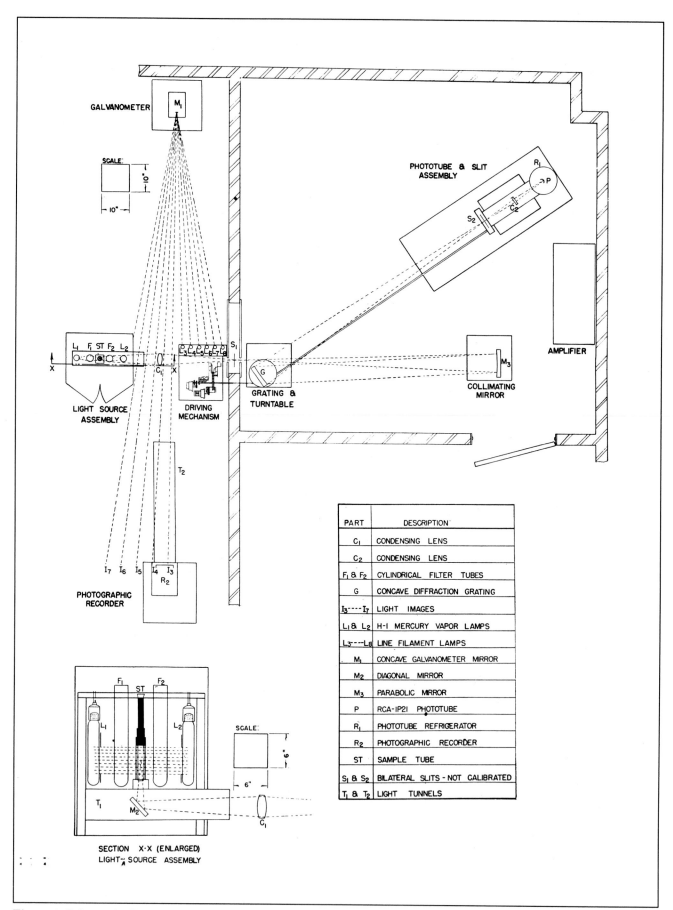

Figure 1. Schematic diagram of spectrograph

The Raman sample tube as shown in Figure 2 was designed for this apparatus, so that an additional filter solution could be placed in the light path and constant-temperature water circulated around the tube. The arrangement shown has the advantage over previously used separate filter holders and cooling systems in that there are fewer glass-to-air surfaces, resulting in less surface reflection losses. The elimination of ring seals at the bottom of the tube allows the sample to be irradiated as far down as the plane window. The all-glass construction facilitates cleaning. (The sample tubes were made by the Pyrocell Manufacturing Co., 207–211 East 84th St., New York 28, N.Y.)

Distilled water is circulated through the outer jacket of the tube from a constant-temperature supply by means of a small centrifugal pump.

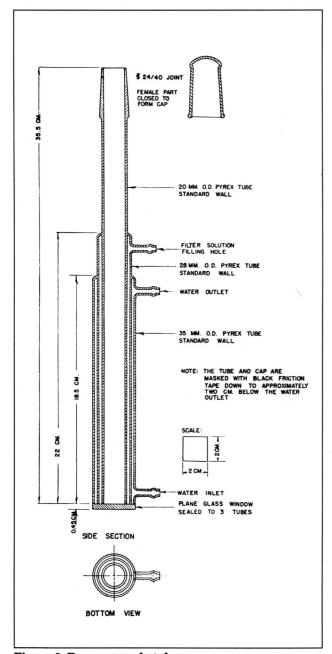

Figure 2. Raman sample tube

The inner jacket has one opening through which a filter solution can be added to remove as completely as possible the mercury continuum between 4400 and 4700 Å. A saturated solution of praseodymium ammonium nitrate has proved to be most efficient for this purpose. Since this salt was rather difficult to purify, a commercial mixture of approximately 50% praseodymium, 30% neodymium, and 20% lanthanum salts was tried and found satisfactory (supplied by Lindsay Light and Chemical Co., West Chicago, Ill.).

The sample tubes shown in Figure 2 require a 35-ml. sample; however, when only smaller amounts of material are available a tube of somewhat similar design is used in which glass wedges are cemented into the tube to fill part of the useless volume. Samples as small as 12 ml. have been used in these modified tubes.

Immediately below the sample tube a slide containing a Polaroid disk may be inserted. Either the parallel or the perpendicular component of the Raman lines can be selected for recording by rotation of the Polaroid. The method of determining depolarization factors from the records of the two components has been reported in the literature (21).

Spectrograph. The grating mounting is a stigmatic type similar to that described by Meggers and Burns (16) except that the grating is mounted on a turntable and an exit slit is used to make the instrument a monochromator. The diffraction grating was ruled at Johns Hopkins University on a Pyrex mirror. The grating has a ruled area of 7 by 3.25 inches with 15,000 lines per inch and a radius of curvature of 15 feet.

The grating and its support bracket are mounted on a precision turntable which can be turned by means of a motor drive, so that the spectrum passes by the exit slit at a rate of 11 Å. per minute. A suitable gear mechanism is provided for a small revolution counter, which indicates directly the wave length in Ångströms, and for a switch mechanism which places fiducial marks on the record at 5 and 25 Å. intervals. The 5 Å. marks are placed on the record by flashing a light in front of the recorder, so that the entire slit is illuminated and a line approximately equal to the width of the slit appears on the finished record. The darker 25 Å. marks are recorded by a brighter light in front of the slit. For this purpose a two-filament automobile lamp serves very well, one filament being used for the 5 Å. marks and two for the 25 Å. marks.

The collimating mirror, M_3, is made from a 25-cm. (10-inch) Pyrex telescope blank ground and polished to a parabola having a focal length of 94.5 inches. The mirror is front-surface aluminized. The mounting bracket is adjustable, so that the mirror may be properly aligned and focused.

The entrance and exit slits of the spectrometer are bilateral and adjustable. Both slits are set at 0.6-mm. opening which gives a resolution of about 15 cm.$^{-1}$ for the spectrograph. The exit slit assembly may be removed and a plate holder inserted for photographic recording if desired.

A simple one-element condensing lens, C_2, serves to focus the light passing through the slit onto the sensitive surface of the phototube.

Detector. A specially designed refrigerator compartment, R_1, is used to house the phototube assembly, so that the unit can be kept at dry ice temperature while in use. This refrigeration is necessary to reduce the fluctuations in the thermal dark current of the tube. The phototube, an RCA 1P21 multiplier type, and the voltage divider supplying the tube are enclosed in an airtight case in the center of the refrigerator. A light tunnel, containing two glass windows to minimize heat transfer, allows light focused on lens C_2 to fall on the photosensitive surface of the tube.

For all the Raman work which has been done with this detector a potential of about 110 volts per stage has been used. This is supplied from a set of 24 radio B-batteries in series with a voltage divider.

Amplifier and Recorder. The signal from the phototube is passed to a one-stage direct current amplifier employing an RH-507 amplifier-electrometer tube. The grid bias and the plate and filament currents are supplied from Willard low discharge wet storage batteries. These batteries provide an exceptionally steady output which, together with proper shielding, good insulation, and drying of the amplifier case make the unit extremely stable for long periods of time.

The plate current from the amplifier is led to an opposing potentiometer circuit which cancels the steady plate current and leaves the variable signal to be fed to the recording galvanometer. A variable shunt across the galvanometer allows the sensitivity to be changed.

The recorder automatically feeds the photographic paper from a 250-foot roll past the recording slit, on which the galvanometer light images are focused, and into a receiver. The entire unit is light-tight, provision having been made for shearing off the record and closing the receiver so that the photographic paper can be removed to the darkroom for processing.

The system of multiple lights and a concave galvanometer mirror for use with the photographic recorder allows deflections of approximately 30 inches to be recorded on paper only 6 inches wide (19).

ANALYTICAL PROCEDURE

Treatment of Sample. The method of obtaining the Raman spectra requires that the samples examined be relatively free of dust, turbidity, colored material, and fluorescing impurities. The presence of an excess of any of these may make the spectra obtained unsuited for analytical work. Their removal is usually necessary.

Dust and suspended matter in a sample cause an increase in background by reason of the Tyndall effect, the suspended material scattering the light of the mercury continuum. The random movements of the particles cause fluctuations in the amount of this scattered light with resulting errors in the Raman line intensity measurements, particularly with recording instruments.

Fluorescent impurities are often present in hydrocarbon samples and may originate from oxidation of the hydrocarbon; contamination by stopcock lubricants, rubber, material extracted from corks and plastic bottle caps; and countless other sources. The usual effect of these is to increase the amount of continuous radiation entering the spectrograph. Since the random fluctuations in the output of the phototube increase

with increasing amounts of light, strongly fluorescing samples will cause random fluctuations which can entirely mask the rather weak Raman radiations. This can make the record useless for analytical work.

Samples which are colored in themselves or which contain colored impurities where the absorbing region lies in the wavelength region of their Raman spectrum do not lend themselves to analysis by the Raman technique. If either the exciting light or the Raman spectrum is absorbed there will be a decrease in the intensity of the Raman lines and any quantitative comparison with external standards becomes difficult. Some work done in these laboratories has shown that a sample which has a transmittance between 4300 and 4700 Å. of 99% through a 1-cm. path will have Raman lines only 90% as intense as those of a sample whose transmittance is 100%.

Several methods of treating samples to remove these materials have been described (11, 14). The one which appeared most satisfactory for hydrocarbon analysis was a simple distillation where the distillate was collected directly in the Raman tube. In the case of mixtures, the distillations must be carried to dryness in order to avoid any effects of fractionation and consequent changes in sample composition. In almost all the cases encountered, the dust and the fluorescent and colored contaminants were removed by this simple distillation; however, some samples of aromatic hydrocarbons, which had been obtained from hydrocarbon mixtures by extraction with aniline, remained colored even after this treatment. It was found that a distillation in which the vapor was passed through hot activated silica gel cleaned the samples sufficiently to obtain satisfactory Raman spectra. Known samples treated in this manner showed no detectable change in composition due to selective adsorption.

Production of Spectrograms. The usual procedure in recording the spectrum of a sample is to place in the apparatus a tube containing pure carbon tetrachloride. The short section of the spectrum which includes the $\Delta\nu = 459$ cm.$^{-1}$ line of this material is next scanned for intensity calibration purposes. The tube containing the sample for analysis is then inserted in the instrument and the spectrum from 1700 to 150 cm.$^{-1}$ (4725 to 4385 Å.) recorded. A second carbon tetrachloride calibration mark is recorded for a check of the first calibration and if, after processing the recording, the two standard deflections of carbon tetrachloride, taken before and after the determination of the spectrum of the sample, are found to agree within 2 or 3%, the spectrogram of the sample is used for the analytical work.

Analysis of Records. For qualitative analytical work the number of possible compounds which must be looked for in an unknown sample may be considerably narrowed down by a knowledge of the source of the sample, the boiling point range, refractive index, bromine number, etc. The analysis then usually involves only the direct visual comparison of a few spectrograms of known compounds with the spectrogram of the unknown. An alternate method lies in calculating the wave number shifts of the Raman lines of the unknown sample and comparing these data with the tables prepared for the pure compounds, similar to those given in this paper.

For quantitative work a base line curve, as shown in Figure 3, must be drawn on the spectrogram. Since the intensities of the Raman lines are directly proportional to the galvanometer deflections, and hence proportional to the heights of the lines above the base line on the spectrogram, the various lines suitable for analytical purposes are measured. The scattering coefficients or the ratios of the heights of these lines to the height of the $\Delta\nu = 459$ cm.$^{-1}$ line of carbon tetrachloride are

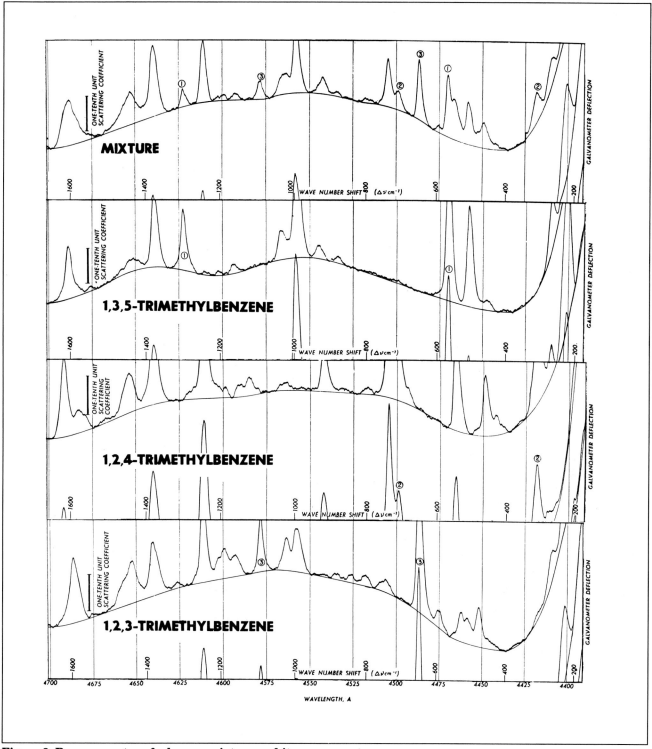

Figure 3. Raman spectra of a known mixture and its components

| | Composition, volume % | |
	Known	Determined
1,3,5-Trimethylbenzene	33.3	33.6
1,2,4-Trimethylbenzene	33.3	33.4
1,2,3-Trimethylbenzene	33.4	33.0
Total	**100.0**	**100.0**

then calculated. The scattering coefficients of the corresponding lines of the pure compounds are next obtained in the same manner. The concentration of the compound present for which the analysis is being made ordinarily is the ratio of the scattering coefficient for one of these lines in the mixture to that of the corresponding line in the pure material.

For most hydrocarbon mixtures, particularly those in which the components are all of the same molecular type, there is a direct proportionality between the scattering coefficient and the volume fraction of the compound present. For mixtures of dissimilar types, such as aromatics and paraffins, there may be deviations from this direct proportionality and additional calibrations may be necessary before accurate analyses can be made.

In determining the percentage of each component present it is desirable to choose positions on the trace where only single substances have moderately strong lines. If the number of components in a mixture is sufficiently small and the spectrum of each component contains several lines, it is usually possible to find more than one peak in the clear for each substance. Figure 3 shows the spectrum of a mixture and the spectra of the pure hydrocarbons in it. The various Raman lines used in the analysis for each of the components have been marked with circled numbers.

If the values obtained from the various peaks for one compound do not check each other sufficiently well, the spectra of the components believed to be present should be rechecked to determine whether some interference has been overlooked or whether the sample is colored. The position of the base line curve should also be checked to see if a slight change in its location could account for the discrepancy. When values are found for all the substances which have peaks in the clear, these may be used to apply corrections on peaks where a component of known percentage may be interfering with a component whose percentage is still unknown. For complex mixtures it may be necessary to apply several corrections to a peak in order to obtain a value for one of the components which cannot be found simply. When corrections are applied to a peak, they must not be a substantial part of its height or the error caused by uncertainty in the base line location will be prohibitive. This doubt in the position of the base line makes it nearly impossible to use simultaneous equations for peaks common to several components and still obtain good results. As a final check on the values obtained, the total should equal 100%. If the total is over this, the base line should be checked to see if it has been drawn too low. If the total is less than 100%, the sample should be checked for color, the base line should be checked to see if it is drawn too high, and, finally, the qualitative analyses should he checked to see that no components have been overlooked.

ANALYTICAL RESULTS

To test the reliability of the method of analysis several known mixtures of hydrocarbons were analyzed. The analyst was not given any information about the samples except the approximate boiling point range and the knowledge that he had at his disposal the spectra of the pure components from which the blends were prepared. The selection of the components for these blends, although dictated somewhat by the availability of the materials used, was such that the samples were in general similar to the fractions which might be obtained from a fractional distillation—that is, they consisted of a mixture of close-boiling materials. The mixtures examined were blends of aromatics, paraffins, naphthenes and aromatics, naphthenes and paraffins, and paraffins and aromatics.

The results of some typical analyses are given in Table I. The percentage error in the analyses, based on the total sample, varies somewhat from compound to compound and is largest in mixtures where the components have similar spectra and in mixtures which have a large number of compounds present. In the former case the analytical difficulties are encountered in overlapping lines, while in the latter the principal difficulty is the uncertainty of the base line location. In general, the Raman analyses have been found to be correct to within 2 percentage units.

CORRELATION OF RAMAN DATA AND MOLECULAR STRUCTURE

Since the Raman spectrum of a material bears a direct relationship to the characteristic frequencies of the various parts of a molecule, the careful examination of the spectrum of a compound should provide information on its molecular configuration. Such knowledge is useful in the study of petroleum fractions where certain types of compounds, which have either not been prepared or are not available for study in their pure state, are to be identified. As additional improvements are made in distillation, extraction, and other separational processes for the higher boiling naphtha and gas oil fractions these correlations, together with those based on the infrared spectra (3), should be invaluable in studying the composition of petroleum fractions.

Of specific interest in the analysis of hydrocarbon mixtures are certain correlations between the Raman spectra and molecular structure which have been made in the course of this work, as well as those which have been published by Kohlrausch, Pongratz, Reitz, and their coworkers (15) and by many others (13). These are summarized in the following paragraphs.

Aliphatic Olefins. Compounds having a C=C bond show a strong Raman line between 1600 and 1685 cm.$^{-1}$, the exact position depending on the configuration of the rest of the molecule. In general, the depolarization factor, ρ_n, for this line is low—namely, between 0.15 and 0.3.

In compounds of the type $CH_2{=}CHR$, and $CH_2{=}CRR'$, where R is any hydrocarbon radical, the strong Raman line lies between about 1640 and 1655 cm.$^{-1}$, while in compounds of the type $HRC{=}CHR'$, $RR'C{=}CHR''$, and $RR'C{=}CR''R'''$ this frequency is shifted to about 1660 to 1685 cm.$^{-1}$. Compounds of these latter three types may show geometrical isomerism and in these cases the frequency is greater by at least 15 units for the trans- than for the cis-isomers.

Compounds which have the structure $RHC{=}CHR'$, where a hydrogen atom is present on both of the carbons which are connected by the double bond, show a Raman line between

Table I. Analysis of Known Hydrocarbon Mixtures

Compound	Boiling point, corrected to 760 mm. of Hg, °C	Known composition, volume %, A	Determined composition, volume %, B	Difference, A–B	Compound	Boiling point, corrected to 760 mm. of Hg, °C	Known composition, volume %, A	Determined composition, volume %, B	Difference, A–B
1,2,3-Trimethylbenzene	176.1	50.0	50.1	−0.1	1,2-Dimethylbenzene	144.4	13.3	13.8	−0.5
1,3,5-Trimethylbenzene	164.7	50.0	50.2	−0.2	Isopropylbenzene	152.4	16.7	16.1	0.6
		100.0	100.3		n-Propylbenzene	159.2	16.7	15.3	1.4
					1-Methyl-3-ethylbenzene	161.3	13.3	14.6	−1.3
1-Methyl-4-isopropylbenzene	177.1	43.4	42.7	0.7	1-Methyl-4-ethylbenzene	162.0	3.3	4.2	−0.9
1,2,3-Trimethylbenzene	176.1	56.6	56.7	−0.1	2-Cyclopentylbutane	154.4	36.7	36.7	0.0
		100.0	99.4				100.0	100.7	
1-Methyl-3-ethylbenzene	161.3	13.9	16.1	−2.2	1-Methyl-2-ethylbenzene	165.2	3.9	3.8	0.1
1-Methyl-4-ethylbenzene	162.0	13.4	15.6	−2.2	1,2,4-Trimethylbenzene	169.2	5.4	5.2	0.2
n-Propylbenzene	159.2	72.7	72.2	0.5	1,3,5-Trimethylbenzene	164.7	15.1	15.9	−0.8
		100.0	103.9		Paraffin-napthene mixture[a]	....	75.6	74.5	1.1
1,2,3-Trimethylbenzene	176.1	33.3	33.0	0.3			100.0	99.4	
1,2,4-Trimethylbenzene	169.2	33.3	33.4	−0.1	n-Propylbenzene	159.2	6.1	5.2	0.9
1,3,5-Trimethylbenzene	164.7	33.4	33.6	−0.2	1-Methyl-3-ethylbenzene	161.3	8.9	9.5	−0.6
		100.0	100.0		1-Methyl-4-ethylbenzene	162.0	1.6	2.3	−0.7
1-Methyl-3-ethylbenzene	161.3	10.6	9.1	1.5	1,3,5-Trimethylbenzene	164.7	2.4	2.3	0.1
1-Methyl-4-ethylbenzene	162.0	10.2	8.5	1.7	Paraffin-naphthene mixture[a]	....	81.0	82.0	−1.0
n-Propylbenzene	159.2	65.3	70.3	−5.0			100.0	101.3	
1,3,5-Trimethylbenzene	164.7	13.9	14.1	−0.2	2-Methylpentane	60.3	30.8	32.6	−1.8
		100.0	102.0		3-Methylpentane	63.3	30.8	29.5	1.3
1,3,5-Trimethylbenzene	164.7	35.2	34.0	1.2	n-Hexane	68.7	38.4	37.9	0.5
n-Propylbenzene	159.2	15.6	15.6	0.0			100.0	100.0	
1-Methyl-2-ethylbenzene	165.2	17.2	16.2	1.0	3-Methylpentane	63.3	30.8	31.6	0.8
1,2,4-Trimethylbenzene	169.2	13.7	12.7	1.0	n-Hexane	68.7	38.4	38.5	−0.1
tert-Butylbenzene	169.1	18.3	19.3	−1.0	Methylcyclopentane	71.8	30.8	30.7	0.1
		100.0	97.8				100.0	100.8	
Ethylbenzene	136.2	50.0	50.1	−0.1	n-Hexane	68.7	38.4	36.2	2.2
2,2,5-Trimethylhexane	124.1	50.0	50.4	−0.4	Methylcyclopentane	71.8	30.8	31.5	−0.7
		100.0	100.5		Cyclohexane	80.7	30.8	33.7	−2.9
							100.0	101.4	

[a] Petroleum fraction having a boiling point of approximately 160 °C. This mixture had been extensively extracted with 98% sulfuric acid to remove aromatic hydrocarbons.

1420 and 1428 cm.$^{-1}$. Pentene-1 and 2-methyl-1-butene are exceptions. This also holds in the cases of the diolefins studied.

In general, normal mono-olefins and some diolefins have strong lines between 1290 and 1300 cm.$^{-1}$. The depolarization factor of the line varies from about 0.2 to 0.25 for the mono-olefins while for diolefins it is 0.3 or greater.

Mononuclear Aromatics. The Raman spectra of the aromatic hydrocarbons have, in general, lines which are quite intense and sharp, as compared with the rather wide bands of comparatively low intensity obtained for many of the paraffins and olefins. All the aromatics have in common one or two rather strong lines near 1600 cm.$^{-1}$, believed by many investigators to correspond to the 1650 cm.$^{-1}$ line of the C=C group found in olefins. In addition, all substituted aromatics, where a methyl group is attached directly to the ring, show a rather strong line between 1373 and 1393 cm.$^{-1}$ (scattering coefficient = 0.05 to 0.25). Also, in common with all substituted aromatics, where the substituting group contains the CH_3 group, is a Raman line between 1430 and 1460 cm.$^{-1}$.

Monosubstituted Aromatics. All monosubstituted aromatics retain some of the strong Raman lines of benzene with but slight shifts of frequency. The lines most characteristic of this group are at about 617, 1001, 1030, and 1200 cm.$^{-1}$. The line at 614 to 620 cm.$^{-1}$ (scattering coefficient greater than 0.1, depolarization factor = 0.65 to 0.88) is easily distinguished from lines of other compounds in this range by its comparatively high intensity and by its depolarization factor. The Raman line at 999 to 1006 cm.$^{-1}$ is very intense (scattering coefficient = 0.55 to 0.9, depolarization factor = 0.1 to 0.2) but polysubstituted compounds where the substitution is in the 1,3-position also have lines in this region which cannot be distinguished from the lines characteristic of monosubstitution. The Raman line at 1025 to 1035 cm.$^{-1}$ for monosubstituted aromatics (scattering coefficient = 0.1 to 0.28, depolarization factor = 0.1 to 0.2) occurs with strong lines for 1,2-disubstituted molecules in the region of 1030 to 1050 cm.$^{-1}$ although, in general, the scattering coefficients for these disubstituted compounds are somewhat greater than for the lines of the monosubstituted type.

The Raman line for monosubstituted compounds at 1183 to 1208 cm.$^{-1}$ (scattering coefficient = 0.11 to 0.3) lies in the same region as that characteristic of 1,4-disubstituted molecules.

Disubstituted Aromatics. Characteristic of 1,2-disubstitution is a strong Raman line between 1030 and 1050 cm.$^{-1}$ (scattering coefficient = 0.23 to 0.45) and in approximately the same range as a line of the monosubstituted molecule. Between 1313 and 1330 cm.$^{-1}$ all polysubstituted aromatics where substitution is in the 1,2 position have a weak Raman line (scattering coefficient = 0.015 to 0.11). The 1600 cm.$^{-1}$ Raman line common to all aromatics is generally split into a pair of lines approximately 20 to 25 cm.$^{-1}$ apart for 1,2-disubstituted aromatics.

The 1,3-disubstitution is characterized by a rather strong Raman line (scattering coefficient = 0.1 to 0.18, depolarization factor = 0.7 to 0.9) between 638 and 641 cm.$^{-1}$ as well as a strong line between 1187 and 1200 cm.$^{-1}$ (scattering coefficient = 0.15 to 0.35). The latter line occurs in the same region as one of the monosubstituted type.

Trisubstituted Aromatics. 1,2,3- and 1,2,4-trisubstituted aromatic hydrocarbons show strong Raman lines between 465 and 480 cm.$^{-1}$. The 1,2,3- lines lie between 479 and 482 cm.$^{-1}$ (scattering coefficient = 0.1, depolarization factor = 0.64 to 0.76) while those of the 1,2,4- type lie between 465 and 476 cm.$^{-1}$ (scattering coefficient = 0.08 to 0.22, depolarization factor = 0.3 to 0.5).

A Raman line at 990 to 995 cm.$^{-1}$ with a high intensity (scattering coefficient = 0.45 to 0.7) occurs only for the 1,3,5-trisubstitution. The 1,3-disubstituted and 1,2,3-trisubstituted molecules may have lines in this region but can generally be distinguished from the 1,3,5- lines either because of their lesser intensity or because they are located closer to 1000 cm.$^{-1}$.

In the 1,2,3- and the 1,3,5-trisubstituted and probably in all the hexasubstituted aromatics, the 1600 cm.$^{-1}$ line is not split into a pair as in the case of the 1,2-disubstituted compounds.

Alkyl Cyclopentanes. Cyclopentane, its monosubstituted, its 1,1- and 1,2-disubstituted, and its 1,1,2-trisubstituted compounds can be recognized by their characteristic line between 884 and 899 cm.$^{-1}$. The intensity of this line decreases as the length of the side chain increases. The depolarization factor is low.

The 1,3-disubstituted and the 1,1,3- and 1,2,3-trisubstituted compounds do not show a strong line in this region.

All the cyclopentanes have a common line between 1450 and 1470 cm.$^{-1}$. In general, this line lies below 1460 cm.$^{-1}$ except when the substitution is in the 1,3-position.

CATALOG OF SPECTRA

The usefulness of any spectrographic method of analysis depends in part on the availability of a set of reference spectra of the pure materials which are apt to be present in the sample under examination. For qualitative purposes the spectra of the unknown and known materials can be compared visually. For quantitative analytical work accurate values of the wave number shifts and scattering coefficients must be known. In the present study on hydrocarbon mixtures the spectra of a large number of relatively pure hydrocarbons have been measured and assembled. These are presented here both as reproductions of the original records obtained with the recording spectrograph and as tables for quantitative analytical purposes.

Such spectra enable the spectroscopist to foresee the difficulties attending a particular analysis and possibly, with the aid of similar infrared and ultraviolet spectrograms now being distributed by the American Petroleum Institute (1), to select the best method for making the analysis.

In common with all other spectroscopic methods of analysis, the reproducibility and accuracy of the Raman procedure are dependent upon certain instrument constants. Accordingly, for the best quantitative work each spectrograph must be calibrated with a complete set of the known pure hydrocarbons which will be found in the samples to be analyzed.

The limitations which govern the applicability of the data concern only the intensity values. The wave number shifts and depolarization factors should, of course, be independent of the instrument used. The main reason for the discrepancy between intensities measured on different instruments is the combined effect of the variation of the degree of polarization of the Raman lines and the difference between the various instruments in transmitting the two polarized components of unpolarized light. The number of reflecting surfaces and their inclination to the path of the light through the spectrograph determine the fraction of each kind of polarized light which will be transmitted. For the instrument described here the ratio of the amount of the parallel polarized component of unpolarized light to that of the perpendicularly polarized component is 0.9, while for an older prism instrument (22) the ratio is 0.3. This difference in the transmittances causes a considerable difference in the scattering coefficients of the various lines in the spectrum of carbon tetrachloride. The scattering coefficient of the $\Delta\nu = 313$ cm.$^{-1}$ (depolarized) line is 0.82 for the grating instrument and 0.66 for the prism instrument.

A second factor which may cause variations in intensities among different instruments is the combined effect of the spectrograph resolution and the Raman line width. In order that enough light flux may be obtained at the phototube when recording Raman spectra photoelectrically it is necessary to open the slits more than is sometimes required for photographic work. Although most Raman lines have a considerable width and may be wider than the slits, some are also narrower and their shape and apparent intensity on the recording will vary with the slit width.

All the spectra presented here were obtained using the 4358 Å. mercury line as the exciting frequency, and since the 4347 and 4339 Å. mercury lines could not be removed some very weak Raman lines due to them are shown in the records. The intensities of these triply excited Raman lines for the 4358, 4347, and 4339 Å. mercury lines are in the ratio of about 1:1/15:1/30, respectively. Since the main utility of these spectra will be for analytical work and since these triply excited lines occur only for strong Raman lines, they have been included in the tabular data given here. The wave number shifts have been calculated as though they originated from the 4358 Å. mercury line.

The values recorded for the Raman frequencies in the tabular data are correct to within ±5 cm.$^{-1}$. Experimental data have been reported exclusively rather than the values corrected to agree with the averages reported by other investigators. The values for the depolarization factors listed in the tabular data are believed to be most accurate where the Raman lines are isolated and somewhat less accurate where the lines are relatively close together. The latter values are, however, still useful for the assignment of molecular vibrations.

The spectra presented are divided into groups according to molecular structure, each group being then arranged in order of molecular weight and of increasing complexity of structure. The groups are: paraffins, olefins and diolefins, naphthenes (including alkylcyclopentanes and alkylcyclohexanes), aromatics, and miscellaneous hydrocarbons.

The indexes of the spectra, given in Tables II to VIII, list in addition to the name and the spectrum number the physical properties of the compounds examined, the best literature data on the properties (2, 6, 7), the sources, and, when known, the purities of the hydrocarbons. In many cases the purities have not been separately determined. However, the physical properties and the methods of preparation indicate that the purities are 98 mole % or higher.

Table II. Spectra Numbers and Properties of Pure Paraffin Hydrocarbons

Name of compound	Spectrum no.	Boiling point at 760 mm. of Hg, °C Determined	Literature values [a]	Refractive index at 20 °C., n_D^{20} Determined	Literature values [a]	Estimated purity, mole % [b]	Source of compound [c]
5-carbon atom							
n-Pentane	1	36.1	36.07	1.3577	1.3575	....	A
2-Methylbutane	2	27.9	27.85	1.3538	1.3537	....	A
6-carbon atom							
n-Hexane	3	68.7	68.74	1.3749	1.3749	....	A
2-Methylpentane	4	60.25	60.27	1.3713	1.3715	....	A
3-Methylpentane	5	63.15	63.28	1.3764	1.3765	....	A
2,2-Dimethylbutane	6	49.65	49.74	1.3687	1.3688	....	A
2,3-Dimethylbutane	7	57.95	57.99	1.3750	1.3750	....	A
7-carbon atom							
n-Heptane	8	98.4	98.43	1.3877	1.3876	....	A
2-Methylhexane	9	....	90.05	....	1.3849	99.77 ± 0.07	D
3-Methylhexane	10	91.85	91.95	1.3887	1.3887	....	A
3-Ethylpentane	11	93.55	93.47	1.3934	1.3934	....	C
2,2-Dimethylpentane	12	79.1	79.21	1.3821	1.3822	99.31	A
2,3-Dimethylpentane	13	89.75	89.79	1.3916	1.3920	....	A
2,4-Dimethylpentane	14	80.4	80.51	1.3815	1.3815	98.66	A
3,3-Dimethylpentane	15	86.1	86.07	1.3909	1.3909	....	C
2,2,3-Trimethylbutane	16	80.9	80.87	1.3895	1.3895	99.7	E
8-carbon atom							
n-Octane	17	125.6	125.67	1.3976	1.3975	....	C
2-Methylheptane	18	117.75	117.65	1.3952	1.3950	....	C
3-Methylheptane	19	118.89	118.93	1.3985	1.3985	....	C
4-Methylheptane	20	117.63	117.71	1.3978	1.3979	....	C
3-Ethylhexane	21	118.55	118.54	1.4017	1.4016	....	C
2,2-Dimethylhexane	22	106.9	106.84	1.3937	1.3935	....	C
2,3-Dimethylhexane	23	115.6	115.61	1.4011	1.4013	....	A
2,4-Dimethylhexane	24	109.35	109.43	1.3953	1.3953	....	D
2,5-Dimethylhexane	25	109.2	109.1	1.3926	1.3925	....	D
3,3-Dimethylhexane	26	111.95	111.97	1.4003	1.4001	....	C
3,4-Dimethylhexane	27	117.45	117.73	1.4039	1.4042	....	C
2-Methyl-3-ethylpentane	28	115.75	115.65	1.4040	1.4040	....	C
3-Methyl-3-ethylpentane	29	118.28	118.26	1.4078	1.4078	....	C
2,2,3-Trimethylpentane	30	109.85	109.84	1.4029	1.4030	....	A
2,2,4-Trimethylpentane	31	99.23	99.24	1.3913	1.3915	....	C
2,3,3-Trimethylpentane	32	114.7	114.76	1.4072	1.4075	....	A
2,3,4-Trimethylpentane	33	113.45	113.47	1.4043	1.4042	99.38	A
9-carbon atom							
n-Nonane	34	150.8	150.80	1.4053	1.4055	....	A
2-Methyloctane	35	....	143.26	1.4035	1.4031	....	B
3-Methyloctane	36	....	144.18	....	1.4062	....	B
4-Methyloctane	37	....	142.48	....	1.4061	....	B
3-Ethylheptane	38	....	143.0	1.4095	1.4092	....	B
4-Ethylheptane	39	....	141.2	....	1.4109	....	B
2,2-Dimethylheptane	40	....	130.5	1.4032	1.402	....	B
3,3-Dimethylheptane	41	....	137.3	1.4090	1.4085	....	B
3,4-Dimethylheptane	42	140.7	140.5	1.4115	1.4108	....	B
3,5-Dimethylheptane	43	136.0	136.0	1.4067	1.407	....	B
4,4-Dimethylheptane	44	135.2	138.0	1.4076	1.408	....	B
2-Methyl-3-ethylhexane	45	138.0	139.0	1.4106	1.411	....	B
2-Methyl-4-ethylhexane	46	133.8	136.0	1.4063	1.407	....	B
3-Methyl-3-ethylhexane	47	140.6	143.0	1.4142	1.415	....	B
3-Methyl-4-ethylhexane	48	140.4	143.0	1.4134	1.416	....	B
2,2,4-Trimethylhexane	49	126.6	126.5	1.4034	1.4033	....	B
2,2,5-Trimethylhexane	50	124.1	124.09	1.3996	1.3996	....	A
2,3,3-Trimethylhexane	51	....	138.0	1.4143	1.4143	....	B
2,3,4-Trimethylhexane	52	139.1	140.0	1.4144	1.415	....	B
2,3,5-Trimethylhexane	53	....	131.37	....	1.4060	....	B
2,4,4-Trimethylhexane	54	130.45	131	1.4072	1.4075	....	B
3,3,4-Trimethylhexane	55	139.9	139	1.4178	1.4178	....	B
2,2-Dimethyl-3-ethylpentane	56	133.6	133.83	1.4125	1.4123	....	B
2,4-Dimethyl-3-ethylpentane	57	136.6	136.73	1.4138	1.4137	....	B
2,2,3,4-Tetramethylpentane	58	133.0	133.01	1.4146	1.4146	....	B
2,2,4,4-Tetramethylpentane	59	....	122.28	1.4072	1.4068	....	B
2,3,3,4-Tetramethylpentane	60	141.2	141.54	1.4222	1.4220	....	B

Table II, continued

Name of compound	Spectrum no.	Boiling point at 760 mm. of Hg, °C Determined	Literature values[a]	Refractive index at 20 °C., n_D^{20} Determined	Literature values[a]	Estimated purity, mole %[b]	Source of compound[c]
10-carbon atom							
n-Decane	61	174.05	174.0[d]	1.4119	1.4114[d]	....	A
2,2,6-Trimethylheptane	62	148.2	148.93[d]	1.4059	1.4078[d]	....	B
2,3,6-Trimethylheptane	63	156.0	155.2[d]	1.4122	1.4130[d]	....	B
2,2,3,3-Tetramethylhexane	64	159.8	159.0[d]	1.4281	1.4264[d]	....	B
2,2,3,4-Tetramethylhexane	65	154.3	156.5[e]	1.4226	1.4224[e]	....	B
2,2,3,5-Tetramethylhexane	66	148.4	....	1.4142	....	....	B
2,2,4,5-Tetramethylhexane	67	148.2	....	1.4132	1.4133[d]	....	B
3,3,4,4-Tetramethylhexane	68	170.9	165.5[d]	1.4379	1.4340[d]	....	B
11-carbon atom							
n-Undecane	69	195.9	195.8[d]	1.4174	1.4173[d]	....	A
2,2,4,6-Tetramethylheptane	70	161.9	162.0[d]	1.4127	1.4127[d]	....	B
12-carbon atom							
n-Dodecane	71	216.26	216.26[d]	1.4217	1.4216[d]	....	B
2,2,3,5,6-Pentamethylheptane	72	188.8	....	1.4283	....	....	B
2,2,4,6,6-Pentamethylheptane	73	177.8	177.2[d]	1.4189	1.4191[d]	....	B
13-carbon atom							
n-Tridecane	74	106.8 at 10 mm.	236.5[d]	1.4256	....	....	B
14-carbon atom							
n-Tetradecane	75	121.1 at 10 mm.	253.5[d]	1.4290	1.4289[d]	....	B
7-Methyltridecane	76	115.3 at 10 mm.	....	1.4291	....	....	B

[a] All physical properties except those marked [d] and [e] are from (2). Values from (2) are given only to nearest 0.01 °C. in boiling point and 0.0001 in refractive index.
[b] Purities listed were determined by freezing point measurements. It is believed that all other materials were 98 mole % pure or higher.
[c] Source of compounds: **A.** Petroleum Refining Laboratory, School of Chemistry and Physics, Pennsylvania State College. **B.** Organic Research Laboratory, School of Chemistry and Physics, Pennsylvania State College. **C.** American Petroleum Institute Research Project 45 at Ohio State University. **D.** American Petroleum Institute Research Project 6 at National Bureau of Standards. **E.** Esso Laboratories, Standard Oil Development Co. **F.** Anglo-Iranian Oil Co., Sunbury-on-Thames, England.
[d] (6).
[e] (7).

Table III. Spectra Numbers and Properties of Pure Olefin Hydrocarbons

Name of compound	Spectrum no.	Boiling point at 760 mm. of Hg, °C Determined	Literature values[a]	Refractive index at 20 °C., n_D^{20} Determined	Literature values[a]	Estimated purity, mole %[b]	Source of compound[c]
Olefins							
5-carbon atom							
1-Pentene	77	30.0	29.97	1.3718	1.3714	....	A
cis-2-Pentene	78	....	37.1	....	1.3820	99.55 ± 0.15	D
trans-2-Pentene	79	....	36.36	....	1.3793	99.91 ± 0.05	D
2-Methyl-1-butene	80	31.05	31.10	1.3776	1.3778	....	A
2-Methyl-2-butene	81	38.6	38.53	1.3874	1.3874	....	A
6-carbon atom							
2,3-Dimethyl-1-butene	82	55.6	55.64	1.3902	1.3904	....	A
3,3-Dimethyl-1-butene	83	....	41.24	....	1.3760	99.7	B
7-carbon atom							
1-Heptene	84	93.65	93.3	1.3998	1.3994	....	A
8-carbon atom							
1-Octane	85	121.25	121.27	1.4088	1.4088	....	A
2,3,3-Trimethyl-1-pentene	86	118.5	108.	1.4188	1.418	....	B
2,3,4-Trimethyl-1-pentene	87	117.15	108.	1.4136	1.415	....	B
2,3,4-Trimethyl-2-pentene	88	116.5	116.26	1.4274	1.4275	....	B
2,4,4-Trimethyl-1-pentene	89	101.55	101.44	1.4096	1.4086	....	B
2,4,4-Trimethyl-2-pentene	90	103.9	104.91	1.4176	1.4160	....	B
3,3,4-Trimethyl-1-pentene	91	....	105.	1.4144	1.414	....	B
3-Methyl-2-isopropyl-1-butene	92	113.55	104.	1.4085	1.409	....	B
3,3-Dimethyl-2-ethyl-1-butene	93	117.2	110.0	1.4159	1.416	....	B
9-carbon atom							
3,3-Dimethyl-2-isopropyl-1-butene	94	121.6	121.6[d]	1.4168	1.4174[d]	....	B
2,3,3,4-Tetramethyl-1-pentene	95	134.55	132.6[d]	1.4303	1.4305[d]	....	B
10-carbon atom							
2,2,6-Trimethyl-1-heptene	96	150.4	....	1.4202	....	....	B
2,4,4,5-Trimethyl-1-hexene	97	158.1	....	1.4350	....	....	B
Diolefins							
5-carbon atom							
2-Methyl-1,3-butadiene	98	....	34.08	1.4218	1.4216	....	A
7-carbon atom							
2-Methyl-1,5-hexadiene	99	88.85	88.1[d]	1.4187	1.4184[d]	....	B
9-carbon atom							
2,3,3,4-Tetramethyl-1,4-pentadiene	100	127.7	....	1.4402	....	....	B

[a] All physical properties except those marked [d] are from (2). Values from (2) are given only to nearest 0.01 °C. in boiling point and 0.0001 in refractive index.
[b] Purities listed were determined by freezing point measurements. It is believed that all other materials were 98 mole % pure or higher.
[c] Source of compounds: **A.** Petroleum Refining Laboratory, School of Chemistry and Physics, Pennsylvania State College. **B.** Organic Research Laboratory, School of Chemistry and Physics, Pennsylvania State College. **C.** American Petroleum Institute Research Project 45 at Ohio State University. **D.** American Petroleum Institute Research Project 6 at National Bureau of Standards. **E.** Esso Laboratories, Standard Oil Development Co. **F.** Anglo-Iranian Oil Co., Sunbury-on-Thames, England.
[d] (6).

Table IV. Spectra Numbers and Properties of Pure Naphthene Hydrocarbons–Alkylcyclopentanes

Name of compound	Spectrum no.	Boiling point at 760 mm. of Hg, °C		Refractive index at 20 °C., n_D^{20}		Estimated purity, mole %[b]	Source of compound[c]
		Determined	Literature values [a]	Determined	Literature values [a]		
5-carbon atom							
Cyclopentane	101	49.2	49.26	1.4065	1.4065	99.9+	A
6-carbon atom							
Methylcyclopentane	102	71.8	71.81	1.4098	1.4097	99.63	A
7-carbon atom							
Ethylcyclopentane	103	103.45	103.45	1.4197	1.4198	97.1	A
1,1-Dimethylcyclopentane	104	....	87.5	1.4137	1.4135	....	B
cis-1,2-Dimethylcyclopentane	105	99.1	99.25	1.4200	1.4221	....	B
trans-1,2,-Dimethylcyclopentane	106	91.85	91.85	1.4418	1.4119	....	A
cis-1,3-Dimethylcyclopentane	107	90.5 at 725 mm.	....	1.4081 at 25 °C.	....	....	D
trans-1,3-Dimethylcyclopentane	108	89.4 at 725 mm.	90.97	1.4065 at 25 °C.	1.4088	99.4 ± 0.12	D
8-carbon atom							
n-Propylcyclopentane	109	131.0	130.8[d]	1.4263	1.4266[d]	....	A
Isopropylcyclopentane	110	126.4	126.4[d]	1.4260	1.4260[d]	....	A
1-Methyl-1-ethylcyclopentane	111	121.45	....	1.4269	....	....	B
cis-1-Methyl-3-ethylcyclopentane	112	121.0	....	1.4202	....	....	B
1,1,2-Trimethylcyclopentane	113	113.7	114.0[d]	1.4228	1.4238[d]	....	B
1,1,3-Trimethylcyclopentane	114	104.9	115–16[d]	1.4111	1.4223[d]	....	C
cis,cis,cis-1,2,3-Trimethylcyclopentane	115	122.8	....	1.4263	....	....	B
cis,cis,trans-1,2,3-Trimethylcyclopentane	116	117.2	....	1.4218	....	....	B
cis,trans,cis-1,2,3-Trimethylcyclopentane	117	109.9	....	1.4133	....	....	B
cis,cis,trans-1,2,4-Trimethylcyclopentane	118	116.9	....	1.4183	....	....	B
cis,trans,cis-1,2,4-Trimethylcyclopentane	119	109.0	....	1.4103	....	....	B
9-carbon atom							
2-Cyclopentylbutane	120	154.4	154.6[d]	1.4360	1.4361[d]	....	A
10-carbon atom							
2-Cyclopentylpentane	121	176.5	177.5[d]	1.4393	1.4438[d]	....	A
12-carbon atom							
2-Cyclopentylheptane	122	219.0	....	1.4450	....	....	A

[a] All physical properties except those marked [d] are from (2). Values from (2) are given only to nearest 0.01 °C. in boiling point and 0.0001 in refractive index.
[b] Purities listed were determined by freezing point measurements. It is believed that all other materials were 98 mole % pure or higher.
[c] Source of compounds: **A.** Petroleum Refining Laboratory, School of Chemistry and Physics, Pennsylvania State College. **B.** Organic Research Laboratory, School of Chemistry and Physics, Pennsylvania State College. **C.** American Petroleum Institute Research Project 45 at Ohio State University. **D.** American Petroleum Institute Research Project 6 at National Bureau of Standards. **E.** Esso Laboratories, Standard Oil Development Co. **F.** Anglo-Iranian Oil Co., Sunbury-on-Thames, England.
[d] (6).

Table V. Spectra Numbers and Properties of Pure Naphthene Hydrocarbons–Alkylcyclohexanes

Name of compound	Spectrum no.	Boiling point at 760 mm. of Hg, °C		Refractive index at 20 °C., n_D^{20}		Estimated purity, mole %[b]	Source of compound[c]
		Determined	Literature values [a]	Determined	Literature values [a]		
6-carbon atom							
Cyclohexane	123	80.8	80.74	1.4263	1.4262	....	A
7-carbon atom							
Methylcyclohexane	124	100.8	100.94	1.4231	1.4231	....	A
8-carbon atom							
Ethylcyclohexane	125	131.7	131.79	1.4330	1.4330	....	A
1,1-Dimethylcyclohexane	126	....	119.50	....	1.4289	99.81 ± 0.03	D
cis-1,2-Diemethylcyclohexane	127	129.65	129.73	1.4360	1.4360	....	A
trans-1,2-Dimethylcyclohexane	128	123.35	123.42	1.4270	1.4270	....	A
cis-1,4-Dimethylcyclohexane	129	124.3	124.32	1.4296	1.4297	....	A
trans-1,4-Dimethylcyclohexane	130	119.3	119.35	1.4209	1.4209	....	A
9-carbon atom							
n-Propylcyclohexane	131	156.6	154.9–155.0[d]	1.4370	1.4370[d]	....	A
Isopropylcyclohexane	132	154.4	154.4[d]	1.4410	1.4408[d]	....	A

[a] All physical properties except those marked [d] are from (2). Values from (2) are given only to nearest 0.01 °C. in boiling point and 0.0001 in refractive index.
[b] Purities listed were determined by freezing point measurements. It is believed that all other materials were 98 mole % pure or higher.
[c] Source of compounds: **A.** Petroleum Refining Laboratory, School of Chemistry and Physics, Pennsylvania State College. **B.** Organic Research Laboratory, School of Chemistry and Physics, Pennsylvania State College. **C.** American Petroleum Institute Research Project 45 at Ohio State University. **D.** American Petroleum Institute Research Project 6 at National Bureau of Standards. **E.** Esso Laboratories, Standard Oil Development Co. **F.** Anglo-Iranian Oil Co., Sunbury-on-Thames, England.
[d] (6).

Table VI. Spectra Numbers and Properties of Pure Cyclo-olefin Hydrocarbons

Name of compound	Spectrum no.	Boiling point at 760 mm. of Hg, °C Determined	Boiling point at 760 mm. of Hg, °C Literature values [a]	Refractive index at 20 °C., n_D^{20} Determined	Refractive index at 20 °C., n_D^{20} Literature values [a]	Estimated purity, mole %[b]	Source of compound[c]
Methylenecyclobutane	133	42.2	41.9[d]	1.4210	1.4204[d]	....	B
1-Methyl-1-cyclopentene	134	75.6	75.85[d]	1.4319	....	....	B
3-Methyl-1-cyclopentene	135	65.2	....	1.4215	1.4248[d]	....	B
cis-3,4-Dimethyl-1-cyclopentene	136	....	....	1.4300	....	....	B
1,2,3-Trimethyl-1-cyclopentene	137	121.9	121.0[d]	1.4457	1.4445[d]	....	B
2,3,3-Trimethyl-1-cyclopentene	138	110.65	108.5[e]	1.4345	1.4324[d]	....	B
2,3,4-Trimethyl-1-cyclopentene	139	112.3	....	1.4345	....	....	B
Cyclohexene	140	....	83.19[d]	1.4464	1.4467[d]	....	A

[a] All physical properties except those marked [d] and [e] are from (2). Values from (2) are given only to nearest 0.01 °C. in boiling point and 0.0001 in refractive index.
[b] Purities listed were determined by freezing point measurements. It is believed that all other materials were 98 mole % pure or higher.
[c] Source of compounds: **A.** Petroleum Refining Laboratory, School of Chemistry and Physics, Pennsylvania State College. **B.** Organic Research Laboratory, School of Chemistry and Physics, Pennsylvania State College. **C.** American Petroleum Institute Research Project 45 at Ohio State University. **D.** American Petroleum Institute Research Project 6 at National Bureau of Standards. **E.** Esso Laboratories, Standard Oil Development Co. **F.** Anglo-Iranian Oil Co., Sunbury-on-Thames, England.
[d] (6).
[e] (7).

Table VII. Spectra Numbers and Properties of Pure Aromatic Hydrocarbons

Name of compound	Spectrum no.	Boiling point at 760 mm. of Hg, °C Determined	Boiling point at 760 mm. of Hg, °C Literature values [a]	Refractive index at 20 °C., n_D^{20} Determined	Refractive index at 20 °C., n_D^{20} Literature values [a]	Estimated purity, mole %[b]	Source of compound[c]
6-carbon atom							
Benzene	141	80.1	80.10	1.5012	1.5011	....	A
7-carbon atom							
Methylbenzene (toluene)	142	110.65	110.63	1.4969	1.4969	....	A
8-carbon atom							
Ethylbenzene	143	136.25	136.19	1.4959	1.4958	....	A
1,2-Dimethylbenzene (o-xylene)	144	144.4	144.42	1.5053	1.5052	....	A
1,3-Diemethylbenzene (m-xylene)	145	139.15	139.10	1.4972	1.4972	....	E
1,4-Dimethylbenzene (p-xylene)	146	138.4	138.35	1.4958	1.4958	....	A
9-carbon atom							
n-Propylbenzene	147	159.25	159.22	1.4919	1.4920	....	A
Isopropylbenzene	148	152.4	152.40	1.4910	1.4913	....	A
1-Methyl-2-ethylbenzene	149	165.15	165.15	1.5042	1.5044	99.1	A
1-Methyl-3-ethylbenzene	150	161.4	161.30	1.4965	1.4965	95.8	A
1-Methyl-4-ethylbenzene	151	161.95	162.05	1.4948	1.4950	95.3	A
1,2,3-Trimethylbenzene	152	176.1	176.15	1.5140	1.5139	....	A
1,2,4-Trimethylbenzene	153	169.2	169.25	1.5049	1.5048	....	A
1,3,5-Trimethylbenzene	154	164.7	164.70	1.4992	1.4991	....	A
10-carbon atom							
n-Butylbenzene	155	183.1	183.28	1.4900	1.4900	....	A
Isobutylbenzene	156	....	172.80	....	1.4865	99.87 ± 0.09	D
sec-Butylbenzene	157	173.15	173.30	1.4900	1.4902	....	A
tert-Butylbenzene	158	169.1	169.10	1.4926	1.4927	....	A
1-Methyl-2-isopropylbenzene	159	178.35	178.3	1.5006	1.5006	99.9	F
1-Methyl-3-isopropylbenzene	160	175.20	175.2	1.4930	1.4930	99.96	F
1-Methyl-4-isopropylbenzene	161	177.15	177.10	1.4905	1.4909	....	A
1,2-Dimethyl-3-ethylbenzene	162	193.80	193.91	1.5117	1.5117	99.6	F
1,2-Dimethyl-4-ethylbenzene	163	189.55	189.75	1.5032	1.5031	99.6	F
1,3-Dimethyl-2-ethylbenzene	164	189.95	190.01	1.5107	1.5107	99.84	F
1,3-Dimethyl-4-ethylbenzene	165	188.45	188.41	1.5039	1.5038	99.95	F
1,3-Dimethyl-5-ethylbenzene	166	183.65	183.75	1.4981	1.4981	99.93	F
1,4-Dimethyl-2-ethylbenzene	167	186.45	186.91	1.5043	1.5043	99.8	F
1,2-Diethylbenzene	168	183.30	184.5	1.5034	1.5034	99.85	F
1,3-Diethylbenzene	169	181.2	181.14	1.4953	1.4955	99.4	A
1,4-Diethylbenzene	170	183.60	183.75	1.4947	1.4947	99.65	F

[a] Physical properties from (2) are given only to nearest 0.01 °C. in boiling point and 0.0001 in refractive index.
[b] Purities listed were determined by freezing point measurements. It is believed that all other materials were 98 mole % pure or higher.
[c] Source of compounds: **A.** Petroleum Refining Laboratory, School of Chemistry and Physics, Pennsylvania State College. **B.** Organic Research Laboratory, School of Chemistry and Physics, Pennsylvania State College. **C.** American Petroleum Institute Research Project 45 at Ohio State University. **D.** American Petroleum Institute Research Project 6 at National Bureau of Standards. **E.** Esso Laboratories, Standard Oil Development Co. **F.** Anglo-Iranian Oil Co., Sunbury-on-Thames, England.

Table VIII. Spectra Numbers and Properties of Miscellaneous Pure Hydrocarbons

| Name of compound | Spectrum no. | Boiling point at 760 mm. of Hg, °C | | Refractive index at 20 °C., n_D^{20} | | Estimated purity, mole %[b] | Source of compound[c] |
		Determined	Literature values[a]	Determined	Literature values[a]		
Indene	171	182.57	181.8–182.3[d]	1.5733	1.5764[d]	....	A
Hydrindene	172	177.85	177.5–178.5[d]	1.5383	1.5383[d]	....	A

[a] All physical properties except those marked [d] are from (2). Values from (2) are given only to nearest 0.01 °C. in boiling point and 0.0001 in refractive index.
[b] Purities listed were determined by freezing point measurements. It is believed that all other materials were 98 mole % pure or higher.
[c] Source of compounds: **A.** Petroleum Refining Laboratory, School of Chemistry and Physics, Pennsylvania State College. **B.** Organic Research Laboratory, School of Chemistry and Physics, Pennsylvania State College. **C.** American Petroleum Institute Research Project 45 at Ohio State University. **D.** American Petroleum Institute Research Project 6 at National Bureau of Standards. **E.** Esso Laboratories, Standard Oil Development Co. **F.** Anglo-Iranian Oil Co., Sunbury-on-Thames, England.
[d] (6).

ACKNOWLEDGMENT

The authors are indebted to the Esso Laboratories of the Standard Oil Development Company for financial assistance and particularly to S. C. Fulton and W. J. Sweeney of this company for their help and encouragement. They are grateful to J. K. Wood and other members of the Petroleum Refining Laboratory for their assistance. For some of the hydrocarbons used in this work the authors wish to thank F. C. Whitmore, N. C. Cook, and R. W. Schiessler, and the members of their research staffs in this school; S. F. Birch of the Anglo-Iranian Oil Company; C. E. Boord and K. W. Greenlee of The Ohio State University; and F. D. Rossini of the National Bureau of Standards.

LITERATURE CITED

(1) American Petroleum Institute Research Project 44, National Bureau of Standards, Catalogs of Infrared and Ultraviolet Spectrograms (June 30, 1945).
(2) Ibid., "Selected Values of Properties of Hydrocarbons," Tables 1a to 14a (June 30, 1945).
(3) Barnes, R. B., Liddel, U., and Williams, V. Z., Ind. Eng. Chem., Anal. Ed., 15, 659 (1943).
(4) Brady, L. J., Oil Gas J., 43, No. 14, 87 (1944).
(5) Brattain, R. R., Rasmussen, R. S., and Cravath, A. M., J. Applied Phys., 14, 418 (1943).
(6) Doss, M. P., "Physical Constants of the Principal Hydrocarbons," 4th ed., New York, Texas Co., 1943.
(7) Egloff, G., "Physical Constants of Hydrocarbons," Vols. I to III, New York, Reinhold Publishing Corp., 1900.
(8) Fry, D. L., Nusbaum, R. E., and Randall, H. M., J. Applied Phys., 17, 150 (1946).
(9) Fulton, S. C., and Heigl, J. J., Instruments, 20, 35 (1947).
(10) Glockler, G., Rev. Modern Phys., 15, 111 (1943).
(11) Goubeau, J., in "Physikalische Methoden der analytischen Chemie," by W. Böttger, Leipzig, Akademische Verlagsgesellschaft, H., 1939.
(12) Herzberg, G., "Infrared and Raman Spectra of Polyatomic Molecules," New York, D. Van Nostrand Co., 1945.
(13) Hibben, J. H., "Raman Effect and Its Chemical Applications," A.C.S. Monograph 80, New York, Reinhold Publishing Corp., 1939.
(14) Kohlrausch, K.W.F., "Der Smekal-Raman-Effekt," Berlin, Julius Springer, 1931; "Der Smekal-Raman-Effekt, Ergänzungsband," Berlin, Julius Springer, 1938.
(15) Kohlrausch, K.W.F., et al., papers appearing in Monatsh., Vol. 60 to present.
(16) Meggers, W. F., and Burns, K., Natl. Bur. Standards, Sci. Papers, 18, 185 (1922).
(17) Naylor, W. H., J. Inst. Petroleum, 30, 256 (1944).
(18) Nielsen, J. R., Oil Gas J., 40, No. 37, 34 (1942).
(19) Pfister, R. J., and Rank, D. H., J. Optical Soc. Am., 32, 397 (1942).
(20) Rank, D. H., Pfister, R. J., and Coleman, P. D., Ibid., 32, 390 (1942).
(21) Rank, D. H., Pfister, R. J., and Grimm, H. H., Ibid., 33, 31 (1943).
(22) Rank, D. H., Scott, R. W., and Fenske, M. R., Ind. Eng. Chem., Anal. Ed., 14, 816 (1942).
(23) Rank, D. H., and Wiegand, R. V., J. Optical Soc. Am., 36, 325 (1946).
(24) Schlesman, C. H., and Hochgesang, F. P., Oil Gas J., 42, No. 36, 41 (1944).
(25) Stamm, R. F., Ind. Eng. Chem., Anal. Ed., 17, 318 (1945).
(26) Sutherland, G.B.B.M., "Infrared and Raman Spectra," London, Methuen and Co., 1935.
(27) Sweeney, W. J., Ind. Eng. Chem., Anal. Ed., 16, 723 (1944).

Note: The original article was followed on pages 712 to 765 by "data in the form of graphs and tables on the Raman spectra of 172 hydrocarbons. Indexes for these graphs and tables are given in Tables II to VIII."

Received May 23, 1947.

Reprinted from *Anal. Chem.* **1947**, *19*, 700–11.

Everything stated in this paper is as true today as it was in 1948. We still teach statistics to undergraduate chemistry majors and nonmajors as well as to graduate students. What would probably surprise Elving and Mellon is the increased amount of time devoted to the topics: several weeks of instruction for undergraduate courses, and about half of a semester for graduate-level courses. One major change from 1948 to today is the virtual disappearance of nomographs. Today calculators and spreadsheets can easily perform most of the calculations that were considered too difficult and instead put into nomograph form. Most students today have never seen a nomograph, let alone used one.

Fred Lytle
Purdue University

Teaching Students How to Evaluate Data

PHILIP J. ELVING AND M. G. MELLON
Purdue University, Lafayette, Ind.

The general topic of what can be taught undergraduates in the beginning course in quantitative analysis is discussed as well as what can be taught graduate students concerning the concepts of precision and accuracy. It is believed sufficient to acquaint undergraduate students with the concepts of the arithmetical mean and the standard deviation, and the relations derived regarding the limits of variation of an observed average as well as the concepts of confidence limits and quality control charts. Although undergraduate students cannot be expected individually to perform sufficient experiments to apply statistical concepts with any degree of validity, it is possible to give them data on which they can base statistical operations, in order to familiarize themselves with the mechanics involved. Stress should be placed on the students acquiring an introduction to the basic concepts which can be used to evaluate the validity of data whether obtained in the chemical laboratory or elsewhere.

The very idea of introducing more subjects into an already overcrowded curriculum fills the teacher of chemistry with dread. However, the subject of statistics is so important in industry and is becoming so increasingly important in the development of analytical and testing methods, that some introduction to the subject should be given in undergraduate courses. Furthermore, in graduate courses in analytical chemistry, sufficient material on the application of statistics should be introduced so that prospective research chemists will realize that they can shorten their research work to an appreciable extent by the proper design of experiments and the proper use of the data obtained from their experiments. The present authors do not believe that sufficient material could be introduced in the usual courses in analytical chemistry to give the students sufficient background to handle statistical techniques competently, or even fully to understand statistical implications. However, the students can be made to realize the importance of the statistical approach and they can be taught a few items of elementary statistical manipulation.

At present, introductory textbooks in quantitative analysis usually content themselves with little beyond the discussion of the terms, accuracy and precision. In their laboratory work the students in the usual beginning quantitative analysis course merely calculate the average of two or three results. This represents, in many cases, the total introduction to the theory of errors and statistical methods given to undergraduates. The question is: what should be taught to these students? What is said below is merely personal opinion concerning what might be taught to undergraduates in the limited time available. At best, this material will have to be included in an hour or two of lecture, supplemented, perhaps, by the assignment of several exercises to be done by the students in the application of statistics to problems.

BASIC CONCEPTS

There is probably no real argument concerning the inclusion in the beginning course in quantitative analysis of an ample consideration of the factors which influence the reliability of a determination, the proper use of significant figures, and some understanding of the nature of deviation from the most probable value or from the average value. Among the allied factors that might be considered are the general nature of experimental error, and the factors which will cause one person using the same method on similar portions of the same substance to obtain replicate results that do not coincide; the latter feature perplexes students. Most students

2855-8/94/0068$08.00/0   ©1994 American Chemical Society   *Milestones in Analytical Chemistry*

have difficulty in realizing why precise results are not in themselves a guarantee of accuracy, or why results differing by several tenths of 1% may yield an accurate average.

One of the difficulties which persist through the undergraduate curriculum into the graduate training is a lack of understanding of significant figures. It is not uncommon to see the results calculated by graduate students in connection with problems or with experimental data expressed to far more significant figures than are warranted by either the data available or the precision and accuracy of the experimental measurements.

In their presentation of material to students, the present authors have availed themselves of the published work of the American Society for Testing Materials (*1*), and of individual authors such as Mitchell (*2*), Moran (*3*), and Wernimont (*4*). While the material presented here is familiar to all with any statistical background, it may be of value to other teachers of analytical chemistry.

MEAN AND STANDARD DEVIATION

The students should be introduced to the normal frequency distribution curve as shown in Figure 1. In this familiar hump-shaped or bell-shaped curve, the number of items of a given value is plotted against the value. A vivid illustration of the validity of the normal curve in practice and of the influence of experimental results on the shape of the curve can be seen in the data obtained in the analysis of a sample of cement for silica and calcium by 182 analysts. (Figures 2 and 3 are reproduced from photographs given to the authors by the late Father Francis W. Power of Fordham University. The data were apparently taken from a cooperative study of cement analysis; the published source of the data, if any, could not be located.)

Figure 2 shows the results obtained for silica. Here the data are in good agreement, and result in a tall, slender curve. The statistical nature of the data as compared to the normal curve is well emphasized by the results tailing to less than 19% and to more than 25%. Incidentally, such a curve is a good morale builder for students in indicating that even experienced analysts do not always get perfect results. Figure 3 shows the data obtained on calcium by the same group. Here the data form a curve which is less tall and less slender.

The graphical arrangements of Figures 2 and 3 enable the instructor to present to the students the concept of the normal curve and its changing shape in accordance with the nature of the data obtained as regards accuracy and precision.

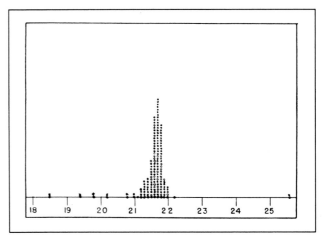

Figure 2

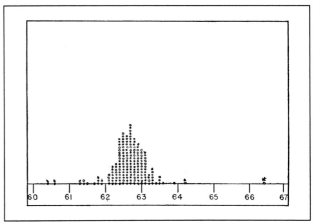

Figure 3

A fundamental introduction of the students to statistical tools can then be obtained by acquainting them with two basic statistical concepts characteristic of the distribution of the experimental data: (1) a measure of the average or central tendency of data such as the usual arithmetic mean or average, and (2) an index to the spread or dispersion of the values obtained about the central value mentioned such as the standard deviation. These values might be defined as follows:

Arithmetic mean or average
$$X = \frac{\Sigma X_i}{n}$$

Standard deviation
$$\sigma = \sqrt{\frac{\Sigma(X_i - \overline{X})^2}{n}} = \sqrt{\frac{\Sigma X_i^2}{n} - \overline{X}^2}$$

$$= \frac{1}{n}\sqrt{n\Sigma X^n - (\Sigma X_i)^n}$$

where *n* is the number of values in the set and X_i is an individual value representing all values from X_1 to X_n.

The summation limits have been omitted, as they are readily defined and explained by the instructor; their inclusion in the formulas, as in the one for average deviation subsequently presented, often confuses beginning students.

This treatment is in agreement with the recommendations of the manual on presentation of data issued by the American

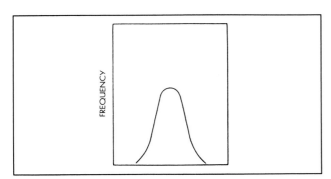

Figure 1

Society for Testing Materials (*1*). In this manual it is recommended that, "given a set of *n* observations of a single variable obtained under the same essential conditions," there be presented "as a minimum, the average, the standard deviation, and the number of observations."

Because the average deviation,

$$\frac{\sum_{i=1}^{n} |X_i - \overline{X}|}{n}$$

is frequently found in chemical literature, it should be defined.

As students are usually interested in applying statistical concepts to small numbers of data, they are introduced to the following modified standard deviation; the role of $(n - 1)$ as a correction factor is indicated.

$$\sigma = \sqrt{\frac{\Sigma(X_i - X)^2}{n - 1}}$$

RANGE

Having introduced the students to the two functions of the mean and the standard deviation, it will then be possible to indicate to them how these functions can be used to estimate percentage of the total number of a group of experimental values which will lie in any stated interval about the average or central value. This can be done by acquainting them with two nomographic charts which are given in the A.S.T.M. manual.

Using the nomograph shown in Figure 4, the minimum fraction or percentage of the results which lie within a given range can be easily calculated. This chart is based on Tchebycheff's theorem. Its use is based only on knowing the average and the standard deviation, and on selecting some value of *t*. For example, for four observations—i.e., *n* = 4—it can be seen that all four observations will probably fall within the region of the average plus or minus twice the standard deviation— i.e, select *t* = 2. This follows from the chart, where, for *t* = 2, less than 25% of the measurements will be outside the limits specified. The nomograph of Figure 4 allows one to state without reservation that, on the presentation of *n*, $\overline{X}$, and σ, more than $(1 - 1/t^2)$ of the total number of observations lie within the range, $\overline{X} \pm t\sigma$.

Further, to calculate the approximate percentage of the total number of observations within given limits, as contrasted to minimum percentage within these limits, use is made of the nomograph shown in Figure 5. This nomographic chart is on the distribution law integral. By means of this chart, it is possible to calculate the percentage of the total number of observations lying within any given symmetrical range about the average

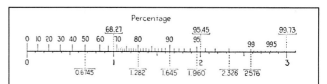

Figure 5

value. For *t* = 2, 95% of the observed results will be within the range of the mean plus or minus twice the standard deviation.

In addition to the average, standard deviation, and the number of observations, use of this chart requires that the data be obtained under controlled conditions. The latter condition is probably fulfilled under careful analytical practice.

Thus, it would be possible to have the students take a group of values obtained under normal laboratory conditions for analyzing a given sample and to calculate what percentage of the total number of observations will fall within, let us say, plus or minus twice the average deviation from the mean.

Furthermore, it is then possible to indicate to students how they might be able to calculate the confidence range or the range within which the true objective average, $\overline{X}'$, may be expected to lie with a given statistical probability. This can be done using the average and standard deviation, and the following relationship:

Computation of Limits for Objective Average, $\overline{X}'$:
$$\overline{X} \pm a\sigma$$
in P_s per cent of observations
$n < 25$, use nomograph of n, a, P_s
$n > 25$, $a = t/\sqrt{n - 3}$

For large numbers of measurements, the formula involving *t* may be used to calculate *a*; for small numbers of data down to four, the nomograph shown in Figure 6 can be used. Thus, for four measurements on a sample, determining one constituent, the objective average will fall within the limits of the average ± 1.9 times the standard deviation 95% of the time.

In presenting such material to students it should be emphasized that the objective average, $\overline{X}'$, of the observed measurements may not be identical with the true or most probable value, $\overline{X}'_T$, of the quantity measured, owing to systematic or constant errors. Such errors include incorrect gravimetric factors, erroneous titrant standardizations, and certain apparatus faults.

After consideration of the topics mentioned, the subject of control charts can be discussed. By this time, the students will have enough background to understand the meaning of control limits and a state of control.

During the current semester at Purdue, an interesting experiment was performed in the second semester part of the first year course in quantitative analysis. The group of approximately forty students was divided into two sections. The members of each section analyzed the same sample. The sample used was a mixture of sodium chloride, potassium chloride, and silica. The determinations were (1) material insoluble in water, (2) chloride by precipitation as silver chloride and weighing, and (3) calculation of the sodium and po-

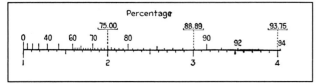

Figure 4

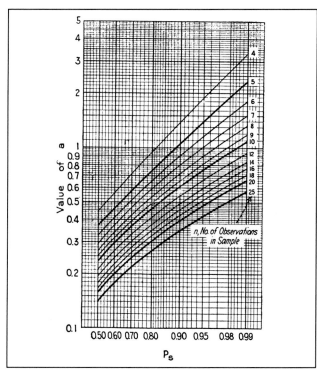

Figure 6

tassium contents using the experimental data obtained. The results of the various determinations for each of the two samples were then plotted in the form of control charts. The data obtained for the silica content of the two samples, which had the same percentage of silica but differing percentages of sodium chloride and potassium chloride, are given in Table I.

In the same course, the students are given the results obtained over a period of years for a sample used in the first semester of the quantitative analysis course. They are asked to calculate the arithmetical mean or average, the average deviation, and the standard deviation. The problem of rejection of observed values—an important problem to students in "quant"—is discussed.

In the basic course in analytical chemistry for graduate students, various indexes of precision used in the literature in addition to those mentioned are discussed. Sets of research data are exhibited and various items such as the mean, average and standard deviations, and variance are calculated. The meanings of the various types of reliability statements that can be made are analyzed. In the seminar, in which all graduate students majoring in analytical chemistry, participate, six hour-sessions were devoted to a discussion of the material in the A.S.T.M. manual (1) and Wernimont's (4) paper on the use of control charts in the analytical laboratory.

ACKNOWLEDGMENT

The authors wish to thank the American Society for Testing Materials for permission to reproduce in Figures 4, 5, and 6 material from (1).

LITERATURE CITED

(1) Am. Soc. Testing Materials, "A.S.T.M. Manual on the Presentation of Data," Philadelphia, 1945.

Table I. Silica Content Calculations for Control Charts

Test no.	X_1	X_2	X_3	Ex	$\bar{X}$	R
			Sample 1			
1	10.06	10.09	10.05	30.20	10.07	0.04
2	9.90	10.01	10.13	30.04	10.01	0.23
3			Results rejected			
4	9.80	9.81	9.99	29.60	9.87	0.18
5	9.96	10.12	10.32	30.40	10.13	0.36
6	10.26	10.28	10.36	30.90	10.30	0.10
7	10.22	10.11	....	20.33	10.16[a]	....
8	10.31	10.32	10.32	30.95	10.32	0.01
9	9.88	9.97	10.32	30.17	10.06	0.44
10	10.03	10.06	10.12	30.21	10.07	0.09
11	9.83	10.08	10.10	29.93	9.98	0.27
12	9.56	9.58	9.59	28.73	9.58	0.03
13	9.74	10.01	10.02	29.77	9.92	0.28
14	10.20	10.01	10.09	30.30	10.10	0.19
15	10.05	10.07	10.10	30.22	10.07	0.05
16	9.82	10.02	10.10	29.94	9.98	0.28
17	10.10	10.14	10.31	30.55	10.18	0.21
18	10.09	10.13	10.31	30.53	10.18	0.22
19	10.06	10.10	10.12	30.28	10.09	0.06
20	9.86	9.85	10.10	29.84	9.95	0.25
21	10.09	10.11	10.23	30.43	10.14	0.14
22	9.96	9.96	9.96	29.88	9.96	0.00
23	10.11	10.14	10.14	30.39	10.13	0.03
24	10.37	10.47	10.53	31.37	10.46	0.16
25	10.36	10.41	10.44	31.21	10.40	0.08
26	10.08	10.10	10.15	30.33	10.11	0.07
27	9.50	10.22	10.15	29.87	9.96	0.72
28	9.67	9.90	10.41	29.98	9.96	0.74
			Sum	806.35		5.23
		$n = 80$	Av.	10.08		
			Sample 2			
1	9.94	9.96	9.97	29.87	9.96	0.03
2	10.06	10.14	0.22	30.42	10.14	0.16
3	10.03	10.06	10.08	30.17	10.06	0.05
4			Results rejected			
5	9.34	10.17	10.26	29.77	9.92	0.92
6	9.45	10.03	10.23	29.71	9.90	0.78
7	9.98	9.99	10.02	29.99	10.00	0.03
8	10.06	10.06	10.08	30.20	10.07	0.02
9	10.37	10.13	10.10	30.60	10.20	0.27
10	9.76	10.02	10.11	29.89	9.96	0.35
11	10.04	10.20	10.00	30.24	10.08	0.20
12	10.05	10.19	10.28	30.52	10.17	0.23
13	10.05	10.09	10.14	30.18	10.06	0.09
14	10.08	10.10	10.26	30.44	10.15	0.18
15	10.06	10.07	10.08	30.21	10.07	0.02
16	9.98	10.14	10.32	30.44	10.15	0.34
			Sum	452.65		3.67
		$n = 45$	Av.	10.06		

[a] Av. of 2.

Preliminary control limits for ranges of 3 using all data:

$$\bar{R} = \frac{8.90}{41} = 0.217; \text{ limits } = 0 \text{ to } (2.574 \times 0.217) = 0 \text{ to } 0.56$$

Final control limits for ranges of 3 after discarding all values above 0.7.

$$\bar{R} = \frac{5.74}{37} = 0.155; \text{ limits } = 0 \text{ to } (2.574 \times 0.155) = 0 \text{ to } 0.40$$

Control limits for averages:
Sample 1. $10.07 \pm (1.023 \times 0.155) = 10.07 \pm 0.16 = 9.91$ to 10.23
Sample 2. $10.06 \pm (1.023 \times 0.155) = 10.06 \pm 0.16 = 9.91$ to 10.22

(2) Mitchell, J. A., *Anal. Chem.*, **19**, 961–7 (1947).
(3) Moran, J., *Ind. Eng. Chem.*, *Anal. Ed.*, **18**, 280–4 (1946).
(4) Wernimont, G., *Ibid.*, **18**, 587–92 (1946).

Received September 10, 1948.

Reprinted from *Anal. Chem.* **1948**, *20*, 1140–43.

The development of searchable libraries for spectral identification of unknowns, like much science, consists of key papers with a hidden background involving people, organizations, and effort. In the late 1940s many scientists were interested in a rapid search system for powder diffraction data. Punched cards, used for a variety of purposes such as bibliographic recording and search with "knitting needle" approaches, seemed a perfect medium. Matthews' recording system used the three strongest lines of knowns, in three of the six combinations, to serve as a primary filter in search strategies. The library, which then had fewer than 4000 members, was built around a hashing-type approach involving "banks" that limited search sizes to perhaps 300 cards. The approach merged with efforts being pioneered at Penn State by Wheeler Davey. An evolutionary product called Termitrex followed. Various coalesced or related professional standards groups carried the concept forward through IBM punched cards, magnetic tapes, and CD-ROM technologies. The current International Center for Diffraction Data library now contains more than 60,000 compounds.

Raymond E. Dessy
Virginia Polytechnic Institute and State University

Punched Card Code for X-Ray Diffraction Powder Data

F.W. MATTHEWS
Canadian Industries Limited, McMasterville, Quebec, Canada

A punched card is described which would enable a rapid and exhaustive search to be made of powder x-ray diffraction data for the identification of crystalline chemical compounds. This search could be based on the most intense lines of the x-ray diffraction pattern or on one intense line of the diffraction pattern and the elemental chemical composition of the substance.

The identification of a chemical compound by the use of x-ray diffraction powder patterns is based on a thesis of Hull (6) "that every crystalline substance gives a diffraction pattern; and that the same substance always gives the same pattern." It remained to be shown by Hanawalt and Rinn (5) that these patterns were sufficiently different to become the basis of a practical method of analysis. These authors described a method of tabulating powder patterns in a manner suitable for routine chemical identification. This scheme used a large ledger which was unsuitable for reproduction and general distribution. As an alternative, a card file of these data, using basically the same scheme, was published by the American Society for Testing Materials (1). This provides an expandable file which is suitable for indexing a comparatively small number of data. The present file, which lists about 4000 substances, has already proved somewhat unwieldy.

If data on three times this number of substances were considered, the size of the file would be such that a search would be very difficult. Alternative solutions to this problem, including the use of punched cards, were discussed in a previous paper (7). The present paper presents a revised punched card code which would facilitate the search of a file of powder diffraction data.

The method of Hanawalt and Rinn (5) used the three most intense (strongest) lines of the powder diffraction pattern as the index lines. The use of three lines is required because variations in x-ray technique and texture of the sample of a given substance cause variations in the relative intensities of the lines of the diffraction pattern. In the ten years this method has been in general use, the use of three index lines has proved necessary and effective for searching purposes.

In searching a file of data for the identification of a powder diffraction pattern, it is usual to start with two diffraction lines of the pattern. Ideally these would be the strongest and second strongest lines of the pattern, but because of the variation in relative intensity of the lines, and the danger that strong lines may be missed in the case of mixtures, an index to be used for an exhaustive search should list the six combinations of the three strongest lines. The card file of diffraction data published by the A.S.T.M provided three cards for each entry, placed in the index in positions determined by three combinations of the strongest lines—1st, 2nd; 2nd, 1st; 3rd, 1st. The combination 1st, 3rd, which has a relatively high probability of occurrence, and the two combinations 2nd, 3rd and 3rd, 2nd, which become important if the first line were missed, would have required more cards in the index.

A preferred arrangement for the routine search of the card file was described in the foreword to the original set of diffraction data index cards (1). This divided the cards into groups or blocks, and each of the three cards for a given substance was placed in the block determined by the three strongest lines of the pattern. Within the block, the cards were arranged in order of the next strongest line of the substance. This placed the cards in the three most probable places, in the card index.

When a punched card system of indexing these data was considered, it was at first thought that the use of the three cards could be eliminated by coding each of the three lines on a single card. This is possible but would have the great disadvantage that a search of the whole file (say 10,000 cards) would have to be made each time a search was attempted. A greatly improved system would be to retain the three-card system, coding the three lines on each of the three cards, and then filing the cards in blocks as described above. With this arrangement, any of the three strongest lines can be made the basis of an exhaustive search by examining one block (approximately 300 cards for 10,000 substances) or at the most two blocks, if the line in question were near the division between two blocks. This would apply equally well if one started with one line or with any two of the three strongest lines. The fact that punched cards can be sorted mechanically greatly facilitates such a search. Because all cards would carry all the data, there would be no necessity, as in the present card index (1), to refer to a master or "first" card.

CHEMICAL DATA

The use of chemical composition as a guide to the identification of powder patterns has been emphasized by several publications. Frevel (4) described a scheme for the listing of powder data by chemical composition. No determinative tables of this type have been published, possibly because of the difficulty of expanding such tables as new data become available.

With the punched card described above, the chemical composition may be coded in such a manner that it may be used in a search of the diffraction data. This would make possible a search based on the following information: If the unknown is a compound of magnesium having a strong powder line at

3.03 Å all index cards having data on compounds containing magnesium can be mechanically sorted from the block 3.00 to 3.05. This accomplishes an exhaustive search which otherwise would have required special tables or have been extremely laborious. In a similar manner information on specific subjects such as organic compounds, metals and alloys, minerals, etc., can be readily sorted from the block by use of a code, punched in the card.

THE PUNCHED CARD

The card presented in Figure 1 is designed around the present 3×5 inch card published by the American Society for Testing Materials (1).

The 3×5 inch area in the center of the card gives crystallographic data on the substance sodium chloride. The 0.5-inch margin carries a double row of holes for punched card coding. The coding of the card is described starting at the upper left-hand corner and proceeding clockwise around the card. On the first card, the block code is determined by the first line; the range 2.80 to 2.85 was numbered 40 in a previous description (7). This is coded 4 in the tens field (4 deep punch) and zero in the units (no punch). The second line is coded 1 in the units field (1 deep) and 9 in the tenths field (7 and 2 shallow) and +7 in the hundredths field. This number has one zero (01.99). It is the first card. The third line is coded 1.62. The chemical composition code for sodium is 1 deep 3 shallow; for chloride 11 deep and 3 shallow. The areas a, b, c, and d are free for coding other data on the cards.

The margin on these cards could be increased to provide a 4×6 inch card with 0.5-inch margins, punched with a double

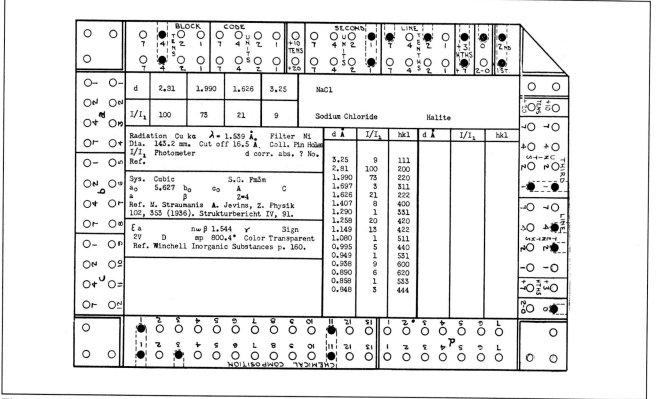

Figure 1. X-Ray Diffraction Data on Punched Card

row of holes and printed to give cards similar to that shown in Figure 1. This type of card is sold under the trade name Keysort by the McBee Company, Athens, Ohio, and under the trade name Cope-Chat, by the Copeland Chatterton Company in Canada and England.

The code should be such that it may be readily used in a wide variety of chemical and crystallographic investigations and a part of the card should be left for individual use. The code suggested would use three of each card in the set, designated as first, second, and third cards. On the first card the order of the lines would be coded in this order: strongest, second strongest; on the second card: second strongest, strongest; and on the third card: third strongest, strongest. The designation of the card could be indicated by punches in the upper right-hand corner, by a narrow colored border, or by the color on the card stock.

POWDER LINE CODE

On the first card the strongest line would determine the block. Provision is made for 99 blocks (the Hanawalt index used 78 blocks). The range of these blocks should be made so that an approximately equal number of cards would fall in each. Normally the cards would be kept in the block. The first line code would enable the cards to be sorted into blocks and would give a proof of file, by which misplaced cards could be readily traced. The order within the block would normally be that of the second line. Searches based on third line, chemical composition, innermost line, or some other feature would destroy this order. When such a search is complete, however, the cards should be mechanically sorted back to the second-line sequence for refiling.

A general rule for the card file may be stated as follows: If the data are approached from one point of view more frequently than another, the cards should be kept in that order, and re-sorted to that order for filing. The attitude that punched cards can be sorted mechanically, and therefore no attempt should be made to keep them in any order, is often, in the end, more time-consuming.

Papers on punched card technique by Casey, Bailey, and Cox (2, 3) describe a variety of coding methods. The numerical code used here is the 7-4-2-1 condensed code, in which any of the digits from 1 to 9 can be represented by one of the four digits or a combination of two. When a single row of holes is used, this code is not selective—i.e., 7's cannot be selected from 8 and 9, which are codes 7 + 1 and 7 + 2. The code is made selective with a double row of holes, by punching the digit deep when the digits 7, 4, 2, or 1 are intended. Digits formed by a combination of two are punched shallow. When used in this manner, any digit can be selected from the others. The second and third lines are coded in the actual value in Ångströms, so that it is unnecessary to consult a table. (With the first-line code, the use of a table is no disadvantage, as the field is not ordinarily used for searching.)

Powder lines in Ångström values may be expressed as four-figure numbers—tens, units, tenths, and hundredths. Because the index lines of a powder pattern rarely have values greater than 20, no provision is made for a tens field other

than +10 and +20. The units and tenths follow the 7-4-2-1 condensed code and in the hundredths field the cards are coded +3 and +7 when values in this range are given. In this code no provision is made for coding zero and in order to separate readily the 2.0 from the 12.0 and from digits other than zero in the tenths field it is necessary to code the number of zeros in the number. For this purpose zeros in the tens field have to be considered—i.e., 2.0 is considered as having 2 zeros: (02.0). Zeros in the hundredths field are not considered.

With a direct code of the type described, the range selected from a block can be as wide or as narrow as the circumstances indicate—i.e., on a search at 3.50, the cards from 3.0 to 4.0 may be selected, or only those of the narrow range 3.50 to 3.53.

Some consideration should be given to limiting the range of index lines by a rule of the following type. The index lines of a pattern are the three strongest lines between the d values of 1 and 10 Ångströms. This would avoid the difficulty that the range of d values recorded varies with camera construction and radiation employed. This is particularly true of focusing cameras. It would also limit the range over which a spectrometer should be run for identification purposes. The coding of data would also be simplified by such a rule.

CHEMICAL COMPOSITION

A number of schemes for coding chemical composition have been considered (7). The scheme suggested here divides the atomic table into thirteen chemically related groups of elements, each of which is given a number which is coded with a deep punch. Within the group, elements are given a second number which is coded with a shallow punch. A suggested code is given in Tables I and II. As each compound contains a number of elements (present average approximately 3.5) which must be coded in this field, there will be some overlapping where a particular element is selected by use of this code. For instance, when compounds of calcium are selected, in addition to all the cards having data on compounds containing calcium, other cards will follow. The deeply punched code number for calcium, however, assures that these additional cards will contain elements chemically related to calcium and will form only a small proportion of the cards sorted from the block. This difficulty could be overcome, in part, by extending the code to a larger number of groups (from 13 to 20 groups) or by using a three-number code for each element (one deep punch and two shallow). In its present form, it is felt that this difficulty would not for practical purposes be serious. Certain arbitrary rules could be set up to simplify this code—e.g., oxygen is coded in oxides only, carbon is coded in inorganic compounds only. If in a laboratory some elements were of particular interest and the overlapping mentioned above is undesirable, these elements could be given a direct code in the spare parts of the card. In the spare parts of the card organic compounds, metal-organic compounds, alloys, minerals, etc., could also be given a direct code. Other codes based on melting point, optical properties, innermost lines of the diffraction pattern, or other easily measured determinative property of a substance could be developed.

Table I. Suggested Code

Element	Code	Element	Code	Element	Code	Element	Code
Li	1-2	Fe	5-2	N	9-4	Tm	12-3
Na	1-3	Co	5-3	P	9-5	Yb	12-4
K	1-4	Ni	5-4	As	9-6	Lu	12-5
Rb	1-5	Cu	5-6	Sb	9-7		
Cs	1-6	Ag	5-7	Bi	9-8		
		Au	5-8				
				O	10-9	Ac	13-6
Be	2-8	Ra	6-9	S	10-11	Th	13-7
Mg	2-9	Rh	6-10	Se	10-12	Pa	13-8
Ca	2-10	Pd	6-11	Te	10-13	U	13-9
Sr	2-11	Os	6-12	Po	10-1	Np	13-10
Ba	2-12	Ir	6-13			Pu	13-11
Ra	2-13	Pt	6-1	F	11-2	Am	13-12
				Cl	11-3	Cm	13-1
B	3-1	C	7-2	Br	11-4		
Al	3-2	Si	7-3	I	11-5		
Sc	3-4	Ti	7-4				
Y	3-5	Zr	7-5				
Ga	3-6	Hf	7-6	La	12-6		
In	3-7			Ce	12-7		
Tl	3-8	V	8-7	Pr	12-8		
		Cr	8-9	Nd	12-9		
Zn	4-9	Mn	8-10	Il	12-10		
Cd	4-10	Cb	8-11	Sm	12-11		
Hg	4-11	Mo	8-12	Eu	12-13		
Ge	4-12	Ti	8-13	Gd	12-1		
Sn	4-13	Ta	8-1	Tb	12-1		
Pb	4-1	W	8-2	Dy	12-2		
		Re	8-3	Ho	12-2		
				Er	12-3		

Table II. Suggested Code

Element	Symbol	Code	Element	Symbol	Code
Aluminum	Al	3-2	Neodymium	Nd	12-9
Americium	Am	13-12	Neptunium	Np	13-10
Antimony	Sb	9-7	Nickel	Ni	5-4
Arsenic	As	9-6	Nitrogen	N	9-4
Barium	Ba	2-12	Osmium	Os	6-12
Beryllium	Be	2-8	Oxygen	O	10-9
Bismuth	Bi	9-8	Palladium	Pd	6-11
Boron	B	3-1	Phosphorus	P	9-5
Bromine	Br	11-4	Platinum	Pt	6-1
Cadmium	Cd	4-10	Plutonium	Pu	13-1
Calcium	Ca	2-10	Polonium	Po	10-1
Carbon	C	7-2	Potassium	K	1-4
Cerium	Ce	12-7	Praseodymium	Pr	12-8
Cesium	Cs	1-6	Protactinium	Pa	13-8
Chlorine	Cl	11-3	Radium	Ra	2-13
Chromium	Cr	8-9	Rhenium	Re	8-3
Cobalt	Co	5-3	Rhodium	Rh	6-10
Columbium	Cb	8-11	Rubidium	Rb	1-5
Copper	Cu	5-6	Ruthenium	Ru	6-9
Curium	Cm	13-1	Samarium	Sm	12-11
Dysprosium	Dy	12-2	Scandium	Sc	3-4
Erbium	Er	12-3	Selenium	Se	10-12
Europium	Eu	12-13	Silicon	Si	7-3
Fluorine	F	11-2	Silver	Ag	5-7
Gadolinium	Gd	12-1	Sodium	Na	1-3
Gallium	Ga	3-6	Strontium	Sr	2-11
Germanium	Ge	4-12	Sulfur	S	10-11
Gold	Au	5-11	Tantalum	Ta	8-1
Hafnium	Hf	7-6	Tellurium	Te	10-13
Holmium	Ho	12-2	Terbium	Tb	12-1
Illinium	Il	12-10	Thallium	Tl	3-8
Indium	In	3-7	Thorium	Th	13-7
Iodine	I	11-5	Thulium	Tm	12-3
Iridium	Ir	6-13	Tin	Sn	4-13
Iron	Fe	5-2	Titanium	Ti	8-13
Lanthanum	La	12-6	Tungsten	W	8-2
Lead	Pb	4-1	Uranium	U	13-9
Lithium	Li	1-2	Vanadium	V	8-7
Lutecium	Lu	12-5	Ytterbium	Yb	12-4
Magnesium	Mg	2-9	Yttrium	Y	3-5
Manganese	Mn	8-10	Zinc	Zn	4-9
Mercury	Hg	4-11	Zirconium	Zr	7-5
Molybdenum	Mo	8-12			

CONCLUSION

The publication of powder diffraction data on punched cards of the type described would increase the cost of publication, but this increase should be justified if the usefulness of the index were increased by making the data more available. The proposed code is only one of many that could be developed; the individual requirements of workers would necessitate variations of the method employed, and many of these variations would make use of the spare parts of the card.

ACKNOWLEDGMENT

The writer wishes to thank A. F. Kirkpatrick of the American Cyanamid Research Laboratories for the suggestions made regarding the coding employed and J. W. Bryers of the McBee Company for his interest and help in the design of the card.

LITERATURE CITED

(1) Am. Soc. Testing Materials, Philadelphia, Pa., "X-Ray Diffraction Data Index," 1942.
(2) Casey, R. S., Bailey, C. F., and Cox, G. J., *J. Chem. Education*, **23**, 495 (1946).
(3) Cox, G. H., Casey, R. S., and Bailey, C. F., *Ibid.*, **24**, 65 (1947).
(4) Frevel, L. K., *Ind. Eng. Chem., Anal. Ed.*, **16**, 209 (1944).
(5) Hanawalt, J. D., and Rinn, H. W., *Ibid.*, **8**, 244 (1936).
(6) Hull, A. W., *J. Am. Chem. Soc.*, **41**, 1168 (1919).
(7) Matthews, F. W., and McIntosh A. O., *Can. Chem. Process Inds.*, **31**, 63 (1947).

Received February 3, 1949.

Reprinted from *Anal. Chem.* 1949, 21, 1172–75.

The 1950s

Redefining the Horizons of Analytical Chemistry

It was a time of optimism and promise and change—a decade that encompassed technological developments that are commonplace today. This was the era that gave birth to color TV, satellites in space and, of course, rock and roll and Elvis.

Progress in science was wide-ranging, and analytical chemistry was no exception. World War II provided the stage for a variety of scientific and engineering advances that found their way into industry and academia. At the start of the decade, the importance of the analytical work done in conjunction with the Manhattan Project was not clear. Although much of the information had not been declassified or revealed, activation analysis was viewed as potentially of the greatest importance for the development of qualitative and quantitative analysis, and with it the analytical techniques of nuclear chemistry.

Defining analytical chemistry

How did analytical chemists of the day describe their discipline? Phillip J. Elving, a faculty member at what was then The Pennsylvania State College, suggested that this had been a relatively straightforward task during the preceding 20–30 years. Analytical chemists generally undertook inorganic, mineral-type analysis; the substantial progress that had been made in biochemistry and in the analysis of organic materials did not involve analytical chemists. Most research in analytical science in the academic setting involved gravimetry and titrimetry, although scientists such as N. H. Furman, I. M. Kolthoff, M. G. Mellon, and Hobart Willard were beginning to expand this scope.

With the new decade came the opportunity to reexamine the direction of analytical chemistry. Previously existing boundaries blurred, and the horizons of the discipline broadened. Elving suggested a new definition, one that is remarkably valid today. He viewed analytical chemistry as a science involving "all techniques and methods for obtaining information regarding the composition, identity, purity, and constitution of samples of matter in terms of the kind, quantity, and groupings of atoms and molecules, as well as the determination of those physical properties and behavior that can be correlated with these objectives."

For the practicing analyst, the 1950s were a time of robust developments. In the separations arena, thin-layer chromatography was born from Kirschner, Miller, and Keller's work in using adsorbent-coated glass strips to separate terpenes in citrus fruits. Study of GC theory contined as Marcel J. E. Golay was able to recognize an analogy between a GC column and an electrical circuit and to derive a mathematical theory that predicted results in accord with experiment. Despite the useful conclusions, he preferred a more physical theory of GC, a desire that eventually led him to the theoretical and experimental demonstration of capillary columns.

In spectroscopy, investigations of NMR led to James N. Shoolery's demonstration of the usefulness of the technique for chemical applications. Kiers, Britt, and Wentworth laid down the basic tenets of phosphorescence theory and instrumentation and its usefulness for the analysis of organic mixtures. Also during the decade, Gilbert, Hawes, and Beckman described the Beckman flame photometer, a commercial instrument that found quick acceptance in the marketplace and expanded the technique of flame spectrophotometry.

Another noteworthy investigation undertaken during this period was the idea of combining a mass spectrometer with a gas chromatograph. R. S. Gohlke used a TOF instrument with

---

1950
North Korea invades South Korea
United Nations building in New York completed
Glenn Seaborg discovers californium
World population reaches approximately 2.3 billion
Cyclamate, an artificial sweetener, is introduced

1951
The African Queen and *An American in Paris* premiere
Moviegoers don special polarizing glasses to view 3D movies
Nobel Prize for Chemistry goes to Edwin McMillan and Glenn Seaborg for the discovery of plutonium
Color television debuts in the U.S.
Chrysler introduces automobile power steering

1952
King George VI of England dies
The Power of Positive Thinking (Norman Vincent Peale) is published
Felix Bloch and Edward Purcell win the Nobel Prize in Physics for their work on magnetic fields in atomic nuclei
"Piltdown Man" is revealed as a fraud
Sony develops the pocket-sized transistor radio

1953
Korean armistice is signed on July 27
Queen Elizabeth II is crowned
Michelin and Pirelli introduce radial ply tires
Joseph Stalin dies
USSR explodes the hydrogen bomb

1954
Jonas Salk begins inoculating students in Pittsburgh with antipolio serum
Nobel Prize in Chemistry goes to Linus Pauling for his study of chemical bonds
Nautilus, the first atomic-powered submarine, is commissioned
TV dinners are introduced in the U.S.
Lord of the Flies (William Golding) *Under Milk Wood* (Dylan Thomas), and *The Lord of the Rings* (J.R.R. Tolkein) are published

photographic recording and concluded that the technique showed great promise for the analysis of volatile mixtures.

Several instrumental components, such as fast, sensitive strip-chart recorders, sensitive IR detectors, detectors to count radioactivity, magnetic stirrers, and thermistors and transistors, became available commercially, and the first digital minicomputer was developed for laboratory use.

A variety of spectrometers, including benchtop recording UV–vis, IR, multichannel emission, NMR, and spark source mass spectrometers, also emerged. Equally important were nonspectroscopic instruments such as single-pan balances, potentiostats, automated liquid–liquid extractors, and gas–liquid chromatographs.

Researchers also made advances in anodic stripping voltammetry, radioimmunoassay, and voltammetry with solid electrodes— techniques for which commercial instruments were unavailable.

Analytical professors of this decade faced a problem that exists today: how to incorporate all the material their students needed in a limited amount of time. Topics such as instruments and instrumentation, quantitative analysis based on organic functional groups, statistics and experimental design, separations, and kinetic methods demanded inclusion. At the same time, developments in other areas of chemistry threatened to decrease the time that was being allotted to the study of analytical chemistry.

Some analytical courses survived, but the situation forced professors to be increasingly selective in the topics they covered. Other analytical topics were incorporated into physical or organic classes. Trends in laboratory instruction included the use of semi-micro techniques and ion-exchange resins as well as the streamlining of Kjeldahl techniques.

Another aspect that reflected the expansive growth of analytical chemistry was the increasing number of conferences devoted to subjects of interest to analysts. Industry supported the idea of having its employees attend these meetings. It recognized that advances in analysis in terms of speed and precision would make it possible to introduce new large-scale continuous processes, better products, and economies in manufacturing. And, according to Journal Editor Walter Murphy, "it also realizes that new and revolutionary methods of analysis

can develop from but one source—increases in research of a fundamental nature."

Important meetings of the day included the Symposia on Analytical Chemistry at Louisiana State University, the annual Microchemical Symposia, the annual Division of Analytical Chemistry Summer Symposia, and the Pittsburgh Conference on Analytical Chemistry and Applied Spectroscopy.

The Pittsburgh Conference

The meeting known today as Pittcon had its genesis during this decade. Its pedigree includes spectroscopists in the Pittsburgh area who organized the 1940 Conference on Applied Spectroscopy; the Society of Analytical Chemists of Pittsburgh, which organized its first symposium in 1945; and the Spectroscopy Society of Pittsburgh, which was formed in 1946. The latter two groups joined forces in 1949 and, one year later, staged the first Pittsburgh Conference.

1957 Pittsburgh Conference Exposition

Held in Pittsburgh's William Penn Hotel, the meeting featured technical sessions on analytical chemistry, IR spectroscopy, emission spectroscopy, UV absorption spectroscopy, and specialized techniques. Presenters dealt with subjects such as the determination of fluoride, electrode problems in electrometric measurements, and recent applications of X-ray diffraction in the analytical field.

This successful conference started the rich tradition that Pittcon enjoys this year as it celebrates its 45th anniversary. One hallmark has been the ongoing efforts of volunteer organizations to maintain the quality of programming while offering improvements to serve conferees and inaugurating ways to recognize contributions to the community of analytical scientists.

The three-day conference held in 1951 featured a full session devoted to testimonial for Keivin Burns, a pioneer in emission spectroscopy, as well as an employment service and an exhibit of books published during the previous five years on topics of analytical interest. The exposition included 20 exhibitors who presented a complete assortment of instruments and apparatus for the laboratory. Attendance exceeded 950 by the second day.

By 1953 the conference had expanded to an entire week, and the exposition gained importance. The list of exhibitors and products displayed in 1954 included some familiar names: a

1955

U.S. Air Force Academy opens in Colorado
"Rock Around the Clock," "Davy Crockett," and "The Yellow Rose of Texas" are popular songs
Artificial diamonds for industrial use are produced in the U.S.
Field ion microscope is developed by E. W. Müller

1956

Neutrino is produced at Los Alamos
Eisenhower is re-elected President; Nixon is elected Vice President
"Blue Suede Shoes," "Hound Dog," and "Don't Be Cruel" are popular songs
Sabin develops oral polio vaccine
John McCarthy develops the computer language Lisp

1957

Common Market is established in Europe
West Side Story is produced
USSR launches the first Earth satellites, Sputnik I and II
New York Giants move to San Francisco; Brooklyn Dodgers head to Los Angeles

1958

Van Allen radiation belts around the Earth are discovered
Venus is observed at radio wavelengths
Guggenheim Museum opens in New York
U.S. satellite Explorer I is launched from Cape Canaveral
U.S. launches first moon rocket that travels 79,000 miles but fails to reach its target

1959

Hawaii becomes the 50th state
Nobel Prize for Chemistry goes to Jaroslav Heyrovsky for the development of polarography
St. Lawrence Seaway opens
First commercial Xerox copier is introduced
Anatomy of a Murder, La Dolce Vita, and *Ben Hur* premiere

Leeds & Northrup recording spectrometer for multiplier phototube detection over the range of 2100–7000 Å, a double-beam continuous recording unit for DU-type monochromators from Fisher Scientific for wavelength ranges of 215–375 mm and 350–750 mm, a Beckman Model DR from Arthur H. Thomas, a Model 21-610 mass spectrometer from Consolidated Engineering, and a dual-grating spectrograph from Bausch & Lomb.

As the decade progressed, the number of papers presented and the number of firms exhibiting products increased. With this expansion came international recognition of the meeting as one of the preeminent analytical gatherings and certainly the place to view the latest manufacturers' offerings. By 1959 the conference attracted more than 2800 attendees and 75 exhibitors, and 160 papers were presented in the technical sessions.

The Journal

Analytical Chemistry began the decade under the direction of Editor Walter Murphy. He was aided by Associate Editor Larry Hallett, who became Science Editor in 1953 and Editor in 1956.

Throughout the 1950s the Editors made a series of changes. A monthly New Products section was started in January 1950, in an attempt to provide readers with a constant source from which they could obtain the information; previously these announcements appeared in a variety of places and were sometimes noticed by interested analysts only by chance. In keeping with this desire to serve readers interested in new instruments and equipment, an advertised product guide was added as part of the review issue starting in April 1955. An A-page feature called Report for Management debuted in November 1955 as a forum for discussing the role of the analyst in industry and subjects of general interest. Finding time to keep up to date with the literature was as much a concern in the 1950s as it is today. The Editor's solution was to give readers a quick survey of the issue's technical papers in a format known as Briefs, which first appeared in January 1957. These interpretive summaries were prepared by a practicing analyst and were designed so they could be cut out and filed on cards for future reference.

The success of the Journal was apparent from its circulation of approximately 24,000 and total of more than 2100 pages published in 1959.

As the decade of the 1950s drew to a close, analytical chemists had redefined their discipline to reflect the changes brought on by the postwar world. This process left them well positioned and eager to meet the challenges of the information era of the 1960s.

LOUISE VORESS

This paper opened an era of rapid expansion in applications of flame spectrophotometry, especially in the United States. It did so not by reporting "breakthrough" science but by describing a convenient, well-engineered, affordable commercial instrument. It established the principle that academic research, although necessary, is not a sufficient base for popular acceptance of a technique.

S. R. Koirtyohann
University of Missouri

Beckman Flame Spectrophotometer

P. T. GILBERT, JR., R. C. HAWES, AND A. O. BECKMAN
National Technical Laboratories, South Pasadena, Calif.

A new flame spectrophotometer, now commercially available, is characterized by a simple detachable atomizer handling samples of 1 ml. or less, a heated spray chamber for evaporating the spray, a versatile burner for oxygen-gas or other flames, and construction of the atomizer-burner unit as an attachment for the Beckman quartz spectrophotometer. Individual measurements can be made in very rapid sequence, and the precision of such measurements is a few tenths per cent of full scale. Interference effects and methods of circumventing them are discussed. Flame spectra and detection limits are given for several dozen elements, and excitation characteristics and methods of analysis are illustrated for a few cases.

Flame spectrophotometry, as a quantitative analytical technique, has only recently aroused widespread interest in this country. Although flame spectra have been used nearly 100 years for the qualitative identification of elements, the literature discloses little work on the use of flame spectra for quantitative determinations prior to 1929, when Lundegårdh (7) published his first treatise on the method. A number of European papers on flame spectrophotometry have appeared since that time, but in this country, it was not until 1939 (6) that any work was published, and only within the past 3 years has any apparatus been described which differs significantly from that of Lundegårdh.

Lundegårdh and most of his followers employed a spectrographic technique. The solution to be analyzed was sprayed into a flame placed in front of a spectrograph. After exposure, the photographic plates were developed and the optical densities of various spectral lines recorded thereon were measured. With the aid of suitable calibration data the optical densities could be correlated with chemical concentrations with relative accuracies of a few per cent.

The photographic step introduced delay and inconvenience and also limited the accuracy of the method to the reproducibility of photographic emulsions. To overcome the disadvantages of the spectrographic method, direct-reading flame photometers were described as early as 1935. Instruments of this type were made by at least three manufacturers in Germany prior to the war. In 1945 the first direct-reading flame photometer was described in this country (1). Since that time many papers have appeared, describing applications and apparatus. Judging from the recent literature, it appears that the apparatus heretofore commercially available leaves much to be desired with respect to convenience and speed of operation, accuracy of measurement, and freedom from spectral interference in multicomponent solutions. To overcome these shortcomings the instrument described in this paper was designed. It has high optical resolving power with attendant freedom from the interference resulting from unresolved overlapping spectral bands. It provides high photometric accuracy. It is simple and fast in operation. Determinations can be made in a few minutes, and no cleaning is required between samples. A single drop of sample is consumed in making a reading, and detectable concentrations may be as low as a few parts per 10^8 for the alkali metals and somewhat higher for other elements. The instrument has already been used for the determination of twenty elements and undoubtedly can be used for many more.

Briefly, the features which distinguish the present instrument, described in greater detail below, from other direct-reading flame photometers, include (1) a one-piece, high-suction, concentric atomizer requiring no rinsing and providing exceptionally constant and low rate of consumption of sample; (2) a heated spray chamber, which completely evaporates the

2855-8/94/0081$08.00/0 © 1994 American Chemical Society

spray, enhances luminous intensity, and improves stability of performance; (3) use of an oxygen–natural gas flame; and (4) construction of the atomizer-burner unit as a separate attachment to be used in conjunction with the Beckman quartz spectrophotometer whose utility for other purposes is not interfered with by its use as a flame spectrophotometer.

DESCRIPTION OF BECKMAN FLAME SPECTROPHOTOMETER

A satisfactory flame spectrophotometer must have adequate resolving power to differentiate between the spectral emission of the element being determined and the emissions of any interfering substances that may be present. For general utility and precise measurements an essential part of a flame spectrophotometer is necessarily a monochromator which provides a continuous selection of wave lengths with resolving power sufficient to separate completely most of the easily excited emission lines, and freedom from scattered radiation sufficient to minimize interferences. These requirements are provided by the monochromator of the Beckman Model DU spectrophotometer (4). As there are many hundreds of these spectrophotometers already in use throughout the world, it was decided to design a flame spectrophotometer which could take advantage of this monochromator and the sensitive and accurate photometric equipment of these instruments. In a paper presented in November 1947 before the Soil Science Society of America, Rogers (9) told of his use of the Model DU instrument with an air-acetylene flame for the routine determination of potassium and calcium in soils. Others also have used the instrument for emission photometry.

General construction of the Beckman flame spectrophotometer accessory is shown in Figure 1.

It consists essentially of a burner and a spray atomizer, with associated gages and regulators for air, gas, and oxygen. The components are assembled in a unit which provides a mounting for the Model DU spectrophotometer. The control box on the left contains the pressure gages, control valves for air, gas, and oxygen, and a pressure regulator and filter for the air. The water-cooled burner (Figure 5) is designed for natural gas and oxygen and produces a broad steady flame which does not require precise optical alignment with the monochromator. The fuel is burned at many small ports, which are located around larger holes through which the air stream containing the atomized sample is introduced. The products of combustion pass upward through a water-cooled chimney. The sample is contained in the 5-ml. beaker at the extreme right (Figures 1 and 6). The capillary inlet tube of the atomizer (Figure 6) which dips into the sample has its inlet orifice sharply tapered to about 70 microns to prevent entrance of solid particles that might clog the capillary. With air at 10 to 30 pounds per square

inch pressure flowing through the atomizer, a suction of about 1 to 3 meters of water is created. The effect of change in level of the sample is, therefore, entirely negligible. The atomizer and spray chamber are made of borosilicate glass and the burner is constructed of nonreactive materials to avoid contamination. The spray tip, which is concentric with the air nozzle, has an internal diameter of about 100 microns and produces a very fine and constant spray which evaporates completely before striking the walls of the spray chamber. Condensation on the walls of the spray chamber is prevented by an electric heater wound on the spray chamber body, and the solute passes completely into the flame in the form of dry microscopic particles. The burner operates normally with natural gas at a pressure of 1 to 3 inches of water and oxygen at a pressure of 15 to 50 inches. It consumes about 15 cubic feet of oxygen per hour.

Figure 2 shows the flame photometer with the Model DU spectrophotometer in place. After the initial installation of a

Figure 1. Flame Spectrophotometer Accessory

Figure 2. Complete Flame Spectrophotometer

10,000-megohm phototube load resistor to increase the sensitivity fivefold, it is merely necessary to detach the lamp housing from the Model DU instrument to change from absorption to emission spectrophotometry.

FLAME CHARACTERISTICS

The oxygen-gas flame was selected because of its high temperature and convenience. An air-gas flame, because of its lower temperature, is suitable only for the determination of elements which are easily excited, such as the alkali metals. Oxygen-acetylene and oxygen-gas produce much hotter flames and correspondingly greater excitation. The former is somewhat superior in this respect to the latter, but this advantage is offset by the higher level of background illumination and the more stringent requirements imposed upon the design of a burner which will produce a steady, quiet flame with an oxygen-acetylene mixture without danger of flashing back. Mixtures of oxygen and artificial gas containing much hydrogen and carbon monoxide also flash back rather readily because of the high flame propagation velocity. Satisfactory fuels having relatively low flame velocity include natural gas and bottled propane and butane (cf. Table I, B).

Dilution of the flame by the air used to spray the sample reduces both the background illumination and the intensity of metallic excitation, and the ratio of spray to flame must be taken into account in the design of a flame photometer. Whereas tank nitrogen can be used instead of air for the atomizer, its use entails a not inconsiderable loss of sensitivity and increases oxygen requirements of the flame. On the other hand, oxygen can be used advantageously to atomize the sample. The flame is thereby altered in the direction of greater background intensity, but at the same time the intensity of metallic emission is roughly doubled. Whereas the present instrument is designed for an oxygen–natural gas flame with air for spraying, it can be used equally well with the following flames when oxygen is used for atomizing: natural gas, with very little additional oxygen; air-acetylene; artificial gas with a little air; and hydrogen alone.

The hydrogen flame is particularly desirable because it eliminates one of the required gases (it replaces compressed air and piped or bottled gas with a tank of hydrogen, the oxygen tank being connected to the atomizer instead of to the burner directly); because it cannot flash back, damage the burner, or produce soot; because it is perfectly quiet and remarkably steady (see Table I, C); and because it provides the greatest ratio of metallic emission to flame background over large regions of the spectrum. Whereas the hydrogen is consumed at a rate of about 40 cu. feet per hour in a typical instrument, and oxygen at about 25 cu. feet per hour, both gases can be conserved by turning them down to very low pressures between runs, without extinguishing the flame.

It is felt that the versatility of the instrument in being equally adaptable, without alteration, to any of a variety of gaseous media is a real advantage.

Typical flame background spectra are shown in Figure 3. Relative photocurrents are plotted against wave length for the natural gas–oxygen flame (atomizer air being turned off) and for the hydrogen flame with atomizer oxygen flowing. With slit widths of 0.1 and 0.2 mm. the OH spectrum in the ultraviolet and the molecular water bands in the infrared are partially resolved. The sharp peaks at 306 to 315 mμ interfere with very few metallic lines, and the same is true of the water peaks near 960 mμ.

It is essential that the flame burn steadily without undue care in the adjustment of controls. The burner has been carefully designed to achieve this result. Gas flow is adjusted with a needle valve, pressure being read on a liquid manometer. No gas pressure regulator is provided, because regulation was found to be superfluous under proper operating conditions where the luminosity is at a broad maximum or plateau with respect to gas and oxygen pressure. Figure 4, for example, shows relative photocurrents for the lithium line at 671 mμ as a function of oxygen pressure for various gas pressures. The broad maxima are clearly evident. Other substances give similar results. Pressure adjustments are not critical and fluctuations of pressure have not been troublesome. Suitable pressures are readily found experimentally by setting the wave-length scale for a particular element and adjusting the gas and oxygen pressures until maximum brightness is obtained. Thereafter the adjustments remain fixed and can be preset on future occasions.

With the hydrogen flame similar conditions prevail. A broad maximum in the luminosity curve will be found by adjusting either the hydrogen or the atomizer oxygen pressure. Near the optimal adjustment a change of 1% in the oxygen pressure affects the intensity of sodium light by 0.05 to 0.1% and a change of 1% in the hydrogen pressure affects it by about 0.1%. Actual maxima with respect to both hydrogen and

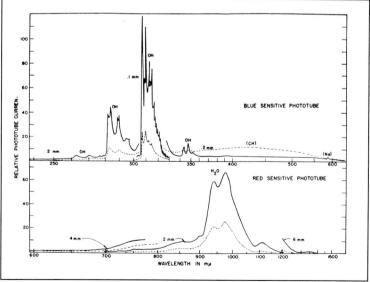

Figure 3. Flame Background

Emission of natural gas–oxygen flame with atomizing air off (---) and hydrogen flame with atomizing oxygen on (—). Slit widths shown as parameters. Molecules responsible for band peaks indicated

Table I. Examples of Data for Various Types of Determinations

A. Rapid determination of sodium in water (4 to 6 readings per minute). Natural gas, heated spray chamber. Readings taken in triplets, on first standard, unknown, and second standard, successively

Standard, 42.3 P.P.M. Na	Unknown	Standard, 92.0 P.P.M. Na
70.3	76.1	108.3
70.5	76.8	107.4
70.1	75.6	107.1
70.2	76.5	107.6
70.4	76.2	...
Av. 70.3	76.2	107.6

Standard errors of ratios = 0.4, 0.5%
Graphical interpolation (Figure 7) yields 49.1 p.p.m. Na for unknown
Probable error of result = 0.2% or 0.1 p.p.m.

B. Readings on sodium in tap water, using oxygen-butane flame

First Series		Second Series	
100.1	100.4	101.5	101.3
100.7	100.7	101.5	101.7
100.8	101.5	101.2	101.6
101.5	100.5	101.4	101.2
101.5	101.4	101.8	101.4
101.2	101.0	101.6	101.5
Av.	101.0		101.5
Standard error %	0.40		0.19

C. Determination of sodium in water, using hydrogen flame and oxygen atomizer

	Blank (Water)	Standard, 50 P.P.M.	Standard, 100 P.P.M.	Unknown
	0.5	68.0	98.5	105.6
		68.0	98.6	105.2
		67.9	98.5	105.5
		68.0	98.5	105.8
		68.0	98.5	105.4
	0.5	67.8	98.4	105.4
Av.	0.5	67.95	98.5	105.5
Net reading		67.45	98.0	105.0
Ratios			1.453	1.072
Ratios for gas flame			1.458	1.072

Sodium found by graphical extrapolation (Figure 7) 114.2 p.p.m.

D. Determination of traces of sodium in distilled water. Part of longer sequence of measurements. Instrument at maximum sensitivity; shutter left open continuously, dark current not adjusted. Hypothetical sodium-free water (which was not available) would depress flame background by 0.2 unit, as determined by measurments at adjacent wave lengths

Water Sample	Readings			Net Reading Corr. for Background	Corrected for Drift	Na Found, P.P.M.[a]
	Flame	Sample	Flame			
A	5.9	30.4	6.1	24.6	24.6	0.163
B	6.4	18.6	6.3	12.5	12.5	0.083
B	6.8	19.3	6.8	12.7	12.8	0.085
A	...	45.7	7.3	38.6	38.8	[b]
A	7.7	31.8	7.8	24.2	24.4	0.162
A	8.7	32.7	8.7	24.2	24.4	0.162

[a] Calculated from calibration curve which was found linear at these concentrations.
[b] Obviously contaminated by accidental Na.

E. Determination of potassium in brine containing 19.58% K as KCl by gravimetric assay. Standard prepared to approximate sample, containing KCl, borax, Na_2CO_3, and NaCl. Natural gas–oxygen flame, unheated spray chamber, propyl alcohol mixed with sample and standard

KCl Found, %	
First sample	Second sample
19.73	19.67
19.47	19.63
19.39	19.65
19.59	
19.57	
Av. 19.55	19.65
General av., % 19.59	

F. Determination of calcium in water, in presence of sodium. Instrument at high sensitivity. Heated spray chamber. Wave length 554 mμ. Both sample and standards contained 102 p.p.m. Na

Water Blank	Standard, No Ca	Sample	Standard, 1.00 P.P.M. Ca	Ca Found, P.P.M.
9.2	11.4	13.0	15.1	0.43
9.5	12.1	13.4	15.6	0.37
8.6	11.4	13.3	15.2	0.50
9.7	12.0	13.7	15.7	0.46
			Av. 0.44	

G. Determination of copper in methyl violet (measurements made by Marshall Odeen). Isopropyl alcohol used as solvent. Sample contained 1.75% (w/v) of dye; standard contained 5 p.p.m. Cu (w/v). Blank, pure isopropyl alcohol. Wave length 324.8 mμ. Natural gas–oxygen flame adjusted to give same reading with blank as with nothing, in order to eliminate possible background effects

Blank	Standard	Sample	Net Standard	Net Sample
99.8	104.6	101.1	4.8	1.3
100.0	104.8	100.9	4.8	0.9
100.2	104.9	101.0	4.7	0.8
100.1	105.1	101.4	5.0	1.3
100.5	105.4	101.3	4.9	0.8
100.2	104.7	101.2	4.5	1.0
Av. 100.1	104.9	101.1	4.8	1.0

Copper found 0.006% of methyl violet

oxygen cannot be achieved simultaneously, and the above figures represent a compromise adjustment. With good commercial regulators on the tanks, changes greater than those mentioned are unlikely during an analysis.

The pressures which produce maximum brightness vary somewhat, in a gas flame, depending upon the element, its concentration, and the wave length used for measurement, but the variation usually is of slight significance, for it is found in practice that the same settings can be used for practically all determinations. Failure to operate at maximum brightness does not directly affect the accuracy of analysis. It is desirable to operate at or near the maximum brightness because the narrowest spectral band widths can be used (important when nearby lines of other elements may cause inter-

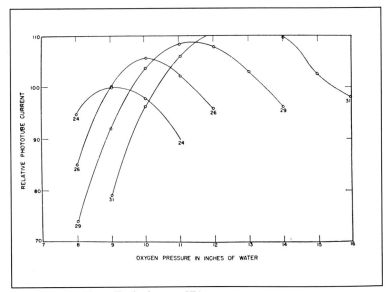

Figure 4. Lithium Emission at 671 mμ
Parameters are gas pressure, m.m. of isopropyl alcohol manometer fluid

ference) and because at maximum brightness the flame burns most steadily, with minimum flicker and least disturbance by fluctuations in gas and oxygen pressures.

The construction of the burner used is shown in Figure 5.

ATOMIZER DESIGN

The most exacting problem in flame spectrophotometer design is presented by the atomizer. This deceptively simple little device must introduce the sample into the flame at the maximum stable and reproducible rate. It must remain unattacked by corrosive solutions and be rugged and easily cleaned. It should preferably be easily altered or interchanged to accommodate samples of varying viscosity and surface tension. The atomizer and spray chamber used are shown schematically in Figure 6.

Experience with reflux-type atomizers, which permit the spray to hit the walls of the chamber and return to the sample, demonstrated that drifts are virtually unavoidable and of such magnitude as to set a severe limit on instrument accuracy and convenience. The drifts result from temperature and concentration changes caused by progressive evaporation, either of the sample or of wash water in air prehumidifying and washing towers. It is also difficult to prevent sample film on the walls from reaching the flame, and to avoid contaminating subsequent samples. Sprayers of designs which permit the spray to hit the walls of the chamber without returning to the sample consume excessive quantities of sample or tend to be erratic, or both.

The solution to all these problems was found in an atomizer in which the spray dries completely before the air stream is deflected from a free straight path. As there is no precipitation or condensation on the walls of the chamber, no cleaning is required when changing samples. The spray chamber is swept completely clean within a few seconds after removing a sample. Readings on different samples can therefore be made

very rapidly (3 to 6 samples per minute; cf. Table I, A). Furthermore, the volume consumed per reading may be as small as one drop (0.05 ml.), while the total volume may be as small as 0.2 ml. A sample of 2 to 5 ml. is usually sufficient for accurate determinations of several constituents.

The adopted atomizer design shown in Figure 6 has a high suction pressure differential (over 100 cm. of water head), making the sample introduction rate relatively independent of height of solution in the sample cup. It is also easy to overcome plugging, because the limiting orifice is made by breaking off a drawn glass tube tip of rapidly diminishing cross section, located at the inlet end of the tube. The heavy-walled air nozzle surrounds and protects the spray tip, so that by merely stopping the air nozzle with the finger air is forced out of the orifice in reverse to blow it out.

Changes in viscosity and surface tension affect the performance of any atomizer. The effects of differences in viscosity and surface tension of samples often can be reduced by operating at higher dilutions to bring these properties of the solution nearer those of the solvent, within the limits set by the sensitivity of the instrument. Here the high photometric sensitivity and continuously adjustable slit of the Model DU spectrophotometer are especially advantageous, for the controls permit a millionfold range in full-scale luminosity. The most intense radiation that can be measured exceeds the faintest detectable by 10^7 to 10^9, depending on the sharpness of the spectral line or band. Thus determinations of major and trace constituents or of easily and difficultly excited spectra may often be made at a single sample strength, or, for routine analyses, conditions where disturbing effects are minimized may be selected.

A dry spray atomizer is difficult to construct so that it delivers an adequate concentration to the flame. This requirement may be satisfied by lowering the surface tension of

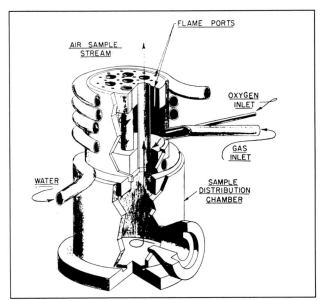

Figure 5. Cut-Away View of Burner

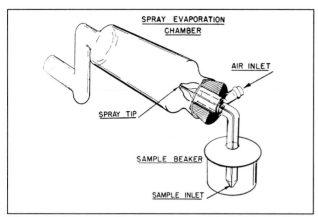

Figure 6. Atomizer Construction and Relation to Spray Chamber

samples by carefully controlled additions of about 20% of propanol or other organic solvent (cf. Table I, E), but after extensive trials it was decided instead to ensure complete evaporation by heating the spray chamber. The 70-watt heater employed can be left connected to the power line while changing samples or adjusting the instrument, and it is not critical as to voltage. Its use makes atomizer design and adjustment much less critical in attempts to obtain maximum spray rate without objectionable fluctuations in flame brilliance.

Solutions in organic solvents of low viscosity can often be used directly in atomizers (cf. Table I, G). When oxygen is used for spraying, one should be certain that explosive mixtures are not formed in the spray chamber. Other organic materials can be prepared by customary wet- or dry-ashing procedures. Typical atomizers exhibit a reproducibility (standard error) of 0.4% or better for individual readings and 0.1% for the average of several readings (see Table I, A, B, and C). As readings may be taken rapidly, averaging several readings is not time-consuming and when interspersed with control and blank readings offers insurance against errors due to drift or excessive pressure fluctuations. The photometric precision of the spectrophotometer is 0.1%, so that an analytical accuracy of 0.2% of the amount present in the sample is entirely feasible for many elements. This approximately equals the analytical accuracy of absorption measurements with the same instrument, and is more than tenfold better than the best spectrographic accuracy.

INTERFERENCE EFFECTS

Interferences often are the most troublesome aspect of flame photometry, and are the more hazardous to analytical accuracy, because they may exist unsuspected. The etiology of six distinct types of interference and procedures for evaluating or minimizing them are discussed in the numbered paragraphs below.

A few experimental studies of interferences with the Beckman flame spectrophotometer have been published, and others are expected to appear soon. A detailed procedure for the accurate determination of sodium and calcium in water has been presented (5), wherein the extent of interferences of types 1 and 5, below, is described for the ranges 0 to 200

p.p.m. of sodium and 0 to 20 p.p.m. of calcium and the method of circumventing them is shown.

Mosher *et al.* (*8*) describe the determination of sodium and potassium in plasma and urine, including effects of instrumental adjustments and of air and oxygen pressures. Their standard solutions included sodium, potassium, calcium, and magnesium chlorides, ammonium phosphate, glucose, urea, gelatin, cholesterol, and alcohol. They show the results of recovery studies on sodium and potassium in plasma, in which the errors averaged 1 to 2%, and of viscosity studies involving gelatin. Sodium in the concentrations dealt with enhanced the potassium light by 8%.

Brown *et al.* (*3*) discuss the determination of calcium and magnesium in leaves. Influences of sodium (with and without a didymium filter), potassium, and acetic and nitric acids on the calcium light are shown. Linear calibration curves were obtained with magnesium at 371 mμ and calcium at 626 mμ. They report that routine analyses could be completed in 2 minutes per metal and that comparison of the photometric with chemical analyses showed standard deviations of 3 or 4%. High and low contents of potassium, sodium, and chloride in the leaves had no important effect. In connection with magnesium, a hydrogen flame has been found to give considerably improved sensitivity, especially at 371 mμ, where convenient instrumental settings will permit of reading easily to 1 p.p.m. of magnesium. Interferences by sodium, potassium, and calcium are also decidedly smaller than in the natural gas flame, and it is considered feasible to determine magnesium routinely, using the hydrogen flame, in biological fluids.

Variations of the apparent or actual emission of radiant energy by excited atoms or by molecules or radicals containing atoms of the species sought may arise in the following ways.

1. **Such mechanisms as quenching, sensitization, reabsorption (reversal), and interference with combustion processes,** resulting in altered flame temperature, may cause alteration of the energy in an emitted line or band by other substances in the flame. In the instrument described here, whereby observations are made in relative terms, we are concerned only with the effects resulting from disparities between the composition of samples and comparison standards. It has been found that instrumental operating conditions can be kept sufficiently uniform during the short time required for a series of observations to make instrumental drifts and random fluctuations practically negligible, although, as in any well-designed device, they set the ultimate limits on available accuracy.

Many of the effects described below have been improperly ascribed to direct interference by one substance on the emission intensity of another. However, after all other sources of interference have been carefully examined and eliminated or allowed for, there sometimes remain substantial direct interference effects. Fortunately, their magnitude is often approximately proportional to the concentration of the interfering substance, and their relative importance can therefore be reduced by merely diluting the solution to the point where other sources of error, such as flame background intensity, become of equal importance. Where this expedient does not render them negligible it often reduces them to such an extent that they are easily compensated by determination of the concentration of the interfering substance (where feasible, by flame photometry on the same sample), and allowed for by computation or by preparation of a more representative comparison standard.

When working with samples that are not of a routine nature, and interference effects are suspected or feared, a simple and practical procedure is to compare the apparent concentration of the element in question in the undiluted sample with that in a portion diluted to half its original concentration, using a suitable pair of standards having a concentration ratio of 2 to 1. In the absence of interference, the second value will be exactly half the first. If this is not the case, a first approximation to the corrected reading on the second sample will be twice the second reading minus half the first. Further dilutions, if they do not entail loss of full-scale precision, can improve the certainty of the interference estimate.

A hypothetical example of this procedure would be the following:

Let Solution I contain 100 p.p.m. of metal A and 100 p.p.m. of metal B, which at this concentration increases the intensity of emission of A by 10%. Furthermore, assume that the intensity of emission of A is directly proportional to its concentration in this range, and that the absolute increment of emission by A due to the presence of B is proportional to the product of their concentrations, which as indicted above is commonly approximately true.

Let Solution II, the standard, contain 100 p.p.m. of A only. Comparing I against II in the flame photometer, I will appear to contain 110 P.P.M. of A. Now with the purpose of detecting interference, dilute Solutions I and II with equal volumes of solvent, forming Solution III (50 p.p.m. of A and 50 p.p m. of B) and Solution IV (50 p.p.m. of A). At 50 p.p.m., B will enhance the emission of A by only 5%. Therefore, Solution III, compared against Solution IV (known to contain 50 p.p.m. of A), will appear to contain 5% more than 50 p.p.m., or 52.5 p.p.m. The fact that 52.5 is less than half of 110 indicates that an interference is present. To arrive at a corrected value for the concentration of A in III, take $2 \times 52.5 - (1/2 \times 110)$, which is 50. Because the simplified assumptions applying to interferences were true in this hypothetical case, the result is exactly correct.

The following is an actual case involving a common interference, that of sodium with calcium:

Solution	Na, P.P.M.	Ca, P.P.M.	Ca, Reading	Apparent Ca
I	100	10	48.2	9.64
II	0	10	50.0	(Standard)
III	50	5	24.6	4.92
IV	0	5	25.0	(Standard)

Fortunately the calcium light is exactly proportional to concentration (Figure 9) but sodium exercises a complex type of interference (see paragraph 5 below) upon calcium. However, the fact that the diluted sample appears to contain more than half as much calcium as the undiluted sample leads one to apply the rule for finding a first approximation: $2 \times 4.92 -$

$(1/2 \times 9.64) = 5.02$, and so the original sample, in the absence of other information, would be surmised to contain 10.04 p.p.m. of calcium. The true value, if greater accuracy is required, could be obtained in this case by measuring the sodium (which is not interfered with by calcium) and then employing a standard containing the same sodium content as found in the sample. The same anions should be employed in the solutions under comparison, inasmuch as differences in emission due to different anions cannot be assumed *a priori* to be negligible.

An example of the determination of a trace of calcium in the presence of 200 times as much sodium is given in Table I, F.

2. **Energy at other wave lengths** than those intended to be measured may reach the photodetector. It is too obvious to require further elaboration that the monochromator must be capable of excluding emission at wave lengths adjacent to or remote from the selected band by substances uncontrolled in the comparison standards. Naturally, this problem is simplified if the photomeasuring equipment is sufficiently sensitive to permit the monochromator to be used at slit widths approaching those at which the resolution is limited only by the optical design of the monochromator. In the ideal monochromator this limit is proportional to the dispersion of the dispersing element (prism or grating) and is set by diffraction effects at the slits. In realizable instruments optical aberrations and slit image curvature (which can be matched by a practicable slit mechanism at only one wave length) introduce additional deviations which add to the Rayleigh limit slit widths. The design features of the Beckman Model DU monochromator have been described (4).

In flame photometry a higher premium must be set on photodetection sensitivity as compared with absorption photometry because line emission is confined practically to single wave lengths. The line energy passed by the monochromator is therefore proportional to the first power of the slit width, whereas background emission and scattered radiation are proportional to the second power. The advantages in interference problems of being able to work at high dilution and the use of a high temperature flame with its associated inherent high background radiance emphasize the desirability of high detection sensitivity. These are the reasons for modification of the photodetection system by increasing the phototube load resistor fivefold to 10,000 megohms, despite some loss of response speed and a slight decrease in linearity and stability.

3. **Direct interference by (nonbackground) radiation** can occur when a desired line falls within a molecular emission band. When emission bands actually overlap, interference cannot be obviated by increased resolution. However, instrumental versatility will often provide an easy solution. An example would be a sample containing barium and lanthanum. The barium emission at 515 to 550 mμ would be convenient to use were it not for the lanthanum (oxide) bands which interfere. Conversely, unresolved barium bands interfere with the lanthanum band at 563 mμ. Here the solution lies in selection of other spectral regions where mutual interference does not occur. The lanthanum peak at 798 mμ and the barium emission at 830 or 873 mμ are much less subject to mutual interference and are more sensitive than the green bands.

Even if overlap cannot be avoided, it will usually be found that multicomponent analyses by methods similar to those used in infrared absorption spectrophotometry (2), where interference is the rule rather than the exception, will prove usefully convenient and accurate.

4. **Intense adjacent line emission** can also cause direct interference. Sometimes an extremely brilliant atomic emission may interfere with the measurement of a neighboring line, either through Doppler broadening or because scattered radiation tends to be concentrated near the line image. An example is found in the determination of boron in borax. The boron band at 548 mμ cannot be used because of the overwhelming brilliance of the sodium D doublet from solutions sufficiently concentrated to permit determining boron to 1% of its amount. The boron bands at 521 and 495 mμ, however, lie well away from the dazzling sodium light and, although weaker than the 548 mμ band, are sufficiently bright for accurate determination of boron.

These procedures for avoiding interference errors by selecting wave lengths free from interference emphasize the inadequacy of filter photometers which are limited to a few, fixed, broad pass bands. The ability to select any desired narrow wave-length band over a wide range is of prime importance. A large proportion of useful wave lengths are shorter than 400 mμ, while the near infrared is particularly useful for the determination of such metals as rubidium, cesium, barium, lanthanum, and strontium.

5. **A general increase of flame background at all wave lengths** may occur, and is common with materials containing sodium and potassium. It extends over most of the spectrum, but is not due to scattered light, as may be shown by interposition of a didymium filter, which cuts out the sodium line almost completely but leaves unaltered the background increase due to sodium at other wave lengths. This effect may be of such magnitude as to require a careful preliminary study, in the event that small amounts of any element are to be determined in the presence of considerable alkali metal and dilution has already been carried to its practical limit. Even magnesium at 285 mμ is subject to this influence by sodium; a sodium concentration of 0.1% will increase the background light at 285 mμ by an amount equivalent to 10 or 20 p.p.m. of magnesium under conditions which in the absence of sodium would permit readings to be made to 1 p.p.m. or less. Similarly, the determination of calcium in the presence of 100 times as much sodium necessitates rather large corrections for both background change and depression of the calcium light proper—effects which though of opposite sign do not by any means cancel each other (see Table I, F). Much potassium in a sample likewise makes difficult the detection of traces of lithium or rubidium, but a careful scanning of the region in the immediate vicinity of the lithium or rubidium lines will reveal the presence of any actual emission by these metals. In practice, such background corrections can be either measured or obviated by use of a blank containing the proper amount of interfering metal but none of the metal being determined.

6. **Spray rate alteration**, by uncontrolled constituents or by conditions (especially the temperature), which affect the hydrodynamics of the atomizer, is a common and frequently misinterpreted cause of interference. Both the viscosity and the surface tension of the solution are of first importance in determining the rate of introduction of the sample into the flame by a given sprayer. It was a study of these critical variables which led to the design of the sprayer used in the present instrument, with its high-suction, orifice-limited flow and dry spray.

STANDARDIZATION

Internal versus External Standardization. The instrument may be standardized with either internal or external standards. The advantages of the latter, discussed below, led to efforts to design an instrument capable of convenient and accurate use with external standards.

As an internal standard a suitable element normally absent from the sample, such as lithium, is added in definite amount to the sample solution. Comparison of the relative emissions of the internal standard and of the element being determined gives a measure of the concentration of the latter. Internal standards tend to compensate for variations in viscosity, surface tension, and perhaps other variables. They have the disadvantage, however, that unsuspected errors may arise from their use. If the test sample should actually contain some of the internal reference element, a corresponding error would result. More serious is the fact that, contrary to the implications of some reported investigations, the emissions of the internal standard element and of the element being measured are usually influenced differently by variations in flame temperature and often by the presence of other components in the sample as well. The use of a foreign element for reference consequently may be misleading. Internal standards, therefore, while often useful, should be employed only after careful tests to determine the relevant facts.

The preferable way of standardizing a flame spectrophotometer ordinarily is to prepare standard solutions which approximate the test samples in composition, and to standardize the instrument with the element that is being determined. In many cases a standard solution containing only the element being determined suffices. In other cases, particularly where interferences are encountered, it may be desirable to use standard solutions containing all the important constituents of the unknown in approximately the correct concentrations (cf. Table I, E and F). Disturbing effects often can be eliminated by working at sufficiently high dilutions. Where this step is insufficient or undesirable, self-compensating standards should be prepared. It often suffices to make a preliminary flame analysis of the unknown and then add to the standard solutions appropriate concentrations of other elements shown in the preliminary analysis. This often involves less effort than preparing an internal standard, particularly in routine determinations, and has the advantage that compensation for disturbing factors usually is much more complete than with an internal standard. As mentioned under Interference Effects, the comparison of diluted and undiluted samples against appropriate standards is an excellent precaution when sample composition is unknown or variable, and usually allows arithmetical correction for interferences of unknown origin.

Flame Background. The flame background must be subtracted from all readings. With many metals, such as the alkalies, the background value may be entirely negligible. In other cases it may be sufficiently small to permit use of a rough average value for a series of measurements. Wherever the background represents an appreciable fraction of the reading, as with such feebly luminous metals as tin and lead, the background value preferably should be obtained by measurement with pure solvent or, better, with a blank solution prepared in the same manner as the sample (cf. Table I, D, F, and G).

Linearity. It is characteristic of flame spectra that the curve of brightness versus concentration usually is substantially linear for low concentrations, with departures from linearity at higher concentrations. Typical calibration curves are shown in Figures 7, 8, and 9. Figure 7 shows the relative photocurrents for sodium in different ranges of concentration. The curves presented are the actual working curves employed in sodium determinations such as those of Table I. In practice, a family of curves is used, covering the rather small range of fluctuating conditions which would prevent one single curve from being applicable at all times. For purposes of mathematical interpolation, an equation can often be accurately fitted to the working curve. For sodium, the region from 50 to 250 p.p.m. accurately fits an equation of the form $c = A + Br^2$, where c is the concentration of sodium and r is the photometer reading, A and B being constants whose ratio is nearly fixed for a given atomizer and flame adjustment but whose absolute magnitudes depend on slit width, etc. Between 0 and 100 p.p.m., an equation of the form $r = Ac - bc^2$ was found to fit the curve very accurately. In the case of lithium (not shown), an equation of the form $c = r/(A + Br + Cr^2)$ was found to fit exactly from 0 to 250 p.p.m.

Figure 8 shows data for potassium, including the effect of flame composition (as defined by oxygen pressure for a given gas pressure) upon the curvature of the graph. This effect is especially pronounced in the case of potassium; with sodium it is observable but considerably smaller—i.e., a set of sodium calibration curves drawn between the same initial and terminal points for various flame compositions nearly coincide. The unusual S-curvature of the potassium curve is due to the fact that the peak luminosity of the potassium emission is found at lower oxygen pressures for smaller concentrations of potassium. Thus, with a particular burner, at 100 p.p.m. potassium the maximum intensity was obtained at about 5 or 6 inches of oxygen; at 200 p.p.m., at 7 inches; at 300 p.p.m., at 9 inches; and at 500 p.p.m., at about 10 inches. In other words, the smaller the potassium content of a flame, the cooler it must be to produce maximum excitation. With sodium the effect is of the same sign but much smaller: the optimal oxygen pressure shifts from 9.5 inches at 5 p.p.m. to 12 inches at 1000 p.p.m. It is interesting to note that, over a thousandfold range, the graph of optimal oxygen pressure versus the logarithm of sodium concentration would be nearly linear. In the light of this phenomenon, it may be understood that, if for each concentration of potassium the corresponding optimal oxygen pressure were employed, the calibration curve would not only lose its inflection but would become nearly straight. Even without readjustment of the oxygen, in practical analyses linear interpolation may be resorted to without serious error over moderate ranges.

Figure 9 shows the strikingly linear relation obtained with calcium at its oxide band peak in the green. This circumstance permits highly accurate determinations to be made at all practical concentrations by simple comparison against a single standard, in the absence of interferences. Such graphs obtained at different emission wave lengths of the same metal may differ in curvature. Whereas the curve at 427 mμ for calcium has not been studied by the authors, in the case of sodium they have found that the weak emission at 330 mμ gives a practically straight graph up to several hundred parts per million. This wave length is therefore sometimes valuable for a preliminary analysis of a sodium solution, although because of the feebleness of the line, it is not suitable for precise determinations.

For most purposes the intercomparison technique for which the instrument is peculiarly suited yields sufficiently accurate results without the aid of a calibration curve. The latter is employed to show the permissible spread between sample and standard, and may be invoked if the desired precision warrants it. Sodium is an exception in that, as shown in Table I,

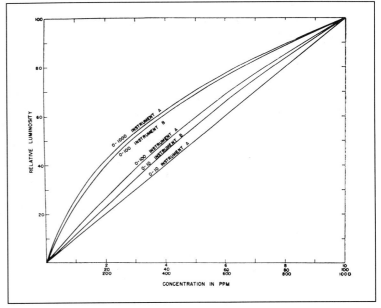

Figure 7. Calibration Curves for Sodium at 589 mμ

Curves for range 0–1000, 0–100, and 0–10 p.p.m. are drawn between same initial and terminal points to show improvement in linearity at low concentrations. Two sets are given: for an earlier instrument, *A*, with unheated spray chamber and with propyl alcohol added to the Na solutions; and for a later and somewhat more sensitive instrument, *B*, with heated spray chamber, using purely aqueous solutions. Curves for *B* are applicable to determinations shown in Table I, A, C, and D

A and C, a curve such as that of Figure 7 is almost always required except at the lowest concentrations.

SENSITIVITY FOR VARIOUS ELEMENTS

Time has not permitted an exhaustive study of the potentialities of the instrument, but at least preliminary information is available on 43 elements, listed in Table II. To the right of the name of each element is given the detection sensitivity, which is the least concentration at which the element can be detected easily by comparison against a suitable blank. In general, it represents the least concentration required to give a reading of 0.5% of full scale and equal to a galvanometer needle response of 1.5 scale divisions, under suitable adjustments of the various controls. It is hardly possible to generalize about the latter because of the widely varying conditions encountered. In the absence of serious interferences, a decidedly smaller concentration than that listed can usually be detected, and even in the presence of interferences, careful technique may yield measurements accurate to a fraction of the detection sensitivity. The

attainable limit depends, however, upon the varying sensitivities of individual atomizers and spectrophotometers.

Wave lengths of flame emission are given in millimicrons. Opposite each wave length is given its intensity, in units obtained by dividing into 100 the detection sensitivity in parts per million for that particular wave length. The asterisk signifies an oxide band which may or may not cover a broad range of wave lengths but has its maximum at the wave length noted.

All intensity values are only approximate, because of the variability of external conditions, properties of the solution, and instrumental adjustments. Most of the intensities have been measured at National Technical Laboratories, using an oxygen–natural gas flame, but some of the data (for cadmium, gadolinium, gallium, gold, indium, mercury, palladium, rhodium, rubidium, ruthenium, silver, strontium, thallium, and yttrium) have been taken or recalculated from the work of Lundegårdh and others, using other kinds of flame, and have not been verified by the authors.

SUMMARY

It is believed that the Beckman flame spectrophotometer offers a combination of features not hitherto available. Of particular significance are the following:

Available accuracy of a few tenths of 1% of the amount present. Emission measurements equal in analytical accuracy to absorption measurements.

High resolving power, affording relative freedom from optical interference effects.

High sensitivity which permits analysis to be made at trace concentrations, or at high dilution to minimize emission interferences.

Speed and convenience. No cleaning between samples. Several readings per minute.

Small samples. Only a single drop of solution is consumed per reading.

Hot flame. Excites more than half the elements.

Low initial and operating cost in comparison with other high resolution emission equipment.

It is believed that these features will greatly extend the field of utility for flame spectrophotometry. On the one hand, the improved quantitative accuracy permits flame spectrophotometry to be used for many analytical determinations where more time-consuming volumetric or gravimetric methods have heretofore been required. On the other hand, the extension of flame technique to embrace additional elements enables the analyst to replace arc spectrography with the more accurate, more convenient, and less expensive flame spectrophotometry.

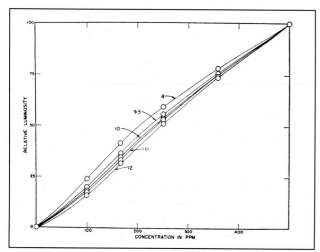

Figure 8. Calibration Curves for Potassium at 769 mμ

Drawn between same initial and terminal points to reveal differences of shape. Gas pressure constant; oxygen pressure in inches of water shown as parameters

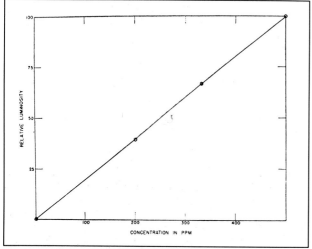

Figure 9. Calibration Curve for Calcium at 554 mμ

ACKNOWLEDGMENT

Thanks are due G. L. Locher, who worked out the design of the burner and performed preliminary experiments on reflux-type atomizers.

LITERATURE CITED

(1) Barnes, R. B., Richardson, D., Berry, J. W., and Hood, R. L., *Ind. Eng. Chem., Anal. Ed.*, **17**, 605 (1945).
(2) Brattain, R. R., Rasmussen, R. S., and Cravath, A. M., *J. Applied Phys.*, **14**, 418 (1943).

Table II. Flame Spectra of Elements

Aluminum

No emission

Antimony

No emission

Barium, 1 p.p.m.

490–570, including

520*	20
550*	20
745*	100
830*	50
873*	50

Boron, 5 p.p.m.

345	0.3
454*	2
473*	5
495*	10
521*	15
548*	20

Cadmium, 500 p.p.m.

326.1	0.2

Calcium, 0.3 p.p.m.

422.7	100
554*	200
603*	100
624*	300
648*	100

Cesium, 0.1 p.p.m.

455.5	1
459.3	0.5
621.3	
672.3	
852.1	1000
894.4	

Chromium, 3 p.p.m.

357.9	30
359.3	30
360.5	30
425.4	20

Chromium (*contd.*)

427.5	20
429.0	20
520.6	10
527*	10
539*	10
544*	10
559*	10
564*	10
582*	10
609*	20
645*	30
685*	20
850*	3

Cobalt, 5 p.p.m.

238.9	
304.4	1
350.2	20
352.7	20
412.1	1

Copper, 1 p.p.m.

324.8	100
327.4	90
512*	5

Dysprosium

Determinable

Europium

Determinable

Gadolinium

451.4*	
461.4*	
569.6	

Gallium, 1 p.p.m.

403.3	
417.2	100

Gold, 50 p.p.m.

267.6	2
365.2*	
397.5*	0.1

Indium, 1 p.p.m.

303.9	
325.6	
410.2	
451.1	100

Iron, 10 p.p.m.

344.1	2
372.0	7
373.6	10
386.0	7

Lanthanum

438.4*	
443.3*	
544*	
563*	
714*	
745*	
798*	
860*	

Lead, 300 p.p.m.

283.3	0.1
364.0	0.1
368.3	0.2
405.8	0.3

Lithium, 0.05 p.p.m.

460.3	0.1
670.8	2000

Magnesium, 10 p.p.m.

285.2	10
370.8	10
382.9 } 283.2 }	8

Manganese, 1 p.p.m.

279.8	1
403.4	100
510*	10
541*	30
561*	70

Mercury, 50 p.p.m.

253.6	2

Neodymium

Determinable

Nickel, 3 p.p.m.

299.4	1
300.3	5
303.8	2
324.3	1
331.6	2
337.0	7
338.1	5
339.3	10
341.5	15
342.4	4
343.4 } 343.7 }	10
344.6	10
345.8 } 346.2 }	20
347.3	5
348.4	2
349.3	15
350.1	3
351.0 } 351.5 }	20
352.4	30
354.9	1
356.6 } 357.2 }	10
359.8	4
361.0	8
361.9	15
367.4	1
373.7	1
377.6	3

Nickel (*contd.*)

378.4	3
380.7	3
385.8	6
515*	3

Palladium, 50 p.p.m.

340.5	2
363.5	2

Phosphorus

No emission

Platinum

Determinable

Potassium, 0.05 p.p.m.

344.6	0.5
404.4 } 404.7 }	7
766.5 } 769.9 }	2000

Praseodymium

Several good bands

Rhodium

332.3	
339.7	
343.5	
352.8	
358.3	
359.6	
369.1	
369.6	
449.2	

Rubidium, 0.1 p.p.m.

420.2	3
421.6	1
780.0	1000
794.8	

Ruthenium, 30 p.p.m.

372.7	3
378.6	3

Samarium

Determinable

Scandium

Determinable

Selenium, 5000 p.p.m.

359*	0.02

Silver, 2 p.p.m.

328.1	7
338.3	50
520.9	

Sodium, 0.01 p.p.m.

330.2	7
589.0 } 589.6 }	10,000

Strontium, 0.5 p.p.m.

407.8	10
421.6	10
460.7	200
870*	

Thallium, 1 p.p.m.

276.8	0.1
351.9	2
377.6	80
535.0	10

Tin, 500 p.p.m.

320–370, including

326.2	0.2
332*	0.2
348*	
358*	
452.5	0.1

Yttrium

464.4	
467.6	
486.0	

Zinc

No emission

(3) Brown, J. G., Lilleland, O., and Jackson, R. K., *Proc. Am. Soc. Hort. Sci.*, **52**, 1 (1948).

(4) Cary, H. H., and Beckman, A. O., *J. Optical Soc. Am.*, **31**, 682 (1941).

(5) Gilbert, P. T., Jr., Natl. Technical Laboratories, *Bull.* **TP 1248-1** (1948).

(6) Griggs, M. A., *Science*, **89**, 134 (1939).

(7) Lundegårdh, H., "Die quantitative Spektralanalyse der Elemente," Jena, Gustav Fischer, Part I, 1929, Part II, 1934.

(8) Mosher, R. E., Boyle, A. J., Bird, E. J., Jacobson, S. D., Batchelor, T. M., Iseri, L. T., and Myers, G. B., *Am. J. Clin. Path.*, **19**, 461 (1949).

(9) Rogers, L. H., *Proc. Soil Sci. Soc. Am.*, **12**, 124 (1947).

Received January 18, 1949. Presented before the Third Analytical Symposium, Analytical Division, Pittsburgh Section, American Chemical Society, Pittsburgh, Pa., February 1948.

Reprinted from *Anal. Chem.* **1950**, *22*, 772–80.

Although Meinhard and Hall (*Anal. Chem.* **1949**, *21*, 185) first used a binding agent for greater layer stability in their separation of inorganic ions on aluminum oxide-coated glass microscope slides by circular "drop chromatography," Kirchner and co-workers conceived the idea to separate terpenes on layers of adsorbent bound to glass strips. They developed these "chromatostrips" in a closed tank in a manner analogous to ascending paper chromatography. This paper, a milestone in the history of TLC, was the first of nine published between 1951 and 1957 that reported the discovery, modification, and application of TLC essentially as it is practiced today.

Joseph Sherma
Lafayette College

Separation and Identification of Some Terpenes by a New Chromatographic Technique

J. G. KIRCHNER, JOHN M. MILLER, AND G. J. KELLER
Bureau of Agricultural and Industrial Chemistry, U.S. Department of Agriculture, Pasadena, Calif.

A chromatographic method for separating terpenes was developed for the determination of the volatile flavoring constituents of citrus fruit. This new technique in organic chromatography has been introduced by using adsorbent-coated glass strips in a manner analogous to paper chromatography. After the mixture has been spotted near one end of the strip, the chromatogram was developed with the aid of capillary attraction by dipping in a suitable solvent. The solvent was then evaporated from the strip and the various zones were indicated by spraying with suitable reagents. On spraying with a fluorescein solution and exposing to bromine vapor, compounds which absorb bromine faster than the fluorescein show up as yellow spots on a pink background. Very unreactive compounds can be located by spraying with a concentrated sulfuric-nitric acid mixture and heating to cause charring of the compounds. The technique can be applied to other types of compounds and is a rapid method of checking solvents and adsorbents for use on larger chromatographic columns. The term "chromatostrip" has been suggested for application to these adsorbent-coated glass strips.

In the course of determining the volatile flavoring constituents of citrus fruit, the authors desired to develop a chromatographic method for the purification and identification of terpenes. The small amounts of oil obtainable from the fruit made this almost imperative, and in addition it was desirable, because of the nature of the terpenes, to eliminate any form of heat treatment.

Although a considerable number of papers (*16, 17*) have been published on the chromatography of terpenes, very little has been done on the chromatography of the simpler terpenes, possibly because of the lack of a suitable indicating reagent. The work on the simpler terpenes, and most of the work on the sesquiterpenes, has been done by arbitrary separation of fractions of the eluting solvent. Winterstein and Stein (*15*), Carlsohn and Müller (*3*), and Späth and Kainrath (*13*) separated a few of the simpler terpenes by chromatography.

The advent of paper chromatography has seen a great increase in the use of chromatographic techniques. Applications of paper chromatography are limited and it was soon evident that ordinary filter paper was unsuitable for chromatographing terpenes. Impregnation of filter paper with various adsorbents has been used to increase the adsorbing strength of the paper both for organic chromatography (*1, 4, 5, 8*) and for inorganic analysis (*6, 7*), but is limited by the relatively small number of adsorbents that can be used.

In contrast to paper chromatography, column chromatography has a distinct disadvantage in work with colorless compounds, such as the terpenes, because of the difficulty in locating the zones. Fluorescent compounds can be located with ultraviolet light and the method of Sease (*12*) can be used to advantage for many ultraviolet absorbing compounds.

The authors desired to combine the advantages of paper and column chromatography in order to obtain a rapid chromatographic method to which zone-indicating developers could be easily applied. The method of Meinhard and Hall (*9*) on the radial surface chromatography of inorganic ions formed the basis of the present work. Their method was modified by coating the adsorbent, mixed with a binder, on suitable glass strips; the strips were activated and then developed in a manner similar to paper strips in test tubes as used by Flood (*6*) and by Rockland and Dunn (*11*). In order to make the method as universally applicable as possible, Sease's (*12*) idea of mixing two fluorescent inorganic materials—for example, zinc cadmium sulfide and zinc silicate—to the adsorbent was incorporated advantageously.

Because not all of the terpenes absorb in the ultraviolet region and are thus not adaptable to the Sease technique, and because reagents were needed to indicate certain specific functional groups, it was necessary to develop a new series of tests for locating compounds on the chromatograms.

For this new technique, the authors propose the name "chromatostrip."

EXPERIMENTAL

Preparation of Strips. Originally the adsorbent mixture was patterned after that of Meinhard and Hall (9) who used 6.2 grams of adsorbent, 3.5 grams of Celite and 0.5 gram of starch, all of which were heated with 18 ml. of distilled water until the starch had coagulated and the mixture had formed a thick paste. The paste was then triturated with water to a consistency just thin enough to spread on the glass strips. Satisfactory strips could be prepared without the addition of filter aid; this proved desirable because the strips containing Celite were weaker in adsorptive power than the strips containing only adsorbent and binder. Therefore, the mixture contained 19 grams of adsorbent, 1 gram of Amioca starch, and (as fluorescing agent described later) 0.15 gram of zinc silicate and 0.15 gram of zinc cadmium sulfide.

Preparation of suitable strips involved careful attention to the details of preparation. The major difficulties were the cracking of the adsorbent and a surface that was too soft. To solve the difficulties, the following procedure was used:

The specified amounts of material, thoroughly blended while dry, were mixed with 36 ml. of distilled water in a 250-ml. beaker. The slurry was then heated on a water bath held at 85° C. with constant stirring until it thickened (1.75 minutes) and was then held at this temperature for 30 seconds longer with stirring. The beaker was removed from the bath and 2 to 7 ml. of water were added immediately to form a thin paste. The mixture was then spread on glass strips 0.5 by 5.25 inches. (Longer strips can be used to increase the resolution. Two-dimensional chromatography has been employed by coating sheets of glass and developing in the same manner as filter paper.) It is necessary to obtain a smooth surface both for writing and for ease of detection of spots; this can be accomplished by coating the glass while it is held between two glass guides 0.02 inch higher than the glass strip. The strips were then dried in a forced-draft oven at 105° C. for 15 minutes. This procedure resulted in a strip approximately 0.02 inch thick, with a minimum of cracks and a surface hard enough to write on with a blunt pencil. These strips would not crumble under mild handling.

Chromatostrips which were not dried prior to use in a uniform fashion exhibited a marked difference in R_F value (R_F = ratio of distance traveled by a spot to the distance traveled by the solvent). Limonene chromatographed with hexane on strips dried in a desiccator over phosphorus pentoxide at 65 mm. of mercury for 0.5 hour had an R_F of 0.8, whereas this oil had an R_F of 0.4 when dried at 3 mm. of mercury over phosphorus pentoxide for the same length of time. Exposure of strips to atmospheric conditions for short periods of time also increased the R_F of some of the samples. Thus, it was necessary to desiccate the strips in a standard manner and to limit the time in which they were exposed to the atmosphere before use.

The strips were placed in a desiccator over powdered potassium hydroxide and evacuated to 3 mm. of mercury. (Reasons for this desiccant are explained later.) If placed in the desiccator while still warm from the oven, the strips reached equilibrium in 30 minutes. Before the strips were removed, it was essential that the vacuum be broken only with dry air. For this reason, a tube packed with Ascarite was used to admit air to the desiccator. The strips could not be used after exposure to the atmosphere for periods longer than 10 minutes. Following these precautions, it was possible to obtain reproducible R_F values.

For the spot test with a sulfuric-nitric acid mixture, it was necessary to eliminate the starch as a binder because of its reaction with the hot acid mixture. Plaster of Paris (20%) was substituted as the binding agent. In preparing these strips, a small quantity of the dry mixture (enough for two strips) was mixed with sufficient water to make a thin paste, and was immediately spread on the glass strips. These strips were dried at 75° C. and were used as soon as they had cooled.

Method of Chromatography. The sample to be chromatographed was placed as a small dot near the bottom of the strip and pencil lines were made on the adsorbent to indicate the original position of the sample and the desired length of solvent travel. The strip was then placed in a test tube which contained 1.5 ml. of fresh solvent. The strip was removed when the solvent reached the desired height (10 cm.), and the solvent was allowed to evaporate from the strip before the qualitative tests were applied.

Solvents. With silicic acid-coated strips, a search was made for suitable solvents to be used in chromatographing terpenes. The solvents tried were divided into four general classes. (Earlier experiments had shown that the strips could not be placed directly in a solvent containing water, because the adsorbent beneath the surface of the liquid tended to slide off the glass strip. Subsequently, this difficulty was eliminated by placing the strip on a wad of cotton saturated with the solvent.) These four classes are described, together with the solvents falling in each class, as follows:

1. Those that carried all oils to the top of the column (ethyl alcohol, dioxane, diethyl ether, acetone, 1-nitropropane, pyridine, ethyl acetate, methanol)

2. Those that did not move the majority of the oils (hexane, petroleum ether, carbon tetrachloride, carbon disulfide)

3. Those that moved the oils a reasonable distance (chloroform, benzene)

4. Those that interfered with the fluorescein-bromine test (tetrahydrofuran, diacetone alcohol, amylene, ethyl triethoxy silane)

Various mixtures of group 1 with 2 and 3 were investigated, and 15% ethyl acetate in hexane was selected as one of the best solvents for this work.

Adsorbents. Of the numerous adsorbents tested (Table I), silicic acid proved to be the best for terpenes. Merck's reagent

Table I. Characteristics of Chromatostrips Made with Various Absorbents

Adsorbent Coating	Physical Characteristics of Strip	Resolution of Oils
Magnesium oxide	Soft	None
Alumina	Excellent	Good
Alumina + silicic acid	Excellent	Good
Calcium hydroxide	Soft, crumbly	None
Starch	Good	None
Dicalcium phosphate	Fair	Some resolution
Bentonite	Good	Oils decomposed
Calcium carbonate	Good	Slight resolution
Magnesium carbonate	Fair	Slight resolution
Filtrol	Good	Separation, but oils decomposed
Filtrol X202	Good	Separation, but oils decomposed
Filtrol, Neutral E	Good	Separation, but oils decomposed
Anex	Fair	None
Florisil	Good	Fair separation
Talc	Good	Slight reolution
Silicic acid	Excellent	Excellent

grade, which had been sifted to pass a 100-mesh sieve, was used in the preparation of the strips for the terpene work.

Color Tests. A wide variety of types of compounds are encountered in the examination of terpenes, so that it is necessary to have several qualitative tests to locate the spots formed on the chromatograms. It is also useful to have, in addition, a variety of tests for specific functional groups. The following tests were worked out and used in the task of locating compounds.

Fluorescein-Bromine. The principle of adding bromine to unsaturated linkages was used in locating a large number of compounds with ethylenic-type double bonds.

The completed chromatogram from which the solvent had evaporated was sprayed with a solution of 0.05% fluorescein in water. The strip was then exposed to bromine vapor by blowing gently across the top of a bottle of bromine (a large excess of bromine should be avoided). The fluorescein reacted with bromine to form the red dye, eosin.

Wherever a material was present which could react with bromine, such as an ethylenic double bond, the fluorescein retained its normal yellow color. The location of these compounds was then readily apparent as yellow spots on a pink background (Figure 1). Amounts of some material as small as 1 microgram were detected with certainty on chromatograms with fluorescein-bromine. Table II reveals the types of compounds which can be detected.

Certain precautions must be observed in performing this test. Unless sufficient water is present, the pink color of eosin

does not develop uniformly. For this reason, the dilute solution of fluorescein is necessary in order to build up the amount of water on the strip. If desired, a 0.1% solution of fluorescein may be used by first spraying with water. The presence of alkali enhances the red color, while acid prevents its formation. Strips desiccated with phosphorus pentoxide apparently absorbed enough acid vapors to depress the formation of the red color; this difficulty was eliminated by using powdered potassium hydroxide as the desiccant.

Fluorescence. Sease's (*12*) method of fluorescent columns was incorporated by adding zinc cadmium sulfide and zinc silicate to the mixture of starch and adsorbent prior to adding the water. The dried strips gave a bright fluorescence; certain types of compounds could be detected as dark spots when viewed under ultraviolet light (Figure 2). As suggested by Sease, the source of ultraviolet light was important in viewing these chromatograms. Unless ultraviolet light of short wave length was used, no spots were discernible. A Mineralight

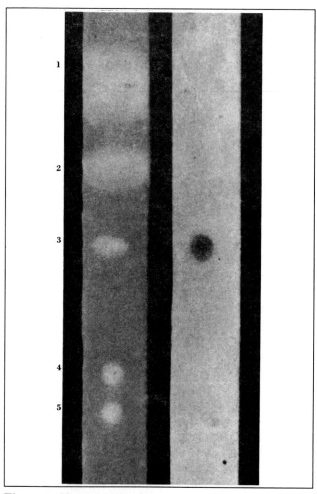

Figure 1. Chromatographs

Left. Mixture of five terpenes as shown by fluorescein-bromine test (from top to bottom)
1. α-Pinene
2. Limonene
3. Terpinyl acetate
4. α-Terpineol
5. Geraniol

Right. Location of cinnamaldehyde with *o*-dianisidine reagent

(Short wave Model SL 2537) gave light of satisfactory wave length for this work. Table II indicates the types of compounds detected by this means.

o-Dianisidine. Aldehydes can be detected as colored compounds with o-dianisidine (14). Cinnamaldehyde was detected by spraying with a solution of o-dianisidine in glacial acetic acid.

Sulfuric Acid. Certain compounds resisted all means of detection because of the absence of reactive groups within the molecule. To detect these compounds, concentrated sulfuric acid was sprayed on the developed chromatogram. This reagent was used in locating 1,8-cineol. Table II indicates the color reactions and types of compounds detected by this reagent.

Special equipment was needed to spray a reagent tag as corrosive as concentrated sulfuric acid. The spraying was done with an all-glass sprayer built in the laboratory (Figure 3). A small stainless steel booth for spraying was constructed to protect the hood, and the sprayer was manipulated behind a glass window.

Bromocresol Green. Ramsey and Patterson (10) have developed a method for the detection of acids on silica gel columns by incorporating bromocresol green in the adsorbent. Acids were detected on the strips by spraying with a solution of 0.3% bromocresol green in 80% by volume methanol, to which had been added 8 drops of 30% sodium hydroxide per 100 ml. The acid appeared as yellow spots on a green background.

Sulfuric-Nitric Acid Mixture. Camphor was so unreactive that it was necessary to develop a special test to indicate the location of this compound. This was accomplished by spraying with concentrated sulfuric acid to which 5% concentrated nitric acid had been added. The chromatograms were then heated on a strip of glass cloth, face down, on top of a hot plate which was turned to full heat and registered approximately 500° C. After the acid fumes had ceased coming off, the glass strip was carefully lifted from the adsorbent and the latter was turned over by means of the glass cloth. The compound locations were observed as black spots on a white background.

Table II. Reactivity of Compounds with Various Zone-Indicating Tests

Compound	Formula	Ultra-violet	Fluorescein-Bromine	Sulfuric (Concd.)	Sulfuric-Nitric (Concd.)
Limonene		−	+	+ Brown	+
α-Pinene		−	+	+ Brown	+
Pulegone		+	+	+ Yellow	+
Camphene		−	+	+ Brown	+
Geraniol		−	+	+ Purple	+
Carvone		+	+	+ Pink	+
p-Cymene		+	−	−	+
α-Terpineol		−	+	+ Green	+
Nopol		−	+	+ Green	+
1,8-Cineol		−	−	+ Green	+
Cinnamaldehyde		+	+	−	+
n-Capric acid	$CH_3(CH_2)_8COOH$	−	−	−	+
Terpinyl acetate		−	+	+ Brown	+
Camphor		−	−	−	+

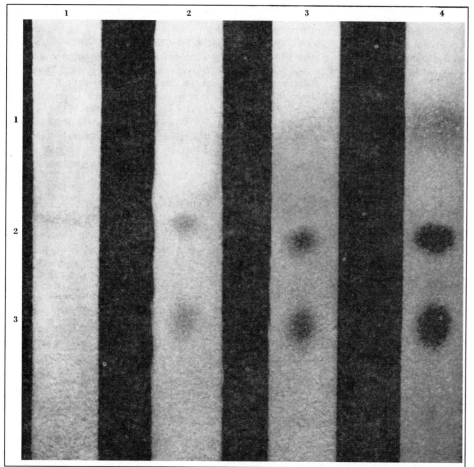

Figure 2. Separation of *p*-Cymene, Pulegone, and Cinnamaldehyde as Shown under Ultraviolet Light on Fluorescent Strips

(From top to bottom)
1. *p*-Cymene
2. Pulegone
3. Cinnamaldehyde

(From left to right)
1. Blank (spot due to traces of impurity in solvent not removed by distillation
2. *p*-Cymene, 143γ, pulegone, 3.3γ, cinnamaldehyde, 3.3γ

3. *p*-Cymene, 358γ, pulegone, 8.2γ, cinnamaldehyde, 9.1γ
4. *p*-Cymene, 1.4 mg., pulegone, 32.8γ, cinnamaldehyde 36.4γ

Chromatography of Terpenes. Fifty-eight samples of different commercial oils were chromatographed with five solvents on fluorescent silicic acid-coated strips. By this series of tests it was hoped to test the applicability of the technique and to select a variety of samples of good purity on which sensitivity and R_F values could be determined. Of the 58 samples chromatographed, two were considered to be sufficiently pure; the other 56 oils gave chromatograms on which 2 to 9 spots were apparent when viewed under ultraviolet light and treated with fluorescein and bromine.

Fourteen compounds were then purified in this laboratory by various techniques. The method of purification and an indication of the purity of these samples are given in Table III. All fractional distillations were made with a Podbielniak Hypercal column. Many different types of compounds were selected in order to indicate the wide range of usefulness of both the technique and the various spot-indicating reagents.

Each of the pure samples was chromatographed on fluorescent silicic acid strips. Five solvents were used to characterize the oil: hexane (boiling point 65° to 69° C.), carbon tetrachloride,

chloroform, benzene, and ethyl acetate in hexane (15% by volume). Sensitivity to the various color tests was determined by measuring the sample applied to the adsorbent, pure or in a suitable solvent, from a calibrated capillary pipet. (The ordinary mercury piston-type microburet could not be used because the oils in contact with the mercury caused the mercury to fall out of the capillary. A calibrated capillary tube fitted with a syringe to control the column of liquid by air displacement was used.) The quantity of sample applied was reduced in systematic fashion until the chromatogram failed to give the color test in question. The last detectable concentration was called the sensitivity limit. On each set of five chromatograms, a control chromatogram of limonene chromatographed with hexane was run in order to make sure the strips were dried properly and to afford a reference for comparing R_F values. A minimum of five samples in each solvent was run to determine the R_F value. Samples diluted with ethyl alcohol tend to have their position on the chromatogram distorted by the alcohol. This phenomenon was not observed when hexane was used as the diluent. On some chromatograms in which the concentration of the oil was high, the R_F was higher than on similar chromatograms with smaller amounts of oil. This was presumed to be caused by a dilution of the ascending solvent by the sample. When the concentration of the sample was high, this dilution was apparently enough to cause a change in the characteristics of the solvent. The results of the sensitivity determinations and the R_F of the terpenes in the various solvents are given in Table IV.

DISCUSSION

By using the technique described herein, a sample of oil can be rapidly chromatographed and many conclusions concerning purity and adulteration can be established at once. Because a large number of terpene samples have been investigated and the great majority of them (56 out of 58) have been found impure by examination of chromatograms in five solvents, much of the tedious physical and chemical examination of the sample can be eliminated. The oils examined gave no sign of decomposing or isomerizing on silicic acid. Silicic acid is an excellent ad-

Table III. Preparation of Pure Compounds

Limonene. By fractionation from grapefruit peel oil. B.p. at 8 mm.: observed, 54 °C., literature, 53.35 °C.

α-Pinene. By fractionation of a commercial sample. B.p. at 30 mm.: observed, 60.5 °C.; literature, 60.5 °C.

Pulegone. By fraction from oil of pennyroyal. B.p. at 7 mm.: observed, 86.5 ° to 89 °C.; literature 87.4 °C.

Camphene. Commercial sample. M.p.: observed, 45 ° to 49 °C.; literature 49 ° to 52 °C.

Carvone. Purified by forming H_2S addition product and then steam distilling. M.p.: observed, 219 ° to 224 °C.; literature, 222 ° to 224 °C.

Geraniol. By forming $CaCl_2$ addition product from commercial sample and them steam distilling

p-Cymene. Comercial sample (Paragon)

Cineol. Fractionated from oil of Cajeput (b.p. at 9.5 mm.: observed 56.4 °C.; literature 54.2 °C.), and then further purified by recovery from resorcinol addition product

α-Terpineol. Recrystallized commercial product from hexane, m.p. 34–34.5 °C.; literature 35 °C.

Nopol. Fractionated, b.p. 92.6 °C. at 4 mm.; literature 71 °C. at 1 mm., 98 °C. at 5 mm.

Terpinyl acetate. Prepared from α-terpineol and acetic anhydride by method of Boulez (2)

Cinnamaldehyde. Commercial sample (Paragon) by recovery from sodium bisulfite addition product

n-Capric acid. Recrystallized from acetone, m.p. 33–34 °C.; literature 31.3 °C.

Camphor. Sample used for molecular weight determinations, m.p. 178.8 °C.; literature 178.8 °C.

---

sorbent for this work, because it resolves compounds of very similar type, as illustrated in Table IV.

The identification of terpenes should be facilitated by application of the principle of characteristic R_F values to the chromatographed samples. In identifying constituents of natural products where the amounts of fractions concerned may be small, the volume of material necessary to secure identification can be substantially reduced. Amounts of oil as small as 0.5 microgram have been detected. Some of the color tests reported here have useful implications for indicating structure—for instance, from Table II it is evident that the fluorescein-bromine test reveals the possible presence of ethylenic-type double bonds. Other indications of structure are obtained from a study of Table II, and many other tests can be applied which have specificity for certain types of structures.

As a chromatographic method aside from the advantages to terpene chemistry, many features can be pointed out.

The chromatostrip combines some of the advantages of paper and column chromatography. It adds the rapidity (a strip may be run in 0.5 hour) and the ease of spot development of paper chromatography to the wide range of adsorbents of column chromatography.

The method can be used for rapidly checking solvents and adsorbents for larger columns, because the results obtained are comparable. Although small conventional columns can be used for the same purpose, the proposed method is more rapid inasmuch as one man can easily run a set of 40 strips in an hour's time.

More drastic reagents can be applied to locate compounds than are permissible with paper strips.

"Wet tails" (irregular flow of solvent up side of paper due to touching of the glass wall) and similar annoyances of the flexible paper are avoided.

It is a microchromatographic method and is much more convenient than a packed column for micro work.

The use of concentrated sulfuric acid and concentrated sulfuric-nitric acid mixture for elucidating positions of organic materials on a chromatogram eliminates the uncertainty usually associated with less drastic revealing agents. These reagents have been used to indicate traces of impurities in some terpenes which were thought to be pure by reason of physical constants and chromatography with other compound-indicating tests.

The spray gun used for all of the work except for the sulfuric acid was an artist's air brush.

As shown by Meinhard and Hall (9), the completed chromatograms could be stripped off on Scotch tape and pasted on suitable cards for filing and reference purposes. This was not possible for strips sprayed with cold sulfuric acid, but the chromatograms which were heated after spraying with acid could be saved in the same manner.

ACKNOWLEDGMENT

The authors are indebted to Richard Course of this laboratory for assistance in fractionating the oils.

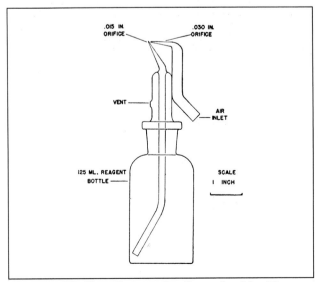

Figure 3. All-Glass Sprayer for Corrosive Liquids

Table IV. Chromatography of Pure Compounds

RF Values

Compound	Hexane	Carbon tetrachloride	Chloroform	Benzene	15% ethyl acetate in hexane	Control, limonene in hexane	Limit of Sensitivity, γ	Color Test
Limonene	0.41	0.37	0.93	0.96	0.66	. . .	37	Fluorescein-bromine
α-Pinene	0.83	0.89	0.95	0.96	0.83	0.50	37	Fluorescein-bromine
Pulegone	0.01	0.01	0.09	0.07	0.49	0.51	4	Ultraviolet and fluorescein-bromine
Camphene	0.74	0.82	0.92	0.94	0.79	0.47	200[a]	Fluorescein-bromine
Geraniol	0.00	0.00	0.05	0.05	0.21	0.46	1.5	Fluorescein-bromine
Carvone	0.00	0.01	0.07	0.04	0.45	0.47	0.4	Ultraviolet
							8.0	Fluorescein-bromine
p-Cymene	0.38	0.56	0.94	0.95	0.60	0.43	100[b]	Ultraviolet
α-Terpineol	0.00	0.00	0.05	0.03	0.24	0.43	4.0	Fluorescein-bromine
Nopol	0.00	0.00	0.11	0.06	0.27	0.53	1.0	Fluorescein-bromine
1,8-Cineol	0.01	0.02	0.12	0.06	0.48	0.48	0.6	Concd. sulfuric
Cinnamaldehyde	0.00	0.00	0.09	0.06	0.31	0.48	0.3	o-Dianisidine
Terpinylacetate	0.00	0.00	0.26	0.25	0.50	0.50	1.0	Fluorescein-bromine
n-Capric acid	0.00	0.00	0.07	0.07	0.43	0.55	4.0	Bromocresol green
Camphor	0.00	0.00	0.28	0.22	0.56	0.67[c]	0.8	Concd. sulfuric-nitric

[a] Sensitivity of detection 15γ with sulfuric-nitric acid.
[b] Sensitivity of detection 30γ with sulfuric-nitric acid.
[c] Comound run on strips with plaster of Paris instead of starch as a binder; thus, RF value for limonene control has a different value.

LITERATURE CITED

(1) Boldingh, J., *Experientia*, **4**, 270 (1948).
(2) Boulez, V., *Bull. soc. chim. Belges*, **1**, 117 (1907).
(3) Carlsohn, H., and Müller, G., *Ber.*, **71**, 858 (1938).
(4) Datta, S. P., and Overell, B. G., *Biochem. J.*, **44**, XLIII (1949).
(5) Datta, S. P., Overell, B. G., and Stack-Dunne, M., *Nature*, **164**, 673 (1949).
(6) Flood, H., *Z. anal. Chem.*, **120**, 327 (1940).
(7) Hopf, P. P., *J. Chem. Soc.*, **1940**, 785.
(8) Kirchner, J. G., and Keller, G. J., *J. Amer. Chem. Soc.*, **72**, 1867 (1950).
(9) Meinhard, J. E., and Hall, N. F., *Anal. Chem.*, **21**, 185 (1949).
(10) Ramsey, L. L., and Patterson, W. L., *J. Assoc. Offic. Agr. Chemists*, **31**, 139 (1948).
(11) Rockland, L. B., and Dunn, M. S., *Science*, **109**, 539 (1949).
(12) Sease, J. W., *J. Am. Chem. Soc.*, **70**, 3630 (1948).
(13) Späth, E., and Kainrath, P., *Ber.*, **70**, 2272 (1937).
(14) Wasicky, R., and Frehden, O., *Microchim. Acta*, **1**, 55 (1937).
(15) Winterstein, A., and Stein, G., *Z. physiol. Chem.*, **220**, 247 (1933).
(16) Zechmeister, L., "Progress in Chromatography, 1938–1947," London, Chapman and Hall, 1949; New York, John Wiley & Sons.
(17) Zechmeister, L., and Cholnoky, L., "Principles and Practice of Chromatography," New York, John Wiley & Sons, 1941.

Received June 26, 1950. Presented before the Division of Analytical Chemistry at the 118th Meeting of the American Chemical Society, Chicago, Ill.

Reprinted from *Anal. Chem.* **1951**, *23*, 420–25.

This paper was published just eight years after the first reports of the detection of nuclear resonance effects in bulk matter. Although chemical shift effects, spin–spin splitting of multiplets, and the dependence of resonance intensity on nuclei concentration were known, it wasn't clear whether NMR would be a particularly useful technique to chemists. Here Shoolery presents some of the earliest examples of NMR applications based on chemical shift and spin–spin coupling effects as well as resonance intensities. The article is noteworthy in that it identifies what remain some of the most widely used applications of high-resolution NMR spectroscopy.

Dallas L. Rabenstein
University of California–Riverside

Nuclear Magnetic Resonance Spectroscopy

JAMES N. SHOOLERY
Varian Associates, Palo Alto, Calif.

Fundamental concepts of nuclear magnetic resonance are discussed and the basic equation relating magnetic field and radio-frequency is derived. The types of apparatus in general use are described. Applications which can be accomplished using fields of moderate homogeneity are quantitative determinations of the number of nuclei of a given type, illustrated by the measurement of moisture. Nuclear magnetic resonance spectroscopy in extremely homogeneous fields is taken up and the various effects which are encountered are explained. Applications to research are found in molecular electron distribution, in structure determination, in organic group identification, and in general studies of the structural features of organic compounds, fluorocarbons, silicones, boranes, and compounds of phosphorus. High-resolution spectra illustrate some of these applications.

The fundamental concept of nuclear magnetic resonance is that an alternating magnetic field can induce transitions between the Zeeman levels of a nucleus placed in a fixed magnetic field. The experiment differs from other spectroscopic techniques in two ways:

1. The separation of the energy levels and consequently the frequency of the absorbed energy can be varied by changing the value of the field. The absolute value of the frequency is not a property of either the nucleus or the molecule being studied, but must satisfy the relation

$$\omega = \gamma H \tag{1}$$

2. The alternating magnetic field is generated by an oscillating current in a coil and is coherent in time, as contrasted with the incoherent oscillating fields of much higher frequency, associated with the electric or magnetic component of electromagnetic radiation. Equation 1 can be derived very simply by the following method. The energy of interaction of the nucleus with a magnetic field is given by

$$W = -\vec{\mu} \cdot \vec{H} \tag{2}$$

Now the nucleus has a spin angular momentum, I, associated with it which points along the same axis as $\vec{\mu}$. Only $2I + 1$ orientations in the magnetic field are allowed by quantum theory. These are characterized by the quantum number m, running from $m = -I \cdots O \cdots + I$.

Now

$$\vec{\mu} \cdot \vec{H} = \mu H \cos \theta = \frac{m}{I} \mu H \tag{3}$$

Only transitions for which $\Delta m = \pm 1$ give detectable absorption, and the oscillating field must therefore be at right angles to the fixed field. Then

$$\Delta W = \frac{\omega h}{2\pi} = \frac{\mu H}{I} \tag{4}$$

$$\omega = \frac{2\pi\mu}{Ih} H = \gamma H$$

γ is usually called the gyromagnetic ratio and is a property of the particular nucleus being studied. Other authors (*2, 4, 8*) have treated the fundamental equations of nuclear magnetic resonance in great detail.

γ differs markedly and in an unpredictable manner from one nucleus to another. Some nuclei, for which $I = 0$, have neither spin nor magnetic moment and cannot be detected.

2855-8/94/0099$08.00/0 © 1994 American Chemical Society

The signal voltage obtained from an ensemble of nuclei depends upon $\gamma^{11/4}$; therefore, the signals from some nuclei are very much stronger than others. Light hydrogen, fluorine-19, phosphorus-31, aluminum-27, and lithium-7 are in this category, along with several rare isotopes.

The relation $\omega = \gamma H$ applies to the field at the nucleus, not the applied field. The applied field is altered inside a sample of material, owing to the bulk magnetic susceptibility. The field inside the electron shell of each atom is altered further by the contribution of the orbital motions of the electrons. Both of these effects result in small shifts, which are very significant.

In an actual physical system, transitions are equally probable in both directions—from higher to lower energy and from lower energy to higher energy. Only the fact that the lower levels are more populated according to a Boltzman distribution gives us net signal different from zero. Since $\Delta W < < kT$, the population difference is only about 10 p.p.m. Also, induced transitions tend to diminish this difference, resulting in smaller signals. Only by keeping the oscillating field small enough to avoid depopulating the lower nuclear spin state more rapidly than natural relaxation mechanisms can re-establish equilibrium can we avoid this saturation effect.

APPARATUS

Apparatus with which nuclear magnetic resonance studies can be conducted has developed along two lines (3, 7, 13-16). Starting with the work in 1946 of Felix Bloch and E. M. Purcell, co-winners of the 1952 Nobel Prize in Physics, various groups have further developed these basic designs. Bloch's nuclear induction experiment requires, first, a magnet, which may be either an electromagnet or a permanent magnet; a generator of radio-frequency energy, usually called a transmitter; a device called a probe with which to apply this energy to the nuclei in the form of an oscillating magnetic field and to receive a response; and a receiver with which to amplify the response sufficiently to display it on some sort of indicator. Such a system is shown in Figure 1. This system measures the energy coupled from one pair of coils into a third coil at right angles to the first pair by the sample of nuclei contained within the coils. In order to observe nuclear resonances, the applied magnetic field is varied through resonance by changing the current in a pair of sweep coils. Other systems differ from Figure 1 in that they combine the function of the transmitter and receiver and detect changes in the level of oscillation in a coil wrapped around the sample being investigated.

The lower limit of detectability in terms of the number of nuclei per cubic centimeter of the sample is, at present, around 10^{16} nuclei for the most favorable ones, ranging up to 10^{20} for other nuclei with spin 1/2. In view of the rather wide range of line widths encountered in nuclear magnetic resonance work, no general guide concerning sensitivity can be formulated, except, perhaps, that liquids give generally much higher amplitude signals than solids or gases.

When very weak lines are encountered, the thermal noise present in the receiver limits sensitivity. The usual way to minimize this difficulty is to modulate the absorption at an audio-frequency and to use a very narrow band width defined

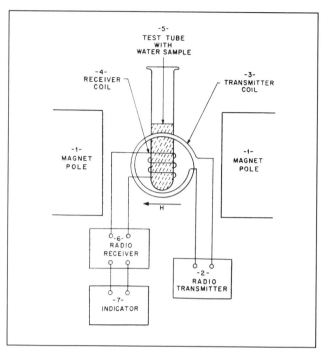

Figure 1. Block Diagram of Nuclear Induction Spectrometer

by the time constant of a coherent detector. One must then sweep through the line slowly enough to maintain its frequency components within the band. This is very time-consuming but necessary for quantitative work with weak resonances. Recently in England (17, 20), it has been demonstrated that the time-consuming slow sweep can be avoided by sweeping rapidly over several thousand gauss with high radio-frequency power and a wide band. Instead of the steady-state nuclear resonance signal, one gets the full polarization of the nuclei for an instant, which makes up for the increased noise due to the wide band. There is a finite limit on the polarization, however, while the noise may be reduced indefinitely by narrowing the band.

Figure 2,a shows a sweep of 500 gauss made in about 20 seconds. The hydrogen and fluorine signals from the compound iso-heptafluoropropane were recorded, and the ratio of fluorine to hydrogen can be determined rapidly. This sort of experiment could be broadened into a set of quantitative analysis schemes, where different sweep ranges, or different frequencies, would respond to several nuclei at a time. It would not be completely general because of the failure of some nuclei to respond. It would be most useful for nonionic substances, such as the fluorocarbons, for which present analytical techniques are extremely laborious.

APPLICATIONS

A more specialized application of quantitative analysis is the measurement of moisture. In a moist solid, only the hydrogen nuclei in the water molecules give sharp, intense signals, the signal from hydrogen nuclei in the solids being broadened by interactions between themselves. Several workers, including Shaw, Elsken, and Kunsman (18) at the Western Regional Research Laboratory, Albany, Calif., and

Rollwitz and O'Meara (*11, 12*) at the Southwest Research Institute, San Antonio, Tex., have demonstrated the feasibility of nuclear magnetic resonance determination of moisture in agricultural products in a rapid and nondestructive way. If apparatus is constructed to take a statistically large sample, say 50 to 100 grams, and provision is made to control the density, then, by integrating the output of the receiver, one can get accurate values of the moisture content in these materials. It is not unreasonable to expect that the measurement will be put on process streams to monitor moisture or some other component continuously.

Perhaps the most promising application of nuclear magnetic resonance is high resolution spectroscopy (*1, 5, 6, 9, 10*), based on the fact that in liquids the resonances are extremely narrow—enough so that the small shifts in field at the nuclei due to contributions from electronic orbital motions result in a splitting of the resonance into a spectrum. When atoms form different types of bonds, the electron cloud surrounding their nucleus is modified, and the result is a characteristic shift in the field compared to some chosen standard reference bond, such as, for example, the OH bond in water for protons or the CF bonds in perfluorocyclobutane for fluorine. These shifts due to the electron orbital motions are linearly field-dependent; therefore, it is advantageous to work at high fields, from 7000 to 10,000 gauss. In order to compare work done at different field strengths, it is useful to divide the shift, measured from a reference compound, by the field, and to express the result as a shift in parts per million. To measure such shifts, it is necessary to have a field which is uniform over the dimensions of the sample and stable during the time of measurement to the order of 1 part in 10,000,000 to 100,000,000.

In Figure 2,*a* a trace is presented showing the hydrogen and fluorine resonances separated by 450 gauss. Figure 2,*b* shows the same two signals with the magnetic field scale increased 500 times. A double peak appears for the hydrogen, and two peaks appear in the fluorine region, corresponding to the CF_3 groups, and the —C—F group. Now, if the scale is increased to 6000 times the original, as in Figure 2,*c*, the double proton peak is seen to consist of two pairs of seven overlapping sharp peaks, and the fluorine signals also exhibit multiplet structure. The origin, zero field, is about one mile to the left of the paper in Figure 2,*c*.

If the experiment were repeated at a different field and frequency, the two groups would have moved relative to each other, but the multiplet splitting would remain unchanged. It is field-independent and results from the splitting of the magnetic energy levels into sublevels because of an indirect interaction between the nuclear magnetic moments through the electron cloud comprising the chemical bond. Each group of *n* identical nuclei gives a resultant $2nI + 1$ components in the multiplet resulting from the interaction with a neighboring chemically different nucleus, where I is the spin of the identical nuclei. The magnitude of the coupling and spacing of the multiplet depends on the number and type of bonds separating the interacting nuclei. This theory is sufficient to explain the general features of the observed fine structure.

The information gained concerning electron distributions in molecules is invaluable in a theoretical sense. Present theories can be checked and, in some cases, modified to agree with the information becoming available for the first time by this powerful new tool. For example, the concept of electronegativity is a well-established one for measuring the electron-attracting power of an element or a group. Suppose an

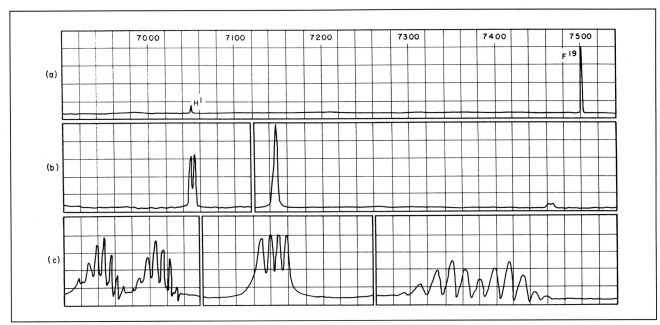

Figure 2. Magnetic Resonance Trace

a. Nuclear magnetic resonance signals from iso-C_3F_7H with sweep from 7000 to 7500 gauss
b. *Left.* Proton signals. *Right.* F^{19} signals. Resolution increased by factor of 500
c. *Left.* Proton signals. Center. F^{19} signals from CF_3 groups. *Right.* F^{19} signals from — CFH group. Resolution increased by factor of 6000.
Each square represents 3 milligauss

electronegative atom is introduced into a molecule such as ethane, in which all the protons are equivalent. If one chooses to introduce oxygen, the resulting molecule is ethyl alcohol. The proton attached to oxygen loses the most electrons; consequently, it is less shielded and the resonance appears at the lowest applied field. The CH_2 gives up fewer electrons, and the CH_3 is almost completely insulated from the oxygen. As the CH_2 and CH_3 bond types—i.e., the orbitals used in bonding—are similar, the splitting of these groups is a measure of the electrons given up by the CH_2 groups to the oxygen, or of the electronegativity of the oxygen. Table I shows the splitting in the ethyl group for a series of substituted ethanes and the result of correlating this with the electronegativity of the substituent (19).

The application of high-resolution nuclear magnetic resonance to structure determination appears to be promising. Figure 3 shows the signals from several fluorocarbons. The shifts recorded in parts per million from the resonance in the reference compound, perfluorocyclobutane, in this case. Three typical group positions can be noted—namely, CF_2 at 0.0 to –15, CF_3 at –50 to –55, and CF_2O at –48. The concept of group positions can be made more general, since the variety of bonding situations which can be resolved is not limited to these particular examples. In the case of fluorine-19 resonances in fluorocarbons, the highly ionic nature of the C-F bonds in fluorocarbons results in the general division of the spectrum into four fairly narrow but widely separated regions, corresponding to competition for the available electrons from each carbon atom between one, two, three, or four atoms which are electronegative compared to carbon. Thus, $-CF_3$, $-CF_2O$, and $-CF_2Cl$ fall in one region,

$$\diagdown_{\diagdown}^{\diagup} CF_2, \quad \diagdown_{\diagdown}^{\diagup} CF\text{-O}, \quad \text{and} \quad \diagdown_{\diagdown}^{\diagup} CFCl \quad \text{fall in another, while } CF_4 \text{ and } \diagdown_{\diagdown}^{\diagup} C\text{-F}$$

typify the other two regions. The effect of atoms with electronegativities more nearly like that of carbon but forming different bond types is less clear and causes some overlapping of the regions. In most cases, however, the distribution of fluorine in a fluorocarbon can be fairly well determined.

In the case of proton resonances, the similarity of the hy-

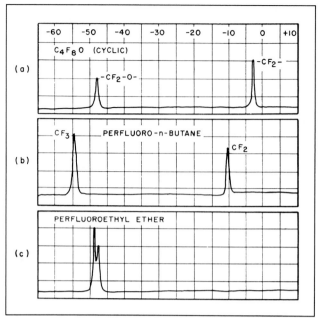

Figure 3. Signals from Fluorocarbons

a. Nuclear magnetic resonance signals from F^{19} nuclei in five-membered ring compound, C_4F_8O. CF_2 groups adjacent to oxygen appear at left, while those adjacent only to other CF_2 groups appear at right
b. Spectrum of n-C_4F_{10}
c. Spectrum of $(C_2F_5)_2O$. Presence of oxygen has shifted CF_3 slightly toward higher fields

drogen and carbon electronegativities seems to lead to equal importance of ionic and covalent bond character effects. Thus, while the effect of electronegativities has been demonstrated in an earlier paragraph, the effect of double bonds or resonant systems cannot be overlooked. Olefinic and aromatic hydrogen atoms possess fairly well-characterized magnetic shielding values readily resolved from each other and from aliphatic protons.

Both fluorine-19 and hydrogen-1 spectra frequently show striking spin-spin multiplets similar to those shown in Figure 2,b and c. Analysis of these multiplets leads to identification and location of groups with respect to their neighbors and is an invaluable aid in assigning structural formulas, where this is possible. Certain groups, such as ethyl, isopropyl, and $\diagdown_{\diagup} C \diagdown_{\diagdown H}^{\diagup F}$ as well as many others give easily recognized multiplet patterns.

Phosphorus-31 signals are also adequate for high-resolution work in liquid samples, and studies of the phosphorus-31 high-resolution nuclear magnetic resonance spectra should prove fruitful.

Silicones and boranes represent still other possible avenues of investigation. The proton spectra of these compounds are already revealing, and the silicon-29 and boron-11 spectra, particularly the latter, are promising possibilities.

Methods of standardizing and classifying spectra are being worked out at the present time. The American Petroleum Institute Research Project 44 is studying schemes for nuclear magnetic resonance classification, and its recommendation is expected to be issued as soon as reasonably complete agreement can be reached as to the most suitable form.

Table I. Chemical Shifts in Substituted Ethanes

Compound	δrel.	$\dfrac{2.1 + 0.45}{\delta \text{rel.}}$	Pauling Electronegativity
C_2H_6	0.0	2.1	2.1
C_2H_5SH	1.2	2.64	2.5
C_2H_5I	1.3	2.68	2.5
C_2H_5Br	1.6	2.82	2.8
$C_2H_5NH_2$	1.7	2.86	3.0
C_2H_5Cl	2.0	3.00	3.0
C_2H_5OH	2.4	3.18	3.5
$C_2H_5OCOCH_3$	3.0	3.45	3.5

$$\delta \text{rel.} = 10^6 \times (H_{CH_3} - H_{CH_2})/H_{CH_3}$$

Intensive study of both the theoretical aspects and practical applications of nuclear magnetic resonance high-resolution spectroscopy as well as work with broad resonances is going on both in the universities and in industrial laboratories. Progress is being made in improving resolution and stability to a degree not believed possible only a short time ago. Spectra of many compounds are being cataloged and classified. Methods of working with liquefied gases and melted solids are already well established. As the technique advances, many new applications of this exciting new spectroscopic method of investigation, to both theoretical and applied problems, should appear.

LITERATURE CITED

(1) Arnold, J. T., Dharmatti, S.S., and Packard, M.E., *J. Chem. Phys.*, **19**, 507 (1951).
(2) Bloch, F., *Phys. Rev.*, **70**, 460 (1946).
(3) Bloch, F., Hansen, W. W., and Packard, M.E., *Ibid.*, **70**, 474 (1946).
(4) Bloembergen, N., Purcell, E. M., and Pound, R. V., *Ibid.*, **73**, 679 (1948).
(5) Gutowsky, H. S., and Hoffman, C. J., *J. Chem. Phys.*, **19**, 1259 (1951).
(6) Gutowsky, H. S., McCall, D. W., McGarvey, B. R., and Meyer, L. H., *J. Am. Chem. Soc.*, **74**, 4809 (1952).
(7) Gutowsky, H. S., Meyer, L. H., and McClure, R. E., *Rev. Sci. Instr.*, **24**, 644 (1953).
(8) Jacobsohn, B. A., and Wangsness, R. K., *Phys. Rev.*, **73**, 942 (1948).
(9) Meyer, L. H., and Gutowsky, H. S., *J. Phys. Chem.*, **57**, 481 (1953).
(10) Meyer, L. H., Saika, A., and Gutowsky, H. S., *J. Am. Chem. Soc.*, **75**, 4567 (1953).
(11) O'Meara. J. P., and Rollwitz, W. L., "Determination of Moisture by Nuclear Magnetic Resonance, Experimental Results on Products of the Corn Wet-Milling Industry." Divisions of Carbohydrate and Analytical Chemistry, Symposium on Analytical Methods and Instrumentation Applied to Sugars and Other Carbohydrates, 124th Meeting, Am. Chem. Soc., Chicago, Ill.
(12) *Ibid.*, "Theory and Design Considerations for a Practical Instrument."
(13) Packard, M. E., *Rev. Sci. Instr.*, **19**, 439 (1948).
(14) Pound, R. V., and Knight, W. D., *Ibid.*, **21**, 219 (1951).
(15) Proctor, W. G., *Phys. Rev.*, **79**, 35 (1950).
(16) Purcell, E. M., Torrey, H. C., and Pound, R. V., *Ibid.*, **69**, 37 (1946).
(17) Ross, I. M., and Johnson, F. B., *Nature*, **167**, 286 (1951).
(18) Shaw, T. M., Elsken, R. H., and Kunsman, C. H., *J. Assoc. Offic. Agr. Chemists*, **36**, 1070 (1953).
(19) Shoolery, J. N., *J. Chem. Phys.*, **21**, 1899 (1953).
(20) Taylor, K., *Nature*, **172**, 722 (1953).

Received for review February 6, 1954. Accepted June 11, 1954. Presented at the Regional Conclave, American Chemical Society, New Orleans, La., December 10 to 12, 1953.

Reprinted from *Anal. Chem.* **1954**, 1400–03.

The seminal paper by Kiers, Britt, and Wentworth popularized phosphorescence spectrometry for the analysis of organics. Phosphorescence theory, instrumentation, and spectral and temporal resolution of mixtures of organic species were eloquently described. This classic article was clearly responsible for the resurgence of low- and room-temperature analytical phosphorimetry during the 1970s and 1980s.

James D. Winefordner
University of Florida

Phosphorimetry
A New Method of Analysis

R. J. KEIRS, R. D. BRITT, JR.[1], AND W. E. WENTWORTH
Department of Chemistry, Florida State University, Tallahassee, Fla.

Phosphorimetry is a means of chemical analysis based upon the nature and intensity of the phosphorescent light emitted by an appropriately excited molecule. Many organic molecules containing multiple bonds, when in a rigid glass formed at low temperature by solutions of the material in suitable solvents, phosphoresce if excited by radiant energy of suitable frequency. Each phosphorescence is unique in regard to its frequency, lifetime, quantum yield, and vibrational pattern and such properties are used for qualitative identification. The correlation of intensity with concentration can serve as a basis for quantitative measurement. Mixtures are analyzed by the use of a resolution phosphoroscope. The method has been applied to the determination of several compounds and their mixtures, three of which are described.

The phosphorescence emissions from over 200 compounds have been reported by molecular spectroscopists interested in the elucidation of the energy schemes of molecules. About 90 such emissions have been tabulated by Lewis and Kasha (11). The use of phosphorescence emission spectra for identification of substances was first suggested by Lewis and Kasha; however, this subject has been developed very little in the past decade. More recently the idea of using a resolution phosphoroscope (proposed by M. Kasha of this laboratory) greatly extended the analytical potentialities of phosphorescence. A theoretical analysis of the method is to be published.

Phosphorescence is not characteristic of a specific class of

[1] Present address, Savannah River Plant, E. I. du Pont de Nemours & Co., Inc., Augusta, Ga.

compounds, but a prime condition for its observation is high viscosity. Substances that phosphoresce may be divided into two classes, based upon the mechanism by which their phosphorescences are produced (11). In the first group are mineral, or crystal, phosphors (15). In this case the individual molecule is not phosphorescent, but the ability to emit an afterglow is associated with the return of an electron to an impurity site in the crystal, following ionization through the process of photon absorption. As this type of phosphorescence cannot be ascribed to a definite substance, phosphorescences of this class were not considered in this investigation. In the second class the emission is attributed to a definite molecular species, whether the substance is in a pure crystalline state, absorbed on a suitable surface, or dissolved in a suitable rigid solution. It is with such rigid transparent solutions that this investigation was concerned.

Phosphorescence is produced in molecules by the absorption of radiant energy of a frequency within the normal absorption band of the molecule—usually in the ultraviolet region of the spectrum. Two of the many possible end results are fluorescence and phosphorescence produced by the excitation of an appropriate electron energy level of the molecule. These two processes may be shown schematically by diagram of molecular electron energy level (Figure 1), similar to that given by Jablonski (8) for dyes.

Molecules in an electron ground state level, G, on absorption of radiant energy, arrow 1, are for a period less than 10^{-12} second (11) in a vibrational level of an upper excited singlet state, S_2, S_3, etc. A singlet state in an unsaturated molecule is characterized by a pair of unsaturation (π) electrons having paired (antiparallel) spins. From one of the upper singlet levels a process may occur in which the absorbed energy is given up in steps that may involve a

2855-8/94/0104$08.00/0 © 1994 American Chemical Society
Milestones in Analytical Chemistry

nonradiative transition (dashed arrow), arrow 2, to the lowest excited singlet level, S_1, followed by a radiative transition, arrow 3, to one of the vibrational levels of the ground electron level, G. This latter transition ($S_1 \rightarrow G$), which is a singlet $\rightarrow$ singlet transition, represents a fluorescence band of the molecule (10, 11).

By a radiationless route, arrow 4, some of the excited molecules can pass from their lowest excited singlet state, S_1, into a vibrational level of a metastable state, T_1. This is the phosphorescent state or triplet state described by Lewis and Kasha (11). A triplet state in an unsaturated molecule is characterized by a pair of unsaturation electrons having unpaired (parallel) spins. Transition from the lowest triplet state to a vibrational level of the ground state, arrow 5, represents a phosphorescence band of the molecule. The transition probability of the triplet $\rightarrow$ singlet transition governs the lifetime of the metastable triplet state.

Because the energy level of the lowest triplet state of such molecules lies below the excited singlet levels, the phosphorescent emissions are of longer wave length than the absorptions; hence the phosphorescences of most molecules lie in the convenient visible or near infrared spectral region.

Several authors (9, 11, 14) have pointed out that a low-lying triplet state can be present in any molecule having a multiple bond—possibly some others. As the phosphorescence emissions from over 200 compounds have been studied thus far, a fertile field is open for analytical exploitation.

In order to observe the triplet $\rightarrow$ singlet transition in an appropriate organic molecule, it is necessary in most instances to have the substance in a dense phase or medium. The reason given (14) for this is that the direct excitation to the triplet state is much too weak, or improbable, to allow a sufficient population of this state to observe phosphorescence; hence, a more devious route must be depended upon. First, excitation to an upper singlet state must be accomplished, followed by an interconversion to a vibrational level of the lowest triplet state. Then the removal of vibrational energy of the triplet state through collisions must result to permit the population of the lowest metastable triplet state, followed, finally, by a radiative transition (phosphorescence) to a vibrational level of the ground state.

The role of the rigid media is explained (14) to be such as to remove triplet state vibrational energy but not remove the triplet state electron energy by thermal deactivation to cause the molecule to dissipate its energy by a nonradiative process (arrow 6, Figure 1). In order for thermal deactivation to occur, electron energy must be converted into kinetic energy of surrounding molecules. Such a process is inefficient compared to the removal of vibrational energy from the same molecule. The momentum transfer from an electron to a molecule is inefficient because

of the large mass difference. However, the extremely high collision rate in liquids overcomes this relative inefficiency and results in no observable phosphorescence because of the complete thermal deactivation of the triplet state.

Phosphorescence has been observed from some substances when in fluid media (3), but, as stated by Kasha (9), this is possible only if the mean lifetime is intrinsically short (order of 10^{-3} second or less).

Each phosphorescence is unique for each molecule and as proposed by Lewis and Kasha (11) may be used to identify phosphors in mixtures. Rybab, Lochet, and Rousset (18) have used the α-phosphorescence (12) as a means of identifying some aromatic amino acids. Each phosphorescence is characteristic of the electron structure of the molecule as a whole and not that of specific groups in the molecule (11, 12). It is characterized by four parameters:

1. The mean lifetime, r, which is the length of time as-

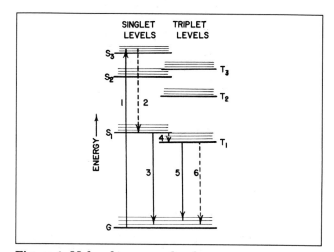

Figure 1. Molecular energy levels

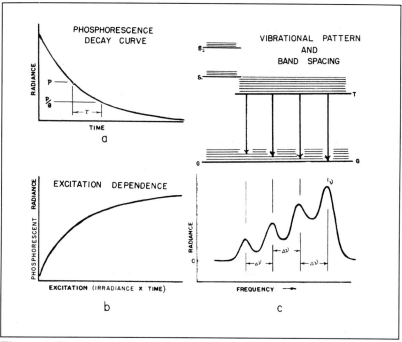

Figure 2. Characteristics of phosphorescence

sociated with the decrease in radiant power from P to P/e, (a, Figure 2).

2. The frequency of the vibrationless transition, υ—i.e., the highest frequency band of the phosphorescence characterized by the transition from the lowest triplet level to the lowest vibrational or ground level (Figure 2,c).

3. The vibrational pattern or band spacing, $\Delta\upsilon$ (Figure 2), which represents the spacing of the vibrational levels of the ground electron level.

4. The quantum efficiency, Φ, the ratio of the quanta of light phosphoresced to the quanta absorbed.

FACTORS AFFECTING PHOSPHORESCENT EMISSIONS

Lewis, Lipkin, and Magel (12) found the lifetime to be nearly independent of temperature for fluorescein dissolved in boric acid, whereas Pyatnitskiĭ and Vinokurova (17) reported that the phosphorescence lifetime of biphenyl varied from 1 to 0.4 second when temperature ranged from 90 ° to 130 °K. Pyatnitskiĭ (16) reported invariant lifetimes for phthalic benzoic acids in the same temperature range. Sveshnikov (19) reported a deviation from the exponential decay law for time intervals that were long compared to r, and involved a thousandfold variation in intensity. He also reported a dependence of r upon intensity of the exciting light. Both phenomena exhibit an apparent decrease in r during the time in which a high concentration of molecules exists in the phosphorescent state. Kasha (9) reports that Calvin explained this as due to magnetic quenching resulting from the interaction of excited triplet-state molecules with the inhomogeneous magnetic fields produced by the relative high concentration of neighboring triplet-state molecules. Sveshnikov and Petrov (20) have reported that the nature of the solvent may affect the lifetime.

Lewis, Lipkin, and Magel (12) have reported that when a liquid is cooled, the intensity of phosphorescence increases as viscosity increases. If there were no thermal quenching in the solid state, both the intensity and the lifetime of phosphorescence should be independent of temperature, and they have shown this to be the case. They also report another indication that collisional deactivation does not alter the triple-state lifetime in rigid media: that the decay curves (phosphorescent radiance vs. time, Figure 2,a) are exponential. They state further that a complete lack of collisional process cannot be shown by such curves, but if as little as 10% of the decay process is a second-order thermal deactivation, it will be detected in the curves. Their deviations have always been traced to phosphorescent impurities and, in fact, they have used the exponential form of the decay curve as a criterion of phosphorescent purity.

The parameter of quantum yield of phosphorescence is too difficult to measure for use in qualitative identification, but it is desired to have as high a value as possible in quantitative phosphorimetry, to increase precision and to extend the range to lower concentrations.

In addition to these qualitative parameters, the radiant power of the phosphorescence is a function of several variables:

The radiant power of the excitation light (12) (Figure 2,b)
The length of time of excitation

The concentration of the phosphor or other substance that may absorb radiation of the frequency used to excite the phosphor or that may absorb any of the phosphorescent emission

The interval of time following cessation of excitation and photometry

Under conditions of the analysis any parameter that may affect the lifetime or the quantum yield.

The Beer-Lambert law cannot be applied directly to this type of absorption-emission, because the light energy absorbed and not that transmitted is of essence. Danckwortt (6) has applied a modified form of the law to fluorometry, and a similar approach must be made in phosphorimetry.

$$P = P_0\, e^{-\epsilon cl}$$

From the Beer-Lambert law above, the fraction of radiant power transmitted by a given medium is given by $T = e^{-\epsilon cl}$. In fluorometry and phosphorimetry the fraction of radiant power absorbed, A, is of value as a function related directly to concentration, where $A = (1 - T) = 1 - e^{-\epsilon cl}$.

Hence the phosphorescent radiant power, P_p, may be expressed as $P_p = K'\,(1 - e^{-\epsilon cl})$. This equation may be differentiated and expanded (13) to a form

$$P_p = K'' \times (2.30\ \epsilon cl - 2.65\ \epsilon^2 c^2 l^2 + 2.04\ \epsilon^3 c^3 l^3 - \ldots)$$

in which the higher order terms may be neglected if the ϵcl value is less than 0.01. Then the phosphorescent radiant power may be related directly to the concentration:

$$P_p = Kc$$

These considerations make it apparent that in a quantitative application, standardization of excitation conditions, resolution time of the phosphoroscope, solvents, temperature, and concentration limits must be observed.

APPARATUS

The essential major pieces of apparatus are shown diagrammatically in Figure 3: an excitation source, resolution phosphoroscope, sample container, and receptor to detect the phosphorescent emission.

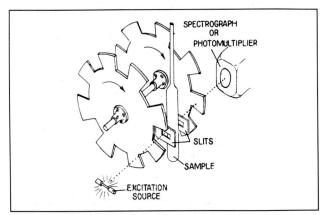

Figure 3. Resolution phosphoroscope

Two excitation sources were independently employed. One was a General Electric 100-watt, A-H4, medium pressure mercury arc (5), with the outer glass jacket removed, which gave a mercury line spectrum superimposed upon an ultraviolet continuum. The second was a General Electric, 1000-watt, A-H6 water-cooled, quartz-jacketed high pressure mercury arc, which yielded a stronger continuum to about 2270 A., together with a broadened line spectrum. The A-H6 was enclosed in a water-cooled compartment. Both sources were positioned relative to the sample in such a manner that quartz collimating and condensing lenses as well as optical filters could be employed.

The phosphoroscope employed was of a modified Becquerel type (1) and consisted of two slotted disks mounted to a common shaft and driven at a selected speed by a synchronous motor (Figure 3). The disks were so arranged that openings in one blade were in line with uncut portions of the second blade, to prevent excitation emission from gaining access to the spectrograph or photometer. This allowed a sample on the optical axis between the rotating blades to be illuminated (excited to phosphorescence), followed by a period of darkness, during which time its phosphorescence was allowed to pass the second disk and thence to the spectrograph or photometer.

The resolution time of such an arrangement is the length of time between the cutoff of excitation to the sample and the clearing of the optical path by the second blade to allow the phosphorescent emission to pass. This time is a function of the motor speed, size and spacing of the cuts in the disks, relative radial position of the disks to each other, and the size of the slits.

Two sets of blades were constructed for the preliminary investigation. The first set, L-10, consisted of two identical 11-inch disks cut from 14-gage aluminum stock with 10 equally spaced slots 1 1/2 inches deep. The slots subtended 9° of arc; the unopened portions 27°. The second set, L-2, had two equally spaced slots per disk with each slot subtending 58° of arc.

The phosphoroscope serves two functions: First, it allows the study of the phosphorescence free from any excitation emission and, secondly, it affords a unique way of resolving simple mixtures of phosphorescences based upon their mean lifetimes. For example, if two compounds of different phosphorescence lifetimes are present in a mixture, the proper choice of disks and motor speeds may give a resolution time which will permit the shorter lived compound to decay to a degree too weak to be detected and the only measurable emission would be from the longer lived component.

In another case where the two lifetimes are nearly equal, resolution may possibly be gained by a spectroscopic resolution of the phosphorescent emission or by a selection of excitation frequency that will excite one phosphor and not the other. In a more difficult case where one phosphorescence cannot be completely isolated, relative intensity values of two sets of conditions, one set optimum for one compound and the second set optimum for the other, may be determined and the concentration found through the solution of two simultaneous equations from appropriate experimental data.

The three motors employed rotated at 1800, 300, and 60 r.p.m. Hence with proper choice of blades and motor a desired resolving time can be chosen—viz., with the L-10 disks and the 1800 r.p.m. motor theoretical resolving times from 0 to 1.5×10^{-3} second could be obtained by suitable positioning of the disks on the shaft. In practice, however, this range in resolving time is limited by the height of the slits employed and is represented in the present equipment by something less than 8° of arc. With the L-10 disk assembly available, a precise setting was not afforded in such a limited angular range and so the blades were symmetrically arranged and permanently fixed on the shaft.

Sample tubes were of the shape shown in Figure 3. They were made from clear fused quartz, 20 mm. in outside diameter in the lower portion and 11 inches high. The samples were placed in a fused quartz Dewar flask that was held securely in place by an upright metal cylinder that supported the two slits shown in Figure 3. The sample tube was centered in the Dewar by means of a plastic washer positioned in the bottom of the Dewar and a hole in a cork that stoppered the Dewar. The samples were maintained at 77 °K. by keeping them submersed in the quartz Dewar filled with liquid nitrogen. A gentle stream of dry air entering the bottom of the metal cylinder and leaving at the slits kept the outside of the Dewar free of condensation during a run.

Either a Hilger glass spectrograph with a linear dispersion of 14 A. per mm. at 4200 A. and an aperture of f. 14 was used to record the phosphorescent spectra, or an Aminco (American Instrument Co.) 10-210 photomultiplier photometer was used to measure relative phosphorescent radiances.

A photomultiplier photometer is preferred to a spectrograph on the basis of time, but does not give a record of the total emission from which wave length maxima and band spacing may be determined. However, plans include the use of a grating monochromator between sample and photomultiplier photometer, so that such data may be readily ascertained. An ultraviolet monochromator between the excitation source and sample should allow better excitation frequency selection, which will permit in an ideal case the excitation of one phosphor in a mixture and not the other, or in the usual case, where mixtures are concerned, selection of optimum conditions of excitation and resolution times to determine the components of a mixture quantitatively.

RESOLUTION OF MIXTURES OF BENZALDEHYDE, BENZOPHENONE, AND 4-NITROBIPHENYL

The three compounds—benzaldehyde, benzophenone, and 4-nitrobiphenyl—were chosen on the basis of their phosphorescence properties, not with the intent to develop a method of analysis for such a mixture, but rather to test the potentialities of the method with mixtures which required both phosphoroscopic and spectroscopic resolution.

Purified samples were prepared and dissolved in a mixed solvent, EPA (11), consisting of purified ethyl ether, isopentane, and ethyl alcohol in a volume ratio of 5:5:2. Such solutions when cooled to liquid nitrogen temperature gave clear transparent glasses. The purity of the solvent is important.

Some phosphorescence can be observed in old or unpurified solvents, and the presence of small amounts of water may cause the supercooled solution to crack and become opaque.

Spectra of the three pure components in varying concentrations were recorded on Kodak III F plates. The plates were developed for 5 minutes in Eastman D-19 developer, followed by standard fixing, washing, and drying. Relative phosphorescent radiant flux values were determined from photographic densities by means of plate calibration (4). A standard benzophenone solution served as a constant radiant energy source for plate calibration for the wave length region used for benzophenone and benzaldehyde. A solution of 4-nitrobiphenyl was used for plate calibration in its wave length region.

The mean lifetime of 4-nitrobiphenyl has been reported as 0.080 second (14); those of benzaldehyde and benzophenone, approximately 0.006 second (11, 14).

The phosphorescent properties of this mixture were such that the 4-nitrobiphenyl could be resolved spectroscopically or phosphoroscopically (on the basis of its slower decay rate). Benzaldehyde and benzophenone were resolved spectroscopically. Their mean lifetimes are too similar to permit resolution phosphoroscopically with present equipment. The spectral distribution of each pure emission is plotted in Figure 4. These emissions are somewhat atypical of phosphorescent emissions. Usually the lowest wave length (highest frequency) band is the most intense (see Figure 2,c). These bands are a little more diffuse than most.

Plate calibration data were obtained by exposing a sample of the phosphorescent substances at various times. A log-step sector (4) was found impractical because of the excessively long exposure times required when the phosphorescent source was moved away from the spectrograph slit to accommodate the log-step sector mechanism. Instead, successive exposures of varying times were made and relying upon the validity of the reciprocity law for photographic emulsions:

$$\text{Exposure} = \text{intensity} \times \text{time}$$

plate calibration curves were plotted from absorbance values evaluated with a microdensitometer. These data when plotted as shown in Figure 5,a gave a means of determining relative intensity values of a source of similar frequency from its photographic density that was recorded on that plate. Density *vs.* relative intensity was plotted rather than the usual density *vs.* log relative intensity, in that an extrapolation to zero intensity is possible and thereby provides a more convenient way of evaluating data in the lower exposure region (2). The working curves for the three pure components were prepared from the spectrograms produced. Photographic density values

were evaluated and from the appropriate plate calibration curve the relative intensity, *I*, that caused that photographic blackening was determined and plotted against molar concentration to give the working curves shown in Figure 5,b, c, d.

Two working curves are shown for benzaldehyde and benzophenone. Two each were needed to resolve a mixture of these two substances spectrophotometrically. Wave length λ_1, 413 mµ, is a maximum emission region for benzophenone and a near minimum emission region for benzaldehyde (see Figure 4); wave length λ_2, 427 mµ, is a region of maximum intensity for benzaldehyde and a near minimum for benzophenone.

The working curve for 4-nitrobiphenyl was not a straight line.

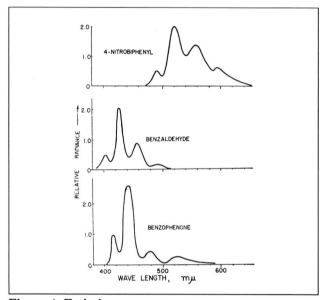

Figure 4. Emission curves

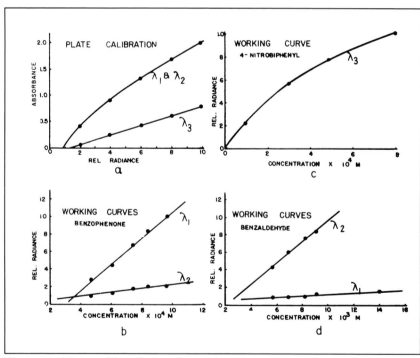

Figure 5. Plate calibration and working curves

There may be two reasons for the deviation from a linear intensity-concentration relationship. First, appreciable photodecomposition occurred. A sample that gave a relative intensity value of 1.07 for the first 15-minute exposure gave 0.77 for the second 15-minute interval, and 0.60 for the third 15-minute interval, indicating considerable photochemical decomposition on exposure to the excitation light. Secondly, the ϵcl value may have been too high. The extinction coefficient, ϵ, of 4-nitrobiphenyl calculated at 3650 A. (the principal exciting line of the mercury arc when copper sulfate filter is used), however, gives an ϵcl value of about 0.01 at the concentration used. But other lines, notably 3350 A. and 3150 A., may have an effect in addition to the 3650 A. line on the amount of light absorbed, even though they were much weaker than the 3650 A. line. Also, the ϵcl value of 0.01 is based on room temperature data and the absorption peaks for the 4-nitrobiphenyl may sharpen and ϵ may increase considerably when conditions change from room to liquid nitrogen temperature.

In order to hold ϵcl values for the mixtures of these compounds within the limits described, a concentration of $10^{-4}M$ for 4-nitrobiphenyl was employed, which required 2-hour exposure times to give applicable plate blackening. Because of the excessive and impractical exposures required as well as serious errors due to decomposition of the sample upon such long exposures it was decided to attempt only a qualitative resolution phosphoroscopically of the 4-nitrobiphenyl and do a quantitative study with more practical substances.

Figure 6 is a print of plate taken on a Hilger spectrograph to show qualitatively the spectroscopic and phosphoroscopic resolution possibilities of such a mixture.

Spectrum A is a record produced by a 15-minute exposure with 1-mm. slit of the phosphorescence of a $5 \times 10^{-3}M$ benzaldehyde solution taken with L-10 blades driven at 1800 r.p.m.

Spectrum B, obtained similarly, is of a $5 \times 10^{-4}M$ benzophenone solution.

Spectrum C is from a $5 \times 10^{-4}M$ 4-nitrobiphenyl solution taken under similar conditions, except that a 2-mm. slit was used.

Spectrum D is the record from a mixture of the three substances taken under conditions of C, while E is the record of the same mixture when the L-2 blades driven at 60 r.p.m. were employed. The exposure time in the latter case was 30 minutes. No trace of the benzaldehyde or the benzophenone spectrum is seen in E, even with a 30-minute exposure.

The phosphorescent components of a mixture such as benzaldehyde and benzophenone may be determined spectrophotometrically. Even though their spectra overlap, a maximum and a minimum of benzophenone, occurring at approximately 413 and 427 mμ, respectively, coincide almost exactly with a minimum and maximum of benzaldehyde at these two wave lengths. Because the radiant powers of phosphorescence of pure benzaldehyde and pure benzophenone are proportional to concentration (Figure 5, b, d) and upon the assumption that the radiances of the two substances in a mixture are additive, the following pair of equations was used to determine the concentrations of the two phosphors in a mixture:

$$P_1 = (k_1C_a + a) + (k_2C_b + b) \tag{1}$$

$$P_2 = (k_3C_a + c) + (k_4C_b + d) \tag{2}$$

where P_1 is the phosphorescent radiant power of the mixture found experimentally at 413 mμ, while P_2 is the radiant power of the mixture found at 427 mμ. k_1 and k_3 are the values of the slopes obtained from the working curves of benzaldehyde at 413 mμ, λ_1, and 427 mμ, λ_2, respectively (Figure 5); a and c are the values of the intercepts of these lines. k_2 and k_4 are the values of the slopes obtained from the working curves of benzophenone at 413 mμ, λ_1, and 427 mμ, λ_2, respectively. C_a and C_b are the molar concentrations of benzaldehyde and benzophenone, respectively, in the mixture. The terms $(k_1C_a + a)$ and $(k_3C_b + c)$ are the slope-intercept values for the intensities of pure benzaldehyde solutions at λ_1 and λ_2, while the terms $(k_2C_b + b)$ and $(k_4C_b + d)$ are the slope intercept values of pure benzophenone at λ_1 and λ_2 respectively. Simultaneous solution of the two equations yields concentration values for the two components.

Samples were prepared for establishing working curves of benzaldehyde in the range 5.0×10^{-4} to $1.0 \times 10^{-3}M$, and from $5.0 \times 10^{-5}M$ to $1.0 \times 10^{-4}M$ for benzophenone. Two mixtures were prepared and run in triplicate. Each replicate, together with exposures for plate calibration and working curves was recorded on separate photographic plates. The slit width used in all runs was 2 mm. with an exposure time of 10 minutes. An 1800 r.p.m. synchronous motor was used to drive the ten-slot, L-10 phosphoroscope. Table I lists the results of this test together with the relative per cent error.

PHOSPHOROSCOPIC RESOLUTION OF ACETOPHENONE AND BENZOPHENONE

A resolution based upon the mean lifetimes of the phosphorescent states was attempted, using acetophenone and benzophenone as the phosphorescent components of a mixture. The uncorrected mean lifetimes of these two compounds have been evaluated by Gilmore, Gibson, and McClure (7) as 0.008 second for acetophenone and 0.006 second for benzophenone. Here also, the compounds were chosen on a basis of their phosphorescent properties, to test the potentialities of

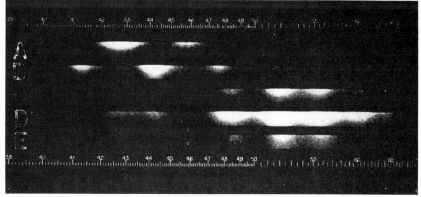

Figure 6. Print of spectrographic plate

Table I. Results of Benzaldehyde-Benzophenone Mixtures

Replica	Mixture	Determined Concn. $\times 10^4\ M$	Actual Concn. $\times 10^4\ M$	% Error
		Results for Benzaldehyde		
1	1	6.0	5.6	7.1
	2	5.9	5.6	5.4
2	1	6.1	5.6	8.9
	2	6.1	5.6	8.9
3	1	6.3	5.6	12.5
	2	6.0	5.6	7.1
		Results for Benzophenone		
		$\times 10^4\ M$	$\times 10^5\ M$	
1	2	5.1	4.7	8.5
	1	7.0	7.5	6.7
2	1	7.2	7.5	4.0
	2	5.0	4.7	6.4
3	1	7.2	7.5	4.0
	2	5.1	4.7	8.5

the method and not necessarily to develop a determination for the specific compounds.

The apparatus and procedure as described were used, except that the spectrograph was replaced by an Aminco photomultiplier photometer with a Type 931-A photomultiplier tube to measure phosphorescent radiant energies. The indicating circuit in the photometer was modified to include auxiliary resistors and capacitors to give the necessary time constant in the circuit to prevent the undesirable "kick" of the meter hand, as each pulse of radiant energy activated the photomultiplier tube when the slow-speed motor was employed to drive the phosphoroscope. Excitation was accomplished with the use of an AH-4 mercury arc filtered by 7 cm. of water and 1.7 cm. of naphthalene in cyclohexane (7.54 grams per liter) The latter filter removed the unwanted region of the mercury spectrum below 316 mμ (10). A dark blue, 5.40-mm. Corning No. 5030 glass filter was used before the photomultiplier to reduce background radiation that develops seemingly from ice formation in the liquid nitrogen surrounding the sample and from a very feeble emission from the solvent.

The phosphoroscope was operated under two sets of conditions to obtain the necessary data for the determination of the mixture. Condition A utilized the L-10 blades and 1800 r.p.m. motor to give a resolution time of approximately 0.001 second. The working curves based on data obtained under these conditions are shown in curves A, Figure 7. Condition B utilized the L-2 blades and 300 r.p.m. motor to give a resolution time of approximately 0.02 second. The data obtained under this condition yielded curves B in Figure 7. Each curve represents

the average of triplicate runs. The data for each of the trials are tabulated in Table II corrected for the indicated solvent background.

Prepared mixtures were run using the described procedure, with results listed in Table III. The determined values for each component were obtained as above, but with intercept values of zero. The relative radiant energy values (photometer scale readings), corrected for background and sample tube size, listed for the mixture are the average values of triplicate determinations. The average error is approximately 10%, excluding the obviously poor acetophenone result of mixture III, where its concentration was in the exceedingly low range of $3.5 \times 10^{-6} M$.

DETERMINATION OF DIPHENYLAMINE AND TRIPHENYLAMINE BY SELECTIVE EXCITATION

To test the resolution potentiality of the method based upon the selective excitation of the components of a mixture, diphenylamine and triphenylamine were used. The mean lifetimes and phosphorescent spectra are such (11) that either a phosphoroscopic or a spectroscopic resolution would probably be gained, but the ultraviolet absorption spectra of the two, although similar, were just different enough to offer a challenge for the method on the basis of selective excitation.

Figure 8 shows the similar absorption characteristics of the two compounds. Curve D represents diphenylamine, while T represents triphenylamine for the wave length region where absorption may give rise to phosphorescence. EPA was used as solvent. Inspection of the curves indicates the most probable wave length region that would permit resolution of these two would be in the 325- to

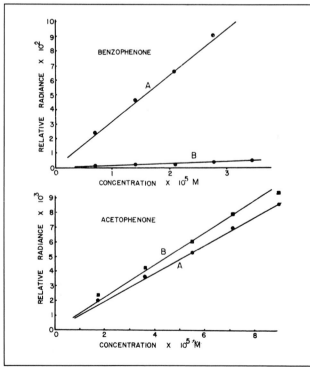

Figure 7. Working curves

350-mμ range. Here the absorption, and hence the excitation, would be small for the diphenylamine, but appreciable for triphenylamine. If the excitation source could be filtered to allow the passage of radiation greater than 325 mμ, but highly absorb radiation with shorter wave length, it would be possible to excite the triphenylamine more strongly than the diphenylamine. Kasha (10) has reported several selective filters for the ultraviolet region, two of which have been used here. The broken curve labeled 1 represents the rather sharp cutoff in the 310-mμ region due to a solution of potassium acid phthalate. Curve 2 is an approximation of the cutoff due to a solution of naphthalene. Neither dashed curve is quantitative and they are included only to help visualize the way in which resolution is gained.

The following changes were made in the apparatus:

The L-10 blades driven by an 1800 r.p.m. motor were used in the two sets of conditions used for resolution. Condition A, under which both components were strongly excited, utilized a 7-cm. water filter and a 1.7-cm. aqueous solution of potassium acid phthalate (2.94 grams per liter) to filter the excitation emission from an A-H4 lamp. Curves A in Figure 9 were obtained from data obtained in this way. Condition B, which affords a stronger excitation of triphenylamine than diphenylamine, utilized a 1.7-cm. cyclohexane solution of naphthalene (7.5 grams per liter) to replace the phthalate filter used in Condition A. Curves B in Figure 9 were obtained in this way.

The data from which these curves were obtained as well as the data from prepared mixtures are listed in Table IV. The concentrations of the unknowns were determined as described above. The average error of the six mixtures determined is approximately 8%.

CONCLUSIONS

These experiments indicate that certain organic compounds can be determined by analysis of the phosphorescent emissions excited in the compounds under appropriate condi-

Table II. Trial Data for Establishing Working Curves

Concn. ×10^5 M	Condition A, Relative Radiance × 10^3			Condition B, Relative Radiance × 10^4		
	Run 1	Run 2	Run 3	Run 1	Run 2	Run 3
	Benzophenone					
3.48	98.3	101.1	99.8	42.7	44.8	42.8
2.78	88.3	89.6	93.6	41.3	36.6	36.1
2.09	65.1	65.0	67.1	29.3	29.4	26.7
1.39	46.1	45.7	48.2	20.5	19.8	19.0
0.70	23.0	23.6	24.2	11.0	10.5	9.7
(Solvent)	(0.32)	(0.22)	(0.37)	(4.0)	(2.7)	(2.1)
	Acetophenone					
	Relative Radiance × 10^4					
8.99	79.0	83.6	86.2	90.0	91.4	94.3
7.19	65.6	68.8	73.7	76.6	76.3	79.9
5.39	51.5	52.0	55.1	59.8	59.8	62.0
3.60	34.2	38.9	38.9	42.6	42.5	44.7
1.80	19.7	20.2	20.3	23.5	23.4	24.5
(Solvent)	(3.5)	(3.2)	(4.3)	(4.2)	(2.8)	(2.1)

Table III. Acetophenone-Benzophenone Mixture

Mix	Av. Relative Radiance × 10^3		Concn. Benzophenone × 10^5 M		Concn. Acetophenone × 10^5 M	
	Condition A	Condition B	Determined	Actual	Determined	Actual
I	25.4	31.6	0.77	0.70	2.02	1.80
II	97.2	5.19	3.14	3.48	0.98	0.90
III	22.3	9.13	0.48	0.35	8.22	8.99

Table IV. Preparation of Working Curves

Concn. ×10^6 M	Condition A, Relative Radiance × 10^2			Condition B, Relative Radiance × 10^4		
	Run 1	Run 2	Run 3	Run 1	Run 2	Run 3
	Diphenylamine					
8.56	63.2	63.0	62.1	43.2	45.0	43.9
6.85	49.4	50.1	49.4	32.0	35.0	34.1
5.14	40.8	37.7	38.6	25.6	25.6	25.5
3.42	27.0	29.9	26.3	17.6	17.9	16.9
1.71	15.7	14.1	13.6	8.7	9.3	8.4
(Solvent)	(1.0)	(0.1)	(0.1)	(0.9)	(1.0)	(1.0)
	Triphenylamine					
				Relative Radiance × 10^3		
2.85	91.0	90.4	88.8	78.7	77.0	75.0
2.28	74.5	72.1	74.5	60.9	62.0	61.5
1.71	60.4	45.6	55.1	49.7	45.6	44.6
1.14	40.6	37.9	36.9	33.4	31.4	31.3
0.57	21.1	18.1	19.3	15.9	15.7	15.7
(Solvent)	(0.0)	(0.0)	(0.0)	(0.1)	(0.1)	(0.1)

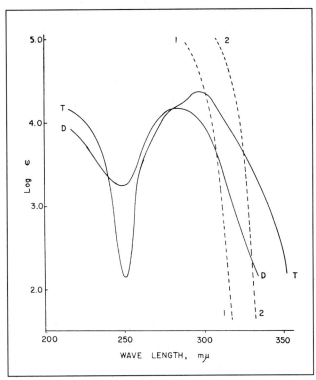

Figure 8. Absorption curves

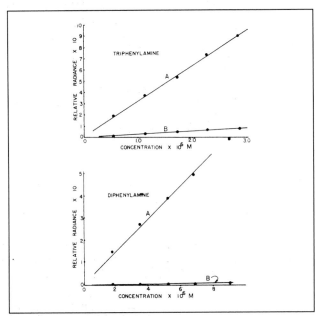

Figure 9. Working curves

tions. Where the method has been challenged by choosing compounds of very similar absorption or phosphorescent characteristics, initial results have been gratifying. The average relative error found upon determining the concentrations of known solutions by this method was about 10%, excluding an unfavorable result involving benzophenone in a mixture with acetophenone, where the concentration was at the exceedingly low value of $3.5 \times 10^{-6}M$. The data in Table V substantiate the statement that the precision determined by running replica samples is much higher than the accuracy obtained in these initial runs, indicating that the accuracy can be improved by controlling experimental parameters more carefully. To this end a calibration of the photomultiplier photometer is indicated, as well as selection of a suitable phosphorescent substance that will serve as a standard to duplicate excitation and photometry conditions from day to day. Mixtures containing more than two phosphorescent substances can be analyzed without physical separation by the proper choice of excitation frequency, resolution time, and emission frequency. The method is currently being applied in areas where its high sensitivity, low limit of detectability, and resolution are advantageous.

ACKNOWLEDGMENT

The authors wish to thank Michael Kasha for his proposal of and assistance in this investigation, and the Office of Ordnance Research, U.S. Army, which has supported this work.

LITERATURE CITED

(1) Becquerel, E., *Ann. chim. et phys.* **27**, 539 (1871).
(2) Boltz, D. F., "Selected Topics in Modern Instrumental Analysis," pp. 202–9, Prentice-Hall, New York, 1952.
(3) Boudin, S., *J. chim. phys.* **27**, 286 (1930).
(4) Brode, W. R., "Chemical Spectroscopy," 2nd ed., pp. 100–14, J. Wiley, New York, 1943.
(5) Buttolph, L. J., *J. Opt. Soc. Amer.* **29**, 124 (1939).
(6) Danckwortt, P. W., "Lumineszenz—Analysis in filtrierten ultravioletten Licht," 4th ed., p. 7, J. W. Edwards, Ann Arbor, Mich., 1944.
(7) Gilmore, E. H., Gibson, G. E., McClure, D. S., *J. Chem. Phys.* **20**, 829 (1952).
(8) Jablonski, A., *Z. Physik.* **94**, 38 (1935).
(9) Kasha, M., *Chem Revs.* **41**, 401 (1947).
(10) Kasha, M., *J. Opt. Soc. Amer.* **38**, 929 (1948).
(11) Lewis, G. N., Kasha, M., *J. Am. Chem. Soc.* **66**, 2100 (1944).
(12) Lewis, G. N., Lipkin, D., Magel, T. T., *Ibid.*, **63**, 3005 (1941).
(13) Lothian, G. T., *J. Soc. Chem. Ind.* **61**, 58 (1942).
(14) McClure, D. S., *J. Chem. Phys.* **17**, 905 (1949).
(15) Pringsheim, P., "Fluorescence and Phosphorescence," Chap. VII, Interscience, New York, 1949.
(16) Pyatnitskiĭ, B. A., *Doklady Akad. Nauk S.S.S.R.* **68**, 281 (1949).
(17) Pyatnitskiĭ, B. A., Vinokurova, T. P., *Ibid.*, **68**, 483 (1949).
(18) Rybab, B., Lochet, R., Rousset A., *Compt. rend.* **241**, 1278 (1955).
(19) Sveshnikov, B. Ya., *Compt. rend. acad. sci. U.R.S.S.* **51**, 429 (1946).
(20) Sveshnikov, B. Ya., Petrov, A. A., *Doklady Akad. Nauk S.S.S.R.* **71**, 461 (1950).

Received for review April 10, 1956. Accepted November 7, 1956.

Reprinted from *Anal. Chem.* **1957**, *29*, 202–09.

Table V. Diphenylamine-Triphenylamine Mixtures

Mix	Av. Relative Radiance × 10³		Concn. Diphenylamine × 10⁶M		Concn. Triphenylamine × 10⁶M	
	Condition A	Condition B	Determined	Actual	Determined	Actual
I	330	16.6	1.98	1.71	0.58	0.57
II	962	75.8	1.00	0.86	2.78	2.85
III	705	12.4	8.18	8.56	0.30	0.28

Golay had the rare ability to bring insights from one field into an apparently disparate one. By seeing an analogy between a GC column and this electrical circuit, he developed a mathematical theory that led to several deductions in accord with experiment. Although not widely appreciated at the time, the significance of the paper arose from Golay's dissatisfaction with this purely mathematical theory and his desire for a more physical theory of GC. This led first to his capillary bundle model of a packed GC column and then to a theoretical description of true wall-coated open-tubular columns, for which Golay is best remembered by chromatographers.

David C. Locke
Queens College, The City University of New York

Vapor Phase Chromatography and the Telegrapher's Equation

MARCEL J. E. GOLAY
The Perkin-Elmer Corp., Norwalk, Conn.

A review is made of the mathematical relationship between the electrical transmission line and the vapor phase chromatography column, when linearity and the absence of diffusion within the mobile phase of the column are assumed, and it is shown that the sample concentration within the mobile or static phase is the analog, not of voltage or current in the transmission line, but of a linear combination of these. The basic chromatographic solution is simpler than the transmission line solution, and may be extended easily and with good approximation to include diffusion within the mobile phase. Several conclusions, such as the existence of an optimum carrier gas flow rate, are derived from the simple theory established.

The purpose of any useful theoretical discussion of the kinetics of gas chromatography should be the prediction of the cause and effect relationship among the parameters at the experimenter's disposal, such as column length, density of packing, and flow rate on the one hand, and the experimental results of interest, such as components separation on the other.

The characteristic feature of chromatography, which must be taken into account in any theory of this process, is a constant departure, a constant upsetting of equilibrium conditions. The assumptions made to establish a simple theory are the essential assumption of linearity, and the further assumption that a sufficient approximation is obtained by treating separately the effect of diffusion between the mobile and the static phase at any point along the column, and the effect of diffusion within the mobile phase along the length of the column. The theory discussed here refers to the kinetics of the chromatographic process at a given temperature, and the

more physical-chemical problem of the effect of temperatures intervenes only to the extent that the several characteristics of the sample and of the column, such as pressure, capacity, and diffusion, which are treated as constant parameters, can be strongly dependent upon the temperature at which the column operates. It must be recognized that the use of relatively large pressures at the column entrance constitutes a departure from linearity, for the treatment of which the discussion presented here should be elaborated.

With the assumption of linearity, the kinematical problem of gas chromatography, in which a small amount of a volatile compound is injected in a carrier gas at the column entrance, is mathematically similar to the problem presented by the older form of chromatography, when a steady stream of sample was inserted in the column, and the delayed and separate arrivals of the several components were observed as a succession of fronts, at the column exit. It is similar also to the problem of heat exchange between a flowing medium and a stationary medium, or, for that matter, to a whole family of countercurrent problems.

There is a certain amount of literature available on this subject (2–4, 7) from which the results shown here could have been derived, but the preferred method used here was to derive these results directly from the known treatment of the telegrapher's equation, which has been masterfully handled by Reimann (6) and by Carson (1), to mention two names only, and which constitutes one of the great classics of communication theory.

DISCUSSION

The assumption of linearity makes it possible to represent a fractionating column by the electrical analog shown in Figure 1. The volume entrained by the mobile phase is represented by

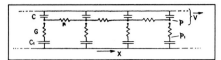

Figure 1. Electrical analog of fractionating column

p, p_1. Mobile and static partial pressures
  G. Diffusion between phases
  v. Flow rate

the upper row of continuously distributed capacitances, C, which travels from left to right at velocity v. The, resistance, R, represents the resistance to the static diffusion which would take place along the column even when v is zero; this resistance will be assumed infinite at first, in order not to introduce a higher order term in the differential equation.

The equivalent free volume of the static phase is represented by the lower row of continuously distributed capacitances C_1, which are connected in turn through the leakance paths, G, to the successive capacitances of the mobile phase. In the case of gas chromatography, these capacitances represent the ability of the carrier gas and of the packing material of the column to hold a certain concentration of the several components of the sample. This concentration can be represented for the mobile phase by the product Cp, in which p designates the partial pressure of a given component in the mobile phase. Likewise, the concentration of the static phase can be represented by the product C_1p_1, in which p_1 represents the partial pressure with which the static phase is in equilibrium.

Two equations can be obtained immediately by means of this model. The first is the conservation equation, which states that the rate of change with time of the integrated concentration to the right of any given abscissa x is equal to the rate of flow, Cvp, of this component across the abscissa, x. The second equation states that the rate of increase of the concentration in the static phase is proportional to the differential pressure, $p - p_1$ and to the leakance, G. These two equations are:

$$\frac{\partial}{\partial t}\left(C\int_x^\infty p\,dx + C_1\int_x^\infty p_1\,dx \right) = Cvp \qquad (1)$$

$$G(p - p_1) = C_1\frac{\partial p_1}{\partial t} \qquad (2)$$

Differentiation of Equation 1 with respect to x yields:

$$C\frac{\partial p}{\partial t} + C_1\frac{\partial p_1}{\partial t} + Cv\frac{\partial p}{\partial x} = 0 \qquad (3)$$

Combining 2 and 3, we obtain a single differential equation, which is valid for both p and p_1:

$$\left[\frac{\partial^2}{\partial t^2} + G\left(\frac{1}{C} + \frac{1}{C_1}\right)\frac{\partial}{\partial t} + \frac{Gv}{C_1}\frac{\partial}{\partial x} + v\frac{\partial^2}{\partial x\partial t} \right](p, p_1) = 0 \qquad (4)$$

New independent variables, u and z, are introduced:

$$u = \frac{1}{v}\left(\sqrt{\frac{C_1}{C}} + \sqrt{\frac{C}{C_1}} \right)x - \sqrt{\frac{C}{C_1}}\,t \qquad (5)$$

$$z = \frac{1}{v}\left(\sqrt{\frac{C_1}{C}} - \sqrt{\frac{C}{C_1}} \right)x + \sqrt{\frac{C}{C_1}}\,t \qquad (6)$$

Their substitution in 4 transforms the latter into a simplified form of the telegrapher's equation:

$$\left(\frac{\partial^2}{\partial z^2} + 2\alpha\frac{\partial}{\partial z} - \frac{\partial^2}{\partial u^2} \right)(\overset{\cdot}{p}, p_1) = 0 \qquad (7)$$

where

$$\alpha = \frac{G}{\sqrt{C_1 C}} \qquad (8)$$

With the new independent variables, u and z, p and p_1 are related by the two equations:

$$p = p_1 + \frac{1}{\alpha}\left(-\frac{\partial p_1}{\partial u} + \frac{\partial p_1}{\partial z} \right) \qquad (9)$$

$$p_1 = p + \frac{1}{\alpha}\left(\frac{\partial p}{\partial u} + \frac{\partial p}{\partial z} \right) \qquad (10)$$

The telegrapher's equation (7) without the α term has the solution:

$$p, p_1 = f(u \pm z) \qquad (11)$$

which represents waves propagating with velocity unity to the right or to the left, and the term in $\partial/\partial z$ imposes the condition that any wave front propagated with velocity unity to the right or to the left decreases exponentially with time, while leaving a diffused trail behind.

The physical meaning of the change of variable introduced by Equations 5 and 6 is that the real column has been replaced by a virtual column in which the mobile phase travels with velocity unity to the right, the formerly immobile static phase travels with velocity unity to the left, and any inserted sample would remain immobile between the two phases if α were infinite, would travel to the right with velocity unity if injected in the mobile phase and if α were zero, and diffuses about the abscissa of injection when α has a finite value.

It is important to note that Relations 9 and 10 are not the relations which exist between current I and voltage V in a transmission line, but are instead the relations which would exist between $V + I$ and $V - I$ in a transmission line of surge impedance unity with series resistance but without shunt leakance, for p and p_1 are analogous to the two quantities just written.

The standard solutions for the telegrapher's equation are usually written for the application of a constant voltage, at the beginning of a semiinfinite line at time zero. Differentiation with respect to time of this standard solution and the application of symmetry consideration give immediately the

114

solution for the voltage and current when a half unit charge is applied at $u = 0$ and $z = 0$ on an infinite line. They are:

$$V_1 = \frac{1}{4} e^{-\alpha z} \left[\frac{z}{\sqrt{z^2 - u^2}} I_1 \left(\alpha \sqrt{z^2 - u^2} \right) + \right.$$

$$\left. I_0 \left(\alpha \sqrt{z^2 - u^2} \right) + \delta (u - z) + \delta (u + z) \right] \quad (12)$$

$$I_1 = \frac{1}{4} e^{-\alpha z} \left[\frac{u}{\sqrt{z^2 - u^2}} I_1 \left(\alpha \sqrt{z^2 - u^2} \right) + \right.$$

$$\left. \delta (u - z) - \delta (u + z) \right] \quad (13)$$

for $|u| < z$ and zero beyond. (I_0 and I_1 are the Bessel functions of an imaginary argument of order 0 and 1, and δ is defined by $\int_{-\epsilon}^{+\epsilon} \delta(x) dx = 1$ for a vanishingly small ϵ.) When a half unit charge is caused to travel from left to right at $u = 0$ and $z = 0$, the current along the resulting voltage and current along the line are obtained similarly and are:

$$V_2 = \frac{1}{4} e^{-\alpha z} \left[\frac{u}{\sqrt{z^2 - u^2}} I_1 \left(\alpha \sqrt{z^2 - u^2} \right) + \right.$$

$$\left. \delta (u - z) + \delta (u + z) \right] \quad (14)$$

$$I_2 = \frac{1}{4} e^{-\alpha z} \left[\frac{z}{\sqrt{z^2 - u^2}} I_1 \left(\alpha \sqrt{z^2 - u^2} \right) - \right.$$

$$\left. I_0 \left(\alpha \sqrt{z^2 - u^2} \right) + \delta (u - z) + \delta (u + z) \right] \quad (15)$$

Since p is analogous to $V + I$, the insertion of a unit p charge in the mobile phase, and of no charge in the static phase, is equivalent to the insertion of a half charge of voltage and a half charge of current in the transmission line, and the resulting expressions for p and p_1 are obtained as indicated below:

$$p = V_1 + I_1 + V_2 + I_2 = e^{-\alpha z} \left[\frac{\alpha (z + u)}{2 \sqrt{z^2 - u^2}} I_1 \right.$$

$$\left. \left(\alpha \sqrt{z^2 - u^2} \right) + \delta (z - u) \right] \quad (16)$$

$$p_1 = V_1 - I_1 + V_2 - I_2 =$$

$$\frac{\alpha}{2} e^{-\alpha z} I_0 \left(\alpha \sqrt{z^2 - u^2} \right) \quad (17)$$

for $|u| < z$, $p = p_1 = 0$ for $|u| > z$

It can be verified immediately that the solutions given by Equations 16 and 17 satisfy 9 and 10 and hence the differential Equation 7. (The δ function in Equation 16 takes the discontinuous character of 17 when forming p with 9, and cancels itself out in the right member of 10.) Furthermore, these solutions satisfy the initial conditions of insertion of a unit charge in the mobile phase, and none in the static phase, and it

is remarkable that these solutions are actually simpler than the corresponding solutions for the electrical problem.

Figure 2 plots Equations 16 and 17 as functions of u for $z = 1$ and $\alpha = 1/2$, 2, and 8, respectively. This is mathematically

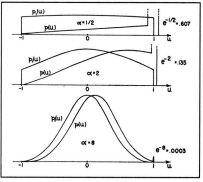

Figure 2. Plots of $p(u)$ and $p_1(u)$ at $z = 1$

equivalent to plotting these functions for the same value of α and increasing values of the virtual time z and changing the scale of u to keep the curves within the confines of the margins. These plots thus give an insight into the diffusion process as time increases.

In Figure 2 (upper), the larger fraction of the initial charge is still contained within the large spike at the right, which represents the δ function. In its progress along the static phase, this charge has left a trail on the static phase, shown by the nearly horizontal p_1 curve. The mobile phase has, in its turn, received a small charge from the static phase, as indicated by the nearly uniformly rising p curve.

In Figure 2 (center), most of the original charge has already diffused, and the p and p_1 curves are beginning to exhibit the shape of an error curve, which is fully developed in Figure 2 (lower), where the undiffused portion has dwindled to 0.03% of the original charge. This last figure underlines the difference of interests of the communication engineer, who amplifies the small information carrying spike at the right and sends it along in the next transmission line section, and of the chemical engineer, who is interested solely in the diffused component exhibited by the error curve, because different values of C_1 cause the different components to separate more and more from each other.

When the initial charge has diffused, use can be made of the first term of the asymptotic expansion of I_1:

$$I_1(x) \cong \frac{e^x}{\sqrt{2\pi x}} \quad (18)$$

and Equation 17 can be transformed into a sufficiently approximate expression for p:

$$p \cong \frac{1}{2} \sqrt{\frac{\alpha}{2\pi z}} e^{-\frac{\alpha u^2}{2z}} \quad (19)$$

The expression Equation 19 is symmetrical in u about the origin 1 because it refers to the virtual column used so far for the sake of mathematical simplicity. It would be also symmetrical in x for a column in which $C_1 = C$. In actual columns we have, in general, $C_1 >> C$; for a study of the general character of our solution in terms of real column length and real time we may neglect the term containing $\sqrt{\frac{C}{C_1}}\, x$ in Equations 5 and 6 and with a simple change of scale in x and t, we may write:

$$u = \xi - \tau \qquad (20)$$

$$z = \xi + \tau \qquad (21)$$

Substitution in Equation 19 yields:

$$p = \frac{1}{2}\sqrt{\frac{\alpha}{2\pi(\xi + \tau)}}\; e^{-\frac{\alpha(\xi - \tau)^2}{2(\xi + \tau)}} \qquad (22)$$

This expression gives the same plot in ξ for a given τ and in τ for a given ξ, and this plot is characterized by a steeper ascent than descent for increasing ξ or τ. This is as expected for ξ, because a large static capacity causes the pressure curve to trail off more in the direction of flow. For this reason, we might expect a slow rise and a faster decay of the elution curve, p, as a function of τ at a given column length ξ. The explanation of the seeming paradox, that the opposite is the case, is given by the circumstance that diffusion continues to take place during elution, so that the descent of p with time at the column exit is actually less steep than the ascent.

The theoretical finding that elution curves have a steeper ascent than descent agrees qualitatively with the elution curves obtained for the more volatile components, which are the first to appear at the column exit. The circumstance that the opposite occurs sometimes for the less volatile components may be interpreted as a departure from linearity for the latter.

Actually, the value of α in Equation 22 is so large in most practical cases that little departure from symmetry should be expected for the elution curves, and large departures in either direction suggest departures from linearity.

One of the most important parameters in gas chromatography is the relative "band width" of the several components—i.e., the ratio of the time of passage of a component to the time interval between adjacent components.

A measure for the band width of a component can be obtained from the δu interval at both ends of which the exponent of Equation 19 is unity:

$$\delta u = 2\sqrt{\frac{2z}{\alpha}} \qquad (23)$$

The band width can be obtained in real time by making the change of variables defined by Equations 5 and 6, and substituting for α its value given by 8:

$$\delta t = 4C_1 \sqrt{\frac{x_0}{CvG}} \qquad (24)$$

where x_0 designates the length of the column.

The flow speed of any component is obtained by setting $u = 0$ in Equation 5 and is:

$$v_s = \left(\frac{x}{t}\right)_{u=0} = \frac{Cv}{C + C_1} \qquad (25)$$

and the excess of the transport time, t, of any component over the transport time, t_a, of a component without affinity for the fixed phase is given by:

$$t - t_a = \frac{C_1 x_0}{Cv} \qquad (26)$$

(The theory developed here is based on the assumption of linearity, and the effects of gas compressibility are not taken into account.)

The ratio $\dfrac{\delta \tau}{t - t_a}$ may thus be considered as a measure of the relative band width due to dynamic diffusion alone, under the assumption made so far that R is infinite, as otherwise we would have been led to a third-order differential equation. A sufficiently good approximation can be obtained for the effect of a finite R by computing the spreading due to R alone within a column in which the flow rate is assumed to be zero, and then computing the elution time of the component thus spread. The differential equation for this case is the simpler telegraph cable equation:

$$(C + C_1)R\frac{\partial p}{\partial t} = \frac{\partial^2 p}{\partial x^2} \qquad (27)$$

for the establishment of which it is permitted to assume G to be infinite, because a small value of G will demand a correspondingly small value of the optimum flow rate, as will be seen shortly. The solution of Equation 27 for a sample of value unity injected at time $t = 0$ is:

$$p = \sqrt{\frac{R(C + C_1)}{4\pi t}}\; e^{-\frac{R(C + C_1)x^2}{4t}} \qquad (28)$$

A measure for the column length over which the sample has diffused statically after a time interval t can be obtained as the x interval at the ends of which the exponent in Equation 28 is unity:

$$\delta x = 4\sqrt{\frac{t}{R(C + C_1)}} \qquad (29)$$

We substitute now for t the time of travel of the component in the column, $\dfrac{x_0(C + C_2)}{VC}$ and obtain the band width δt_s, due to the static diffusion alone, by dividing δx by the flow rate of this component:

$$\delta t_s = 4\frac{C + C_1}{V^{3/2}C^{3/2}}\sqrt{\frac{x_0}{R}} \qquad (30)$$

The combined effect of the kinetic and static diffusions can be obtained by making the convolution of the elution curves which would be obtained for each individual effect alone. As the convolution of two error functions of area unity each is an error function, the second moment (moment of inertia) of which is the sum of the second moments of the two generating error functions, we obtain the combined kinetic and static diffusion time as $\sqrt{\delta t^2 + \delta t_s^2}$. The relative band width:

$$\delta B = \frac{\sqrt{\delta t^2 + \delta t_s^2}}{\dfrac{C_1 x_0}{Cv}} \tag{31}$$

has a minimum value which occurs when the flow rate has the optimum value:

$$v_{\text{opt.}} = \frac{C + C_1}{CC_1} \sqrt{\frac{G}{R}} \tag{32}$$

for which case we have:

$$B_{\text{opt.}} = 4 \sqrt{\frac{2(C + C_1)}{C_1 x_0}} \left(\frac{1}{RG}\right)^{1/4} \tag{33}$$

The concept of the height of the equivalent theoretical plate (HETP), h, and the associated concept of the number of theoretical plates in a column of length x_0, $\dfrac{x_0}{h}$, are intimately connected with the equivalent transmission line parameters utilized in this discussion.

Consider first the linear extent, within the column, of the "band" corresponding to one component. After its travel through the column, the square of this linear extent, Δx^2, measured between the e^{-1} points of the band, will be given by the square of the band width measured in seconds, $\delta t^2 + \delta t_s^2$, multiplied by the square of the component velocity, $Cv^2/(C + C_1)$. Setting v_{opt} for v, we obtain from Equations 24, 30, and 32:

$$\Delta x^2 = (\delta t^2 + \delta t_s^2) \left(\frac{C}{C + C_1} v_{\text{opt}}\right)^2 =$$
$$32 \frac{C}{C + C_1} \frac{x_0}{\sqrt{RG}} \tag{34}$$

Consider next a partition column similar to that illustrated by Figure 1, but differing from it in two important respects:

1. R and G are both infinite.
2. C and C_1 are no longer continuously distributed; instead they are lumped every interval h along the column.

As the mobile phase moves, or rather skips along the fixed phase, at the rate of v/h condensers per second, any unit charge inserted at any epoch of time on one of the $C - C_1$ pairs will be redistributed, after n skips, in accordance with the terms of the binomial expansion of:

$$\left(\frac{C_1}{C + C_1} + \frac{C}{C + C_1}\right)^n \tag{35}$$

The largest term of this expansion will be the $\dfrac{C}{C + C_1} n - th$ one, and the term y terms distant from this largest term will have a value which can be easily obtained by an application of Sterling's formula $\left(a! \cong \left(\dfrac{a}{e}\right)^a \sqrt{2\pi a}\right)$:

$$\frac{C + C_1}{\sqrt{2\pi n\ CC_1}}\ e^{-\frac{1}{2}\frac{(C + C_1)^2 y^2}{nCC_1}} \tag{36}$$

The number Δy of terms clustered around the central maximum term, and greater than e^{-1} times that central term, will be given by:

$$\Delta y^2 = \frac{8n\ CC_1}{(C + C_1)^2} \tag{37}$$

The width in real length of the band thus developed after n skips will be simply:

$$\Delta x = \Delta y . h \tag{38}$$

The number n of skips required for the band maximum to travel the length of the column, x_0—i.e., to travel along $\dfrac{x_0}{h}$ condensers of the fixed phase—is:

$$n = \frac{C + C_1}{C} \frac{x_0}{h} \tag{39}$$

The last three equations yield:

$$\Delta x^2 = \Delta y^2 h^2 = 8 \frac{C_1}{C + C_1} x_0 h \tag{40}$$

Comparison of Equations 34 and 40 gives immediately the HETP of the column:

$$h = \frac{4}{\sqrt{RG}} \tag{41}$$

The number of theoretical plates, n_0, can be expressed as follows by means of Equations 33 and 41:

$$n_0 = \frac{x_0}{h} = \frac{x_0 \sqrt{RG}}{h} = 8 \frac{C + C_1}{C} \frac{1}{(\delta B_{\text{opt}})^2} \tag{42}$$

In terms of observable quantities, δB_{opt} can be expressed as the ratio of the band width between the e^{-1} points, over the time $t - t_a$, and the ratio $\dfrac{C + C_1}{C_1}$ is simply the ratio $\dfrac{t}{t - t_a}$. In practice, it is more convenient to measure the band width at half height, δt_0, which is $\sqrt{lge^2}$ smaller than the band width between e^{-1} points, and Equation 42 may be written entirely in terms of conveniently measured quantities:

$$n_0 = 8lge^2 \frac{t(t - ta)}{\delta t_0^2} = 5.54 \frac{t(t - t_0)}{\delta t_0^2} \qquad (43)$$

If $t - t_a$ were replaced by t in Equation 43, that expression would then be equivalent to the formula given by Phillips (5).

CONCLUSIONS

The flow rate of any component is proportional to the flow rate of the carrier gas (Equation 25.)

When the condition of linearity is met, the elution curves in gas chromatography have nearly the symmetrical shape of error functions.

There is an optimum flow rate for maximum band separation, and the maximum band separation obtained with the optimum flow rate is nearly independent of the mobile and static capacities, C and C_1, when $C_1 >> C$—i.e., when $t >> t_a$.

The maximum band separation is proportional to the square root of the column's length and to the fourth root of the diffusion parameters, G and R.

For a given component, the number of theoretical plates of a column is proportional to the product of the transport time of that component times the time interval between the elution of an inert component and that component, and inversely proportional to the square of the band width of that component.

LITERATURE CITED

(1) Carson, J. R., "Electric Circuit Theory and the Operational Calculus," pp. 100–3, McGraw-Hill, New York, 1926.
(2) Goldstein, S., *Proc. Roy. Soc. London* **A219**, 151 (1953).
(3) Goldstein, S., *Quart. J. Mech. Appl. Math.* **4**, 129 (1951).
(4) Hiester, N. K., Vermeulen, T., *J. Chem. Phys.* **16**, 1087 (1948).
(5) Phillips, Courtenay, "Gas Chromatography," p. 15, Academic Press, New York, 1956.
(6) Reimann-Weber, "Die Partielle Differentialgleichungen der Mathematischen Physik.," Vol. **2**, pp. 306–14, Vierneg and S. Braunschweig, 1901.
(7) Schumann, T. E. W., *J. Franklin Inst.* **208**, 405 (1929).

Received for review June 2, 1956. Accepted February 18, 1957. Division of Analytical Chemistry, Symposium on Vapor Phase Chromatography, 129th Meeting ACS, Dallas, Tex., April 1956.

Reprinted from *Anal. Chem.* **1957**, *29*, 928–32.

This pioneering demonstration of the potential of GC/MS foreshadowed the maturation of the technique to become what is today, in terms of volume of applications, the predominant method in MS. Because this work preceded the general introduction of computers into the analytical chemistry laboratory, the use of a time-of-flight mass spectrometer combined with photographic recording was essential. Nevertheless, the conclusion that the GC/MS combination represents an analytical tool of "near ultimate power" for volatile mixtures showed that Gohlke clearly foresaw future developments.

Charles L. Wilkins
University of California–Riverside

Time-of-Flight Mass Spectrometry and Gas-Liquid Partition Chromatography

R. S. GOHLKE
Spectroscopy Laboratory, The Dow Chemical Co., Midland, Mich.

The direct combination of a time-of-flight mass spectrometer, which scans a mass range from m/e = 1 to m/e = 6000 at the rate of 2000 times per second, with a gas-liquid partition chromatographic apparatus results in an instrument capable of rapidly and completely characterizing organic chemical mixtures boiling below 350 °C.

Mass spectrometry, long used in the petroleum and more recently in the chemical industry (*11, 12, 20*), is one of the finest single tools available for the analysis of volatile chemical mixtures. It is a rapid, precise method, uses sample sizes on the order of a few milligrams, and is often capable of identifying single components even if previously obtained standard mass spectra are not available for comparison purposes.

Because mass spectra result from the rupture of the chemical bonds in a molecule, the presence of various functional groups in the molecule usually will produce predictably distinctive mass spectra. Moreover, differences between the mass spectra of various types of chemical compounds are of such a magnitude that unknown spectra can be identified from correlations of the known behavior of similar compounds. These correlations of mass spectra, which are based on empirical evidence, have been made for the substituted aromatic hydrocarbons (*14*), halogenated compounds (*10*), alcohols (*4*), acids (*13*), esters (*13*), aldehydes (*5*), and a number of other compound types [for a complete list see (*12*)]. They are very valuable in the identification and analysis of unknown components for which detailed reference spectra are unavailable.

The interpretation and calculations involved in the analysis of mixtures containing more than 10 to 15 components are either rather laborious and time-consuming or

impossible by mass spectrometry, unless high speed computers are available to perform the necessary mathematical operations.

This disadvantage of the use of mass spectrometry for the analysis of complex mixtures is not shared by gas-liquid partition chromatography. In addition to the usual ability of gas-liquid partition chromatography to separate complex mixtures completely into their component parts, it also is a rapid, precise method, and the instrumentation cost is low (usually between $1500 and $4000 per complete installation).

Several books discuss the theory and practices of gas-liquid partition chromatography (*2, 8, 16*).

The applications to which this technique has been applied are rather diverse—analysis of petroleum hydrocarbons (*3, 9, 19*), determination of allyl sulfides in onion juice (*1*), fluorocarbon separations (*15, 18*), and rare gas separations (*6*), to mention only a few.

The major deterrent to the use of gas-liquid partition chromatography as an aid in the analysis of nonroutine samples is the fact that the separated components cannot be absolutely identified from the chromatogram itself. There are two common means of overcoming this difficulty: to compare the retention volume of the unknown component with that of known compounds on two or more chromatographic columns containing different liquid substrates (*17*), or to collect the component as it leaves the column and subsequently identify the collected material by another method.

If the first method is to be effective, one must have data available regarding the behavior of the members of various homologous series on columns of different liquid substrates, at various operating temperatures. The experience of this laboratory has been that for identification of the components present in nonroutine samples, the data which must be ob-

The 1950s   2855-8/94/0119$08.00/0   © 1994 American Chemical Society   **119**

tained to use the comparative retention volume system of identification are so extensive as essentially to preclude use of this method.

The techniques of infrared spectrometry and mass spectrometry have been widely used for the second identification procedure, but several difficulties are attendant on these or any other subsequently applied identification procedures. The devices generally used for the collection of the chromatographic samples for infrared use involve bubbling the effluent gas from the column through a suitable solvent, or passing the effluent gas through a trap cooled in liquid air, with subsequent sample transfer to an infrared microcell for identification. These techniques always demand constant supervision of the fraction collection device and often involve the use of toxic, flammable solvents or fragile cold traps of diminutive size.

The chromatographic samples for mass spectrometry are invariably collected in liquid air-cooled cold traps. The use of cold traps for sample collection involves a high degree of skill, if samples free of atmospheric contaminants and previously eluted components from the chromatographic column are desired. It is almost always desirable to use the samples as soon after collection as possible, to avoid the difficulties encountered in the storage of minute samples for any length of time.

The greatest disadvantage to fraction collection is the time required to identify the collected fractions subsequently, particularly when mass spectrometers of conventional design are used, as it requires 20 to 40 minutes to obtain the spectrum and prepare the instrument for the next sample. This would mean that if one had a chromatogram containing 10 peaks whose identity was desired, it would require 3 to 8 hours of mass spectrometer instrument time to obtain the mass spectra of the fractions, in addition to the time required to collect them.

Infrared spectrometry for the identification is not significantly superior to conventional mass spectrometry. The time required to scan the infrared spectrum is somewhat less (12 to 15 minutes) than for mass spectrometry, but the small size of the chromatographic fractions usually requires the use of a liquid microcell (volume less than 1 drop) as a sample container. The proper use of a microcell makes the adjustment of certain instrument operating conditions (amplifier gain, slit width) desirable—and these adjustments often cannot be reliably performed by the personnel operating the instrument. A recent promising development for component identification by infrared involves collection of the gaseous samples, directly in micro gas absorption cells of the multiple traversal type (21).

Application of infrared spectrometry for direct qualitative analysis of the effluent stream, however, suffers the disadvantages that the absorption spectra of heated gaseous samples are less sharp than those of solutions at room temperature and the very fast scanning required results in loss of detail.

The problem of chromatogram component identity was solved with the recent marketing of a mass spectrometer manufactured by the Bendix Aviation Corp., which is capable of directly and continuously monitoring the effluent stream from the chromatographic column. The mass spectrum of the effluent vapor is presented on an oscilloscope 2000 times per second, thus completely eliminating the need for collection, or any other manipulation, of the components represented by the peaks on a chromatogram.

EXPERIMENTAL APPARATUS

Gas-Liquid Partition Chromatography. The apparatus consists of a gas-liquid partition chromatography apparatus which has a time-of-flight mass spectrometer connected to the system at a point between the column exit and the thermal conductivity cell, as shown schematically in Figure 1.

The gas chromatography portion of the instrument is of essentially conventional design, except that four chromatographic columns are enclosed in an insulated oven instead of the more widely used single column. The use of four columns connected in parallel, each complete with sample injection port, permits the separation of samples requiring different chromatographic columns without requiring the time-consuming act of physically changing columns every time this is desired. The exit ends of the four columns terminate in a block shown in detail in Figure 2. From this block a single line leads to the mass spectrometer and another single line leads to the separately enclosed Gow-Mac thermal conductivity cell used as the detector. The use of separately enclosed and heated column and conductivity cell chambers permits operation of the conductivity cell at a temperature independent of the column temperature. The temperature of the thermal conductivity cell is usually kept somewhat higher than the temperature of the columns to minimize the tendency for high boiling components of a sample to condense in the cell.

The design details of gas chromatography equipment appear to be somewhat a matter of personal preference, and only the barest description is given here.

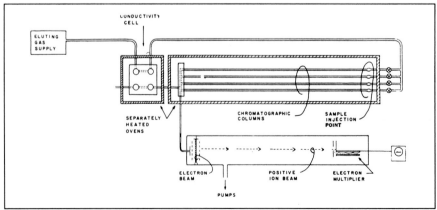

Figure 1. Schematic view of method of attachment of time-of-flight mass spectrometer to gas chromatographic apparatus

The chromatographic oven is a double-walled $17 \times 17 \times 6$ inch box built of magnesium, with the 1-inch interstice between the double walls packed with glass wool for insulating purposes. A 600-watt cartridge-type heater is sufficient to raise the temperature of the oven interior to approximately 200 °C. Because accurate reproduction of component residence time was not necessary for the application, no attempt was made to regulate the temperature of the oven carefully. In use, the 600-watt heater is connected to a variable autotransformer, which is supplied with a constant voltage input. The temperature of the oven interior is therefore a balance between the heat input from the heater and the radiated heat loss of the oven exterior and, in practice, may vary as much as ±5 °C. from a constant value.

One arm of each of four copper tubing T fittings extends through the oven wall, is sealed with a silicone rubber disk, and is the point at which samples are introduced via a hypodermic syringe. Each column is equipped with an eluting gas shutoff valve mounted outside the oven (Figure 1).

The oven enclosing the thermal conductivity cell is identical to the column oven, except for somewhat smaller dimensions and a 300-watt heater. The thermal conductivity cell is mounted with the filaments vertical, the 300-watt heater is strapped directly to the cell block, and the temperature of the cell is measured by an iron-constantan couple inserted into a hole drilled into the cell block. The thermal conductivity cell used is a Gow-Mac Instrument Co. product (TE-III geometry, stainless steel block, 9225 filaments). The thermal conductivity cell is wired as a conventional Wheatstone bridge and is supplied by a 150-ma. constant current power supply shown schematically in Figure 3.

Drifts in cell balance as observed by recorder base line due to temperature variations are on the order of 0.05 to 0.10 mv. over a 24-hour period at 200° C. Again, no attempt at precise thermoregulation was made.

Mass Spectrometry. The spectrometer used to monitor the effluent vapor of the chromatographic column is a time-of-flight mass spectrometer manufactured by the Bendix Aviation Corp., 3130 Wasson Road, Cincinnati, Ohio, Type 12-100. The theory of operation of time-of-flight mass spectrometry has been adequately described (22). The instrument was modified by replacing the supplied oil diffusion pump and water cooled baffle with a mercury diffusion pump, a Freon 22 refrigerated baffle (about −20° C. baffle interior temperature), and a liquid air trap. These modifications reduced the instrument background spectrum to an unobjectionable level for even the most critical work. The following features make this instrument admirably suited to a gas chromatographic application.

High rate of scan—2000 (or 10,000 to 20,000, optionally) new complete mass spectra per second are produced.

Resolution of adjacent mass units is complete to about mass 200 and adjacent mass resolution is usable to about mass 450.

The sensitivity of the mass spec-

trometer is such that spectra of perfect quality are presented if the maximum concentration of the chromatographic peak exceeds 0.2 mv., with the thermal conductivity detector and power supply described. The absolute sensitivity of the instrument is such that a partial pressure of 1×10^{-8} mm. of mercury of argon in the ion source will produce one recorded argon ion per instrument cycle (operating at 10,000 cycles per second) (7).

The mass spectra are presented on a Tektronix Type 541 oscilloscope equipped with a Tektronix Type 53154K plug-in preamplifier. The mass range presented on the oscilloscope screen is easily and continuously variable from a single mass unit to 6000 mass units.

EXPERIMENTAL

The mass spectra produced by the time-of-flight mass spectrometer can be made directly comparable to spectra produced by mass spectrometers of the conventional electrostatic or electromagnetic scanning design.

Thus, existing files of mass spectra can be directly compared to the time-of-flight spectra for identification purposes.

Table I compares portions of the mass spectrum of vinyl chloride as obtained on the time-of-flight mass spectrometer and on a Consolidated Electrodynamics Corp. 21-103 mass spectrometer. The difference in the relative intensity of the peaks in the time-of-flight spectra is due to the fact that it is difficult to measure the height of the ion peaks precisely in the time-of-flight mass spectrometer and is of no concern when the spectra are used solely for identification purposes.

If necessary, however, peak heights can be accurately determined by the use of pulse-counting equipment (at a scan-

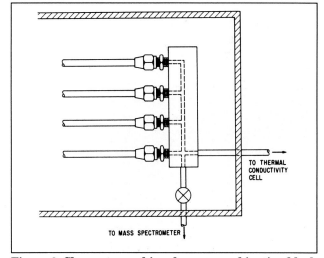

Figure 2. Chromatographic column recombination block

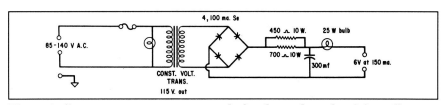

Figure 3. Constant current power supply for thermal conductivity cell

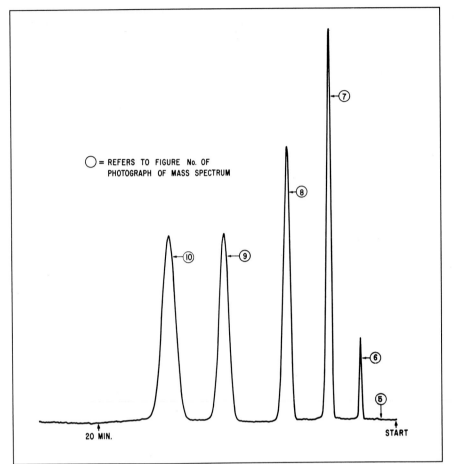

Figure 4. Sample chromatogram of acetone, benzene, toluene, ethylbenzene, and styrene

The sample was chromatographed on a 10 foot × 1/4 inch column of Tide detergent at a temperature of 138 °C. with a helium inlet pressure of 10 psig.

Within the chromatogram:
○ = REFERS TO FIGURE No. OF PHOTOGRAPH OF MASS SPECTRUM

20 MIN.

START

Table I. Relative Ion Intensity

m/e	CEC No. 21-103	Bendix TOF No. 12-100
12	4.66	5
25	17.4	18
26	43.3	42
27	132	140
35	8.97	10
47	4.74	5
48	2.06	2
59	1.89	2
60	6.26	6
61	9.00	9
62	100	100
63	4.82	5
64	31.61	32
65	0.67	. . .

As it would serve no useful purpose to present a chromatogram containing 30 peaks and the corresponding photographs of the mass spectra, the chromatogram of a single synthetic mixture is used to illustrate the manner in which the combined gas-liquid chromatography-mass spectrometry technique is used.

The chromatogram of a mixture of acetone, benzene, toluene, ethylbenzene, and styrene is shown in Figure 4. Photographs of the mass spectrum appearing on the oscilloscope screen at the points indicated are shown in Figures 5 to 10.

The lag between the appearance of the peak on the chromatographic recorder and its mass spectrum on the oscilloscope screen is on the order of 1 second. The height of ion peaks in the mass spectrum rise and fall in intensity at the same rate as the component peak rises and falls in concentration, as indicated by the conductivity cell recorder. Single chromatographic peaks containing two or three components can usually be successfully resolved by a careful examination of several mass spectra obtained at various times during the development of the chromatographic peak.

In the photograph (Figure 11) of the complete ethylbenzene mass spectrum (trace 1) the ion peaks are so close

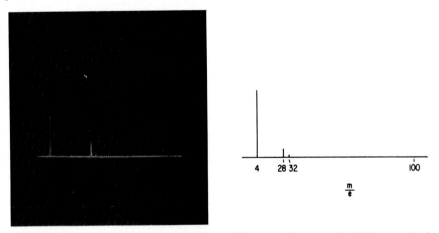

Figure 5. The mass spectrum of helium with a small amount of nitrogen (m/e 28) and oxygen (m/e 32) impurity present

ning rate of 10,000 spectra per second), but this has been found unnecessary for the gas chromatography applications of the instrument in this laboratory. An analog output system has become available recently, which provides meter indications and recordings of several peaks simultaneously, either directly or as ratios of each other, plus the ability to record spectra in the conventional manner on strip charts (7).

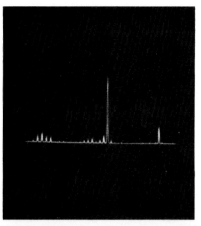

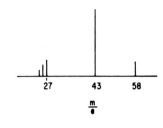

Figure 6. Mass spectrum of acetone (first chromatographic peak)

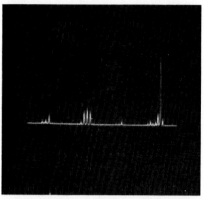

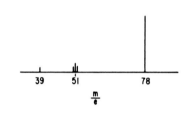

Figure 7. Mass spectrum of benzene (second chromatographic peak)

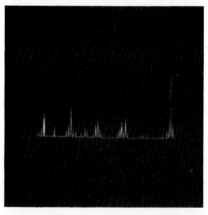

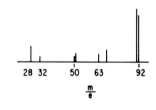

Figure 8. Mass spectrum of toluene (third chromatographic peak)

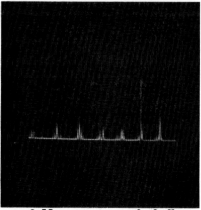

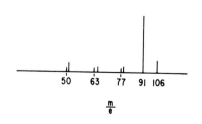

Figure 9. Mass spectrum of ethylbenzene (fourth chromatographic peak)

together that accurate mass determination can be somewhat doubtful. The use of an auxiliary 10-position mass range switch makes it possible to present magnified overlapping portions of the complete spectrum as separate exposures in the same photograph. The ion mass position in traces 2, 3, and 4 is easily determined by placing a precalibrated mass ruler upon the photographed trace. The four traces comprising Figure 11 were photographed in a total of 3 seconds.

The tap switch has 10 positions, giving precisely controlled mass ranges from mass 10 to mass 300, in increments comprising about 36 mass units. About mass 300, the mass spectrometer controls are used manually in the normal operating manner.

Any portion of the mass spectrum may be viewed at will, the adjustments involved are extremely simple to perform, and the results of adjustment may be continuously viewed on the oscilloscope screen.

The adjacent mass resolution of the time-of-flight instrument is sufficient for any conceivable gas-chromatographic applications. Figure 12 shows the appearance of the spectrum of mercury. The resolution between adjacent masses is, for all practical purposes, complete. Beyond mass 200, adjacent mass resolution gradually is lost, until in the mass range of 600 to 700 no adjacent mass resolution is obtained (Figure 13). Figure 13 is a portion of the mass spectrum of a perfluorinated kerosene showing the ion peaks from this compound at masses 617, 631, and 643. No trace of resolution between these and the corresponding carbon-13–containing fragments at 618, 632, and 644 is observed.

A very finely controllable needle valve placed in the line between the gas-liquid partition chromatography apparatus and the mass spectrometer acts as the mass spectrometer leak and allows the amount of sample being drawn into the mass spectrometer to be varied at will. During normal operation the pressure in the gas-liquid partition chromatography apparatus is 760 mm. of mercury and the pres-

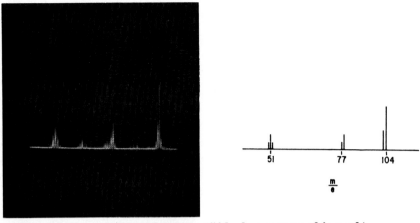

Figure 10. Mass spectrum of styrene (fifth chromatographic peak)

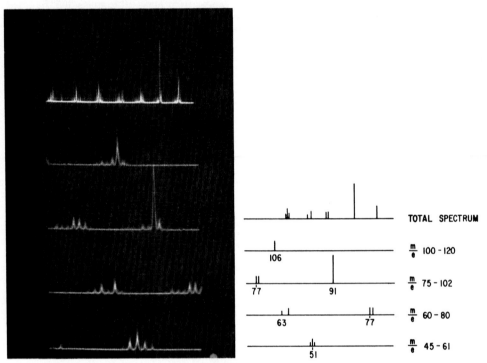

Figure 11. Mass spectrum of ethylbenzene

The top trace is the complete spectrum and the four lower traces are expanded portions of the top trace. The tap switch used for the lower traces has an additional seven mass ranges not used here

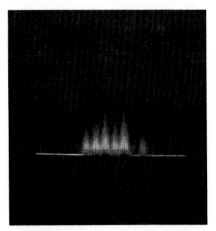

Figure 12. Mass spectrum of mercury

sure in the mass spectrometer ion source is adjusted to about 1×10^{-4} mm. of mercury by means of the needle valve. Under these conditions, the mass spectrometer is consuming the effluent vapor from the chromatographic column at a rate of about 0.003 cc. per second—before this vapor reaches the conductivity cell.

With a flow rate of 60 cc. per second this loss is equivalent to 0.3% of the stream, which is not considered serious enough to affect the chromatogram to a point where quantitative determinations of the components present based on the area of the chromatogram peak become inaccurate.

Any quantitative determination of the components is therefore performed in the usual manner—integration of the peak area and adjustment of this area by predetermined constants to relate the area per cent directly to mole or weight per cent.

SUMMARY

The technique of combined mass spectrometry and gas-liquid partition chromatography has been used in this laboratory for the characterization of samples whose great complexity makes normal analytical procedures prohibitive from the standpoint of time and analytical cost, and has resulted in an analytical tool of near ultimate power when applied to chemical mixtures boiling below 350 °C. to 760 mm.

ACKNOWLEDGMENT

The author is indebted to V. J. Caldecourt, W. J. Felmlee, and E. D. Ruby of this laboratory, whose contributions markedly aided the development of this technique.

LITERATURE CITED

(1) Carson, J. E., Wong, F. F., Division of Analytical Chemistry, Symposium on Advances in Gas Chromatography, 132nd Meeting, ACS, New York, N. Y., September 1957.
(2) Desty, D. H., "Vapour Phase Chromatography," Butterworths Scientific Publications, London, 1957.
(3) Eggersten, F. T., Groennings, S., *Anal. Chem.* **30**, 20 (1958).
(4) Friedel, R. A., Schultz, J. L., Sharkey, A. G., Jr., *Ibid.*, **28**, 926 (1956).
(5) Gilpin, J. A., McLafferty, F. W., *Ibid.*, **29**, 990 (1957).

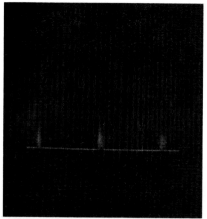

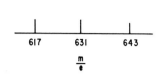

Figure 13. Portion of mass spectrum of a perfluorinated kerosene, illustrating appearance of ion peaks in the *m/e* 600 region

(6) Greene, S. A., Division of Analytical Chemistry, Symposium on Advances in Gas Chromatography, 132nd Meeting, ACS, New York, N. Y., September 1957.

(7) Harrington, D., Bendix Aviation Corp., private communication.

(8) Keulemans, A. I. M., "Gas Chromatography," Reinhold, New York, 1957.

(9) Knight, H. S., *Anal. Chem.* **30**, 9 (1958).

(10) McLafferty, F. W., ASTM E-14 Conference on Mass Spectrometry, San Francisco, Calif., May 1955.

(11) McLafferty, F. W., *Anal. Chem.* **28**, 306 (1956).

(12) McLafferty, F. W., *Appl. Spectroscopy* **11**, 148 (1957).

(13) McLafferty, F. W., Gohlke, R. S., ASTM E-14 Conference on Mass Spectrometry, San Francisco, Calif., May 1955.

(14) Meyerson, S., *Appl. Spectroscopy* **9**, 120 (1955).

(15) Percival, W. C., *Anal. Chem.* **29**, 20 (1957).

(16) Phillips. C., "Gas Chromatography," Butterworths Scientific Publications, London, 1956.

(17) Ray, N. H., *J. Appl. Chem.* **4, 21** (1954).

(18) Reed, T. M., III, *Anal. Chem.* **30**, 221 (1958).

(19) Simmons, M. C., Snyder, L. R., *Ibid.*, **30**, 32 (1958).

(20) Waldron, J. D., *Metropoliton-Vickers Gazette* July 1956.

(21) White, J. U., Alpert, N. L., Weiner, S., Ward, W. M., Galloway, W. S., Pittsburgh Conference on Applied Spectroscopy and Analytical Chemistry, Pittsburgh, Pa., March 1958.

(22) Wiley, W. C., McLaren, I. H., *Rev. Sci. Instr.* **26**, 1150 (1955).

Received for review May 31, 1958. Accepted November 10, 1959. Presented in part before the Division of Analytical Chemistry, 132nd Meeting, ACS, New York, N. Y., September 1957, Pittsburgh Conference on Analytical Chemistry and Applied Spectroscopy, 1958, and Symposium on Gas Chromatography, National Institutes of Health, Bethesda, Md., 1958.

Reprinted from *Anal. Chem.* **1959**, *31*, 535–41.

The 1960s

Birth of the Electronic Era

When we think of the 1960s, among the things that come to mind are the race to put a man on the moon, the Vietnam War, flower children, the civil rights movement, miniskirts, and the Beatles. But the sixties were also a time when analytical chemists were making large strides in many areas of measurement science.

This was a decade of advances in the Journal and at the Pittsburgh Conference. The year 1960 marked the 32nd volume of the Journal, Larry Hallett's sixth year as Editor, and the 11th Pittsburgh Conference.

The science

During the 1960s the transition from traditional wet chemistry to instrumental methods of analysis continued. The major instrumental advances affecting analytical chemists were the introduction of semiconductor devices to replace transistors and vacuum tubes, the commercial availability of minicomputers, and the introduction of continuous-wave and pulsed lasers. These advances, along with the promotion of

1960	First laser is developed Willard Libby wins the Nobel Prize in Chemistry for radiocarbon dating Brezhnev becomes Premier of the USSR "Let's Do the Twist" is a popular song
1961	Lawrencium (element 103) is produced First "letter" in genetic code is determined Berlin Wall is constructed *West Side Story* wins the Academy Award
1962	*Silent Spring* (Rachel Carson) is published Cuban missile crisis occurs John Glenn is the first American astronaut to orbit Earth Linus Pauling wins the Nobel Peace Prize
1963	John F. Kennedy is assassinated Cassette tape is invented Carl Sagan finds ATP in a mixture of chemicals indicative of early Earth Civil rights riots break out in Birmingham, AL

1964	American involvement in the Vietnam War increases
	    with the Tonkin Resolution
	Verrazano Bridge opens
	Beatles hit the popular music charts with "I Want to
	    Hold Your Hand"
	Dorothy M.C. Hodgkin wins the Nobel Prize in
	    Chemistry for the structural analysis of vitamin B12

modular electronic instrumentation by Howard Malmstadt and Christie Enke, allowed analytical chemists to readily develop their own instruments and to become more involved with fundamental aspects of measurements.

One area of active research was atomic absorption spectrometry. Most of the previous work in this field, which had been done in the 1950s, was aimed at developing new methodologies and applications. Development of the nitrogen oxide–acetylene flame improved detection for elements (such as Mo and Be) that had been difficult to determine in the acetylene–air flames previously used. Graphite furnace atomization, developed by Boris L'vov in the early 1960s, made it possible to determine small amounts of sample, and a flurry of activity followed. Numerous applications were developed, especially after commercial instruments became available in 1970.

The use of an inductively coupled plasma for atomic emission spectrometric excitation was described by S. Greenfield in 1964 and by Velmer Fassel in 1965. The Grimm discharge for the atomic emission analysis of solids, although developed in 1967, was not widely used until the 1970s.

J. D. Winefordner and his co-workers at the University of Florida developed atomic fluorescence spectrometry as an analytical tool in the mid-1960s. Winefordner's group built on the work of C. Th. J. Alkemade, who had mentioned the possibility of AFS in 1963.

Molecular spectroscopy research centered on fluorescence and phosphorescence spectrometries. Although numerous papers were published on UV–vis, IR, and Raman spectroscopies, most were new applications. It wasn't until the 1970s that Fourier and Hadamard transform methods, along with diode lasers, were used to make significant improvements in traditional UV-vis, IR, and Raman spectroscopies.

Mass spectrometry in the 1960s was aimed primarily at analysis methods that would be considered primitive now. Work by Fred McLafferty, Klaus Biemann, Maurice Bursey, and John Beynon, to name a few, enabled researchers to determine the specific mass and structure of analytes of interest. Spark source MS was being developed as an analytical tool by George Morrison, who later become Editor of the Journal.

Research in kinetic methods of analysis and automated clinical methods was also performed. Groups led by Howard Malmstadt, Harry Pardue, Charles Reilley, and George Guilbault, among others, laid the groundwork for automated clinical analysis and for the sophisticated stopped-flow, temperature and pressure jump, and relaxation methods used today.

Much progress was made in the chromatographic arena.

Liquid chromatography was first used as an analytical technique in the early 1900s (although it underwent a dormant period and was not "rediscovered" until the 1930s), and gas chromatography was developed in the 1950s. Applications of these techniques continued to be developed in the 1960s, and it was also a time for refinement of the theory involved in the chromatographic process and for development of detectors based on fluorescence, electrochemistry, and ionization. Work by J. Calvin Giddings, John Knox, R.P.W. Scott, and Georges Guiochon, among others, led to a more complete understanding of the processes taking place during chromatographic separations, and detectors developed by J. E. Lovelock, S. R. Lipsky, and others increased the selectivity and sensitivity of chromatographic methods.

Capillary GC was first described by Marcel Golay in the late 1950s and, although the results were impressive, progress during the 1960s was slow. Work was aimed at solving the problems of sample injection and column technology, but it was not until the 1970s that capillary GC enjoyed a renaissance with improved column technology and instrumentation.

Marcel Jules Edouard Golay

The use of supercritical fluids in chromatography was reported in 1962; the apparatus had no pump, however, and allowed only visual detection of colored solute bands directly on a glass column. The technique did not become a practical analysis method until 1968, when Karayannis et al. published a paper describing an instrument for what was then called "hyperpressure gas chromatography."

The 1960s also marked the beginnings of hyphenated techniques. The first GC/IR systems were described in 1964, with the back-to-back publication of papers by Bartz and Ruhl of the Dow Chemical Co. and by Wilks and Brown of Wilks Scientific. Both groups independently developed their own GC/IR interface, but the Bartz and Ruhl paper is generally considered to take precedence because it was submitted for review five weeks before that of Wilks and Brown.

With the advent of minicomputers, analytical chemists finally had the necessary tools to handle large amounts of data. GC/MS was first described in the 1950s, but it wasn't until the 1960s, when computer systems were developed to handle the large volume of data produced, that these systems gained widespread acceptance. Twenty-five years later, it is rare to find any type of

instrumental analysis system without an attached computer!

Computers were also used to recognize spectral features and smooth noisy data, resulting in the birth of the field known now as chemometrics. In 1964 Abraham Savitsky and Marcel Golay of Perkin Elmer published a landmark paper describing a smoothing algorithm for reduction of random noise from spectra. The digital smoothing filter they described is still used extensively and is the predecessor of the many digital filters used today.

Electrochemistry, the mainstay of analytical chemistry in the 1940s and 1950s, continued as an area of great activity in the 1960s; work progressed in potentiometry, polarography, and ion-selective electrodes. Nicholson and Shain's classic 1964 paper describing a numerical method for the solution of boundary value problems for a variety of reactions that take place at the electrode surface marked the beginning of a new era in what was then called stationary electrode polarography. Today this technique is known as linear scan, or cyclic, voltammetry.

Ion-selective electrodes were developed by Garry Rechnitz, Martin Frant, James Ross, Erno Pungor, George Guilbault, and Richard Buck, among others. In 1966 Frant and Ross of Orion Research described a fluoride electrode that helped lead to the acceptance of ion-selective electrodes in the early years of commercial development.

The theory and technique of studying small-solution thicknesses chronopotentiometrically was first described by Fred Anson in 1963, and in 1965 Charles Reilley expanded the practice of thin-layer electrochemistry by introducing a simple, versatile apparatus.

Although biotechnology didn't attract widespread interest until the 1980s, the groundwork for genetic engineering and the human genome project was laid in the 1960s. In 1966 R. Bruce Merrifield of the Rockefeller University published a paper in *Analytical Chemistry* in which he described an instrument for automated peptide synthesis. This work was the beginning of research for which he was awarded the Nobel Prize in Chemistry in 1984.

The Pittsburgh Conference

The 1960 Pittsburgh Conference was held at the Penn–Sheraton Hotel in Pittsburgh and featured 23 technical sessions on topics

1965
BASIC is developed
Measles vaccine becomes available
First Gemini space flight is launched
Blackout leaves 30 million people in the northeast
    U.S. without power

1966
Indira Ghandi becomes Prime Minister of India
Miniskirts become fashionable
Edwin Aldrin walks in space from Gemini 12
Man of La Mancha is a hit on Broadway

1967
Computer keyboards are used
Six-Day War between Israel and Arab nations takes
    place
First human heart transplant is achieved
Synthetic DNA is produced at Stanford University

1968
Soviets invade Czechoslovakia
First Apollo space flight is launched
Riots disrupt the Democratic Convention in Chicago
Watson writes *The Double Helix*

1969
Fermilab is founded
Neil Armstrong walks on the moon
Woodstock attracts more than 300,000 music fans
Scanning electron microscopy is used after 15 years of
    development

such as nucleonics, emission spectroscopy, GC, and IR spectroscopy. More than 100 exhibitors, among them such familiar names as Applied Research Laboratories, Bausch & Lomb, Beckman Instruments, Fisher Scientific, Jarrell-Ash, Mettler, Perkin Elmer, Sadtler Research Labs, and Varian Associates, participated in the Exposition of Modern Laboratory Equipment.

In 1968, because of unresolved labor–management problems in Pittsburgh, the Pittcon organizers moved the conference to Cleveland's Sheraton Hotel. For the first time, the exhibition was held at a remote site: the Public Auditorium about half a mile from the hotel. That year there were 160 exhibitors (down from 180 the year before), but by 1969 Pittcon consisted of 40 sessions and more than 230 exhibitors.

The Journal

In 1960 a subscription to *Analytical Chemistry* cost only $4.00. Each monthly issue contained a "Report for Analytical Chemists," featuring topics such as the reorganization of Esso Research, new approaches to teaching analytical chemistry, and the state of the art of atomic absorption spectrometry; a news section, which included meeting coverage and award announcements; a column by Editor Larry Hallett, discussing various issues of the day; and an Instrumentation column by contributing editor Ralph Müller.

Each issue also included a profile of the "Laboratory of the Month," which highlighted modern labs in companies such as Shell Development Co., Allied Chemical Corp., and Schering Corp. Advertisements featured instruments such as polarographs and spectrometers as well as equipment such as balances, glassware, pH meters, and chemicals.

As the decade progressed, changes were made to the look of the Journal. In 1963 the annual Buyer's Guide and the Review issue were split into separate issues, although both were still published in April. In 1966 Larry Hallett retired as Editor and was succeeded by Herbert Laitinen, the first academic researcher to serve as Editor while remaining a full-time university professor. In 1967, in deference to the information age, the Journal added the "ANCHAM" codon to the cover. Among those serving as Managing Editor during the 1960s were Virginia Stewart Jackhellm, now ACS ombudsman, and John K Crum, currently Executive Director of the ACS.

The progress made during the beginning of the electronic era in the 1960s set the stage for the explosive growth of the information age in the 1970s.

MARY WARNER

Milestones in Analytical Chemistry

This classic, elegant article marked the beginning of a new era in what was then called stationary electrode polarography but is now known as linear scan, or cyclic, voltammetry. It provided details of a numerical method (the use of integral equations) for the solution of boundary value problems for a variety of reactions that take place at the electrode surface. It also provided a firm, quantitative foundation for a great deal of work in electroanalytical chemistry, and it is one of the most cited papers in the area.

Robert A. Osteryoung
North Carolina State University

Theory of Stationary Electrode Polarography: Single Scan and Cyclic Methods Applied to Reversible, Irreversible, and Kinetic Systems

RICHARD S. NICHOLSON and IRVING SHAIN
Chemistry Department, University of Wisconsin, Madison, Wis.

The theory of stationary electrode polarography for both single scan and cyclic triangular wave experiments has been extended to systems in which preceding, following, or catalytic (cyclic) chemical reactions are coupled with reversible or irreversible charge transfers. A numerical method was developed for solving the integral equations obtained from the boundary value problems, and extensive data were calculated which permit construction of stationary electrode polarograms from theory. Correlations of kinetic and experimental parameters made it possible to develop diagnostic criteria so that unknown systems can be characterized by studying the variation of peak current, half-peak potential, or ratio of anodic to cathodic peak currents as a function of rate of voltage scan.

Stationary electrode polarography (6) (voltammetry with linearly varying potential) has found wide application in analysis and in the investigation of electrolysis mechanisms. For analysis, the method is more sensitive and faster than polarography with the dropping mercury electrode (37) and, when used with stripping analysis, can be extended to trace determinations (7, 17, 45). In studying the mechanism of electrode reactions, the use of stationary electrodes with a cyclic potential scan makes it possible to investigate the products of the electrode reaction and detect electroactive intermediates (10, 11, 18). Furthermore, the time scale for the method can be varied over an extremely wide range, and both relatively slow and fairly rapid reactions can be studied with a single technique. Various electrodes have been used in these studies, but the most important applications have involved the hanging mercury drop electrode [reviewed by Kemula (16) and Riha (36)] and the dropping mercury electrode [reviewed by Vogel (50)].

Since the first application of the method by Matheson and Nichols (21), numerous investigators have contributed to the theory of stationary electrode polarography. The first were Randles (29) and Sevcik (44) who considered the single scan method for a reversible reaction taking place at a plane electrode. The theory was extended to totally irreversible charge transfer reactions by Delahay (4), and later Matsuda and Ayabe (23) rederived the Randles-Sevcik reversible theory, the Delahay irreversible theory, and then extended the treatment to the intermediate quasi-reversible case. Other workers also have considered both reversible (13, 22, 32, 35) and totally irreversible (12, 32) reactions taking place at plane electrodes.

In addition, the theory of the single scan method has been extended to reversible reactions taking place at cylindrical electrodes (25) and at spherical electrodes (9, 31, 32, 35). Totally irreversible reactions taking place at spherical electrodes (8, 32) also have been discussed. Further contributions to the theory have included systems in which the products of the electrode reaction are deposited on an inert electrode (2); the reverse reaction, involving the dissolution of a deposited film (26); and systems involving multi-electron consecutive reactions, where the individual steps take place at different potentials (14, 15).

Even in the cases involving reversible reactions at plane electrodes, the theoretical treatment is relatively difficult, ultimately requiring some sort of numerical analysis. Because of this, the more complicated cases in which homogeneous chemical reactions are coupled to the charge transfer reaction have received little attention. Saveant and Vianello developed the theory for the catalytic mechanism (39), the preceding chemical reaction (38, 41), and also have discussed the case involving a very rapid reaction following the charge transfer (40). Reinmuth (32) briefly discussed the theory for a system in which a first order chemical reaction follows the charge transfer.

The mathematical complexity also has prevented extensive study of the cyclic triangular wave methods, in spite of the value of this approach. Sevcik (44) qualitatively discussed the method for reversible reactions at a plane electrode under steady state conditions—i.e., after many cycles when no further changes in the concentration distributions take place in the solution from one cycle to the next. Later Matsuda (22) presented the complete theory for this multisweep cyclic triangular wave method, for a reversible reaction at a plane electrode. The only other contributions to the theory of cyclic methods were those of Gokhshtein (13) (reversible reactions), Koutecky (19) (reversible and quasi-reversible reactions), and Weber (51) (catalytic reactions). A generalized function of time was included in each of these derivations, which thus could be extended to cyclic triangular-wave voltammetry. In the last two papers, however, the case actually considered was for a cyclic step-functional potential variation.

Because of the increased interest in stationary electrode polarography, it has become important to extend the theory to include additional kinetic cases. Furthermore, many of the recent applications of cyclic triangular wave voltammetry have involved only the first few cycles, rather than the steady state multisweep experiments. Therefore, a general approach was sought, which could be applied to all these cases. In considering the mathematical approaches of other authors, at least three have been used previously: applications of Laplace transform techniques, direct numerical solution using finite difference techniques, and conversion of the boundary value problem to an integral equation.

The first approach is the most elegant, but is applicable only to the simplest case of a reversible charge transfer reaction (2, 19, 31, 35, 44), and also to the catalytic reaction (51). Even in these cases, definite integrals arise which can only be evaluated numerically.

The second approach (8, 9, 25, 29) is the least useful of the three, because functional relations which may exist between the experimental parameters are usually embodied in extensive numerical tabulations and are often missed. Thus, the results may depend on an extremely large number of variables. This is particularly so in the more complicated cases involving coupled chemical reactions, which may require the direct simultaneous solution of three partial differential equations together with three initial and six boundary conditions.

The third method possesses the advantages of the first, and yet is more generally applicable. Several methods can be used to convert the boundary value problem to an integral equation (33), and at least two methods of solving the resulting integral equations have been used. The series solution proposed by Reinmuth (32) is very straightforward, but only in cases involving totally irreversible charge transfer does it provide a series which is properly convergent over the entire potential range. [This approach to obtaining series solutions is essentially the same as used by Smutek (47) for irreversible polarographic waves. Series solutions of the same form can also be obtained directly from the differential equations, as was shown recently by Buck (3)]. Reinmuth has outlined a method for evaluating these series in regions where they are

divergent (33), but attempts to use that approach in this laboratory (with a Bendix Model G-15 digital computer) produced erratic results.

The methods most frequently used for solving the integral equations have been numerical (4, 12, 13, 22, 23, 26, 39) and an adaptation of the approach suggested by Gokhshtein (13) was used in this work. The method which was developed is generally applicable to all of the cases mentioned above, all additional first order kinetic cases of interest, and both single scan and cyclic triangular wave experiments. Except for reversible, irreversible, and catalytic reactions, the treatment is limited to plane electrodes because of the marked increase in complexity of the theory for most of the kinetic cases if an attempt is made to account for spherical diffusion rigorously. Conditions under which derivations for plane electrodes can be used for other geometries have been discussed by Berzins and Delahay (2). An empirical approach to making approximate corrections for the spherical contribution to the current will be described elsewhere. The cases considered here all involve reductions for the first charge transfer step, but extension to oxidations is obvious.

To present a logical discussion, each of the kinetic cases was compared to the corresponding reversible or irreversible reaction which would take place without the kinetic complication. Thus, it was necessary to include in this work a substantial discussion of these two cases, in spite of the extensive previous work. However, this makes it possible to discuss the numerical method proposed here in terms of the simplest possible case for clarity, and at the same time summarizes the widely scattered previous work in a form which is most convenient for comparison of experimental results with theory.

I. REVERSIBLE CHARGE TRANSFER

Boundary Value Problem. For a reversible reduction of an oxidized species O to a reduced species R,

$$O + ne \rightleftarrows R \qquad (\text{I})$$

taking place at a plane electrode, the boundary value problem for stationary electrode polarography is

$$\frac{\partial C_O}{\partial t} = D_O \frac{\partial^2 C_O}{\partial x^2} \qquad (1)$$

$$\frac{\partial C_R}{\partial t} = D_R \frac{\partial^2 C_R}{\partial x^2} \qquad (2)$$

$t = 0, x \geq 0$:

$$C_O = C_O{}^*; \; C_R = C_R{}^* (\sim 0) \qquad (3)$$

$$t \geq 0, \text{x} \to \infty: C_O \to C_O{}^*; \; C_R \to 0 \qquad (4)$$

$t > 0, x = 0$:

$$D_O \left(\frac{\partial C_O}{\partial x}\right) = -D_R \left(\frac{\partial C_R}{\partial x}\right) \qquad (5a)$$

$$C_O/C_R = \exp\left[(nF/RT)(E - E^\circ)\right] \tag{5b}$$

where C_O and C_R are the concentrations of substances O and R, x is the distance from the electrode, t is the time, C_O^* and C_R^* are the bulk concentrations of substances O and R, D_O and D_R, are the diffusion coefficients, n is the number of electrons, E is the potential of the electrode, E° is the formal electrode potential, and R, T, and F have their usual significance. The applicability of the Fick diffusion equations and the initial and boundary conditions has been discussed by Reinmuth (*33*).

For the case of stationary electrode polarography, the potential in Equation 5b is a function of time, given by the relations

$$0 < t \leqslant \lambda \quad E = E_i - vt \tag{6a}$$

$$\lambda \leqslant t: \qquad E = E_i - 2v\lambda + vt \tag{6b}$$

where E_i is the initial potential, v is the rate of potential scan, and λ is the time at which the scan is reversed (Figure 1).

Equations 6a and 6b can be substituted into Equation 5b to obtain the boundary condition in an abridged form:

$$C_O/C_R = \theta S_\lambda(t) \tag{7}$$

where

$$\theta = \exp\left[(nF/RT)(E_i - E^\circ)\right] \tag{8}$$

$$S\lambda(t) = \begin{cases} e^{-at} & \text{for } t \leqslant \lambda \\ e^{at - 2a\lambda} & \text{for } t \geqslant \lambda \end{cases} \tag{9}$$

and

$$a = nFv/RT \tag{10}$$

If t is always less than λ, then Equation 7 reduces to

$$C_O/C_R = \theta e^{-at} \tag{11}$$

which is the same boundary condition that has been used previously for theoretical studies of the single scan method for a reversible charge transfer.

The direct use of the Laplace transform to solve this boundary value problem is precluded by the form of Equation 7. However, the differential equations can be converted into integral equations by taking the Laplace transform of Equations 1 to 4, solving for the transform of the surface concentrations in terms of the transform of the surface fluxes, and then applying the convolution theorem (*33*):

$$C_O(0,t) = C_O^* - \frac{1}{\sqrt{\pi D_O}} \int_0^t \frac{f(\tau)d\tau}{\sqrt{t - \tau}} \tag{12}$$

$$C_R(0,t) = \frac{1}{\sqrt{\pi D_R}} \int_0^t \frac{f(\tau)d\tau}{\sqrt{t - \tau}} \tag{13}$$

where

$$f(t) = D_O \left(\frac{\partial C_O}{\partial x}\right)_{x=0} = i/nFA \tag{14}$$

The boundary condition of Equation 7 now can be combined with Equations 12 and 13, to eliminate the concentration terms and obtain a single integral equation, which has as its solution the flux of substance O at the electrode surface:

$$\int_0^t \frac{f(\tau)d\tau}{\sqrt{t - \tau}} = \frac{C_O^* \sqrt{\pi D_O}}{1 + \gamma\theta S_\lambda(t)} \tag{15}$$

where

$$\gamma = \sqrt{D_O/D_R} \tag{16}$$

Referring to Equation 10, it can be noted that the term at is dimensionless

$$at = nFvt/RT = (nF/RT)(E_i - E) \tag{17}$$

and is proportional to the potential. Since the ultimate goal is to calculate current-potential curves rather than current-time curves, it is useful to make all calculations with respect to at rather than t. This can be accomplished by a change in variable

$$\tau = z/a \tag{18}$$

$$f(t) = g(at) \tag{19}$$

and Equation 15 becomes

$$\int_0^{at} \frac{g(z)dz}{\sqrt{a}\sqrt{at - z}} = \frac{C_O^* \sqrt{\pi D_O}}{1 + \gamma\theta S_{a\lambda}(at)} \tag{20}$$

This integral equation can be made dimensionless (especially important if numerical methods are used) by the substitution

$$g(at) = C_O^* \sqrt{\pi D_O a}\ \chi(at) \tag{21}$$

and the final form of the integral equation is:

$$\int_0^{at} \frac{\chi(z)dz}{\sqrt{at - z}} = \frac{1}{1 + \gamma\theta S_{a\lambda}(at)} \tag{22}$$

The solution to Equation 22 provides values of $\chi(at)$ as a function of at, for a given value of $\gamma\theta$. From Equations 5b and 7, the values of at are related to the potential by

$$E = E^\circ - (RT/nF) \ln \gamma +$$
$$\qquad (RT/nF)\left[\ln \gamma\theta + \ln S_{a\lambda}(at)\right] \tag{23a}$$

or

$$(E - E_{1/2})n = (RT/F)\left[\ln \gamma\theta + \ln S_{a\lambda}(at)\right] \tag{23b}$$

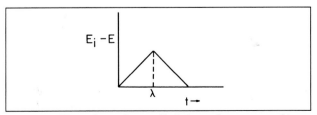

Figure 1. Wave form for cyclic triangular wave voltammetry

where $E_{1/2}$ is the polarographic half wave potential

$$E_{1/2} = E° + (RT/nF) \ln \sqrt{D_R/D_O} \qquad (24)$$

Thus, values of $\chi(at)$ can be regarded as values of $\chi[(E - E_{1/2})n]$, and will ultimately furnish values of the current as a function of potential (Equations 14, 19, 21):

$$i = nFAC_O^* \sqrt{\pi D_O a}\ \chi(at) \qquad (25)$$

The values of $\chi(at)$ are independent of the actual value of $\gamma\theta$ selected, provided in $\gamma\theta$ is larger than perhaps 6, and a formal proof has been given by Reinmuth (35). This corresponds to the usual experimental procedure of selecting an initial potential anodic of the foot of the wave, and in effect reduces the number of variables involved by one.

Numerical Solution. Although Equation 22 has been solved in several ways, only the numerical approaches appeared to be readily applicable to the cyclic experiment. The technique developed here involves dividing the range of integration from $at = 0$ to $at = M$ into N equally spaced subintervals by a change of variable,

$$z = \delta v \qquad (26)$$

and the definition

$$n = at/\delta \qquad (27)$$

Here, δ is the length of the subinterval ($\delta = M/N$), and n is a serial number of the subinterval. Thus, Equation 23 becomes

$$\sqrt{\delta} \int_0^n \frac{\chi(\delta v)\, dv}{\sqrt{n - v}} = \frac{1}{1 + \gamma\theta S_{\delta\lambda}(\delta n)} \qquad (28)$$

where n varies from 0 to M in N integral steps. The point of singularity ($n = v$) in the kernel in Equation 28 can be removed through an integration by parts to obtain

$$\int_0^n \frac{\chi(\delta v)dv}{\sqrt{n - v}} = 2\left[\chi(0) \sqrt{n} + \int_0^n \sqrt{n - v}\, d[\chi(\delta v)] \right] \qquad (29)$$

The integral on the right-hand side of Equation 29 is a Riemann-Stieltjes integral, which can be replaced by its corresponding finite sum (1). Eliminating the special points $i = 0$ and $i = n$ from the summation, one obtains

$$\int_0^n \frac{\chi(\delta v)dv}{\sqrt{n - v}} = 2\left[\chi(1) \sqrt{n} + \sum_{i=1}^{n-1} \sqrt{n-i}\,[\chi(i + 1) - \chi(i)] \right] \qquad (30)$$

and substituting this result in Equation 28,

$$2\sqrt{\delta} \left[\chi(1) \sqrt{n} + \sum_{i=1}^{n-1} \sqrt{n - i}\, [\chi(i + 1) - \chi(i)] \right] = \frac{1}{1 + \gamma\theta S_{\delta\lambda}(\delta n)} \qquad (31)$$

Equation 31 defines N algebraic equations in the unknown function $\chi(n)$, where each nth equation involves the previous $n - 1$ unknowns. These equations are then solved successively for the values of $\chi(at)$—i.e., $\chi(\delta n)$. When $\delta n \leq \delta\lambda$, the function $S_{\delta\lambda}(\delta n)$ is exp $(-\delta n)$, and when the point in the calculations is reached where $\delta n > \delta\lambda$, the function is replaced by exp $(\delta n - 2\delta\lambda)$. In this way, single scan or cyclic current-potential curves can be calculated easily, and the extension to multicycles is obvious.

Analytical Solution. It is also possible to obtain an analytical solution to Equation 22. It is an Abel integral equation (48), and the solution can be written directly as

$$\chi(at) = \frac{L(0)}{\pi \sqrt{at}} + \frac{1}{\pi} \int_0^{at} \frac{1}{\sqrt{at - z}} \left[\frac{dL(at)}{d(at)} \right]_{at=z} dz \qquad (32)$$

where $L(at)$ represents the right-hand side of Equation 22. Performing the differentiation indicated (using Equation 11 as the boundary condition) the exact solution is

$$\chi(at) = \frac{1}{\pi \sqrt{at}\,(1 + \gamma\theta)} + \frac{1}{4\pi} \int_0^{at} \frac{dz}{\sqrt{at - z}\ \cosh^2\left(\dfrac{\ln \gamma\theta - z}{2} \right)} \qquad (33)$$

Equation 33 has been given previously by Matsuda and Ayabe (23), and by Gokhshtein (13). If it is assumed that $\gamma\theta$ is large, Equation 33 reduces to the result obtained by Sevcik (44) and by Reinmuth (35), but such an assumption is not required in this case. Although the definite integral in Equation 33 cannot be evaluated in closed form, several numerical methods (such as the Euler-Maclaurin summation formula or Simpson's rule) can be used, provided the singularity at $at = z$ is first removed by a change of variable (23) or an integration by parts (35).

Series Solution. If only the single scan method is considered, Equation 22 can be solved in series form, and although the resulting series does not properly converge for all potentials, the form of the results obtained is very useful for comparison of the limiting cases obtained in the kinetic systems.

One possible approach is to expand the right-hand side of Equation 22 as an exponential power series in at, as done by Sevcik (44), but this cannot be done for most cases involving coupled chemical reactions. Thus, Reinmuth's approach (32, 33) is more general, and the final result is

$$\chi(at) = \frac{1}{\sqrt{\pi}} \sum_{j=1}^{\infty} (-1)^{i+1} \sqrt{j} \times$$

$$\exp\left[(-jnF/RT)(E - E_{1/2})\right] \quad (34)$$

Single Scan Method. In every case, the solution of Equation 22 ultimately requires numerical evaluation which in the past has been carried out with varying accuracy. This has led to some uncertainty in the literature regarding the location of the peak potential with respect to $E_{1/2}$, the height of the peak, the significance of the half-peak potential, etc. For this reason, the calculations were carried out (using the numerical solution, Equation 31) on an IBM 704 digital computer to obtain accurate values of $\chi(at)$ as a function of potential (Table I). Since the hanging mercury drop electrode is frequently used in analytical work, there is also listed a function based on Reinmuth's equation (35), from which currents at a spherical electrode can be calculated. The calculations were made using a value of $\delta = 0.01$, with $\ln \gamma\theta = 6.5$, and are accurate to ±0.001. The temperature was assumed to be 25° C., and values for some other temperature can be obtained by multiplying the potential by the factor $(273.16 + T)/298.16$ (see Equation 23). The factor $\sqrt{\pi}$ was included in the tabulation for convenience in making comparisons with previous work.

The reversible stationary electrode polarogram for a plane electrode exhibits a maximum value of $i_p/(nFA \sqrt{D_{Oa}} C_O{}^*) = 0.4463$ at a potential $28.50/n$ millivolts cathodic of $E_{1/2}$,— i.e.,

$$(E_p - E°)n + (RT/F) \ln \gamma = -28.50 \pm 0.05 \text{ mv.} \quad (35)$$

or

$$E_p = E_{1/2} - (1.109 \pm 0.002)(RT/nF) \quad (36)$$

Actually, the peak of a reversible stationary electrode polarogram is fairly broad, extending over a range of several millivolts if values of $\chi(at)$ are determined to about 1%. Thus, it is sometimes convenient to use the half-peak potential as a reference point (24), although this has no direct thermodynamic significance. The half-peak potential precedes $E_{1/2}$ by $28.0/n$ mv., or

$$E_{p/2} = E_{1/2} + 1.09(RT/nF) \quad (37)$$

The $E_{1/2}$ value can be estimated from a reversible stationary electrode polarogram from the fact that it occurs at a point 85.17% of the way up the wave.

Calculations based on Equation 33 and also Equation 34 (at least for potentials anodic of $E_{1/2}$) agree exactly with those in Table I, and these data can be used to construct accurate theoretical stationary electrode polarograms.

Cyclic Triangular Wave Method. For the first cycle, the cathodic portion of the polarogram is, of course, the same as described above for the single scan method. However, the height and position of the anodic portion of the polarogram will depend on the switching potential, E_λ, being used.

Polarograms for various values of $(E_\lambda - E_{1/2})n$ were calculated from Equation 31, and, provided that the switching po-

tential is not less than about $35/n$ mv. past the cathodic peak, the curves all have the same relative shape. (For switching potentials close to the peak, the shape of the anodic curve is very dependent on the switching potential, but this is not a usual experimental condition, and will not be considered further.) Typical cyclic polarograms are shown in Figure 2. By using for the base line the cathodic curve which would have been obtained if there had been no change in direction of potential scan, all of the anodic curves are the same, independent of switching potential, and identical in height and shape to the cathodic wave. Thus, when the anodic peak height is measured to the extension of the cathodic curve, the ratio of anodic to cathodic peak currents is unity, independent of the switching potential. This behavior can be used as an important diagnostic criterion to demonstrate the absence (or unimportance) of the various coupled chemical reactions. Of those considered in this paper, only the catalytic case (which can easily be distinguished from the reversible case by other behavior) gives a constant value of unity for the ratio of anodic to cathodic peak height on varying the switching potential. Experimentally, the cathodic base line can be obtained by extending a single scan cathodic sweep beyond the selected switching potential or, if another reaction interferes, by stop-

Table I. Current Functions $\sqrt{\pi}\chi(at)$ for Reversible Charge Transfer (Case I)

$(E-E_{1/2})n$ mv.	$\sqrt{\pi}\chi(at)$	$\phi(at)$	$(E-E_{1/2})n$ mv.	$\sqrt{\pi}\chi(at)$	$\phi(at)$
120	0.009	0.008	−5	0.400	0.548
100	0.020	0.019	10	0.418	0.596
80	0.042	0.041	15	0.432	0.641
60	0.084	0.087	20	0.441	0.685
50	0.117	0.124	25	0.445	0.725
45	0.138	0.146	−28.50	0.4463	0.7516
40	0.160	0.173	30	0.446	0.763
35	0.185	0.208	35	0.443	0.796
30	0.211	0.236	40	0.438	0.826
25	0.240	0.273	50	0.421	0.875
20	0.269	0.314	−60	0.399	0.912
15	0.298	0.357	80	0.353	0.957
10	0.328	0.403	100	0.312	0.980
5	0.355	0.451	120	0.280	0.991
0	0.380	0.499	150	0.245	0.997

To calculate the current:

(1) $i = i(\text{plane}) + i(\text{spherical correction})$.

(2) $= nFA \sqrt{aD_O}C_O{}^* \sqrt{\pi}\chi(at) + nFAD_OC_O{}^*(1/r_o)\phi(at)$

(3) $= 602\ n^{3/2} A \sqrt{D_O v}\ C_O{}^* [\sqrt{\pi}\chi(at) + 0.160\ (\sqrt{D_O}/r_o) \sqrt{nv})\phi(at)]$, amperes.

Units for (3) are: A, sq. cm.; D_O, sq. cm./sec.; v, volt/sec.; $C_O{}^*$, moles/liter; r_o, cm.

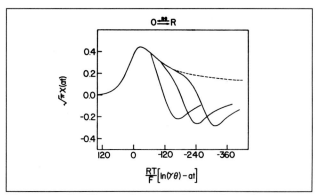

Figure 2. Cyclic stationary electrode polarograms (Case I)

Switching potentials correspond to $(E_{1/2} - E_\lambda)n$ of 64, 105, and 141 mv. for anodic scans

ping the scan at some convenient potential past the peak and recording the constant potential current-time curve (using appropriate corrections for charging current). The latter method of obtaining the base line has been proposed for analytical purposes by Reinmuth (34).

The position of the anodic wave on the potential axis is a function of the switching potential, especially for small values of $(E_\lambda - E_{1/2})n$. This results from the fact that on the cathodic scan, the surface concentration of substance R does not quite equal C_O^* at potentials close to the peak and, if the anodic scan is started under these conditions, $C_R(0, t)$ is slightly less than $C_O(0, t)$ was at the corresponding potentials for the cathodic scan. This causes an anodic shift in the wave, which decreases as the switching potential is made more cathodic. The behavior is summarized in Table II.

The transition from the single cycle to the multicycle triangular wave method involves a gradual realignment of concentration gradients at the same relative potentials on successive cycles, which, in turn, causes a gradual change in the shape of both the anodic and cathodic curves. After about 50 cycles (44), further changes in shape are very slow, and essentially a steady state cyclic curve is obtained. The relations between the cathodic and anodic peak potentials, peak

Table II. Anodic Peak Potential as a Function of Switching Potential for Reversible Charge Transfer (Case I)

$(E_{1/2} - E_\lambda)n$ mv.	$[E_p(\text{anodic}) - E_{1/2}]n$ mv.
−65	34.4
70	33.7
75	33.3
80	32.9
−100	32.0
150	30.7
200	29.8
300	29.3

heights, and switching potentials for the steady state case are given by Matsuda (22). Generally, unless a reacting system is being studied, there is little point in investigating the intermediate cyclic scans beyond the first and before the steady state is reached. Thus, although the individual curves for any number of cycles can be calculated from Equation 31, only the first few have been tabulated for an arbitrarily selected value of E_λ (27) and copies of these data are available on request.

II. IRREVERSIBLE CHARGE TRANSFER

Boundary Value Problem. For the case of a totally irreversible reduction taking place at a plane electrode

$$O + ne \overset{k}{\rightarrow} R \qquad (II)$$

the boundary value problem for stationary electrode polarography is similar to the reversible case and Equations 1, 3, 4, and 5a are applicable, except the terms involving substance R are not used. However, Equation 5b is replaced by

$$t > 0, x = 0; D_O\left(\frac{\partial C_O}{\partial x}\right) = k\,C_O \qquad (38)$$

where

$$k = k_s \exp[(-\alpha n_a F/RT)(E - E°)] \qquad (39)$$

and the other terms have their usual significance (5). In addition, only Equation 6a is required to describe the potential variation, since there is no anodic current on the reverse scan. Thus, the last boundary condition can be written

$$f(t) = D_O\left(\frac{\partial C_O}{\partial x}\right) = C_O k_i e^{bt} \qquad (40)$$

where

$$k_i = k_s \exp[(-\alpha n_a FR/T)(E_i - E°)] \qquad (41)$$

$$b = \alpha n_a Fv/RT \qquad (42)$$

Here k_i is the rate constant at the initial potential, and b is analogous to its counterpart a for the reversible case. Using exactly the same methods as above, this boundary value problem can be converted to a single integral equation, first given by Delahay (4)

$$1 - \int_0^{bt} \frac{\chi(z)dz}{\sqrt{bt - z}} = (e^{u - bt})\chi(bt) \qquad (43)$$

where

$$e^u = \sqrt{\pi D_O b}/k_i = (\sqrt{\pi D_O b}/k_s) \times$$
$$\exp[(\alpha n_a F/RT)(E_i - E°)] \qquad (44)$$

For a particular value of u, the solution to Equation 43 provides values of $\chi(bt)$ as a function of bt, which in turn are related to the potential by

$$bt = (\alpha n_a F/RT)(E_i - E) \tag{45}$$

Provided u is greater than about 7 (which corresponds to selecting an initial potential anodic of the foot of the wave) the values of $\chi(bt)$ are independent of u (4). With respect to the initial potential, however, the entire wave shifts along the potential axis as a function of u. Experimentally, the initial potential is a convenient reference point, since E° is seldom known for a totally irreversible system, and data calculated for a particular value of u can be used for any initial potential by arbitrarily shifting the potential axis. For tabulation of data, however, it is more convenient to define a potential axis which will be independent of an arbitrary function such as u, and this can be done by utilizing the relation

$$(E - E^\circ)\alpha n_a + (RT/F)\ln(\sqrt{\pi D_O b}/k_s) =$$
$$(RT/F)(u - bt) \tag{46}$$

Thus, values of $\chi(bt)$ can be used to calculate the current

$$i = nFAC_O{}^* \sqrt{\pi D_O b}\, \chi(bt) \tag{47}$$

as a function of the potential, which in turn is defined by the left-hand side of Equation 46. Equation 43 is a Volterra integral equation of the second kind, and was evaluated numerically by Delahay (4) and also by Matsuda and Ayabe (23). The numerical approach described in this paper also is applicable. In addition, a series solution has been reported (32)

$$x(bt) = \frac{1}{\sqrt{\pi}} \sum_{j=1}^{\infty} \left(-1 \right)^{j+1} \frac{\sqrt{(\pi)^j}}{\sqrt{(j-1)!}} \times$$
$$\exp\left[\left(-\frac{j\alpha n_a F}{RT} \right) \left(E - E^\circ + \frac{RT}{\alpha n_a F} \ln \frac{\sqrt{\pi D_O b}}{k_s} \right) \right] \tag{48}$$

The series in Equation 48 is strictly applicable only for values of $bt \geq 4$—i.e., potentials about $100/\alpha n_a$ millivolts cathodic of the initial potential. This is not a serious restriction, and the series is properly convergent over the entire range of interest. Although the series converges very slowly near the peak of the wave, Equation 48 is probably the best way to calculate the current. Since precise values of $\chi(bt)$ have not been published previously, Equation 43 was evaluated using the numerical method and, in addition, values of $\chi(bt)$ were calculated from Equation 48. The results (Table III) were identical and agreed well with the less accurate data previously presented by Delahay (4) and Matsuda and Ayabe (23). These values of $\chi(bt)$ were calculated using $\delta = 0.01$, with $u = 7.0$, and are accurate to ± 0.001.

Experimental Correlations. The relation between the peak potential and the other experimental parameters can be derived from Table III

$$(E_p - E^\circ)\,\alpha n_a + (RT/F)\ln\sqrt{\pi D_O b}/k_s =$$
$$-5.34 \text{ mv.} \tag{49}$$

which can be rearranged to obtain

$$E_p = E^\circ - (RT/\alpha n_a F)(0.780 + \ln\sqrt{D_O b} - \ln k_s) \tag{50}$$

This was first derived by Delahay (4, 6) (but note typographical error) and later by Matsuda and Ayabe (23). The half-peak potential also can be used as a reference point, and from Table III

$$(E_{p/2} - E^\circ)\,\alpha n_a + (RT/F)$$
$$\ln\sqrt{\pi D_O b}/k_s = 42.36 \text{ mv.} \tag{51}$$

Thus,

$$E_p - E_{p/2} = -1.857(RT/\alpha n_a F) \tag{52}$$

Table III. Current Functions
$\sqrt{\pi}\chi(bt)$ for Irreversible Charge Transfer (Case II)

Potential,[a] mv.	$\sqrt{\pi}\chi(bt)$	$\phi(bt)$	Potential,[a] mv.	$\sqrt{\pi}\chi(bt)$	$\phi(bt)$
160	0.003	0	15	0.437	0.323
140	0.008		10	0.462	0.396
120	0.016		5	0.480	0.482
110	0.024		0	0.492	0.600
100	0.035		−5	0.496	0.685
90	0.050	0	−5.34	0.4958	0.694
80	0.073	0.004	10	0.493	0.755
70	0.104	0.010	15	0.485	0.823
60	0.145	0.021	20	0.472	0.895
50	0.199	0.042	25	0.457	0.952
40	0.264	0.083	−30	0.441	0.992
35	0.300	0.115	35	0.423	1.00
30	0.337	0.154	40	0.406	
25	0.372	0.199	50	0.374	
20	0.406	0.253	70	0.323	

[a] The potential scale is $(E - E^\circ)\alpha n_a + (RT/F)\ln\sqrt{\pi D_O b}/k_s$ The initial potential for any value of u can be obtained from $(E - E_i)\alpha n_a = (E - E^\circ)\alpha n_a\ (RT/F)(u - \ln\sqrt{\pi D_O b}/k_s)$.

To calculate the current:
(1) $i = i(\text{plane}) + i(\text{spherical correction})$
(2) $= nFA\sqrt{bD_O}C_O{}^*\sqrt{\pi}\chi(bt) + nFAD_O C_O{}^*(1/r_o)\phi(bt)$
(3) $= 602n(\alpha n_a)^{1/2}A\sqrt{D_O v}C_O{}^*[\sqrt{\pi}\chi(bt) + 0.160(\sqrt{D_O}/r_o\sqrt{\alpha n_a v})\phi(bt)]$

Units for (3) are same as Table I.

From Equations 50 and 52, both the peak potential and the half-peak potential are functions of the rate of potential scan

$$(E_{p/2})_2 - (E_{p/2})_1 = (E_p)_2 - (E_p)_1 =$$
$$(RT/\alpha n_a F) \ln \sqrt{v_i/v_2} \qquad (53)$$

and thus, for a totally irreversible wave, there is a cathodic shift in peak potential or half-peak potential of about $30/\alpha n_a$ millivolts for each tenfold increase in the rate of potential scan.

An alternate form of Equation 50 also can be derived by combining Equations 40, 44, and 45 with the value of $\chi(bt)$ at the peak, to obtain

$$(C_O)_p/C_O{}^* = \chi(bt)_p \times \exp\left[(-\alpha n_a F/RT)(E_p - E^\circ +\right.$$
$$\left.\frac{RT}{\alpha n_a F} \ln \sqrt{\pi D_O b}/k_s)\right] = 0.227 \qquad (54)$$

This result does not depend on the other experimental parameters (such as v) and Equation 54 can be solved for the surface concentration of substance O at the peak. This can be substituted into the Eyring equation for a totally irreversible reaction (5) to obtain a result first derived by Gokhshtein (12) (with a slightly higher constant):

$$i_p = 0.227 \; nFAC_O{}^* K_s \times$$
$$\exp\left[(-\alpha n_a F/RT)(E_p - E^\circ)\right] \qquad (55)$$

Thus, a plot of $\ln(i_p)$ vs. $E_p - E^\circ$ (or $E_{p/2} - E^\circ$) for different scan rates would be a straight line with a slope proportional to αn_a, and an intercept proportional to k_s. This appears to be an extremely convenient method of determining the kinetic parameters for this case, although the scan rate would have to be varied over several orders of magnitude.

Still another approach to obtaining kinetic information from stationary electrode polarograms was described by Reinmuth (30) who showed that for an irreversible reaction, the current flowing at the foot of the wave is independent of the rate of voltage scan. This same conclusion can be drawn by combining Equations 46 to 48, and considering the first few terms of the series.

$$i = nFAC_O{}^* \sqrt{\pi D_O b}\,[e^{bt-u} \sqrt{\pi}\, e^{2(bt-u)} + \ldots] \qquad (56)$$

The current will be independent of v [note definition of exp (u), Equation 44] when the second term is small compared to the first, or to the 5% error level, when $\sqrt{\pi} \exp(bt-u) \le 0.05$. This condition holds for values of $\sqrt{\pi}\chi(bt)$ less than about 0.05, or about 10% of the peak value. Thus, the range of applicability of this useful criterion of irreversibility does not extend as high along the wave as implied by Reinmuth (30). For those cases where the second term can be dropped, however, Equation 56 reduces to

$$i = nFAC_O{}^* k_s \times \exp\left[(-\alpha n_a F/RT)(E - E_i)\right] \qquad (57)$$

and offers a simple way of obtaining kinetic data.

Spherical Electrodes. For an irreversible reaction taking place at a spherical electrode, Reinmuth was able to derive a series solution which is convergent over the entire potential range of interest (32). Unfortunately, it is not possible to separate the spherical correction term from the expression for the plane electrode as for the reversible system (Case I). Since the series converges very slowly at potentials near the peak, making the use of a computer almost mandatory, an alternate means of expressing the spherical contribution to the current was sought. A large number of curves were calculated, and the spherical correction—i.e., the difference between the current obtained at spherical and plane electrodes of the same area under identical conditions—was plotted as a function of the dimensionless parameter $\sqrt{D_O}/(r_0 \sqrt{b})$ where r_0 is the radius of the electrode. The plot was linear (to better than 1%) for values of $\sqrt{D_O}/(r_0 \sqrt{b})$ less than 0.1. Since this includes all values which are of practical use, an irreversible spherical correction term, $\phi(bt)$ was evaluated which can be used just as the analogous term for the reversible case. These values of $\phi(bt)$ are listed in Table III, and values of the current calculated in this way agree well with the considerably less convenient data presented previously (8).

COUPLED CHEMICAL REACTIONS

If a homogeneous chemical reaction is coupled to the charge transfer reaction, stationary electrode polarography provides an extremely powerful method of investigating the kinetic parameters. Several of the important kinetic systems are discussed in this work, including those which involve first order (or pseudo first order) preceding, following, or catalytic chemical reactions. Several other cases, including the chemical reaction coupled between two charge transfer reactions, have also been considered, and will be presented elsewhere.

For each of the kinetic cases, the boundary value problem was formulated in a manner similar to Case I or Case II (depending on the nature of the charge transfer reaction) but modified to reflect the kinetic complication. These boundary value problems are presented in Table IV. In each case the rate constant is first order or pseudo first order—e.g., in Cases VII and VIII it was assumed that $C_Z \gg C_O$, and the rate constant which was used in the calculations was $k_f (= k_f' C_Z)$. Using the same procedure as outlined in Equations 12 through 22, each of these boundary value problems was converted to a single integral equation. However, in all the kinetic cases except Case VI, at least one of the differential equations involved two concentration variables. Thus, in order to simplify the problem, the usual changes in variable were made (20).

The integral equation obtained for each case is presented in Table V. All terms in these equations have been defined previously, except K, which is the equilibrium constant for the chemical reaction, and l which is the sum of the rate constants $(k_f + k_b)$. Each of these integral equations was then solved numerically using the approach described in Equations 26 to 31. Only two different kernels are involved in all the integral equations, and once the procedures were worked out for one case, they could be extended readily to the other cases.

Table IV. Boundary Value Problems for Stationary Electrode Polarography with Coupled Chemical Reactions

	Reaction	Diffusion Equations	Initial Conditions $t=0$, $x \geq 0$	Boundary Conditions $t>0$, $x \to \infty$	Boundary Conditions $t>0$, $x=0$
III	$Z \underset{k_b}{\overset{k_f}{\rightleftharpoons}} O$ $O + ne \rightleftharpoons R$	$\frac{\partial C_Z}{\partial t} = D_Z \frac{\partial^2 C_Z}{\partial x^2} - k_f C_Z + k_b C_O$ $\frac{\partial C_O}{\partial t} = D \frac{\partial^2 C_O}{\partial x^2} + k_f C_Z - k_b C_O$ $\frac{\partial C_R}{\partial t} = D_R \frac{\partial^2 C_R}{\partial x^2}$	$C_O/C_Z = K$ $C_O + C_Z = C^*$ $C_R = C_R^* \;(\approx 0)$	$C_O/C_Z \to K$ $C_O + C_Z \to C^*$ $C_R \to 0$	$D_Z \frac{\partial C_Z}{\partial x} = 0$ $D_O \frac{\partial C_O}{\partial x} = -D_R \frac{\partial C_R}{\partial x}$ $C_O/C_R = \Theta S_\lambda(t)$
IV	$Z \underset{k_b}{\overset{k_f}{\rightleftharpoons}} O$ $O + ne \overset{k}{\rightarrow} R$	Same as III (a)	Same as III (a)	Same as III (a)	$D_Z \frac{\partial C_Z}{\partial x} = 0$ $D_O \frac{\partial C_O}{\partial x} = k C_O = k_i C_O \exp(bt)$
V	$O + ne \rightleftharpoons R$ $R \underset{k_b}{\overset{k_f}{\rightleftharpoons}} Z$	$\frac{\partial C_O}{\partial t} = D_O \frac{\partial^2 C_O}{\partial x^2}$ $\frac{\partial C_R}{\partial t} = D_R \frac{\partial^2 C_R}{\partial x^2} - k_f C_R + k_b C_Z$ $\frac{\partial C_Z}{\partial t} = D_Z \frac{\partial^2 C_Z}{\partial x^2} + k_f C_R - k_b C_Z$	$C_O = C_O^*$ $C_R = C_R^* \;(\approx 0)$ $C_Z = K C_R^* \;(\approx 0)$	$C_O \to C_O^*$ $C_R \to 0$ $C_Z \to 0$	$D_O \frac{\partial C_O}{\partial x} = -D_R \frac{\partial C_R}{\partial x}$ $D_Z \frac{\partial C_Z}{\partial x} = 0$ $C_O/C_R = \Theta S_\lambda(t)$
VI	$O + ne \rightleftharpoons R$ $R \overset{k_f}{\rightarrow} Z$	$\frac{\partial C_O}{\partial t} = D_O \frac{\partial^2 C_O}{\partial x^2}$ $\frac{\partial C_R}{\partial t} = D_R \frac{\partial^2 C_R}{\partial x^2} - k_f C_R$	Same as V (b)	Same as V (b)	Same as V (b)
VII	$O + ne \rightleftharpoons R$ $R + Z \overset{k_f'}{\rightarrow} O$	$\frac{\partial C_O}{\partial t} = D_O \frac{\partial^2 C_O}{\partial x^2} + k_f C_R$ $\frac{\partial C_R}{\partial t} = D_R \frac{\partial^2 C_R}{\partial x^2} - k_f C_R$	Same as Equation 3	Same as Equation 4	Same as Equations 5a and 7
VIII	$O + ne \overset{k}{\rightarrow} R$ $R + Z \overset{k_f'}{\rightarrow} O$	Same as VII	Same as Equation 3 (a)	Same as Equation 4 (a)	Same as Equation 40

[a] Since the charge transfer is totally irreversible, those equations involving substance R are not used.
[b] Since the chemical reaction is irreversible, the equations involving substance Z are not used.

These numerical results (presented in the various tables) provided values of the current functions $\chi(at)$ or $\chi(bt)$ which could be related to potential by Equations 23 and 24 for reversible charge transfers, and Equations 45 and 46 for irreversible charge transfers.

In each case, the integral equation was made dimensionless by the same substitution (Equation 21) and, as a result, the current always can be calculated from the current function $\chi(at)$ or $\chi(bt)$ merely by multiplication by the term $nFAC_O^* \sqrt{\pi D_O a}$ for reversible charge transfers, or by $nFAC_O^* \sqrt{\pi D_O b}$ for irreversible charge transfers, as in Equations 25 and 47. [Note that for the preceding chemical reaction (Cases III and IV), the stoichiometric concentration C^* appears in these equations rather than the equilibrium concentration C_O^*.]

For the cases involving an irreversible chemical reaction, no additional restrictions—other than that of selecting an initial potential anodic of the foot of the wave—were introduced in the derivation. For the three cases involving reversible chemical reactions (Cases III, IV, and V), however, a simplify-

ing assumption was made in order to reduce the number of variables. In these cases, the integrals which arise corresponding to that in Equation 28 are of the form

$$(\sqrt{\delta}/K) \int_0^n \frac{\exp[\delta\psi](n - v)]\chi(\delta v)\,dv}{\sqrt{n-v}}$$

for preceding chemical reactions and

$$K\sqrt{\delta} \int_0^n \frac{\exp[\delta\psi](n - v)]\chi(\delta v)\,dv}{\sqrt{n - v}}$$

for succeeding chemical reactions. Here ψ is l/a for reversible charge transfer and l/b for irreversible charge transfer. With these integrals, the integration by parts (used to remove the point of singularity) produces terms of the form

$$(\sqrt{\pi}/K \sqrt{\psi}) \operatorname{erf} \sqrt{\delta\psi(n - v)}$$

for a preceding chemical reaction and

III $\qquad 1 - \int_0^{at} \dfrac{\chi(z)dz}{\sqrt{at - z}} = \dfrac{1+K}{K} \, {}_{\ominus}S_{a\lambda}(at) \int_0^{at} \dfrac{\chi(z)dz}{\sqrt{at - z}} + \dfrac{1}{K} \int_0^{at} \dfrac{e^{-(1/a)(at-z)}\chi(z)dz}{\sqrt{at - z}}$ $\qquad$ (58)

IV $\qquad 1 - \int_0^{bt} \dfrac{\chi(z)dz}{\sqrt{bt - z}} = \dfrac{1+K}{K} e^{u-bt} \chi(bt) + \dfrac{1}{K} \int_0^{bt} \dfrac{e^{-(1/b)(bt-z)}\chi(z)dz}{\sqrt{bt - z}}$ $\qquad$ (59)

V $\qquad 1 - \int_0^{at} \dfrac{\chi(z)dz}{\sqrt{at - z}} = \dfrac{1}{1+K} \, {}_{\ominus}S_{a\lambda}(at) \int_0^{at} \dfrac{\chi(z)dz}{\sqrt{at - z}} + \dfrac{K}{1+K} \, {}_{\ominus}S_{a\lambda}(at) \int_0^{at} \dfrac{e^{-(1/a)(at-z)}\chi(z)dz}{\sqrt{at - z}}$ $\qquad$ (60)

VI $\qquad 1 - \int_0^{at} \dfrac{\chi(z)dz}{\sqrt{at - z}} = {}_{\gamma\ominus}S_{a\lambda}(at) \int_0^{at} \dfrac{e^{-(k_f/a)(at-z)}\chi(z)dz}{\sqrt{at - z}}$ $\qquad$ (61)

VII $\qquad 1 - \int_0^{at} \dfrac{e^{-(k_f/a)(at-z)}\chi(z)dz}{\sqrt{at - z}} = {}_{\ominus}S_{a\lambda}(at) \int_0^{at} \dfrac{e^{-(k_f/a)(at-z)}\chi(z)dz}{\sqrt{at - z}}$ $\qquad$ (62)

VIII $\qquad 1 - \int_0^{bt} \dfrac{e^{-(k_f/b)(bt-z)}\chi(z)dz}{\sqrt{bt - z}} = e^{u-bt} \chi(bt)$ $\qquad$ (63)

$$(K\sqrt{\pi}/\sqrt{\psi}) \, \mathrm{erf} \sqrt{\delta\psi(n - v)}$$

for a succeeding chemical reaction. Although the numerical evaluation could have been carried out retaining all these terms, considerable simplification resulted in assuming that $\delta\psi (n - v) \geqslant 4$. This assumption makes it possible to consider the erf terms as unity, but excludes the cases involving small ψ from the numerical data—i.e., cases where the chemical reaction has no effect on the charge transfer.

These restrictions, once noted, are relatively unimportant and will be discussed in connection with the individual cases. The simplification which results, however, is extremely important. In some previous treatments (41) involving reversible chemical reactions, it has been necessary to select a specific value of the equilibrium constant in order to calculate a set of data for a range of rate constants. In this case, for example, a table comparable to Table VII would be required for each value of K likely to be of interest. However, the simplification discussed above makes it possible for the effect of the equilibrium constant on the shape of the polarograms to be separated from its effect on the location of the curve on the potential axis. This can be seen in Equations 58 and 60 where the effect on the potential can be handled by defining a new potential axis in terms of $(E — E_{1/2})n — (RT/F) \ln K/(1 + K)$ for Case III and $(E — E_{1/2})n — (RT/F) \ln (1 + K)$ for Case V. For Case IV (Equation 59) the new potential axis is defined in terms of $(E—E°) \alpha \, n_a + (RT/F) \ln \sqrt{\pi D_O b/k_s} - (RT/F) \ln K/(1+K)$. The effect of K on the shape of the polarograms is included in the numerical data. This approach places no restrictions on the value of the equilibrium constant, and the only exception to the applicability of the numerical data is that stated above: the sum of the rate

constants cannot be small compared to a.

The values of $\chi(at)$ and $\chi(bt)$ were calculated using $\delta = 0.02$, and are accurate to $+0.000$, -0.002—i.e., if any error is present, the values tend to be slightly low. The use of a larger value of δ also introduces a small error in the potential and, for each case where the kinetic case reduces to Case I, the waves are shifted cathodic by a few tenths of a millivolt in comparison to Table I.

A series solution also was obtained for each kinetic case (Table VI). As noted previously, only in those cases involving irreversible charge transfer was it possible to obtain series solutions which converged properly over the entire potential range of interest. However, even when unsuitable for calculation of theoretical current voltage curves, the series solutions were extremely useful in correlation of the experimental and kinetic parameters. This was particularly true in those cases in which the series was of the same type as obtained for Cases I or II, since then the form of the solution was known, and correlations between the potential and kinetic parameters could be obtained from the exponential terms.

III. CHEMICAL REACTION PRECEDING A REVERSIBLE CHARGE TRANSFER

A large group of coupled chemical reactions involves cases in which the electroactive species is produced by a homogeneous first-order chemical reaction preceding a reversible charge transfer

$$Z \underset{k_b}{\overset{k_f}{\rightleftarrows}} O$$

$$O + ne \rightleftarrows R \qquad \text{(III)}$$

Only the case in which the chemical reaction is reversible is

Table VI. Series Solutions for Stationary Electrode Polarography with Coupled Chemical Reactions

$$\text{III} \qquad \chi(at) = \frac{1}{\sqrt{\pi}} \sum_{j=1}^{\infty} (-1)^{j+1} \left[\sqrt{j} \prod_{i=1}^{j-1} \left(1 + \frac{\sqrt{i}}{K\sqrt{(l/a)+i}} \right) \right] \exp\left[-\frac{jnF}{RT}\left(E - E_{1/2} - \frac{RT}{nF}\ln\frac{K}{1+K} \right) \right] \tag{64}$$

$$\text{IV} \qquad \chi(bt) = \frac{1}{\sqrt{\pi}} \sum_{j=1}^{\infty} (-1)^{j+1} \left[\frac{(\sqrt{\pi})^j}{\sqrt{(j-1)!}} \prod_{i=1}^{j-1} \left(1 + \frac{\sqrt{i}}{K\sqrt{(l/b)+i}} \right) \right] \exp\left[-\frac{j\alpha n_a F}{RT}\left(E - E^0 + \frac{RT}{\alpha n_a F}\ln\frac{\sqrt{\pi Db}}{k_s} - \frac{RT}{\alpha n_a F}\ln\frac{K}{1+K} \right) \right] \tag{65}$$

$$\text{V} \qquad \chi(at) = \frac{1}{\sqrt{\pi}} \sum_{j=1}^{\infty} (-1)^{j+1} \left[\sqrt{j} \Big/ \prod_{i=1}^{j} \left(1 + \frac{K\sqrt{i}}{\sqrt{(l/a)+i}} \right) \right] \exp\left[-\frac{jnF}{RT}\left(E - E_{1/2} - \frac{RT}{nF}\ln(1+K) \right) \right] \tag{66}$$

$$\text{VI} \qquad \chi(at) = \frac{1}{\sqrt{\pi}} \sum_{j=1}^{\infty} (-1)^{j+1} \left[\frac{1}{\sqrt{(j-1)!}} \prod_{i=1}^{j} \sqrt{(k_f/a)+i} \right] \exp\left[-\frac{jnF}{RT}(E - E_{1/2}) \right] \tag{67}$$

$$\text{VII} \qquad \chi(at) = \frac{1}{\sqrt{\pi}} \sum_{j=1}^{\infty} (-1)^{j+1} \sqrt{(k_f/a)+j} \; \exp\left[-\frac{jnF}{RT}(E - E_{1/2}) \right] \tag{68}$$

$$\text{VIII} \qquad \chi(bt) = \frac{1}{\sqrt{\pi}} \sum_{j=1}^{\infty} (-1)^{j+1} \left[(\sqrt{\pi})^j \Big/ \prod_{i=1}^{j-1} \sqrt{(k_f/b)+i} \right] \exp\left[-\frac{j\alpha n_a F}{RT}\left(E - E^0 + \frac{RT}{\alpha n_a F}\ln\frac{\sqrt{\pi Db}}{k_s} \right) \right] \tag{69}$$

relevant; however, there are no restrictions on possible values of the equilibrium constant.

This case has been discussed (for the single scan method only) by Saveant and Vianello (38, 41) and, in addition, a series solution has been presented (46).

Qualitatively, the effect of a preceding chemical reaction on the cathodic scan depends on several factors, and three distinct limiting cases can be recognized. First, if l/a is very small, the experiment is over before significant conversion of Z to O can take place. Under such circumstances, the curve obtained is the same shape as Case I, appears at the normal potential for the uncomplicated reduction of O to R, but the magnitude of the current is proportional to the equilibrium concentration of substance O in the bulk of the solution, rather than its stoichiometric concentration C^*. This result can be obtained from Equation 64, which, if l/a is small, reduces to

$$\chi(at) = \frac{1}{\sqrt{\pi}} \sum_{j=1}^{\infty} (-1)^{j+1} \sqrt{j} \left(\frac{K}{1+K} \right) \times \exp\left[(-jnF/RT)(E - E_{1/2}) \right] \tag{70}$$

This series is the same as Equation 34 for the reversible case, except for the factor $K/(K+1)$. If K is large (equilibrium favoring O) a normal stationary electrode polarogram is obtained. On the other hand, if K is small, the current is determined by the equilibrium concentration $C_O^* = [K/(K+1)]C^*$. This is the one correlation which was not obtained from the numerical solution, because of the simplifying assumption (that l/a is not very small) which was made in the derivation. This restriction is not particularly serious, however, because the option of

varying the rate of potential scan lends great versatility to the method. For example, if low values of l/a are encountered experimentally, and reasonable values of l are involved, one could merely reduce the rate of voltage scan to a region where the numerical data are applicable. On the other hand, if l is low and a has already been reduced to the lower practical limit, Equation 70 could be used directly to calculate the theoretical current potential curve. The converse approach, of using large enough rates of voltage scan to ensure that this condition holds, has been used by Papoff to determine equilibrium constants (28).

The two other limiting cases can be obtained from Equation 64 by assuming first that l/a is large, so that $l/a + i \approx l/a$, and then considering either large or small values of $\sqrt{a}/K\sqrt{l}$. If $\sqrt{a}/K\sqrt{l}$ is small, Equation 64 reduces to

$$\chi(at) = \frac{1}{\sqrt{\pi}} \sum_{j=1}^{\infty} (-1)^{j+1} \times \sqrt{j} \exp\left[(-jnF/RT) \left(E - E_{1/2} - \frac{RT}{nF}\ln\frac{K}{1+K} \right) \right] \tag{71}$$

which is the same as Equation 34 for the reversible case, except that the wave appears at a potential determined by the equilibrium constant. At the other limit, for large values of $\sqrt{a}/K\sqrt{l}$, Equation 64 reduces to

$$\chi(at) = \frac{1}{\sqrt{\pi}} \sum_{j=1}^{\infty} (-1)^{j+1} \sqrt{j!} \; K \sqrt{l/a} \times \exp\left[-\frac{jnF}{RT}\left(E - E_{1/2} - \right. \right.$$

$$\frac{RT}{nF} \ln \frac{K}{1+K} - \frac{RT}{nF} \ln \frac{\sqrt{a}}{K\sqrt{l}}\bigg)\bigg] \quad (72)$$

This series does not correspond to one of the previously encountered cases, and it cannot be characterized merely by inspection. Nevertheless the appearance of the term $1/\sqrt{a}$ in each term of the sum indicates that the magnitude of the current will be independent of the rate of potential scan.

Single Scan Method. The characteristics of the stationary electrode polarogram between these two limiting cases were obtained from the numerical calculations. Typical cyclic polarograms calculated for several values of $\sqrt{a}/K\sqrt{l}$ are shown in Figure 3. The curve for $\sqrt{a}K\sqrt{l}$ equal to zero corresponds to the reversible case and, as the chemical step becomes more important, the curves become more drawn out. It should be noted that in Figure 3 the ordinate is $\chi(at)$, and not the current. Thus, for the situation where an increase in $\sqrt{a}/K\sqrt{l}$ corresponds to an increase in a, the current function $\chi(at)$ decreases, but the actual current increases, since the factor $\sqrt{a}$ appears in Equation 25 for the current. In the limit for large values of a, the decrease in $\chi(at)$ exactly balances this factor $\sqrt{a}$ in Equation 25, so that the current becomes independent of a, as indicated in Equation 72.

Values of the rate constants can be obtained from the cathodic stationary electrode polarograms (provided the equilibrium constant is known) by comparing the experimental curves with theoretical plots such as in Figure 3. Data for construction of accurate theoretical polarograms are presented in Table VII.

Since this is a rather cumbersome way of handling the data, an alternate approach may be more convenient. A large number of theoretical curves were calculated for various values of $\sqrt{a}/K\sqrt{l}$, in addition to those listed in Table VII. From these data, a working curve was constructed for the ratio i_k/i_d of the peak current obtained in a kinetic case to that

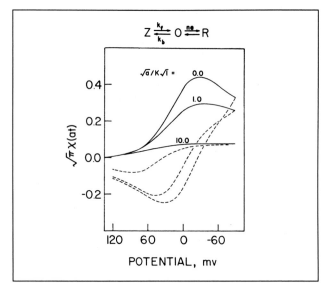

Figure 3. Cyclic stationary electrode polarograms (Case III)

Potential scale is $(E - E_{1/2})n - (RT/F)\ln K/(1+K)$

Table VII. Current Functions $\sqrt{\pi}\chi(at)$ for a Chemical Reaction Preceding a Reversible Charge Transfer (Case III)

Potential[a]	$\sqrt{a}/K\sqrt{l}$						
	0.2	0.5	1.0	1.5	3.0	6.0	10.0
120	0.009	0.009	0.009	0.009	0.009	0.009	0.008
100	0.019	0.019	0.019	0.019	0.018	0.017	0.015
80	0.041	0.040	0.039	0.038	0.035	0.031	0.027
60	0.081	0.080	0.075	0.072	0.063	0.051	0.041
50	0.113	0.108	0.100	0.094	0.080	0.062	0.049
45	0.132	0.125	0.116	0.108	0.089	0.068	0.052
40	0.152	0.144	0.131	0.121	0.099	0.074	0.055
35	0.174	0.164	0.149	0.135	0.109	0.079	0.059
30	0.199	0.184	0.164	0.150	0.118	0.084	0.062
25	0.224	0.206	0.183	0.164	0.127	0.089	0.064
20	0.249	0.228	0.199	0.178	0.136	0.093	0.067
15	0.275	0.249	0.216	0.191	0.144	0.098	0.069
10	0.301	0.270	0.232	0.204	0.151	0.101	0.071
5	0.324	0.289	0.246	0.215	0.158	0.104	0.072
0	0.345	0.307	0.259	0.225	0.163	0.107	0.074
−5	0.364	0.321	0.271	0.234	0.168	0.109	0.075
10	0.379	0.334	0.280	0.241	0.173	0.111	0.076
15	0.391	0.344	0.288	0.247	0.176	0.113	0.077
20	0.399	0.351	0.293	0.252	0.179	0.114	0.077
25	0.404	0.355	0.297	0.255	0.181	0.115	0.078
−30	0.406	0.358	0.299	0.257	0.182	0.116	0.078
35	0.405	0.358	0.300	0.258	0.183	0.116	0.079
40	0.402	0.357	0.300	0.258	0.183	0.117	0.079
45	0.397	0.353	0.298	0.258	0.183	0.117	0.079
50	0.390	0.349	0.296	0.256	0.183	0.117	0.079
−60	0.373	0.338	0.289	0.252	0.181	0.116	0.079
80	0.337	0.310	0.272	0.240	0.176	0.115	0.078
100	0.301	0.284	0.253	0.227	0.170	0.113	0.077
120	0.273	0.260	0.236	0.214	0.164	0.110	0.076
140	0.250	0.240	0.222	0.203	0.158	0.108	0.075
$E_{p/2}$, mv.	+29.3	31.3	34.4	37.5	44.2	53.4	62.2

[a] Potential scale is $(E - E_{1/2})n - (RT/F) \ln K/(1+K)$.

expected for an uncomplicated diffusion controlled case. This procedure makes it unnecessary to have the value of several of the experimental parameters such as the diffusion coefficient, electrode area, etc. The working curve for this case (Figure 4) fits the empirical equation

$$i_k/i_d = \frac{1}{1.02 + 0.471 \sqrt{a/K}\sqrt{l}} \tag{73}$$

to about 1% (except for very low values of $\sqrt{a/K}\sqrt{l}$, where the error does not exceed 1.5%). More precise values of the ratio i_k/i_d as a function of $\sqrt{a/K}\sqrt{l}$ have been tabulated (27) and, if required for construction of a more exact working curve, are available on request.

As indicated from the series solutions, the potential at which the wave appears is independent of $\sqrt{a/K}\sqrt{l}$ for small values (Equation 71) while the wave shifts anodic by about $60/n$ mv. for each tenfold increase in $\sqrt{a/K}\sqrt{l}$ for large values (Equation 72). For intermediate values, the shift in potential can be obtained from the numerical data. Since the wave is drawn out for large values of $\sqrt{a/K}\sqrt{l}$ in this case, it is more convenient to consider shifts in half peak potential rather than peak potential, and this behavior (Figure 5) can be used to obtain kinetic data.

Cyclic Triangular Wave Method. As shown in Figure 3, the anodic portion of a cyclic stationary electrode polarogram is not affected quite as much by the preceding chemical reaction as is the cathodic portion. For example, for $\sqrt{a/K}\sqrt{l}$ equal to 3.0, the cathodic curve is quite flat, but since the anodic portion involves a chemical reaction after charge transfer, it has many of the characteristics of Case V, and is still markedly peak shaped. For these curves, a switching potential $(E_\lambda - E_{1/2})n - (RT/F)\ln(K/(1+K))$ of -90 mv. was selected. The shape of the anodic curves is a function of the switching potential and, thus, it is not practical to compile a table of the anodic current function, although such data are available on request (27). In general, the height of the anodic peak (as measured to the extension of the cathodic curve) increases as the switching potential is made more cathodic.

The anodic current is fairly insensitive to changes in $\sqrt{a/K}\sqrt{l}$ for values larger than about 5, and thus the range of applicability of this correlation is about the same as for the cathodic current (Figures 3 and 4). However, for those cases where a value of i_d cannot be obtained for use with the working curve of Figure 4, the kinetic parameters can be determined by relating $\sqrt{a/K}\sqrt{l}$ to the ratio of the anodic to cathodic peak currents. A working curve for this correlation is shown in Figure 6. Data for construction of a more accurate working curve are available on request (27).

IV. CHEMICAL REACTION PRECEDING AN IRREVERSIBLE CHARGE TRANSFER

The case in which a first order chemical reaction precedes an irreversible charge transfer

$$Z \underset{k_b}{\overset{k_f}{\rightleftarrows}} O$$

$$O + ne \overset{k}{\rightarrow} R \tag{IV}$$

has not been discussed previously. The stationary electrode po-

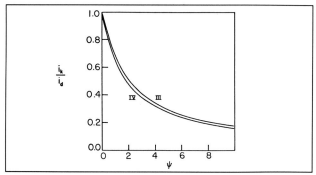

Figure 4. Working curve, ratio of kinetic peak current to diffusion controlled peak current

Case III: $\psi = \sqrt{a/K}\sqrt{l}$
Case IV: $\psi = \sqrt{b/K}\sqrt{l}$

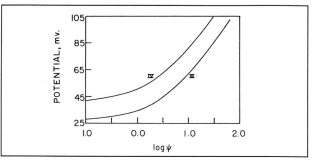

Figure 5. Variation of half-peak potential as a function of the kinetic parameters

Case III: $\psi = \sqrt{a/K}\sqrt{l}$; potential scale is $(E_{p/2} - E_{1/2})n - (RT/F)\ln K/(1+K)$. Case IV: $\psi = \sqrt{b/K}\sqrt{l}$; potential scale is $(E_{p/2} - E^\circ)\alpha n_a - (RT/F)\ln K/(1+K) + (RT/F)\ln \sqrt{\pi Db}/k_3$

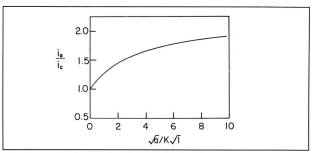

Figure 6. Ratio of anodic to cathodic peak current as a function of kinetic parameters, Case III

larograms are qualitatively similar to Case III, except, of course, no anodic current is observed in the cyclic triangular wave experiment and, further, the curves are even more drawn out because of the effect of the electron transfer coefficient, α. Thus, although the main effect of the preceding chemical reaction is the same as Case III—i.e., to decrease the current (compared to the irreversible charge transfer without chemical complication)—the detailed characteristics of the two cases are markedly different, and effects of the rate controlled charge transfer can be separated from the rate controlled chemical reaction by quantitative evaluation of the stationary electrode polarograms.

As in Case III, three distinct limiting cases can be considered depending on the kinetic parameter $\sqrt{b/K}\sqrt{l}$. First, if

l/b is small, the curve is the same shape as Case II, its potential is unaffected by the kinetic complication, and the magnitude of the current is a function of the equilibrium concentration of substance O. Thus, for small values of l/b, Equation 65 reduces to

$$\chi(bt) = \frac{1}{\sqrt{\pi}} \sum_{j=1}^{\infty} (-1)^{j+1} \frac{(\sqrt{\pi})^j}{\sqrt{(j-1)!}} \times \left(\frac{K}{1+K}\right) \exp\left[-\frac{j\alpha n_a F}{RT} (E - E^\circ + \frac{RT}{\alpha n_a F} \ln \sqrt{\pi D b}/k_s)\right] \quad (74)$$

which is the same as Equation 48 for the uncomplicated irreversible case except for the factor $K/(1 + K)$. As in Case III, this is the correlation which is not included in the numerical data because of the simplifying assumptions made.

The two other limiting cases are obtained by assuming first that l/b is large, and then considering the series for both small and large values of $\sqrt{b}/K\sqrt{l}$. If $\sqrt{b}/K\sqrt{l}$ is small, Equation 65 reduces to

$$\chi(bt) = \frac{1}{\sqrt{\pi}} \sum_{j=1}^{\infty} (-1)^{j+1} \frac{\sqrt{(\pi)^j}}{\sqrt{(j-1)!}} \times \exp\left[-\frac{j\alpha n_a F}{RT} \left(E - E^\circ + \frac{RT}{\alpha n_a F} \ln \frac{\sqrt{\pi D b}}{k_s} - \frac{RT}{\alpha n_a F} \ln \frac{K}{1+K}\right)\right] \quad (75)$$

which is the same as Equation 48 for the irreversible case, except that the potential at which the wave appears has been shifted by the equilibrium constant.

At the other limit, for large values of $\sqrt{b}/K\sqrt{l}$, Equation 65 reduces to

$$\chi(bt) = \frac{1}{\sqrt{\pi}} \sum_{j=1}^{\infty} (-1)^{j+1} (\sqrt{\pi})^j K \sqrt{l/b} \times \exp\left[-\frac{j\alpha n_a F}{RT} \left(E - E^\circ + \frac{RT}{\alpha n_a F} \ln \frac{\sqrt{\pi D b}}{k_s} - \frac{RT}{\alpha n_a F} \ln \frac{K}{1+K} + \frac{RT}{\alpha n_a F} \ln \frac{K\sqrt{\pi b}}{\sqrt{l}}\right)\right] \quad (76)$$

At this limit, a peak is no longer observed, and both the potential of the wave and magnitude of the current are independent of b. In addition, Equation 76 can be written in a closed form valid for large values of $\sqrt{b}/K\sqrt{l}$

$$i = (nFAC^* \sqrt{D} K \sqrt{l})/1 + \exp\left[\frac{\alpha n_a F}{RT} (E - E^\circ + \right.$$

$$\left. \frac{RT}{\alpha n_a F} \ln \frac{\sqrt{\pi D b}}{k_s} - \frac{RT}{\alpha n_a F} \ln \frac{K}{1+K} + \frac{RT}{\alpha n_a F} \ln \frac{K\sqrt{\pi b}}{\sqrt{l}}\right)\right] \quad (77)$$

This is the equation for an S-shaped current-voltage curve, and at cathodic potentials the current is directly proportional to the term $K(k_f + k_b)^{1/2}$.

Between these last two limiting cases, the characteristics of the stationary electrode polarogram can be obtained from the numerical solution to Equation 59. Alternatively, theoretical current-potential curves could be calculated from the series solution, Equation 65, which converges properly over the entire potential range. Typical polarograms are given in Figure 7, and data for construction of accurate curves are listed in Table VIII.

As in Case III, the kinetic parameters can be obtained from a direct comparison of experimental and theoretical polarograms, or from working curves of the ratio i_k/i_d of the kinetic peak current to the diffusion controlled peak current for an irreversible charge transfer (Figure 4). The working curve for this case was found to fit the empirical equation

$$i_k/i_d = \frac{1}{1.02 + 0.531 \sqrt{b}/K \sqrt{l}} \quad (78)$$

The applicability of Equation 78 is about the same as Equation 73.

As indicated from the series solutions, the potential at which the peak appears differs from Case III. The potential does not depend on $\sqrt{b}/K\sqrt{l}$ for small values of the kinetic parameter (Equation 75), while the potential shifts anodic by about $60/\alpha n_a$ mv. for each tenfold increase for large values of $\sqrt{b}/K\sqrt{l}$ (Equations 76 and 77). This behavior, which also can be used to obtain kinetic data, is contrasted with Case III in Figure 5.

V. CHARGE TRANSFER FOLLOWED BY A REVERSIBLE CHEMICAL REACTION

The case in which a reversible chemical reaction follows a reversible charge transfer

$$O + ne \rightleftarrows R$$

$$R \overset{k_f}{\underset{k_b}{\rightleftarrows}} Z \quad (V)$$

includes a fairly large group of organic electrode reactions. Such reactions have been studied by numerous workers using a variety of electrochemical techniques. However, the theory for stationary electrode polarography has not previously been presented.

If the charge transfer is irreversible, a succeeding chemical reaction will have no effect on the stationary electrode polarogram. The chemical reaction can still be studied if either sub-

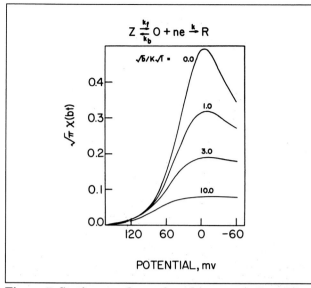

Figure 7. Stationary electrode polarograms, Case IV

Potential scale is $(E - E°)\alpha n_a + (RT/F) \ln \sqrt{\pi Db/k_s} - (RT/F) \ln K/(1 + K)$

stance R or Z is electroactive at some other potential but such cases will not be considered here, and are probably more easily handled by other techniques such as step functional controlled potential electrolysis (43).

Depending on the magnitude of the kinetic parameters, three limiting cases can be distinguished. First, if the rate of the chemical reaction is very fast, the system will be in equilibrium at all times, and the only effect will be an anodic displacement of the wave along the potential axis. This result can be obtained from Equation 66, which for large values of l/a reduces to

$$\chi(at = \frac{1}{\sqrt{\pi}} \sum_{j=1}^{\infty} (-1)^{j+1} \sqrt{j} \times$$

$$\exp\left[- \frac{jnF}{RT} \left(E - E_{1/2} - \frac{RT}{nF} \ln (1 + K)\right)\right] \quad (79)$$

This series is the same as Equation 34 for the reversible case, except for the term $(RT/nF) \ln(1 + K)$ in the exponential, which reflects the anodic displacement of the wave.

The second limiting case to be considered is that in which the chemical reaction is very slow, so that essentially no chemical reaction takes place during the experiment. Under such conditions the curve should again be the normal reversible shape, but should appear at its normal potential. Thus, if l/a is small, Equation 66 reduces exactly to Equation 34. This is the limiting case which is not included in the numerical solution because of the simplifying assumption. Unfortunately this case is important experimentally, particularly when the equilibrium constant is large. Under these circumstances, however, the chemical reaction can be considered irreversible ($k_f \gg k_b$), and an alternate approach to the theory is available. This will be treated as Case VI, below.

The third limiting case occurs when l/a is large and the ki-

Table VIII. Current Functions $\sqrt{\pi}\chi(bt)$ for a Chemical Reaction Preceding an Irreversible Charge Transfer (Case IV)

Potential[a]	$\sqrt{b}/K \sqrt{l}$						
	0.2	0.5	1.0	1.5	3.0	6.0	10.0
160	0.003	0.003	0.003	0.003	0.003	0.003	0.003
140	0.007	0.007	0.007	0.007	0.007	0.007	0.007
120	0.016	0.016	0.016	0.016	0.015	0.015	0.014
110	0.024	0.024	0.023	0.023	0.022	0.021	0.019
100	0.035	0.034	0.034	0.033	0.031	0.029	0.026
90	0.050	0.049	0.048	0.047	0.044	0.039	0.033
80	0.070	0.070	0.067	0.065	0.059	0.050	0.042
70	0.102	0.099	0.094	0.090	0.079	0.063	0.050
60	0.140	0.134	0.126	0.117	0.100	0.076	0.058
50	0.190	0.179	0.164	0.151	0.122	0.088	0.065
40	0.248	0.230	0.205	0.185	0.143	0.099	0.070
35	0.280	0.257	0.226	0.201	0.152	0.103	0.072
30	0.312	0.282	0.244	0.216	0.161	0.107	0.074
25	0.343	0.307	0.263	0.230	0.168	0.110	0.076
20	0.370	0.330	0.279	0.241	0.174	0.112	0.077
15	0.395	0.349	0.292	0.251	0.179	0.115	0.078
10	0.414	0.364	0.302	0.260	0.183	0.116	0.079
5	0.430	0.375	0.310	0.265	0.186	0.117	0.079
0	0.440	0.382	0.315	0.269	0.188	0.118	0.080
−5	0.444	0.385	0.318	0.271	0.189	0.119	0.080
−10	0.443	0.386	0.318	0.272	0.189	0.119	0.080
15	0.438	0.383	0.317	0.271	0.189	0.119	0.080
20	0.430	0.378	0.314	0.269	0.189	0.119	0.080
25	0.419	0.371	0.310	0.267	0.188	0.119	0.080
30	0.407	0.362	0.306	0.263	0.187	0.118	0.080
−35	0.394	0.354	0.301	0.260	0.186	0.118	0.080
40	0.381	0.345	0.295	0.257	0.184	0.117	0.079
50	0.355	0.327	0.283	0.248	0.180	0.116	0.079
60	0.333	0.309	0.272	0.240	0.177	0.115	0.078
70	0.313	0.294	0.261	0.233	0.174	0.114	0.078
$E_{p/2}$, mv.	+44.2	47.3	51.4	54.5	62.2	71.9	82.2

[a] Potential scale is $(E - E°)\alpha n_a - (RT/F) \ln K/(1 + K) + (RT/F) \ln \sqrt{\pi Db/k_s}$

netic parameter $K \sqrt{a/l}$ also is large. Under these conditions Equation 66 reduces to

$$\chi(at) = \frac{1}{\sqrt{\pi}} \sum_{j=1}^{\infty} (-1)^{j+1} \frac{(\sqrt{\pi})^j}{\sqrt{(j-1)!}} \times$$

$$\exp\left[-\frac{jnF}{RT}\left(E - E_{1/2} - \frac{RT}{nF}\ln(1 + K) + \frac{RT}{nF}\ln K\sqrt{\pi a/l}\,\right)\right] \quad (80)$$

which is the same form as the irreversible series, Equation 48. Thus, a stationary electrode polarogram is obtained which is the same shape as an irreversible curve, $\chi(at)$ at the peak will equal 0.496 (the value of $\chi(bt)$ in Case II) and, using an approach similar to Equations 49 and 50, it can be shown that

$$E_p = E_{1/2} - (RT/nF)[0.780 + \ln K\sqrt{a/l} - \ln(1 + K)] \quad (81)$$

In these limiting cases, the peak potential will shift cathodic by about $60/n$ mv. for a tenfold increase in $K\sqrt{a/l}$.

Single Scan Method. Typical stationary electrode polarograms calculated from Equation 60 are shown in Figure 8 and the data are listed in Table IX. However, the cathodic current function $\chi(at)$ varies only from 0.446 to 0.496 with a variation of three orders of magnitude in $K\sqrt{a/l}$, and thus is not particularly useful for kinetic measurements.

The variation of peak potential with $K\sqrt{a/l}$, is probably more useful for characterizing the kinetic parameters from the cathodic scan. For small values of $K\sqrt{a/l}$, the curves approach reversible behavior (Equation 79) and the potential is independent of $K\sqrt{a/l}$. At the other limit (Equations 80 and 81) the peak potential shifts cathodically by about $60/n$ mv. for a tenfold increase in $K\sqrt{a/l}$. For intermediate values of $K\sqrt{a/l}$, the variation in peak potential can be obtained from Table IX, and a working curve can be constructed as in Figure 9. Extrapolation of the straight line segments of the data indicate that Equation 79 will hold whenever $K\sqrt{a/l}$ is less than about 0.05, and that Equation 80 (or 81) will hold when $K\sqrt{a/l}$ is larger than about 5.0.

Cyclic Triangular Wave Method. As expected from the mechanism, the anodic portion of the cyclic stationary electrode polarogram is very sensitive to the kinetic parameters (Figure 8). Unfortunately, the peak height also is a function of

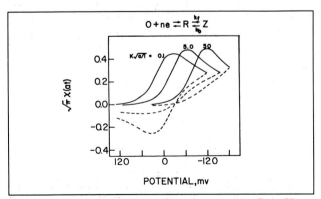

Figure 8. Stationary electrode polarograms, Case V
Potential scale is $(E - E_{1/2})n - (RT/F)\ln(1 + K)$

the switching potential, and thus it is not convenient to compile tables of $\chi(at)$ for the anodic portion of the scan, although such data are available on request (27). In order to use the anodic current for kinetic measurements, therefore, it was necessary to select arbitrary switching potentials and prepare working curves of the ratio i_p (anodic) $/i_p$(cathodic) as a function of $K\sqrt{a/l}$. The curve for a switching potential of $(E_\lambda - E_{1/2})n - (RT/F)\ln(1 + K)$ equal to -90 mv. is shown in Figure 10.

VI. CHARGE TRANSFER FOLLOWED BY AN IRREVERSIBLE CHEMICAL REACTION

As indicated above, an alternate approach to the theory of stationary electrode polarography for succeeding chemical reactions—which permits correlations for small values of l/a—involves considering the chemical reaction to be irreversible

$$O + ne \rightleftarrows R$$
$$R \xrightarrow{k_f} Z \qquad\qquad \text{VI}$$

Except for a brief presentation of a series solution (32) the theory of stationary electrode polarography for this case has not been presented previously.

Qualitatively, the behavior is very similar to Case V, except that neither the equilibrium constant nor k_b is included in the kinetic parameter. Thus, for small values of k_f/a, the chemical reaction has little effect, and a reversible stationary electrode polarogram is observed at its normal potential. This result can be obtained from the series solution, since when k_f/a is small Equation 67 reduces to Equation 34. This limiting case is precisely the one not included in the numerical calculations of Case V, but it is included here.

At the other limit, as k_f/a becomes large, Equation 67 reduces to

$$\chi(at) = \frac{1}{\sqrt{\pi}}\sum_{j=1}^{\infty}(-1)^{j+1}\frac{(\sqrt{\pi})^j}{\sqrt{(j-1)!}} \times$$
$$\exp\left[-\frac{jnF}{RT}\left(E - E_{1/2} - \frac{RT}{nF}\ln\sqrt{k_f/a\pi}\,\right)\right] \quad (82)$$

which is the same form as the series for the irreversible case (Equation 48). In this case it can be shown that

$$E_p = E_{1/2} - (RT/nF) \times (0.780 - \ln\sqrt{k_f/a}) \quad (83)$$

Thus, it can be seen that this limiting case is the same one as the third listed under Case V, except that it is approached from the opposite direction. For Equation 83 an increase in the kinetic parameter k_f/a causes an anodic shift in the wave, while for Equation 81 an increase in the kinetic parameter $K\sqrt{a/l}$ causes a cathodic shift in the wave. This is because the term $\ln(1 + K)$ in Equation 81 has already shifted the wave anodic by an amount corresponding to the maximum possible for complete equilibrium.

Table IX. Current Functions $\sqrt{\pi}\chi(at)$ for Charge Transfer Followed by a Reversible Chemical Reaction (Case V)

Potential[a]	$K\sqrt{a/l}$							
	0.1	0.25	0.55	1.0	2.0	3.25	5.0	10.0
120	0.008	0.007	0.006	0.005	0.003	0.002	0.002	0.001
100	0.018	0.016	0.013	0.010	0.007	0.005	0.003	0.002
80	0.037	0.034	0.028	0.021	0.015	0.010	0.007	0.004
60	0.079	0.070	0.057	0.046	0.031	0.022	0.016	0.009
50	0.108	0.097	0.082	0.066	0.045	0.032	0.023	0.013
45	0.128	0.115	0.097	0.079	0.054	0.039	0.028	0.016
40	0.149	0.135	0.114	0.093	0.065	0.047	0.034	0.019
35	0.173	0.157	0.134	0.110	0.077	0.056	0.041	0.022
30	0.200	0.182	0.158	0.129	0.092	0.068	0.049	0.028
25	0.227	0.210	0.183	0.152	0.110	0.081	0.059	0.033
20	0.256	0.239	0.210	0.177	0.129	0.096	0.070	0.040
15	0.286	0.269	0.240	0.205	0.151	0.114	0.084	0.048
10	0.316	0.301	0.272	0.235	0.177	0.135	0.099	0.058
5	0.342	0.330	0.303	0.265	0.205	0.159	0.118	0.069
0	0.372	0.359	0.334	0.300	0.236	0.184	0.141	0.083
−5	0.395	0.386	0.364	0.332	0.268	0.213	0.166	0.100
10	0.414	0.408	0.390	0.363	0.302	0.245	0.192	0.118
15	0.430	0.426	0.414	0.391	0.336	0.278	0.222	0.138
20	0.440	0.439	0.433	0.416	0.368	0.313	0.254	0.163
25	0.447	0.448	0.446	0.436	0.398	0.347	0.288	0.189
−30	0.449	0.452	0.455	0.452	0.424	0.380	0.322	0.220
35	0.447	0.452	0.459	0.461	0.446	0.410	0.357	0.252
40	0.443	0.449	0.458	0.465	0.461	0.436	0.392	0.287
45	0.435	0.442	0.453	0.464	0.470	0.455	0.420	0.323
50	0.426	0.433	0.446	0.459	0.474	0.469	0.445	0.358
−55	0.416	0.423	0.436	0.451	0.472	0.478	0.464	0.391
60	0.405	0.412	0.425	0.441	0.446	0.479	0.476	0.422
65	0.393	0.400	0.412	0.429	0.456	0.476	0.483	0.447
70	0.381	0.387	0.400	0.416	0.444	0.467	0.483	0.468
75	0.369	0.376	0.387	0.403	0.431	0.456	0.477	0.480
−80	0.357	0.364	0.375	0.389	0.417	0.443	0.468	0.487
85	0.346	0.352	0.362	0.376	0.402	0.429	0.455	0.487
90	0.336	0.341	0.351	0.363	0.389	0.414	0.440	0.482
95	0.326	0.330	0.339	0.351	0.375	0.399	0.425	0.471
100	0.316	0.320	0.328	0.340	0.362	0.383	0.409	0.458
−110			0.309	0.319	0.337	0.356	0.378	0.427
120				0.316	0.333	0.352	0.395	
130						0.312	0.329	0.366
140							0.308	0.340
E_p, mv.	−30.3	−32.4	−36.5	−41.6	−50.9	−59.1	−67.3	−82.7

[a] Potential scale is $(E - E_{1/2})\, n - (RT/F) \ln (1 + K)$.

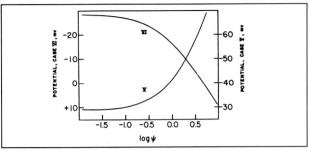

Figure 9. Variation of peak potential as a function of kinetic parameters

Case V: $\psi = K\sqrt{a/l}$; the potential scale is $(E_p - E_{1/2})n - (RT/F) \ln(1 + K)$

Case VI: $\psi = k_F/a$; the potential scale is $(E_p - E_{1/2})n$

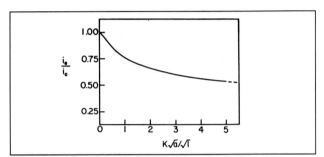

Figure 10. Ratio of anodic to cathodic peak current as a function of the kinetic parameters, Case V

Single Scan Method. Stationary electrode polarograms calculated from Equation 61 are shown in Figure 11, and the data are presented in Table X. As in Case V, the cathodic currents vary by only about 10% for a variation in k_f/a of about three orders of magnitude, and thus are not useful for kinetic measurements. Again, the variation in peak potential is probably more useful. For small values of k_f/a, the peak potential is independent of k_f/a, while for large values, an anodic shift of about $30/n$ mv. for a tenfold increase in k_f/a is observed. This behavior, and also the behavior for intermediate values of k_f/a are included in Figure 9, and the limits of applicability of Equations 34 and 82 (or 36 and 83) can be estimated as before.

Cyclic Triangular Wave Method. As in Case V, measurements on the anodic portion of the cyclic triangular wave are most suitable for kinetic measurements (Figure 11), but as before the anodic current function depends on the switching potential. In this case, however, an extremely simple way of handling the data was found. A large number of single cycle theoretical curves were calculated varying both k_f/a and the switching potential, and for a constant value of the parameter $k_f\tau$ (where τ is the time in seconds from $E_{1/2}$ to E_λ), the ratio of anodic to cathodic peak currents was found to be constant. (As before, the anodic peak current is measured to the extension of the cathodic curve.) Thus a working curve could be constructed for the ratio of peak currents i_a/i_c as a function of $k_f\tau$ (Figure 12) and, if $E_{1/2}$ is known, a rate constant can be calculated from a single cyclic curve.

By using faster scan rates (small k_f/a), $E_{1/2}$ also can be obtained experimentally. For accurate work, a large scale plot of Figure 12 would be required, and the data are presented in Table XI.

Qualitatively, from Figure 12, it is very difficult to measure the anodic scan for values of $k_f\tau$ much greater than 1.6. Conversely, a value of $k_f\tau$ much less than 0.02 cannot be distinguished from the reversible case. The method is very convenient, since the ratio of peak heights is independent of such experimental parameters as the electrode area and diffusion coefficient.

VII. CATALYTIC REACTION WITH REVERSIBLE CHARGE TRANSFER

For an irreversible catalytic reaction following a reversible charge transfer,

$$O + ne \rightleftarrows R$$
$$\overset{k'_f}{R + Z \rightarrow O} \qquad (VII)$$

the theory of stationary electrode polarography (single sweep method) has been treated by Saveant and Vianello (39) who also applied the method in an experimental study (38). Since additional correlations can be obtained by considering alternate approaches to the problem, this case will be reviewed briefly. The cyclic triangular wave method has not been treated previously.

Qualitatively, the effect of the chemical reaction on the cathodic portion of the cyclic wave would be an increase in the maximum current. This can be seen by considering the two limiting cases. First, if k_f/a is small, Equation 68 reduces directly to Equation 34, and a reversible stationary electrode polarogram is obtained. At the other limit, for large values of k_f/a, Equation 68 reduces to

$$\chi(at) = \frac{1}{\sqrt{\pi}} \sum_{j=1}^{\infty} (-1)^{j+1} \sqrt{k_f/a} \times$$

$$\exp\left[-\frac{jnF}{RT}(E - E_{1/2})\right] \qquad (83)$$

which indicates that the current is directly proportional to $\sqrt{k_f}$ and independent of the rate of voltage scan, since $\sqrt{a}$ is a coefficient in Equation 25.

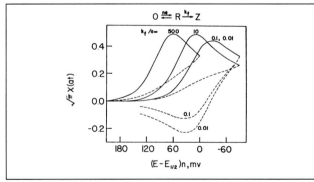

Figure 11. Stationary electrode polarograms, Case VI

Table X. Current Functions $\sqrt{\pi}\chi(at)$ for Charge Transfer Followed by an Irreversible Chemical Reaction (Case VI)

Potential[a]	k_f/a						
	0.05	0.2	0.5	1.0	1.6	4.0	10.0
160	0.003	0.003	0.003	0.003	0.003	0.004	0.006
150	0.003	0.003	0.004	0.004	0.005	0.006	0.009
140	0.004	0.005	0.005	0.006	0.007	0.009	0.020
130	0.006	0.007	0.008	0.009	0.010	0.014	0.020
120	0.009	0.010	0.011	0.013	0.015	0.020	0.030
110	0.014	0.015	0.016	0.019	0.022	0.030	0.044
100	0.020	0.021	0.024	0.027	0.032	0.043	0.062
90	0.029	0.031	0.035	0.040	0.046	0.062	0.090
80	0.042	0.045	0.050	0.057	0.065	0.089	0.126
70	0.161	0.065	0.072	0.083	0.093	0.124	0.175
65	0.073	0.077	0.085	0.098	0.110	0.147	0.203
60	0.086	0.092	0.101	0.116	0.130	0.173	0.234
55	0.102	0.108	0.120	0.136	0.153	0.200	0.268
50	0.120	0.128	0.141	0.160	0.177	0.230	0.303
45	0.140	0.148	0.163	0.185	0.206	0.263	0.338
40	0.163	0.172	0.189	0.213	0.236	0.297	0.373
35	0.188	0.199	0.218	0.244	0.269	0.332	0.406
30	0.216	0.227	0.248	0.276	0.302	0.367	0.435
25	0.244	0.256	0.278	0.308	0.335	0.398	0.459
20	0.273	0.288	0.310	0.340	0.367	0.427	0.477
15	0.303	0.317	0.341	0.371	0.396	0.451	0.488
10	0.333	0.347	0.371	0.400	0.423	0.469	0.491
5	0.359	0.374	0.396	0.424	0.445	0.480	0.489
0	0.384	0.398	0.419	0.443	0.461	0.486	0.481
−5	0.405	0.418	0.437	0.458	0.472	0.485	0.469
−10	0.423	0.434	0.451	0.467	0.476	0.480	0.455
15	0.435	0.445	0.459	0.471	0.476	0.469	0.439
20	0.444	0.452	0.462	0.470	0.471	0.456	0.422
25	0.448	0.454	0.461	0.464	0.461	0.440	0.405
30	0.448	0.452	0.457	0.456	0.450	0.424	0.389
−35	0.445	0.448	0.448	0.444	0.436	0.407	0.374
40	0.439	0.440	0.438	0.431	0.420	0.392	0.360
45	0.431	0.430	0.426	0.417	0.406	0.377	0.346
50	0.421	0.419	0.413	0.403	0.391	0.362	0.334
60	0.399	0.395	0.387	0.375	0.363	0.337	0.312
−70	0.375	0.369	0.361	0.349			
80	0.351	0.346	0.337	0.325			
90	0.330	0.325					
100	0.311	0.305					
E_p, mv.	−27.7	−25.2	−21.1	−16.4	−11.8	−1.5	+9.8

[a] Potential scale is $(E - E_{1/2})n$.

Equation 83 can be written in the form

$$i = \frac{nFA\sqrt{Dk_f}\,C_O^*}{1 + \exp\left[\dfrac{nF}{RT}(E - E_{1/2})\right]} \qquad (84)$$

which provides a closed form solution describing the entire wave. Under these conditions, no peak is obtained and, for large values of k_f/a, the potential at which exactly one-half of the catalytic limiting current flows is the polarographic $E_{1/2}$. In addition, for very cathodic potentials, Equation 84 reduces to a form derived by Saveant and Vianello (39)

$$i = nFAC_O^*\sqrt{Dk_f} \qquad (85)$$

which indicates, just as Equation 83, that the limiting current for large values of k_f/a is independent of scan rate.

Single Scan Method. Since Equation 62 is an Abel integral equation, the solution can be written directly in a manner similar to Case I, Equation 32. This is the approach used by Saveant and Vianello (39). Calculations carried out using the numerical method agreed exactly with the results of Saveant and Vianello (for the cathodic portion of the scan) and typical curves are shown in Figure 13. However, the values of $\chi(at)$ presented by Saveant and Vianello do not cover the rising portion and peak of the stationary electrode polarogram at small enough potential intervals to permit convenient calculation of accurate theoretical curves. Thus, additional data are presented in Table XII.

The cathodic current function $\chi(at)$ can be correlated with the kinetic parameter k_f/a by comparison of experimental and theoretical polarograms. However, it is probably more convenient to use a working curve in which the ratio of the catalytic peak current to the reversible peak current is plotted as a function of $(k_f/a)^{1/2}$ (Figure 14). For values of k_f/a larger than about 1.0, the plot is essentially linear, which defines the range of applicability of Equations 84 and 85. For values of k_f/a less than about 0.06, the peak ratio of i_k/i_d is fairly insensitive to changes in k_f/a. The method appears very convenient, but for accurate work with small values of k_f/a a large scale plot of Figure 14 would be required, and the necessary data can be obtained from Tables I and XII.

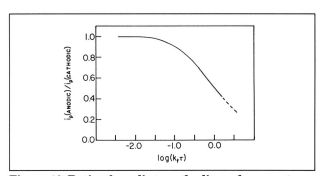

Figure 12. Ratio of anodic to cathodic peak current as a function of $k_f\tau$, Case VI

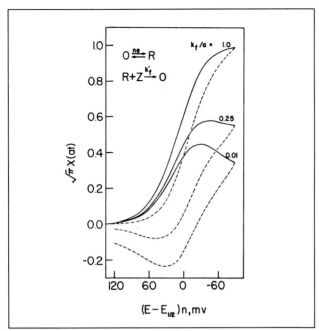

Figure 13. Stationary electrode polarograms, Case VII

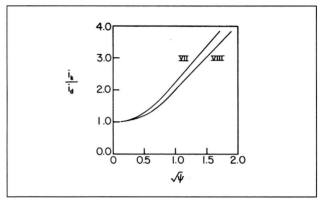

Figure 14. Ratio of kinetic peak current to diffusion controlled peak current

Case VII: $\psi = k_f/a$
Case VIII: $\psi = k_f/b$

The variation in peak potential with changes in rate of potential scan also can be used to obtain kinetic data. For low values of k_f/a, the peak potential is independent of k_f/a, and is constant at $28.5/n$ mv. cathodic of $E_{1/2}$ as in the reversible case. As k_f/a increases, the peak potential shifts cathodically by about $60/n$ mv. for a tenfold increase in k_f/a but, simultaneously, the peak becomes quite broad and, for values of k_f/a larger than about 1.0, no peak is observed. Thus, it is more useful to correlate half-peak potentials with k_f/a [as suggested by Saveant (38)] since this correlation can be extended to the region where no peak is observed. As predicted from Equation 84, when k_f/a is larger than about 10, the potential at which the current is equal to half of the limiting current is independent of variations of k_f/a, and is equal to the polarographic half-wave potential. The behavior for intermediate values of k_f/a is shown in Figure 15, and the data for an accurate plot can be obtained from Table XII.

Spherical Electrodes. For the catalytic case, Weber (51) was able to derive a spherical correction term in closed form, which is the same as the one derived by Reinmuth for the reversible case (35). Thus, the value of $\phi(at)$ listed in Table I can be used directly with the values of $\chi(at)$ in Table XII to calculate the current at a spherical electrode.

Cyclic Triangular Wave Method. Provided a switching potential is selected at least $35/n$ mv. cathodic of the peak potential, the anodic curve (as measured to the extension of the cathodic curve) is the same shape as the cathodic curve, independent of both the switching potential and k_f/a. The ratio of the cathodic peak current to the anodic peak current is unity, exactly as in the reversible case. At large values of k_f/a—e.g., larger than about 1.0—where no cathodic peak is observed, anodic peaks are not observed either and, on the anodic scan, the current simply returns to zero at potentials corresponding to the foot of the cathodic wave.

Since the anodic portion of the scan has exactly the same properties as the cathodic, no additional quantitative kinetic information can be obtained.

VIII. CATALYTIC REACTION WITH IRREVERSIBLE CHARGE TRANSFER

For catalytic systems in which both the charge transfer reaction and the chemical reaction are irreversible,

$$O + ne \xrightarrow{k} R$$
$$R + Z \xrightarrow{k'_f} O \tag{VIII}$$

the theory of stationary electrode polarography has not been considered previously. Qualitatively, one would expect that

Table XI. Ratio of Anodic to Cathodic Peak Currrents, Case VI

$k_f \tau$	i_a/i_c
0.004	1.00
0.023	0.986
0.035	0.967
0.066	0.937
0.105	0.900
0.195	0.828
0.350	0.727
0.525	0.641
0.550	0.628
0.778	0.551
1.050	0.486
1.168	0.466
1.557	0.415

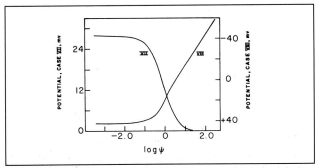

Figure 15. Variation of half-peak potential as a function of kinetic parameters

Case VII: $\psi = k_f/a$; the potential scale is $(E_{p/2} - E_{1/2})n$
Case VIII: $\psi = k_f/b$; the potential scale is $(E_{p/2} - E^\circ)\alpha n_a + (RTF)$ $\ln \sqrt{\pi Db/k_s}$

the polarograms would be similar to the catalytic case with reversible charge transfer, except that the curves would be lower and more spread out on the potential axis. In addition, no anodic current would be observed in a cyclic triangular wave experiment.

Again, two limiting cases can be considered. For small values of k_f/b, Equation 69 reduces to Equation 48 and an irreversible stationary electrode polarogram is obtained. For large values of k_f/b, Equation 69 reduces to

$$\chi(bt) = \frac{1}{\sqrt{\pi}} \sum_{j=1}^{\infty} (-1)^{j+1} (\sqrt{\pi})^j \sqrt{k_f/b} \times$$
$$\exp\left[-\frac{j\alpha n_a F}{RT}\left(E - E^\circ + \frac{RT}{\alpha n_a F} \times \right.\right.$$
$$\left.\left. \ln \frac{\sqrt{\pi Db}}{k_s} + \frac{RT}{\alpha n_a F} \ln \sqrt{\frac{k_f}{\pi a}} \right)\right] \quad (86)$$

and, analogous to Case VII under these conditions, the current is directly proportional to $\sqrt{k_f}$ and is independent of b.

Similarly, a closed form solution describing the entire wave can be obtained by rearranging Equation 86

$$i = nFAC_O^* \sqrt{Dl_f}/1 + \exp\left[\frac{\alpha n_a F}{RT}\left(E - E^\circ + \right.\right.$$
$$\left.\left. \frac{RT}{\alpha n_a F} \ln \frac{\sqrt{\pi Db}}{k_s} + \frac{RT}{\alpha n_a F} \ln \sqrt{\frac{k_s}{\pi a}} \right)\right] \quad (87)$$

Typical curves for several values of k_f/b are shown in Figure 16 and the data are listed in Table XIII. These curves qualitatively are similar to those of Figure 14 except that $\chi(bt)$, k_f/b, and the potential axis all depend on αn_a.

For quantitative characterization of the kinetics, the data can be treated in a manner similar to the catalytic reaction coupled to a reversible charge transfer. In this case, however, the ratio of the catalytic current to the irreversible current for

the system is the parameter used. The working curve for this case is included in Figure 14. As in the previous case, the ratio i_k/i_d is fairly insensitive to changes in k_f/b less than about 0.06, and the points required for a large scale plot of the working curve can be obtained from Tables III and XIII.

The variation of the half-peak potential with changes in k_f/b also can be used to characterize the system. A working curve for this method is shown in Figure 15, and the data can be obtained from Table XIII.

DIAGNOSTIC CRITERIA

In each of the kinetic cases, the effect of the chemical reaction will depend on its rate, as compared with the time required to perform the experiment. Taking Case V (irreversible succeeding chemical reaction) as an example, if a very rapid reaction is involved in experiments with very slow scan rates, the stationary electrode polarogram will reflect the characteristics of the chemical step almost entirely. On the other hand, if the rate of voltage scan is rapid compared to the rate of the reaction, the curves are identical to those for the corresponding uncomplicated charge transfer.

Similar observations can be extended to the other kinetic cases and are borne out quantitatively in the theory presented here—i.e., in every kinetic case, the ratio of the rate constant to the rate of voltage scan appears in the kinetic parameter. This, in turn, makes it possible to use these relations to define diagnostic criteria for the investigation of unknown systems.

Although there are innumerable correlations which could be made, perhaps those which are most useful experimentally are the variation—with changes in rate of voltage scan—of the cathodic peak current, the cathodic peak (or half-peak) potential, and the ratio of the anodic to cathodic peak currents. These diagnostic relations are most useful for qualitative characterization of unknown systems, since only trends in the experimental behavior are required. Therefore, the correlations in Figures 17, 18, and 19 have all been calculated for an arbitrary value of rate constant (and equilibrium constant, if necessary) so that the behavior could be compared readily on the same diagram.

Among the relations to be noted in Figure 17 is that by plotting the quantity $i_p/nFA\sqrt{Da}C_O^*$ as a function of the rate of voltage scan, the effect of v on the diffusion process can be separated from its effect on the kinetics. Thus, for the uncomplicated charge transfer reactions (Cases I and II) horizontal straight lines are obtained. The behavior for each kinetic case approaches one of these straight lines when the rate of voltage scan is such that the chemical reaction cannot proceed significantly before the experiment is over. Experimentally, this correlation is extremely easy to obtain, since it is only necessary to plot $i_p/v^{1/2}$ vs. v.

The rate at which the wave shifts along the potential axis as the rate of voltage scan is varied (Figure 18) is also useful for investigation of unknown systems. Because several of the cases do not exhibit peaks under some conditions, the half-peak potential was used in this correlation. However, these curves apply as well to the peak potential for Cases I, II, V, and VI.

Table XII. Current Functions $\sqrt{\pi}\chi(at)$ for a Catalytic Reaction with Reversible Charge Transfer (Case VII)

Potential[a]	k_f/a								
	0.04	0.1	0.2	0.4	0.6	1.0	1.78	3.16	10.0
120	0.009	0.010	0.010	0.011	0.012	0.013	0.015	0.019	0.030
100	0.020	0.021	0.021	0.023	0.025	0.028	0.033	0.040	0.066
80	0.042	0.043	0.045	0.049	0.052	0.059	0.069	0.086	0.139
60	0.086	0.088	0.093	0.100	0.108	0.121	0.144	0.176	0.289
50	0.120	0.123	0.129	0.140	0.150	0.170	0.201	0.249	0.409
45	0.140	0.145	0.152	0.165	0.178	0.201	0.239	0.294	0.482
40	0.163	0.168	0.177	0.193	0.207	0.234	0.279	0.345	0.567
35	0.189	0.195	0.205	0.224	0.242	0.273	0.326	0.403	0.665
30	0.216	0.224	0.236	0.258	0.278	0.315	0.378	0.467	0.773
25	0.245	0.254	0.267	0.294	0.318	0.361	0.433	0.539	0.894
20	0.275	0.285	0.301	0.331	0.359	0.409	0.493	0.614	1.022
15	0.306	0.318	0.337	0.371	0.403	0.461	0.558	0.695	1.162
10	0.336	0.349	0.370	0.410	0.447	0.512	0.623	0.782	1.310
5	0.364	0.380	0.404	0.449	0.491	0.566	0.690	0.867	1.459
0	0.391	0.408	0.436	0.487	0.534	0.617	0.756	0.955	1.614
−5	0.414	0.434	0.465	0.522	0.574	0.668	0.821	1.042	1.769
10	0.432	0.455	0.489	0.552	0.611	0.715	0.883	1.124	1.919
15	0.448	0.472	0.510	0.580	0.644	0.757	0.942	1.204	2.061
20	0.459	0.485	0.527	0.604	0.673	0.796	0.996	1.278	2.197
25	0.465	0.494	0.540	0.622	0.697	0.829	1.044	1.345	2.322
−30	0.468	0.499	0.548	0.638	0.719	0.861	1.088	1.046	2.436
35	0.467	0.500	0.553	0.649	0.735	0.885	1.126	1.462	2.540
40	0.463	0.499	0.556	0.658	0.749	0.907	1.159	1.510	2.633
45	0.457	0.495	0.555	0.663	0.759	0.924	1.188	1.552	2.713
50	0.450	0.490	0.553	0.666	0.766	0.939	1.211	1.587	2.782
−60	0.431	0.476	0.545	0.668	0.776	0.961	1.250	1.644	2.894
80	0.390	0.442	0.522	0.662	0.782	0.984	1.295	1.715	3.034
100	0.354	0.413	0.502	0.653	0.781	0.994	1.315	1.749	3.102
120	0.326	0.390	0.486	0.646	0.779	0.997	1.326	1.765	3.134
140	0.305	0.374	0.474	0.641	0.777	0.999	1.330	1.772	3.149
$E_{p/2}$, mv.	+27.3	25.7	23.5	19.6	16.3	11.2	6.7	3.8	1.0

[a] Potential scale is $(E - E_{1/2})n$.

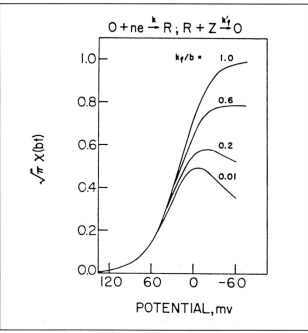

Figure 16. Stationary electrode polarograms, Case VIII

Potential scale is $(E - E°)\alpha n_a + (RT/F) \ln \sqrt{\pi D b/k_3}$

versible charge transfer reaction is involved, it is always possible to obtain αn_a from the shift in peak potential, provided the rate of voltage scan is in the proper range. For reversible charge transfer, other correlations are possible involving variation in anodic peak current or peak potential with switching potential. However, these correlations can be defined more easily in connection with experimental results on a particular kinetic case. Several studies involving applications of the theory presented here are now in progress, and these, with the additional correlations, will be presented in the future.

Although various electrochemical methods have been developed recently which can be used for similar measurements on kinetic systems, stationary electrode polarography appears to be particularly convenient to use. From the experimental point of view, the time scale of the experiment can be varied from conventional polarographic rates of voltage scan of a few millivolts per second (where convection ultimately sets the lower limit of scan rate) to perhaps several thousand volts per second (where charging current and adsorption phenomena become important). Furthermore, this extremely wide range of scan rates can be obtained with ease if instruments based on operational amplifiers are available (42, 49). Thus, the use of the diagnostic criteria is particularly simple, and stationary electrode polarography is a very powerful method for studying electrochemical kinetics.

ACKNOWLEDGMENT

The numerical calculations were carried out at the Midwest Universities Research Association, and the help of M. R. Storm in making available the IBM 704 computer is gratefully acknowledged. During the academic year 1962-63, R. S. Nicholson held the Dow Chemical Fellowship at the University of Wisconsin.

In Figure 19, which is relevant only for reversible charge transfers, the ratio of anodic peak current to the cathodic peak current is unity for the reversible (Case I) and catalytic (Case VII) systems only, and this serves as a quick test for the presence of kinetic complications.

In some of the individual cases, additional experimental correlations are particularly useful. For example, if an irre-

Table XIII. Current Functions $\sqrt{\pi}\chi(bt)$ for a Catalytic Reaction with Irreversible Charge Transfer (Case VIII)

Potential[a]	$k_f\,b$								
	0.04	0.1	0.2	0.4	0.6	1.0	1.78	3.16	10.0
160	0.004	0.004	0.004	0.004	0.004	0.004	0.004	0.004	0.004
140	0.008	0.008	0.008	0.008	0.008	0.008	0.008	0.008	0.008
120	0.016	0.016	0.016	0.016	0.016	0.016	0.016	0.016	0.016
110	0.024	0.024	0.024	0.024	0.024	0.024	0.024	0.024	0.024
100	0.035	0.035	0.035	0.035	0.035	0.035	0.035	0.036	0.036
90	0.050	0.050	0.051	0.051	0.051	0.051	0.051	0.052	0.052
80	0.072	0.073	0.073	0.073	0.074	0.074	0.075	0.075	0.076
70	0.104	0.105	0.105	0.106	0.107	0.108	0.109	0.110	0.113
60	0.145	0.147	0.147	0.148	0.150	0.152	0.155	0.157	0.162
50	0.198	0.200	0.201	0.205	0.208	0.213	0.218	0.224	0.234
40	0.264	0.267	0.271	0.278	0.283	0.291	0.302	0.313	0.334
35	0.301	0.305	0.311	0.320	0.327	0.339	0.354	0.370	0.399
30	0.334	0.344	0.349	0.362	0.372	0.388	0.409	0.430	0.471
25	0.376	0.383	0.392	0.408	0.422	0.444	0.473	0.500	0.558
20	0.410	0.418	0.431	0.453	0.470	0.498	0.534	0.574	0.650
15	0.443	0.454	0.470	0.500	0.521	0.556	0.600	0.656	0.761
10	0.469	0.483	0.503	0.538	0.567	0.612	0.673	0.740	0.876
5	0.490	0.506	0.532	0.575	0.610	0.666	0.742	0.827	1.002
0	0.504	0.524	0.555	0.608	0.651	0.719	0.813	0.920	1.145
−5	0.511	0.534	0.571	0.633	0.685	0.766	0.878	1.007	1.288
−10	0.511	0.539	0.581	0.653	0.713	0.809	0.941	1.097	1.443
15	0.506	0.538	0.586	0.667	0.735	0.844	0.998	1.178	1.595
20	0.497	0.532	0.585	0.676	0.753	0.875	1.049	1.258	1.755
25	0.485	0.523	0.581	0.681	0.765	0.900	1.094	1.328	1.903
30	0.470	0.512	0.575	0.683	0.774	0.921	1.133	1.394	2.053
−35	0.456	0.500	0.568	0.683	0.780	0.937	1.166	1.450	2.186
40	0.440	0.487	0.559	0.681	0.783	9.951	1.195	1.502	2.317
50	0.411	0.463	0.541	0.674	0.786	0.969	1.238	1.583	2.538
60	0.386	0.442	0.525	0.667	0.786	0.980	1.269	1.643	2.710
70	0.366	0.425	0.512	0.661	0.785	0.988	1.289	1.684	2.840
−80	0.348	0.409	0.501	0.655	0.783	0.992	1.303	1.714	2.936
100	0.320	0.386	0.484	0.646	0.780	0.997	1.320	1.748	3.056
120	0.300	0.371	0.473	0.641	0.778	0.999	1.328	1.764	3.112
140			0.466	0.638	0.776	0.999	1.331	1.772	3.139
160			0.461	0.636	0.776	1.000	1.332	1.776	3.152
$E_{p/2}$ mv.	+41.1	39.6	37.0	32.4	27.7	19.8	10.3	1.3	−14.4

[a] Potential scale is $(E - E^\circ)\,\alpha n_a + (RT/F)\ln\sqrt{\pi Db/ks}$.

LITERATURE CITED

(1) Apostol, T. M., "Mathematical Analysis," p. 200, Addison Wesley, Reading, Mass., 1957.
(2) Berzins, T., Delahay, P., *J. Am. Chem. Soc.* **75,** 555 (1953).
(3) Buck, R. P., *Anal. Chem.* **36,** 947 (1964).
(4) Delahay, P., *J. Am. Chem. Soc.* **75,** 1190 (1953).
(5) Delahay, P., "New Instrumental Methods in Electrochemistry," Chap. 3, Interscience, New York, 1954.
(6) *Ibid.*, Chap. 6.
(7) DeMars, R. D., Shain, I., *Anal. Chem.* **29,** 1825 (1957).
(8) DeMars, R. D., Shain, I., *J. Am. Chem. Soc.* **81,** 2654 (1959).
(9) Frankenthal, R. P., Shain, I., *Ibid.*, **78,** 2969 (1956).

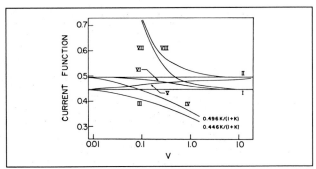

Figure 17. Variation of peak current functions with rate of voltage scan

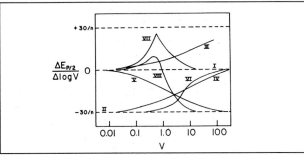

Figure 18. Rate of shift of potential as a function of scan rate

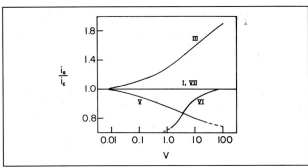

Figure 19. Ratio of anodic to cathodic peak currents as a function of rate of voltage scan

(10) Galus, Z., Adams, R. N., *J. Phys. Chem.* **67**, 862 (1963).
(11) Galus, Z., Lee, H. Y., Adams, R. N., *J. Electroanal. Chem.* **5**, 17 (1963).
(12) Gokhshtein, A. Y., Gokhshtein, Y. P., *Doklady Akad. Nauk SSSR* **131**, 601 (1960).
(13) Gokhshtein, Y. P., *Ibid.*, **126**, 598 (1959).
(14) Gokhshtein, Y. P., Gokhshtein, A. Y., in "Advances in Polarography, " I. S. Longmuir, ed., **Vol. II,** p. 465, Pergamon Press, New York, 1960.
(15) Gokhshtein, Y. P., Gokhshtein, A. Y., *Doklady Akad. Nauk SSSR* **128,** 985 (1959).
(16) Kemula, W., in "Advances in Polarography," I. S. Longmuir, ed., **Vol. I,** p. 105, Pergamon Press, New York, 1960.

(17) Kemula, W., Kublik, Z., *Anal. Chim. Acta* **18,** 104 (1958).
(18) Kemula, W., Kublik, Z., *Bull. Acad. Polon. Sci., Ser. Sci. Chim., Geol. et Geograph.* **6,** 661 (1958).
(19) Koutecky, J., *Collection Czechoslov. Chem. Communs.* **21,** 433 (1956).
(20) Koutecky, J., Brdicka, R., *Ibid.,* **12, 337** (1947).
(21) Matheson, L. A., Nichols, N., *Trans. Electrochem. Soc.* **73,** 193 (1938).
(22) Matsuda, H., *Z. Elektrochem.* **61,** 489 (1957).
(23) Matsuda, H., Ayabe, Y., *Ibid.,* **59,** 494 (1955).
(24) Mueller, T. R., Adams, R. N., *Anal. Chim. Acta* **25,** 482 (1961).
(25) Nicholson, M. M., *J. Am. Chem. Soc.* **76,** 2539 (1954).
(26) Nicholson, M. M., *Ibid.,* **79,** 7 (1957).
(27) Nicholson, R. S., Ph.D. Thesis, University of Wisconsin, 1964.
(28) Papoff, P., *J. Am. Chem. Soc.* **81,** 3254 (1959).
(29) Randles, J.E.B., *Trans. Faraday Soc.* **44,** 327 (1948).
(30) Reinmuth, W. H., *Anal. Chem.* **32,** 1891 (1960).
(31) Reinmuth, W. H., *Ibid.,* **33,** 185 (1961).
(32) *Ibid.,* p. 1793.
(33) Reinmuth, W. H., *Ibid.,* **34,** 1446 (1962).
(34) Reinmuth, W. H., Columbia University, New York, N. Y., unpublished data, 1964.
(35) Reinmuth, W. H., *J. Am. Chem. Soc.* **79,** 6358 (1957).
(36) Riha, J., in "Progress in Polarography," P. Zuman, ed., **Vol. II,** Chap. 17, Interscience, New York, 1962.
(37) Ross, J. W., DeMars, R. D., Shain, I., *Anal. Chem.* **28,** 1768 (1956).
(38) Saveant, J. M., Ecole Normale Superieure, Paris, private communication, 1963.
(39) Saveant, J. M., Vianello, E., in "Advances in Polarography," I. S. Longmuir, ed., **Vol. I,** p. 367, Pergamon Press, New York, 1960.
(40) Saveant, J. M., Vianello, E., *Compt. Rend.* **256,** 2597 (1963).
(41) Saveant, J. M., Vianello, E., *Electrochim. Acta* **8,** 905 (1963).
(42) Schwarz, W. M., Shain, I., *Anal. Chem.* **35,** 1770 (1963).
(43) Schwarz, W. M., Shain, I., Division of Physical Chemistry, 142nd National Meeting, ACS, Atlantic City, N. J., September 1962.
(44) Sevcik, A., *Collection Czechoslov. Chem. Communs.* **13,** 349 (1948).
(45) Shain, I., in "Treatise on Analytical Chemistry," Kolthoff and Elving, eds., **Part I,** Sec. D-2, Chap. 50, Interscience, New York, 1963.
(46) Shain, I., Southeast Regional Meeting, ACS, New Orleans, December 1961.
(47) Smutek, M., *Collection Czechoslov. Chem. Communs.* **20,** 247 (1955).
(48) Tricomi, F. G., "Integral Equations," p. 39, Interscience, New York, 1957.
(49) Underkofler, W. L., Shain, I., *Anal. Chem.* **35,** 1778 (1963).
(50) Vogel, J., in "Progress in Polarography," P. Zuman, ed., **Vol. II,** Chap. 20, Interscience, New York, 1962.
(51) Weber, J., *Collection Czechoslov. Chem. Communs.* **24,** 1770 (1959).

Received for review November 5, 1963. Accepted December 30, 1963. This work was supported in part by funds received from the U. S. Atomic Energy Commission under Contract No. AT(11-1)-1083. Other support was received from the National Science Foundation under Grant No. G15741.

Reprinted from *Anal. Chem.* **1964,** *36,* 706–23.

Savitsky and Golay reported an engagingly simple yet robust method for smoothing noisy data. Their new method for S/N enhancement, one of the first applications of computer-based data analysis in analytical chemistry, was quickly embraced by researchers in this and many other fields. Sometime later it was discovered that the tables contained many errors, and corrected tables are provided in later papers (*Anal. Chem.* **1972**, *44*, 1906, and *Anal. Chem.* **1978**, *50*, 1383). The Savitsky–Golay filter continues to be used extensively today.

Steven D. Brown
University of Delaware

Smoothing and Differentiation of Data by Simplified Least Squares Procedures

ABRAHAM SAVITZKY and MARCEL J. E. GOLAY
The Perkin-Elmer Corp., Norwalk, Conn.

In attempting to analyze, on digital computers, data from basically continuous physical experiments, numerical methods of performing familiar operations must be developed. The operations of differentiation and filtering are especially important both as an end in themselves, and as a prelude to further treatment of the data. Numerical counterparts of analog devices that perform these operations, such as *RC* filters, are often considered. However, the method of least squares may be used without additional computational complexity and with considerable improvement in the information obtained. The least squares calculations may be carried out in the computer by convolution of the data points with properly chosen sets of integers. These sets of integers and their normalizing factors are described and their use is illustrated in spectroscopic applications. The computer programs required are relatively simple. Two examples are presented as subroutines in the FORTRAN language.

The primary output of any experiment in which quantitative information is to be extracted is information which measures the phenomenon under observation. Superimposed upon and indistinguishable from this information are random errors which, regardless of their source, are characteristically described as noise. Of fundamental importance to the experimenter is the removal of as much of this noise as possible without, at the same time, unduly degrading the underlying information.

In much experimental work, the information may be obtained in the form of a two-column table of numbers, *A vs. B*. Such a table is typically the result of digitizing a spectrum or digitizing other kinds of results obtained during the course of an experiment. If plotted, this table of numbers would give the familiar graphs of %T *vs.* wavelength, pH *vs.* volume of titrant, polarographic current *vs.* applied voltage, NMR or ESR spectrum, or chromatographic elution curve, etc. This paper is concerned with computational methods for the removal of the random noise from such information, and with the simple evaluation of the first few derivatives of the information with respect to the graph abscissa.

The bases for the methods to be discussed have been reported previously, mostly in the mathematical literature (*4, 6, 8, 9*). The objective here is to present specific methods for handling current problems in the processing of such tables of analytical data. The methods apply as well to the desk calculator, or to simple paper and pencil operations for small amounts of data, as they do to the digital computer for large amounts of data, since their major utility is to simplify and speed up the processing of data.

There are two important restrictions on the way in which the points in the table may be obtained. First, the points must be at a fixed, uniform interval in the chosen abscissa. If the independent variable is time, as in chromatography or NMR spectra with linear time sweep, each data point must obtained at the same time interval from each preceding point. If it is a spectrum, the intervals may be every drum division or every 0.1 wavenumber, etc. Second, the curves formed by graphing the points must be continuous and more or less smooth—as in the various examples listed above.

ALTERNATIVE METHODS

One of the simplest ways to smooth fluctuating data is by a moving average. In this procedure one takes a fixed number of points, adds their ordinates together, and divides by the number of points to obtain the average ordinate at the center abscissa of the group. Next, the point at one end of the group is dropped, the next point at the other end added, and the process is repeated.

Figure 1 illustrates how the moving average might be obtained. While there is a much simpler way to compute the moving average than the particular one described, the following description is correct and can be extended to more sophisticated methods as will be seen shortly. This description is based on the concept of a convolute and of a convolution function. The set of numbers at the right are the data or ordinate values, those at the left, the abscissa information. The outlined block in the center may be considered to be a separate sheet of paper on which are written a new set of abscissa numbers, ranging from -2 thru zero to $+2$. The C's at the right represent the convoluting integers. For the moving average each C is numerically equal to one. To perform a convolution of the ordinate numbers in the table of data with a set of convoluting integers, C_i, each number in the block is multiplied by the corresponding number in the table of data, the resulting products are added, and this sum is divided by five. The set of ones is the convoluting function, and the number by which we divide, in this case, 5, is the normalizing factor. To get the next point in the moving average, the center block is slid down one line and the process repeated.

The concept of convolution can be generalized beyond the simple moving average. In the general case the C's represent any set of convoluting integers. There is an associated normalizing or scaling factor. The procedure is to multiply C_{-2} times the number opposite it, then C_{-1} by its number, etc., sum the results, divide by the normalizing factor, if appropriate, and the result is the desired function evaluated at the point indicated by C_0. For the next point, we move the set of convoluting integers down and repeat, etc. The mathematical description of this process is:

$$Y_j^* = \frac{\sum\limits_{i=-m}^{i=m} C_i Y_{j+i}}{N}$$

The index j represents the running index of the ordinate data in the original data table.

For the moving average, each C_i is equal to one and N is the number of convoluting integers. However, for many types of data, the set of all 1's, which yields the average, is not particularly useful. For example, on going through a sharp peak, the average would tend to degrade the end of the peak. There are other types of smoothing functions which might be used, and a few of these are indicated in Figure 2.

Figure 2A illustrates the set where all values have the same weight over the interval—essentially the moving average.

The function in Figure 2B is an exponential set which simulates the familiar RC analog time constant—i.e., the most recent point is given the greatest weight, and each preceding point gets a lesser weight determined by the law of exponential decay. *Future* points have no influence. Such a function treats *future* and *past* points differently and so will obviously

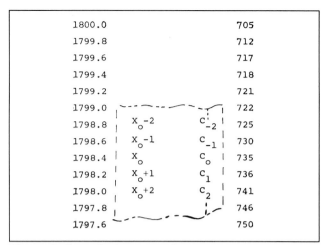

Figure 1. Convolution operation

Abscissa points at left, tabular data at right. In box area the convolution integers, c_i. Operation is the multiplication of the data points by the corresponding C_i, summation of the resulting products, and division by a normalizer, resulting in a single convolute at the point X_0. The box is then moved down one line, and the process repeated.

introduce a unidirectional distortion into the numerical results, as does the RC filter in an actual instrument.

When dealing with sets of numbers in hand, and not an actual run on an instrument where the data is emerging in serial order, it is possible to look ahead as well as behind. Then we can convolute with a function that treats *past* and *future* on an equal basis, such as the function in Figure 2D. Here the most weight is given to the central point, and points on either side of the center are symmetrically weighed exponentially. This function acts like an idealized lead-lag network, which is not practical to make with resistors, capacitors, and so on.

The usual spectrum from a spectrophotometer is the resultant of two convolutions of the actual spectrum of the material, first with a function representing the slit function of the instrument, which is much like the triangular convolute shown in Figure 2C, and then this first convolute spectrum is further convoluted with a function representing the time constant of the instrument. The triangular convoluting function could in many cases yield results not significantly different from the symmetrical exponential function.

Figure 3 illustrates the way in which each of these functions would act on a typical set of spectroscopic data. Curve 3A is replotted directly from the instrumental data. It is a single sharp band recorded under conditions which yield a reasonable noise level. The isolated point just to the right of the band has the value of 666 on the scale of zero to 1000 corresponding to approximately 0 to 100% transmittance. This point is introduced to illustrate the effect on these operations of a single point which has a gross error. The numbers along the bottom are the digital value at the lowest point of the plot, and one may consider that the peak goes down to 34.2% transmittance. The base line at the top is at about 79%.

Curve 3B is a nine-point moving average of the data. As expected, the peak is considerably shortened by this process. Especially interesting is the step introduced by the isolated error. In effect, it has the shape of the boxlike convolute in

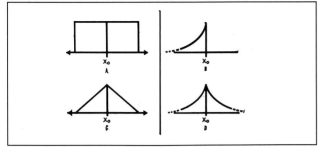

Figure 2. Various convolute functions

A. Moving average. B. Exponential function. C. Symmetrical triangular function, representing idealized spectrometer slit function. D. Symmetrical exponential function

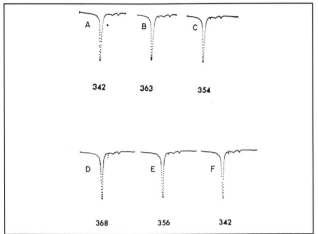

Figure 3. Spectral band convoluted by the various 9-point functions

The number at the bottom of each peak refers to the lowest recorded point and is a measure of the ability to retain the shape of the peak A. Raw data with single isolated error point. B. Moving average. C. Triangular function. D. Normal exponential function. E. Symmetrical exponential function. F. Least squares smoothing function

Figure 2*A*, which is exactly what one would expect from the convoluting process (*3*).

Curve 3*C* is for a triangular function which obviously forces both the peak itself and the isolated error into a triangular mold.

Curve 3*D* is the result of convoluting with the numerical equivalent of a conventional *RC* exponential time constant filter using only five points. The peak is not only shortened, but is also shifted to the right by one data point, or 0.002 micron and the isolated data point is asymmetric in the same manner. The convolution with a symmetrical lead-lag exponential, as in Figure 3*E*, does not distort the peak but does still reduce its intensity.

Note that while all of these functions have had the desired effect of reducing the noise level, they are clearly undesirable because of the accompanying degradation of the peak intensity.

METHOD OF LEAST SQUARES

The convoluting functions discussed so far are rather simple and do not extract as much information as is possible. The experimenter, if presented with a plot of the data points, would tend to draw through these points a line which best fits them. Numerically, this can also be done, provided one can adequately define what is meant by best fit. The most common criterion is that of least squares which may be simply stated as follows:

A set of points is to be fitted to some curve—for example, the curve $a_3x^3 + a_2x^2 + a_1x + a_0 = y$. The *a*'s are to be selected such that when each abscissa point is substituted into this equation, the square of the differences between the computed numbers, *y*, and the observed numbers is a minimum for the total of the observations used in determining the coefficients. All of the error is assumed to be in the ordinate and none in the abscissa.

Consider the block of seven data points enclosed by the left bracket in Figure 4. If these fall along a curve that can be described approximately by the equation shown, then there are specific procedures—which are described in most books on numerical analysis—to find the *a*'s. One then substitutes back into the resulting equation the abscissa at the central point indicated by the circle. The value which is obtained by this procedure is the best value at that point based on the least squares criterion, on the function which was chosen, and on the group of points examined.

This procedure can be repeated for each group of seven points, dropping one at the left and picking up one at the right each time. A somewhat later block is indicated at the right. In the usual case, there is found a different set of coefficients for each group of seven points. Even with a high-speed computer this is a tedious proposition at best.

Note, however, that finding the *a* coefficients is required only as a means for determining the final best value at just one point, the central point of the set. A careful study of the least squares procedure using these constraints, leads to the derivation of a set of integers which provide a weighting function. With this set of integers the central point can be evaluated by the convoluting procedure discussed above. This procedure is exactly equivalent to the least squares. It is not approximate.

The derivation is presented in Appendix I. For either a cubic or a quadratic function, the set of integers is the same, and

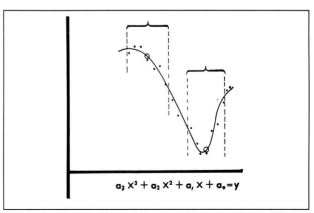

Figure 4. Representation of a 7-point moving polynomial smooth

the set for up to 25 points is shown in Table I of Appendix II with the appropriate normalizing factors. A most instructive exercise is to tabulate a simple function such as $y = x^3$ over any interval, apply these smoothing convolutes and compare these new values with the original. The answers will be found to be exact.

In Figure 3F this least squares convoluting procedure has been applied to the data of Figure 3A, using a 9-point cubic convolute. The value at the peak and the shape of the peak are essentially undistorted. As always, the isolated point assumes the shape of the convoluting function. The FORTRAN language computer program for performing this operation is presented in Program I of Appendix III.

Going beyond simple curve fitting, one can find in the literature on numerical analysis a variety of least squares procedures for determining the first derivative. These procedures are usually based on interpolation formulas and are for data at any arbitrary interval. Again, if we restrict ourselves to evaluating the function only at the center point of a set of equally spaced observations, then there exist sets of convoluting integers for the first derivative as well. (These actually evaluate the derivative of the least squares best function.)

A complete set of tables for derivatives up to the fifth order for polynomials up to the fifth degree, using from 5 to 25 points, is presented in Appendix II. These are more than adequate for most work, since, if the points are taken sufficiently close together, then practically any smooth curve will look more or less like a quadratic in the vicinity of a peak, or like a cubic in the vicinity of a shoulder. More complete tables can be found in the statistical literature (2, 4, 6, 9). Program II of Appendix III shows the use of these tables to obtain the coefficients of a polynomial for finding the precise center of an infrared band.

The shapes of the 9-point convolutes for a few of the functions are illustrated in Figure 5. Of special interest is the linear relation of the first derivative convolute for a quadratic. This is quite unique operationally because in processing a table of data, only one multiplication is necessary for each convolution. The remainder of the points are found from the set calculated for the previous point by simple subtraction. In Figure 6B is shown the first derivative of the spectrum in Figure 6A, obtained using a 9-point convolute.

The derivatives are useful in cases such as our methods of band finding on a computer (7), in studies of derivative spec-

tra, in derivative thermogravimetric analysis, derivative polarography, etc.

CONCLUSIONS

With the increase in the application of computers to the analysis of digitized data, the convolution methods described are certain to gain wider usage. With these methods, the sole function of the computer is to act as a filter to smooth the noise fluctuations and hopefully to introduce no distortions into the recorded data (3).

This problem of distortion is difficult to assess. In any of the curves of Figure 3, there remain small fluctuations in the data. Are these fluctuations real, or, as is more likely, are they just a low frequency component of the noise level which could not be smoothed? The question cannot be answered by taking just the data from a single run. However, if one were to take more than one run, average these and then smooth, or smooth and then average, the computer-plus-instrument system could decide, since even low frequency noise will not recur in exactly the same place in different runs. Computer time is most efficiently used if the averaging is done prior to smoothing.

Recent work (1) has shown the utility of simple averaging of a large number of runs in the enhancement of signal-to-noise ratios. The use of combined smoothing and averaging can considerably reduce the instrument time required, throwing the burden onto the computer which operates in a wholly different time domain. A characteristic of both procedures is that the noise is reduced approximately as the square root of the number of points used. This is illustrated for the smoothing case by Figure 7. At the upper left is the raw data, at the upper right a 5-point smooth, lower left 9 points, and lower right 17 points. If it is desired to improve the signal-to-noise ratio by a factor of 10, simple averaging would require a total of 100 runs. Similar improvement could be achieved by making only 16 runs plus a 9-point least squares smooth (average of 16 runs $\cong 4 \times$ improvement, and 9-point smooth $\cong 3 \times$ improvement) or only 4 runs plus a 25-point smooth (average of 4 runs $\cong 2 \times$ improvement plus 25-point smooth $\cong 5 \times$ improvement). The distribution between the number of runs required and the number of points which may be used for the smoothing is a function of the experimental curve under examination. The minimum distortion will occur when the polynomial accurately describes the analytical data, and will deviate as the polynomial departs from the true curve. The best results are obtained when the data are digitized at high densities—i.e., points very close together—and the number of points used in the convolute is chosen to be small enough so that no more than one inflection in the observed data is included in any convolution interval. Our results should be compared with those achievable using conventional instrument filtering. In a sense, we are substituting idealized filters and filter-networks for electronic hardware such as resistors, condensers, servos, etc. If one examines the time relationships, it probably takes longer to get the information using the digitizer and then the computer than with the analogous electronic network. There is, very definitely, the advantage in

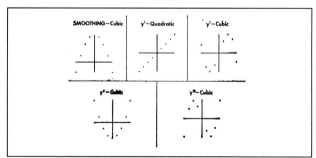

Figure 5. Nine-point convoluting functions (orthogonal polynomial) for smoothing and first, second, and third derivative

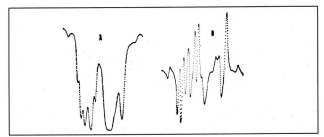

Figure 6. Seventeen-point first derivative convolute

A. Original spectrum. *B.* First derivative spectrum

the computer of being able to vary the processing completely unfettered by the practical restriction of real circuitry and servo loops. Note too, that this processing can be done after the fact of data collection, and indeed several different procedures may be applied in order to assess the optimum. This is a real advantage in itself, and provides ample justification for use of the computer solely as a noise filter. However, the greatest utility of these methods comes in the pretreatment of data to be further processed, as in our bandfinding procedures, in any curvefitting operations, in quantitative analyses, etc.

This type of data processing, so far as computers are concerned, requires a relatively small amount of programming and relatively little use of the computer memory or of the computer's processing capability. Therefore, even accounting-type computers, such as the IBM 1401 can be used to process data in this way. Furthermore, on such computers there is generally a high-speed line printer which can be turned into a relatively crude point plotter. On each line an X can be placed at the appropriate position to 1% of value, and the actual value is printed at the edge of the paper. The rate can be on the order of 10 lines, or points, per second.

ACKNOWLEDGMENT

The authors appreciate the assistance of Harrison J. M. Kinyua in the recompilation and checking of the convolute tables, of Robert Bernard of the Perkin-Elmer Scientific Computing Facility for his advice and assistance in program development, and of Joel Stutman, University of Maryland, for his advice on Appendix I.

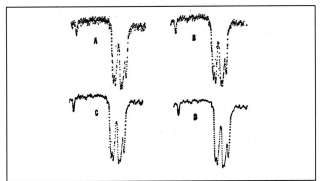

Figure 7. Square root relation between number of points and degree of smoothing

A. Raw data. *B.* Five-point smooth. *C.* Nine-point smooth. *D.* Seventeen-point smooth

APPENDIX I

The general problem is formulated as follows:

A set of $2m + 1$ consecutive values are to be used in the determination of the best mean square fit through these values of a polynomial of degree n (n less than $2m + 1$). This polynomial is of the form:

$$f_i = \sum_{k=0}^{k=n} b_{nk} i^k = I$$

$$b_{n0} + b_{n1} i + b_{n2} i^2 + \ldots + b_{nn} i^n \qquad \text{Ia}$$

The derivatives of this polynomial are

$$\frac{df_i}{di} = b_{n1} + 2b_{n2}i + 3b_{n3}i^2 + \ldots + nb_{nn}i^{n-1} \qquad \text{Ib}$$

$$\frac{d^2 f_i}{di^2} = 2b_{n2} + 3 \times 2b_{n3}i + \ldots + (n-1)nb_{nn}i^{n-2} \qquad \text{Ic}$$

$$\frac{d^n f_i}{di^n} = n! b_{nn} \qquad \text{Id}$$

Note that, in the coordinate system being considered, the value of i ranges from $-m$ to $+m$, and that $i = 0$ at the central point of the set of $2m + 1$ values. Hence, the value of the sth derivative at that point is given by:

$$\left(\frac{d^s f_i}{di^s} \right)_{i=0} = s! b_{ns} = a_{ns} \qquad \text{II}$$

where

$$f_0 = b_{n0} = a_{n0} \qquad \text{IIa}$$

$$\frac{df_0}{di} = b_{n1} = a_{n1} \qquad \text{IIb}$$

$$\frac{d^2 f_0}{di^2} = 2b_{n2} = a_{n2} \qquad \text{IIc}$$

The least squares criterion requires that the sum of the squares of the differences between the observed values, y_i, and the calculated, f_i, be a minimum over the interval being considered.

$$\frac{\partial}{\partial b_{nk}} \left[\sum_{i=-m}^{i=m} (f_i - y_i)^2 \right] = 0 \qquad \text{III}$$

Minimizing with respect to b_{n0}, we have

$$\frac{\partial}{\partial b_{n0}} \left[\sum_{i=-m}^{i=m} b_{n0} + b_{n1}i + \ldots + (b_{nn}i^n - y_i)^2 \right] = \qquad \text{IIIa}$$

$$2 \sum_{i=-m}^{i=m} (b_{n0} + b_{n1}i + \ldots + b_{nn}i^n - y_i) = 0 \qquad \text{IIIa}$$

and with respect to b_{ni}, we have

$$\frac{\partial}{\partial b_{ni}} \left[\sum_{i-m}^{i=m} (b_{no} + b_{ni} + \ldots + b_{nn}i^n - y_i)^2 \right] =$$

$$2 \sum_{i=-m}^{i=m} (b_{n0} + b_{n1}i + \ldots + b_{nn}i^n - y_i)\, i = 0 \qquad \text{IIIb}$$

and with respect to the general b_{nr}, we obtain

$$2 \sum_{i=-m}^{i=m} \left[\left(\sum_{k=0}^{k=n} b_{nk}i^k \right) - y_i \right] i^r = 0 \qquad \text{IIIc}$$

or

$$\sum_{i=-m}^{i=m} \sum_{k=0}^{k=n} b_{nk}i^{k+r} = \sum_{i=-m}^{i=m} y_i i^r \qquad \text{IVa}$$

where r is the index representing the equation number which runs from 0 to n (there are $n + 1$ equations). The summation indexes on the left side may be interchanged—i.e.,

$$\sum_{i=-m}^{i=m} \sum_{k=0}^{k=n} b_{nk}i^{k+r} = \sum_{k=0}^{k=n} \sum_{i=-m}^{i=m} b_{nk}i^{k+r} \qquad \text{IVb}$$

and finally, since b_{nk} is independent of i,

$$\sum_{k=0}^{k=n} b_{nk} \sum_{i=-m}^{i=m} i^{k+r} = \sum_{i=-m}^{i=m} y_i i^r = F_k \qquad \text{Va}$$

or

$$\sum_{k=0}^{k=n} b_{nk} S_{r+k} = F_k \qquad \text{Vb}$$

where

$$S_{r+k} = \sum_{i=-m}^{i=m} i^{r+k} \qquad \text{Vc}$$

and

$$F_k = \sum_{i=-m}^{i=m} i^k y_i \qquad \text{Vd}$$

Note that $S_{r+k} = 0$ for odd values of $r + k$. Since S_{r+k} exists for even values of $r + k$ only, *the set of $n + 1$ equations can be sepa-*

rated into two sets, one for even values of k and one for odd values. Thus, for a 5th degree polynomial, where $n = 5$

$$S_0 b_{50} + S_2 b_{52} + S_4 b_{54} = F_0$$
$$S_2 b_{50} + S_4 b_{52} + S_6 b_{54} = F_2 \qquad \text{VIa}$$
$$S_4 b_{50} + S_6 b_{52} + S_8 b_{54} = F_4$$

which can be used to solve for b_{50}, b_{52}, and b_{54}, while

$$S_2 b_{51} + S_4 b_{53} + S_6 b_{55} = F_1$$
$$S_4 b_{51} + S_6 b_{53} + S_8 b_{55} = F_3 \qquad \text{VIb}$$
$$S_6 b_{51} + S_8 b_{53} + S_{10} b_{55} = F_5$$

which can be used to solve for b_{51}, b_{53}, and b_{55}. The set of equations in VIa has the same form for $n = 4$, so that $b_{40} = b_{50}$, $b_{42} = b_{52}$, and $b_{44} = b_{54}$ while the set VIb has the same form for $n = 6$, so that $b_{51} = b_{61}$, $b_{53} = b_{63}$, and $b_{55} = b_{65}$. In other words, $b_{ns} = b_{n+1,s}$ for n and s both odd. For example, to determine the third derivative for the best fit to a curve of third (or fourth) order, we would have:

$$S_2 b_{31} + S_4 b_{33} = F_1$$
$$S_4 b_{31} + S_6 b_{33} = F_3$$

from which $b_{33} = \dfrac{S_2 F_3 - S_4 F_1}{S_2 S_6 - S_4^2}$

When, for instance, $m = 4$ ($2m + 1 = 9$ points), we have from Vc that

$$S_2 = 60, \quad S_4 = 708, \quad S_6 = 9780$$

and

$$b_{33} = \frac{60 F_3 - 708 F_1}{60\,(9780) - (708)^2} = \frac{F_3 - 7 F_1}{7128}$$

which reduces to:

$$\frac{-14y_{-4} + 7y_{-3} + 13y_{-2} + 9y_{-1} + 0y_0 - 9y_1 - 13y_2 - 3y_3 + 14y_4}{1188}$$

The coefficients of y_i constitute the convoluting integers (Table VIII) for the third derivative of a cubic polynomial determined from a least squares fit to 9 points. Since the value of a_{33} is $3! b_{33}$, the denominator in the above expression must be divided by 6 to get the normalizer of 198 found in Table III.

In all of the above derivations, it has been assumed that the sampling interval is the same as the absolute abscissa interval—i.e., $\Delta x = 1$. If not, the value of Δx must be included in the normalization procedure. Hence, to evaluate the sth derivative at the central point of a set of m values, based on an nth degree polynomial fit, we must evaluate

$$a_{nsm} = s! b_{nsm} = \frac{\displaystyle\sum_{i=-m}^{i=m} C_{ism} y_i}{\Delta x^s N_{sm}} \qquad \text{VII}$$

TABLE I

CONVOLUTES	SMOOTHING			QUADRATIC	CUBIC		A20	A30			
POINTS	25	23	21	19	17	15	13	11	9	7	5
-12	-253										
-11	-138	-42									
-10	-33	-21	-171								
-09	62	-2	-76	-136							
-08	147	15	9	-51	-21						
-07	222	30	84	24	-6	-78					
-06	287	43	149	89	7	-13	-11				
-05	322	54	204	144	18	42	0	-36			
-04	387	63	249	189	27	87	9	9	-21		
-03	422	70	284	224	34	122	16	44	14	-2	
-02	447	75	309	249	39	147	21	69	39	3	-3
-01	462	78	324	264	42	162	24	84	54	6	12
00	467	79	329	269	43	167	25	89	59	7	17
01	462	78	324	264	42	162	24	84	54	6	12
02	447	75	309	249	39	147	21	69	39	3	-3
03	422	70	284	224	34	122	16	44	14	-2	
04	387	63	249	189	27	87	9	9	-21		
05	322	54	204	144	18	42	0	-36			
06	287	43	149	89	7	-13	-11				
07	222	30	84	24	-6	-78					
08	147	15	9	-51	-21						
09	62	-2	-76	-136							
10	-33	-21	-171								
11	-138	-42									
12	-253										
NORM	5175	8059	3059	2261	323	1105	143	429	231	21	35

TABLE II

CONVOLUTES	SMOOTHING			QUARTIC	QUINTIC		A40	A50			
POINTS	25	23	21	19	17	15	13	11	9	7	5
-12	1265										
-11	-345	285									
-10	-1122	-114	11628								
-09	-1255	-285	-6460	340							
-08	-915	-285	-13005	-255	195						
-07	-255	-165	-11220	-420	-195	2145					
-06	590	30	-3940	-290	-260	-2860	110				
-05	1503	261	6378	18	-117	-2937	-198	18			
-04	2385	495	17655	405	135	-165	-160	-45	15		
-03	3155	705	28190	790	415	3755	110	-10	-55	5	
-02	3750	870	36660	1110	660	7500	390	60	30	-30	
-01	4125	975	42120	1320	825	10125	600	120	135	75	
00	4253	1011	44003	1393	883	11053	677	143	179	131	
01	4125	975	42120	1320	825	10125	600	120	135	75	
02	3750	870	36660	1110	660	7500	390	60	30	-30	
03	3155	705	28190	790	415	3755	110	-10	-55	5	
04	2385	495	17655	405	135	-165	-160	-45	15		
05	1503	261	6378	18	-117	-2937	-198	18			
06	590	30	-3940	-290	-260	-2860	110				
07	-255	-165	-11220	-420	-195	2145					
08	-915	-285	-13005	-255	195						
09	-1255	-285	-6460	340							
10	-1122	-114	11628								
11	-345	285									
12	1265										
NORM	30015	6555	260015	7429	4199	46189	2431	429	429	231	

TABLE III

CONVOLUTES   1ST DERIVATIVE   QUADRATIC   A21

POINTS	25	23	21	19	17	15	13	11	9	7	5
-12	-12										
-11	-11	-11									
-10	-10	-10	-10								
-09	-9	-9	-9	-9							
-08	-8	-8	-8	-8	-8						
-07	-7	-7	-7	-7	-7	-7					
-06	-6	-6	-6	-6	-6	-6	-6				
-05	-5	-5	-5	-5	-5	-5	-5	-5			
-04	-4	-4	-4	-4	-4	-4	-4	-4	-4		
-03	-3	-3	-3	-3	-3	-3	-3	-3	-3	-3	
-02	-2	-2	-2	-2	-2	-2	-2	-2	-2	-2	-2
-01	-1	-1	-1	-1	-1	-1	-1	-1	-1	-1	-1
00	0	0	0	0	0	0	0	0	0	0	0
01	1	1	1	1	1	1	1	1	1	1	1
02	2	2	2	2	2	2	2	2	2	2	2
03	3	3	3	3	3	3	3	3	3	3	
04	4	4	4	4	4	4	4	4	4		
05	5	5	5	5	5	5	5	5			
06	6	6	6	6	6	6	6				
07	7	7	7	7	7	7					
08	8	8	8	8	8						
09	9	9	9	9							
10	10	10	10								
11	11	11									
12	12										
NORM	1300	1012	770	570	408	280	182	110	60	28	10

TABLE IV

CONVOLUTES   1ST DERIVATIVE   CUBIC   QUARTIC   A31   A41

POINTS	25	23	21	19	17	15	13	11	9	7	5
-12	30866										
-11	8602	3938									
-10	-8525	815	84075								
-09	-20982	-1518	10032	6936							
-08	-29236	-3140	-43284	68	748						
-07	-33754	-4130	-78176	-4648	-98	12922					
-06	-35003	-4567	-96947	-7481	-643	-4121	1133				
-05	-33450	-4530	-101900	-8700	-930	-14150	-660	300			
-04	-29562	-4098	-95338	-8574	-1002	-18334	-1578	-294	86		
-03	-23806	-3350	-79564	-8179	-902	-17842	-1796	-532	-142	22	
-02	-16649	-2365	-56881	-5363	-673	-13843	-1489	-503	-193	-67	1
-01	-8558	-1222	-29592	-2816	-358	-7506	-832	-296	-126	-58	-8
00	0	0	0	0	0	0	0	0	0	0	0
01	8558	1222	29592	2816	358	7506	832	296	129	58	8
02	16649	2365	56881	5363	673	13843	1489	503	193	67	-1
03	23806	3350	79504	8179	902	17842	1796	532	142	-22	
04	29562	4098	95338	8574	1002	18334	1578	294	-86		
05	33450	4530	101900	8700	930	14150	660	-300			
06	35003	4567	96947	7481	643	4121	-1133				
07	33754	4130	78176	4648	98	-12922					
08	29236	3140	43284	-68	-748						
09	20982	1518	-10032	-6936							
10	8525	-815	84075								
11	-8602	-3938									
12	-30866										
NORM	1776060	197340	3634092	255816	23256	334152	24024	5148	1188	252	12

TABLE V

POINTS	CONVOLUTES 25	1ST DERIVATIVE 23	21	QUINTIC 19	17	SEXIC 15	A51 13	A61 11	9	7	5
-12	-6356625										
-11	-11820675	-357045									
-10	-15593141	-654687	-15977364								
-09	-17062146	-840937	-28754154	-332684							
-08	-15896511	-878634	-35613829	-583549	-23945						
-07	-12139321	-752859	-34807914	-686099	-40483	-175125					
-06	-6301491	-478349	-26040033	-604484	-43973	-279975	-31380				
-05	544668	-106911	-10949942	-348823	-32306	-266401	-45741	-3084			
-04	6671883	265164	6402438	9473	-8671	-130506	-33511	-3776	-5758		
-03	9604353	489687	19052988	322378	16679	65229	-12	-1244	-4538	-90	
-02	6024183	359157	16649358	349928	24661	169819	27093	2166	2762	18	
-01	-8322182	-400653	-15033066	-255102	-14404	-78351	-14647	-573	-508	-2	
00	0	0	0	0	0	0	0	0	0	0	0
01	8322182	400653	15033066	255102	14404	78351	14647	573	508	2	
02	-6024183	-359157	-16649358	-349928	-24661	-169819	-27093	-2166	-2762	-18	
03	-9604353	-489687	-19052988	-322378	-16679	-65229	12	1244	4538	90	
04	-6671883	-265164	-6402438	-9473	8671	130506	33511	3776	5758		
05	-544668	106911	10949942	348823	32306	266401	45741	3084			
06	6301491	478349	26040033	604484	43973	279975	31380				
07	12139321	752859	34807914	686099	40483	175125					
08	15896511	878634	35613829	583549	23945						
09	17062146	840937	28754154	332684							
10	15593141	654687	15977364								
11	11820675	357045									
12	6356625										
NORM	7153575	312455	5311735	81719	41990	20995	2431	143	143	1	

TABLE VI

POINTS	CONVOLUTES 25	2ND DERIVATIVE 23	21	QUADRATIC 19	CUBIC 17	15	A22 13	A32 11	9	7	5
-12	92										
-11	69	77									
-10	48	56	190								
-09	29	37	133	51							
-08	12	20	82	34	40						
-07	-3	5	37	19	25	91					
-06	-16	-8	-2	6	12	52	22				
-05	-27	-19	-35	-5	1	19	11	15			
-04	-36	-28	-62	-14	-8	-8	2	6	28		
-03	-43	-35	-83	-21	-15	-29	-5	-1	7	5	
-02	-48	-40	-98	-26	-20	-48	-10	-6	-8	0	2
-01	-51	-43	-107	-29	-23	-53	-13	-9	-17	-3	-1
00	-52	-44	-110	-30	-24	-56	-14	-10	-20	-4	-2
01	-51	-43	-107	-29	-23	-53	-13	-9	-17	-3	-1
02	-48	-40	-98	-26	-20	-48	-10	-6	-8	0	2
03	-43	-35	-83	-21	-15	-29	-5	-1	7	5	
04	-36	-28	-62	-14	-8	-8	2	6	28		
05	-27	-19	-35	-5	1	19	11	15			
06	-16	-8	-2	6	12	52	22				
07	-3	5	37	19	25	91					
08	12	20	82	34	40						
09	29	37	133	51							
10	48	56	190								
11	69	77									
12	92										
NORM	26910	17710	33649	6783	3876	6188	1001	429	462	42	7.

TABLE VII

CONVOLUTES	2ND DERIVATIVE			QUARTIC		QUINTIC	A42	A52			
POINTS	25	23	21	19	17	15	13	11	9	7	5
-12	-429594										
-11	31119	-346731									
-10	298155	61845	-37791								
-09	413409	281979	11628	-96084							
-08	414786	358530	35802	45084	-121524						
-07	336201	331635	41412	105444	82251	-93093					
-06	207579	236709	34353	109071	153387	88803	-72963				
-05	54855	104445	19734	76830	137085	133485	98010	-10530			
-04	-100026	-39186	1878	26376	71592	95568	115632	20358	-4158		
-03	-239109	-172935	-15678	-27846	-11799	19737	53262	17082	12243	-117	
-02	-348429	-280275	-30183	-74601	-88749	-59253	-32043	117	4983	603	-3
-01	-418011	-349401	-39672	-105864	-141873	-116577	-99528	-15912	-6963	-171	48
00	-441870	-373230	-42966	-116820	-160740	-137340	-124740	-22230	-12210	-630	-90
01	-418011	-349401	-39672	-105864	-141873	-116577	-99528	-15912	-6963	-171	48
02	-348429	-280275	-30183	-74601	-88749	-59253	-32043	117	4983	603	-3
03	-239109	-172935	-15678	-27846	-11799	19737	53262	17082	12243	-117	
04	-100026	-39186	1878	26376	71592	95568	115632	20358	-4158		
05	54855	104445	19734	76830	137085	133485	98010	-10530			
06	207579	236709	34353	109071	153387	88803	-72963				
07	336201	331635	41412	105444	82251	-93093					
08	414786	358530	38802	45084	121524						
09	413409	281979	11628	-96084							
10	298155	61845	-37791								
11	31119	-346731									
12	429594										
NORM	4292145	2812095	245157	490314	478686	277134	160446	16731	4719	99	3

TABLE VIII

CONVOLUTES	3RD DERIVATIVE			CUBIC	QUARTIC		A33	A43			
POINTS	25	23	21	19	17	15	13	11	9	7	5
-12	-506										
-11	-253	-77									
-10	-55	-35	-285								
-09	93	-3	-114	-204							
-08	196	20	12	-68	-28						
-07	259	35	98	28	-7	-91					
-06	287	43	149	89	7	-13	-11				
-05	285	45	170	120	15	35	0	-30			
-04	258	42	166	126	18	58	6	6	-14		
-03	211	35	142	112	17	61	8	22	7	-1	
-02	149	25	103	83	13	49	7	23	13	1	-1
-01	77	13	54	44	7	27	4	14	9	1	2
00	0	0	0	0	0	0	0	0	0		0
01	-77	-13	-54	-44	7	-27	-4	-14	-9	-1	-2
02	-149	-25	-103	-83	13	-49	-7	-23	-13	-1	1
03	-211	-35	-142	-112	17	-61	-8	-22	-7	1	
04	-258	-42	-166	-126	18	-58	-6	-6	14		
05	-285	-45	-170	-120	15	-35	0	30			
06	-287	-43	-149	-89	7	13	11				
07	-259	-35	-98	-28	-7	91					
08	-196	-20	-12	68	28						
09	-93	3	114	204							
10	55	35	285								
11	253	77									
12	506										
NORM	296010	32890	86526	42636	3876	7956	572	858	198	6	2

TABLE IX

CONVOLUTES	3RD DERIVATIVE		QUINTIC		SEXIC		A53	A63			
POINTS	25	23	21	19	17	15	13	11	9	7	5
-12	118745										
-11	217640	23699									
-10	279101	42704	425412								
-09	290076	52959	749372	317655							
-08	244311	51684	887137	1113240	4915						
-07	144616	38013	787382	1231500	8020	93135					
-06	5131	13632	448909	932760	7975	141320	11260				
-05	-146408	-16583	-62644	259740	4380	113065	15250	1580			
-04	-266403	-43928	-598094	-589080	-1755	3800	8165	1700	2295		
-03	-293128	-55233	-908004	-1220520	-7540	-150665	-6870	-55	1280	65	
-02	-144463	-32224	-625974	-1007760	-7735	-260680	-16335	-2010	-2285	-40	
-01	284372	49115	748068	948600	5720	-169295	7150	645	500	5	
00	0	0	0	0	0	0	0	0	0	0	
01	-284372	-49115	-748068	-948600	-5720	-169295	-7150	-645	-500	-5	
02	144463	32224	625974	1007760	7735	260680	16335	2010	2285	40	
03	293128	55233	908004	1220520	7540	150665	6870	55	-1280	-65	
04	266403	43928	598094	589080	1755	-3800	-8165	-1700	-2295		
05	146408	16583	62644	-259740	-4380	-113065	-15250	-1580			
06	-5131	-13632	-448909	-932760	-7975	-141320	-11260				
07	-144616	-38013	-787382	-1231500	-8020	-93135					
08	-244311	-51684	-887137	-1113240	-4915						
09	-290076	-52959	-749372	-317655							
10	-279101	-42704	-425412								
11	-217640	-23699									
12	-118745										
NORM	5722860	749892	4249388	4247012	16796	2144809	9724	572	286	2	

TABLE X

CONVOLUTES	4TH DERIVATIVE			QUARTIC		QUINTIC	A44	A54			
POINTS	25	23	21	19	17	15	13	11	9	7	5
-12	858										
-11	803	858									
-10	643	793	594								
-09	393	605	540	396							
-08	78	315	385	352	36						
-07	-267	-42	150	227	31						
-06	-597	-417	-130	42	17	621	84				
-05	-857	-747	-406	-168	-3	251	64	6			
-04	-982	-955	-615	-354	-24	-249	11	4	18		
-03	-897	-950	-680	-453	-39	-704	-54	-1	9	6	
-02	-517	-627	-510	-388	-39	-869	-96	-6	-11	1	
-01	253	133	0	-68	-13	-429	-66	-6	-21	-7	
00	1518	1463	969	612	52	1001	99	6	14	-3	
01	253	133	0	-68	-13	-429	-66	-6	-21	-7	
02	-517	-627	-510	-388	-39	-869	-96	-6	-11	1	
03	-897	-950	-680	-453	-39	-704	-54	-1	9	6	
04	-982	-955	-615	-354	-24	-249	11	4	18		
05	-857	-747	-406	-168	-3	251	64	6			
06	-597	-417	-130	42	17	621	84				
07	-267	-42	150	227	31	756					
08	78	315	385	352	36						
09	393	605	540	396							
10	643	793	594								
11	803	858									
12	858										
NORM	1430715	937365	408595	163438	8398	92378	4862	143	143	11	

TABLE XI

POINTS	CONVOLUTES 25	5TH DERIVATIVE 23	21	QUINTIC 19	SEXIC 17	15	A55 13	A65 11	9	7	5
−12	−275										
−11	−500	−65									
−10	−631	−116	−1404								
−09	−636	−141	−2444	−44							
−08	−501	−132	−2819	−74	−55						
−07	−236	−87	−2354	−79	−88	−675					
−06	119	−12	−1063	−54	−83	−1000	−20				
−05	488	77	788	−3	−36	−751	−26	−4			
−04	753	152	2618	58	39	44	−11	−4	−9		
−03	748	171	3468	98	104	979	18	1	−4	−5	
−02	253	76	1938	68	91	1144	33	6	11	4	
−01	−1012	−209	−3876	−102	−104	−1001	−22	−3	−4	−1	
00	0	0	0	0	0	0	0	0	0	0	
01	1012	209	3876	102	104	1001	22	3	4	1	
02	−253	−76	−1938	−68	−91	−1144	−33	−6	−11	−4	
03	−748	−171	−3468	−98	−104	−979	−18	−1	4	5	
04	−753	−152	−2618	−58	−39	−44	11	4	9		
05	−488	−77	−788	3	36	751	26	4			
06	−119	12	1063	54	83	1000	20				
07	236	87	2354	79	88	675					
08	501	132	2819	74	55						
09	636	141	2444	44							
10	631	116	1404								
11	500	65									
12	275										
NORM	1300650	170430	1931540	29716	16796	83980	884	52	26	2	

Note that since $\Delta x^0 = 1$, the interval is of no concern in the case of smoothing.

Repeated Convolution. The process of convolution can be repeated if desired. For example, one might wish to further smooth a set of previously smoothed points, or to obtain the derivative only after the raw data has been smoothed. Thus, if we convolute using p points the first time, and m points the second,

$$(f)_{s_1 s_1} = a_{n_2 s_2} m = \frac{\sum_{i=-m}^{i=m} C_{i s_2 m} \, a_{n_1 s_1 p}}{\Delta x^{(s_2)} N_{s_2 m}} \qquad \text{VIII}$$

$$= \frac{\sum_{i=-m}^{i=m} C_{i s_2 m} \sum_{j=-p}^{j=p} C_{j s_1 p} F_j}{\Delta x^{(s_1 + s_2)} N_{s_2 m} N_{s_1 p}} \qquad \text{IX}$$

$$= \frac{\sum_{i=-m}^{i=m} \sum_{j=-p}^{j=p} C_{i s_2 m} C_{j s_1 p} F_{i+j}}{\Delta x^{(s_1 + s_2)} N_{s_2 m} N_{s_1 p}} \qquad \text{X}$$

thus:

$$(f)_{s_1 s_2} = \frac{\sum_{h=-(m+p)}^{h=(m+p)} d_h y_h}{N_h} \qquad \text{XI}$$

where

$$h = i + j$$

$$N_h = \Delta x^{(s_1 s_2)} N_{s_1 p} N_{s_2 m}$$

$$d_h = \sum_{i=-m}^{i=m} \sum_{j=-p}^{j=p} c_i c_j$$

Equation XI shows that one need not go through the convolution procedure twice, but can do a single convolution, using $2(m+p)+1$ points, and a table of new integers formed by combining the c's.

For the case where a cubic smooth is to be followed by obtaining the quadratic first derivative using $m = 2$ and $p = 2$:

$C_i = -2, -1, 0, 1, 2, N_i = 10$
$C_j = -3, 12, 17, 12, -3, N_j = 35$
$d_{-4} = C_{-2} C'_{-2} = 6$
$d_{-3} = C_{-2} C_{-1} + C'_{-1} C_{-2} = -36 + 3 = -33$

$$d_{-2} = C_{-2}C'_0 + C_0C'_2 + C_{-1}C'_{-1} = -34 - 12 = -46$$
$$d_{-1} = C_{-2}C'_1 + C_1C'_{-2} + C_{-1}C'_0 + C_0C'_{-1} = -24 - 3 - 17 = -44$$
$$d_0 = C_{-2}C'_2 + C_2C'_{-2} + C_1C'_{-1} - C_{-1}C'_1 = 0$$
$$d_1 = \text{By symmetry} = 44$$
$$d_2 = 46$$
$$d_3 = 33$$
$$d_4 = -6$$
$$N_h = 350\ \Delta x$$

APPENDIX II

The following eleven tables contain the convoluting integers for smoothing (zeroth derivative) through the fifth derivative for polynomials of degree two through five. They are in the form of tables of A_{ij}, where i is the degree of the polynomial and j is the order of the derivative. Thus, to obtain the third derivative over 17 points, assuming a fourth degree polynomial (A_{43}), one would use the integers in the column headed 17 of Table VIII.

APPENDIX III. COMPUTER PROGRAMS

The programming of today's high speed digital computers is still an art rather than a science. Different programmers presented with the same problem will, in general, write quite different programs to satisfactorily accomplish the calcula-

tion. Two programs are presented here as examples of the techniques discussed in this paper. They are written in the FORTRAN language, since this is one of the most widespread of the computer programming languages. Programmers using other languages should be able to follow the logic quite readily and make the appropriate translations. Each is written as a subroutine for incorporation into a larger program as required.

Program 1 is a 9-point least squares smooth of spectroscopic data. The raw data has previously been stored by the main program in the region NDATA.

Lines 1 through 25 are explanatory and housekeeping to set up the initial conditions. The 9-point array NP contains the current set of data points to be smoothed. The main loop consists of lines 26 through 35. The inner loop, lines 28 through 30, moves the previous set of points up one position. The next point is added by line 31. In lines 32 and 33 the convoluting integers are multiplied by the corresponding data, and the products summed. In line 34, the sum is divided by the normalizing constant and the resulting smoothed point is stored.

Program 2 computes the precise peak position and intensity of a set of points which is known to contain a spectroscopic peak. In order to approximate a Lorentz contour (5), values are converted to absorbance and the reciprocals are

---

PROGRAM 1

```
*         SUBROUTINE SMOOTH -9 POINT                                          1
C                                                                            2
          SUBROUTINE SMOOTH (N,NDATA,M,MDATA)                                3
C                                                                            4
C         INPUTS                                                             5
C             N        NUMBER OF RAW DATA POINTS                             6
C             NDATA    ARRAY OF N RAW DATA POINTS STORED IN MAIN PROGRAM     7
C                      DUMMY DIMENSION                                       8
C         OUTPUTS                                                            9
C             M        NUMBER OF SMOOTHED DATA POINTS = N-8                  10
C             MDATA    ARRAY OF M SMOOTHED POINTS STORED IN MAIN PROGRAM     11
C                      MAY BE SAME REGION IN MAIN PROGRAM AS NDATA           12
C                      DUMMY DIMENSION                                       13
C                                                                            14
          DIMENSION NDATA(1000),MDATA(1000),NP(9)                           15
C                                                                            16
C         INITIALIZATION SEGMENT                                            17
C                                                                            18
          M=N-8                                                             19
          DO 10 I=2,9                                                       20
          J=I-1                                                             21
    10    NP(I) = NDATA(J)                                                  22
C                                                                            23
C         SMOOTHING LOOP                                                    24
C                                                                            25
          DO 200 I=1,M                                                      26
          J=I+8                                                             27
          DO 11 K=1,8                                                       28
          KA = K+1                                                          29
    11    NP(K)=NP(KA)                                                      30
          NP(9) = NDATA(J)                                                  31
          NSUM=59*NP(5)+54*(NP(4)+NP(6))+39*(NP(3)+NP(7))+14*(NP(2)+NP(8))- 32
          121*(NP(1)+NP(9))                                                 33
          MDATA(I) = NSUM/231                                               34
    200   CONTINUE                                                          35
C                                                                            36
    9999  RETURN                                                            37
C                                                                            38
*         END                                                               39
```

PROGRAM 2

```
*       SUBROUTINE CENTER      LORENZ
C                                                                        2
        SUBROUTINE CENTER(NPOINT,X,Y,TRANS,DENS)                         3
C                                                                        4
C       COMPUTATION OF PRECISE PEAK POSITION AND INTENSITY USING 9 POINTS 5
C       TO A QUADRATIC,Y=A20+A21X+A22XSQ, X=(-A21/2A22). IN ORDER TO      6
C       APPROXIMATE A LORENTZ CONTOUR, VALUES ARE CONVERTED TO ABSORBANCE 7
C       AND THE RECIPROCALS ARE USED IN DETERMINING THE COEFFICIENTS BY   8
C       ORTHOGONAL POLYNOMIALS                                            9
C                                                                       10
        DIMENSION NPOINT(25),DNS(9)                                     11
     60 DO 61 I=1,9                                                      12
        IP=I+8                                                          13
     61 DNS(I)=1./ALOG10F(1000./NPOINT(IP))                             14
        P4=DNS(1)+DNS(9)                                                15
        P3=DNS(2)+DNS(8)                                                16
        P2=DNS(3)+DNS(7)                                                17
        P1=DNS(4)+DNS(6)                                                18
        A20=(-21.*P4)+(14.*P3)+(39.*P2)+(54.*P1)+(59.*DNS(5))           19
        A20=A20/231.                                                    20
        A21=(4.*(DNS(9)-DNS(1)))+(3.*(DNS(8)-DNS(2)))+(2.*(DNS(7)-DNS(3))) 21
       1+(DNS(6)-DNS(4))                                                22
        A21=A21/60.                                                     23
        A22=(28.*P4)+(7.*P3)-(8.*P2)-(17.*P1)-(20.*DNS(5))              24
        A22=A22/924.                                                    25
        X=(-A21/(2.0*A22))                                              26
        Y=A20+X*(A21+X*A22)                                             27
        DENS=1.0/Y                                                      28
        TRANS=1000./(10.0**DENS)                                        29
   1000 RETURN                                                          30
*       END                                                             31
```

used in determining the coefficients of a polynomial (6) having the form:

$$y = a_{20} + a_{21}x + a_{22x}{}^2$$

The center is at the point where $x - a_{21}/2a_{22}$.

The data points to be used are points 9 through 17 stored in the array NPOINT and may have any value from 30 through 999. In the loop lines 12 through 14, each of these points is converted to absorbance, the reciprocal taken, and the result stored in the array DNS.

Since for a_{20} and a_{22}, the convolute function is symmetric about the origin, forming the sums $P4$ through $P1$ in lines 15 through 18 shortens the computation. The first constant, a_{20} is found in line 19, using the values of the 9-point convolute from Table I, and normalized in line 20.

The value of a_{21} is computed in lines 21–22 using the convolute from Table III (first derivative-quadratic). Note that this is an antisymmetric function. Table VI furnished the constants for the computation of a_{22} in line 24. Note that the normalizing factor of 924 is 2! times the value given in the table. X is computed in line 26 and the corresponding value of y is computed in line 27 by substituting the appropriate values

into the polynomial. The absorbance or optical density (DENS) is, of course, the reciprocal of y (line 28).

LITERATURE CITED

(1) Allen, L. C., Johnson, L. F., *J. Am. Chem. Soc.* **85**, 2668 (1963).
(2) Anderson, R. L., Houseman, E. F., "Tables of Orthogonal Polynominal Values Extended to $N = 104$," Agricultural Experiment Station, Iowa State College of Agriculture and Mechanical Arts, Ames, Iowa, 1942.
(3) Cole, L., *Science* **125**, 874 (1957).
(4) Guest, P. G., "Numerical Methods of Curve Fitting," p. 349 ff., University Press, Cambridge, England, 1961.
(5) Jones, R. N., Seshadri, K.D., Hopkins, J.W., *Can. Jour. Chem.* **40**, 334 (1962).
(6) Kerawala, S. M., *Indian J. Phys.* **15**, 241 (1941).
(7) Savitzky, A, *Anal. Chem.* **33**, 25A (Dec. 1961).
(8) Whitaker, S., Pigford, R.L., *Ind. Eng. Chem.* **52**, 185 (1960).
(9) Whittaker, E., Robinson, G., "The Calculus of Observations," p. 291 ff., Blackie and Son, Ltd., London & Glasgow, 1948.

Received for review February 10, 1964. Accepted April 17, 1964. Presented in part at the 15th Annual Summer Symposium on Analytical Chemistry, University of Maryland, College Park, Md., June 14, 1962.

Reprinted from *Anal. Chem.* **1964**, *36*, 1627–39.

This article is one of the early harbingers of HPLC. Written in 1964, it explains the origin of the major advantage of the modern LC method, yet to come, over the already well-established GC method. For both methods, the critical pressure below which a separation cannot be achieved increases rapidly with decreasing particle size; at the same time, the analysis time decreases. Although experimental conditions have evolved considerably in the past 30 years (the particles used in LC are much smaller and the inlet pressure is much higher than was conceivable then), the conclusions remain valid.

Georges Guiochon
University of Tennessee and Oak Ridge National Laboratory

Comparison of the Theoretical Limit of Separating Ability in Gas and Liquid Chromatography

J. CALVIN GIDDINGS
Department of Chemistry, University of Utah, Salt Lake City, Utah

It is established that the theoretical limit of separating ability in chromatography is a function of relative selectivity, zone migration rates, and the number of theoretical plates. Only the latter is expected to show a significant difference between gas and liquid chromatography. The main limitation to acquiring an unlimited number of plates is the pressure drop available to the system. For a given pressure drop, equations are derived showing the maximum number of plates which can be possibly achieved in single columns. Liquid chromatography, owing mainly to the difference in diffusivities, shows a 100- or 1000-fold advantage in maximum plates. These concepts are related to the critical pressure, where the latter is the minimum pressure leading to a specified separation.

Gas and liquid chromatography are two powerful and complementary techniques for chemical analysis. Their theoretical basis is nearly identical. Despite this, a wide schism has evolved in the manner of their scientific growth. Few real efforts have been made to compare the performance and potential performance of one technique with the other. This void makes it more difficult than necessary for the analyst to choose the proper alternative when faced with a new separations problem. The present paper is intended to explore one aspect of the potential of the two methods. This aspect is concerned with the theoretical limit of separability. A subsequent paper will deal with the comparative speed of analysis.

The theoretical limit of separating ability may be difficult to realize fully. Such a limit may require the use of extreme operating conditions and may require an inordinate amount of time for separation. Nonetheless the theoretical limit

should be approachable within a factor of two or so, a rather small factor compared to the order-of-magnitude differences involved. The theoretical limit may not be needed for ordinary separations. However it does give an indication of the present scope of chromatography and the possible approaches to extending this scope.

Once the best possible column materials and conditions have been established for a given problem, it seems obvious that the degree of separation can be increased to any desired level by adding to the length of the column. One soon reaches a length of such magnitude, however, that optimum flow can no longer be maintained within the column. This problem can be temporarily skirted by increasing the pressure drop through the column. Ultimately one reaches a point where the equipment will stand no further pressure increases. Thus the ultimate limit of separability may well be determined by the maximum pressure limitations of the equipment. It is true that column length can itself be a troublesome barrier if excessive values are required. However, the use of long columns is largely a matter of inconvenience, and length is therefore probably not such a basic limitation as pressure drop in most systems. The limiting role of pressure has also been considered by Knox in connection with the ultimate speed of separation (4). The recent development of high-pressure laboratory systems, particularly by Hamilton (3) (up to 600 p.s.i.), gives an empirical indication of the importance of pressure limitations. These considerations lead us, in this paper, to consider pressure drop as the most critical limiting factor to ultimate separability.

FACTORS INVOLVED IN SEPARATION

The success of any separation must hinge, first, on the differential movement of zone centers, and second, on the zones remaining sufficiently compact during migration that they do not spread into one another despite the disengagement of centers. The most commonly used measure of these requirements for separation is the resolution, Rs. This quantity can be defined by

$$Rs = \Delta z / 4\sigma \qquad (1)$$

where Δz is the distance between zone centers (indicating the differential movement) and σ is the mean standard deviation in zone width, or roughly the quarter-width of the zone (indicating solute mixing owing to zone spreading). For close lying zones the widths are usually of the same approximate magnitude. A resolution of unity indicates that the separation is adequate for most purposes. This value indicates that the zones overlap only in their outer edges where the concentration has dropped to about 10% of its maximum value.

The resolution of neighboring zones obviously increases with column length, L, and with the relative selectivity exhibited by the column toward the pair of components. (The latter is given by $\Delta K/K$ where K is the mean thermodynamic distribution coefficient and ΔK is the difference in K from one component to the other.) It also increases as the column plate height, H, is reduced. (Plate height is the usual measure of zone spreading or σ, being given by $H = \sigma^2/L$.) It can be shown that these various terms contribute to resolution as follows

$$Rs = (L/16H)^{1/2}(\Delta K/K)(1 - R) \qquad (2)$$

The R value, similar but not identical to R_f (3), is the ratio of the zone velocity to that of the mobile fluid. The ratio L/H can be written as N, the number of theoretical plates in the column. Thus

$$Rs = (N/16)^{1/2}(\Delta K/K)(1 - R) \qquad (3)$$

In comparing gas and liquid chromatography, it is necessary to consider the maximum potential magnitude of the various terms in this equation.

The maximum relative selectivity, $\Delta K/K$, will vary greatly depending on the similarity in molecular structure for the pair under consideration. However, for a given pair it is logical to expect that gas and liquid chromatography would show about the same maximum capacity for $\Delta K/K$. Selectivity, after all, reduces to the problem of molecular interactions and the degree to which these can be made different for two similar solutes. The difference depends on factors such as polarity, hydrogen bonding, polarization, etc., and probably can be enhanced to nearly an equal degree in gas and liquid chromatographic systems. (It is assumed throughout, of course, that the solutes are volatile so that a legitimate comparison can be made.)

The maximum value of $(1 - R)$ is unity. This value is approached as R goes to zero. In practice R must be kept above zero to get a finite migration rate and a reasonably short analysis time. As a compromise between speed and resolution, R may best be operated in the range from 0.1 to 0.5. Even with $R = 0.5$, the value of $(1 - R)$ is within a factor of 2 of its maximum. This is not a large variation compared to other factors involved in resolution. Furthermore, there is no significant difference on this matter between gas and liquid chromatography.

The only other factor involved in the resolution equation, the number of theoretical plates, N, becomes decisive in view of the fact that $\Delta K/K$ and $(1 - R)$ are roughly equal under the best of conditions. Thus to a good approximation we may regard the maximum achievable N as the best criterion for the ultimate separation potential of a chromatographic system.

MAXIMUM NUMBER OF THEORETICAL PLATES

In view of the importance to separation of the maximum plate number, N, an equation will be derived for the purpose of relating N to the limiting pressure drop. For the present it will be assumed that the column under consideration is uniform in its properties throughout its length. The obvious modification needed to account for the high compressibility of gases will be discussed later.

The pressure gradient existing in a chromatographic column is proportional to the mean flow velocity, v, and the viscosity, η, of the carrier fluid. It increases as the particle size, d_p, of the packing is made smaller. In quantitative form the pressure gradient is

$$\Delta p / L = 2\varphi \eta v / d_p^2 \qquad (4)$$

where Δp is the pressure drop across the column of length L (the ratio, $\Delta p/L$, being the pressure gradient). The quantity is a structural factor (1) (indicating the tightness of packing) of approximate magnitude 3×10^2. The right hand side of Equation 4 must be modified by a quadratic term in velocity if turbulence exists, but this is not of much concern in the present case.

The number of theoretical plates in a column is given by $N = L/H$. Upon solving for L from Equation 4 and substituting it into this expression, the plate number is found to be

$$N = d_p^2 \Delta p / 2\sigma \eta v H \qquad (5)$$

The nature of the plate height, H, to be used in this equation has been intensively investigated in recent years. For our purposes it is sufficient to represent this complex quantity by

$$H = 2\gamma D_m / v + g(v) \qquad (6)$$

where D_m is the binary diffusion coefficient for solute in the mobile phase, γ is a constant of order 0.6, and $g(v)$ is a complex function of flow velocity, v, whose most important characteristic is

$$dg(v)/dv > 0 \qquad (7)$$

for all real values of v. This simply means that $g(v)$, accounting for nonequilibrium and eddy-diffusion phenomena, is an increasing function of v throughout the full operating range. The first term, $2\gamma D_m/v$, accounts for longitudinal molecular diffusion within the column. (The value of γ may exceed unity, especially in liquid chromatography, due to longitudinal diffusion in the stationary phase. Ordinarily, however, γ remains within a narrow range which is about the same for gas and liquid systems.)

When the above expression for H is substituted into Equation 5, we have

$$N = \frac{d_p^2 \Delta p}{2\varphi\eta \ [2\gamma D_m + vg(v)]} \qquad (8)$$

This equation shows, for a fixed pressure drop, that N decreases as v increases due to the $vg(\gamma)$ term in the denominator. The theoretical limit of N is found as v approaches zero, viz.,

$$N_{lim} = \frac{d_p^2 \Delta p}{2\varphi\gamma\eta D_m} \qquad (9)$$

Quite significantly, this equation contains no terms related to the nonequilibrium (or mass-transfer) and eddy-diffusion contributions to the plate height. It is a limit to be approached strictly at very low velocities where the latter effects are negligible.

Equation 9 shows that the maximum achievable N is proportional to the pressure drop available to the system. It is also proportional to the square of particle size and inversely proportional to the carrier's viscosity and diffusivity.

In the comparison of gas and liquid chromatography it can be assumed, as a first approximation, that the maximum pressure drop is the same and that the largest practical particle size is comparable. Since φ and γ are fixed structural factors, the ultimate separating potential of the two methods depends mainly on the product of viscosity and diffusivity, $\eta D_m \eta$. Thus

$$\frac{N_{lim} \ (gas)}{N_{lim} \ (liq.)} = \frac{\eta_l D_l}{\eta_g D_g} \qquad (10)$$

where subscripts l and g stand for liquid and gas, respectively. The viscosity of liquids is roughly 100 times larger than that of gases. Diffusivity in liquids is nearly 10^5 times smaller than that in gases. Hence the theoretical limit to the number of plates is roughly 1000 times larger in liquid than in gas chromatography; viz.,

$$\frac{N_{lim} \ (gas)}{N_{lim} \ (liq.)} \sim \frac{1}{1000} \qquad (11)$$

(Since D_g decreases as pressure is increased, this ratio may be increased to 1/100 or more at high inlet pressures.) Thus for potential separating ability without regard for speed, liquid chromatography shows a distinct basic advantage over gas chromatography. The advantage originates with the slow diffusivity found in liquids. The 1000-fold advantage in N becomes a 30-fold advantage in actual resolution, Rs, since the dependence in Equation 3 is of the square root type.

Capillary columns in gas chromatography make a closer approach to classical liquid chromatography because the structural constant φ equals only 16 for open tubes (1) (d_p must be equated to the tube diameter for this comparison).

This brings N_{lim} for capillary gas chromatography within a 50- to 100-fold margin of liquid chromatography. This would revert back to the 1000-fold margin if capillaries were used in liquid chromatography. There is no basic reason why this cannot be done although the advantages, except as one approaches the theoretical limit, may be minor.

The order of magnitude of N_{lim} can be easily calculated by substituting the appropriate quantities, in consistent units, into Equation 9. As an example for liquid chromatography we may assume d_p = 0.05 cm. (fairly large 30- to 40-mesh range), Δp = 10^7 dynes per sq. cm. (~ 10 atm.), η = 10^{-2} poise, D_m = 3×10^{-5} sq. cm. per second, γ = 0.6, and γ = 300. This gives a value much larger than any yet reported—i.e., $N_{lim} \sim 10^8$.

THE CRITICAL PRESSURE

The idea that there is a critical inlet pressure, below which a separation can not be achieved by any manipulation of column length or flow velocity, was introduced by the author in relationship to gas chromatography (2). The foregoing arguments show that the concept is a general one. Thus Equation 9 gives the maximum number of plates to be achieved with a pressure drop of Δp. The quantity Δp may be regarded as the critical pressure (or pressure drop) for $N = N_{lim}$, since any reduction in Δp would mean that this number of plates could no longer be achieved.

If one wishes to make the greatest use of a limited pressure drop, it is desirable to operate the column outlet as close as possible to a vacuum (a limit is imposed here by the boiling of liquids under reduced pressures). While an absolute vacuum is not necessary, we will assume here that the outlet pressure is negligible compared to the inlet pressure. Thus we may replace Δp in Equation 9 by the critical inlet pressure, p_c. This leads to the following equation for the critical pressure

$$p_c = 4\varphi\gamma\eta D_m N/d_p^2 \qquad (12)$$

where N has replaced N_{lim} since we are now expressing limiting pressures rather than limiting plate numbers. This equation shows that if N plates are needed for a certain separation on a given kind of column (fixed d_p, D_m, η), an inlet pressure equal to or greater than p_c must be available. Once again the demands on light chromatography are relatively small due to the slight magnitude of the ηD_m product.

The nature of Equation 12 for gas chromatography is altered by the strong dependence of the gaseous diffusion coefficient, $D_g = D_m$, on pressure.

This can be expressed by

$$D_g = D_g'/p \qquad (13)$$

where D_g' is the value of the diffusion coefficient at unit pressure. The mean value of D_g for use in Equation 12 can be obtained by using the length–average pressure, $\bar{p}$, in place of p. Under vacuum outlet conditions, with p_c as the inlet pressure, $\bar{p}$ is found to equal $2p_c/3$. Consequently

$$D_g = 3D_g'/2p_c \qquad (14)$$

Upon substituting this back into Equation 12 we obtain

$$p_c' \quad (6\varphi\gamma\eta \, D_g'N/d_p^2)^{1/2} \tag{15}$$

Although the gradients existing in gas chromatography have been rather loosely treated here, this equation differs from the earlier rigorous form only by the small numerical constant of $\sqrt{2/3}$.

The development given above indicates the limits associated with single chromatographic columns. One can imagine column segments joined together by pumps of low dead volume such that each segment experiences the maximum possible pressure drop. Under these circumstances the arguments given above would apply to the individual segments rather than to the column as a whole.

LITERATURE CITED

(1) Giddings, J. C., *Anal. Chem.* **34**, *314 (1962)*.
(2) Giddings, J. C., Stewart, G. H., Ruoff, A. L., *J. Chromatog.* **3**, 239 (1960).
(3) Hamilton, P. B., *Anal. Chem.* **32**, 1779 (1960).
(4) Knox, J. H., *J. Chem. Soc.* 1961, p. 433.

Received for review April 1, 1964. Accepted June 1, 1964. Work supported by a research grant from the National Science Foundation.

Reprinted from *Anal. Chem.* **1964**, *36*, 1890–92.

The field of hyphenated techniques using IR spectrometry was born with the back-to-back publication of two articles on GC-IR interfaces in 1964. This article by Bartz and Ruhl takes precedence because it was received for review five weeks earlier than a similar paper by Wilks and Brown! Although the sensitivity of modern instruments is almost a million times better than that of the system described here, the nature of the interface is very similar to that of many contemporary instruments.

Peter R. Griffiths
University of Idaho

Rapid Scanning Infrared–Gas Chromatography Instrument

A. M. BARTZ and H. D. RUHL

The Dow Chemical Co., Chemical Physics Research Laboratory, Midland, Mich.

This instrument was designed to utilize the ability of gas chromatography (GC) to physically separate a multicomponent chemical sample into its individual components and the ability of infrared spectroscopy to specifically identify reasonably pure compounds. The effluent from a GC column (vaporized sample plus helium) is passed through a heated light pipe which serves as an infrared absorption cell with a large optical path length-to-volume ratio. The infrared absorption spectrum of the vapor sample is obtained by using two single-beam grating spectrometers in parallel. One spectrometer covers the range 2.5 to 7 microns while the other scans from 6.5 to 16 microns. By using two spectrometers, a high chopping rate, and fast recorders, a complete spectrum may be obtained in 16 seconds comparable in quality to a normal 12-minute scan by commercial spectrometers. The high scanning speed is necessary if an IR spectrum is desired of each successive GC peak of a multicomponent sample.

Previous techniques of combining infrared and gas chromatography have required condensation of the GC sample (effluent) and obtaining an IR spectrum of the condensed sample by use of either a microcell or by attenuated total reflectance. Both methods are tedious, slow, and rather difficult since extremely small quantities of condensed sample are involved.

An instrument was needed which would obtain an infrared spectrum of the GC sample directly, without condensation, and do it quickly so that closely spaced (in time) successive GC peaks could be readily identified. There were two problems to be solved: design a satisfactory sample cell and obtain a good IR spectrum in 30 seconds or less preferably using a

standard room temperature IR detector. The infrared absorption cell had to satisfy the following requirements: small sample (10 to 20 ml., such that GC peaks of 10-second halfwidth at a 1-ml.-per-second flow rate would fill the cell), large path length to volume ratio for maximum sensitivity, operable at 250 °C., short sample flushing time, good optical transmittance, and good corrosion resistance.

A cell satisfying these requirements is a rectangular light pipe. In this cell the inside walls are made highly reflecting to propagate the light through the cell by multiple reflection. The cell used in obtaining the spectra has inside dimensions of $1/8 \times 1/2 \times 12$ inches (volume of 12 ml.). Any metal with good corrosion resistance and capable of being polished to a highly reflecting finish may serve as the body material. For the particular optical arrangement we used, the 12-inch geometric length gave an effective optical path length of 16 inches. The transmittance is about 40% and is essentially constant with wavelength in the 2.5- to 16-micron range. The sample flushing time is 2 seconds using a nitrogen purge of 25 ml. per second (2 p.s.i.).

The maximum scanning rate in an infrared spectrometer is ultimately limited by the detector response time. The detector, which sees modulated or chopped radiation, must receive a certain minimum number of chops (bits of information) per wavelength interval (determined by the required resolution) to present a good profile of the spectrum being scanned. Rapid scanning therefore requires a high chopping rate; unfortunately room temperature IR detectors have decreasing sensitivity for increasing chopping rates thereby imposing a limit to scan rate.

A performance goal in designing this instrument was that it should have resolution at least equal to that of a bench-top prism spectrometer and be able to obtain spectra of GC peaks

30 seconds apart. Since the spectra were to be used solely for qualitative purposes the wavy I_0 of single beam spectrometers was acceptable. The problem of reducing the scan time to less than 30 seconds was solved in the following manner: the radiation from a single infrared source, after passing through the light pipe is flickered with a reflecting chopper to two different single beam grating spectrometers. One spectrometer covers the range 2.5 to 7 microns while the other concurrently covers the range 6.5 to 16 microns. Since chopping is necessary for each spectrometer normally, this method halves the detector response requirements at no cost in energy. It does, however, require single beam operation. The use of two spectrometers in parallel further reduces scan time in that each spectrometer uses only one grating in first order—no time is spent switching gratings.

EXPERIMENTAL

Apparatus. Figure 1 indicates the optical path of the infrared radiation. The radiation from the Nernst glower, S, is focused on the entrance end of the IR cell by means of spherical mirror, $M1$. The image is transferred through the pipe by multiple reflection and is directed onto spherical mirror, $M3$, by plane mirror, $M2$. The rotating chopper reflects the beam from $M3$ onto spherical mirror, $M8$, one half of the time and allows the light to proceed to spherical mirror, $M4$, when it is not reflecting the beam. Mirrors, $M8$ and $M4$, fill out the entrance slits of the long wavelength and short wavelength spectrometer, respectively. The light from the entrance slit is dispersed by the grating of each spectrometer and focused on the exit slit which permits radiation of a very narrow wavelength interval to pass and be focused onto the thermocouple detector. Each spectrometer uses two transmission filters to remove higher order light. These are flipped in and out at the

appropriate time by an air piston-lever arrangement, the air pistons being controlled by cam actuated valves. The signal from each detector is amplified, synchronously demodulated, and recorded by a dual channel Sanborn recorder.

Scanning action is accomplished by rotating the gratings by means of cams and cam follower arms. The slits are programmed to give nearly constant energy by using cams to rotate an arm attached to a spindle which acts as the pivot point of a lever. The slit jaws are fastened to bars attached (through flexible connections) to the lever arm. The wavelength scanning and slit program cams are mechanically coupled and driven by a six-speed motor.

The sample handling arrangement utilizes three solenoid valves which are controlled by a panel switch. The valves have three conditions of operation: purge, fill, and seal, described as follows:

Purge: GC effluent passes through 1 to vent, N_2 passes through 2, through the IR cell, past 3 to vent.

Fill: GC effluent passes by 1, by 2 to the IR cell and by 3 to vent. N_2 is blocked at 2.

Seal: GC effluent passes through 1 to vent. N_2 is blocked at 2, 3 is shut off, thereby sealing any gas in the IR cell.

This arrangement allows the operator to select the desired GC peaks for spectra (by-passing others if he wishes) and keep the sample in the cell at a fixed concentration for the scan and for possible rescans under different spectrometer or recorder conditions.

Figure 1 illustrates the 12-inch cell in place with the top half of the magnesium heater block removed. The cell windows are NaCl sealed with silicone rubber adhesive. The heater block is heated by 12 symmetrically spaced, 100-watt cartridge heaters. The cell temperature is monitored by a thermocouple with a dial indicator.

The overall instrumental arrangement is as indicated in Figure 1 with the infrared source compartment, IR cell heater block, spectrometer case, and the GC equipment mounted on a table 46×60 inches. A $25- \times 56$-inch relay rack houses the IR amplifiers, power supply, Sanborn recorder, and a control panel for the spectrometers and the sample cell. Operation of the instrument is simplified by using microswitches actuated by cams which are mechanically coupled to the scanning cams to automatically stop the recorder chart advance at the end of the useful scan and stop the scan when the grating and slit program cams are in a ready (for rescan) position. During the reset portion of the scanning cam rotation, the transmission filters are also reset. The dry air used to drive the filter flippers also serves as a purge to remove atmospheric H_2O from the spectrometer case. Scanning action is initiated by a

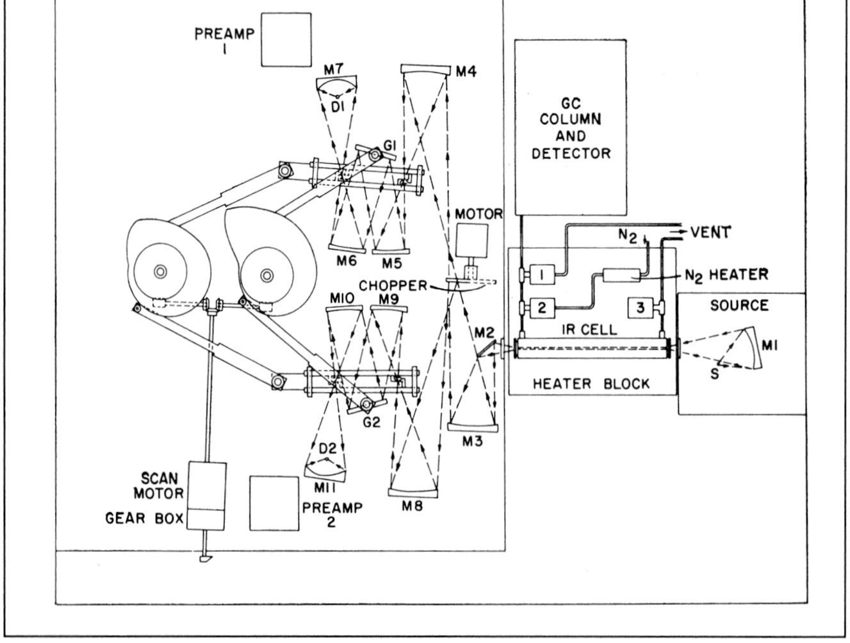

Figure 1. Optical path of infrared radiation

push-button switch. Scanning speed is selected by a six-position knob and spectral frequency is indicated by a counter whose reading is linearly proportional to the cm.$^{-1}$ position of each spectrometer.

RESULTS

Each half of the spectrum is normally recorded on a chart, 50 mm. (the transmittance scale) by 80 mm. (the linear wavenumber scale). The charts are printed adjacent to each other on a single roll. Wavenumber identification of unknown absorption bands is made by overlaying a transparent calibration chart, readable to about ±0.1 mm. or ±3 cm.$^{-1}$ for the

short wave spectrometer and ±1.2 cm.$^{-1}$ for the long wave spectrometer. Reproducibility error is small relative to the reading error. Figure 2 shows a scan of polystyrene film with the two halves of each spectrum rearranged to look more familiar for comparison with the standard spectrum on top. The spectrum on top was obtained with a laboratory prism spectrometer with a normal 12-minute scan time. The spectrum in the middle was obtained in 16 seconds and demonstrates comparable resolution to the prism instrument. Increasing scan time to 80 seconds noticeably improves the resolution as evidenced by the bands in the 3.5-micron region. There is no further improvement by increasing the scan time addition-

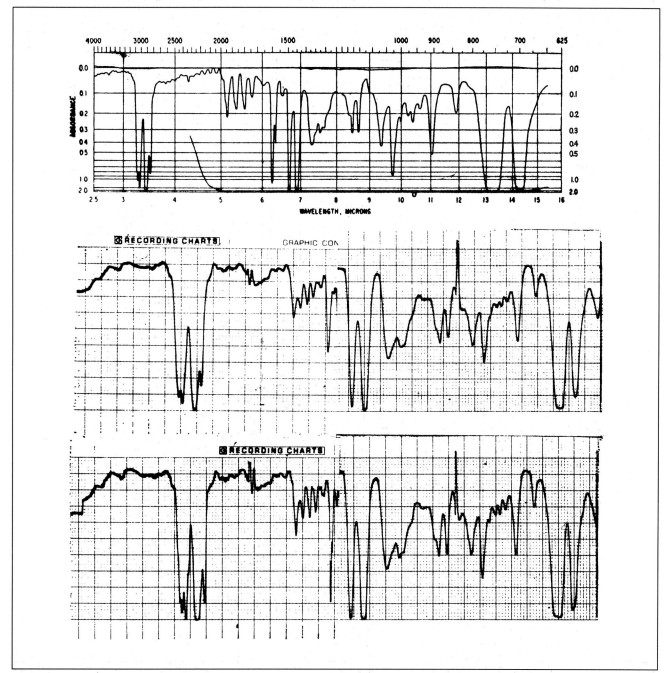

Figure 2. Top: standard spectrum of polystyrene film run on lab. prism spectrometer. Middle: polystyrene film run on this instrument in 16 seconds. Bottom: polystyrene film run on this instrument in 80 seconds

ally. At a chopping rate of 24 c.p.s. the 16-second scan represents a spectral profile made from 384 bits of information or about 6 cm.$^{-1}$ per bit for the short wavelength half of the spectrum.

Figure 3 shows a spectrum (bottom) made from a 10-µl. sample of 5% acetone in toluene with the acetone chromatographically separated from the toluene, captured in the IR cell, and scanned in 16 seconds. The library reference standard spectrum (top) is a scan of acetone vapor obtained with a 10-meter path cell. Again the two halves of the spectrum have been rearranged to look more familiar to demonstrate that identification of the GC peak may readily be made by comparing its IR spectrum with a standard.

This instrument was designed for sizable components, 0.5 mg. or more; if increased sensitivity is needed, a longer cell would have to be used. For this cell, at a carrier flow rate of 1 ml. per second, 12 seconds is required to fill the cell, 16 seconds to scan, 4 seconds for resetting and purge, for a total of 32 seconds. This is the minimum time required between adjacent GC peaks.

Figure 4 shows spectra (rearranged) of three alcohols. In each case the GC sample was a 10-µl. load of a 10% mixture of alcohol in toluene and scanned in 16 seconds. Identification by comparison with library standards is clearly possible.

Figure 5 shows spectra (normal chart presentation) of some chlorobenzenes. This also indicates that complete GC separation of peaks is not required to obtain quite pure samples for IR identification, if the volume of the IR cell is small relative to the available sample volume. By filling the IR cell from the leading edge of peak 1 (taking a narrow slice out of 1) very little contamination from peak 2 is found since bands due to o-dichlorobenzene (peak 2) do not appear in the spectrum of peak 1 (identified as p-dichlorobenzene). The p-dichlorobenzene is an 8% component again using a 10-µl. load and 16-second scan time. The spectrum on the right is included to demonstrate that relatively high-boilers may be easily separated chromatographically and identified by their IR vapor spectrum. 1,2,3,5-Tetrachlorobenzene has a boiling point of 246 °C.

Our experience with this instrument has indicated it requires some familiarization because the spectra are of vapor samples and are split into halves, but the advantage of obtaining an IR spectrum in 16 seconds to allow specific identification of each GC peak in a complicated sample makes it all worthwhile.

ACKNOWLEDGMENT

The authors are grateful to L. Gould and his associates in the Mechanical Fabrication and Development Laboratory of

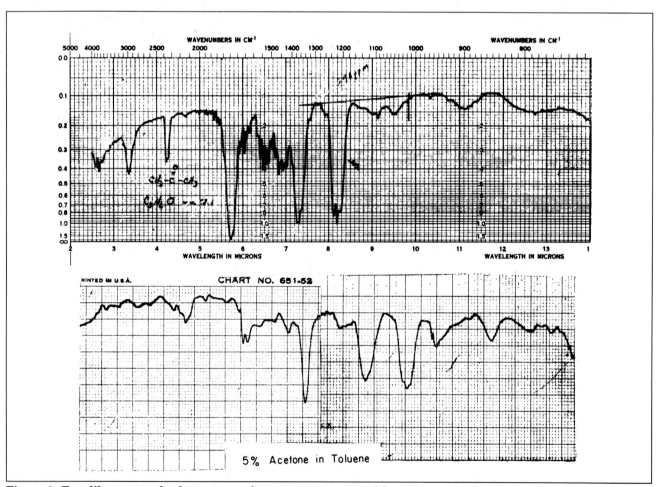

Figure 3. Top: library standard spectrum of acetone vapor run with a 10-meter cell. Bottom: spectrum of the acetone peak of a GC separation of a 5% acetone in toluene mixture using a 10-µl. sample

The Dow Chemical Co. for valuable assistance in the design and construction of the light pipe cell. The contributions of C. Pratt, L. Westover, L. Herscher, W. Felmlee, and D. Erley (of this laboratory) are also very much appreciated.

Received for review April 1, 1964. Accepted June 24, 1964. Pittsburgh Conference on Analytical Chemistry and Applied Spectroscopy, Pittsburgh, Pa., 1964.

Reprinted from *Anal. Chem.* **1964**, *36*, 1892–96.

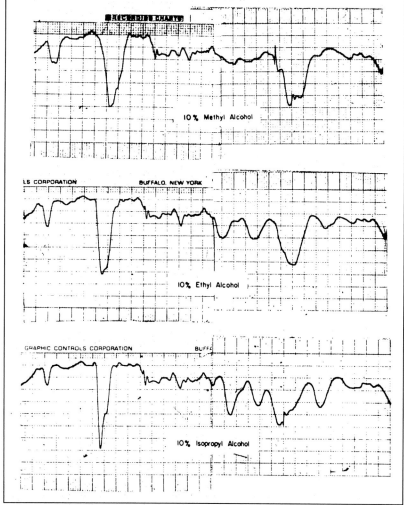

Figure 4. IR vapor spectra of several alcohols which were separated chromatographically from toluene

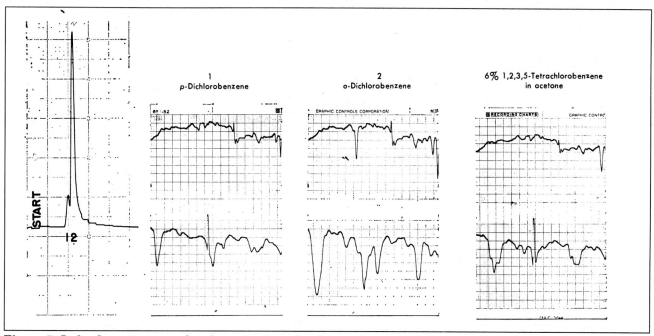

Figure 5. Left: chromatogram showing two unresolved components. Middle: IR spectra of each peak permitting identification. Right: IR vapor spectrum of chromatographic cut of 246° C. boiler

This scientific communication is one of the most important in analytical plasma spectroscopy. It pointed the way toward analytically useful ICPs, with the introduction of aerosol-containing analyte by an independent argon stream in a nebulizer tube. The introduction of coolant gas via laminar flow led to plasma stability which, in turn, led to reasonable experimental precision. These two features boded well for commercial production and for subsequent analytical uses. Even today, ICP torches are similar in design to those described here.

Joseph Caruso
University of Cincinnati

Induction-Coupled Plasma Spectrometric Excitation Source

Richard H. Wendt and Velmer A. Fassel
Institute for Atomic Research and Department of Chemistry, Iowa State University, Ames, Iowa

Sir: During the past two years we have employed various experimental configurations for forming plasmas of the inductively coupled type (9, 10) and have evaluated their potentialities as a source for the excitation of atomic and molecular spectra. One of these configurations was particularly useful as a practical source for analytical spectrometry. In this configuration, an induction coupled plasma is maintained by a high-frequency, axial magnetic field in a laminar flow of argon at atmospheric pressure. No electrodes are in contact with this discharge as opposed to the capacitively coupled plasma—i.e., high-frequency torch or radio-frequency discharge (6, 7)—and d.c. plasma jet. When the discharge gas in the induction-coupled plasma is pure argon, a temperature of about 16,000° K. is obtained at or near thermal equilibrium (9, 10). To contain a discharge of this temperature without wall contamination, a carefully controlled, laminar flow of cold argon surrounds the plasma. Atomic spectra are obtained by introducing ultrasonically generated aerosols into the argon flow which supports the plasma.

The recent appearance of a paper by Greenfield, Jones, and Berry (3) describing a similar source has prompted this communication on our independent observations. Our experimental arrangement for maintaining the plasma and for introducing an aerosol into the discharge is considerably different from the Greenfield design and possesses definite advantages for analytical applications.

The laminar flow of gas employed to support our plasma differs distinctly from other published induction-coupled designs. Single-tube versions described by Reed (9, 10), Cannon (2), Mironer and Hushfar (8), and Kana'an and co-workers (1, 4) have employed tangential gas inlets to form vortex flows of high velocity at the walls and of low velocity in the center. The resulting recirculation of the hot gas was considered essential for operation of these designs. In the laminar-flow plasma described in this communication, there is no recirculation. In fact, a pure argon plasma has been operated on flow rates as high as 4.6 liters/minute in the plasma tube (which corresponds to a velocity of 38 cm./second) without any decrease in stability or intensity.

Tangential inlets are also used in both tubes of the dual-tube designs of Greenfield, Jones, and Berry (3) and Lepel (Lepel High Frequency Laboratories, Woodside, N.Y.), but their designs possess several inherent disadvantages in comparison to the laminar-flow configuration. First, a vortex flow has more turbulence than a laminar flow and this turbulence may decrease the stability of the discharge. This loss of stability is drastically enhanced by minor distortions in the walls of the tubes, whereas our plasma has been operated successfully with the coolant tube purposely bent and kinked. Second, the addition of aerosols of solutions or powders to a tangential flow of gas tends to cause them or their vapors to be thrown against the inner wall of the coolant tube, thus decreasing the transmittance of the tube and devitrifying the quartz. Because the coolant flow in our design is laminar and mixes only slightly with the hot vapors in the core of the discharge, the coolant tube remains clean and useful throughout months of continuous use. The Forrest plasma torch (Forrest Electronics Corp., Las Vegas, Nev.) is the only other dual-tube, laminar-flow design to our knowledge, but no data have been published on its performance.

As shown in Figure 1, an ultrasonic atomizing system, similar in design to that of West and Hume (11), is used to produce an aerosol of the sample solution with an average droplet diameter of about 5 microns (5). These droplets are carried into the discharge by an independent stream of argon via a small central tube.

2855-8/94/0178$08.00/0   ©1994 American Chemical Society   *Milestones in Analytical Chemistry*

The ultrasonic aerosol generator allows the introduction of solutions of virtually any sample concentration (so long as the viscosity is not too high) and of any degree of acidity or basicity. Thus, absolute detection limits are greatly improved. Organic solvents may also be used if the sample container is made of an appropriate material.

Although variations in viscosity, surface tension, and sample depth will slightly affect the production rate and droplet size of the aerosol, the primary factor controlling the rate of addition of aerosol to the plasma is the argon flow rate, which may easily be reproduced.

EXPERIMENTAL FACILITIES

Plasma. Power supply: Lepel High Frequency Laboratories, Woodside, N.Y. Model T-5-3-MC-J-S generator; 3.4-mc. frequency; 5-kw. nominal output.

Coil: Lepel pancake-concentrator type, 5 turn.

Coolant tube: clear fused quartz, 22-mm. i.d., 24-mm. o.d.; 22 cm. total length, extending 11 cm. beyond coil.

Plasma tube: clear fused quartz, 16-mm. i.d., 18-mm. o.d.; 12.5 cm. total length, terminating 17 mm. below top of concentrator ring; centered within coolant tube by four "feet" located 3 cm. from open end.

Aerosol tube: borosilicate glass, 5-mm. i.d., 7-mm. o.d., terminating 4 mm. below end of plasma tube.

Base: brass; double O-ring seals on each quartz tube; optional screen to ensure laminar flow.

Flow rates to discharge: coolant, 22 liters/minute of Ar; plasma, 0.4 liter/minute of Ar; aerosol, 0.5 liter/minute of Ar which carries 0.12 ml. solution/minute.

Ignition: graphite rod, not grounded; lowered into high field region until plasma is formed, then withdrawn.

Aerosol Generator. Power supply: Siemens (Siemed, Inc., Hinsdale, Ill.) Sonostat 631; 12-watt output from transducer; 870-kc. frequency.

Lens: acrylic plastic (Plexiglass), plano-concave; 7.5-cm. focal length; attached to transducer with Eastman 910 cement.

Sample container: inner glass joint, ẟ 50/50, with 0.010-inch Plexiglass bottom; 10- to 25-ml. capacity.

Aerosol chamber: borosilicate glass tube, 28-mm. o.d., 7 inches long, sealed into ẟ 50/50 outer glass joint.

Spectrograph. Jarrell-Ash, Boston, Mass., 1.5-meter Wadsworth Model 78.

OBSERVATIONS

The induction-coupled plasma has the general appearance of a bright flame with three regions or zones. The core or first region, as outlined in Figure 2, is centered at the top of the concentrator ring, and is about 8 mm. in diameter, 25 mm. long, nontransparent, and brilliant. The

core fades into the second region which is about 16 mm. in diameter and about 75 mm. long (depending to a great degree on the length of the coolant tube and whether aerosol is being added). The second region is also bright but slightly transparent. The third region or tailflame is distinctly separated from and extends about 15 cm. above the tip of the second region. When the plasma is supported by pure argon, the tailflame is barely visible because the strong argon emission lines are outside the range of sensitivity of the eye. The tailflame assumes typical flame colors when solutions are added to the plasma.

The radiation from the core includes an intense continuum extending from about 3000 to 5000 A. which apparently arises from recombination and perhaps cyclotron radiation. The intensity of the continuum is sharply reduced in the second region and is negligible in the tailflame. A fairly well developed spectrum of neutral argon is emitted from the core and second region but only two Ar II lines (4806 and 7589 A.) have been observed and these were weak even from the center of the core. The stronger Ar I lines are weak but still detectable in the tailflame more than 7 cm. above the tip of the second region. Because these lines have excitation potentials of 13.0 to 14.5 e.v., a significant concentration of argon atoms in the metastable state probably exists in the tailflame.

When an aqueous aerosol is introduced into the discharge, the over-all intensity of the plasma is reduced somewhat and the expected hydrogen Balmer-series lines and the 3064-A. OH band system appear in the spectrum. The hydrogen lines are strong and broad in the core, but become sharper and weaker in the second region, and disappear entirely before the

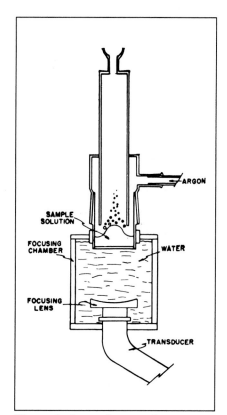

Figure 1. Aerosol generator

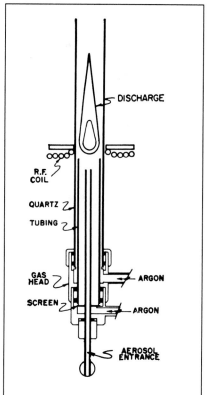

Figure 2. Plasma torch

tailflame begins. The OH band head at 3064 A. is of moderate and relatively constant intensity throughout all three regions. The other OH band heads are weak or not visible in a normal exposure of the emission spectrum.

The portion of the tailflame which extends beyond the coolant tube emits band systems of O_2, N_2, NH, and N_2^+ in addition to OH bands. When the coolant tube is short—e.g., Greenfield, Jones, and Berry—the second region also extends into the atmosphere and these band systems, especially N_2^+, become considerably more intense. The discharge outside the tube, especially the tailflame, wanders slightly because of air currents and cannot be expected to have a stability as high as the discharge within the tube.

Neutral atom lines of most elements exhibit their best line to background ratio approximately 9 cm. above the core. High energy neutral atom lines and most ion lines exhibit their best ratio at 4.5 cm. above the core although for many elements the ratio is not significantly different.

Typical detection limits for a representative list of elements in aqueous solutions are given in Table I. The detection limit is defined here as that minimum concentration in solution which emits a detectable spectral line observable over the background.

The observation that useful lines of most elements can be detected at trace concentration levels indicates a wide range of analytical applications of this discharge. The use of aqueous solutions rather than organic solvents eliminates the band systems of C_2 and CH from the emission spectra of the discharge. The background spectrum is thus low, especially from within the coolant tube. In addition, the discharge offers a wide range of excitation energy for special applications. Greenfield, Jones, and Berry (3) have confirmed our expectations that matrix effects are negligible. Although only a convenience advantage, the discharge emits considerably less audible noise than a Beckman burner or a constricted d.c. arc plasma jet. Our observations indicate that this combination of plasma and aerosol generator is a practical and versatile source for analytical spectrometry. We are now exploring further analytical applications and will discuss these in a more complete communication.

LITERATURE CITED

(1) Beguin, C. P., Kana'an, A. S., Margrave, J. L., *Endeavor* **23** (89), 55 (1964).
(2) Cannon, H. R., "A Study of an Induction-Coupled Plasma Operating at 400 Kilocycles," M.S. thesis, Air Force Institute of Technology, Wright-Patterson AFB, Ohio, GA/Phys/62-2, **AD-286404** (1962).
(3) Greenfield, S., Jones, I. Ll., Berry, C. T., *Analyst* **89**, 713 (1964).
(4) Kana'an, A. S., "Studies at High Temperature," Ph.D. thesis, University of Wisconsin, Madison, Wis., 1963.

Table I. Detection Limits of Elements

Element	Line, A.	Detection limit, µg./ml. 4.5 cm.	Detection limit, µg./ml. 9.0 cm.
Al	3961	3	3
As	2780	25	
Ca	4226	0.5	0.2
	3933	0.5	0.8
Cd	3261		20
Cr	3578		0.3
Cu	3247	1.2	0.2
Fe	3719		3
La	4086	50	50
Mg	2852	2	2
Mn	4030	2	1
Ni	3524		1
P	2535	10	
Si	2516	3	
Sn	3034	50	50
Sr	4607	0.09	0.09
Ta	2685	16	
Th	4019	40	40
W	4008	3	3
Zn	4810	30	
Zr	3438	15	15

(5) Lang, R. J., *J. Acous. Soc. Am.* **34**, 6 (1962).
(6) Mavrodineanu, R., Hughes, R. C., *Spectrochem. Acta* **19**, 1309 (1963).
(7) Mavrodineanu, R., Boiteux, H., "Flame Spectroscopy," pp. 51–3, Wiley, New York, 1965.
(8) Mironer, A., Hushfar, H., "Radio Frequency Heating of a Dense, Moving Plasma," presented at AIAA Electric Propulsion Conference, Colorado Springs, Colo. (March 1963); preprint 63045-63, American Institute of Aeronautics and Astronautics, New York, N. Y.
(9) Reed, T. B., *Intern. Sci. Technol.* (6), 42 (1962).
(10) Reed, T. B., *J. Appl. Phys.* **32**, 821 (1961).
(11) West, D. C., Hume, D. N., *Anal. Chem.* **36**, 412 (1964).

Received for review February 24, 1965. Accepted April 1, 1965. Work was performed in the Ames Laboratory of the U.S. Atomic Energy Commission.

Reprinted from *Anal. Chem.* **1965**, *37*, 920–22.

Thin Layer Electrochemical Studies Using Controlled Potential or Controlled Current

DONALD M. OGLESBY,[1] **SVERRE H. OMANG,**[2] and **CHARLES N. REILLEY**

Department of Chemistry, University of North Carolina, Chapel Hill, N. C. 27515

A new, versatile electrode design suitable for thin layer electrochemical studies was developed, and its accuracy and practicality were verified. The use of both platinum-metal and mercury-coated platinum as working electrode materials is discussed. Errors resulting from residual currents were investigated. Rapid thin layer chronopotentiometric studies of the ferri-ferrocyanide couple at a polished platinum electrode gave results in good agreement with theory. Diffusion coefficients were determined using the chronopotentiometric technique. The basic equations describing chronoamperometry in the thin layer are presented and experimentally verified using the Cu(II)-Cu(0) and the ferri-ferrocyanide couples.

Theory and technique of studying small solution thicknesses chronopotentiometrically were introduced by Christensen and Anson (2). These authors later applied the technique to the study of kinetics of hydrolysis of *p*-benzoquinoneimine (3). Hubbard and Anson (5) published thin layer electrode designs which were more versatile than those employed previously. Osteryoung and Anson (8) used thin layer chronopotentiometry to show that adsorbed iodide ion was not oxidizable at the same potential as the solution species.

This work presents an improved thin layer electrode design. The design allows the use of a variety of electrode materials as working electrodes, including the smooth platinum and mercury-coated platinum surfaces investigated here. Solution thicknesses from 1×10^{-3} cm. to 1 cm. are readily obtained. The problem of iR drop between the working and the

reference electrodes is reduced significantly, and the problem of contamination of the solution by oxygen is overcome. The same basic assembly can be used to study a thin layer of solution bounded by one working electrode and an inert barrier; two equipotential working electrodes; or two working electrodes with independent potential and/or current control (Figure 1).

EXPERIMENTAL

The basic thin layer electrode assembly was constructed from a 0- to 2-inch micrometer with attachment (L. S. Starrett Co., Athol, Mass.). A precision thimble (Starrett No. T2211) allowed direct reading to one ten-thousandth of an inch. To modify the micrometer for use as an electrode, the spring loading in the precision thimble was replaced by a friction washer. A platinum disk was silver-soldered onto the face of the micrometer spindle and the platinum tip machined flat and lapped to a mirror finish. The area of the electrode face, calculated from the outside diameter of the micrometer spindle, was 0.278 sq. cm. The Teflon collar (Figure 2 *b*) was pressed onto the micrometer spindle, resulting in a solution-tight seal between the metal and the Teflon. The collar was positioned so that the platinum tip extended approximately 0.001 inch past the face of the collar (Figure 2).

A cup machined from Teflon, designed to hold the sample solution, was mounted on a detachable anvil (Starrett 212). An optically flat glass disk was press-fitted into the bottom of the Teflon cup and rested flat against the face of the detachable anvil (Figure 2 *d*).

A circular piece of platinum gauze, placed around the inside of the cup and attached to a platinum wire lead, constituted the auxiliary electrode. A Lucite cover was placed over the cup to maintain a nitrogen atmosphere over the solution. The cover, constructed in two sections for easy placement and removal, was provided with holes to allow insertion of the arm

---

[1]Present address: Old Dominion College, Norfolk, Va. 23508.
[2]Present address: Central Institute for Industrial Research, Oslo, Norway.

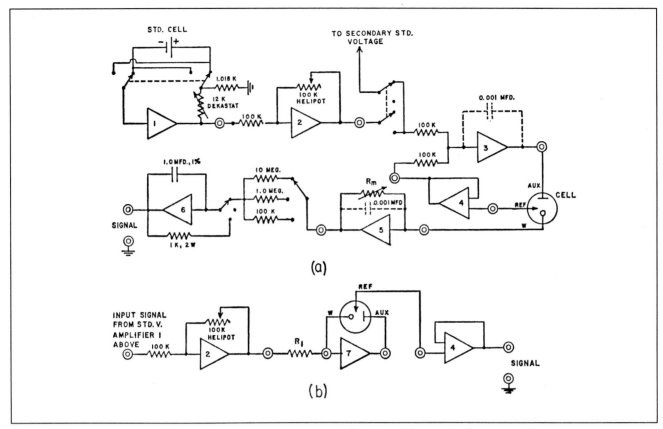

(a)

(b)

Figure 1. Schematic diagram of instrument

1. Standard voltage amplifier
2. Multiplier
3. Control amplifier
4. Follower
5. Current-measuring amplifier
6. Integrator
7. Current-control amplifier

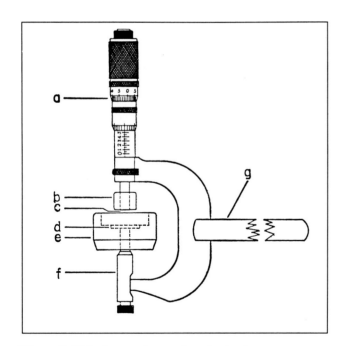

Figure 2. Thin layer micrometer electrode

a. Outer thimble for direct reading to one ten-thousandth of an inch
b. Press-fitted Teflon collar
c. Platinum face of the micrometer spindle
d. Flat glass disk pressed into the Teflon cup against the face of the detachable anvil
e. Cup machined from Teflon
f. Starret No. 212 detachable anvil
g. Stainless steel rod for mounting cell assembly

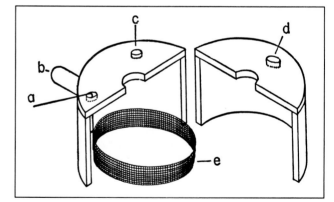

Figure 3. Lucite cover for cup of thin layer electrode

a. Auxillary electrode lead
b. Nitrogen inlet
c. Hole for reference electrode salt bridge
d. Hole for spout of deaeration bottle
e. Platinum gauze auxiliary electrode

of an S.C.E., the delivery tube of the deaeration bottle, and a stream of nitrogen (Figure 3).

The solution was deaerated prior to introduction into the Teflon cup. The deaeration bottle was fitted with a gas dispersion tube, a delivery tube, and a small vent. When the vent was covered by the thumb, the increased nitrogen pressure forced the solution to flow out of the delivery tube. The first few milliliters of solution were discarded to assure freedom from oxygen contamination.

Mercury-Coated Platinum Electrode. For the mercury-coated platinum electrode assembly, a Starrett T2F1 0- to 2-inch micrometer was employed. A platinum tip and Teflon collar were fixed to the spindle of the micrometer in the same manner as described for the platinum thin layer electrode. The platinum was coated with mercury by successively polishing with rouge and dipping into mercury, a technique similar to that used by Moros (7). A flat spatula, covered with a piece of filter paper impregnated with a fine grade of lens rouge, was used for polishing.

After thorough amalgamation, the tip was washed with a stream of mercury from a polyethylene wash bottle. Thorough coating of the surface was essential to obtaining reproducible runs. Excess mercury was removed by suction through a fine capillary connected to an aspirator. The mercury-coated surface could be used repeatedly unless the potential of the electrode was allowed to become sufficiently negative for appreciable hydrogen evolution or sufficiently positive for oxidation of the mercury surface.

Normally, the electrode surface was washed with mercury and the excess removed with the capillary before each run. The mercury-coated platinum electrode was placed in a nitrogen atmosphere as soon as possible after preparation to minimize oxidation of the surface by oxygen. When not in use, the electrode surface was covered with a nitrogen-filled balloon.

An S.C.E. was used as the reference electrode for both the controlled-current and controlled-potential studies. Philbrick UPA-2 and chopper-stabilized K2-W operational amplifiers, powered by a Philbrick R-300 power supply, with appropriate circuitry, were used to obtain the desired instrumentation. The input and feedback circuits are shown in Figure 1. A Sargent Model SR recorder was used for recording the integrated current-time curves and the potential-time curves for times longer than about 5 seconds. For the faster potential-time curves, a Sanborn Model 151-100A single channel recorder with a Model 150-400 drive amplifier and power supply and a Model 15-1800 stabilized d. c. preamplifier were used.

All solutions were prepared from J. T. Baker reagent grade chemicals. The $1.0 \times 10^{-3}M$ potassium ferri-cyanide was $0.5M$ in sodium sulfate supporting electrolyte. The $1.040 \times 10^{-3}M$ cupric nitrate solution was $1.0M$ in potassium nitrate supporting electrolyte.

After deaeration of the solution, the cup was filled from the deaeration bottle. The desired solution thickness was established by turning the thimble of the micrometer to the appropriate setting, based on a previous calibration. Electrical contact with the working electrode surface was made by attaching a lead to the body of the micrometer.

DISCUSSION

An important aspect of using thin-layer electrodes is accurate knowledge of the effective thickness of the solution (2). With the micrometer electrode different solution thicknesses are readily obtained. However, it is necessary first to establish the micrometer setting at which the solution thickness is zero. One method of finding the effective solution thickness is the following: using chronopotentiometry, a plot of $i\tau$ as a function of micrometer setting will yield an intercept at $i\tau$ equal to zero, which will represent a solution thickness of zero (2).

$$l = i\tau/nFAC°$$

For chronoamperometry, the analogous plot of Q, the coulombs of electricity passed, as a function of micrometer setting yields the zero-thickness setting at $Q = 0$. Once this is done for a given electrode, one may accurately and directly set any desired solution thickness from less than 1.0×10^{-3} cm. up to the depth of the solution in the cup. The screw of the micrometer is precisely machined so the reproducibility of the settings is usually limited by the precision with which the worker can read the scale. The calibration curves for the platinum thin-layer electrode are shown in Figure 4.

The intercepts were calculated by the least-squares method. The intercept based on the data from controlled potential electrolysis with integrated current was a micrometer reading of 0.15688, and the intercept based on controlled current data was at a reading of 0.15683, a difference of 0.00005 inch. This difference is small and consistent. The reproduc-

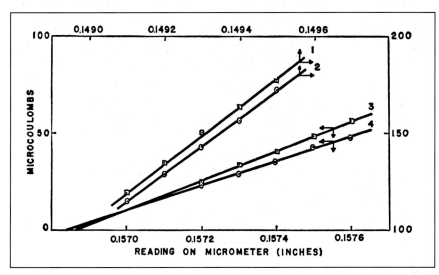

Figure 4. Calibration curves

mM Cu(NO$_3$)$_2$ in 1.0M KNO$_3$, Hg-coated Pt electrode
1. Chronopotentiometric data, $i = 5$ μa., intercept 0.14829 cm.
2. Chronoamperometric data, potential stepped from +0.15 to −0.20 volt; intercept, 0.14831 cm.

mM K$_3$Fe(CN)$_6$ in 0.5M Na$_2$SO$_4$, Pt electrode
3. Chronoamperometric data, potential stepped from −0.15 to −0.30 volt; intercept 0.15688 cm.
4. Chronopotentiometric data, $i = 2.5$ μa., intercept 0.15683 cm.

ibility of data at a given micrometer setting, using a given technique—e.g., controlled potential or controlled current—was $\pm 2 \times 10^{-5}$ inch (it should be noted that the two lines are different).

Figure 4 also shows the calibration curve for the mercury-coated platinum electrode, based on the data in Table I. Setting this electrode involves reproducing the mercury coating on the platinum face. Several runs representing renewal of the mercury surface by the procedure already given are summarized in Table I. A representative set of integrated current-time curves is shown in Figure 5.

The useful negative potential range is much greater with the mercury-coated platinum than with the polished platinum surface. Hydrogen evolution does not occur until approximately -1.3 volts $vs.$ S.C.E., as may be seen from Figure 6. In the study of adsorption and electrochemical processes, irreversible on platinum, by the thin-layer technique, the unique

properties of mercury as an electrode material can be utilized.

An important portion of the residual current in thin layer electrolysis is caused by the edge effect—diffusion of the electroactive species into the thin layer from the perimeter of the volume element. This cylindrical edge is exposed to semi-infinite diffusion from the bulk of the solution, causing a residual current proportional to the concentration of all species electrolyzed at the potential of the working electrode. Complete elimination of this edge effect poses a difficult design problem because a solution contact must be made with the reference and auxiliary electrodes.

Contributions to the current from reduction of surface metal oxides or catalytic reduction of the solvent will be more important in thin layer than in semi-infinite techniques. A typical thin layer of solution ($l = 2 \times 10^{-3}$ cm., $A = 0.3$ sq. cm., $C^{\circ} = 10^{-3}M$) contains only 6×10^{-10} mole of electroactive species, equivalent to approximately three monolayers on the

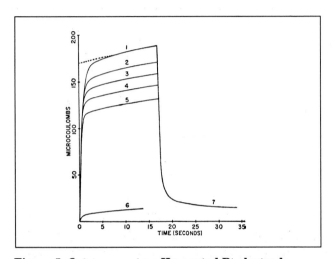

Figure 5. $Q\text{-}t$ curves at an Hg-coated Pt electrode

mM Cu^{+2} in $1M$ NO$_3^{-}$, potential stepped from $+0.15$ to -0.20 volt and *vice versa*. **Solution thickness: 1. 3.05×10^{-3}; 2. 2.80×10^{-3}; 3. 2.54×10^{-3}; 4. 2.28×10^{-3}; 5. 2.03×10^{-3} cm.; 6. background curve on $1M$ KNO$_3$; 7. stripping curve**

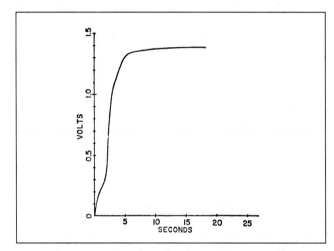

Figure 6. Chronopotentiogram of $1.0M$ KNO$_3$ with the mercury-coated thin layer electrode

$l = 2.28 \times 10^{-3}$ cm.; current $= 5$ µa.; electrode area $= 0.278$ sq. cm.

Table I. Reproducibility of Electrochemical Studies Using Mercury-Coated Electrode

Data taken on $1.040 \times 10^{-3}M$ Cu^{+2} in $1.0M$ KNO$_3$
Electrode area $= 0.278$ sq. cm.
Each average represents six different mercury coatings

Soln. thickness based on calibration curve of Figure 4, cm. $\times 10^{-3}$	Av. chronopot. transition time at 5 µa., sec.	Av. $i\tau$ µcoulombs	Rel. std. dev., %	Av. integrated current from chronoamp., µcoulombs	Rel. std. dev., %
2.03	23.8	119	±1.6	115	±1.7
2.28	27.0	135	±3.1	129	±2.0
2.54	30.0	150	±2.3	143	±2.3
2.80	32.8	164	±1.4	157	±1.6
3.05	35.6	178	±1.3	173	±1.9

electrode surface.

In practice, residual currents less than 1 μa. per sq. cm. have been attained by excluding oxygen, using small solution thicknesses, and minimizing the current contribution from electrolysis of hydrogen ion or formation or reduction of surface metal oxides.

To verify that residual current is in part caused by an edge effect, the Teflon collar was slipped slightly past the electrode face. The micrometer screw was turned to a setting which previously corresponded to a thickness of 2.03×10^{-3} cm. However, at this setting the Teflon collar touched the bottom of the cup, causing an upward force on the spindle of the micrometer. Recalibration showed that this caused the thickness to be 2.75×10^{-1} cm., as a result of backlash in the micrometer screw. The residual current was reduced by about one half, but the iR drop between the reference and working electrodes was increased, causing a shift in the measured reduction potential of Cu^{+2} by about -0.06 volt.

The error caused by residual current depends on the time required for the electrochemical measurement, and controlled current experiments can be carried out at shorter times than those indicated by Hubbard and Anson (5).

A practical lower limit to the length of transition times was not encountered using the electrode design shown in Figure 3. The ill-defined chronopotentiograms with transition times less than 10 seconds reported by Hubbard and Anson (5) might have been caused by the proximity of the working electrode to the auxiliary electrode; this results in poor current distribution over and large resistance drop across the electrode face. The large current required for reduction of the concentrated solutions used made the resistance factor more important. In the present work, transition times as short as 0.2 second were obtained without difficulty, as may be seen from the chronopotentiogram shown in Figure 7.

Shown in Figure 8 is a plot of the transition time *vs.* $1/i$ for different solution thicknesses. The linearity of these plots indicates good agreement with the theoretical relationship:

$$\tau = \frac{nFAC^\circ}{i} - \frac{l^2}{3D} \qquad (1)$$

The transition times were measured graphically by the "method of Kuwana" (9). Equation 1 should hold for times down to about 0.2 second because the exponential term of Equation 2

$$\frac{nFAlC^\circ}{i} - \frac{l^2}{3D} = \tau - \frac{2l^2}{\pi^2 \Delta} \sum_{k=1}^{\infty} \frac{1}{k^2} \exp\left(-\frac{Dk^2\pi^2\tau}{l^2}\right) \qquad (2)$$

is negligible at these times.

Diffusion Coefficients. Because a plot of τ *vs.* $1/i$ has an intercept with the τ axis at $-l^2/3D$ and because l is determined on the basis of the calibration plot of Figure 4, diffusion coefficents can be determined without a knowledge of the solution concentration or the electrode area. The $l^2/3D$ intercepts of the plots shown in Figure 8 give the values shown in Table

III. The data for Figure 8 and the values given in Table III are based on Table II. The average value of 0.45×10^{-5} sq. cm./ second found is about one half that given by Kolthoff and Lingane (6). Application of the thin layer electrochemical method to the study of diffusion coefficients is being further investigated.

Thin Layer Chronoamperometry. Consider an electroactive solution species within the finite boundaries at zero and l, such that $0 < x < l$. Also, consider the electrode surface to be at l. With the initial condition $C_{(x,0)} = C^\circ$ and the boundary conditions $[\partial C_{(x,t)}/\partial x]_{x=0} = 0$ and $C_{(l,t)} = 0$ for chronoamperometry, one may readily derive an expression for $C_{(x,t)}$ by Laplace transforms or by reflection and superposition (4). The expression, previously derived for similar problems in heat conduction, converges most rapidly for small times and is given by (1).

$$C_{(x,t)} = C^\circ \sum_{n=0}^{\infty} (-1)^n \times$$
$$\left[\text{erfc} \frac{(2n+1)\,l - x}{2\sqrt{Dt}} + \text{erfc} \frac{(2n+1)\,l + x}{2\sqrt{Dt}} \right] \qquad (3)$$

Another form of this equation, which converges more rapidly for solutions at long times, is given by:

$$C_{(x,t)} = \frac{4C^\circ}{\pi} \times \sum_{n=0}^{\infty} \frac{(-1)^n}{(2n+1)} \left[\exp\left(-\frac{D(2n+1)^2\pi^2 t}{4l^2}\right) \right] \left[\cos \frac{(2n+1)\,\pi x}{2l} \right] \qquad (4)$$

Equation 3, valid for short times, becomes more unwieldy at longer times and is not a valid solution of the differential equation at the limit as t approaches infinity. Similarly, Equation 4 is not a valid series solution as t approaches zero. This is also true of the chronopotentiometric equations presented by Christensen and Anson (2). Equation 1 of those authors is the series valid at long times, and Equation 2 is valid for short times.

The expression for current at the electrode surface as a function of time may readily be obtained by multiplying $\partial C_{(\partial x,t)}/x$ by $nFAD$ and substituting $l = x$.

$$i_{(l,t)} = \frac{nFAD^{1/2}C^\circ}{\pi^{1/2}t^{1/2}} \times \sum_{k=0}^{\infty} (-1)^k \left[\exp\left(-\frac{k^2l^2}{Dt}\right) - \exp\left(-\frac{(k+1)^2l^2}{Dt}\right) \right] \qquad (5)$$

This equation shows that when $l^2 < Dt$, the current decays rapidly. The use of controlled potential with integrated current allows not only precise control of the potential through-

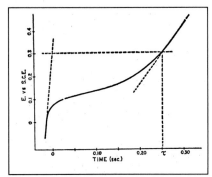

Figure 7. Short-time thin layer chronopotentiogram

Platinum thin-layer electrode; $1.0 \times 10^{-3}M$ $K_3Fe(CN)_6$ in $0.5M$ Na_2SO_4; $l = 1.12 \times 10^{-3}$ cm.; current = 100 μa.; electrode area = 0.278 sq. cm.; $\tau = 0.25$ sec.

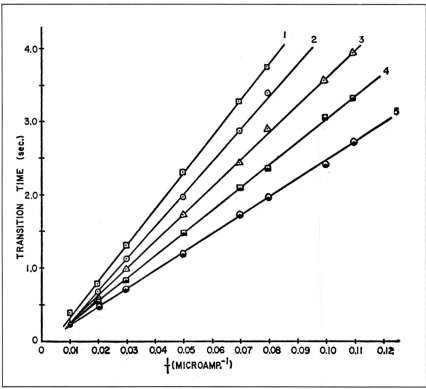

Figure 8. Dependence of τ on $1/i$ at different solution thicknesses

Pt electrode area, 0.278 sq. cm.; mM $K_3Fe(CN)_6$ in $0.5M$ Na_2SO_4.
Solution thickness: 1. 1.88×10^{-3}; 2. 1.62×10^{-3}; 3. 1.37×10^{-3}; 4. 1.12×10^{-3}; 5. 0.864×10^{-3} cm. Although only data to $\tau \cong 4$ sec. are shown, the data included transition times from about 0.25 sec. to about 20 sec. with the same linearity shown here

Table II. Chronopotentiometric Data Taken at Current Densities Such That $l^2/3D$ Term Is Significant

$1.0 \times 10^{-3}M$ $Fe(CN)_6^{-3}$ in $1.0M$ Na_2SO_4; electrode area = 0.278 sq. cm.

$1/i$, μa.$^{-1}$	Solution thickness, cm. $\times 10^{-3}$				
	0.864	1.12	1.37	1.62	1.88
	Transition time, sec.				
0.01	0.20	0.25	0.26	0.25	0.27
0.02	0.45	0.52	0.54	0.66	0.77
0.03	0.70	0.82	0.98	1.10	1.29
0.05	1.18	1.46	1.71	1.96	2.30
0.07	1.72	2.08	2.43	2.86	3.27
0.08	1.96	2.36	2.89	3.38	3.73
0.10	2.40	3.04	3.56	4.23	4.85
0.11	2.72	3.31	3.89	4.72	5.40
Exptl. slope	2.50	3.09	3.70	4.48	5.09
Theoret. slope	2.32	3.00	3.67	4.36	5.04

Table III. Diffusion Coefficient Values Based on $l^2/3D$ Intercepts of Plots in Figure 8

Least-squares method used to calculate intercepts from data given in Table II

Cm. $\times 10^{-3}$	$\dfrac{l^2}{3D}$ (sec.)	D, sq. cm./sec. $\times 10^5$
0.864	0.0502	0.49
1.12	0.0883	0.47
1.37	0.140	0.45
1.62	0.236	0.37
1.88	0.252	0.47
		Av. 0.45

out the experiment but also rapid depletion of the species being studied. A partial correction for residual current and charging of the double layer may be made from an integrated current-time curve on the supporting electrolyte at the same potential used for the electrolysis. Point by point subtraction of this background curve from the i-t curve, obtained from a run with the depolarizer present, yields a measure of the amount of depolarizer present.

Many of the advantages of the thin-layer electrochemical cell have already been pointed out ($2, 3, 5, 8$). One is that, in the times required for many analytical processes, one has homogeneous diffusional mixing of the solution in the volume element being studied. Further studies are under way for applying this principle to the coulometric analysis of ion mixtures by controlled potential with integrated current.

LITERATURE CITED

(1) Carslaw, H. S., Jaeger J. C., "Conduction of Heat in Solids," 2nd ed., p. 96, Oxford University Press, Oxford, 1959.
(2) Christensen, C. R., Anson, F. C. , *Anal. Chem.* **35**, 205 *(1963)*.
(3) *Ibid.*, **36**, 495 (1964).
(4) Crank, J., "The Mathematics of Diffusion," Chap. 10, Oxford University Press, Oxford, 1956.
(5) Hubbard, A. T., Anson, F. C., *Anal. Chem.* **36**, 723 *(1964)*.
(6) Kolthoff, I. M., Lingane, J. J., "Polarography," 2nd ed., Interscience, New York, 1952.
(7) Moros, S. A., *Anal. Chem.* **34**, 1584 *(1962)*.
(8) Osteryoung, R. A., Anson, F. C., *Ibid.*, **36**, 975 (1964).
(9) Russell, C. D., Peterson, J. M., *J. Electroanal. Chem.* **5**, 467 (1963).

Received for review January 25, 1965. Accepted May 7, 1965. Division of Analytical Chemistry, 149th Meeting, ACS, Detroit, Mich., April 1965. Research supported in part by the Advanced Research Projects Agency.

Reprinted from *Anal. Chem.* **1965**, *37*, 1312–16.

Spectral Interferences in Atomic Absorption Spectrometry

S. R. KOIRTYOHANN and **E. E. PICKETT**
Department of Agricultural Chemistry, University of Missouri, Columbia, Mo.

Absorption by molecular species in the flame produces background which must be considered in atomic absorption measurements. The alkali halides absorb in much of the ultraviolet region when the cooler flames are used. In hotter flames, calcium, strontium, and magnesium oxide and hydroxide molecules show appreciable absorption in the same spectral regions in which they emit. Resonance lines of barium, lithium, sodium, and chromium can be interfered with in varying amounts by these spectra in the conventional air-acetylene flame. A method to correct for this background is given.

One frequently encounters statements in the literature on atomic absorption to the effect that no spectral interferences occur in this method (*7, 10*). Zaidel and Korennoi (*11*) reported that there was no interference by strontium on the determination of lithium by absorption even though the SrOH band interferes seriously in the emission flame method. Small light losses that have been encountered in the absence of the test element have been attributed to scattering by solid particles in the flame (*10*).

Because of the well known relationships between absorption and emission, it seems likely that molecular species which emit strongly in the flame should exhibit absorption at the same wavelengths. Billings (*1*) has reported difficulty in the determination of barium by absorption in the presence of large amounts of calcium in the premix air-acetylene flame, and we found background absorption from various salts (*6*) using the long absorption path method suggested by Fuwa and Vallee (*5*).

These apparent contradictions led us to make a more thorough investigation of background absorption as it affects the practice of atomic absorption.

EXPERIMENTAL

Instrumentation. The instrument upon which this work was done has a Jarrell-Ash Model 82,000 monochromator equipped with 25-micron slits (about 0.4A. band pass) and mounted with an optical bar. Standard spectrograph bench riders are used to support the external optics which were arranged as in our earlier work (*6*). The d.c. electronics were assembled from components that were on hand, consisting of a line-operated regulated power supply for the 1P28 photomultiplier, a Keithley Model 620 electrometer-amplifier, and a Heathkit Model EUW-20A servo-recorder. A Beckman hydrogen lamp and a tungsten lamp provided the continuous sources in the ultraviolet and visible regions. The burner for the premix air-acetylene flame was modified from the Hilger H909 atomic absorption attachment and provides an absorption path of about 10 cm.

Procedure. Our arrangement of the burner and tube for the long path atomic absorption measurements and the procedure for obtaining the absorption spectra from solutions of salts have been given (*6*).

The premix flame was set up in the conventional way except that a continuous source replaced the cathode lamp for background measurements. A slightly fuel rich (nonluminous) flame was used throughout.

Our electronics do not discriminate against flame emission which was quite significant in some cases. The true absorbance of a substance in the flame at a given wavelength was calculated from the formula:

$$A = \log \frac{I_0}{I_s - I_e}$$

where I_0 is the source intensity measured through the flame,

I_s is the intensity measured while aspirating the sample, and I_e the intensity of emission measured during sample aspiration with a shutter between the light source and the flame.

RESULTS

Spectra of the more common alkali halides were run by aspirating rather concentrated aqueous solutions into a hydrogen–oxygen flame directed into a 40-cm. Vycor tube. The curves for KI and KBr, which show more structure than most of the others, are given in Figure 1. These spectra and others which were reported earlier (6) are in good general agreement with those published by Müller (8). His spectra were obtained from the alkali halides heated in a furnace. There are differences in the relative intensity of some peaks, but the wavelength of each maximum and minimum agrees within the precision of our measurements. This is true for all except the lithium salts where the deviations are probably caused by the formation of LiOH in the flame (4).

According to Müller the spectra are due to absorption by halide molecules. There can be little doubt that we are observing the same process, perhaps complicated by other reactions in the flame. This provides a background of molecular absorption to complicate atomic absorption determinations in this type of matrix and flame.

The same types of spectra are observed when alkali halides are aspirated into an air–natural gas flame in a conventional premix burner. This has been observed by Willis (10), although he attributes the light losses to scattering.

In the air–acetylene flame or oxygen-hydrogen flame formed by a total consumption atomizer-burner, the alkali halide spectra are very weak and frequently undetectable.

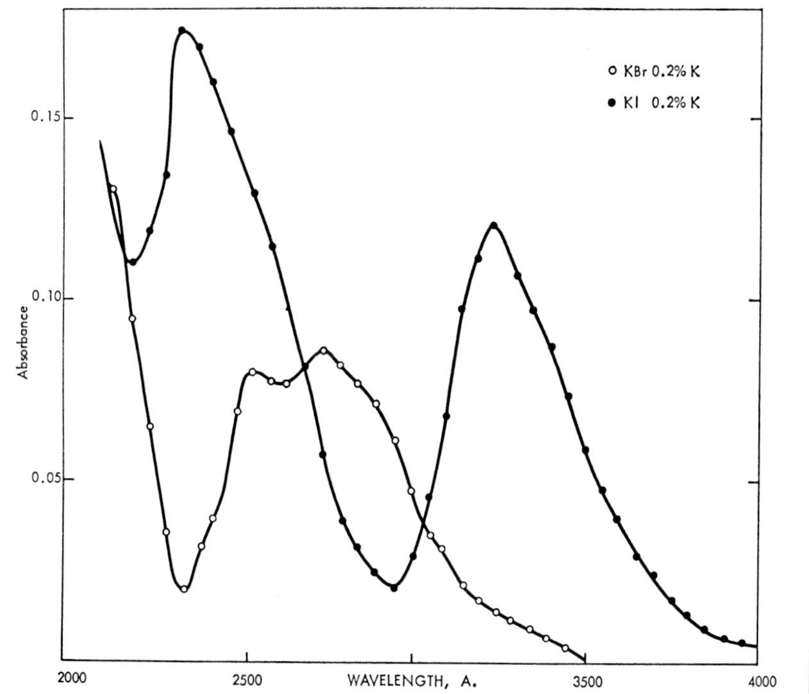

The alkaline earth elements tend to form stable compounds in the higher temperature flames. The intense oxide and hydroxide bands of calcium and strontium are well known in emission flame work. We find that the same spectra can be observed in absorption.

Figure 2 shows the emission response (solid line) and the absorption (dashed line between data points) observed when a solution containing 1% calcium as the chloride was aspirated into the air–acetylene flame. This band overlaps the strongest barium absorption line at 5536 A. The absorption observed from 1% calcium is about what would be expected from 75 p.p.m. of barium (9). For these curves the emission correction was about 1/3 of the total absorption signal.

Measurements on this band were also made by Capacho-Delgado and Sprague (2) using an instrument with a.c. electronics. They obtained a similar spectrum and found that the absorbance by CaOH at 5536 A. was the same using either a continuous source or a barium cathode lamp.

Emission and absorption curves for the SrO bands in the 6400–6900 A. region are shown in Figure 3. The similarity is again obvious and would be even more striking if the emission curve were not distorted by lower detector sensitivity at the longer wavelengths. A small amount of lithium was added to the strontium solution to show the position of the Li 6708 A. line on the emission curve. The absorption from a 1% strontium solution is about the amount expected from 0.3 p.p.m. of lithium (9).

While aspirating a 1% strontium solution containing no detectable lithium, the absorption was checked at 6708 A. with a continuous source and with a monochromatic lithium source obtained from a second flame which was aspirating a 10-p.p.m. lithium solution. No significant difference in absorption was found, indicating that there probably is no unresolved fine structure in the SrO band in this region.

Similar spectra were obtained for the SrOH band in the 6000–6100 A. region, for CaO at 5900–6400 A., and for MgOH, although very much weaker, about 3600–3900 A. The edge of the CaO band includes the sodium absorption lines at 5890 and 5896 A. A solution containing 5% calcium gave an absorbance of 0.01 at 5893 A. This is a weak interference, but one that could be important in cases such as the determination of sodium in very pure calcium salts. The Cr 3579 A. line is in the edge of the MgOH bands. These bands are quite weak but we still observe a few tenths of a per cent absorption from a 1% magnesium solution at this wavelength.

Scattering by particles in the flame has been proposed to explain light losses in the absence of the test element (10). We observed no light losses from any of the

Figure 1. Absorption by potassium halides in the long-path flame

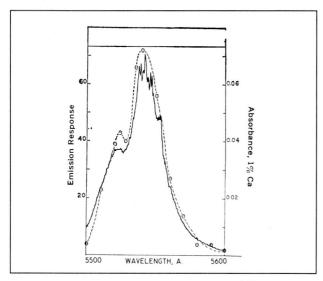

Figure 2. Emission and absorption by CaOH

alkaline earth solutions in spectral regions free of molecular absorption and emission. As a further test for this effect, a solution containing 10% lanthanum as the nitrate was aspirated into the flame. This compound undoubtedly decomposes into refractory oxides and should cause large light losses if scattering is important. No detectable light losses occurred in the 2200–6500 A. region. Similar results were found with a solution containing 2.5% aluminum as the nitrate with the air–acetylene flame and with a 0.4% suspension of finely divided alumina in the long path procedure.

DISCUSSION

The molecular absorption spectra of alkali halides are observed with the long path absorption method, in which the temperature in much of the tube is relatively low, and in the

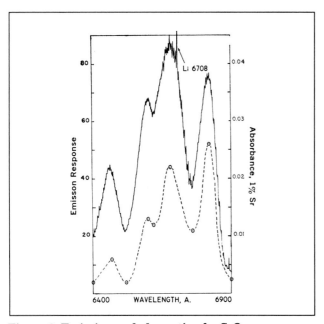

Figure 3. Emission and absorption by SrO

low-temperature air–natural gas flame. These molecules are evidently decomposed almost completely in the hotter air–acetylene or oxygen–hydrogen flames, even when rather concentrated solutions are aspirated.

Spectra from the more stable alkaline earth oxide and hydroxide molecules can be observed in absorption as well as emission in the hotter flame and, if these elements are present in the sample solution, will appear as background for atomic absorption measurements of elements that have resonance lines in the same spectral regions. The intensity of the bands relative to atomic lines is much lower in absorption than in flame emission. This is because the effective band pass is controlled by the width of the atomic line rather than the resolution of the monochromator. The effects would probably be about the same in emission and absorption if a monochromator with resolution equal to the width of the atomic line were used.

These processes explain the difficulty in the determination of barium in the presence of calcium found by Billings (*1*), but contradict the work of Zaidel and Korennoi (*11*) who found no interference by strontium on lithium in absorption. The latter workers were primarily concerned with strontium band emission affecting their absorption measurement, and evidently did not look carefully for an absorption effect.

The CaO absorption in the vicinity of the sodium lines and that of MgOH around Cr 3579 are very weak and are unlikely to be important except in special cases. However, in using commercial instruments which have scale expansion capability of as much as 30-fold, these small absorbances increase in importance by about the same factor.

We were unable to observe light losses due to scattering by particles in the flame even when concentrated solutions of elements forming refractory oxides were used. It is likely that the workers who report observing this effect (*3, 10*) were actually measuring molecular absorption. The "scattering" curves given by Willis (*10*) for NaCl and K_2SO_4 show structure that could hardly be explained by this process. Scattering may occur in some cases but our data indicate that the process is much less important than previously believed.

The fact that molecular interferences are smaller in atomic absorption than in flame emission is partially offset by the increased experimental difficulty in obtaining an appropriate correction. The monochromatic nature of the light source prevents scanning wavelengths in the vicinity of the absorption line. By use of measurements at the wavelength of the resonance line of the test element, absorption by the matrix salts cannot be distinguished from a small impurity of that element in the matrix. Unabsorbed lines from the cathode lamp have been used to measure the absorbance at adjacent wavelengths (*10*), but the possible presence of structure in the background absorption and the fact that frequently no suitably located line is available make this unsatisfactory in many cases.

In the ultraviolet region we have used a continuous source to measure the matrix absorption in the vicinity of resonance lines and thus to correct for it (*6*). The intense flame emission in the visible region has prevented our testing this method there. With a.c. amplification and a modulated source the same methods should apply, in spite of the intense emission, as long as the background does not contain unresolved fine structure at the wavelength in question.

CONCLUSION

The number of specific spectral interferences that we have found is rather small. Freedom from such interferences will continue to be one of the major advantages of the atomic absorption method, but it seems logical to expect that many weak absorption systems will be encountered. Work with very small absorbances using concentrated solutions of complex materials must be done with this possibility in mind.

LITERATURE CITED

(1) Billings, G. K., Ph.D. thesis, Rice University, Houston, Texas, 1963.
(2) Capacho-Delgado, L., Sprague, S., Perkin-Elmer Corp., *Atomic Absorption Newsletter* **4,** 363 (1965).
(3) David, D. J., *Analyst* **86,** 730 (1961).
(4) Dean, J. A., "Flame Photometry," p. 156, McGraw-Hill, New York, 1960.
(5) Fuwa, K., Vallee, B. L., *Anal. Chem.* **35***, 942 (1963).*
(6) Koirtyohann, S. R., Pickett, E. E., *Ibid.,* **37,** 601 (1965).
(7) Manning, D. C., *Atomic Absorption Newsletter* No. 24, Perkin-Elmer Corp. (1964).
(8) Müller, L. A., *Ann. Physik* **82,** 39 (1927).
(9) Slavin, W., Sprague, S., Manning, D. C., *Atomic Absorption Newsletter* No. 18, Perkin-Elmer Corp. (1964).
(10) Willis, J. B., "Methods of Biochemical Analysis," Vol. 11, David Glick, ed., Interscience, New York, 1964.
(11) Zaidel, A. N., Korennoi, E. P., *Opt. Spectro.* **10**, 299 (1961).

Received for review January 7, 1966. Accepted February 14, 1966.

Reprinted from *Anal. Chem.* **1966**, *38*, 585–87.

This paper describes an instrument that allowed, for the first time, the automated chemical synthesis of polypeptides. This technology has had a tremendous impact on chemistry, biology, and medicine by allowing the rapid preparation of polypeptides of up to 100 residues. The significance of this invention was widely recognized, and in 1984 Bruce Merrifield was awarded the Nobel Prize in Chemistry. After two decades of active research, such automated instruments are available from several manufacturers and, in selected cases, the synthesis of long polypeptides—a feat that is nearly impossible with conventional methods—is possible.

W.S. Hancock
Hewlett Packard

Instrument for Automated Synthesis of Peptides

R. B. Merrifield, John Morrow Stewart, and Nils Jernberg
The Rockefeller University, New York, N. Y. 10021

An instrument which can perform automatically all of the operations involved in the stepwise synthesis of peptides by the solid phase method is described in detail. The synthesis of the peptide chain takes place on a solid polymer support and all of the reactions are conducted within a single vessel. The apparatus is composed of two main parts—the reaction vessel and the components required to store, select, and transfer reagents, and the programmer which controls and sequences the operation of the various components. The operation of the instrument and its application to the synthesis of several peptides are described.

Extensive advances in methods of isolation, purification, analysis, and structure determination of peptides and proteins have outdistanced our synthetic achievements in this area. To cope with many of the new problems which have arisen, a greatly accelerated and simplified approach to peptide synthesis was required. Solid phase peptide synthesis was devised (*13*), and developed (*9, 10*) with these objectives as guides. The principles of the method and the special features which make it adaptable to an automated process have been reviewed (*11, 14*), and an apparatus designed for automated peptide synthesis was constructed and briefly discussed (*15*). This article describes in detail the instrument which can perform automatically all of the operations involved in the stepwise synthesis of polypeptides.

GENERAL PRINCIPLES

The method is based on the fact that a peptide chain can be synthesized in a stepwise manner while one end of the chain is covalently attached to an insoluble solid support. During the intermediate synthetic stages the peptide remains in the solid phase and can therefore be manipulated conveniently without significant losses. All of the reactions, including the intermediate purification procedures, are conducted within a single reaction vessel. It is this feature which permits convenient automation of the process. The problem in essence is simply to introduce the proper reagents and solvents into the vessel in the proper sequence at the proper times.

The solid support is a chloromethylated styrene-divinylbenzene copolymer bead. The C-terminal amino acid is coupled as a benzyl ester to the resin and the peptide chain grows one residue at a time by condensation at the amino end with *N*-acylated amino acids. The *tert*-butyloxycarbonyl group has been the protecting group of choice and activation has usually been by the carbodiimide or active ester routes. Since each of the reactions in the synthesis can be modified in a variety of ways it was important to design the apparatus with sufficient flexibility to cope with a wide range of reactions and conditions.

APPARATUS

The apparatus is composed of two main parts, the first being the reaction vessel with the components required to store and select reagents and to transfer them into and out of the vessel, and the second being the programmer which automatically controls and sequences the operation of the various components. A photograph of the complete instrument is shown in Figure 1, a schematic drawing is given in Figure 2, and the wiring diagram of the programmer is shown in Figure 3. Parts for the instrument are listed below.

Programmer. Stepping drum programmer model A-31-EZ-30, Tenor Co., Butler, Wis.

C1. Electrolytic capacitor, 100 mf., 50 volts.

C2. Paper capacitor, 1 mf., 600 volts.

2855-8/94/0192$08.00/0 ©1994 American Chemical Society

D1. Silicon diode rectifier, 5 amperes, 100 piv.; RCA 1N1613.

P. Pilot lamp, neon, Drake HR117.

P1–P3. Pilot lamp, Dialco 812210, with GE 1829 bulbs (28 volts) and green jewels.

Pump. Beckman metering pump model 74603, 0 to 20 ml. per minute, modified as described.

R1–R3. 32-Pole shorting relay, Guardian IR-805-S, 24-volt d.c. coil.

R4. 3-Pole double throw relay, Potter-Brumfield KRP14AG, 115-volt a.c. coil.

R5. Time-delay relay, Amperite 115N010 (thermal, 115 volts, 10-second delay).

R6. Stepping relay, Guardian Rotomite IR-705-12P-24D, 24 volts d.c.

S1, S14. Switch, DPDT, 6 amperes, Cutler-Hammer 8373K7.

S2. Switch, SPST, 6 amperes, Cutler-Hammer 8381K8.

S3, S4, S5. Push-button switch, momentary contact, NO, Arrow 80541E.

S6–S10. Unit control switches, SPDT, center off, Cutler-Hammer 7503K13.

S11. Pump limit switch. Miniature microswitch, SPDT, Micro 1SM1.

S12. Shaker limit switch. Microswitch, SPDT, roller type, Micro BZ-2RM22-A2.

S13. End-of-run microswitch on amino acid selector valve, SPDT, Micro BZ-2RM22-A2, NC contacts used.

S15. Tap switch, rotary, 12-position, nonshorting, Mallory 32112J, on shaft of solvent valve.

S16. Tap switch, rotary, 2-gang, 12-position, nonshorting, Centralab 2005, on shaft of amino acid valve.

Drum Switches. Drum-controlled switches of Tenor programmer. NO contacts of all switches are used except Home switch, of which NC contacts are used.

Solvent Valve; Amino Acid Valve. 12-Position all-Teflon motor-driven rotary selector valves (see text).

Shaker. Motor-driven, to invert reaction vessel (see text).

SV1–SV3. Solenoid-operated valves, miniature diaphragm type, all-Teflon, normally closed, Mace EDV-122, Mace Corporation, San Gabriel, Calif.

T1. Transformer, primary 117 volts, secondary 25 volts, 2 amperes, Stancor P8357.

Timers. Cutler-Hammer 10336H46A. No. 1, 1 min; Nos. 2 and 3, 3 min.; No. 4, 30 min.; No. 5, 300 min.

Z1. Resistor, wire-wound, 200 ohms, 5 watts.

For the operation of the instrument the proper reagents and solvents are selected by the amino acid–and solvent selector valves and are transferred by the metering pump from the reservoirs to the reaction vessel which contains the peptide–resin. After the desired period of mixing by the shaker the solvents, excess reagents, and by-products are removed to the waste flask by vacuum filtration. These basic operations are repeated in a prearranged sequence under the control of the programmer until the synthesis of the desired peptide chain is complete. All parts of the apparatus which come into contact with the solvents and reagents are made of glass or chemically resistant polymers.

Reaction Vessel. A modification of the reaction vessel previously described (13) for use in the manual method has been designed and is shown in Figure 4. It consists of a glass cylinder (total volume 45 ml.) with a coarse grade fritted disk (Corning) at the lower end. The bottom is constructed to leave a minimum of space below the filter and is sealed to a male Luer connector. The top end is fitted with a ST 14/20 female joint which holds a specially ground stopper containing a 1-mm. i.d. tube extending 1.5 cm. into the vessel and ending with a 5-mm. coarse fritted disk. The outer end of the stopper is terminated in a male Luer connector. The vessel is held by a three-fingered, hollow-stemmed clamp (Fisher No. 5-742). Solvent and air lines are attached to the reaction vessel by means of Kel-F female Luer fittings (fittings with attached 0.076-inch i.d. tubing of Teflon are available from Hamilton Co., Whittier, Calif.). The tubing of Teflon passes through the hollow clampshaft to avoid entanglement during the shaking operation. Solvents and reagents are pumped into the bottom of the vessel while air is displaced at the top (air outlet solenoid SV1 open). Solvents are removed at the bottom by vacuum (solvent outlet solenoid SV2 open) while air is drawn in at the top through a tower of Drierite (air inlet solenoid SV3 open). The volume of solvent is adjusted to fill the vessel more than half full but to keep the level below the tip of the air outlet. Thus, when the vessel is inverted all of the inner surface is washed with solvent, and any resin adhering to the walls is brought into contact with the reagents. With solenoid valves SV1 and SV3 closed, an air lock is maintained in the capillary

Figure 1. Apparatus for automated peptide synthesis

tubing which prevents solvent from escaping at the top when the vessel is inverted. This vessel will accommodate 2 to 4 grams of resin. Two larger vessels of 80- and 120-ml. capacity but otherwise of similar design have been constructed for use with approximately 7- and 10-gram batches of resin.

Shaker. This is a device for producing a gentle mixing of the resin and solvents in the reaction vessel. An eccentric drive from a Hurst synchronous motor (Model PC-DA, 10 r.p.m., with clutch and brake) moves a gear through a 90° arc. This gear drives a second, smaller gear through 180°. The latter is attached to the clamp which holds the vessel, and thus repeatedly inverts the vessel to mix its contents during the shaking periods. A cam mounted on the rear end of the clampshaft is so positioned that it activates a microswitch (S12, shaker limit switch) each time the vessel comes to the upright position. This microswitch is energized by the programmer only at the end of each shaking cycle, and then serves to stop the vessel in the vertical position and to step the programmer to the next operation (see description below).

Pump. The Beckman metering pump (0 to 20 ml. per minute) is used for all of the pumping operations. It is modified in three ways. The standard Viton diaphragms are replaced with Teflon-coated diaphragms (No. 70992) to withstand the effects of dimethylformamide and other organic solvents. The holdup volume of the pump is minimized by insertion of threaded Kel-F plugs into the inlet and outlet ports. The holdup is also minimized by stopping the pump each time

at the end of an exhaust stroke. This is done by means of a microswitch (S11, pump limit switch) mounted on the pump backplate so that it is actuated by the lever arm which drives the piston. The total volume pumped is controlled by adjusting both the volume per stroke and the pumping time.

When the pump was turned off high voltage transients were produced. These ruptured the silicon diodes of a bridge rectifier within SV1, which is in parallel with the pump. Addition of the surge filter (C2, Z1) corrected this difficulty.

Solvent Outlet System. Solvents are removed from the reaction vessel by vacuum filtration through the fritted disk at the bottom. The solvent passes out the Teflon tubing through a T connector (machined from Kel-F) to the solenoid valve SV2 and into the 12-liter round-bottomed waste flask. To accelerate the filtration rate and to avoid precipitation of materials and possible obstruction of the waste line, $1\frac{1}{8}$ inch i.d. polyethylene tubing is used for the line between the outlet solenoid (SV2) and the waste flask. A pressure of about 100 mm. is maintained in the waste flask with the laboratory vacuum line. Under these conditions solvent is removed at the rate of approximately 100 ml. per minute. To ensure complete removal of solvents, an excess of time is allowed for each filtration step in the automatic program. Since the Mace diaphragm solenoid valves are not designed to open against a vacuum, satisfactory operation was obtained only by connecting SV2 in the reverse direction—i.e., with the waste flask connected to the "in" port of the valve.

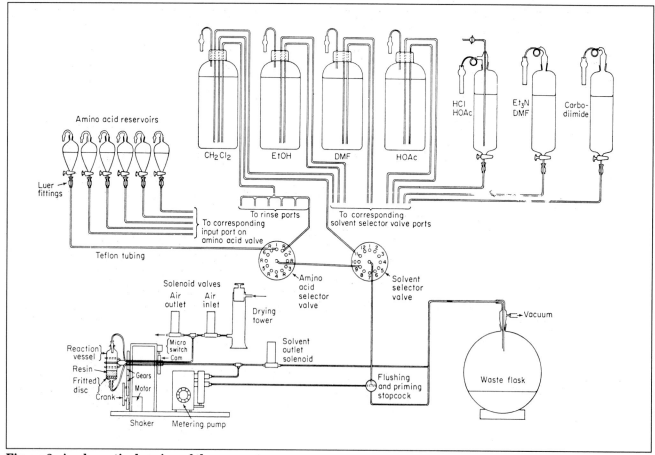

Figure 2. A schematic drawing of the apparatus

Reagent and Solvent Reservoirs. The solvents such as methylene chloride, dimethylformamide, and ethanol are stored in 4-liter brown bottles fitted with No. 38 polyethylene screw caps. The glacial acetic acid is stored in the original commercial 5-lb. bottles to avoid picking up moisture during transfer. Solvent is withdrawn through lines of Teflon tubing (AWG No. 16, standard natural, Pennsylvania Fluorocarbon Co., Clifton Heights, Pa.) which are inserted through tight-fitting holes in the caps and extend to the bottom of the bottles. Two or three such lines from each bottle are then attached to the appropriate inlet ports of the solvent selector valve (Figure 2). An additional line runs to a drying tube containing indicating Drierite to replace the solvents with dry air.

Solutions of *tert*-butyloxycarbonyl amino acids or esters are stored in 60-ml. separatory funnels having Teflon stopcocks. The solutions are protected from moisture by small Drierite-packed drying tubes which are attached to the funnels with ST 14/20 connections. The stems of the funnels are fitted with Luer connectors and attached to Teflon lines (Hamilton) which run to the odd-numbered positions of the amino acid selector valve (Figure 2). The apparatus is fitted with six funnels which handle a 1-day supply of amino acids.

The other three reagents are contained in 1-liter cylindrical separatory funnels as shown in the upper right of Figure 2. To prevent excessive loss of the volatile solvents by diffusion during storage the drying tubes are connected to the funnels

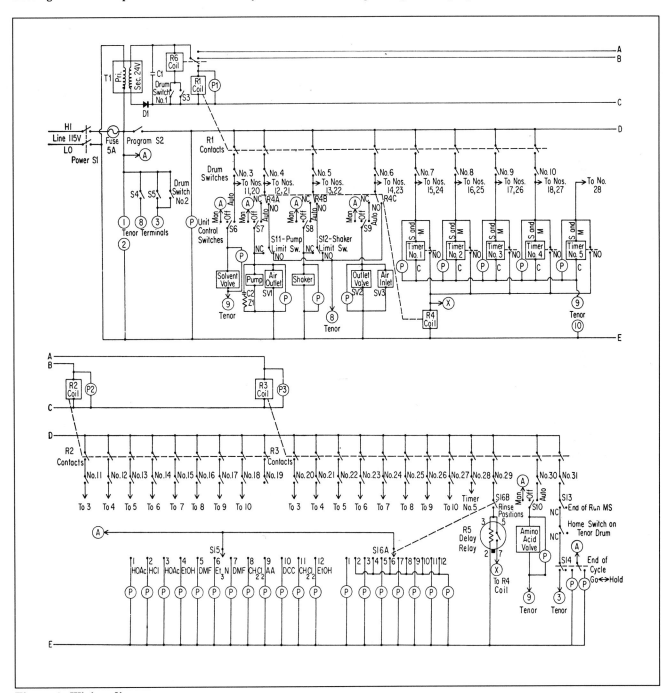

Figure 3. Wiring diagram

by coiled 3-ft. lengths of 0.076-inch i.d. Teflon tubing. These funnels are fitted with Teflon stopcocks and are connected by Luer joints to Teflon outlet lines which run to the solvent selector valve.

Rotary Selector Valves. To select the desired solvents and reagents rotary Teflon selector valves were constructed (Figure 5). This type of valve was chosen because it gives a very sharp cutoff of liquid, and prevents mixing of one reagent with another. With a manifold in which many reagents flow into a common chamber or line there is danger of cross contamination of reagents. Two such valves are necessary, one to select the amino acid derivatives and a second one to select all of the other solvents and reagents.

The valves are constructed from two disks of Teflon with carefully machined faces which are mounted between stainless steel plates and held together under spring pressure. The front plate and disk are stationary and contain a center port and 12 evenly spaced ports around the circumference. The various inlet solvent tubes are connected to the outer ports and the single outlet tube is jointed to the center port. The connections are with a threaded nylon pressure screw and a tapered Teflon ferrule. The rear disk contains a center port and one circumferential port which are joined by a 1.5-mm. hole within the disk. As this disk is turned it connects, one at a time, the 12 inlet ports to the central outlet port. A leak-free seal between the two Teflon disks of the valve is obtained by fitting each orifice in the front disk with a specially machined X-ring of Kel-F (see Figure 5). The pressure spring applies sufficient force to the movable disk and pressure plate to seat the X-rings in their recesses and bring the two Teflon disks into light contact. The valve is advanced one position at a time by a Geneva-type intermittent gear drive mechanism (6) which is operated by a 23-r.p.m. Bodine capacitor motor No. KC1-23RM. This drive lacks the locking feature usually found on Geneva drives and thus allows the disk to be turned manually when the drive stud is not engaged. Spring pressure on the Teflon disks maintains the proper indexing in automatic operation. The rear end of the solvent selector valve shaft turns a 12-point tap switch (S15) which indi-

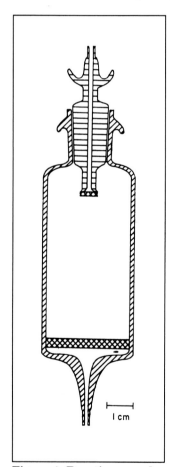

Figure 4. Reaction vessel

cates the position of the valve at any time by pilot lamps on the programmer panel (Figure 1).

The valves are designed to move in only one direction and cannot select solvents at random. Therefore, the 12 inlet solution lines must be connected to the solvent selector valve in the sequence in which they are required during the synthesis (Figure 2). The outlet tube of the amino acid selector valve is connected to one of the outer ports (No. 9) of the solvent selector valve and the outlet tube of the solvent selector valve is connected to the pump through a three-way stopcock. The third arm of the stopcock is connected to the waste flask and is used to flush and prime the solvent and reagent lines.

The amino acid valve also contains 12 inlet ports, with the six amino acid reservoirs being connected to the odd-numbered positions, while alternate ports are connected to a rinse solvent to flush the line between the two valves and thereby prevent contamination by the previous amino acid solution. The six rinse lines are all supplied with methylene chloride from a six-arm glass manifold. Instead of the single-gang tap switch shown in Figure 5 for the solvent selector valve the amino acid valve shaft is fitted with a 12-point, 2-gang tap switch (S16A and B). The first gang (S16A) operates pilot lights to indicate the position of the valve. The odd-numbered positions are connected to the numbered amino acid pilot lights, while the even-numbered positions are all connected in parallel to a single "rinse" pilot (Figure 3). The second gang (S16B) has all of the even-numbered (rinse) positions wired together. These are used in conjunction with the time-delay relay (R5) to perform the rinse function after each amino acid is pumped. Operation of this function is described later. The odd-numbered contacts of S16B are not used. The amino acid valve also bears a microswitch (S13) so positioned that it is actuated by a pin in the Geneva drive plate when the valve is in position 12 (rinse after the sixth amino acid). This end-of-run switch serves to stop the entire instrument after the coupling of the sixth amino acid is completed.

The wiring diagram for the Geneva drive of the solvent and amino acid selector valves is given in Figure 6. The 23-r.p.m. motor shaft bears a cam which actuates a microswitch (Micro BZ 2RM22-A2, roller type, SPDT) once each revolution. The cam is positioned so that the switch is actuated and encloses the NO contacts just as the drive stud is emerging from the slot in the Geneva drive plate after advancing the valve to the next position. The microswitch is actuated only momentarily, as the inertia of the motor carries the cam off the switch roller and restores the NC circuit. At the beginning of the Geneva drive cycle, the motor is energized through the NC contacts of the microswitch and the NC contacts of the relay (Potter-Brumfield KHP17A11). When the motor shaft has made one revolution to advance the valve to the next position, actuation of the microswitch by the cam stops the motor and energizes the relay. The NO contacts of the relay now serve to hold the relay energized and prevent the motor from advancing further until the source of power has been interrupted by the programmer. The return lead from the Geneva drives of the selector valves to the low side of the line passes through terminals 9 and 10 of the Tenor programmer. These terminals are connected to a switch which is opened each time the programmer steps. This allows the valve drive relay to open and makes it possible to operate the same selector valve on two successive steps if desired. This arrangement is necessary because the drum microswitches do not open during stepping when plugs are inserted in two successive positions.

Programmer. The proper sequential operation and timing of the previously described components is controlled by a Tenor stepping drum programmer. This programmer consists of a row of 32 roller-type microswitches positioned under a revolving drum. The drum bears rows of holes around its circumference, one row of 30 holes being positioned above each microswitch. The drum is driven by a Geneva-type drive which advances the drum by steps, 30 steps constituting one revolution of the drum. The desired program is established by the insertion of nylon plugs into the appropriate holes of the drum. These plugs actuate the corresponding microswitches which in turn control the various operating units (pump, shaker, or valves) and timers. The arrangement of the plugs in the diimide program drum is shown in Figure 7; that for the active ester program drum is shown in Figure 8. The two drums are readily interchangeable. Any future changes in the program necessitated by the use of different reagents or amino acid derivatives can be easily accommodated by changing the location of the plugs in the drums. At the end of the selected time for each step the timer furnishes a signal to cause the programmer drum to step to the next position where a new combination of switches will be actuated. The position of the programmer drum can be advanced manually one step at a time by depressing pushbutton switch S4, or continually by depressing S5. All of the operating units of the instrument can also be operated manually (independently of the automatic program) by means of the unit control switches (S6–S10).

One cycle in the automatic solid phase synthesis—i.e., the lengthening of the peptide chain by one amino acid residue—requires nearly 90 steps of the programmer drum (Figures 7 and 8) while the programmer model used provides only 30 steps per revolution. However, three times as many switches are available on the programmer as are needed for peptide synthesis and these switches can therefore be divided into three banks. Each bank of switches is energized in turn with 115-volt power by means of a stepping relay and a multicontact program relay. In this way the drum makes three revolutions during the execution of the complete program and thus makes 90 separate steps available. This drum expansion and

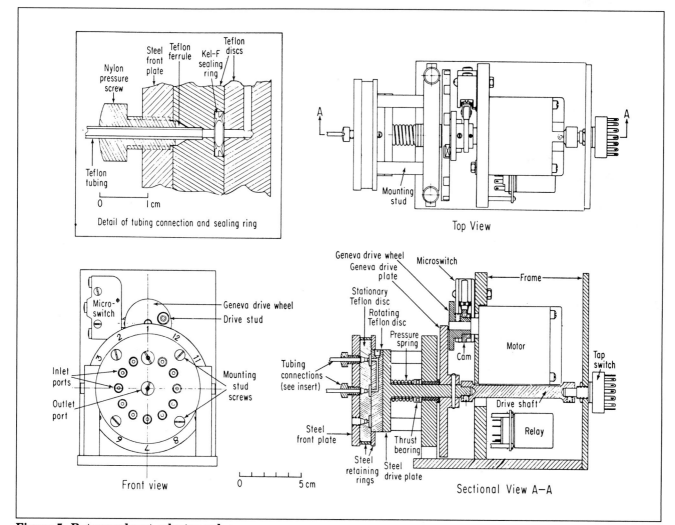

Figure 5. Rotary solvent selector valve

switch bank sequencing is controlled by a 24-volt d.c. system provided by T1, D1, and C1 (see Figure 3). The rotary stepping relay (R6) causes each of the three program relays (R1, R2, R3) to be energized in turn; the pilot lights (P1, P2, P3) indicate to the operator which relay is energized and thus show where in the program the instrument is operating. The Guardian stepping relay used (R6) makes 12 steps per revolution. Since this programmer needs only three steps, every third contact of R6 is wired together, thus converting it to a three-step device (contacts 1, 4, 7, and 10 to R1, contacts 2, 5, 8, and 11 to R2, and contacts 3, 6, 9, and 12 to R3). The stepping relay can be advanced manually by push-button switch S3, or automatically by drum switch No. 1, which is actuated by a plug in the drum at step 30. Drum switch 2 is also actuated at step 30 to advance the drum to step 1. Thus during automatic operation the program begins at step 1 of the drum with R1 and the first bank of drum switches (Nos. 3 to 10) energized. At step 30, R6 advances one step and the program drum makes another revolution with R2 and drum switches 11 to 19 energized. At step 30 on this second revolution R6 steps again and, during the third revolution of the program, drum R3 and drum switches 20 to 31 are energized. When the program is completed at step 26 of the third revolution (Figure 7), further operation is controlled by the position of the end-of-cycle switch, S14. If S14 is in the "hold" position (open) the programmer stops. If S14 is in the "go" position (closed), power is applied to terminal 3 of the Tenor programmer causing it to step continuously back to the "home" position (step 1) of the drum. The second pole of S14 energizes pilot lights which indicate the setting of S14. During this "return to home" phase, R6 is again stepped once, thus returning the instrument to the beginning of the program with R1 and drum switches 3 to 10 energized. The instrument then proceeds to carry out another cycle of the synthesis.

Full details of the wiring of the instrument are given in Figure 3. Connections to the Tenor programmer are indicated by circled numbers which correspond to numbered terminals on the Tenor unit (1 and 2, power for drive motor; 3, drum continuous step; 8, drum single step; 9 and 10, to NC microswitch which opens during each stepping operation). All points marked with circled A's are connected together to furnish power to operating units and pilot lights when the program is not in operation. A circled X indicates the connection of R5 switch to R4 coil.

Timers. All of the operations of the automatic instrument are time controlled. All of the operations except the rinse steps are controlled by one of the five Cutler-Hammer timers. Terminals S (solenoid) and M (motor) of the timers are connected together to the appropriate drum switch, while the terminals C (common) of all the timers are connected together and then to terminal 9 of the Tenor programmer. This use of the step-interrupted return circuit is necessary to reset the timer in those cases where the program requires use of the same timer on two successive steps. When a given timer is running and the preset time elapses, the normally open contacts of the timer-controlled switch close, and line voltage is applied to the coil of R4. Thus the timers do not cause the

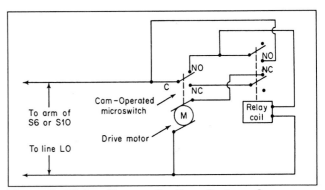

Figure 6. Wiring diagram for the selector valves

Tenor drum to step directly, but rather actuation of R4 causes the power being supplied to an operating unit to be switched to terminal 8 of the Tenor programmer, thus causing the drum to step. When the operating unit in use is the outlet valve, line voltage is switched by R4C directly to Tenor terminal 8, and the drum steps. When the operating unit is either the pump or shaker, however, the drum does not step until the operating unit limit switch (S11 or S12) is actuated—i.e., until the pump piston reaches the end of an exhaust stroke or the shaker brings the vessel to the upright position. In these cases the energization of R4 only brings the proper limit switch (S11 or S12) into the circuit; when the limit switch is actuated, the drum steps.

The rinse steps which flush the amino acid valve and tubing with methylene chloride to prevent cross-contamination of successive amino acid residues—i.e., step 7, bank 3 of the diimide program and step 3, bank 3 of the active ester program—are not timed by one of the Cutler-Hammer timers, but rather by the time-delay relay (R5). When the amino acid pumping step is completed—e.g., step 6, bank 3 of the diimide program—the drum advances to step 7, where drum switches for the pump (No. 21), amino acid selector valve (No. 30), and the rinse switch (No. 29) are actuated. Both the pump and the amino acid valve begin to operate. Both these devices are operated by 20-r.p.m. motors, and the cam on the amino acid valve Geneva drive is positioned so that the actual movement of the valve disk occurs during the second half of the motor rotation. Since the pump always stops at the end of an exhaust stroke, during the first intake stroke of the pump the valve disk will still be stationary and in the amino acid position. During the following exhaust stroke of the pump, the actual movement of the valve disk occurs, advancing it to the rinse position. This synchronization assures that the pump will not be on an intake stroke when the disk is in motion. When the amino acid valve reaches the rinse position the associated tap switch (gang 2, S16B) (see description above) closes and applies power from Tenor drum switch 29 to the delay relay, R5. After 10 seconds, R5 closes and applies power (through point X, Figure 3) to R4 coil, and the Tenor drum steps when the pump reaches the end of its next exhaust stroke. A similar scheme is used to provide a brief rinse after the diimide pumping step, although in this case (step 10, bank 3) the R5 timing operation begins immediately, since the amino acid valve is already in the rinse position, and S16B is closed. During the first pump exhaust stroke the solvent selector valve advances to position 11 to flush the line with methylene chloride.

The solvent and amino acid selector valves do not incorporate a device to signal the programmer drum to step. Therefore these valves are always used in conjunction with a timer and another operating unit (pump, shaker, outlet) which will furnish the necessary stepping signal.

Housing and Interconnections. The programmer module is housed in a standard relay rack cabinet (Premier cabinet rack, DCR 210, $22\frac{3}{4}$ inches high). All of the switches and pilot lights are mounted on a standard $10\frac{1}{2}$ inch high panel (Figure 1) which is anchored to the Tenor programmer by sheet metal brackets. The remainder of the cabinet front is covered with a Lucite panel. A shelf-type chassis attached to the back of the front panel carries the low-voltage power supply (T1, D1, C1), the stepping relay R6, the program relays (R1, R2, R3), relay R4, and the delay relay R5. The Cutler-Hammer timers are mounted in a row on a panel above the back of the Tenor programmer (Figure 1). The operating units are connected to the programmer module by means of multiconductor cables which plug into the rear of the module. Sockets for these plugs are mounted on a small sheet metal chassis attached to the rear of the Tenor unit. The solvent and amino acid selector valves are connected by 18-conductor cables (Belden 8744), while the pump–shaker solenoid valve unit is connected by means of a 9-conductor cable (Belden 8449).

OPERATION OF THE INSTRUMENT

Several preliminary operations are necessary before the synthesis of a peptide can be started. First, the supporting resin containing the C-terminal amino acid of the proposed peptide chain must be prepared and analyzed. This is done as previously described (9, 10) by esterification of a chloromethylated copolymer of styrene and divinyl-benzene with the *tert*-butyloxycarbonyl (*t*-BOC) amino acid (1, 4, 8, 17) (Cyclo Chemical Corp., Los Angeles). The product is freed of very fine particles of resin by flotation in methylene chloride to prevent subsequent clogging of the fritted disks of the reaction vessel. A sample of the vacuum-dried product is hydrolyzed in a 1:1 mixture of dioxane and 12N HCl (9) and the liberated amino acid is measured quantitatively on an amino acid analyzer. The amino acid content is used to calculate the amounts of subsequent amino acid derivatives and dicyclohexylcarbodiimide reagent which will be used in the synthesis. The best range of substitution has been 0.1 to 0.3 mmole per gram. *t*-BOC amino acid–resins are usually prepared in advance and are stored until needed.

The appropriate solvent reservoirs are filled with glacial acetic acid (Mallinckrodt analytical reagent), methylene chloride (dichloromethane, Matheson Coleman & Bell DX 835) and commercial (99.5%) absolute ethanol. N,N-Dimethylformamide (Matheson Coleman & Bell DX 1730) is freed of dimethylamine and formic acid by shaking with barium oxide and distillation under reduced pressure (20). The 1N HCl–acetic acid solution is prepared by adding 700 ml. of glacial acetic acid to the storage separatory funnel and passing in a slow stream of anhydrous hydrogen chloride. Samples are withdrawn at the bottom and titrated for chloride by the Volhard method. This solution, when protected by the long coil of capillary tubing and drying tube, is stable for several weeks without a significant decrease in concentration. The triethy-

Drum step No.	H	1	2	3	4	5	6	7	8	9	10	11	12	13	14	15	16	17	18	19	20	21	22	23	24	25	26	27	28	29	30	31
																					Solvent valve	Pump	Shaker	Outlet	Timer 1	Timer 2	Timer 3	Timer 4	Timer 5	Rinse timer	Amino acid valve	Return home
1	X			X₁		X	X								X	X						X			X							
2				X			X					X				X							X	X								
3					X			X					X				X				X				X							
4					X	X						X₅			X	X						X			X							
5					X			X					X				X				X₉				X	X						
6					X				X					X				X				X			X							
7					X	X									X	X															X	X_R
8					X			X					X				X				X₁₀	X					X					
9					X				X					X				X				X			X							
10				X₂		X	X								X	X					X₁₁	X							X			
11					X		X						X				X					X						X				
12					X				X				X					X				X		X	X							
13					X				X			X₆		X		X	X				X						X					
14					X				X				X	X		X					X				X							
15				X₃		X	X							X			X			X					X	X						
16					X			X				X₇				X	X				X						X					
17					X				X				X				X				X				X							
18					X	X								X			X				X	X										
19					X				X							X	X				X			X								
20					X				X			X				X					X				X							
21					X	X									X				X		X₁₂		X	X								
22				X			X						X			X X					X			X							
23					X			X					X				X				X			X								
24				X₄		X	X							X			X				X		X X									
25					X			X				X₈				X	X				X				X							
26					X				X				X				X				X			X							X_A	
27					X	X									X			X														X
28					X			X					X				X X				X										X
29					X				X				X				X				X										X
30		X	X																													

Figure 7. Drum program for use with dicyclohexylcarbodiimide

The *x* marks indicate the positions of the nylon plugs on the programmer drum. Subscripts 1 to 12 indicate the port on the solvent selector valve which opens when the drum moves to the step shown. Subscripts A and R indicate that the amino acid valve is opened to an amino acid reservoir or to a rinse line, respectively

lamine reagent is prepared by mixing 50 ml. of triethylamine (Matheson Coleman & Bell TX 1200) with 450 ml. of purified dimethylformamide.

The solvent lines from the reservoirs are filled one at a time by turning the solvent selector valve to the corresponding position and applying suction through the flushing and priming stopcock.

The metering pump is calibrated by pumping methylene chloride into a graduated cylinder for a measured period. This operation is carried out under the control of the programmer with timer 2 set for 1 minute. The pump rate (approximately 20 ml. per minute) can be varied by changing the length of stroke of the piston.

The holdup volume of the system between the solvent selector valve and the bottom of the reaction vessel is determined by filling the line with the HCl–acetic acid reagent (port 2), turning the valve to the acetic acid line (port 3) and pumping until all of the HCl has been flushed out. The effluent is titrated for chloride and the holdup volume is calculated. In the system now in use it was 4.2 ml.

The reaction vessel is loaded with a weighed amount of the t-BOC amino acid–resin (2 to 4 grams for the small, 45-ml. capacity vessel). The stopper is lubricated with silicone high vacuum grease and secured in place with springs, and the inlet and outlet lines are attached. In the synthesis three equivalents of each t-BOC amino acid derivative are used per equivalent of the first amino acid on the resin. The calculated quantity of each of the first six amino acids is dissolved in 7 ml. of methylene chloride, filtered if necessary, and placed in the amino acid reservoirs in the proper sequence. Because of poor solubility in methylene chloride, t-BOC-nitro-l-arginine is first dissolved in 2 ml. of dimethylformamide and diluted with 5 ml. of methylene chloride, while t-BOC-im-benzyl-l-histidine is dissolved in 7 ml. of pure dimethylformamide. The t-BOC amino acid p-nitrophenyl esters are dissolved in 16 ml. of pure dimethylformamide. During the automated synthesis the amino acid solutions are pumped completely into the reaction vessel and a precise concentration therefore is not required.

The dicyclohexylcarbodiimide solution, on the other hand, is metered by the pump and the concentration of the reagent must be calculated for each run. Since the holdup volume and the total volume pumped are known the actual volume of diimide solution delivered into the vessel can be calculated. The required quantity of dicyclohexylcarbodiimide is

dissolved in this volume of methylene chloride. The total volume of solution prepared at one time depends on the number of amino acids to be added.

The programmer is set for the run by inserting the proper program drum (diimide or active ester) and stepping it manually (S5) to step 2. Bank 1 of the drum switches is energized by using S3 to step R6 to the desired position. The timers are set as follows: No. 1, 30 seconds; No. 2, 60 seconds; No. 3, 90 seconds; No. 4, 10 minutes; No. 5, 120 minutes. For runs in the larger vessels, settings of timers 1 and 2 must be increased. The amino acid and solvent selector valves are set to position 1. The pump is set for 20 ml. per minute. The unit control switches (S6–S10) are placed in the auto position and the end-of-cycle switch (S14) is placed in the go position. Closing of the program switch (S2) starts the automatic synthesis.

Functioning of a Typical Diimide Cycle. The instrument first washes the resin three times with acetic acid by means of three sets of pumping, shaking, and outlet steps. As described above, the pump always stops at the end of an ex-

Drum step No.	Home	Bank relay	Step drum	Solvent valve (B1)	Pump (B1)	Shaker (B1)	Outlet (B1)	Timer 1 (B1)	Timer 2 (B1)	Timer 3 (B1)	Timer 4 (B1)	Solvent valve (B2)	Pump (B2)	Shaker (B2)	Outlet (B2)	Timer 1 (B2)	Timer 2 (B2)	Timer 3 (B2)	Timer 4 (B2)	(19)	Solvent valve (B3)	Pump (B3)	Shaker (B3)	Outlet (B3)	Timer 1 (B3)	Timer 2 (B3)	Timer 3 (B3)	Timer 4 (B3)	Timer 5 (B3)	Rinse timer	Amino acid valve	Return home
	H	1	2	3	4	5	6	7	8	9	10	11	12	13	14	15	16	17	18	19	20	21	22	23	24	25	26	27	28	29	30	31
1	X			X_1			X	X									X	X				X									X	X	X_R
2				X				X					X				X					X_{10}	X								X		
3					X					X					X			X					X								X		
4						X	X					X_5				X	X					X_{11}				X	X						
5				X				X					X				X						X					X					
6					X					X					X			X						X					X				
7						X	X									X	X									X	X						
8				X				X					X				X						X					X					
9					X					X					X			X						X					X				
10				X_2		X	X									X	X									X	X						
11					X			X					X				X						X					X					
12					X					X					X			X						X					X				
13					X					X		X_6				X	X					X_{12}				X	X						
14					X					X			X				X						X					X					
15				X_3		X	X						X					X					X				X						
16					X			X				X_7		X			X	X					X					X					
17					X					X			X				X						X				X						
18					X	X								X				X						X					X				
19					X			X					X				X	X					X	X									
20					X					X			X				X						X				X						
21					X	X								X				X					X				X					X_A	
22					X			X					X	X				X															X
23					X				X				X				X																X
24				X_4		X	X						X				X																X
25					X			X					X				X	X															X
26					X					X			X				X																X
27					X	X						X_8	X				X																X
28					X			X				X_9	X	X			X																X
29					X					X			X				X																X
30		X	X																														

Figure 8. Drum program for use with active esters
See Figure 7 for explanation

haust stroke to minimize solvent mixing, and the shaker always stops with the vessel in the upright position to make the following filtering (outlet) step possible. During the third of these outlet steps (step 10, bank 1), the solvent valve advances to position 2, and the HCl–acetic acid reagent is then pumped into the vessel. The 30-minute reaction period necessary for complete removal of the *tert*-butyloxycarbonyl protecting group is obtained by use of three successive 10-minute shaking steps.

After this deprotection step the resin is washed three times with acetic acid to remove hydrogen chloride, three times with ethanol to remove acetic acid, and three times with dimethylformamide. A 10-minute shaking period with triethylamine in dimethylformamide serves to neutralize the hydrochloride of the amino acid on the resin, thus liberating the free amine in preparation for coupling with the next protected amino acid. Triethylammonium chloride and excess triethylamine are removed by three washes with dimethylformamide, and three methylene chloride washes then prepare the resin for the coupling step. The *t*-BOC amino acid solution is then pumped into the vessel in a 30-second (timer 1) pumping step; the small amount of air pumped caused no harm. On the next step (rinse), the pump draws one more stroke of air, then three strokes of methylene chloride to flush the amino acid line.

The next step is a 10-minute shaking operation to allow the amino acid to soak into the resin beads. During this step, the solvent valve advances to the diimide (No. 10) position. At the next step, diimide solution is pumped for 30 seconds, and then the rinse step adds one more stroke of diimide solution and three strokes of methylene chloride. The coupling reaction then takes place during a 2-hour (timer 5) shaking cycle. After the coupling reaction, by-products and excess reagents are removed by three washes in methylene chloride and two washes in ethanol.

If the end-of-cycle switch (S14) was set in the hold position, the instrument stops after the third ethanol wash and the resin is left suspended in ethanol. If S14 is in the go position, the programmer returns to the beginning of the program and proceeds to carry out the next cycle of operation. The instrument will continue to operate for approximately 24 hours until the coupling cycle for the sixth amino acid has been completed. Then the end-of-run microswitch (S13) stops the instrument. To continue the run, the amino acid reservoirs are washed (solvents are added to the reservoirs and drawn through the amino acid valve and the solvent valve to the waste flask through the three-way stopcock). The amino acid reservoirs are then refilled with the proper new solutions, the reagent and solvent reservoirs are replenished if necessary, the amino acid valve is set to position 1, and the solvent valve is set to position 12. The programmer is then stepped manually (S5) back to step 1 to start the coupling of the next six amino acid residues.

The functioning of the active ester coupling cycle is similar in most respects to the diimide cycle, but with the following exceptions:

Since the active ester coupling reactions are done in dime-thylformamide, the three methylene chloride washes preceding the coupling reaction are omitted, and instead the resin is washed six times with dimethylformamide. This is accomplished by programming the active ester drum to move the solvent valve past the unused methylene chloride position (No. 8). During the shaking period of the last dimethylformamide wash (step 27, bank 2), the solvent valve is advanced to position 8 (methylene chloride), and during the outlet step (step 28) the valve is advanced again to the amino acid position (No. 9). The instrument is thus ready to pump the amino acid active ester solution (16-ml., 1-minute pumping, timer 2) at step 29.

A 4-hour reaction period for the coupling reaction is provided by programming two successive 120-minute shaking steps.

Since the diimide reagent is not used, the program drum moves the solvent valve past this position without using it. During the coupling reaction the solvent valve is advanced to the diimide position, and during the next step (outlet, step 4) the valve is again stepped to bring it to the methylene chloride position (No. 11) for the washing operations.

The diimide reaction has been used routinely for the introduction of all amino acids except asparagine and glutamine. Since the diimide reagent causes an undesirable side reaction with these amino acids (*5, 7, 16*), they have been employed as their *p*-nitrophenyl esters (*2, 3*). When it is desired to change to the active ester program during the course of a synthetic sequence, the filling of the amino acid reservoirs is arranged so that the last amino acid before the asparagine or glutamine is placed in reservoir 6. This causes the instrument to stop at the point where the drum change must be made. A similar procedure may be followed for the reverse program change, or if only a single asparagine or glutamine is to be introduced, it may be placed in any reservoir and the end-of-cycle switch placed in the hold position.

Resin samples may be removed at any point during the synthesis for hydrolysis and amino acid analysis as described above. This allows the operator to ascertain that the synthesis is proceeding satisfactorily.

When the synthesis of the desired amino acid sequence has been completed, the peptide–resin is removed from the reaction vessel with the aid of ethanol, filtered, and dried. Weight gain of the resin during the synthesis provides an indication of the amount of peptide incorporated. The peptide is cleaved from the resin with HBr–trifluoroacetic acid as previously described (*11*), and subjected to a suitable purification procedure.

Reliability, Maintenance, and Possible Improvements. The automatic instrument described above has been used for over 400 coupling cycles. In general, a high degree of reliability of the instrument has been experienced. Certain precautions should be mentioned, however. Vacuum lines must be kept clear of accumulations of solid triethylammonium chloride which sometimes form from the vapors which pass through the system. The relays should be checked routinely to be sure that contacts are not fouled. The Tenor programmer unit incorporates three relays, one of which arcs badly and tends to transfer contact material; it failed after approximately 30,000 steps. The shaker limit microswitch (S12) has performed satisfactorily for over 1 million cycles, and the pump limit switch (S11) over 200,000 times. These values are in excess of the manufacturer's minimum life expectancy ratings.

While the Beckman pump has shown satisfactory resistance to the wide range of solvents used, some problems have been experienced resulting from failures of the valve actuating mechanism. One such failure of the outlet valve caused excessive cylinder pressure, leakage of HCl–acetic acid, and resultant dissolution of the nylon cylinder-retaining nut. This nylon nut has been replaced by one machined from Kel-F.

Experience in the use of the automatic instrument has suggested some possible improvements. A desirable change would be the incorporation of both the diimide and active ester programs on a single large drum, with a provision for automatic change of program as desired. In areas where power interruptions are common, a safety relay could be incorporated into the circuit to prevent the instrument from continuing upon restoration of power. This would be desirable because of the design of the amino acid and solvent selector valve circuits. These valves advance one position each time they receive power, requiring three seconds for this operation, and then hold until the power is removed. During automatic operation these valves are in the holding phase for considerable lengths of time. If even an instantaneous power failure should occur during such a holding period, upon restoration of power the valve would again advance and become completely out of phase with the remainder of the program. It would be preferable to stop the instrument entirely. An improved instrument embodying these modifications is currently under construction.

Other possible modifications would be the incorporation of other redundant or fail-safe controls. For example, a solvent level control for the reaction vessel to circumvent possible pump malfunction or exhaustion of solvents; interlocking controls to assure that the solvent and amino acid valves are in the proper position at any point in the program; a vacuum sensor in the waste receiver to prevent possible failure of the filtering operation; and a recorder to make a permanent record of the performance of each operation of the automatic cycle for help in trouble-shooting possible failures.

APPLICATIONS

The automated apparatus described here has been successfully utilized for the synthesis of several peptides, among which were bradykinin (15), several analogs of this nonapeptide plasma kinin (18), angiotensinylbradykinin (12), a decapeptide from tobacco mosaic virus protein (19), and insulin (8). There was, in each case, substantial saving of time and ef-

fort in the synthesis of these peptides and the overall yields were better than those usually achieved by conventional techniques. These advantages should become even more important as the synthesis of longer peptides is undertaken.

Since the various peptides which have been prepared have contained most of the common naturally occurring amino acids, it is believed that the method will have rather wide applicability to problems of peptide synthesis. However, several problems concerned with the chemistry of peptide synthesis by the solid phase method remain, and certain amino acids and combinations of amino acids still present difficulties. These questions are under active investigation.

It has already been suggested (12) that the principles of solid phase synthesis should be applicable to the synthesis of other polymers of defined structure. The flexibility which has been incorporated into the design of this instrument is expected to facilitate its application to the automated synthesis of such polymers in addition to the specific application described here for the synthesis of peptides.

LITERATURE CITED

(1) Anderson, G. W., McGregor, A. C., *J. Am. Chem. Soc.* **79**, 6180 (1957).
(2) Bodanszky, M., duVigneaud, V., *Ibid.,* **81**, 5688 (1959).
(3) Bodanszky, M., Sheehan, J. T., *Chem. Ind.* **1964**, p. 1423.
(4) Carpino, L. A., *J. Am. Chem. Soc.* **79,** 98 (1957).
(5) Gish, D. T., Katsoyannis, P. G., Hess, G. P., Stedman, R. J., *Ibid.,* **78**, 5954 (1956).
(6) Jones, F. D., "Ingenious Mechanisms for Designers and Inventors," Vol. 1, p. 69, Industrial Press, New York, 1930.
(7) Kashelikar, D. V., Ressler, C., *J. Am. Chem. Soc.* **86**, 2467 (1964).
(8) McKay, F. C., Albertson, N. F., *Ibid.,* **79**, 4686 (1957).
(9) Marshall, G. R,, Merrifield, R. B., *Biochemistry* **4**, 2394 (1965).
(10) Merrifield, R. B., *Biochemistry* **3**, 1385 (1964).
(11) Merrifield, R. B., *Endeavour* **24**, 3 (1965).
(12) Merrifield, R. B., in "Hypotensive Peptides," E. G. Erdös, N. Back, and F. Sicuteri, eds., Springer Verlag, New York, 1966.
(13) Merrifield, R. B., *J. Am. Chem. Soc.* **85**, 2149 (1963).
(14) Merrifield, R. B., *Science* **150**, 178 (1965).
(15) Merrifield, R. B., Stewart, J. M., *Nature* **207**, 522 (1965).
(16) Paul, R., Kende, D. S., *J. Am. Chem. Soc.* **86**, 741 (1964).
(17) Schwyzer, R., Sieber, P., Kappeler, H., *Helv. Chim. Acta* **42**, 2622 (1959).
(18) Stewart, J. M., Woolley, D. W., in "Hypotensive Peptides," E. G. Erdös, N. Back, and F. Sicuteri, eds., Springer Verlag, New York, 1966.
(19) Stewart, J. M., Young, J. D., Benjamini, E., Shimizu, M., Leung, C. Y., *Federation Proc.* **25**, 653 (1966).
(20) Thomas, A. B., Rochow, E. G., *J. Am. Chem. Soc.* **79**, 1843 (1957).

Received for review September 6, 1966. Accepted October 10, 1966. Work was supported in part by Grant A 1260 from the U. S. Public Health Service.

Reprinted from *Anal. Chem.* **1966**, *38*, 1905–14.

The fluoride electrode developed by Frant and Ross in 1966 (*Science* **1966**, *154*, 1533) was the real success story in the acceptance of ion-selective electrodes in their early years of commercial development. This electrode, however, was not without its limitations: Its performance was influenced by ionic strength, hydroxide ion concentration, and interfering fluoride complexes. This classic article described the use of TISAB, which overcame these limitations through conditions designed to provide a constant ionic strength, suitable pH, and complexation of interfering Al^{3+} and Fe^{3+} ions. It paved the way for widespread use of the electrode.

Gary D. Christian
University of Washington

Use of a Total Ionic Strength Adjustment Buffer for Electrode Determination of Fluoride in Water Supplies

Martin S. Frant and James W. Ross, Jr.
Orion Research Inc., Cambridge, Mass. 02139

A recently announced (*1*) fluoride ion activity electrode has been shown to give close to a theoretical response to fluoride ion activity (*1–3*) and to be suitable for titrations (*2*) and for the determination of fluoride in a variety of different solutions (*4–8*). One of the most obvious applications of the electrode would appear to be monitoring the addition of fluoride ion to public water supplies, and in many cases this can be done quite simply. We, and others (*9, 10*), have found that with calibration curves prepared by adding known amounts of fluoride ion to untreated water, direct determination of fluoride concentration can be done at the 1 mg per liter (henceforth, the abbreviation "ppm" will be used for mg per liter) level with an accuracy of ±5%, provided only that the pH of the water is less than 8.5.

While the electrode has no direct interferences other than hydroxide ion, the difficulty in broadly extending the techniques is that the electrode responds to fluoride ion activity, not concentration. Although it may be that activity will prove to correlate better with biological response than concentration, all additions of fluoride currently are monitored on a concentration basis. A problem exists for those who must express results in concentration units, for activity will differ from concentration by an amount which is a function of the total ionic background of the water, and will differ further if a significant fraction of the fluoride ions are complexed with ions such as Al^{3+} or Fe^{3+}.

A technique is reported which eliminates virtually all of the effect of variation in the ionic composition of the water supply, and makes it possible to use a single calibration curve for a wide range of water samples. It is based on a 1:1 dilution of both standards and samples with a solution which simultaneously performs three functions: (1) the total ionic strength of the water is fixed by adding a much larger level of ions

than that found in drinking water; (2) the solution is buffered in a range which avoids hydroxide interference; and (3) citrate is added to complex Fe^{3+} or Al^{3+}, displacing any bound fluoride.

EXPERIMENTAL

Total Ionic Strength Adjustment Buffer ("TISAB"). To approximately 500 ml of H_2O was added 57 ml of Analytical Reagent grade glacial acetic acid, 58 grams of A.R. grade NaCl, and 0.30 gram of sodium citrate. The solution was titrated to pH 5.0–5.5 using A.R. grade $5M$ NaOH. The solution was cooled, and diluted to make 1 liter.

All fluoride standards were prepared by dilution from Orion Research $0.1M$ Fluoride Activity Standard.

Apparatus. The fluoride ion activity electrode (Orion Model 94-09) was used with the Orion Model 801 digital pH meter and a single-junction reference electrode with an equitransferent filling solution (Orion Model 90-01). All solutions were stirred with a magnetic stirrer.

Procedure. Ten milliliters of a standard 1.0 ppm F− solution were pipetted into a beaker, and 10.0 ml of TISAB were then added. The solution was stirred, and the electrode potential determined after three minutes. The same 1:1 mixing procedure was used with other standards and with unknowns.

RESULTS AND DISCUSSION

The calibration curves for the electrode response to fluoride levels in public water supplies are shown in Figure 1. In the absence of TISAB, the electrode gives a straight line with a slope of 56 mV per decade change from well above 100 ppm to 0.01 ppm. Below 0.01 ppm there is curvature as the electrode approaches the lower limit of detection at about 0.005 ppm. If the fluoride solution is diluted 1:1 with TISAB, the

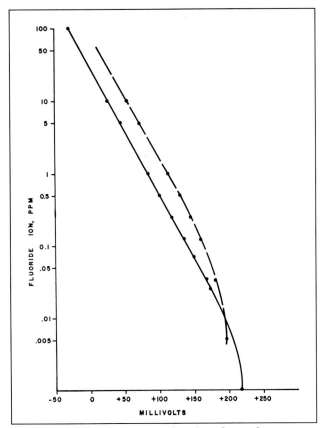

Figure 1. Calibration curves showing electrode response in millivolts *vs.* fluoride ion concentration in ppm

The solid line is the response in a pure NaF solution, the dashed line the response to NaF in a TISAB background

slope increases to 58.3 mV, just slightly less than the theoretical slope of 58.8 at 23 °C, because the ionic strength has been held constant. However, the curvature begins at a higher concentration. Part of this is caused by the dilution factor, and part by a slight suppression of activity in the $2M$ TISAB solution, but there may also be fluoride contamination in the reagents used to prepare the TISAB. This latter possibility seems quite likely for Mesmer (11) recently estimated that he had over 1 ppm fluoride in a $1M$ solution of recrystallized NaCl. However, measurements in TISAB do not lose precision at fluoride levels above 0.1 ppm, and this is adequate for most water supply purposes.

The method was tested by running 26 random amounts of NaF in distilled water, ranging from 0.1 to 5 ppm of F^- ion. The samples were prepared by one operator and run by a second as unknowns. The averages were: known, 0.96 ppm and experimentally measured, 1.00 ppm; the standard deviation between the two was 0.06 ppm.

Because fluoride is often added to water supplies as the fluosilicate ion, known samples were run using Na_2SiF_6 in distilled water and in Cambridge tap water—which has a background of 0.05 ppm of F^- and about 200 ppm of dissolved solids. The means for 11 samples were 0.98 and 1.05 ppm with a standard deviation of 0.07 ppm in distilled water and 0.80 and 0.81 with a standard deviation of 0.04 ppm for 10

Cambridge tap water samples.

Arrangements were then made with a New England state health agency to obtain over 60 water samples which had been analyzed by the SPADNS (12) technique. They use a Technicon AutoAnalyzer which includes a distillation module, and restandardize every fifth sample. The history of the samples, including both source and content, was unknown to us. The samples were not randomly selected, but were chosen to include a larger representation of stations known to have possibly interfering ions present.

The results were summarized by plotting the electrode values *vs.* the SPADNS values for the same samples. Without TISAB, 13 of the 48 samples (or more than 25%) fall more than ±0.05 ppm from the best fit line. With TISAB, all of the samples fall within, or very close to, the ±0.05 ppm limits.

While the combined scatter for both methods is quite low there is a distinct bias between the two sets of data, with the electrode readings running about 10–12% low. The bias is the same with or without TISAB: the best fit lines have almost exactly the same slope.

The bias was not particularly disturbing, for a large scale survey in 1961 on the SPADNS method (with distillation) showed that a single fluoride solution containing 0.68 ppm was analyzed by 50 major laboratories as containing from 0.24 to 1.22 ppm with a standard deviation of 0.09 ppm. We attempted to find the source of bias by rerunning under different code numbers some of the same samples several months apart, but both techniques reproduced their results quite well (see Table I). A fluoride recovery experiment was run on the same samples, and both methods agreed on the added amount. The standards in both laboratories appear to agree, for there is good agreement for the two distilled water samples, and on earlier analyses of sample solutions.

A second, smaller scale test was run in cooperation with the Illinois Department of Public Health. Their Division of Laboratories supplied 24 samples from various sources with fluoride contents ranging from 0.1 to 2.0 ppm.

They had used both the SPADNS procedure including distillation, and had also run replicate analyses on some samples by Mergregian–Maier method. The Illinois mean for the 24 samples by SPADNS was 0.66 ppm; our mean by the TISAB and electrode procedure was 0.65 ppm; the standard deviation between the two methods was 0.088 ppm. In this instance there was no bias between the methods, although they use the same procedure as the New England health agency.

Industrial and waste waters having a much higher and more variable dissolved solids content were also examined. Sixty-two samples of water taken from rivers in various sections of continental United States were supplied by the Division of Pollution Surveillance of the Federal Water Pollution Control Administration, U.S. Department of the Interior. Complete analyses were available, including fluoride by the SPADNS technique. They do not use distillation, but add $BaCl_2$ to those samples which are found to be high in sulfate or phosphate, for these are known to interfere in the SPADNS test.

A comparison of the results by the SPADNS method and

Table I. Recovery of Fluoride by Electrode and SPADNS Procedure

Sample No.	Replicate SPADNS analyses	Replicate electrode analyses	Added F⁻	SPADNS Expected	SPADNS Found	Electrode Expected	Electrode Found
15913	0.61, 0.67	0.60, 0.62	0.643	1.28	1.35	1.25	1.25
15616	0.20, 0.20	0.10, 0.09	0.646	0.85	0.78	0.75	0.70
15098	0.46, 0.48	0.39, 0.40	0.536	1.01	1.00	0.94	0.94
16596	1.08, 1.07	0.90, 0.92	0.461	1.55	1.60	1.38	1.40
15176	1.15, 1.10	0.91, 0.98	0.770	1.90	1.90	1.72	1.75
15894	0.82, 0.80	0.68, 0.72	0.319	1.13	1.13	1.02	1.02
dist H$_2$O	. . .	. . .	0.523	0.52	0.56	0.52	0.51
dist H$_2$O	. . .	. . .	0.949	0.95	0.93	0.95	0.94

the electrode showed that 85% of the values were within ±0.05 ppm, and all of the values fell within ±0.1 ppm. For the two points which were most out of range, one sample had over 1 cm of mud at the bottom, and the other had a considerable growth of algae. For comparison with other techniques, preparation of standards and 62 analyses required two working days by a lab technician without previous electrode experience.

The means for the two methods were SPADNS 0.28 ppm, and electrode with TISAB 0.29 ppm. Here again, there was no bias. The standard deviation between the two methods was 0.06 ppm. In an earlier test with the same agency, TISAB was not used. With a different lot of 27 samples, the SPADNS mean was 0.38 and the electrode 0.28. Without TISAB, the standard deviation was 0.14 ppm.

We believe that the TISAB approach will make possible direct concentration readings for low levels of fluoride in a wide variety of aqueous systems. Aside from the examples cited here, we believe that the same approach can probably be extended in fluoride determinations in urine, or from scrubbers which monitor for air pollution.

ACKNOWLEDGMENT

The generous cooperation of Robert C. Kroner of the Federal Water Pollution Control Administration and Robert M. Scott of the Illinois Department of Health is gratefully acknowledged.

LITERATURE CITED

(1) M. S. Frant and J. W. Ross, Jr., *Science*, **154**, 1553–5 (1966).
(2) J. J. Lingane, *Anal. Chem.*, **39**, 881–7 (1967).
(3) J. B. Andelman, Fortieth Water Pollution Control Federation Conference, New York, October 12, 1967.
(4) B. A. Raby and W. E. Sunderland, *Anal. Chem.*, **39**, 1304–5 (1967).
(5) M. S. Frant, *Planting,* **54**, 702–4 (1967).
(6) R. A. Durst and J. K. Taylor, *Anal. Chem.*, **39**, 1483–5 (1967).
(7) I. A. Elfers, unpublished data, The Robert A. Taft Sanitary Engineering Center, Cincinnati, Ohio, 1967.
(8) M. D. Morris and J. B. Orenberg, *Anal. Chem.*, **39**, 1776–80 (1967).
(9) K. Knowlton, Superintendent, Beverly-Salem Water Department, Beverly, Mass., personal communication, 1967.
(10) T. S. Light, Pittsburgh Conference on Analytical Chemistry and Applied Spectroscopy, March 1967.
(11) R. E. Mesmer, *Anal. Chem.*, **40**, 443 (1968).
(12) "Standard Methods for the Examination of Water and Wastewater," American Public Health Association, New York, 1965, pp. 144–6.

Received for review December 27, 1967. Accepted March 8, 1968.

Reprinted from *Anal. Chem.* **1968**, *40*, 1169–71.

It is hard to imagine a GC/MS instrument without the now-integrated computer-based data system that collects successive mass spectra and aids in identifying the eluting components. This landmark paper describes the first of such continuous spectral recording systems and introduces the then-visionary concepts of the computer-controlled mass spectrometer, calculated total intensity chromatograms, automatic background subtraction, individual *m/z* value chromatograms, and automated matching of unknown spectra against an online spectral library.

Chris Enke
Michigan State University

Mass Spectrometer–Computer System Particularly Suited for Gas Chromatography of Complex Mixtures

Ronald A. Hites[1] and Klaus Biemann
Department of Chemistry, Massachusetts Institute of Technology, Cambridge, Mass. 02139

A digital recording technique for low resolution fast scanning mass spectrometers employing a medium size on-line computer is discussed. The speed with which data are taken and the large spectra storage capacity of the system make it particularly suited for recording mass spectra of gas chromatographic effluents. Some of the features of this system are that spectra are recorded continuously regardless of the emergence of gas chromatographic fractions; peak center and intensity calculations proceed while the spectrum is being scanned; secondary storage on magnetic disks allows space for a practically unlimited number of spectra; the computer controls the scanning function of the mass spectrometer; the spectra are correlated with the chromatogram by a plot of total intensity (calculated by the computer) *vs.* spectrum index number; and all spectra are presented in digital form (mass-intensity tables and/or plots) suitable for further processing such as correcting for background or searching standard files of spectra.

U se of a gas chromatograph as a sophisticated sample introduction system for a mass spectrometer has gained wide acceptance because of its great utility and power as an analytical tool. This utility, however, is somewhat limited by the large volume of data that can be produced.

In the past it has been common practice to record a mass spectrum during the emergence of each gas-chromatographic fraction that seemed to be of interest, as well as a few spectra between fractions to obtain scans representative of the background of the entire system. Because gas chromatograms of

[1] N.I.H. predoctoral fellow, 1966–68.

mixtures encountered in a wide variety of research or routine problems generally result in 10–100 fractions, and sometimes even more, the number of mass spectra recorded within a very short time (5 minutes to 2 hours) is proportionally large. Conversion of all these raw spectra (oscillograph traces) into an interpretable form which requires at least the identification of all mass numbers, if not conversion to a mass *vs.* intensity plot or table, is quite time consuming in comparison to the actual recording time (1–3 seconds).

Utilization of the recorded spectra can hardly keep pace with their accumulation, even when spectra are recorded selectively. This selective process not only places the burden of constant decision (to record or not to record) on the operator but also leads to the loss of important information from those parts of the gas chromatogram that are allowed to pass without recording the mass spectrum. Thus, minor or partly unresolved components may easily escape detection. This risk would be eliminated if spectra were recorded continuously, but that would produce an unmanageably large number of spectra—a few hundred per gas chromatogram. Furthermore, at this stage more automated approaches to the identification of spectra would have to be utilized and these techniques require data in digital (*m/e vs.* intensity) form.

As an aid in alleviating this burden, we have previously reported a computer compatible recording system for fast scanning single focusing mass spectrometers employing digital magnetic tape (*1*). The output of the electron multiplier amplifier was digitized at a rate of 3000 samples per second and this information, written on magnetic tape, was processed at a local batch processing computer installation. Processing of the tape consisted of finding peak centers, converting them to masses (based on an external calibration of the time–mass relationship) and finally plotting the normalized spectra. However, batch processing proved somewhat inconvenient, and with the subsequent installation of a medium-sized digital computer in our laboratory, we had the opportunity to test a more elegant and potentially more useful approach—real time data acquisition.

The basic feature of this on-line system is the continuous scanning and recording of mass spectra at four-second intervals. The amplified analog output of the electron multiplier is read by the internal analog-to-digital converter (resolution = 14 bits plus sign) of the computer, peak centers are located while the data is being read, and finally, when the scan is complete (3 seconds), the peak position (in arbitrary time units) and intensity data are transferred to the computer's magnetic disks. Spectra are recorded continuously during the entire period of the gas chromatogram regardless of the emergence of peaks. The ability to record all possible spectra during the chromatogram eliminates the need to constantly decide during the experiment whether or not to record a given spectrum and also eliminates the risk of not scanning a spectrum, which later (during the evaluation of the data) turns out to be important.

When accumulating these large numbers of mass spectra (450 spectra for a 30-minute chromatogram), the need for a method for precisely correlating each scan with the chromatogram soon became apparent. It was realized that if the raw data points (approximately 8000 of them) are summed, the result is directly proportional to the unresolved ion beam current as measured by the beam monitor of the mass spectrometer. If each spectrum is indexed serially, a plot of this index *vs.* summed intensity closely resembles the gas chromatogram (see Figures 2 and 3). In fact, with most magnetically scanning mass spectrometers, the beam monitor signal fluctuates during each scan, making it impossible to obtain a smooth recording resembling a gas chromatogram. (This disturbance is caused by the interactions of the magnetic field with the secondary electrons generated when a positive ion strikes the beam monitor electrode.)

From the plot of summed intensities, one is able to note the relative intensity of various spectra, and thus evaluate the quality of the spectra—i.e., relative combination of compound and background, mainly column bleed. The absolute intensities of the spectra are not apparent in the final form since all intensities are normalized to the most abundant peak in the particular spectrum. However, this plot is used primarily to correlate the spectrum index with the chromatogram. For example, one can select from this plot the indexes representing those intense spectra in gas chromatographic fractions that are of the most interest. By entering these indexes using the keyboard, the computer will search the stored data to find the corresponding spectra and present them in plotted or tabular form at the discretion of the user.

It is the rapid processing ability of the computer and the large data storage capacity of the magnetic disks that allow one to record great numbers of spectra. In addition, one is aided by another feature of an on-line computer—its ability to actually control the mass spectrometer. Since in the GC–MS–computer system described here, the reproducibility of the repetitive scanning (cycling) of the magnetic field is of

critical importance for the reliable conversion of peak positions in time units to masses, we use the internal timers of the computer to trigger the repetitive switching of the magnetic field scanner. The necessary subroutines are part of the software package supplied by the computer manufacturer. The reproducibility of the repetitive magnetic scan is improved because computer controlled cycling is more accurate than the previously used (*1*) electro-mechanical relay supplied with the mass spectrometer's scan circuit. As a consequence, the frequency of correction during the time-to-mass conversion is greatly reduced.

EXPERIMENTAL

Figure 1 represents a schematic diagram of the gas chromatograph–mass spectrometer–computer system. The effluent of a Varian Aerograph Model 600 gas chromatograph, equipped with a flame ionization detector and linear temperature programmer, passes through a splitter that diverts about one third of the sample to the flame detector. The remaining two thirds of the effluent are carried to a fritted glass helium separator (*2*) by a 12-inch length of 0.010 inch i.d. stainless steel capillary tubing. This tubing is uniformly heated with heating tape since even a minute cold spot leads to loss of higher boiling components by condensation. The capillary connects to a high temperature diaphragm valve used to isolate the gas chromatograph and the mass spectrometer when the two instruments are to be operated independently from each other. The valve and helium separator are housed in a heated oven. Table I summarizes the operating conditions of the GC–MS instrument.

The mass spectrometer employed is a Hitachi–Perkin-Elmer RMU6-D, a 90° sector, magnetic scanning instrument. The magnet scanner is repetitively started and stopped under control of the internal timers of the computer. For gas chromatographic work, the field is swept from mass 20 to 600 in 3 seconds and allowed to collapse for 1 second, with this cycle repeating continuously. Thus, since magnet hysteresis effects have been eliminated after a few scans, a given mass is always

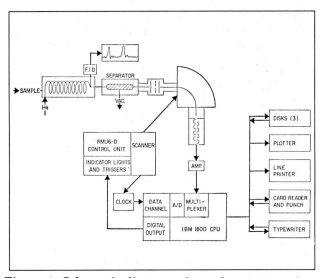

Figure 1. Schematic diagram of gas chromatograph–mass spectrometer–computer system

Table I. Operating Conditions of the Gas Chromatograph–Mass Spectrometer Instrument

GC column	5′ × 1/8″, 1% SE-30 on Chromosorb W, glass
GC temp	Injector: 290 °C Oven: programmed 100 to 360 °C at 10°/min
He flow	25 ml/min through column
Split to F.I.D.	35% at 200 °C (GC oven)
Separator temp	200 °C
M.S. pressure	7×10^{-6} torr
M.S. resolution	1:500
M.S. scan rate	3 sec for $M/e = 20$ to $M/e = 600$, repeated in 4-sec cycles
Acceleration potential	1800 V
Electron energy	70 eV
Ionizing current	60×10^{-6} A
Ion source temp	250 °C

swept across the collector at the same time relative to the start of the scan.

The computer (see Figure 1) is an IBM 1800 system, a medium sized computer incorporating special features designed for data acquisition and process control. The configuration available in our laboratory has a cycle time of 2 μsec, divide time of 45 μsec, word length of 16 bits, and 32,767 words of core storage. It is equipped with various input–output devices: three magnetic disks (12,000 words each), typewriter and keyboard, card reader and punch, line printer, and incremental plotter. The interface to the mass spectrometer is simply a pulse generator that operates only when the mass spectrometer is scanning but not when the field is collapsing to the starting point. The pulse generator is set for 350-μsec intervals. This rate was chosen to produce about 3000 digital samples per second, the rate used with the magnetic tape system for reasons discussed previously (1). These pulses are sensed by the computer through a data channel. Each pulse causes the analog input feature of the computer to accomplish the following without interrupting the normal sequence of instructions: it determines which multiplexer point to read; where in core storage the result is placed (both specified by the programmer); and then converts the analog signal to digital form using an analog-to-digital converter which is an integral part of this computer. In this way, the data are taken under external control while the program concerns itself with further processing as outlined below.

PROGRAMMING

When preparing to record spectra of gas chromatographic effluents, a punched card is read into the computer which causes the program (stored on one of the magnetic disks) to load and begin execution. The date and an experiment code are entered into the keyboard to be used along with the spectrum index number for later identification of the spectra. The

computer now starts cycling the magnetic field scanner and the sample is injected into the gas chromatograph. After the apparent pressure surge due to the solvent has decreased, the user turns on the filament and the computer is set in the read mode using a switch located at the mass spectrometer. From this point on, the experiment continues automatically to the end of the chromatogram.

As mentioned above, the output of the electron multiplier amplifier is digitized every 350 μsec and the result stored. While being read, the data are processed in the following manner. Whenever five data points have been accumulated, they are smoothed using a simple convolution technique (3). As each point is produced, it is examined to see whether it could be the beginning of a peak, defined as a point that is a given number of units greater than its predecessor. If the next data point is again higher, the peak center calculation begins. The peak centers are found by the centroid method; the center is the ordinate point at which the area is halved. The highest abscissa value in the peak is taken as the intensity. Since this peak finding proceeds while the spectrum is being scanned, peak center calculations (in terms of clock-pulses and intensities) are complete a few milliseconds after the end of the scan. These peak heights and centers are transferred to the magnetic disks along with an index number indicating the scan. The raw data points, which have been smoothed, remain in core storage and are now summed to give a value corresponding to "total ionization" which is stored in core memory along with the scan index.

The sequence of these processes is indicated to the user by signal lights on the mass spectrometer console. Since the above operations must be complete within one four-second scan cycle, the programs for the peak finding procedure, the disk operations, and the intensity summation are all written in IBM 1800 assembler language rather than Fortran.

The above cycle of events is continuously repeated during

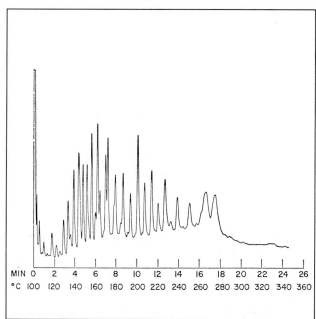

Figure 2. Flame ionization detector chromatogram of methyl esters of a naturally occurring acidic extract

the course of the chromatogram and after a 30-minute chromatogram, for example, about 450 spectra are stored on the disks. When the chromatogram is complete, the user actuates a sense switch on the computer, which causes the program to terminate the read mode and to plot the summed intensities *vs.* the spectrum index number. Such a plot is shown in Figure 3. It serves to correlate the spectrum index with the flame ionization detector recording (Figure 2). Either all of the spectra or specified ones can be processed further, a choice that is available to a user who wishes to examine the spectra of certain GC peaks that appear to be of greater interest than others. Processing consists of conversion of the peak heights and positions stored on the disk in terms of time units into a table of mass *vs.* intensity. This is accomplished using an external standard in the same manner as with the previously described recording system employing digital magnetic tape (*1*).

RESULTS

The flame ionization detector chromatogram of a naturally occurring acidic extract which has been converted to methyl esters is shown in Figure 2. This recording was obtained simultaneously with the recording of the mass spectra. The corresponding summed intensity plot is shown in Figure 3 (note the similarity of the two figures). The apparent increase · in the abundance of components of long retention time in the summed intensity plot as compared to the F.I.D. record is due to the thermal expansion of that part of the metal capillary (connecting the gas chromatograph with the mass spectrometer) that lies in the temperature programmed GC oven.

This total ionization plot indicates that about 320 spectra were stored on the disks. Since one disk can be replaced while data are being written on the other, the capacity is thus in principle unlimited. The area from scan 95 to 160 has been replotted in Figure 4 in an enlarged scale. Normally,

the entire plot appears in this form; the condensed form (Figure 3) is presented here only for the benefit of the reader. Each X-mark represents the summed intensities of a stored spectrum. The spectra marked with arrows are plotted (by the computer) in Figure 5. It can be seen that spectrum No. 134 corresponds to a mixture of the two components represented by the adjacent spectra (Nos. 133 and 135). Close examination of spectrum No. 133 reveals that it compares very favorably with the published spectrum of methyl 4,8,12-trimethyloctadecanoate (*4*), and was finally interpreted to be that of methyl 4,8,12,16-tetramethylheptadecanoate. Spectrum 135 is that of methyl nonadecanoate. This example illustrates that it is indeed possible to correctly identify a minor component which manifests itself only as a small shoulder on a gas chromatographic peak. Processing of the spectra corresponding to all significant fractions of this chromatogram revealed that the sample is a mixture of methyl esters of straight chain and polymethyl substituted fatty acids ranging up to C_{30}. A discussion of the significance of these spectra, which would go beyond the subject of this paper, will be published elsewhere.

Generating the spectra in this computer-compatible form opens the way to many types of further processing such as searching files of authentic spectra, resolving spectra of simple mixtures into those of the individual components, subtracting background, etc. The identifications mentioned above have indeed involved such techniques (*5*).

While the utility of this system was discussed above in the context of gas chromatography, it is also well suited for recording the spectra of samples introduced directly into the ion source. Fractionation and decomposition can be easily detected; the need for establishing a constant vapor pressure over a long period of time is eliminated by the fast scanning speed; and the sensitivity is not sacrificed due to the fast response and low noise of the computer system.

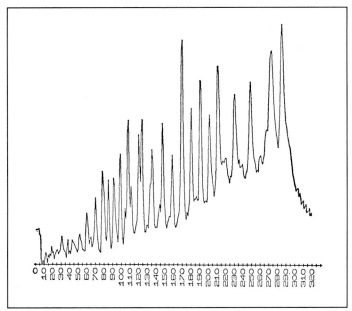

Figure 3. Plot of total ionization (ordinate) *vs.* mass spectrum index number corresponding to chromatogram of Figure 2

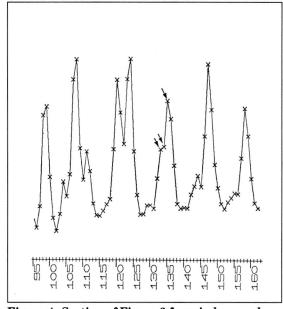

Figure 4. Section of Figure 3 from index numbers 95 to 160 plotted in expanded scale

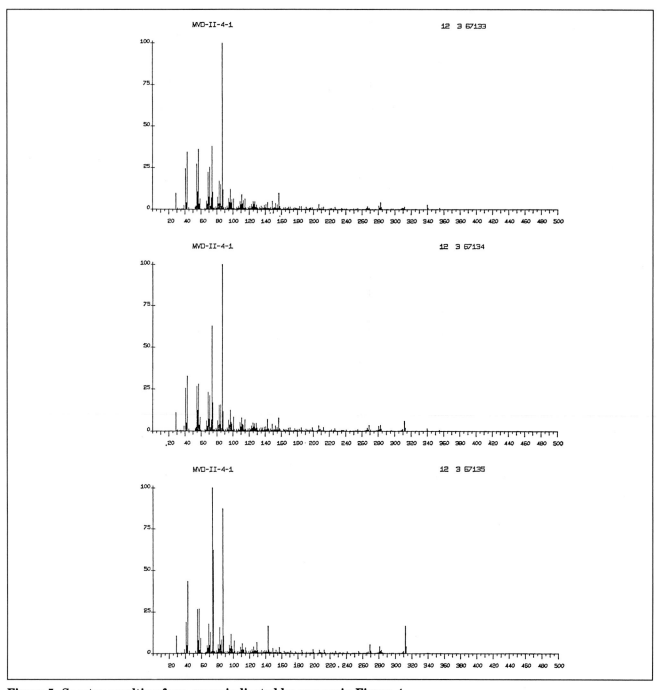

Figure 5. Spectra resulting from scans indicated by arrows in Figure 4

Top: Scan number = 133   Middle: Scan number = 134   Bottom: Scan number = 135

ACKNOWLEDGMENT

The authors are indebted to Mrs. Vivian Beecher and Donald Sordillo for their help with the programming and computer interfacing, to Dr. Milica V. Djuricic for supplying the sample discussed, and to Robert C. Murphy for help with the gas chromatography.

LITERATURE CITED

(1) R. A. Hites and K. Biemann, *Anal. Chem.*, **39**, 965 (1967).
(2) J. T. Watson and K. Biemann, *Anal. Chem.*, **37**, 844 (1965).
(3) A. Savitzsky and M. J. E. Golay, *Anal. Chem.*, **36**, 1627 (1964).
(4) R. Ryhage and E. Stenhagen, *Arkiv Kemi,* **15**, 333 (1960).
(5) R. A. Hites and K. Biemann, International Conf. on Mass Spectrometry, Berlin, September 1967.

Received for review February 28, 1968. Accepted May 6, 1968. Work supported by research grants from the Facilities Research Division of the National Institutes of Health (FR-00317) and the National Aeronautics and Space Administration (NGR-22-009-102).

Reprinted from *Anal. Chem.* **1968**, *40*, 1217–21.

The invention of supercritical fluid chromatography is attributed to Klesper, Corwin, and Turner (*J. Org. Chem.* **1962**, *27*, 700), whose original apparatus had no pump and allowed only visual detection of colored solute bands directly on a glass column. This article, although published after other groups had begun research in the field, represents the first report of a practical instrument by the inventor. This work relied on CCl_2F_2 as a mobile phase (an unlikely prospect today) and focused on separating nonvolatile compounds (still the focus for many SFC practitioners). Many features of this instrument, such as the detector–pressure regulator configuration, are still used in modern SFC instruments.

T. L. Chester
The Procter & Gamble Company

Apparatus and Materials for Hyperpressure Gas Chromatography of Nonvolatile Compounds

Nicholas M. Karayannis,[1] Alsoph H. Corwin, Earl W. Baker,[2] Ernst Klesper,[3] and Joseph A. Walter
Department of Chemistry, The Johns Hopkins University, Baltimore, Md.

Introduction of the method of gas-liquid chromatography (*1*) literally revolutionized the art of separation of materials possessing sufficient vapor pressure to permit its use. In practice, however, a wide variety of materials of limited volatility have resisted efforts to purify them by this means. In particular, organic compounds of moderate molecular weights with polar groups attached, most inorganic compounds (*2*), and ionic organic compounds, such as amino acids (*3*), etc., have resisted efforts at separation without functional group modification.

In these laboratories, especial interest has attached to the purification of porphyrins. The potential advantages of the use of gas chromatography in this application are obvious but the limited volatility of the materials made the method impractical. Efforts were made to bring about the volatilization of porphyrins by raising the column temperature and by operating under vacuum. Partial success was achieved with etioporphyrin II at temperatures in excess of 250 °C, using the lowest pressure available with an oil pump. Under these conditions, however, more porphyrin was decomposed than was volatilized. After numerous unsuccessful efforts to volatilize porphyrins for gas chromatographic purification, our search for a new approach to the problem of gas chromatography of materials of low volatility was successful. A preliminary publication by Klesper, Corwin, and Turner (*4*) reported the separation of a metalloporphyrin mixture by high pressure gas chromatography above the critical temperature of the carrier gas, which was dichlorodifluoromethane. The method was suggested by the fact that liquids, for thermodynamic reasons, show higher vapor tension when under pressure from insolu-

ble gases (*5*). It had been observed further that inorganic and organic solids (*6*, *7*), including a derivative of chlorophyll (*8*), are soluble in various solvents—*e.g.*, carbon dioxide, sulfur dioxide, carbon disulfide, ammonia, methanol, ethanol, and ether—under pressure and above the critical temperature of the solvent. It had also been found that the quantity of solute in the gas phase is frequently far greater than can be accounted for by normal volatility (*9*, *10*), while the effect of pressure on the amount of solid dispersed is much larger than that expected from the normal increase of vapor pressure because of external pressure.

Klesper, Corwin, and Turner reported that dichlorodifluoromethane, c.t. 111.5 °C (*11*), is a solvent, under high pressure, for porphyrins. It is superior to other less chlorinated freons and nitrogen (*4*). Dichlorodifluoromethane has the additional advantages of noninflammability and favorable corrosion characteristics with many metals and alloys (*12*). The pressures applied during the preliminary explorations (*4*) were in the range of 800–2300 psi.

Since then work on high pressure gas chromatography has been reported but, with one exception (*13*, *14*), in quite different directions. Thus, Kobayashi *et al.* (*15–17*) have applied pressures up to 2000 psi for the evaluation of physical constants. Myers and Giddings (*18*, *19*) have made a combined high pressure–small particle ($<1\mu$) approach to gas chromatography, at pressures up to 2500 psi and, in collaboration with Manwaring (*20*), a study of the effect of turbulent flow in gas chromatography with high-speed, high-pressure equipment up to 170 atm. Their most recent instrumentation works with inlet pressures up to 2000 atm (*21*). Sie, van Beersum, and Rijnders (*13*), working with CO_2 as carrier gas, at 30–40 °C and up to 80 atm, concluded that, with increasing pressure, a pronounced drop of the partition coefficient is

[1]Drexel Institute of Technology, Philadelphia, Pa.
[2]Mellon Institute, Pittsburgh, Pa.
[3]Institut für Makromolekulare Chemie, Freiburg, W. Germany.

achieved, permitting separations to be carried out at lower temperatures than usual. Sie and Rijnders (*14*) have also reported recently on the efficacy of supercritical fluids in securing increased volatility and improved separations. Their apparatus, however, lacks the capability of pumping to increase pressures and so to secure the maximum advantage of the method.

Our original apparatus utilized for the gas chromatography of metalloporphyrins (*4*) was not supplied with satisfactory devices for pressure or flow rate control. In order to extend our studies on the development of gas chromatography under high pressure as an analytical or preparative method for nonvolatile organic, organometallic, and inorganic compounds, a better experimental apparatus was needed. A joint effort was undertaken by the Perkin-Elmer Corporation (Norwalk, Conn.) and this laboratory, which resulted in the construction of the hyperpressure gas chromatograph by Perkin-Elmer.

Our explorations with hyperpressure gas chromatography show that this field is not a mere extension of conventional gas chromatography. Success of the method depends upon interaction between the carrier gas and the material being chromatographed of a type not present in conventional operations. Most conventional liquid phases are unsatisfactory because of the high solvent power of the system. Conventional detection methods are mostly inapplicable because of the high density of the carrier gas. In addition, the limitations of the development of the general technology of manipulations at high pressures pose special handling problems.

In spite of these difficulties, substantial progress has been made toward practical gas chromatographic operations under high pressure conditions. In the present paper, the current apparatus is described. Our findings concerning the suitability of various column packing materials are included and a survey is presented of the types of chelates which can be volatilized successfully in the apparatus for the separation of metals. Detailed information on the conditions permitting separations to be achieved with each of these types of chelating agent will be presented in subsequent papers.

APPARATUS AND MODIFICATIONS

Description of the Apparatus. *Pumping System.* The pumping system (Pressure Products Industries, Hatboro, Pa.) is shown schematically in the lower portion of Figure 1. The air pump supplies pressures up to 2800 psi and the screw pump smoothes and meters the pressure for an additional 100–300 psi. The output pressure is controlled by a backpressure regulator (Skyvalve Inc., Syracuse, N. Y.) at the instrument exit before the flowmeter. Safety devices include a diaphragm vent valve, rupture disks, check valves and filters, and a steel enclosure.

Gas Chromatographic System. This is shown schematically in the upper portion of Figure l. A vaporizer heats the incoming CCl_2F_2 above the critical temperature. The injection system allows sample introduction while the system is at elevated temperature and pressure. All components between the vaporizer and the trap are constructed of titanium. It was found

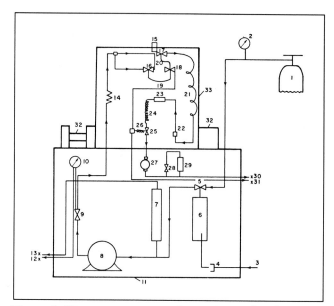

Figure 1. Hyperpressure gas chromatograph, schematic

1, Freon cylinder 35 °C.; *2, 10,* gauges; *3,* air supply; *4,* air pressure regulator; *5, 9, 25, 28,* valves; *6,* air pump; *7,* accumulator; *8,* screw pump; *11,* pumping system enclosure (steel); *12, 13, 30, 31,* vents; *14,* vaporizer; *15,* injection port; *16, 17, 18,* injection system valves; *19,* 6-inch, 1/8 inch i.d., sample loop; *20,* 7-inch, 1/8 inch i.d. tubing, connecting valves *17* and *18; 21,* column; *22,* optical cell; *23,* alumina trap; *24, 26,* capillary restriction tubing; *27,* backpressure regulator; *29,* flowmeter; *32,* filter photometer assembly; *33,* oven enclosure

later that stainless steel can be substituted and is resistant to the chelating action of porphyrins. Chromatographic alumina was selected as the trapping material because it does not allow the migration of porphyrins through it (*4*). After reduction of the pressure, the gas flow is measured by means of a rotameter (Brooks Instrument Division, Hatfield, Pa.).

Detector. The sample detector is a flow-through cell of fused silica scanned by a Hitachi–Perkin-Elmer Model 139 spectrophotometer. The cell is tubing with polished 2-mm flats for the transmission of the light beam. The internal chamber is 2 mm in diameter. The cell is held by spring pressure. Sealing is achieved by two Viton "O" rings. As this cell was sensitive to changes in refractive index of the carrier gas, because of pressure inequalities, a second cell was designed with all optical components flat. This cell was scaled with thin sheet washers made of Teflon (Du Pont). The spectrophotometer may be used with an interference filter photometer for gross determinations or with the instrument monochromator for resolution and sensitivity.

Recorder. A Leeds and Northrup Speedomax H, Model S, AZAR recorder is used.

Operation. In practice, the sample loop, the 7-inch tubing connecting valves 17 and 18 and the column of Figure 1 (items 19, 20, and 21, respectively) were used as a unit to form the chromatographic column. For the introduction of the sample, a borosilicate glass wool plug was impregnated with the sample solution in a suitable solvent. The solvent was evaporated by warming and the plug was introduced at the inlet of the sample loop 19 before starting the apparatus. In some cases, a few crystals of the sample were applied to the plug instead of a solution.

The operating laboratory must be vigorously air conditioned to maintain the ambient temperature of the apparatus below 30 °C to prevent formation of bubbles of CCl_2F_2; otherwise vapor lock prevents filling of the system.

COLUMN PACKING MATERIALS

Solid Supports. Chromosorbs W and P proved suitable for the gas chromatography of porphyrins and metal chelates. Under the same conditions, sharper peaks are obtained with acid-washed and silanized Chromosorb W than with the untreated solid support. KCl crystals can be coated with 5% of a suitable liquid phase and used under high pressure without bleeding. The peaks produced by metal acetylacetonates from KCl columns were considerably sharper than those produced by the same chelates under the same conditions from Chromosorb P columns.

The columns can be unpacked by removing the plugs from both ends and subjecting them to a pressure of *ca.* 2000 psi. The first two components of the chromatographic column were plugged with aluminum wire plugs, and the third component and the alumina trap were plugged with barosilicate glass wool.

Stationary Phases. The problem of the liquid phase proved to be the most severe one, owing to the critical conditions of operation during hyperpressure gas chromatography (HPGC). Packings were, in most cases, prepared by the conventional slurry and evaporation technique. For Carbowax 20M columns, the stationary phase was pretreated by heating for 16 hours at 100 °C, under a vacuum less than 0.01 mm Hg, in order to prevent the formation of formaldehyde and

Table 1. Chelates Giving Incomplete Volatilization[a]

Chelating agent	Metals	Result
8-Quinolinol	Co(II), Ni(II)	Not volatile
	Co(II) (205 °C 2100 psi)	Partially volatile, decomposition
Salicylaldehyde	Co(II), Ni(II), Cu(II)	Not volatile
Salicylaldimine	Ni(II), Cu(II)	Not volatile
Salicylaldoxime	Mn(II), Fe(II), Co(II), Cd(II), V(V), VO(IV), Pd(II), Pb(II), Bi(III)	Not volatile
Dimethylglyoxime	Ni(II), Pd(II)	Not volatile
Thenoyl-trifluoro-acetone	Mn(II), Co(II)	Not volatile
	Ni(II)	Volatile, retained
	Nd(III)	Partially volatile, much residue, not eluted
Dithizone	Ni(II), Cd(II)	Volatile but altered chemically

[a] Except with salicylaldoximates (see Table II), the support was hexamethyldisilazane (HMDS)–treated Chromosorb P and the liquid phase was Epon 1001. Temp., 160 °C; pressures from 1500 to 2000 psi.

formic acid in the column (*22–24*). It was then dissolved in acetone, the predried Chromosorb W was added and, after evaporation of the solvent by mild heating, the mixture was heated for 1 hour at 100 °C, under a vacuum of less than 0.01 mm Hg. Acetone, benzene, or chloroform were usually used as the stationary phase solvents. Versamid 900 was found soluble in a boiling 1:1 mixture of methanol–ethylene dichloride.

Our experience with some liquid phases which are unsatisfactory suggests the application of the method of HPGC to the purification of plastics. It should be noted, in this connection, that the dissolution of polymers under pressure has been reported (*25*). Ucon 50-HB-2000 and Apiezon M were almost completely eluted from the column in a few minutes. Silicone Gum Rubber SE-52 columns bled continuously and considerable amounts of eluted materials were deposited in the apparatus after the column. These residues were difficult to remove. Carbowax 20M and Harflex 370 were partially decomposed. Metalloporphyrins gave broad peaks with these two substances.

Kel-F wax (15 % w/w on Chromosorb W) did not bleed and was suitable for the elution of porphyrin and acetylacetone metal chelates and etioporphyrin II. The alkyl esters of porphyrin carboxylic acids could not be eluted from this column, however. After runs with these esters, the Kel-F wax was altered and did not permit the elution of the metal chelates mentioned above. Bentone 34, coated together with Epon 1001 resin on Chromosorb P, gave rather broad but symmetrical peaks with etioporphyrins II and III. Silicone Nitrile Gum Rubber XE-60, Versamid 900, and Epon 1001 and 1009 resins are suitable for porphyrins and metal chelates and can be used for chromatographic separations. Silicone XE-60 is negligibly eluted from the column at pressures up to 3100 psi. Versamid 900 on Chromosorb W (20% w/w) bled heavily until the Versamid concentration dropped to 12-13% and then bleeding became negligible. Epon 1001 and 1009 resins did not bleed. The former proved best for the separation of porphyrins and metalloporphyrins. Porphyrin peak widths decrease on the liquid phases in the following order: Carbowax 20M and Harflex 370, Versamid 900, Epon 1009, XE-60, Epon 1001.

PRACTICAL APPLICATIONS

With the apparatus described above, separations were achieved between various types of metalloporphyrins and between various metallic chelates of acetylacetone. These will be reported in detail in separate communications. In addition, the possibilities of separations among metal chelates of seven additional chelating agents were studied. The results are reported herewith. Table I lists a group of metal chelates in which volatilization was incomplete or otherwise unsatisfactory. Table II lists a group of metal chelates which are volatile and eluted from the column.

Combining the information in Tables I and II, a sharp separation of Ni(II) and Cu(II) from many other elements may be achieved by use of the salicylaldoximates. By the use of thenoyltrifluoroacetonates, sharp separations of many elements may be obtained from Mn(II) and Co(II).

Table II. Retention Times of Volatile Chelates[a]

Salicylaldoximates

Chelate	Relative retention time
Ni(II)	1.00
Cu(II)	1.11
Zn(II)[b]	1.03

Retention time for Ni(II) chelate, 3.75 min.

Thenoyltrifluoroacetonates

Sc(III)	1.12
Y(III)	1.15
Eu(III)	1.18
Al(III)	0.97
Zr(IV)	0.98
VO	1.29
Fe(III)	1.03
Co(III)	1.16
Ni(II)	1.31
Pd(II)	1.08
Cu(II)	1.00
Zn(II)	0.97
Th(IV)	1.04
UO$_2$	1.21

Retention time for Cu(II) chelate, 5.625 min.

[a] Operating conditions: Column: 27-inch, 1/8-inch i.d. stainless steel. 2% Kel-F wax on acid washed–HMDS treated Chromosorb W, 80–100 mesh. For salicylaldoximates: pressure: 1000 psi; temp.: 130 °C; CCl$_2$F$_2$ flow: 321 ml/min gas at atmospheric pressure. For thenoyltrifluoroacetonates: pressure: 925 psi; temp.: 130 °C; gas flow: 181 ml/min.
[b] Zn salicylaldoximate is a monohydrate and loses water at 950 °C (26). It decomposes slowly at 148 °C, so that recovery is not complete.

Solvent Action. Our first publication in this field (4) indicated that the effects on volatility achieved were caused by solvent action, not by pressure alone, because high pressures of many gases fail to achieve the volatilization obtained with CCl$_2$F$_2$. The present study confirms this conclusion. Mn(II) thenoyltrifluoroacetonate has a higher vapor pressure than the Fe(III) and Al(III) chelates (27), yet the latter two are volatilized in CCl$_2$F$_2$ and the former is not. The Co(II) chelate sublimes at 0.25 mm at 690 °C., the Al(III) chelate at 125 °C, and that of Fe(III) at 135 °C (27), yet the more volatile Co(II) chelate is not volatilized in CCl$_2$F$_2$ while the latter two are.

This is indicative of the primary solvent effect of the CCl$_2$F$_2$ during volatilization of these compounds under pressure.

ACKNOWLEDGMENT

The authors acknowledge the collaboration of Harold Hill, John Baudean, and Stanley Norem of the Perkin-Elmer Corp., in the design and construction of the hyperpressure gas chromatograph.

LITERATURE CITED

(1) A. T. James and A. J. P. Martin, *Biochem. J.,* **50**, 679 (1952).
(2) R. S. Juvet, Jr., and F. Zado, in "Advances in Chromatography," J. C. Giddings and R. A. Keller, Eds., Vol. 1, Marcel Dekker, New York, 1965, pp 249–307.
(3) C. W. Kehrke and F. Shahrokhi, *Anal. Biochem.,* **15**, 97 (1966).
(4) E. Klesper, A. H. Corwin, and D. A. Turner, *J. Org. Chem.,* **27**, 700 (1962).
(5) F. Pollitzer and E. Strebel, *Z. Physik. Chem.,* **110**, 768 (1924).
(6) H. S. Booth and R. M. Bidwell, *Chem. Revs.,* **44**, 477 (1949).
(7) M. Centnerszwer, *Z. Physik. Chem.,* **46**, 427 (1903).
(8) J. B. Hannay and J. Hogarth, *Chem. News,* **41**, 103 (1880).
(9) A. Smits, *Rec. Trav. Chim.,* **49**, 962 (1930).
(10) C. J. van Nieuwenburg and P. M. van Zon, *Ibid.,* **54**, 129 (1935).
(11) R. F. Bichowsky and W. K. Gilkey, *Ind. Eng. Chem.,* **23**, 366 (1931).
(12) T. Midgley, Jr., and A. L. Henne, *Ibid.,* **22**, 542 (1930).
(13) S. T. Sie, W. van Beersum, and G. W. A. Rijnders, *Separation Sci.,* **1**, 459 (1966).
(14) S. T. Sie and G. W. A. Rijnders, *Anal. Chim. Acta,* **38**, 31 (1967).
(15) F. I. Stalkup and R. Kobayashi, *A.I.Ch.E.J.,* **9**, 121 (1963).
(16) H. B. Gilmer and R. Kobayashi, *Ibid.,* **10**, 797 (1964).
(17) K. T. Koonce, H. A. Deans, and R. Kobayashi, *Ibid.,* **11**, 259 (1965).
(18) M. N. Myers and J. C. Giddings, *Anal. Chem.,* **37**, 1453 (1965).
(19) *Ibid.,* **38**, 294 (1966).
(20) J. C. Giddings, W. A. Manwaring, and M. N. Myers, *Science,* **154**, (no. 3745), 146 (1966).
(21) M. N. Myers and J. C. Giddings, *Separation Sci.,* **1**, 761 (1966).
(22) E. R. Adlard in "Vapour Phase Chromatography," D. H. Desty and C. L. A. Harbourn, Eds., Academic Press, New York, 1957, p 98.
(23) C. Weurman and J. Dhont, *Nature,* **184**, 1480 (1959).
(24) M. E. Kieser and D. J. Sissons, *Ibid.,* **185**, 529 (1960).
(25) P. Ehrlich and E. B. Graham, *J. Polymer Sci.,* **45**, 245 (1960).
(26) P. Lumme, *Suomen Kemistilehti,* **32B**, 261 (1959).
(27) E. W. Berg and J. T. Truemper, *Anal. Chim. Acta,* **32**, 245 (1965).

Received for review November 29, 1967. Accepted June 24, 1968. This work was supported by the National Institutes of Health under Grant No. 2 RO1 GM 11159.

Reprinted from *Anal. Chem.* 1968, *40*, 1736-39.

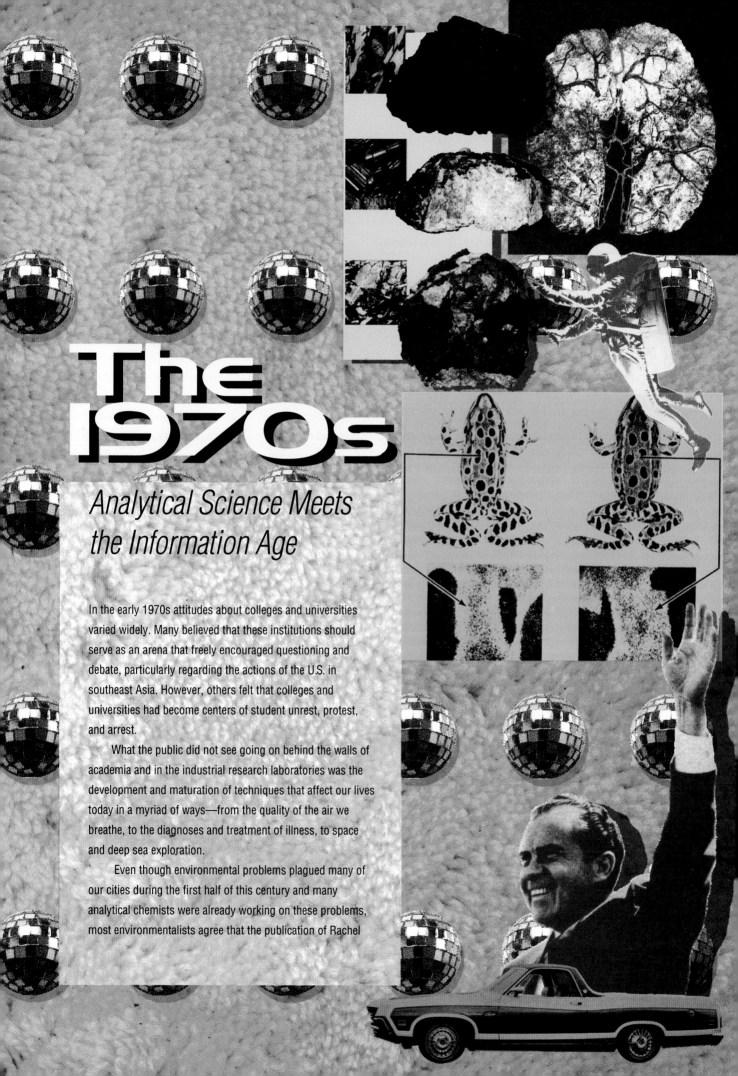

The 1970s

Analytical Science Meets the Information Age

In the early 1970s attitudes about colleges and universities varied widely. Many believed that these institutions should serve as an arena that freely encouraged questioning and debate, particularly regarding the actions of the U.S. in southeast Asia. However, others felt that colleges and universities had become centers of student unrest, protest, and arrest.

What the public did not see going on behind the walls of academia and in the industrial research laboratories was the development and maturation of techniques that affect our lives today in a myriad of ways—from the quality of the air we breathe, to the diagnoses and treatment of illness, to space and deep sea exploration.

Even though environmental problems plagued many of our cities during the first half of this century and many analytical chemists were already working on these problems, most environmentalists agree that the publication of Rachel

Carson's *Silent Spring* (1962) was undoubtedly the single most galvanic event in the history of the environmental movement. Although the first Earth Day (1970) was an occasion for average citizens to express their commitment to the environment, the real teeth to the movement came in the form of 22 separate pieces of legislation passed during the 1970s, including the creation of the Environmental Protection Agency. No decade before or since has been as legislatively fruitful. The criteria for the improvement of the environment were now set down in law.

New techniques

Chemometrics. According to Bruce Kowalski (*Anal. Chem.* **1978**, *50*, 1309 A–1313 A), the 1970s saw the advent of the information age and the creation of a new specialty: chemometrics. By using mathematics and statistics to improve the measurement process, the analytical chemist can get the most out of the data—more analytical bang for your buck, if you will.

One aspect of this new specialty was the transform domain. The success of fast Fourier transform analysis applied to NMR and IR spectroscopies led to experimentation and success in other spectroscopies and electrochemistry. Spectrum and waveform analysis in which techniques such as convolution, deconvolution, correlation, and cross-correlation were used could reduce noise and enhance resolution. Gary Horlick's work on digital data handling demonstrates useful manipulation of atomic emission spectra based on discrete Fourier transformation of the digitized spectrum, multiplication by a weight function in the Fourier domain, and inverse Fourier transformation to yield the final spectrum.

Modeling was performed using linear and nonlinear regression analysis, and univariate and multivariate control strategies increased sensitivity. A program that allowed a computer to examine and learn from a set of spectra was the beginning of pattern recognition.

Electrochemistry. Of all human organs, the brain is undoubtedly the most fascinating. Because the brain is a hubbub of electrical activity, it is only natural that electrochemical methods be used to decipher its secrets. By using electrochemical methods as detectors for LC, researchers developed clinical assays for catecholamines and their metabolites, phenolic substances, pharmaceuticals, enzymes, and endorphins. Researchers used anodic stripping voltammetry to follow catecholamines in vivo, thus demonstrating that on-line monitoring of induced chemical changes can be followed in the living brain.

During this decade, chromatographers modified stationary phases to effect a particular separation. In a similar endeavor, electrochemists sought to modify the surfaces of electrodes to effect a particular electrochemical measurement. In their work, Royce W. Murray and colleagues described their efforts to covalently bond several ligands to the surfaces of tin oxide electrodes, which was a first step toward developing various applications for chemically modified electrodes.

Mass spectrometry. This decade saw advances in many areas of MS. Several theories were proposed to clarify the processes involved in field desorption MS; problems in biochemical and natural products analysis lent themselves very well to field desorption MS, as did inorganic salts and organic molecules. Developments in the analysis of multicomponent mixtures and the use of multilabeled molecular tracers in samples hinted at how field ionization MS might be used in the next decade. Laser desorption MS of biomolecules by P. G. Kistemaker and colleagues demonstrated that the technique has potential for the analysis of nonvolatile and thermally labile compounds.

Advances in instrumentation and application of the Fourier transform methods mentioned earlier gave a boost to ion cyclotron resonance MS as an analytical technique. Christie Enke and Rick Yost introduced the triple quadrupole mass spectrometer, and chemical ionization MS experienced rapid growth during these years, as demonstrated by the work of Donald Schoengold and Burnaby Munson. The GC/MS techniques that would become the workhorse methods in environmental analysis were also developed. C. A. Evans, Jr., and J. P. Pemsler reported on their success with using the ion microprobe mass spectrometer to elucidate the isotopic and compositional gradients in thin films. Instrument manufacturers took advantage of the advances made in the early 1970s in coupling LC to MS.

Separations. LC was a technique that complemented GC and was especially useful for thermally unstable or nonvolatile compounds. However, analytical chemists realized that LC enjoyed certain other advantages.

Because lower temperatures could be used and because mobile and stationary phases allow for selective interaction of molecules, researchers were able to greatly expand this class of techniques. The development of modes of LC and various columns, such as chemically bonded stationary phases, increased the analysts' repertoire. Biochemical analysis alone was enhanced by the development of reversed-phase LC; and

"**Analytical chemistry is what analytical chemists do.**"
Charles N. Reilley

because it was so easy to recover separated fractions, articles on preparative LC soon appeared in the literature. Advances in capillary GC (direct injection, improved column technology, better detectors) contributed to the blossoming of this technique for the analysis of complex environmental and biological samples.

Hamish Small and colleagues used a combination of resins to neutralize the ions of the background electrolyte, in ion-exchange chromatography coupled to conductometric detection, so that only the species of interest is left in the effluent that enters the conductivity cell. J. Calvin Giddings and colleagues' work in sedimentation field-flow fractionation and Takao Tsuda and Milos Novotny's work in microcolumn LC established these techniques as new modes of separation.

Spectroscopy. Even though the optoacoustic effect was known in the late 19th century, it was not until the early 1970s that photoacoustic spectroscopy was reported and used to study insulator, semiconductor, and metallic systems as well as solid and semisolid biological systems; for surface analysis; and for de-excitation studies. The development of resonance ionization spectroscopy gave a sensitive detector for verification of classical physical concepts, for the study of kinetics and diffusion, and for ultra low-level counting.

Tunable dye lasers, although developed in the 1960s, were applied in the 1970s for techniques such as laser-excited atomic fluorescence flame spectrometry, reported by James Winefordner and L. M. Fraser; atmospheric monitoring; and for the development of coherent anti-Stokes Raman spectroscopy. In the realm of trace analysis, Joseph M. Jaklevic and colleagues reported their work using X-ray fluorescence with semiconductor detector spectrometry, and R. Cournoyer and colleagues reported their work using Fourier transform IR spectrometry.

Other techniques. Surface techniques proliferated. Indeed, David Hercules (*Anal. Chem.* **1978**, *50*, 734 A–744 A) observed that "each form of surface spectroscopy has an acronym, an obvious attempt by surface scientists to compete with the Federal Government." X-ray photoelectron spectroscopy was used to determine binding energies for carbon, nitrogen, phosphorus, and sulfur compounds as well as for compounds of some miscellaneous elements. Many techniques were used for bulk analysis at the parts-per-billion level, and parts-per-million sensitivity levels could be obtained by secondary ion MS.

Continuous-flow analysis, demonstrated in the late 1950s,

1975
Jimmy Hoffa disappears
King Faisal of Saudia Arabia is assassinated
Sarah Caldwell becomes the first woman conductor of
the Metropolitan Opera

1976
U.S. celebrates its 200th birthday
Agatha Christie dies
Air Force Academy admits women
Legionnaires' disease kills 29
The Orient Express ends its Istanbul-to-Paris run

1977
Elvis Presley dies
G. Gordon Liddy is released from prison
Star Wars opens
U.S. tests the neutron bomb
New York City experiences a massive blackout

1978
First test tube baby is born in England
Son of Sam receives life imprisonment for six murders
Norman Rockwell dies
U.S. and People's Republic of China establish full
diplomatic relations

1979
U.S. embassy staff is taken hostage in Iran
Jimmy Carter, Menachem Begin, and Anwar Sadat
agree on Mideast peace treaty
Mother Teresa wins the Nobel Peace Prize
Reactor building at Three Mile Island nuclear power
plant is badly contaminated

required air segmentation of the flowing stream. By modifying the technique to inject the sample directly into the carrier stream, Jaromir Ruzicka, Elo Hansen, and others created flow injection analysis. A myriad of applications were developed for plants, water, blood, soil, fertilizers, and pharmaceuticals.

V.V.S. Eswara Dutt and Horacio A. Mottola reported on how continuous kinetic-based determinations using a reagent regeneration cycle can be used to determine a variety of chemical species. All reagents are contained in a single reservoir and are continuously circulated at constant flow through the cell into which an aliquot of the sample is injected.

If there were ever a time when an analytical chemist would drool, it must have been when Apollo 11 returned from the Moon with 22 kg of rocks. The samples were given to 141 principal investigators, who reported their findings at the beginning of 1970. The techniques used on the samples included MS; neutron activation analysis; atomic absorption, emission, and X-ray fluorescence spectrometries; wet chemistry; and GC.

The Pittsburgh Conference

Another institution grew along with the Journal: the Pittsburgh Conference. Throughout the decade the meeting was held in Cleveland, OH; in 1980 it was moved to Atlantic City, NJ. In 1970, 45 sessions were held, more than 300 papers were presented, and more than 240 exhibitors hawked their wares. By the end of the decade, the conference had essentially doubled in size with 84 sessions, 740 papers, and more than 370 exhibitors.

The Journal

During this decade, the Journal was under the stewardship of Herbert H. Laitinen. For a time, John K Crum, currently Executive Director of the ACS, served as Managing Editor.

Most changes in the Journal took place in the A-pages. The first Analytical Approach appeared in April 1974; Focus was first published in June 1979. Early in the decade, *Analytical Chemistry* was essentially black and white; color was supplied only in the advertisements. Redesigns done in 1973 and 1976 contributed to a more modern-looking magazine, and by the end of the decade the publication sported a new logo, type, and layout; some new department names; more color in the feature articles; and more attractive covers.

Between the new look of the magazine, the scientific accomplishments of the decade, and the strength of its readership and leadership, the Journal was positioned to enter the 1980s as a vibrant, indispensable tool for researchers, students, and the discipline itself.
FELICIA WACH

Analysis of Thin Films by Ion Microprobe Mass Spectrometry

C. A. Evans, Jr., and J. P. Pemsler

Ledgemont Laboratory, Kennecott Copper Corp., Lexington, Mass. 02173

The ion microprobe mass spectrometer was used to investigate isotopic and compositional gradients in thin films of both oxides and metals. Continuous recording of intensities, and sputter rates as low as 0.5 Å/sec enabled depth resolutions of the order of 20 Å. Oxygen isotope mixing in duplex $Ta_2{}^{16}O_5/Ta_2{}^{18}O_5$ film was shown to vary with the $Ta_2{}^{18}O_5$ thickness added. Phosphorus gradients in Ta_2O_5 anodized in H_3PO_4 varied with film thickness and H_3PO_4 concentration. Homogeneities in thin films of Ag–Cu and Al–Ge–Nb depended on their modes of preparation. The ion microprobe is concluded to be a powerful tool for examining thin films.

The concept of an ion microprobe mass spectrometer was first described by Herzog and Viehbock (*1*). A sample was bombarded with ions of energy in the KeV range causing surface atoms of the target to be sputtered, a small fraction of them being ionized. These secondary or sputtered ions were extracted into a mass spectrometer and analyzed.

A number of investigators have explored the application of this technique. Anderson (*2*) used positive and negative ions for bombarding the target, and Benninghoven (*3*) investigated the use of positive and negative secondary ions. Castaing and Slodzian (*4*) and Robinson, Liebl, and Andersen (*5*) demonstrated that surface spatial resolution of the order of 1 μ can be obtained. Satkiewicz applied the technique to a variety of materials including metals, minerals (*6*), organic materials (*7*), and thin films (*8*). Relative ionization efficiencies were found to vary by several orders of magnitude from element to element (*2, 3*) so that detection limits vary widely. In favorable cases detectabilities in the ppm range are attainable. The removal rate of atoms from the sample is variable over a wide range—0.5 Å/sec for low ion current densities to 100 Å/sec for high current densities and high energy primary ions. Sputtering rates and ion yields are further influenced by parameters such as the nature of the bombarding ion, ion energy, sample and surface characteristics, and residual gas pressure, so that conditions must be optimized for a particular objective.

THIN FILMS

In addition to their great technological significance, thin films are of fundamental interest because they offer an opportunity to study an almost two-dimensional solid. Many analytical techniques have been used to study the physical, chemical, and structural properties of thin films (*9*). Chemical characterization requires a high degree of spatial resolution in the depth dimension. In a few specific cases, precision anodic stripping (*10*) can achieve depth resolutions of 100 Å. This technique, however, is not broadly applicable. Spark source mass spectroscopy is capable of good sensitivities for almost every element, but the depth resolution is only 1000 Å in the best of cases and more typically about 1 μ (*11–13*).

The ability to control penetration rates and the broad elemental coverage make the ion microprobe particularly suitable to the study of thin films. Slow sputtering rates provide the best depth resolution. A corresponding signal decrease reduces detection limits, but this is generally tolerable as trace detection is not always necessary in thin film analysis. With these capabilities and limitations in mind, the ion microprobe mass spectrometer was evaluated as a technique for the chemical characterization of thin films.

EXPERIMENTAL

The GCA Ion Microprobe Analytical Mass Spectrometer (GCA Technology Division, Bedford, Mass.) was used for this study. A simplified representation of the instrument is shown in Figure 1. A duoplasmatron ion source provides the primary argon ions. These ions are then accelerated toward the sample with up to 15 KeV of kinetic energy. The ion beam diameter and position are determined by the Einzel lens and beam deflection electrodes. Sample bombardment by the primary ions causes sputtering of neutral atoms as well as positive and negative ions. In this study, the positive ions are extracted into a double-focusing mass spectrometer for mass and energy resolution. A 20-stage Al dynode electron multiplier is used to detect the mass resolved secondary ions, and a vibrating reed electrometer and recorder provide a visual and permanent record of the ion intensity. Scanning the magnetic field during sputtering enables recording a mass spectrum of the secondary ions. A concentration *vs.* depth profile of a particular mass can be studied by setting the mass spectrometer to that peak and recording the intensity variations with time.

The following modifications and additions were made to the basic instrument.

Lif-O-Gen (Lif-O-Gen, Inc., Lumberton, N. J.) research grade argon and a two-stage regulator were used to supply argon to the duoplasmatron. This system is preferred over glass flasks in that it supplies a long-term constant pressure of primary gas, enabling more stable operation of the duoplasmatron. Also, the argon supply line is operated at a pressure above ambient, reducing contamination of the gas by inbound leakage.

All three Einzel lens electrodes in the primary optics were grounded, as suggested by Herzog and Satkiewicz (*14*), and the last aperture diameter was reduced to 3 mm. Aberrations due to the Einzel lenses were thereby eliminated.

A Faraday cage was inserted so that retraction of the sample holder allowed the primary ion beam to be monitored. The current was read on a digital panel meter.

The single coarse and fine controls provided for manual adjustment of the magnet current were replaced with a four position peak switcher. The shorting type, rotary switch, and coarse and fine precision potentiometers permitted selection of four preset magnet currents.

Samples were degreased, mounted on the sample holder, inserted into the target chamber, and the chamber evacuated to approximately 5×10^{-7} torr. At this time the primary ion source was started and adjusted to maintain 2.0 μA of 14 keV Ar^+ as measured by the Faraday cage. The primary ion current was monitored and the duoplasmatron adjusted until the ion current remained constant at 2.0 μA for at least 15 minutes. Once the constant current condition was met, the sample was rapidly translated into the primary beam and the desired mass monitored *vs.* time.

RESULTS AND DISCUSSION

Optimization of spatial resolution and accurate calibration of penetration depth as a function of sputtering time requires an accurate knowledge of the magnitude and spatial homogeneity of the primary beam. If ions produced at any given time are to be interpreted as representative of the composition at a particular depth, then the primary ion beam must maintain a constant current density over its entire area. Any nonuniformities in the beam profile would result in some areas being sputtered at different rates than others. The ion sample at any time would then consist of ions from different depths.

The defocused mode of the primary ion optics was, therefore, studied by two techniques. (1) Thin films were vapor-deposited on glass slides and then sputtered. The uniformity of light transmission through the sputtered area was then examined with a microphotometer. (2) A series of Ta_2O_5 films were prepared by controlled anodization of Ta in suitable electrolytes. These films exhibit sharp interference colors, and color variations equivalent to thickness changes of as little as 30 Å can be visually detected. The Ta_2O_5 films were partially sputtered with the primary ion beam and the color uniformity of the craters was examined for depth inhomogeneities.

The instrument as received provided a nominally uniform beam with inhomogeneities greater than ±10%. Excess cratering occurred primarily near the spot edges, and, to a lesser extent, in the center of the sample. Experiments with the focusing apertures and lenses indicated that by grounding the Einzel lens electrodes and using the resulting divergent beam, craters were produced whose depth was homogeneous to within ±50 Å at a total depth of 1000 Å, as determined by examination of the Ta_2O_5 interference colors. In addition to distortions

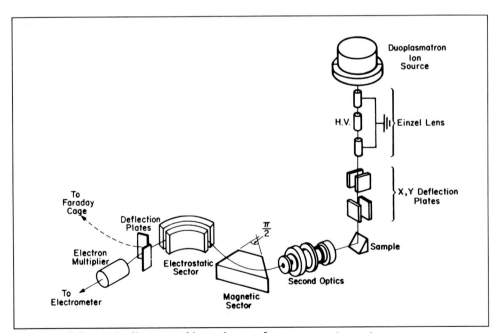

Figure 1. Schematic diagram of ion microprobe mass spectrometer

caused by the primary optics, variations in the electrostatic field at the sample can be caused by sample holding clips. It was necessary to ensure that the sample holding device had no projections upward in the vicinity of the primary ion beam.

Accurate knowledge of penetration depths and sputtering rates is dependent on the constancy and reproducibility of the primary ion current. The as-received instrument had no provision for directly measuring the flux of bombarding ions. Even when duoplasmatron parameters were reset as carefully as possible, factors such as pressure dependence and filament aging resulted in irreproducible sputter conditions. Constancy of the target current was not sufficient to establish a reproducible sputter rate since the target current varies with primary ion current as well as secondary ion and electron emission effects, which in turn depend on surface condition, sample composition, and ion energy. To obtain a positive measure of the primary ion current, a Faraday cage was mounted below the sample holder on the primary beam axis. With the sample moved out of the ion path, the primary ion current was monitored with the Faraday cage and read on a digital panel meter. Primary beam currents can then be monitored and adjusted. With a constant primary ion current, features and character of spatial distribution could be reproduced to about 3% of the actual depth.

An additional instrumental parameter plays a significant role in the development of this method for the study of thin films. Herzog et al. (15) showed that atomic and molecular ions sputtered from a sample have a different distribution with respect to initial kinetic energy. An examination of the secondary ion yield vs. secondary ion kinetic energy shows that the number of molecular ions drops rapidly with increasing kinetic energy, but the yield of atomic ions decreases only slowly at higher energy. Studies involving oxygen isotope tracers are subject to interference since both $^{18}O^+$ and $(^1H_2^{16}O)^+$ occur at mass 18. If the accelerating potential and the electrostatic analyzer voltage are preset to accept only those ions with high initial kinetic energy, then the interference of the molecular species $(^1H_2^{16}O)^+$ will be greatly reduced.

Isotope Gradients in Oxides. The anodic oxidation of tantalum has been studied extensively by various investigators and as such represents an ideal metal-oxide system with which to explore the capabilities of the ion microprobe. If Ta is anodized first in $H_2^{16}O$ and then in $H_2^{18}O$ (or $D_2^{18}O$), then a duplex film $Ta_2^{16}O_5$ and $Ta_2^{18}O_5$ will result. Pringle (16) used

anodic stripping combined with activation analysis to study the ^{18}O–^{16}O intermixing after the duplex anodization procedure. Oxygen isotope intermixing was found to vary as the square root of the added $Ta_2^{18}O_5$ thickness. The distance about the $Ta_2^{18}O_5/Ta_2^{16}O_5$ interface where the % ^{18}O varies by σ (i.e., 84-17% ^{18}O) is defined as d. Pringle's data were fit to the expression

$$d = 3.58 \sqrt{^{18}O \text{ thickness}} \qquad (1)$$

Five samples of Ta anodized first in $H_2^{16}O$ and subsequently in $D_2^{18}O$ were obtained from AECL, Chalk River. The thickness/voltage relation during the anodization of Ta is well established (17) and precise values of the thickness of the $Ta_2^{16}O_5$ and $Ta_2^{18}O_5$ layers could readily be determined. These values are given in columns 2 and 3 of Table I. Samples were sputtered with the homogeneous ion beam and $^{18}O^+$ and $^{16}O^+$ intensities were recorded as functions of time. The depth at any time was determined by the sputter rate, which was calculated by measuring the time necessary for the $^{18}O/^{16}O$ ratio to reach 0.5 of its original value. This represented the time necessary to traverse the $Ta_2^{18}O_5$ thickness as determined by the anodization conditions. Calculated sputtering rates were substantially constant and are given in column 4 of Table I. Results for two of the samples are shown in Figure 2. The spread at the $^{18}O/^{16}O$ interface increases with increasing $Ta_2^{18}O_5$ thickness in agreement with Pringle (16), and a comparison of observed σ values with those calculated from Equation 1 is given in columns 5 and 6 of Table I. The ion microprobe results are somewhat larger than those of Pringle, but agreement between the two techniques is considered quite good. Some additional mixing of oxygen isotopes may have been caused by energy from the primary ion beam.

Impurity Gradients in Oxides. Tantala films resulting from the anodic oxidation of tantalum are sensitive to the inclusion of impurity atoms from the electrolyte. Randall, Bernard, and Wilkinson (18) have shown that appreciable quantities of phosphorus are present in anodic films formed on tantalum in phosphoric acid solutions. Phosphorus gradients in these films were studied by radiotracer techniques and by electrical property measurements. Since the ion microprobe provides a continuous monitoring of concentration as a function of depth, it is well suited to examine impurity gradients in thin films.

Table I. Comparison of Results of Pringle (16) and Ion Microprobe Study of $Ta_2^{18}O_5/Ta_2^{16}O_5$ Interface

Sample	$Ta_2^{16}O_5$, Å	$Ta_2^{18}O_5$, Å	Sputtering rate, Å/sec	1σ distance, Å	
				Pringle (16)	This study
1	4141	298	0.67	62	110
2	3633	825	0.63	103	130
3	3144	965	0.73	111	150
4	2653	1261	0.67	127	170
5	2163	2180	0.66	167	215

Tantalum oxide films formed in phosphate electrolytes of different concentrations were examined and the phosphorus gradients are shown in Figure 3a and b. The $^{31}P^+$ intensity is much higher for films formed in concentrated $14M$ H_3PO_4 than in those formed in $0.9M$ H_3PO_4. In the 300- Å films formed in $0.9M$ H_3PO_4, the phosphorus signal drops rapidly from the oxide/solution interface and falls to zero at the oxide/metal interface. There is evidence of a small plateau in concentration in the 300- Å film formed in concentrated H_3PO_4. The thicker oxide films show definite evidence of phosphorus plateaus within the oxide films. Further details of these experiments will be reported in a separate communication.

Gradients in Metallic Films. The ion microprobe was used to study compositional variations in the metallic film Ag–Cu and the amorphous semimetallic alloy Al–Ge–Nb. Anderson (19) used a spectrophotometric method to monitor the vapor species produced by dc sputtering of the Ag–Cu eutectic alloy. He observed Ag/Cu ratios which varied with time and target temperature. At a target temperature of 80 °C, approximately 40 minutes of sputtering were required to obtain a constant Ag/Cu ratio. Longer presputter times were required at higher target temperatures.

Several films were prepared by sputtering an Ag–Cu eutectic onto a cooled target after long presputter times to achieve steady-state. These films were then examined with the ion microprobe for variations in the Ag/Cu ratio as a function of depth. No substantial variations were found in several samples prepared in this manner.

Thin film deposits of Al–Ge–Nb alloy were prepared by dc sputtering of an Al–Ge alloy button surrounded by an annulus of Nb. This configuration can be expected to lead to inhomogeneities in the ternary alloy film. The intensity versus sputtering time for $^{27}Al^+$ and $^{93}Nb^+$ in the Al–Ge–Nb film is shown in Figure 4. Variations in concentration within the film are apparent.

CONCLUSIONS

The ion microprobe has been shown to be a powerful tool for exploring isotopic and composition gradients in thin films of both metals and insulators. Suitable modifications of the instrument have provided a uniform sputtering rate and crater profile, so that reliable analyses can be obtained as a function of depth into the sample. Continuous recording of intensities and sputter rates as low as 0.5 Å/sec enable depth resolutions of the order of 20 Å.

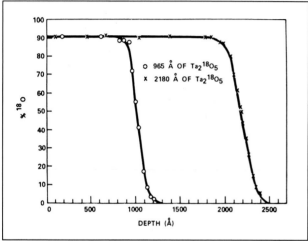

Figure 2. $\%^{18}O^+$ *vs.* depth for Ta_2O_5 anodized in $H_2^{16}O_2$ followed by $D_2^{18}O_2$

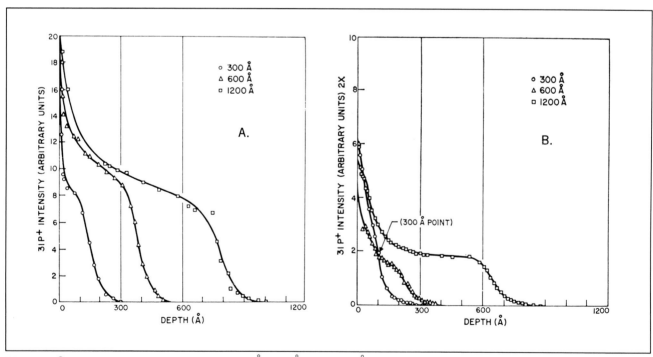

Figure 3. $^3P^+$ intensity vs. film depth for 300 Å, 600 Å, and 1200 Å Ta_2O_5
(a). Anodized in $14M$ H_3PO_4, *(b)*. Anodized in $0.09M$ H_3PO_4

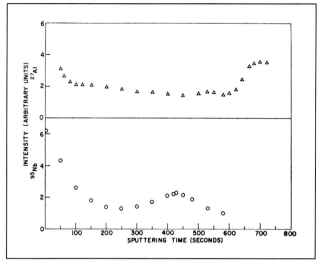

Figure 4. Intensity *vs.* sputtering time for $^{27}Al^+$ and $^{93}Nb^+$ for Al–Ge–Nb thin film

ACKNOWLEDGMENT

The authors thank J.P.S. Pringle for providing the duplex anodized Ta_2O_5; R. E. Pawel for the Ta_2O_5 anodized in H_3PO_4; and K. L. Chopra and M. R. Randlett for the Ag–Cu and Al–Ge–Nb thin films. The comments and suggestions of R.F.K. Herzog and F. W. Satkiewicz are gratefully acknowledged.

REFERENCES

(1) R.F.K. Herzog and F. P. Viehbock, *Phys. Rev.*, **76**, 855 (1949).
(2) C. A. Anderson, *Int. J. Mass Spectrom. Ion Phys.*, **2**, 61 (1969).
(3) A. Benninghoven, *Z. Physik*, **199**, 141 (1967).
(4) R. Castaing and G. Slodzian, *J. Microsc. (Paris)*, **1**, 395 (1962).
(5) C. F. Robinson, H. Liebl, and C. A. Andersen, Third National Electron Microprobe Conference, Chicago, 1968.
(6) R.F.K. Herzog, W. P. Poschenrieder, and F. G. Satkiewicz, Final Report NASA Contract No. NAS 5-9254 (1967).
(7) F. G. Satkiewicz, GCA Technology Div., Bedford, Mass., personal communication, 1969.
(8) F. G. Satkiewicz, Air Force Avionics Laboratory Technical Report TR-69-332, Jan. 1970.
(9) K. L. Chopra, "Thin Film Phenomena," McGraw-Hill Book Company, New York, 1969.
(10) R. E. Pawel and T. S. Lundy, *J. Electrochem. Soc.*, **115**, 233 (1968).
(11) A. J. Ahearn, 1959 Sixth National Symposium on Vacuum Technology, Transactions, New York, Pergamon Press, 1960.
(12) D. L. Malm, Fourteenth Annual Conference on Mass Spectrometry and Allied Topics, Dallas, 1966.
(13) W. M. Hickam and G. G. Sweeney, Eleventh Annual Conference on Mass Spectrometry and Allied Topics, San Francisco, 1963.
(14) R.F.K. Herzog and F. G. Satkiewicz, GCA Technology Div., Bedford, Mass., personal communication, 1969.
(15) R.F.K. Herzog, W. P. Poschenrieder, F. G. Ruedenauer, and F. G. Satkiewicz, Fifteenth Annual Conference on Mass Spectrometry and Allied Topics, Denver, 1967.
(16) J.P.S. Pringle, AECL Chalk River, private communication, 1970.
(17) L. Young, "Anodic Oxide Films," Academic Press, New York, 1961.
(18) J. J. Randall, Jr., W. J. Bernard, and R. R. Wilkinson, *Electrochem. Acta*, **10**, 183 (1964).
(19) G. S. Anderson, *J. Appl. Phys.*, **40**, 2884 (1969).

Received for review March 30, 1970. Accepted May 11, 1970.

Reprinted from *Anal. Chem.* **1970**, *42*, 1060–64.

In this article is the seed of a technique that is almost universally available on GC/MS instruments today. GC/CIMS is the first resort when the molecular weight of an eluting compound cannot be established by routine GC/EIMS because of excessive fragmentation. The technique has since developed differently from the conditions presented herein; now helium is the universal carrier gas and methane is admitted to the MS source separately to form the reagent ion CH_5^+. How quick the push to make practical use of CI, discovered only a few years earlier!

Maurice M. Bursey
University of North Carolina–Chapel Hill

Combination of Gas Chromatography and Chemical Ionization Mass Spectrometry

Donald M. Schoengold[1] and Burnaby Munson
Department of Chemistry, University of Delaware, Newark, Del. 19711

Chemical ionization mass spectrometry (or perhaps "ion-molecule reaction mass spectrometry") is a recently developed technique of analytical interest for the production of mass spectra through gaseous ionic reactions (1, 2). In this technique, primary ions are generally produced by electron impact in a gaseous mixture at pressures as high as a few Torr. The gaseous mixture consists of the reactant gas and the analytical sample, normally in a reactant gas/sample mole ratio of 100 to 1 to 1000 to 1. Because of the large excess of reactant gas virtually all of the primary ions are produced by direct ionization of the reactant gas, not the analytical sample. These primary ions will undergo reactive collisions with the bulk reactant gas and perhaps produce other ions. To be a suitable reactant gas, a compound must produce a set of ions which does not react with the bulk reactant gas to give further products; *i.e.*, the distribution of ions must achieve a substantially constant value as the pressure or reaction time is increased.

To illustrate:

$$CH_4 \xrightarrow{e} \begin{array}{c} CH_4^+ \\ CH_3^+ \end{array} \xrightarrow{CH_4} \begin{array}{c} CH_5^+ \\ CH_2H_5^+ \end{array} \xrightarrow{CH_4} \text{no other ions} \quad (1)$$

That is, the major (90%) primary ions produced by high energy (≥ 35 eV) electrons in methane, CH_4^+ and CH_3^+, react rapidly with the major component, CH_4, to give CH_5^+ and $C_2H_5^+$. On the other hand, CH_5^+ and $C_2H_5^+$ do not react with methane to produce any other ions; therefore, they may react with the small amount of analytical sample to produce a set of ions which is characteristic of the sample. CH_5^+ and $C_2H_5^+$ react by proton and hydride transfer to give $(MW + 1)^+$ and $(MW - 1)^+$ ions which may dissociate further. This distribution of ions is

the chemical ionization (CI) mass spectrum of the sample and does depend on the ions of the reactant gas. CI mass spectra are frequently less complex and easier to interpret than electron impact mass spectra (1, 2).

Application of the technique of CI mass spectrometry to gas chromatography–mass spectrometry (GC-MS) can be used to avoid some of the problems inherent in normal GC-MS work. The effluent from a gas chromatograph can be used directly in CI mass spectrometry if the GC carrier gas is suitable as a CI reactant gas. This technique enables GC-MS to be accomplished without a molecular separator and eliminates the restriction imposed by the separator that the carrier gas be helium.

Recently, two other combinations of chemical ionization mass spectrometry and gas chromatography have been reported (3, 4). Both of these reports concerned quadrupole mass spectrometers and used methane as the carrier gas.

EXPERIMENTAL

The mass spectrometer which was used in these experiments was a Bendix Model 12 T0F instrument modified for high pressure work and operated in the pulsed mode with variable time delay (5, 6). An Aerograph dual column, temperature programmed gas chromatograph was connected to the mass spectrometer by a system of copper tubing and Swagelok fittings. A Nupro valve, Model "M" Cross Pattern Fine Metering Valve, was adjusted to allow sufficient carrier gas and sample into the source of the mass spectrometer to produce the desired pressure, 0.01 to 0.1 Torr. The bulk of the material was exhausted to the atmosphere. Helium and methane were used as carrier gases. The source pressure was varied from 0.01 to 0.1 Torr with ionic residence times as long as 10 μsec. A sim-

[1] Present address, Sun Oil Co., Marcus Hook, Pa.

ple circuit was designed to start the scan of a mass spectrum after each GC peak reached a given height.

In these experiments, the ratio of sample to carrier gas was about 0.02 with 1 μl of liquid sample injected into the gas chromatograph. Because this concentration of sample is higher than has generally been used in CI studies, direct ionization of the sample and secondary reactions of sample ions with sample might occur. Consequently, the material was diluted by adding an increased flow of carrier gas through the reference column to the gas flow from the sample column and cell. This combined sum was the total effluent which was sampled. The sample to carrier gas ratio was less than 0.005 in the final mixture.

RESULTS AND DISCUSSION

The majority of the reported chemical ionization mass spectra have been obtained with methane as the reactant gas. Consequently, methane was used as the carrier-reactant gas. At a source pressure of 0.12 Torr, with ionic residence times as large as 9 μsec, CI spectra were obtained for several compounds. Where comparisons are possible, the spectra are in reasonable agreement with the other data ($1, 2, 7$). Simple aliphatic ketones give primarily $(MW + 1)^+$ and acyl ions; esters give $(MW + 1)^+$, protonated acid ions, and acyl ions; aromatic compounds give primarily $(MW + 1)^+$ ions; and low molecular weight alcohols give $(MW + 1)^+$, $(MW - 1)^+$, and alkyl ions.

Figure 1a shows a simple, well-resolved two-peak chromatogram of methanol and ethanol using methane as the carrier gas. Table I shows the simple chemical ionization mass spectra (CH_4 reactant gas, $m/e \geq 29$) obtained during two time intervals on each peak. Methanol is easily recognized by the $(MW + 1)^+$ and $(MW - 1)^+$ peaks at $m/e = 33$ and 31, which are characteristic of low molecular weight alcohols. Ethanol is recognized by $(MW + 1)^+$ and $(MW - 1)^+$ peaks at $m/e = 47$ and 45. The ions at $m/e = 29, 41$, and 43 which are present in CH_4 alone are useful as mass markers and are disregarded in CI spectra.

The ion currents for the samples are reported relative to the ion current for $C_2H_5^+$ simply for convenience in this Table and to indicate the extent of reaction of CH_5^+ and $C_2H_5^+$ with the alcohols. At present, no quantitative significance can be

attached to the absolute values of ion currents or peak heights. The limits of detectability depend upon the nature of the chemical ionization spectrum as well as the operating parameters of a particular system. We have not yet attempted to optimize these parameters; however, even under these conditions about 0.1 μl of sample could be identified.

There is no intermingling of material within the source of the mass spectrometer for these widely separated chromatographic peaks. The ratios of $(MW + 1)^+$ to $(MW - 1)^+$ ions for the duplicate scans of a chromatographic peak are sufficiently close (1.56 and 1.50 for methanol and 1.00 and 0.87 for ethanol) to allow identification of the material without undue concern about when the spectra are taken during a chromatographic peak.

Figure 1b shows an unresolved peak of acetone and ethanol on a gas chromatograph with methane as the carrier gas. The column conditions and components were chosen for this purpose. The CI mass spectrum of scan 1 was a predominant peak at $m/e = 59$ which is characteristic of acetone. In scan 2, both ethanol and acetone could be identified by $(MW + 1)^+$ ions at $m/e = 47$ and 59. For this and several other composite peaks of simple compounds, the composite nature of the chromatographic peak and the identity of the compounds could be determined by obtaining the chemical ionization spectra (CH_4) twice during a peak. Multicomponent peaks obviously present further problems.

The results obtained with helium as a carrier and reactant gas were pleasantly surprising. From previous data on charge exchange reactions of He^+ ($8, 9$), it was expected that the spectra obtained with He as the reactant gas would be analytically useless because of virtually complete dissociation of the molecular ions produced by charge exchange reactions with He^+. The spectra which were obtained, of which Table II is

Table I. Spectra with CH_4

m/e	CH_3OH		C_2H_5OH	
	Scan 1	Scan 2	Scan 3	Scan 4
29	100	100	100	100
31	16	14	...	...
33	25	21	...	...
41	9	4	5	7
43	6	6	6	7
45	...	...	6	9
47	...	...	6	7

Spectra normalized to largest peak = 100.

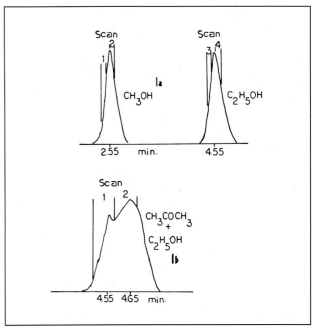

Figure 1. Gas chromatographic peaks for mixtures with methane as carrier-reactant gas

Table II. Spectra with He

CH₃OH

m/e	API No. 282[a]	Charge exchange[b]	This work
32	72	1	38
31	100	4	100
30	8	3	10
29	42	100	70

n-C₃H₇OH

m/e	API No. 284[a]	Charge exchange[c]	This work
60	10	...	12
59	15	...	14
58	5	...	17
57	3	...	10
43	4	21	18
42	13	8	20
41	10	9	23
39	6	34	10
31	100	86	100

[a] American Petroleum Institute Project 44, ref. *10*.
[b] Reference *9*.
[c] Reference *8*.
Spectra normalized to largest peak = 100.

representative, resembled very closely the conventional electron impact mass spectra of the compounds (*10*). The relative abundance of He⁺ is unreliable and not given in the table. The observation that spectra were not very sensitive functions of either the total pressure within the source or the residence time of the ions also suggested that the spectra were not caused by ion-molecule reactions. For comparison with our data on methanol and propanol, the charge exchange spectra obtained with He⁺ and conventional 70 eV electron impact spectra are shown in Table II.

It is obvious that the spectra cannot be produced by charge exchange reactions of He⁺. While there may be some contribution of charge exchange under these conditions of relatively low pressure, the differences in spectra between our work and the other electron impact data are probably attributable to different instruments.

For a mixture of reactant and sample in the sources of the mass spectrometer, two competitive electron impact ionization processes can occur:

$$\text{Reactant} \xrightarrow{e} \text{Ions (He}^+, \text{ or CH}_4^+, \text{ etc.)} \quad (2)$$

$$\text{Sample} \xrightarrow{e} \text{Ions} \quad (3)$$

The relative importance of each of the reactions is determined by the electron impact cross section for ionization. The ioniza-

tion cross section for He is about 0.4×10^{-16} cm²; CH_4, 4.7×10^{-16} cm²; and acetone, 10×10^{-16} cm² (*11, 12*). Since the ratio of ionization cross sections of sample/He is much greater than the ratio of cross sections of sample/CH_4, direct ionization of the samples will be more likely with He as a carrier gas than with CH_4.

The spectra of the same compounds obtained with the two reactant-carrier gases, He and CH_4, are very different from each other as may be seen by comparing the spectra for methanol in Tables I and II. CH_4 as a carrier gas gives ions resulting from proton and hydride transfer and subsequent decomposition, $(MW + 1)^+$, $(MW - 1)^+$, and lower fragment ions. He as a carrier gas gives ions resulting from predominantly electron impact ionization or charge transfer, MW^+, and fragment ions.

The work done with methane on well resolved and composite chromatographic peaks was repeated with helium as the carrier gas and comparable results were obtained.

Further work on these systems which is planned includes studying the reproducibility of the system, the effects of operating parameters, and the possibility of quantitation.

ACKNOWLEDGMENT

Acknowledgment is made to the donors of the Petroleum Research Fund, administered by the American Chemical Society, and to the University of Delaware Research Foundation for partial support of this research. One of us (DMS) is grateful to the Sun Oil Co. for a leave of absence and support during part of this research. The use of this method is covered by patents pending by Frank Field and M. S. B. Munson. This patent is owned by Esso Research and Engineering Company with exclusive manufacturing rights to Scientific Research Instruments Corp., Baltimore, Md.

REFERENCES

(1) M.S.B. Munson and F. H. Field, *J. Amer. Chem. Soc.*, **88**, 2621 (1966).
(2) F. H. Field, *Accounts Chem. Res.*, **1**, 42 (1968).
(3) G. P. Arsenault and J. J. Dolhun, Paper presented at the 18th Conference on Mass Spectrometry, San Francisco, Calif., June 1970.
(4) Marvin L. Vestal, Paper presented at the 18th Conference on Mass Spectrometry, San Francisco, June 1970.
(5) C. D. Miller, T. O. Tiernan, and J. H. Futrell, *Rev. Sci. Instrum.*, **40**, 503 (1969).
(6) C. W. Hand and H. von Weyssenhoff, *Can. J. Chem.*, **42**, 195 (1964).
(7) John Michnowicz, unpublished data from these laboratories.
(8) P. Wilmenius and E. Lindholm, *Ark. Fys.*, **21**, 97 (1962).
(9) E. Pettersson, *Ibid.*, **25**, 181 (1963).
(10) Catalog of Mass Spectral Data, API Research Project 44, Carnegie Institute of Technology, Pittsburgh, Pa.
(11) F. W. Lampe, J. L. Franklin, and F. H. Field, *J. Amer. Chem. Soc.*, **79**, 6129 (1957).
(12) J. A. Beran and L. Kevan, *J. Phys. Chem.*, **73**, 3866 (1969).

Received for review June 23, 1970. Accepted September 14, 1970.

Reprinted from *Anal. Chem.* **1970**, *42*, 1811–13.

The Winefordner group pioneered the analytical application of atomic fluorescence spectrometry to trace element analysis in the mid-1960s. Conceptually, the ideal source for such measurements is a tunable laser. This article represents the first application of a tunable dye laser as a source for atomic fluorescence spectrometry. Although the ideal tunable laser (in terms of linewidth, ease of tunability, wavelength range, and cost) remains elusive even today, the current application of laser-excited atomic fluorescence in glow discharge atom cells now exhibits some of the lowest recorded detection limits in atomic spectrometry and even approaches single-atom detection.

Gary Horlick
University of Alberta

Laser-Excited Atomic Fluorescence Flame Spectrometry

L. M. Fraser and J. D. Winefordner[1]
Department of Chemistry, University of Florida, Gainesville, Fla. 32601

Both line and continuum sources of excitation have been used for atomic fluorescence spectrometry (1, 2). These sources are usually either operated continuously (CW), and the radiation is modulated by mechanical choppers, or is modulated electrically. Mechanical or electrical modulation of exciting radiation and ac (often phase sensitive) detectors) detection of the modulated photodetector signals due to modulated fluorescence is utilized to eliminate measurement of emission signals resulting from the flame gases as well as other constant (dc) signals. However, sinusoidal (or square wave) modulation of source radiation does not usually result in any significant increase in the atomic fluorescence signal-to-noise ratio and, in fact, can even result in a decrease under certain circumstances (3). On the other hand, a stable, repetitively pulsed source of excitation with a small duty cycle, i.e., small ratio of on-to-off-time, could result in a significant increase in the fluorescence signal-to-noise ratio due primarily to the decreased noise; during the short on-time, the signal can be of the same order as the average signal resulting from CW sources, whereas the noise will be small because of the small number of random photodetector pulses due to most noise sources commonly present in atomic fluorescence flame spectrometry, e.g., dark current, flame background, analyte emission, etc. (4).

According to the above discussion, an ideal source would be a high power source with a small duty cycle. Because of practical problems associated with exchanging line sources, the ideal source should also produce a continuum from below 200 nm to about 800 nm. Pulsed high pressure inert gas continuum sources, however, are very expensive, bulky, and rather inconvenient to use. In addition, with high pressure discharge

lamps, stray light can be considerable, pulse duration can be quite long, the spectral radiance per pulse can be quite variable, and the spectral radiance per pulse at pulse repetition rates of 10 Hz or greater can be quite low.

A possible alternative to the pulsed continuum high pressure source is given in the present investigation. A stable, pulsed, tunable dye laser pumped with a N_2 laser is used to excite atomic fluorescence in flames. The dye laser system used here has a peak power of greater than 10 KW at all wavelengths, a pulse repetition rate of about 1–25 Hz, a spectral half width of about 0.1–1 nm, and a pulse half width of about 2–8 nsec. By proper choice of dye and grating angle, any wavelength region (0.1–1 nm wide) between 360 and 650 nm can be selected to excite atomic fluorescence. By use of a fast response multiplier phototube and a boxcar integrator (gated amplifier) capable of aperture (sampling) gate widths of the order of 10 nsec wide, all noises except scatter of laser radiation within the relatively turbulent flame gases can be essentially eliminated. Some initial atomic fluorescence measurements, including detection limits, analytical curves, and spectral resolution, for Al, Ca, Cr, Fe, Ga, In, Mn, Sr, and Ti in either H_2/air or C_2H_2/N_2O flames are reported here.

EXPERIMENTAL

A block diagram of the experimental setup is shown in Figure 1. The specific components of the experimental setup are also designated in Figure 1. Two nebulizer-burner systems were used for the present studies. A modified Jarrell-Ash Tri-flame burner system was used for H_2/air flames. The modified nebulizer-burner system, shown in Figure 2, consists of the Hetco nebulizer-burner, a cylindrical chimney, and a special Alkemade-type burner head (5) containing a matrix of 73

[1] Author to whom reprint requests should be sent.

0.75-mm diameter holes (2×0.5 cm burner area). The H_2/air flame produced with the modified Jarrell-Ash Triflame burner was very laminar compared to total-consumption burner flames. The nebulizer-burner used for C_2H_2/N_2O flame in the present studies was of the capillary type described by Aldous *et al.* (6). This nebulizer-burner utilized a Perkin-Elmer aspiration chamber assembly for the Model 290 atomic absorption spectrometer and a circular (10-mm diameter) burner head containing a bundle of 75 capillaries (0.69-mm i.d.).

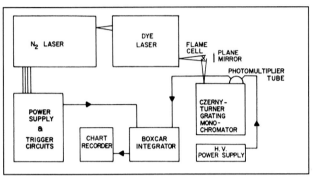

Figure 1. Block diagram of experimental system for laser-excited atomic fluorescence flame spectrometry

Instrumental components are: Model 1000-N_2 laser with power supply and trigger circuits and Dial-a-Line dye laser, AVCO Everett Research Laboratory, Everett, Mass. 02149; RCA 1P28A photomultiplier tube; Model 4-8400 scanning 0.25 m Czerny-Turner grating monochromator, American Instrument Co., Inc. Silver Spring, Md. 20910; Model 160 Boxcar Integrator, Princeton Applied Research Corp., Princeton, N. J. 08540; Servoriter 11 Potentiometric Recorder, Texas Instruments, Inc., Houston, Texas 77006; Model 412B high voltage power supply, John Fluke Manufacturing Co., Inc., Seattle, Wash. 98133; Modified Jarrell-Ash Triflame nebulizer burner (see Figure 2) and capillary burner mounted on Perkin-Elmer chamber-nebulizer (see text); 2-stage regulators on gas cylinders and burner regulator with flow meters for Perkin-Elmer Model 303 atomic absorption flame spectrometer. Perkin-Elmer, Inc., Norwalk, Conn. 06852

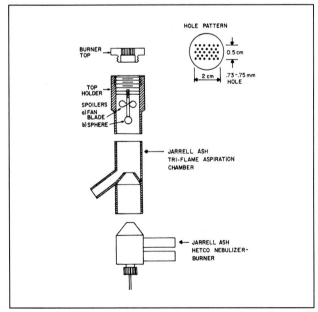

Figure 2. Modified Jarrell-Ash Triflame nebulizer burner for use with H_2/air flame

A simple scan drive system was constructed to vary the grating angle in the dye laser and therefore to allow a continuous variation of the dye laser wavelength over a small wavelength range (say 10–50 Å).

RESULTS AND DISCUSSION

The spectral distribution (obtained by varying the dye laser wavelength) in approximately a 1.0-nm wavelength region surrounding 422.7 nm and the relative magnitudes of signals obtained by spraying water (background), blank, and 1000 µg ml^{-1} of Ca into a H_2/air (15.2 l. min^{-1} H_2 and 7.5 l. min^{-1} air) flame using the modified Jarrell-Ash Triflame burner system (solution flow rate was 4.0 ml min^{-1}) are shown in Figures 3c and 3b, respectively. Also shown in Figure 3a is the spectrometer slit function for the monochromator obtained by scattering dye laser radiation off a Teflon (Du Pont) sheet while varying the dye laser wavelength. It is evident that the spectral distribution of the blank follows the spectrometer slit distribution. Also, it is evident that the

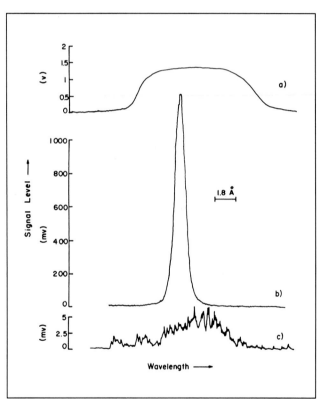

Figure 3. *a*. Spectrometer slit function obtained by wavelength scanning the dye laser radiation scattered from a Teflon sheet used in place of the flames (monochromator wavelength is set at 422.7 nm)

b. Spectral distribution of fluorescence radiation of Ca (1000 µg/ml) obtained by varying the dye laser wavelength over about 1 nm surrounding 422.7 nm (monochromator wavelength set at 422.7 nm). Solution aspiration rate is 4 ml/min. All other experimental conditions are as in Table I

c. Spectral distribution of scattered radiation by the flame gases obtained by varying the dye laser wavelength over about 1 nm surrounding 422.7 nm (monochromator wavelength set at 422.7 nm)

Milestones in Analytical Chemistry

blank signal is many times smaller than the atomic fluorescence signal obtained at 422.7 nm with 1000 µg ml⁻¹ of Ca; the blank signal is probably due to scatter from refractive index variations within the flame (*i.e.*, the same response was observed whether water, 500 ppm of diverse salts, or nothing was being aspirated). Also solvent and solute vaporization are quite complete in both flame–nebulizer–burner systems used in these studies. The blank signal (and noise) is nearly the same at all wavelengths (*i.e.*, for all dyes and all wavelengths) and depends primarily upon the flame gases (*i.e.*, the scatter signal and noise are independent of thermal emission from the flame gases, from the analyte, and from the matrix and are nearly independent of the flow rate of solution into the spray chamber).

The flame background with laser source off and with the laser source on (in both cases, water is being introduced into an C_2H_2/N_2O–6.5 l. min⁻¹ C_2H_2 and 14.5 l. min⁻¹ N_2O) is shown in Figure 4, The noise level with the laser source on is several times greater than with the laser source off. Also the noise level with water being aspirated into the C_2H_2/N_2O flame is very similar to the noise level with water being aspirated into the H_2/air flame; however, it should be noted by comparing Figures 4 and 5 that although the peak-to-peak noises with water being aspirated are similar, the dc offset differs between C_2H_2/N_2 and H_2/air. Of course, the dc offset can be suppressed.

The signal-to-noise ratio obtained for Mn (403.3 nm) at a concentration of 5 µg ml⁻¹ is shown in Figure 5. The limit of detection for Mn is about 0.3 µg ml⁻¹ under similar experimental conditions. Other elements give similar signal-to-noise ratios near the limit of detection.

As long as a sufficiently wide monochromator spectral bandpass is used, it is possible to scan the laser wavelength over a small wavelength region and thereby record the fluorescence spectra of closely-spaced fluorescence lines, *i.e.*, a fluorescence excitation spectrum. In Figure 6 is shown the fluorescence excitation spectra of Cr in the wavelength region of 351.0–361.0 nm. Within this region Cr has a multiplet of 3 lines (357.8, 359.3, and 360.5 nm). The relative magnitudes of

the three lines is modified to some extent by the spectrometer slit function. The resolution of the recorded fluorescence spectra is determined only by the spectral half-width of the laser emission band.

Analytical curves for Ca, Cr, Fe, Ga, In, Mn, and Sr in H_2/air flames and Al and Ti in an C_2H_2/N_2O flame are shown in Figures 7 and 8, respectively. The analytical curves (log-log plots) are linear (slopes of unity) from the detection limits to concentrations at least 10³ greater than the detection limits. The linearity is as expected from theoretical considerations (7). The relative standard deviations of fluorescence signals 10-fold, 100-fold, and 1000-fold above the detection limits are about 8%, 5%, and 3%, respectively.

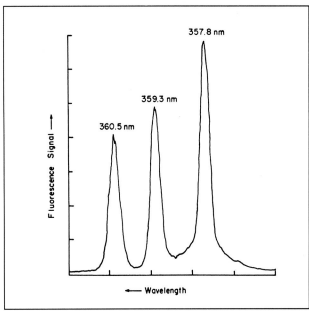

Figure 5. Signal-to-noise ratio for Mn at 5 µg ml⁻¹ and a water blank in an H_2/air flame

Experimental conditions are as in Table I. Note that the noise level for blank with laser on is similar to the noise level with laser on for C_2H_2/N_2O flame shown in Figure 4. Also note the slightly different dc offset for the H_2/air flame in Figure 5 as compared to C_2H_2/N_2O flame in Figure 4

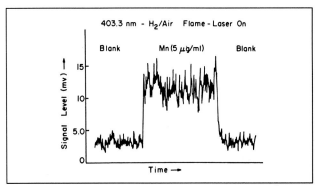

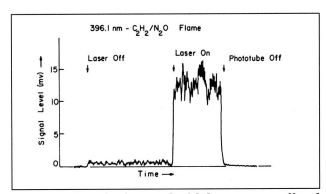

Figure 4. Flame background with laser source off and flame background with laser source on at 3961 Å

In both cases, water is being introduced at a rate of 4.0 ml/min into the Perkin-Elmer chamber. Experimental conditions of boxcar integrator: 10-nsec aperture gate width; 3-nsec aperture time constant; 10-Hz repetition rate

Figure 6. Atomic fluorescence spectra of Cr (1000 µg ml⁻¹) in wavelength range of 357–361 nm

Monochromator conditions: wavelength set at 359 nm; spectral bandwidth about 4 nm. Other experimental conditions are as in Table I

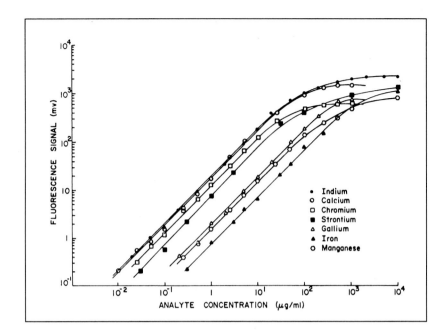

Figure 7. Analytical curves for laser excited atomic fluorescence of Ca, Cr, Fe, Ga, In, Mn, and Sr in H₂/air flames
Experimental conditions are same as in Table I

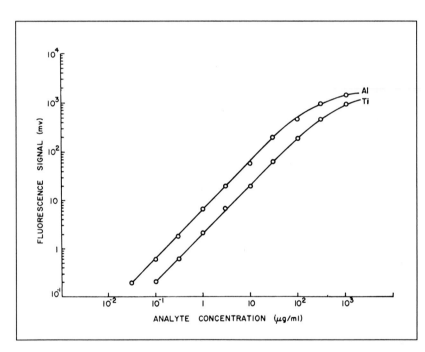

Figure 8. Analytical curves for laser excited atomic fluorescence of Al and Ti in an C₂H₂/N₂O flame
Experimental conditions are same as in Table I

Detection limits (concentrations producing a signal-to-noise ratio of 2) for Al in an C_2H_2/N_2O flame and for Ca, Cr, Fe, Ga, In, Mn, and Sr in H^2/air flame are given in Table I. The pertinent experimental conditions for Figures 7 and 8 and for the detection limits in Table I are given in Table I as footnotes. The laser excited atomic fluorescence detection limits for Al, Ca, Cr, In, and Sr are of the same order of magnitude or better than any previous atomic fluorescence flame spectrometric values obtained with *any* source type. The values for Ga, Fe, and Mn are definitely inferior to previously reported values obtained with line sources; no explanation can be given for these diverse values. The value for Ti is the first reported atomic fluorescence limit of detection. *All* detection limits obtained with laser excitation are lower than previ-

ously reported values obtained with continuum high pressure gas discharge lamps. The laser excited atomic fluorescence limit of detection for Ti is comparable with the best previously reported value by atomic absorption and atomic emission flame spectrometry. The laser excited atomic fluorescence flame spectrometric detection limits for Al, Cr, and In are within about 10-fold and for Ca, Ga, Mn, and SR are within about 100-fold of the best previously reported detection limits by either atomic absorption or atomic emission flame spectrometry (see Table II). The laser excited atomic fluorescence limits of detection could certainly be improved by 10- to 100-fold by a more extensive optimization of experimental conditions.

These initial experimental results indicate possible use of dye lasers as sources of excitation in atomic fluorescence

Table I. Detection Limits for Several Elements by Laser Excited Atomic Fluorescence Flame Spectrometry

Element—line (nm)	Flame conditions (gas flow rates l./min)[a]		Monochromator slit width,[b] nm	Laser dye no.	Detection limit,[c] µg/ml
	N$_2$O–C$_2$H$_2$	Air–H$_2$			
Al—396.1	14.5–6.5	...	1.44	17/66-B	0.03
Ca—422.7	...	7.5–15.2	2.40	17/52	0.01
Cr—359.3	...	7.5–11.8	1.44	17/66-D	0.03
Fe—372.0	...	7.5–11.8	2.40	17/66-D	0.3
Ga—403.2	...	7.5–20.0	1.44	17/66-B	0.3
In—410.4	...	7.5–18.3	2.40	17/66-B	0.01
Mn—403.1	...	7.5–20.0	1.44	17/66-B	0.3
Sr—460.1	...	7.5–15.2	1.44	17/53	0.03
Ti—399.8	14.5–6.5	...	1.44	17/66-B	0.1

[a] All measurements taken at 2.5–3.0 cm above burner top.
[b] Other experimental conditions: photomultiplier voltage −900 V; boxcar integrator conditions: 50-ohm input; >50-MHz bandwidth; dc coupling; 10-nsec aperture gate width; 10-nsec aperture time constant; AVCO dial-a-line laser: 10-Hz repetition rate, 17 kV.
[c] Concentration resulting in signal-to-noise ratio of 2.

Table II. Comparison of Detection Limits by Atomic Flame Spectrometry

Element	AF[a] (laser)	AFL[b] (2)	AFC[b] (7)	AE[c] (2, 7)	AAL[c] (2, 7)
Al	0.03	0.1	...	0.005	0.04
Ca	0.01	0.02	0.1	0.0001	0.0005
Cr	0.03	0.05	10.0	0.005	0.005
Fe	0.3	0.008	1.0	0.05	0.005
Ga	0.3	0.01	5.0	0.01	0.07
In	0.01	0.1	2.0	0.005	0.05
Mn	0.3	0.006	...	0.005	0.002
Sr	0.03	0.03	...	0.0002	0.004
Ti	0.1	...	...	0.02	0.1

[a] This study. These values are obtained under good but not necessarily optimized conditions.
[b] AFL = Atomic fluorescence flame spectrometry excited with narrow line sources. AFC = Atomic fluorescence flame spectrometry excited with continuum sources. Detection limits are best reported values by a number of workers and the specific references can be found in the general references listed after the methods.
[c] AE = Atomic emission flame spectrometry. AAL = Atomic absorption flame spectrometry with narrow line sources. Detection limits are best reported values by a number of workers and the specific references can be found in the general references listed after the methods.

flame (or nonflame) spectrometry. With the dye laser source, it is now possible to obtain low detection limits, freedom from flame background noise, and long linear analytical curves for many elements (most elements can be excited by the dye laser—*i.e.*, relatively few elements have no sensitive spectral absorption lines above 360 nm). By frequency doubling, it may be possible to extend the wavelength range down to nearly 200 nm.

REFERENCES

(1) J. D. Winefordner and T. J. Vickers, *Anal. Chem.*, **42**, 206R (1970).
(2) J. D. Winefordner and R. C. Elser, *Ibid.*, **43** (3), 24A (1971).
(3) J. D. Winefordner, M. L. Parsons, J. M. Mansfield, and W. J. McCarthy, *Ibid.*, **39**, 436 (1967).
(4) J. D. Winefordner, *Accounts Chem. Res.*, **2**, 361 (1969).
(5) T. Hollander, Ph.D. Thesis, University of Utrecht, The Netherlands, 1964.
(6) K. M. Aldous, R. F. Browner, R. M. Dagnall, and T. S. West, *Anal. Chem.*, **42**, 939 (1970).
(7) J. D. Winefordner, V. Svoboda, and L. J. Cline, *CRC Crit. Rev. Anal. Chem.* **1**, 233 (1970).

Received for review April 12, 1971. Accepted June 11, 1971. Research sponsored by AFOSR (AFSC), U.S.A.F. Grant No. 70-1880B.

Reprinted from *Anal. Chem.* **1971**, *43*, 1693–96.

Horlick demonstrated three useful manipulations of atomic emission spectra—each based on discrete Fourier transformation of the digitized spectrum, multiplication by a weight function in the Fourier domain, and inverse Fourier transformation—to yield the final spectrum. The respective weight functions for smoothing, differentiation, and resolution enhancement were a rectangular function, a linear ramp function, and a convolution-derived function. Horlick thus showed that FT methods could be applied to frequency-domain experimental optical spectra, as had previously been shown for optical spectra derived from interferometry.

Alan G. Marshall
Florida State University

Digital Data Handling of Spectra Utilizing Fourier Transformations

Gary Horlick
Department of Chemistry, University of Alberta, Edmonton, Alberta

Utilizing the information available upon Fourier transformation of spectra, several data handling operations are performed, including smoothing, differentiation, and resolution enhancement. These operations are carried out by appropriate simple modifications of the spacial frequency spectrum of the original spectrum. The spacial frequency spectrum is calculated by taking the Fourier transformation of the original spectrum. This calculation and the distribution of information in the spacial frequency spectrum are discussed and illustrated. Then the implementation of the above operations (smoothing, differentiation, and resolution enhancement) by utilization of the spacial frequency information is described. In particular, this approach to spectral smoothing provides an effective way of maximizing the signal-to-noise ratio of a measurement.

One of the major developments in spectrometric measurements in recent years is the acquisition of spectral information in digital form. As technological developments make the so-called small computer more powerful, convenient, and inexpensive, this trend is sure to continue and expand. A major driving force in this development is the desire to perform various types of digital data handling on the digitized spectrum. Typical operations that have been performed on spectra include smoothing (*1*), differentiation (*2, 3*), and resolution enhancement (*4*). A particularly powerful route for performing these and similar operations on spectra is through the utilization of Fourier transformations (*5–9*). In addition, this approach provides unique insight into the implementation and fundamental limitations of these techniques as the manner in which

the available spectral information can be manipulated is clearly revealed.

Data handling of signals utilizing Fourier transformations is not new. In particular, the concepts of smoothing were developed to a high degree of mathematical sophistication during the 1940's by Norbert Wiener and applied to the design of radar receivers by many workers. However, the basic simplicity of data handling utilizing Fourier transformations is often lost in an excess of mathematical equations and in applications to arbitrary waveforms. It is important, in order to effectively utilize these techniques, to know and appreciate in a practical sense the types of information that are obtained upon Fourier transformation of real signals. The next section discusses this in detail for simple flame emission spectra. The main aim is to provide an intuitive feeling for the distribution of the spectral information in the Fourier domain. Smoothing, differentiation, and resolution enhancement of these spectra using Fourier transformations are then discussed and illustrated in the last section.

EXPERIMENTAL

All spectra were measured using a Heath EU-700 monochromator. The spectral bandwidth of this monochromator is approximately 2 Å at a slit width of 100 µm. The signal from the photomultiplier tube was converted to a voltage with a Heath EU-703-71 photometric readout module and the voltage was digitized with a DANA Model 5400/015 DVM. The spectra were sampled at 0.2-Å intervals and the digitized points were punched directly onto cards.

The Fourier transformations were calculated on an IBM 360 computer. All the transforms and plots are 128 points long. Thus the horizontal axis for the optical spectral plots have a normalized length of 25.6 Å. The horizontal axis for the

2855-8/94/0232$08.00/0   ©1994 American Chemical Society   *Milestones in Analytical Chemistry*

spacial frequency plots extends from 0 to 2.5 Å^{-1}. Except where noted, the vertical axes are arbitrary. All the plots were drawn using the standard CALCOMP system.

RESULTS AND DISCUSSION

Information Obtained upon Fourier Transformation of a Spectrum. Fourier transformation is a technique for determining the frequency spectrum of a waveform. For the purposes of this paper, an optical spectrum is the waveform of interest. Carrying out a Fourier transformation on an optical spectrum results in a spacial frequency spectrum. The term "spacial frequency" refers to a frequency in the plane of the paper on which the optical spectrum is plotted and should be carefully distinguished from the term "optical frequency." The spacial frequency spectrum has units that are the reciprocal of those used for the optical spectrum. Thus, if the original spectrum has units of Å, the spacial frequency spectrum has units of Å^{-1}.

Most Fourier transformations are carried out using the so-called Fast Fourier Transform (FFT) (10). This is simply an efficient algorithm for the calculation of the Fourier transformation of a set of points. Several versions of this algorithm have been programmed. The input to a typical program can be a set of real data (i.e., a digitized optical spectrum) or a set of real and imaginary inputs. The output of a typical FFT program consists of two series, the real part of the transform [X(J)] and the imaginary part [Y(J)]. These two outputs can be used to generate two additional series, the amplitude spectrum of the spacial frequencies that make up the original optical spectrum and the phase spectrum of these spacial frequencies. The amplitudes of the spacial frequencies [A(J)] are calculated from the real and imaginary outputs by taking the root sum of squares of the two series, i.e.:

$$A(J) = [X(J)^2 + Y(J)^2]^{1/2} \quad (1)$$

The phases of these spacial frequencies [P(J)] are calculated using the following equation:

$$P(J) = \arctan [Y(J)/X(J)] \quad (2)$$

All these outputs are illustrated in Figure 1 for a single line optical spectrum.

Figure 1A is the original optical spectrum. This is a spectrum of the emission of 0.5 ppm Ca at 4226.7 Å in an O_2–H_2 flame. The real output of the FFT for this input spectrum is shown in Figure 1B. It is simply a damped cosine wave. The frequency of this cosine wave depends on the position of the spectral peak with respect to the origin in the original spectrum, and the functional form of the damping depends on the line shape in the original spectrum (11). The imaginary output (Figure 1C) is a damped sine wave with similar characteristics to the real output.

The amplitude spectrum of the spacial frequencies is shown in Figure 1D. This amplitude spectrum indicates that the original optical spectrum is composed mainly of low frequency spacial frequencies including a relatively large dc level. The amplitudes of the higher frequency spacial frequencies are small but their presence is significant in that it is primarily

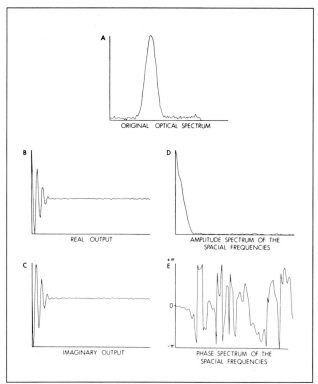

Figure 1. Outputs resulting from the Fourier transformation of a single line optical spectrum. (See text) The phase spectrum (1E) is modulo 2π

these spacial frequencies that make up the noise in the original optical spectrum. This distribution of information in the spacial frequency spectrum becomes intuitively obvious when it is realized that the peak is spread out over several sampled points while the noise occurs from point to point. Thus the information about the peak occurs in a different region of the spacial frequency spectrum than does some of the noise information. Hence the noise information can be discriminated against with respect to signal information. This forms the basis for spectral smoothing operations.

The phase spectrum of the spacial frequencies is shown in Figure 1E. Phase, in this case, simply refers to the phase of the individual spacial frequencies at one specific point on the optical spectrum. The phase is most usefully calculated with respect to a point on the optical spectrum at which the spacial frequencies are in phase. For the spectrum illustrated in Figure 1A, this point is the peak maximum. This, indeed, is the reason for the existence of the peak when the optical spectrum is interpreted as a Fourier summation of spacial frequencies.

It can be seen from the phase spectrum that the phases of the low frequency spacial frequencies (the peak information) are essentially the same—i.e., these spacial frequencies are in phase. A small slope in the phase spectrum in this region indicates that the reference point chosen is not the exact peak maximum or that the peak is slightly asymmetric. After a certain point, the phase spectrum begins to fluctuate essentially at random. This is an indication that the corresponding spacial frequencies are essentially due to noise in the original optical spectrum. In other words, it is unlikely that spacial frequencies resulting from noise in the original spectrum would happen, by chance, to have the same phase as the spacial frequencies resulting from the signal (*i.e.*, the peak) at a specific point in space along the optical spectrum axis (*i.e.*, the peak maximum). Thus the phase spectrum provides additional information about the distribution of signal and noise spacial frequencies.

Fourier transformation is a cyclic operation. The original optical spectrum can be regenerated using the real and imaginary outputs. The specific method will depend on the particular FFT program that is being used. For the operations illustrated in the next section, the optical spectrum was regenerated using the real output as a set of real input data and the resulting real output was the desired optical spectrum. A second approach was to use the real output as the real input and the negative of the imaginary output as the imaginary input to the FFT program (*12*). Again the resulting real output was the desired optical spectrum.

It is important to note that the amplitude spectrum of the spacial frequencies does not contain any information about the position of the spectral peaks in the original optical spectrum. Only the real and imaginary outputs do in the frequency of their oscillations. However, the real and imaginary outputs can be regenerated from the amplitudes of the spacial frequencies using the phase information.

Several data handling operations can be readily carried out on spectra using the information obtained upon Fourier transformation of the spectra. In general, the amplitude spectrum of the spacial frequencies or the real and imaginary outputs are modified before reconstruction of the optical spectrum. The modification typically involves multiplication of the real output by a relatively simple function. This is analogous to convolving the original optical spectrum with the Fourier transformation of the multiplication function. Smoothing, differentiation, and resolution enhancement of spectra using Fourier transformations are discussed and illustrated in the next section. These operations are only representative of the large number of data handling operations that can be performed on spectra using this approach.

SPECIFIC OPERATIONS PERFORMED ON SPECTRA USING FOURIER TRANSFORMATIONS

Smoothing. It is often desirable to smooth a spectrum in order to improve the signal-to-noise ratio. Several approaches to this problem have been discussed in the literature (*1, 6, 7*). One of the most common is to convolve the spectrum with an appropriate weighting function. This convolution can be carried out in a versatile and effective way using Fourier transformations.

It was noted in the last section that some of the noise information in the original optical spectrum appears in a different region of the spacial frequency spectrum than does the signal information. This provides a means of discriminating against the noise with respect to the signal before reconstruction of the optical spectrum and, hence, some of the noise can be filtered out. This is illustrated in Figure 2.

The original spectrum is shown in Figure 2*A*. This is the spectrum of the emission of 0.05 ppm Ca at 4226.7 Å in an O_2–H_2 flame. The real part $[X(J)]$ of its Fourier transformation is shown in Figure 2*B*. The high frequency spacial frequency region (primarily noise information) can be truncated by multiplying $X(J)$ by the simple function shown in Figure 2*C* to generate the modified real output shown in Figure 2*D*. The smoothed optical spectrum that results upon regeneration from the modified real output shows a marked reduction in the noise level (Figure 2*E*). In addition, since the signal information was not truncated, the peak shows little or no broadening. This type of operation is approximately analogous to analog low pass filtering. The smoothing function illustrated in Figure 2*C* is a low pass digital filter for spacial frequencies. The digital filter approach allows for complete control of the cutoff frequency and control or elimination of phase shifts. These are often difficult to control with analog filtering. In addition the peak shifting property of analog RC filters is eliminated as both future and prior information is utilized (*8, 13*).

It is difficult to present a hard and fast rule for determining the extent of the truncation function illustrated in Figure 2*C*. This choice is highly dependent on the specific experimental conditions, the amount of filtering desired, and the degree of signal distortion that can be tolerated. The simplest approach to this problem is an empirical one. The optical spectrum of interest is first measured under conditions that yield a high signal-to-noise ratio. Transformation of this signal will result in a well defined spacial frequency spectrum and the point be-

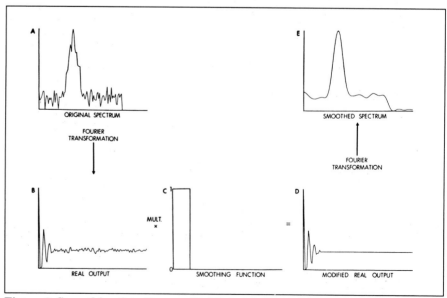

Figure 2. Smoothing function applied to a spectrum using Fourier transformations

yond which little or no signal information is present can be easily established.

The abrupt filter illustrated in Figure 2 may not be the most desirable for certain measurements. If some of the signal information is truncated abruptly, spurious side lobes will result. This is shown in Figure 3A. However, with very noisy spectra it may be desirable to have a low cutoff frequency. This will, in general, necessitate the truncation of some of the higher spacial frequencies that contribute to the peak information and as such the peak will be broadened. In this case, a smoothing function can be used that minimizes side lobes such as a linear truncation (see Figure 3B). The broadening of the peak is often quite acceptable and is simply the standard trade-off between the signal-to-noise ratio and resolution.

Several other smoothing functions can be used such as Gaussians, exponentials, etc., and at this point the question might well be asked, "Is there a smoothing function that will result in an optimum value for the signal-to-noise ratio?" Suffice it to say that a considerable amount of work has been reported in this general area and the characteristics of such a matched filter have been rigorously established (14). When the noise is white, a filter that takes the form of the spacial frequency spectrum of the instrumental line shape function is a close approximation to a matched filter for these signals. This can be determined by calculating the spacial frequency spectrum of a line measured with a very high signal-to-noise ratio. The smoothing function determined in this way is shown in Figure 3C along with the resulting smoothed spectrum. This operation essentially amounts to convolving a noisy line with an essentially noise-free version of itself.

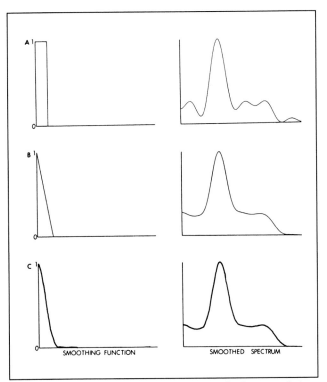

Figure 3. Smoothing functions and the resulting smoothed spectra. The original spectrum is shown in Figure 2A (See text for discussion)

Note that with this filter the line is considerably widened. This is simply a result of the goal that was set for the filter, *i.e.*, maximizing the signal-to-noise ratio. If it is also desirable to preserve, as well as possible, the observed line shape, then a different filter must be designed. Ernst has discussed this problem (8).

The improvement in the signal-to-noise ratio achieved by these smoothing operations can be used as an indication of their effectiveness. Calculated on the basis of peak maximum divided by the root mean square value of the base-line noise over sixteen points of the base line, the improvement in the signal-to-noise ratio achieved by the filter shown in Figure 2 is about 8. It must be emphasized that this number is only valid for this specific spectrum because the improvement is highly dependent on the relative extents and magnitudes of the spacial frequency spectra of the signal and noise information. Thus a statement about the relative bandwidth of the signal and noise information should be made when a signal-to-noise ratio is reported.

This problem can be looked at in a very qualitative way. For the spectrum illustrated in Figure 2A, it could be said that the signal-to-noise ratio is not very good. But it could perhaps more accurately be said that the signal-to-noise ratio is not very good for narrow peaks but that it is not too bad for broad peaks, *i.e.*, peaks composed of low frequency spacial frequencies. The improvement achievable using the smoothing functions attests to this fact. Thus attempts should always be made to limit the noise bandwidth to that just necessary for the accurate transmission of the signal information. This bandwidth control can be accurately achieved with the digital smoothing functions described above.

Differentiation. Differentiation of spectra has often been used to modify the spectral information. The first derivative of a spectral peak is used as an aid in exact peak location (3, 15) and higher derivatives are used for peak sharpening (2, 16). The derivative theorem of Fourier transforms (12) states that if the imaginary part of the Fourier transformation of a function is multiplied by a linear ramp (starting at the origin) the result is the real part of the Fourier transformation of the derivative of the original function. Multiplication by this linear ramp function in the Fourier domain eliminates the dc level and attenuates low frequency spacial frequencies with respect to high frequency spacial frequencies. It simply amounts to a high pass digital filter for spacial frequencies. The accentuation of high frequency spacial frequencies is a well-known characteristic of a differentiation step (17). The accentuation should not be carried out beyond the point at which the signal spacial frequencies disappear or else the resulting derivative will be very noisy. Thus some low pass digital filtering of the spacial frequencies should always be used in conjunction with differentiation. This is directly analogous to techniques used in the design of analog differentiating circuits using active filters. In addition, as with analog differentiation, the quality of the differential is dependent on the high and low frequency cutoffs relative to the signal frequencies.

The effect of such a digital differentiating filter on a spectrum is illustrated in Figure 4. The original spectrum is the sodium doublet as emitted by a sodium hollow cathode lamp and is shown in Figure 4A. The differentiating filter is shown in Figure 4B and it was applied in a manner analogous to that depicted in Figure 2. Note that both high and low pass filtering are readily carried out in one simple multiplication step. The resulting first derivative spectrum is shown in Figure 4C. Higher derivatives can easily be obtained by successive application of this filter.

Resolution Enhancement. Many approaches to resolution enhancement have been discussed in the literature. Common methods have used pseudo-deconvolution (4, 18, 19), special convolving filters (9), and differentiation (16). Resolution enhancement is desirable because the observed spectrum is often not an accurate representation of the real spectrum. The observed spectrum is the result of the convolution of the real spectrum by the resolution function of the spectrometer. It is often the slit width of the spectrometer that determines the width of the resolution function. This convolution distorts both the shape and the width of the real spectral lines and this can limit fundamental interpretation of line shapes and resolution.

The effect that this convolution can have on a simple spectrum is shown in Figure 5. The sodium doublet (5895.92 Å, 5889.95 Å) was measured at two spectral bandwidths, ~1 Å (Figure 5A) and ~4 Å (Figure 5C). The respective spacial frequency spectra are shown in Figures 5B and 5D. A consideration of these spacial frequency spectra provides unique insight into resolution enhancement.

It is obvious that observation of the sodium doublet with the wider spectral bandwidth has altered the spacial frequency spectrum. For the situation illustrated in Figure 5 in which the spectral bandwidth is wider than the spectral line width of the real spectrum, some of the upper spacial frequencies are completely lost. In addition, for all situations, the amplitudes of the lower spacial frequencies are reduced. Essentially all resolution enhancement procedures utilizing a convolution or pseudo-deconvolution approach attempt to restore the spacial frequency spectrum to that of the real spectrum or in this case to that of a spectrum observed with higher resolution.

This general approach to resolution enhancement has been discussed in an excellent paper by Bracewell and Roberts (20). This paper deals with radio astronomy but there is no fundamental difference between scanning the sky with a radio antenna and scanning a spectrum with a slit.

The effectiveness and limitations of this approach to resolution enhancement should now be obvious. The upper spacial frequencies that are lost cannot be recovered. This imposes a fundamental limit as to how far resolution enhancement can be carried out using the information available in the spacial frequency spectrum. Some-

times it is possible, using a priori information about the spacial frequency spectrum, to extrapolate into this region but this is risky at best, as is effectively illustrated in Figure 8 of Bracewell and Roberts (20). Thus resolution enhancement, in general, must rely on restoration of the amplitudes of the lower spacial frequencies still present in the spacial frequency spectrum of the observed optical spectrum.

Resolution enhancement may be necessary or desirable in many experimental situations. For example, it may be necessary in carrying out a series of spectral measurements at low concentrations, to use a wider slit width than is desirable with respect to resolution in order to have sufficient sensitivity. The spectrum shown in Figure 5C can serve as an example of such a measurement. In general, it is possible under more ideal experimental conditions (i.e., higher concentration) to carry out the same measurement with better resolution. The spectrum in Figure 5A can serve as an example of this

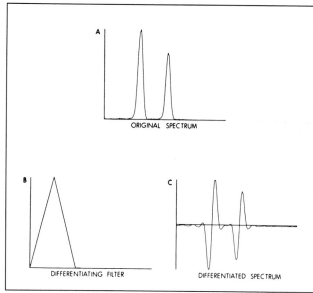

Figure 4. Differentiation of a spectrum. The differentiating filter is applied in a manner analogous to that depicted in Figure 2

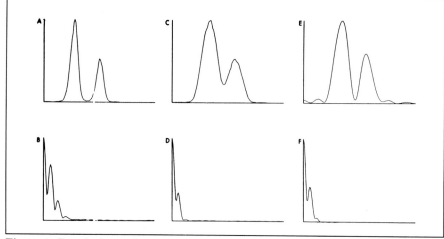

Figure 5. Resolution enhancement of a spectrum. See text for discussion of the figures

measurement. Resolution enhancement of the series of measurements made with lower resolution may now be carried out using the information contained in the higher resolution spectrum. The spacial frequencies of the lower resolution spectra are simply multiplied by an appropriate function to restore their amplitudes to those of the spacial frequencies of the higher resolution spectrum.

For the situation discussed above and illustrated in Figure 5, this function may be determined by dividing the spacial frequency spectrum illustrated in Figure 5B by that illustrated in Figure 5D. The spacial frequency spectra should be normalized and the division can only be carried out to the point where noise begins to dominate the result.

Using this approach, the spacial frequency spectrum in Figure 5D was restored to that shown in Figure 5F. The resolution-enhanced spectrum is shown in Figure 5E. Often side lobes are generated in the resolution-enhanced spectrum because of truncation effects similar to those illustrated in Figure 3A. Side lobe generation can be minimized by using the smoothing techniques discussed earlier.

All the data handling operations have been discussed and illustrated using simple optical spectra. There is no limitation to their application to more complex spectra of any kind except that they be digitized. In addition the operations that can be performed on spectra using Fourier transformations are by no means limited to those discussed here. Other possible operations include cross correlation of spectra and several other approaches to resolution enhancement.

REFERENCES

(1) A. Savitzky and M.J.E. Golay, *Anal. Chem.*, **36**, 1627 (1964).
(2) A. E. Martin, *Spectrochim. Acta*, **14**, 97 (1959).
(3) J. R. Morrey, *Anal. Chem.*, **40**, 905 (1968).
(4) R. N. Jones, R. Venkataragharan, and J. W. Hopkins, *Spectrochim. Acta*, **23A**, 925 (1967).
(5) J. R. Izatt, H. Saki, and W. S. Benidict, *J. Opt. Soc. Amer.*, **59**, 19 (1969).
(6) Mihai Caprini, Sorin Cohn-Sfetcu, and Anca Maria Manof, *IEEE Trans. Audio Electroacoustics*, **AU-18**, 389 (1970).
(7) T. Inouge, T. Harper, and N. C. Rasmussen, *Nucl. Instrum. Methods*, **67**, 125 (1969).
(8) R. R. Ernst, "Advances in Magnetic Resonance, Vol. 2," J. S. Waugh, Ed., Academic Press, New York, N.Y., 1966, p 1.
(9) D. W. Kirmse and A. W. Westerberg, *Anal. Chem.*, **43**, 1035 (1971).
(10) R. S. Singleton, *IEEE Trans. Audio Electroacoustics*, **AU-17**, 166 (1969).
(11) G. Horlick, *Anal. Chem.*, **43**(8), 61 A (1971).
(12) R. Bracewell, "The Fourier Transform and Its Application," McGraw-Hill Book Co., New York, N.Y., 1965.
(13) K. S. Seshadri and R. N. Jones, *Spectrochim. Acta*, **19**, 1013 (1963).
(14) G. L. Turin, *IRE Trans. Information Theory*, **IT-6**, 311 (1960).
(15) J. P. Walters and H. V. Malmstadt, *Appl. Spectrosc.*, **20**, 193 (1966).
(16) L. C. Allen, H. M. Gladney, and S. H. Glarum, *J. Chem. Phys.*, **40**, 3135 (1964).
(17) F. R. Stauffer and H. Sakai, *Appl. Opt.*, **7**, 61 (1968).
(18) W. F. Herget, W. E. Deeds, N. M. Gailar, R. J. Lovell, and A. H. Nielsen, *J. Opt. Soc. Amer.*, **52**, 1113 (1962).
(19) P. A. Jansson, R. H. Hunt, and E. K. Plyler, *Ibid.*, **58**, 1665 (1968).
(20) R. N. Bracewell and J. A. Roberts, *Austr. J. Phys.*, **7**, 616 (1954).

Received for review September 13, 1971. Accepted November 30, 1971. Financial support by the National Research Council of Canada and the University of Alberta is gratefully acknowledged.

Reprinted from *Anal. Chem.* **1972**, *44*, 943–47.

Before the development of semiconductor detectors, X-ray fluorescence had been used primarily for the analysis of geological materials. The analysis was slow and the equipment cumbersome. With the concise, detailed description of the many factors involved in X-ray fluorescence and the description of the newly developed semiconductor detectors presented in the following landmark paper, it became clear that this technique would solve problems in many areas. As the equipment improved and the number of applications increased, it became clear that this paper had already provided insight into the advantages of this new detector, a warning about potential problems, and some approaches to improving accuracy and sensitivity.

Ralph O. Allen
University of Virginia

Trace Element Determination with Semiconductor Detector X-Ray Spectrometers

Robert D. Giauque, Fred S. Goulding, Joseph M. Jaklevic, and Richard H. Pehl
Lawrence Berkeley Laboratory, University of California, Berkeley, Calif. 94720

A method of obtaining high sensitivity and accuracy in X-ray fluorescence analysis using semiconductor detector spectrometers is discussed. Mono-energetic exciting radiation is employed to generate characteristic X-rays from trace elements in thin, uniform specimens. Corrections for absorption effects are determined; enhancement effects are omitted as they are negligible for many thin specimens. A single element thin-film standard is used to calibrate for the X-ray geometry, and theoretical cross sections and fluorescent yield data are employed to relate the X-ray yields for a wide range of elements to the thin-film standard. Various corrections which affect the accuracy of the method are discussed including the method for determining X-ray spectral background. Results obtained in the analyses of biological and geological specimens, and of air particulate filters are reported. Using a single excitation energy, the concentrations of more than fifteen trace elements may be simultaneously determined during a fifteen-minute interval for concentrations of 1 ppm or less. This corresponds to less than 10 ng/cm^2 on air particulate filters.

The analytical technique of X-ray emission spectroscopy depends upon the ability to excite and accurately measure characteristic K and L X-rays emanating from the specimen. Prior to 1966, energy separation was usually achieved by using wavelength dispersive spectrometers. Since the first utilization of semiconductor detectors for X-ray spectrometry (1), major advancements have been made. Progress in electronic design over the past several years has significantly improved the energy resolution and count rate performance of semiconductor detector X-ray spectrometers

(2-4). More recently, the advent of guard-ring detectors (5) has drastically reduced X-ray spectrum background resulting from the degradation of signals due to incomplete charge collection in the detectors. As a result of these improvements, semiconductor detectors are now applicable to many analytical problems, including trace element analyses.

Some of the principal attributes of semiconductor detector spectrometers for energy dispersive analysis are: (1) simultaneous detection of a wide range of energies allowing the intensities of many characteristic X-rays and their respective spectrum backgrounds to be determined together; (2) interelement interferences due to overlapping X-ray lines are clearly shown; (3) compact excitation radiation–specimen–detector geometries are permitted, minimizing the required intensity of excitation radiation; (4) high detection efficiencies over a wide energy range; and (5) no requirements for moving parts or mechanical alignments. The energy resolution capabilities of semiconductor detector spectrometers are more than sufficient for most analytical applications, although X-ray spectrometers using crystals for energy dispersion provide higher energy resolution for radiation of less than approximately 15 keV. Also, higher individual X-ray intensities can be measured with crystal spectrometers since only a narrow radiation energy range is normally imposed on the detection system count rate limitations. However, for most multielement analytical problems, semiconductor detector X-ray spectrometers now permit analyses to be carried out in much shorter time periods and are less complex to operate than are crystal X-ray spectrometers.

In the first part of this paper, factors that affect sensitivities attainable in X-ray fluorescence analysis are discussed. A following section describes the theory of the calibration method employed for multielement analyses. Next, experimental procedures are described and results, including comparisons with results achieved by other methods, are reported. Factors which affect the accuracy of the calibration method are discussed in detail in an appendix.

DISCUSSION OF METHOD

General Technique for Obtaining and Characterizing the X-ray Spectrum. Figure 1 illustrates the technique employed for X-ray fluorescence analysis. Exciting radiation provided by an X-ray tube or radioisotope, either directly or indirectly with secondary targets, impinges upon a specimen. A fraction of these photons, if of sufficient energy, produces vacancies in the inner shells of atoms within the specimen, which in turn can emit characteristic X-rays that are then measured by the detector. In addition to these photoelectric interactions, a portion of the radiation striking the specimen is scattered either coherently (no energy loss) or incoherently (energy loss determined by the Compton process). For many analyses, particularly with low atomic number matrices, the intensity of scattered X-rays can be quite large compared with that of the characteristic X-rays. Figure 2 shows a generalized energy spectrum obtained from a semiconductor detector in such a situation. The exciting radiation in this case is assumed to be monochromatic. The features of interest in the spectrum are:

a) The two high-energy peaks produced by coherent and incoherent scattering of the exciting radiation from the specimen.

b) A low-energy continuum due to specimen scattered X-rays that are in turn scattered out of the detector leaving only a small fraction of the energy in the detector.

c) A general, rather flat, background which arises partially from incomplete charge collection in the detector.

d) The characteristic X-rays from the elements in the specimen.

Factors Determining Sensitivity. The main conditions that limit X-ray fluorescence sensitivities are the X-ray peak intensities and their ratios to the background. Some of the principal factors affecting sensitivities are the selection of the exciting radiation, the geometry employed, and the form and composition of the specimen.

Selection of the Exciting Radiation. Since the ratio of fluorescent to scattered X-ray intensities is often very small, the energy of the scattered X-rays should be sufficiently high that the scattered X-rays will not cause substantial interference with the characteristic X-rays. Because a continuum of such interfering scattered radiation is obtained if conventional X-ray tubes or bremsstrahlung sources are employed to directly provide the exciting radiation, monochromatic exciting radiation is used in our method. Monochromatic exciting radiation can be obtained by employing characteristic X-ray tubes (6), conventional X-ray tubes and secondary targets, radioisotope source-target assemblies (7), or radioisotopes which decay only by electron capture, *i.e.*, ^{109}Cd and ^{55}Fe.

The selection of the exciting radiation energy is strongly influenced by the range of elements to be studied. Maximum sensitivity is achieved by employing an exciting radiation energy slightly greater than the K or L absorption edge energies of the elements to be analyzed, but of sufficient energy that the incoherently scattered radiation does not produce significant overlapping background. Figure 3 is a plot of photoelec-

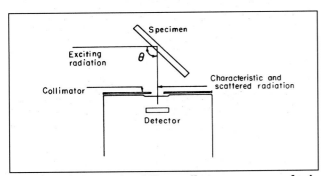

Figure 1. Schematic of X-ray fluorescence analysis technique

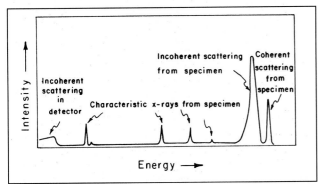

Figure 2. Generalized energy spectrum obtained from a low atomic number matrix with a semiconductor detector. The exciting radiation is assumed to be monochromatic

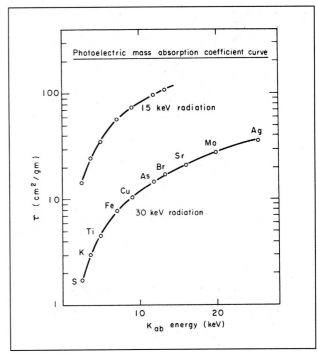

Figure 3. Photoelectric cross section curves for 15 and 30 keV photons

tric cross sections (τ) of various elements *vs.* their K absorption edge energies for two excitation energies. As shown, the elements with K absorption edge energies in the 2–15 keV range photoelectrically absorb 15 keV radiation approximately seven times more efficiently than the 30 keV radiation.

Geometry Considerations. It is important that the excitation–specimen–detector geometry be designed for the maximum practical efficiency so that a large number of counts can be acquired in a short time. Furthermore, it is desirable that the energy loss by the incoherent scattering process be a minimum. The minimum energy loss occurs when the exciting X-rays are scattered at small angles, and the energy loss which is defined by the Compton scattering process increases slightly for $\theta = 90°$ as can be seen from Equation 1

$$E_{\text{incoherent}} = E_{\text{exciting}}/[1 + 0.001957(1-\cos\theta)E_{\text{exciting}}] \quad (1)$$

where the X-ray energies are expressed in keV. However, the intensity of the scattered radiation relative to the fluorescent radiation intensities is a minimum when $\theta = 90°$—a factor of two less than at 180°. Thus, it is desirable to maintain an angle near 90° between the exciting and the detected radiation to yield the highest peak to background ratios.

Form and Composition of Specimen. Absorption effects increase more rapidly for the fluorescent X-rays than for the scattered exciting radiation; hence, sensitivity decreases as the specimen thickness approaches the critical thickness (i.e., the thickness beyond which an increase in thickness does not measurably increase a fluorescent X-ray intensity). Thus, high sensitivity is obtained by using relatively thin specimens. For example, Figure 4 shows the spectra obtained from a NBS SRM 1571 orchard leaves specimen prepared in the form of pellets, 30 and 300 mg/cm² thick. The spectra were obtained in equal times from the same specimen areas by adjusting the X-ray tube current to yield identical total counting rates. The improvement in sensitivity for lower energy X-rays

with the thin specimen is obvious. For most applications, the reduced counting rates realized with thinner specimens can be compensated by employing a more intense exciting radiation. Higher sensitivities are obtained for low atomic number matrices since absorption effects are considerably less than for high atomic number matrices.

Equipment and Characteristics. For the experiments discussed in this paper, a guard-ring detector (5) with pulsed light feedback electronics (4) and a 512 channel pulse height analyzer were employed. The total resolution of the system, FWHM, was 225 eV at 6.4 keV (FeKα X-ray energy) at 5,000 counts/sec using an 8-μsec pulse peaking time. Excitation was provided by a molybdenum transmission X-ray tube (6) with a combined anode plus window thickness of 0.010 cm. The X-ray tube was operated at 42 kV with regulated currents varying from 10–400 μA. The X-ray tube, detector, and specimen changer are shown in Figure 5. The distances between the X-ray tube anode to the specimen, and between the specimen to the detector were approximately 6 and 2.5 cm, respectively. The angles formed by the exciting and the detected emergent radiations with the specimen surface were both near 45°. The total area of the specimens used varied from approximately 2 to 3 cm² depending upon the external X-ray tube collimation employed. Corrections for system dead time resulting either through pile-up rejection or analyzer dead time were made using a gated clock that measured the total system live time (8). The maximum total count rates were 10,000 counts/sec. At this count rate, the system dead time was 50%. The high count rate capabilities permitted the use of intense X-ray excitation so that adequate statistical accuracies could be obtained for the measured fluorescent X-ray intensities in relatively short periods of time.

CALIBRATION METHOD

General Considerations. The use of thin specimens minimizes matrix effects in X-ray fluorescence analysis. The preparation of thin uniform specimens usually eliminates the tedious procedure of either preparing sets of standards simi-

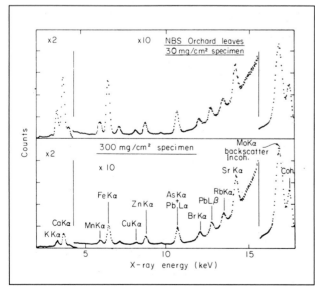

Figure 4. Spectra from NBS orchard leaves specimens of masses 30 and 300 mg/cm²

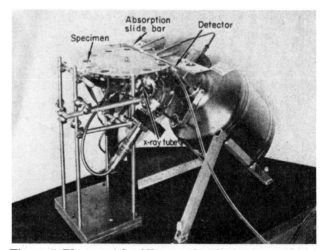

Figure 5. Photograph of X-ray tube, detector, specimen changer, and built-in absorption measurement capabilities

lar to the specimens to be analyzed, or adding internal standards to correct for matrix effects. When employing thin specimens, matrix enhancement effects are normally very small. As will be described later, matrix absorption effects can be determined if the critical thicknesses for the X-ray energies of interest have not been reached. For most analytical applications, the corrections necessary for matrix absorption effects are one to two orders of magnitude higher than for matrix enhancement effects. The use of monochromatic exciting radiation and a single element thin film standard for calibration of the X-ray spectrometer, and the employment of theoretical calculations to standardize for the analyses of many elements have previously been reported (9). Factors converting counts/sec to μg/cm^2 for the elements are obtained by calculating the relative probability of fluorescence excitation and detection of X-ray lines as a function of Z. Absolute normalization of the calibration curve to the X-ray tube intensity is accomplished by measuring an X-ray line intensity from the single element thin film standard of known mass. An elaboration of the technique is presented in this section.

Most problems in X-ray fluorescence analysis deal with the measurement of characteristic K or L X-rays arising from transitions to the inner atomic shell vacancies. These vacancies are created by photoelectric interactions of the atoms with radiation of sufficient energy. A simplified version of some of the basic atomic processes occurring in X-ray fluorescence analysis is shown in Figure 6. K X-rays are usually employed for the analysis of elements up to atomic numbers in the 55 to 60 range, and L X-rays are used for the elements of higher atomic number. For the development of the method, the specimen will be treated initially as being infinitely thin.

The relative ability of the elements to photoelectrically absorb the exciting radiation is determined by their photoelectric cross sections. To ascertain the contribution of a particular energy level to the total photoelectric cross section at the exciting energy, corrections must be made for the fraction of the cross section arising from interactions involving the other energy levels. The total photoelectric cross section for this individual energy level plus all lower energy levels (τ) is multiplied by the quantity $(1-1/J_{K,L})$ where $J_{K,L}$ is the ratio (jump ratio) between the photoelectric mass absorption coefficients at the top and bottom of the absorption edge energy. For absorption occurring in the K shell, the value of τ is the total photoelectric mass absorption coefficient for this exciting energy. However, for L energy levels, the value of τ is obtained

by extrapolation of the curve for the particular energy level to the exciting energy. This procedure compensates for the contribution of energy levels higher than the level of interest.

Only a fraction of the vacancies created in a particular energy level is filled by transitions which give rise to the direct emission of X-rays. Some vacancies are filled by transitions involving the emission of Auger electrons (10). The fraction of vacancies filled by transitions which directly yield X-rays is known as the fluorescence yield value ($\omega_{K,L}$). Figure 7 shows curves of theoretical fluorescence yield values for a wide range of elements for the K and L_{III} energy levels.

Transitions to a particular energy level give rise to the emission of more than one X-ray line since they can originate from several initial energy states. The ratio of the intensity of a particular X-ray line with respect to the total intensity is referred to as the fractional value (f).

For a particular excitation radiation, X-ray excitation curves may be established for X-ray lines from individual energy levels by multiplying the values of the terms τ, $(1-1/J_{K,L})$, $\omega_{K,L}$, and f. The values of each of these terms are reported in the literature (11–15). Figure 8 shows some calculated curves for the excitation of characteristic Kα and Lα X-rays with molybdenum exciting radiation (71% MoKα and 29% MoKβ radiation). Such curves can be determined for any monochromatic exciting radiation. As shown, the relative ability to excite characteristic X-rays with a specific excitation radiation energy drops off rapidly with decreasing absorption edge energy. Even though sensitivity for lower atomic number elements can be enhanced by selecting an excitation radiation of lower energy, sensitivities attainable are

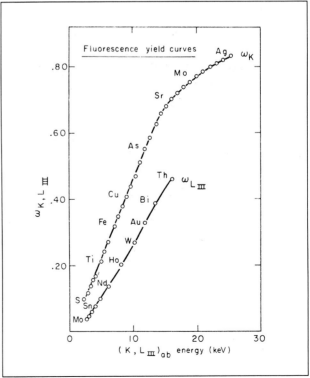

Figure 7. Theoretical fluorescence yield curves for the K and L_{III} energy levels

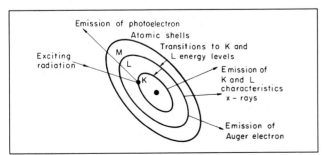

Figure 6. Simplified version of some of the basic atomic processes that can occur in X-ray fluorescence analysis

less than for elements near the middle of the periodic table as a result of low fluorescence yield values and, in most cases, high absorption effects.

Because of geometry considerations, only a small fraction of the X-ray lines emitted may be detected. A small proportion of the X-rays may be attenuated by the air or helium path, if present, and by the detector window. Also, for higher energy X-rays the efficiency (ϵ) of the detector may be less than unity. For any specific energy, the fraction transmitted (T) by the media between the specimen and the detector, and the efficiency of the detector may be calculated with relative ease using mass absorption coefficients listed in the literature. For any given geometry and for constant exciting radiation intensity, the relative ability to excite and detect various X-ray lines from "infinitely" thin specimens may be determined from the ratios of the product:

$$K_j = \tau \left(1 - \frac{1}{J_{K,L}}\right) \omega_{K,L} f T \epsilon \tag{2}$$

where the subscript j refers to the specific element.

Table I lists a comparison of calculated and experimental values of relative excitation and detection efficiencies, K_j, for the system employed, for seven X-ray lines normalized to the CuKα values. Theoretical fluorescence yield values (12, 13) were used to determine the calculated values. The experimental values were determined from the average value obtained from three thin film standards of masses varying from 50 to 150 µg/cm². The standards were prepared by evaporation of the elements onto thin aluminum backings of mass 800 µg/cm². The precisions listed are for one standard deviation. Al-

though the accuracy of the calculations depends on the selection of theoretical values, the actual calibration could be done by using experimental data such as shown in the table.

Since the intensity of an X-ray line from an "infinitely" thin specimen is directly proportional to the concentration, m_j(g/cm²) of the element, the intensities and elemental concentrations may be expressed as:

$$I = I_0 G K_j m_j \tag{3}$$

where I_0 is the exciting radiation intensity and G is a geometric factor.

For constant exciting radiation intensity and for constant geometry, the value of $I_0 G$ can be determined from a single element thin-film standard for which the absorption effects are negligible. In effect, theoretical values of relative excitation and detection efficiencies are calculated. The values are calibrated in units of cm²/g using a convenient element for which a thin-film standard is available.

Corrections for Absorption Effects. Since specimens prepared for analyses are not infinitely thin, corrections must be applied for matrix absorption effects. Two processes of absorption by the specimen occur, one for the exciting radiation, the other for the X-rays from elements within the specimen. The net absorption equals the product of these two absorption effects in the total specimen mass. Integrating these effects over the thickness of the uniform specimen, the correction may be written:

$$\text{Absorption correction} = \frac{1 - e^{-(\mu_1 \csc \phi_1 + \mu_2 \csc \phi_2)m}}{(\mu_1 \csc \phi_1 + \mu_2 \csc \phi_2)m} \tag{4}$$

where μ_1 and μ_2 are the total mass absorption coefficients (cm²/g) of the specimen for the exciting and the characteristic radiation, respectively. ϕ_1 and ϕ_2 are the angles formed by the exciting and characteristic radiation with the specimen surface. m is the mass (g/cm²) of the specimen. Applying this correction for absorption effects to Equation 3 gives

$$I = I_0 G K_j m_j \frac{(1 - e^{-(\mu_1 \csc \phi_1 + \mu_2 \csc \phi_2)m})}{(\mu_1 \csc \phi_1 + \mu_2 \csc \phi_2)m} \tag{5}$$

If the critical thickness for the X-ray energy of interest is not attained, the individual values of μ_1, μ_2, ϕ_1, ϕ_2, and m need not

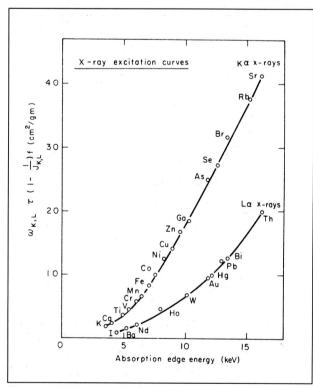

Figure 8. X-Ray excitation curves for molybdenum exciting radiation

Table I. Relative Excitation and Detection Efficiencies (K_j)

Line	Calculated	Determined
CrKα	0.381	0.370 ± 0.011
MnKα	0.450	0.435 ± 0.003
FeKα	0.587	0.559 ± 0.003
NiKα	0.884	0.882 ± 0.007
CuKα	1.000	1.000 ± 0.015
AsKα	1.653	1.660 ± 0.083
SeKα	1.776	1.753 ± 0.057
PbLα	0.804	0.774 ± 0.019

be known for the determination of this correction because the quantity $e^{-(\mu_1 \csc\phi_1 + \mu_2 \csc\phi_2)m}$ may be measured directly since it represents the combined attenuations of the incident and fluorescent X-rays in the total specimen thickness. This is done by measuring the relative X-ray intensity with and without the specimen from a target located at a position adjacent to the back of the specimen as shown in Figure 9. The calculation is expressed:

$$A \equiv \frac{I_T' - I_S}{I_T} = e^{-(\mu_1 \csc\phi_1 + \mu_2 \csc\phi_2)m} \tag{6}$$

where I_S, I_T, and I_T' are the intensities of the X-ray plus background from the specimen alone, the target alone, and the specimen with the target, respectively. The target should be at least of critical thickness for the X-ray energy of interest. To minimize possible absorption correction errors arising from enhancement of the specimen radiation by scattered target radiation, targets that yield a high ratio of scattered to fluorescent radiation should not be used.

Figure 5 shows the assembly employed to make absorption measurements. Many absorption correction measurements may be easily made by positioning the absorption slide bar containing the targets. With this assembly, the targets are located less than 0.15 cm from the specimen. Simultaneous absorption measurements for many X-rays can often be made by using a multielement uniform target. However, to use this procedure, the contribution of enhancement effects between the specimen and the target must be negligible. Conversely, enhancement effects between the specimen and the target may exist for low trace element concentrations and still have a negligible effect on the results since the value of I_S is low compared to the value of I_T'. For thin specimens with similar element concentrations, repetitive absorption measurements often need not be made if the specimens are prepared of approximately equal total mass concentrations (g/cm^2). For some thin specimens, particularly when higher energy X-rays are employed for analyses, absorption effects are negligible.

Preparation of Thin Standards and Specimens. *Preparation of Standards.* Thin-film standards are prepared by evaporation of the elements onto thin aluminum films. The standards are made thin enough that self-absorption effects are negligible, generally around 100 μg/cm^2. Thin standards can also be prepared by absorbing known amounts of the elements in solution on cellulose powder, drying at approximately 80 °C, weighing, pulverizing, and using a fraction of the mixture to press a thin pellet which is then weighed. Ab-

sorption correction measurements must be made on cellulose standards.

Preparation of Specimens. The principal considerations dictating the desired specimen thickness are the matrix effects that will be prevalent and the characteristics of the X-ray spectrometer employed. Specimens should be prepared thin enough that: a) The critical thicknesses for the X-ray energies of interest are not attained. b) Matrix enhancement effects are quite small for the analyses to be performed. c) Count rate limitations are not imposed upon the X-ray detection system when near maximum excitation intensity is employed. d) Total system geometries do not significantly vary with respect to the standard employed for analysis. Nevertheless, specimens should be prepared thick enough that favorable statistics in analysis are obtained in relatively short periods of time.

Biological Specimens. Specimens consisting of biological material such as tissue, blood, plants, etc., are either freeze-dried or oven-dried, pulverized, and subsequently 2.54-cm diameter pellets are pressed at 15,000 psi and weighed. The pellets typically are of mass 30 mg/cm^2 and are approximately 0.03 cm thick. The freeze-drying process not only allows the preparation of thin uniform specimens, but also permits higher trace element sensitivity due to the concentration factor of up to ten and more which can often be obtained.

Rock, Glass, and Pottery Specimens. Finely pulverized rock, glass, or pottery specimens are collected on thin filters using the apparatus shown in Figure 10. While maintaining a vacuum behind a filter on a glass frit, short blasts of air are let into the glass vessel by quick manipulation of the stopcock. This procedure converts the specimen into a fine dust which is

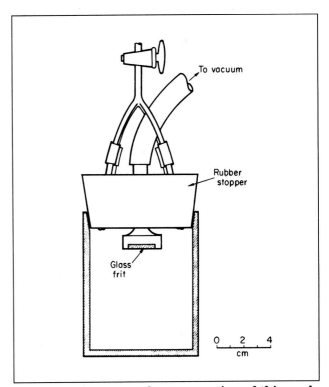

Figure 10. Apparatus for preparation of thin rock, glass, and pottery specimens

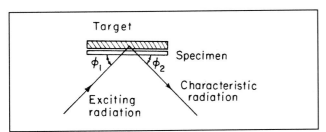

Figure 9. Schematic of method for absorption correction measurements

collected on a 0.8-μ Millipore filter 2.5 cm in diameter and of mass 5 mg/cm². Deposits prepared cover an area of 3 cm² and are of approximate mass 5 mg/cm². Specimens must be pulverized fine enough that particle size will neither affect the deposit nor the analysis. Grinding the specimens to pass through a 325 mesh screen (<44 microns) is sufficient for many analyses.

This preparation procedure has worked well in some cases and been entirely unsatisfactory in other cases. Some of the finely pulverized specimens have fine particles which selectively stick to the side of the vessel and, consequently, unrepresentative specimens are prepared. To overcome this problem, some specimens may be fused with lithium metaborate to form a homogeneous mixture which is then finely pulverized.

Elements in Solution. Elements in solution may be absorbed on cellulose powder, dried at approximately 80 °C, pulverized, and pressed into 2.54-cm diameter pellets. Trace ionic impurities often may be collected on ion-exchange resin-loaded papers (16), but care must be taken to correct for uneven distributions between the front and the back of the papers when lower energy X-rays are employed for analysis. Trace amounts of many elements may be precipitated with a carrier and collected on a filter paper (17).

Air Particulate Filters. The concentrations of many elements present in air collection specimens on filter papers are simply determined by measuring the intensities of the characteristic X-rays since absorption effects can be considered negligible. Absorption by the filter of lower energy X-rays (around 4 keV and less) from particles impacted within the pores of the filter can become important for some analyses. A procedure to compensate for this effect is not known. To obtain high sensitivity in analysis on air pollution specimens, filters of low mass should be used to minimize scattered excitation radiation background.

RESULTS

The following results were obtained with the previously described equipment. Standardization was accomplished in 100 sec using a 101 μg Cu/cm² evaporated thin-film standard. Calculated values for relative excitation and detection efficiencies (K_j) were employed for analyses. Absorption correction measurements were made when necessary and the concentrations of the elements were calculated using Equation 7.

$$\text{ppm}(j) = \frac{c_j}{c_s} \times \frac{i_s}{i} \times \frac{\ln A}{1-A} \times \frac{m_s}{m} \times \frac{1}{K_j} \times 10^6 \quad (7)$$

where c_j and c_s are the characteristic X-ray count rates from element j and the standard. i and i_s are the X-ray tube currents employed for the specimen and the standard. A is defined by Equation 6. m and m_s are the specimen and the standard masses (g/cm²). Because of the count rate limitations of the detection system, absorption correction measurements were usually made at lower tube currents than those employed for obtaining the specimen spectrum.

Biological Specimens. Table II shows the results of the analyses from five separately prepared specimens of standard NBS SRM 1571 orchard leaves. Pellets 2.54 cm in diameter and of mass 30 mg/cm² were prepared. To determine the correction for the moisture content, a weighed amount of the orchard leaves was dried for 24 hours at 90 °C and weighed. Total analysis time, including absorption measurements, was thirty minutes for each specimen. The errors listed are for two standard deviations.

Figure 11 shows the spectrum and results obtained in thirty minutes from a freeze-dried human serum specimen which was pulverized and pressed into a 2.54-cm diameter pellet of mass 30 mg/cm². Since the concentration factor obtained by freeze-drying the specimen was ten, the concentrations of the elements in the original specimen are one-tenth the values listed.

Pottery and Rock Specimens. Table III illustrates a comparison of the results obtained by X-ray fluorescence and LBL neutron activation on a pottery specimen. Thin specimens of mass 5 mg/cm² were prepared on 0.8-μ Millipore filter of mass 5 mg/cm², employing the apparatus shown in Figure 10. The errors listed are for one standard deviation for five separately prepared specimens. The Ti, V, Cr, and Mn results have been corrected for X-ray background from BaL X-rays. For these corrections, the barium concentration of

Table II. Analysis of NBS SRM 1571 Orchard Leaves

	X-Ray fluorescence	NBS
Cr	2.5 ppm ± 1.6	2.3 ppm
Mn	88.6 ppm ± 2.2	91 ppm ± 4
Fe	276 ppm ± 8	300 ppm ± 20
Ni	1.3 ppm ± 0.4	1.3 ppm ± 0.2
Cu	12.6 ppm ± 0.6	12 ppm ± 1
Zn	23.7 ppm ± 0.8	25 ppm ± 3
As	10.6 ppm ± 0.8	14 ppm ± 2
Br	9.3 ppm ± 0.6	10 ppm
Rb	11.0 ppm ± 0.8	12 ppm ± 1
Sr	36.6 ppm ± 1.2	37 ppm
Pb	45.4 ppm ± 2.0	45 ppm ± 3

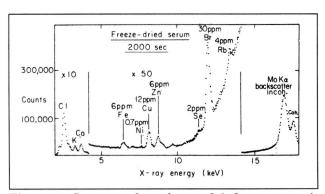

Figure 11. Spectrum from freeze-dried serum specimen. Concentrations listed are for freeze-dried specimen. This preparation gave a concentration factor of ten

712 ppm determined by neutron activation was chosen to calculate the interfering X-ray correction. The barium concentration could be determined by employing exciting radiation of sufficiently high energy to produce BaK X-rays from the specimen. Figure 12 shows a spectrum obtained from one of the specimens. Forty minutes were required to analyze each specimen.

A comparison of the X-ray fluorescence and LBL neutron activation results in the analysis of U.S. Geological Survey Andesite is shown in Table IV. Again, the errors tabulated are for one standard deviation for five separately prepared specimens of mass 5 mg/cm². Total analysis time for each specimen was forty minutes. The neutron activation result of

1210 ppm Ba was used to correct for BaL X-ray line interferences.

The X-ray fluorescence potassium and calcium results are probably low because of absorption by the filter of lower energy X-rays from particles collected within the pores of the filter rather than on the surface. For these analyses, the specimens are assumed to have been collected entirely on the surface of the filter. Also, particle size effects can increase rapidly with decreasing X-ray energy (18, 19), and slight nonuniformities in the thickness of the specimens when large absorption correction measurements are necessary would both yield low results.

Air Particulate Filters. Figure 13 shows the spectrum and results obtained on an air pollution filter of mass 5 mg/cm². This specimen represents three cubic meters of air passed through a filter area of 4 cm² in one hour. The spectrum was taken in 800 seconds and the concentrations listed are in nanograms/cm².

Theoretical Limits of Detection. Theoretical limits of detection with the equipment used are shown in Table V for three separate types of specimens. In this paper the minimum detectable amount is defined as that concentration or quantity which gives a line intensity above background equal to three times the square root of the background for counting times not to exceed 1000 sec including total system dead time. (In our work, approximately 75% of the total peak areas are employed.) This criterion gives a confidence level of 95%. Higher sensitivity for many of these elements can be obtained by employing different characteristic X-ray tubes (6). Detection limits can be further improved by approximately a factor of four if both a double guard-ring detector (5) and a pulsed X-ray tube (20) are employed.

Table III. Analysis of Pottery Specimen

	X-Ray fluorescence	Neutron activation
K	0.95% ± 0.02	1.35% ± 0.04
Ca	0.28% ± 0.01	<1.0%
Ti	0.77% ± 0.01	0.78% ± 0.03
V	176 ppm ± 16	176 ppm ± 16
Cr	107 ppm ± 5	115 ppm ± 4
Mn	31 ppm ± 4	40.9 ppm ± 0.5
Fe	1.05% ± 0.01	1.017% ± 0.012
Ni	277 ppm ± 4	279 ppm ± 20
Cu	56 ppm ± 2	59 ppm ± 8
Zn	61 ppm ± 1	59 ppm ± 8
Ga	46 ppm ± 1	44 ppm ± 5
As	32 ppm ± 1	30.8 ppm ± 2.2
Se	<2 ppm	. . .
Br	<3 ppm	2.3 ppm ± 0.9
Rb	70 ppm ± 2	70.0 ppm ± 6.3
Sr	145 ppm ± 2	145 ppm ± 22
Pb	35 ppm ± 3	

Table IV. Analysis of Andesite

	X-Ray fluorescence	Neutron activation
K	1.79% ± 0.02	2.6% ± 0.3
Ca	3.17% ± 0.04	3.9% ± 0.5
Ti	0.61% ± 0.01	0.59% ± 0.04
V	125 ppm ± 16	157 ppm ± 28
Cr	<20 ppm	8.5 ppm ± 4
Mn	734 ppm ± 11	736 ppm ± 12
Fe	4.61% ± 0.06	4.88% ± 0.06
Ni	20 ppm ± 2	. . .
Cu	69 ppm ± 2	. . .
Zn	87 ppm ± 2	104 ppm ± 9
Ga	23 ppm ± 1	. . .
As	3 ppm ± 1	. . .
Se	<2 ppm	. . .
Br	<2 ppm	. . .
Rb	69 ppm ± 2	91 ppm ± 12
Sr	755 ppm ± 8	1021 ppm ± 276
Pb	38 ppm ± 3	. . .

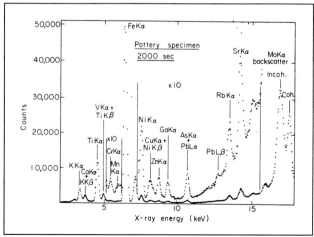

Figure 12. Spectrum from pottery specimen of mass 5 mg/cm²

In this section, factors which affect the accuracy of the method are discussed. Included are considerations for some of the individual terms used to calculate the relative excitation and detection efficiencies for various X-ray lines with the system employed. Also, methods for the accurate determination of X-ray spectra background and for the compensation of overlapping X-rays are described.

Determination of the Values for the Terms $\omega_{K,L}$, τ_j, $J_{K,L}$, ϵ, and f. *Fluorescence Yield, $\omega_{K,L}$.* Theoretical fluorescent yield values (12, 13) are used in the calculation. L X-rays originate from transitions to three separate energy levels. Since possible errors exist in the reported fluorescence yield values for X-rays from transitions to the L_{II} and L_I energy levels due to considerations for Coster-Kronig transitions, an X-ray resulting from a transition to the L_{III} energy level is selected for

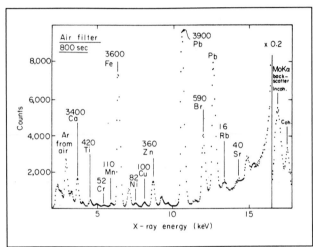

Figure 13. Spectrum from air filter of mass 5 mg/cm².
Concentrations listed in nanograms/cm²

Table V. Theoretical Limits of Detection

Specimen	Biological	SiO₂ on filter of mass 5 mg/cm²	Air particulate on filter of mass 5 mg/cm²
Mass (mg/cm²)	30	5	...
Tube current (μA)	250	400	400
Area analyzed (cm²)	3	2	3
Element and spectral line			
TiKα	3 ppm	10 ppm	15 ng/cm²
CrKα	1	4	9
FeKα	0.6	3	5
CuKα	0.3	1	3
HgLα	0.4	2	5
PbLα	0.4	2	5
BrKα	0.3	1	3
RbKα	0.4	2	6

the calculations; hence, the $\omega_{L(III)}$ values are used. There is generally several percent disagreement between the theoretical and the experimental fluorescent yield values reported in the literature. As previously shown in Table I, the experimental data obtained were consistent with the use of theoretical values in the calculations.

Photoelectric Mass Absorption Coefficient, τ_j. Photoelectric mass absorption coefficients are reported in the literature (11). Figure 14 shows the output of the molybdenum transmission X-ray tube operated at 42 kV. As illustrated, the exciting radiation is treated as two monochromatic X-ray beams corresponding to the MoKα and MoKβ X-ray energies. The net effective photoelectric mass absorption coefficient is determined by multiplying the fraction of each radiation times the photoelectric mass absorption coefficient for each radiation and summing them. As shown, there is a radiation continuum in the range of 25 to 40 keV. The continuum accounts for about 10% of the total exciting radiation. The photoelectric mass absorption coefficients for this continuum are approximately a factor of five less than for the MoK radiation. The net effect on the photoelectric mass absorption coefficients for elements with absorption edge energies below the MoKα energy would be only 2% and essentially would be proportionally the same for all of these elements. Thus, the contribution from this continuum is ignored in the calculations. In addition, a small fraction of the exciting radiation consists of energies just below that of the MoKα X-ray energy. The molybdenum anode and window are poor absorbers of radiation of these energies since they are barely below the MoK ab-

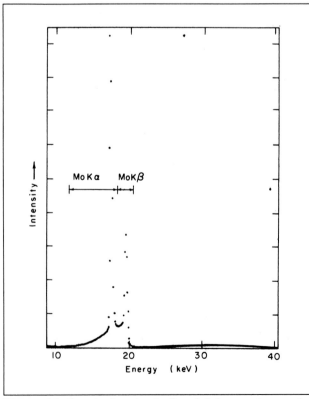

Figure 14. Spectral output of molybdenum transmission X-ray tube operated at 42 kV

sorption edge energy. As illustrated, this radiation is treated as part of the MoKα radiation for analysis. Thus, the photoelectric mass absorption coefficient is appropriately reduced for those elements that have absorption edge energies above part of this radiation which corrects for the ineffective portion of the exciting radiation. To illustrate the magnitude of those corrections, the total photoelectric mass absorption coefficients for strontium and bromine are reduced by 9.7 and 1.8%, respectively. If the exciting radiation had consisted of only MoKα and MoKβ X-rays, and if only the photoelectric mass absorption coefficients for MoKα radiation were used for elements with characteristic X-rays in the 3- to 14-keV range (potassium to strontium K X-rays, and cadmium to bismuth L X-rays), the relative error introduced over this range would have been only 1%.

For L X-rays, the photoelectric mass absorption coefficient for the production of X-rays from transitions to the L_{III} energy level are used in the calculations. Figure 15 shows a plot of the total mass absorption coefficient for lead. The photoelectric cross section due to the L_{III} energy level is determined by extrapolation, as illustrated.

Photoelectric Mass Absorption Coefficient Jump Ratio, $J_{K,L}$. Photoelectric mass absorption coefficient jump ratios are reported in the literature (*11*). For L X-rays, the J_L values corresponding to the L_{III} energy level are employed in the calculations.

Detector Efficiency, ε. The detector efficiency for radiation striking the detector at an angle of 90° near the center of the sensitive region of the detector can be calculated from the detector mass and the photoelectric mass absorption coefficient of the detector as written:

$$\text{Detector Efficiency} = 1 - e^{-\tau \rho t} \tag{8}$$

where τ is the photoelectric mass absorption coefficient (cm^2/g) of the detector for the energy of interest. ρ is the density (g/cm^3) of the detector. t is the thickness (cm) of the detector. For radiation between 4 and 15 keV striking a 5-mm thick silicon detector in the described manner, the efficiency is unity. However, with geometries employed for analyses, a fraction of the radiation strikes the detector at angles of less than 90° and impinges upon the detector near the periphery of the sensitive region. Figure 16 shows the total system geometry employed. The approximate sensitive region of the detector is shaded. To determine the effect of geometry on the detector efficiency, the intensities of characteristic X-rays from reference specimens were measured (1) using the geometry employed for analysis, and (2) with the fine collimator set in place of the exterior of the two detector collimators such that X-rays detected approached the detector in the center of the sensitive region at an angle close to 90°. The reference specimens were prepared of infinite thickness for the radiation of interest so that the geometry would not affect the apparent reference specimen thickness.

Detector efficiency was assumed to be unity for both geometries for lower energy X-rays (around 4 keV), and thus, the effect of the geometry employed on the detector efficiency was determined from the ratio of the intensities with and without the fine collimation. This geometry was found to reduce the

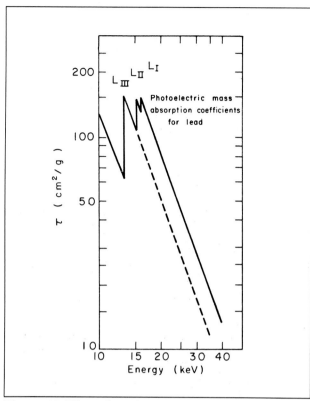

Figure 15. Photoelectric mass absorption coefficient plot for lead

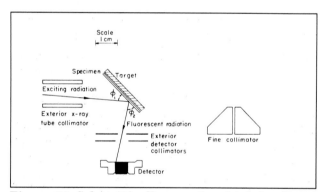

Figure 16. Schematic diagram for obtaining X-ray spectrum and absorption measurements

detector efficiency to 0.79, 0.90, and 0.98 for SrKα (14.14 keV), AsKα (10.53 keV), and CuKα (8.04 keV) X-rays, respectively.

Fraction of X-rays of Interest with Respect to Total X-Rays Emitted from Transitions to a Particular Energy Level, f. Figure 17 presents the most prominent K and L transitions leading to K and L X-ray series lines. All K X-rays originate from transitions to a single energy level. Accurate relative X-ray transition probabilities to the K shell are reported in the literature (*14*). The L X-ray series is more complex and originates from transitions to three energy levels as shown. The literature (*15*) cites approximate relative probabilities for these transitions. Sometimes more than one of the L X-rays from transitions to the L_{III} energy level are not resolved from X-rays originating from transitions to other L energy levels (such as the Lβ₂

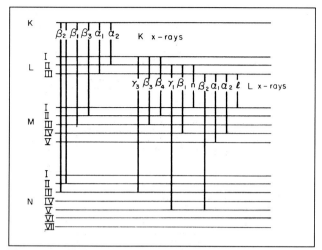

Figure 17. X-Ray energy level diagram

and $L\beta_1$ X-rays from lead). This, in turn, requires approximation as to the fraction of the unresolved X-rays that originate from transitions to the L_{III} energy level. The percentage of errors arising from these approximations is small when the most intense L X-ray lines ($L\alpha$) are employed for the calculations.

Determination of X-Ray Spectra Background from Scattered Exciting Radiation. The use of monochromatic exciting radiation permits the development of accurate background curves which are related to the ratio of the intensities of the coherent and the incoherent scattered radiation. These curves are established from reference specimens of varying effective atomic number. The reference specimens are prepared of mass similar to the specimens to be analyzed, and should not contain elements which have X-rays in the energy range of interest. Using these curves, the total spectrum backgrounds due to scattered exciting radiation can be determined from the intensities of the coherent and incoherent scattered exciting radiation.

As an illustration, five reference specimens of varying cellulose and sulfur contents were prepared to simulate scattered exciting radiation backgrounds that would be obtained with biological specimens. All reference specimens were of mass 30 mg/cm^2; the sulfur contents varied from 0 to 40%. Since the respective ratios of the coherent and incoherent scattering cross sections (for $MoK\alpha$ and $MoK\beta$ radiation) from low atomic number matrices such as these are nearly constant, and since the relative difference in the absorption effects of the scattered radiation is small for similar specimens to be analyzed, only the intensities of the scattered $MoK\alpha$ X-rays were employed to determine spectral line background contributions from the scattered exciting radiation. To further simplify the determination of the background contributions of the coherent and incoherent scattered radiation, all reference specimens were run to yield approximately equal total scattered radiation. Although the coherent and incoherent scattered $MoK\alpha$ X-rays were not completely resolved, fixed energy ranges centered on the peaks were selected that gave approximately the same total net counts for the two peaks

from all five reference specimens. Even though the ratio of the intensities of the coherent to the incoherent scattered radiation varied by a factor of two, accurate background curves were established for the individual X-ray lines and were expressed by the equation for a straight line as written:

$$Bkg_j = (Bkg_{coh} + Bkg_{incoh}) \times \left[\left(\frac{Bkg_{coh}}{Bkg_{coh} + Bkg_{incoh}} \right) \times S_j + B_j \right] \quad (9)$$

where Bkg_{coh} and Bkg_{incoh} are the total counts for the fixed energy ranges for the coherent and incoherent peaks, respectively. S_j is the slope of the background curve for the X-ray line from element j. B_j is the intercept at $Bkg_{coh} = 0$.

Typically, the standard deviations for the points forming the individual curves were in the range of 0.1 to 0.5%. In X-ray fluorescence analysis, the ratio of incoherent to coherent scattered radiation increases as the exciting radiation energy increases, and as the effective atomic number of the matrix decreases.

Determination of Overlapping X-Ray Background. Thin-films are prepared to determine overlapping X-ray lines by establishing relationships between individual X-ray peak shapes and relative intensities. These thin-films are prepared either by dusting the finely pulverized material onto Mylar tape, or by evaporation onto thin backings, such as aluminum or Mylar. The masses of these thin-films are unimportant as long as the relative difference of the absorption of the X-rays is negligible. For analyses, absorption correction measurements are applied to determine the relative effects. Also, from these films relative excitation and detection efficiencies, such as for L X-rays originating from transitions to either the L_{II} or L_I energy levels, are established in relationship to previously calculated values (K_j in Equation 2) for X-rays from transitions to the L_{III} energy level.

X-Rays detected by semiconductor detectors often produce peaks that tail on the low energy side. This tailing effect becomes more pronounced with increasing X-ray energy and can cause overlapping of X-ray lines, particularly when trying to discern X-rays of low intensity adjacent to a high-intensity higher energy peak.

When widely differing counting rates are used, a slight base-line shift can occur in the energy spectrum. Thus, one or more of the intense X-ray lines in the energy spectrum are used to appropriately adjust for any base-line shift that may occur.

Escape peaks and pile-up can be of consideration for trace element analyses when both high and low intensity (differing by more than an order of magnitude) X-rays are present. The energy of escape peaks corresponds to the original X-ray energy minus the energy of the escaping photons, which in the case of silicon detectors are the SiK X-rays. The intensity of the escape peaks, although energy dependent, are more than two orders of magnitude less than the initial X-ray intensity. At high count rates, pile-up (simultaneous detection of more than one X-ray at a time, hence, sums of the X-rays are recorded) can occur even though pile-up rejection is used. At 10,000 counts/sec, the peak intensity due to pile-up is more than two orders of magnitude less than the intensity of the initial radiation. The minor corrections for overlapping X-ray background due to peak tailing, escape peaks, and pile-up are established with relative ease from the thin-films.

Geometry–Absorption Correction Consideration. Figure 16 illustrates the geometrical relationship between the exciting radiation, the specimen, and the detector. The absorption correction for radiation coming from various parts of the specimen changes with ϕ_2. The net effective value of ϕ_2 determines the apparent mass of the specimen. To ascertain whether the net effective value of ϕ_2 changes with increasing correction for absorption effects and, in turn, introduces errors in analyses, thin aluminum foils of mass 8.1 mg/cm^2 were successively stacked and the concentrations of iron present were determined. The analysis of iron was selected because the mass absorption of the aluminum for the FeKα X-rays was a factor of twenty more than that of the aluminum for the exciting MoK radiation. Thus, the absorption correction is almost solely dependent on the absorption of FeKα X-rays. The results obtained are shown in Table VI. It is concluded from these results that for absorption corrections up to a factor of five (the value $[\mu_1 \csc\phi_1 + \mu_2 \csc\phi_2] \, m \cong 5$) errors due to varying effective values of ϕ_2 are negligible.

ACKNOWLEDGMENT

The assembly shown in Figure 5 uniting the X-ray tube, detector, and absorption measurement capabilities was designed and constructed by Hardy Wandesforde. The current regulator for the X-ray tube was designed and fabricated by Don Landis. We want to especially thank William Searles for his untiring effort and assistance in assembling and maintaining the equipment. We are indebted to Frank Asaro and Harry Bowman for providing the neutron activation analyses, and to Karl Scheu for preparing the thin-film standards. We are grateful to Ursula Abed for her comments about the preparation of this paper.

Table VI. Analysis of Aluminum Foil

Mass (mg/cm^2)	Absorption correction	Fe, ppm
8.1	1.61	4,130
16.2	2.45	4,130
24.3	3.38	4,120
32.4	4.35	4,070
40.5	5.30	4,120

REFERENCES

(1) H. R. Bowman, E. K. Hyde, S. G. Thompson, and R. C. Jared, *Science*, **151**, 562 (1966).
(2) F. S. Goulding, J. Walton, and D. F. Malone, *Nucl. Instrum. Methods,* **71,** 273 (1969).
(3) F. S. Goulding, J. T. Walton, and R. H. Pehl, *IEEE Trans. Nucl. Sci.*, **17** (1), 218 (1970).
(4) D. A. Landis, F. S. Goulding, R. H. Pehl, and J. T. Walton, *IEEE Trans. Nucl. Sci.*, **18** (1), 115 (1971).
(5) F. S. Goulding, J. M. Jaklevic, B V. Jarrett, and D. A. Landis, *Advan. X-ray Anal.*, **15**, 470 (1972).

(6) J. M. Jaklevic, R. D. Giauque, D. F. Malone, and W. L. Searles, *Advan. X-ray Anal.*, **15**, 266 (1972).
(7) J. R. Rhodes, in "Energy Dispersion X-ray Analysis," J. C. Russ, Ed., **ASTM STP 485**, 243, American Society for Testing Materials, Philadelphia, Pa., 1971.
(8) D. A. Landis, F. S. Goulding, and B. V. Jarrett, *Nucl. Instrum. Methods*, **101**, 127 (1972).
(9) R. D. Giauque and J. M. Jaklevic, *Advan. X-ray Anal.*, **15**, 164 (1972).
(10) I. Bergstrom, C. Nordling, A. H. Snell, R. W. Wilson, and B. G. Petterson, in "Alpha-, Beta-, and Gamma-Ray Spectroscopy," K. Siegbahn, Ed., Vol. 2, North-Holland Publishing Co., Amsterdam, Netherlands, 1965, p 1523.
(11) W. H. McMaster, N. K. Del Grande, J. H. Mellett, and J. H. Hubbell, "Compilation of X-ray Cross Sections," University of California, Lawrence Livermore Laboratory, *Rept.* **UCRL-50174**, Section II, Revision I (1969).
(12) R. W. Fink, R. C. Jopson, N. Mack, and C. D. Swift, *Rev. Mod. Phys.*, **38**, 513 (1966).
(13) E. J. McGuire, *Phys. Rev.*, **A3**, 587 (1971).
(14) J. S. Hansen, H. U. Freund, and R. W. Fink, *Nucl. Phys.*, **A142**, 604 (1970).
(15) G. G. Johnson, Jr., and E. W. White, "X-ray Emission Wavelengths and keV Tables for Nondiffractive Analysis," **ASTM DS 46**, American Society for Testing Materials, Philadelphia, Pa., 1970.
(16) W. J. Campbell, E. F. Spano, and T. E. Green, *Anal. Chem.*, **38**, 987 (1966).
(17) C. L. Luke, *Anal. Chim. Acta*, **41**, 239 (1968).
(18) F. Claisse, *Spectrochim. Acta*, **25B**, 209 (1970).
(19) D. F. Ball, *Analyst (London)*, **90**, 258 (1965).
(20) J. M. Jaklevic, F. S. Goulding, and D. A. Landis, *IEEE Trans. Nucl. Sci.*, **NS-19**, No. 3, 392 (1972).

Received for review July 28, 1972. Accepted November 17, 1972. Work performed under the auspices of the U.S. Atomic Energy Commission.

Reprinted from *Anal. Chem.* **1973**, *45*, 671–81.

Sedimentation Field-Flow Fractionation

J. Calvin Giddings, Frank J. F. Yang[1], and Marcus N. Myers
Department of Chemistry, University of Utah, Salt Lake City, Utah 84112

Sedimentation field-flow fractionation (SFFF), one member of the field-flow fractionation (FFF) class of techniques, is described and its potential advantages and disadvantages for macromolecular and particle separations are discussed. Theoretical equations for retention and plate height are given along with an analysis of possible disturbances caused by relaxation effects, polydispersity, secondary flow, and sample size effects. The conclusions are tested experimentally using polystyrene latex particles having diameters from 907 Å to 4808 Å. Measured retention parameters are in excellent agreement with theory but a considerable discrepancy exists in plate height results. This is traced to a possible sample overloading effect. Particle fractionation is illustrated and the further potential of the method is discussed.

Sedimentation field-flow fractionation (SFFF) is a name describing a group of separation methods constituting a subclass of general field-flow fractionation (FFF) (*1–6*). In SFFF, sedimentation, induced by gravity or by a centrifuge, is employed to layer molecules or particles over one wall of a flow tube (see Figure 1). These layers are of different thicknesses for particles of different size and density. Solvent flow in the tube—perpendicular in direction to the sedimentation velocity vector—carries each constituent down the channel, but it carries fastest those particles forming the thickest layers and which therefore extend furthest toward the tube center where flow velocity is greatest (Figure 1). One thus gets a differential migration according to layer thickness. As layer thickness is a parameter in the migration rate of var-

ious components, we will deal with its characteristics in the theory section.

In theory, SFFF has several advantages over conventional sedimentation techniques for analytical separations. First of all, SFFF, by its very nature, is an elution technique, with corresponding advantages in sample detection and collection (*1, 6*). Second, it has been shown that for a given field, SFFF can generate separability (as measured by the number of theoretical plates) per unit time or per unit length only two to three lower than the maximum values obtainable from the direct application of an equal field using normal centrifugal methods (*6*). This disadvantage is compensated in part by the fact that one can use, throughout the SFFF run, the more intense fields at the very outer edge of the rotating system. More importantly, one can extend the path length of the separation by coiling the SFFF tube any desired number of times around the outer perimeter of the rotation chamber (*1, 6*). Because of this, the number of plates achievable is theoretically

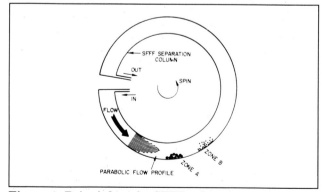

Figure 1. Principles of a SFFF column operating in a centrifuge

Species of zone A have a larger effective mass and are therefore pushed closer to the wall than those of zone B. Because of this, zone A is retained more than zone B and separation occurs

[1]Present address: Department of Chemistry, Oregon State University, Corvallis, Ore., 97331.

unlimited in SFFF, but has a definite ceiling in normal centrifugation. [Our comparison is valid for both kinetic and equilibrium forms of sedimentation, which have been shown to produce roughly equal levels of overall resolution (7).]

The third advantage of SFFF is that no special density gradients or other arrangements need be provided for convective stabilization. The small scale and cross-sectional geometry of the flow channel obviate this problem. (However one must take precautions to avoid secondary flow, as will be explained.)

One obvious disadvantage of SFFF, compared to direct-field sedimentation, is that it is inherently limited to small samples. Scale-up would be difficult and would involve other sacrifices, such as in separation speed.

Another disadvantage, at this point in time, is that the special long-tube capability of SFFF requires the use of a low-volume seal to get samples out of the rotor system. As constructed in our laboratory, this seal was not able to handle rotation speeds greater than 4000 rpm, which provided about 1400g. Longer seal life was assured by not exceeding 500g. Higher speeds would extend the range of applicability to much smaller particles, and would improve both resolution and separation speed. For this reason, the present work is in no sense a description of an ultimate system. It must be regarded as a prototype system, used to demonstrate retention, separation, the role of various parameters, and the applicability of theory.

While SFFF was first envisioned as a special case of FFF (1), it was developed independently as a method in particle separations by Berg, Purcell, and Stewart (8–10). These authors used both gravitational and centrifugal fields in early work on the subject. They employed an ingenious centrifugal device that bypassed the need for complicated rotor seals (10). Flow occurred in a direction parallel to the axis of rotation, thus precluding channels of extraordinary length. With this system, they demonstrated the retention of the bacteriophage, R17, of molecular weight 4×10^6. Work with particles as small as 10^5 mol wt was regarded as possible at $10^5 g$.

THEORY

The flux density of dilute solute in the field direction (normal to flow) within an SFFF tube is (2)

$$J_x = -D\frac{dc}{dx} + Uc \tag{1}$$

By convention in FFF, coordinate x is the distance into the channel measured from the wall at which the solute accumulates. Quantity D is the diffusion coefficient, c is the concentration and U is the field induced drift velocity. For a sedimentation field induced by centrifugation, the latter is (11)

$$U = \frac{F}{f} = -\frac{m(1-\rho/\rho_s)\omega^2 r_0}{f} \tag{2}$$

in which F is the net force acting on the solute particle, f is its friction coefficient, m its mass and ρ_s its density. In some instances it is necessary to replace ρ_s by $1/\bar{v}_s$ where $\bar{v}_s$ is the partial specific volume of the solute. The centrifugal accelera-

tion is $\omega^2 r_0$ and the solvent density is ρ). The minus sign is applied because the sedimentation force is directed *out* from the center of rotation, whereas coordinate x measured the distance *in* from the outer channel wall. (The same sign can be shown to be applicable when inverted densities bring the solute to the inner wall.)

Quality U is virtually constant in any given segment of a SFFF column. First of all, the channel width w is ordinarily small by comparison to the rotational radius r_0. The SFFF columns reported here, for instance, have a width of 0.635 mm, less than 1% of the radius $r_0 = 8$ cm. Values of w significantly larger are not expected because column resolution deteriorates rapidly with increasing w. Another potential influence on U is the variability of ρ, ρ_s, and f across the column width; this variability is also reduced to negligible proportions because of the small magnitude of w. The same argument shows that diffusion coefficient D is essentially constant across the tube width.

The steady-state distribution of solute in a segment of the channel is obtained by setting the net solute flux, J_x, equal to zero. This and the substitution of Equation 2 for U yields the differential equation

$$\frac{d\ln c}{dx} = -\frac{m(1-\rho/\rho_s)\omega^2 r}{fD} \tag{3}$$

where, in view of the discussion above, the right hand side is constant. Intergration provides an equation for the concentration profile over the column width (from $x = 0$ to $x = w$)

$$\frac{c}{c_0} = \exp\left(-\frac{m(1-\rho/\rho_s)\omega^2 r}{fD}x\right) = \exp\left(-\frac{x}{l}\right) \tag{4}$$

where c_0 is the concentration at $x = 0$. This exponential layer of solute has a thickness characterized by the length parameter l, a quantity fully defined by the relationships of Equation 4. From Equation 4

$$l = \frac{fD}{m(1-\rho/\rho_s)\omega^2 r} \tag{5}$$

When $l \ll w$ (column width), l becomes the mean thickness of the solute layer.

An alternate form of Equation 5 is useful. First, in view of the Einstein relationship, $D = kT/f$, quantity fD can be replaced by kT. Also $\omega^2 r$ can be replaced by the general acceleration term G, applicable to laboratory centrifugal or gravitational forces. Finally, we replace $(1-\rho/\rho_s)$ by $\Delta\rho/\rho_s$, where $\Delta\rho = \rho_s - \rho$. These three changes give

$$l = \frac{kT}{mG\Delta\rho/\rho_s} \tag{6}$$

Retention. If we divide Equation 6 by w, we get the dimensionless ratio, $\lambda = l/w$, which is the basic retention parameter of FFF (3)

$$\lambda = \frac{kT}{mGw\Delta\rho/\rho_s} \tag{7}$$

For spherical particles mass, m can be replaced by $\rho_s \pi d^3/6$, where d is the particle diameter. In this case

$$l = \frac{6kT}{\pi d^3 G w \Delta \rho} \tag{8}$$

If parameter λ is known or can be calculated as above, the retention ratio, R = (solute velocity)/(mean carrier velocity), can be predicted using the Equation (3).

$$R = 6\lambda \left[\coth (1/2\lambda) - 2\lambda \right] \tag{9}$$

In the limit of high retention (small R and small λ), R becomes

$$R = 6\lambda \quad \lambda \to 0 \tag{10}$$

This is a useful equation because it describes retention in the region of the greatest practical value (6).

Plate Height. Plate height in SFFF is expected to follow the same equation as in other FFF methods, expressed generally as (3).

$$H = \frac{2D}{R\langle v \rangle} + \chi \frac{w^2 \langle v \rangle}{D} + \Sigma H_i \tag{11}$$

where $\langle v \rangle$ is the mean flow velocity of the solvent. This equation is similar to that describing chromatography, the successive terms accounting for longitudinal diffusion, nonequilibrium, and extraneous factors such as dead volume and relaxation effects. The nonequilibrium coefficient, χ, has the following limits corresponding to no retention and to high retention, respectively (3).

$$\chi = 1/105 \quad \lambda = \infty \tag{12}$$

$$\chi = 24\lambda^3 \quad \lambda \to 0 \tag{13}$$

In practical work, the first term of Equation 11 is negligible because of the sluggish diffusion of the large solute species. Ideally, the third term is also negligible; making it so is one of the prime objectives in developing any of the FFF techniques. This leaves only the nonequilibrium term to determine the ultimate plate height, resolution, and separation speed of SFFF. The limits imposed by this term have been described generally for FFF, and will not be repeated here (6). Our objective, in part, is to test the experimental achievability of plate height described solely by the nonequilibrium term

$$H = \chi w^2 \langle v \rangle / D \tag{14}$$

Relaxation Time. When first injected into a SFFF column, solute particles will be distributed rather uniformly over the cross-section at the head of the column. A finite time is required for the exponential layer to form under the influence of the field. Until this equilibrium layer is formed, any downstream migration will be accompanied by a decreased retention, an increased retention parameter, and an increased plate height. This distortion can be removed by a stop-flow method in which flow is halted after injection for a period of time adequate to establish the exponential distribution.

The relaxation time associated with layer formation is defined generally for FFF (3) as the time required by the field to carry solute particles from the column center, $x = w/2$, to the center of gravity of the equilibrium layer, x_{cg}. Thus, $t_r = [(w/2) - x_{cg}]/U$ where U is the particle velocity. With the aid of Equation 2 and Einstein's relationship, $D = kT/f$, this becomes

$$t_r = \frac{[(w/2) - \chi_{cg}] k T}{m(1 - \rho_0/\rho)\omega^2 r_0 D} \tag{15}$$

An equivalent expression makes this calculable in terms of the FFF retention parameter, λ (3)

$$t_r = \frac{w^2 \lambda}{D} \left[\frac{1}{2} - \lambda + (e^{1/\lambda} - 1)^{-1} \right] \tag{16}$$

which shows the desirability of working with a small column width w. For small particles, where D is relatively large, relaxation is, of course, relatively slow.

Two relaxation periods should be allowed to elapse if stop-flow injection is used. This is illustrated elsewhere (3). However, a simpler expression for an adequate waiting period is clearly

$$t_{\text{stop}} = w/U \tag{17}$$

which is the time needed to traverse the width of the column. Equation 2 can be utilized to get U. It can be shown that $t_{\text{stop}} \geq 2t_r$.

Polydispersity. Unless all the particles in a sample are of identical size, the elution peak for that sample will be broadened by the unequal migration rates of particles at different size. The experimental plate height will increase by an incremental value, H_p, reflecting this broadening. We shall derive an equation for the H_p value of a nonhomogeneous sample of spherical particles using a method analogous to that employed for polymer fractionation by thermal FFF (3).

The variance of a peak in a SFFF column due to polydispersity is

$$\sigma^2 = \left(\frac{dZ}{dd} \right)^2 \sigma_d^2 \tag{18}$$

where Z is the migration distance along the column and σ_d^2 is the variance in particle diameter. The term dZ/dd is best obtained through the series of known conversions: Z to R, R to λ, and λ to d. Thus

$$\sigma^2 = \left(\frac{dZ}{dR} \frac{dR}{d\lambda} \frac{d\lambda}{dd} \right)^2 \sigma_d^2 \tag{19}$$

Since $Z = Rt$, $dZ/dR = t = Z/R$. Here t represents the duration of the run. The term $dR/d\lambda$ is obtainable from Equation 9, although its form is moderately complex. In the limit $\lambda \to 0$, it reduces simply to 6. The term $d\lambda/dd$ can be derived from Equation 8 as $-3\lambda/d$. When we substitute these quantities back into Equation 19, we obtain

$$\sigma^2 = 9Z^2 \left(\frac{d \ln R}{d \ln \lambda} \right)^2 \frac{\sigma_d^2}{d^2} \tag{20}$$

The incremental plate height is $H_p = \sigma^2/Z$, or

$$H_p = 9Z \left(\frac{d \ln R}{d \ln \lambda} \right)^2 \left(\frac{\sigma_d}{d} \right)^2 \tag{21}$$

This quantity, unlike most plate height terms, depends on migration distance Z. At elution, Z is equal to the column length L, and we have

$$H_{\text{p}} = 9L \left(\frac{\text{d} \ln R}{\text{d} \ln \lambda}\right)^2 \left(\frac{\sigma_d}{d}\right)^2 \tag{22}$$

The term $\text{d} \ln R/\text{d} \ln \lambda$ is to be obtained from Equation 9. In the limit of high retention, $\lambda \to 0$, R becomes 6λ and $\text{d} \ln R/\text{d} \ln \lambda$ equals unity. Thus

$$H_{\text{p}} = 9L(\sigma_d/d)^2 \tag{23}$$
$$\lambda \to 0$$

The dimensionless ratio, σ_d/d, is available from the manufacturer for the polystyrene particles employed in this study. In other systems, these equations could be used to determine the degree of heterogeneity from plate height measurements. The principal requirement is that H_{p} be large enough so that it is not obscured by other plate height effects. On this basis, the method would be fairly sensitive. Values of σ_d/d from 0.1 to 0.01 should be measurable on columns with 10 and 1000 intrinsic theoretical plates, respectively. One could take this a step further and derive the detailed size or molecular weight distribution curve. The study of heterogeneity is an important part of ultracentrifugation (12). For many classes of particles and macromolecules, SFFF should be able to accomplish this task with greater resolution and accuracy because of its greater intrinsic resolving power.

Column Geometry and Secondary Flow. Earlier efforts in this laboratory to achieve SFFF retention failed, apparently because the circular cross-section of the columns then employed encouraged secondary flow effects. Secondary flow is a component of flow normal to the column axis of a curved tube induced by the increased centrifugal force experienced by fluid moving in the fast streamlines (13–15). A continuous recirculation of fluid occurs; if the flow is strong enough, it will recirculate the solute, thereby causing a loss of retention and an increase in peak spreading. A rectangular cross-section with a large breadth to width ratio, by contrast, will inhibit secondary flow because of the large frictional drag in the narrow space between long wall surfaces.

Concentration Effects. If the concentration of solute in the zone becomes excessive, significant interactions will occur between individual solute particles and the ideal retention and plate height expressions derived earlier will fail.

Concentration effects may be envisioned as stemming from the compression of the sample into a volume too small to accommodate it. The volume available for sample is defined roughly by the physical space occupied by the zone. This space is most restrictive for the narrow zones of a high-efficiency column. In general, it becomes more restrictive as retention increases because the "layer thickness," l, falls off with increasing retention. The magnitude of the concentration under these circumstances is evaluated as follows.

The concentration of solute under ideal circumstances has a Gaussian dependence along flow coordinate z and an exponential dependence (Equation 4) on field coordinate x. The concentration does not vary along breadth coordinate y. If the variance, σ^2, of the Gaussian is replaced by Z^2/N, where Z is the distance migrated and N is the number of plates generated in migrating to point Z, then the solute concentration can be expressed as

$$c = c_{00} \exp(-z^2 N/2Z^2) \exp(-x/l) \tag{24}$$

where c_{00} is the highest concentration in the zone, found at $x = 0$ and at the center of the Gaussian, $z = 0$. Here, coordinate z is measured from the Gaussian maximum.

The total sample content in the zone is simply

$$m' = \iiint c\, \text{d}x\text{d}y\text{d}z \tag{25}$$

When Equation 24 is substituted into this and the integrals are evaluated (keeping in mind that c is independent of y), we get

$$m' = c_{00}(2\pi/N)^{1/2}\, a\, Z\, l\, [1 - \exp(-1/\lambda)] \tag{26}$$

where a is the column breadth and λ is l/w, as before. This equation can be rearranged to obtain maximum concentration, c_{00}, as a function of injected sample size, m', and the other parameters of Equation 26.

$$c_{00} = \frac{m'}{(2\pi/N)^{1/2} a\, Z\, l\, [1 - \exp(-1/\lambda)]} \tag{27}$$

Equation 27 expresses the maximum value of concentration at migration distance Z, and therefore provides the raw data necessary to determine if solute interaction is of major proportions. Whether, in fact, interactions are significant at these concentration levels must be judged by other criteria related to intermolecular or interparticle forces and statistical distributions.

One can imagine a situation in which (nonlinear) interactions force the zone to spread apart, at which time linearity is essentially restored. Hence, there is a valid argument for substituting the theoretical value of N into Equation 27 instead of the experimental value, since the latter may reflect a situation in which interactions have already done their major damage. Thus if we write $N = Z/H$ and use the equation $H = \psi l^2 v/D$, which is equivalent (2) to Equation 14, we obtain

$$c_{00} = \frac{m'}{(2\pi\psi v/D)^{1/2} a\, Z^{1/2} l^2\, [1 - \exp(-1/\lambda)]} \tag{28}$$

where v is the migration velocity of the zone and ψ is a numerical coefficient determined from nonequilibrium theory (2). In the limit of high retention, $\lambda \to 0$, coefficient ψ approaches 4, thus giving

$$c_{00} = \frac{m'}{(8\pi v/D)^{1/2} a\, Z^{1/2} l^2} \tag{29}$$

Since concentration varies throughout the zone and varies with time at a given point in the zone, it is useful to select a reference concentration to evaluate the probable effect of interactions. A reasonable choice is $c_{00}{}^L$, the value of c_{00} at the end of the column, $Z = L$. This is near the high end of the concentration scale but does not represent the extreme value for the entire experiment because higher c_{00} values occur before the zone has migrated to the end of the column and increased its dilution. Thus c_{00} at $Z = L$ should characterize in some average way the high concentration conditions in the column.

When $Z = L$, the last three equations become, respectively

$$c_{00}{}^L = \frac{m'/V_0}{(2\pi/N)^{1/2}\lambda\,[1 - \exp(-1/\lambda)]} \tag{30}$$

$$c_{00}{}^L = \frac{m'/V_0}{(2\pi\psi v/LD)^{1/2}w\lambda^2\,[1 - \exp(-1/\lambda)]} \tag{31}$$

$$c_{00}{}^L = \frac{m'/V_0}{(8\pi v/LD)^{1/2}w\lambda^2} \tag{32}$$

In these expressions, V_0 is the column void volume. Thus m'/V_0 would be the concentration if the sample were diluted over the entire column volume; the denominator is the correction factor yielding the degree concentration above the m'/V_0 level.

EXPERIMENTAL

The basic apparatus used in this study was an International Equipment Co. refrigerated Model B-20 centrifuge. This was fitted with a special rotary seal designed and built in this laboratory, providing a connection to the stainless steel channel of a FFF column held on the inside wall of an 8-cm radius centrifuge basket (I.E.C. No. 1470). The rotary seal allowed continuous flow to and from the channel without mixing of the inlet and exit streams while the centrifuge was spinning or at rest. The seal consisted of a hard chrome-plated stationary shaft with internal concentric flow channels, and a moving body of stainless steel with the flow streams isolated from each other and the stabilizing bearings by means of double "O" ring seals on graphite-impregnated Teflon spacers. The internal volume in the exit stream portion of the seal (the center flow channel) was kept to a minimum (under 0.1 cm³) to reduce peak spreading. An injection port was placed between the seal and the column inlet enabling sample injections by microsyringe when the centrifuge was stopped.

The column was fabricated from two pieces of 0.0762-cm thick 304 stainless steel sheet with 0.0635-cm stainless steel spacer from which the desired channel shape had been cut. The three pieces were welded together on the outside edges while bent just enough to fit inside the centrifuge basket. The channel thus formed was 45.7 cm long (squared length), 2.54 cm in breadth and 0.0635 cm in width, with the ends tapered to streamline entrance and exit flow. Inlet and outlet tubing were also welded into the system for connection to the rotary seal.

Control of the rotation rate in the centrifuge required a special control unit (designed and built in this laboratory) because the low rotation speeds (500–3000 rpm) were not well regulated by the normal centrifuge control system. Rotation rates were determined with the aid of a digital interval counter (Computer Measurement Co., Model 1189C) which measured the interval between successive passes of a magnet attached to the basket and a fixed coil.

The carrier fluid was degassed at 70 °C and pumped through the system by a Chromatronix Model CMP-1V me-

tering pump. Precautions were taken to avoid bacterial contamination. A Laboratory Data Control Model 1205 UV Monitor, set at 254 mμ was used for detection. Readout was achieved on a Varian Aerograph G2000 recorder.

The monodisperse spherical polystyrene latex beads (Dow Chemical Co.) used as sample materials had particle diameters of 907, 1087, 1756, 2339, 3117, 3570, and 4808 Å. The $\overline{M}w/\overline{M}_n$ ratio varied from 1.0004 to 1.0333. The density was reported as 1.051 ± 0.001 g/cm³ at 25 °C (10). A density value of 1.05 g/ml was used in all calculations in this paper. The beads were prepared as 10% by weight aqueous suspensions containing an emulsifier. These were diluted to 2% before use.

Ribonuclease free sucrose (N.B.C. Research Biochemical Co.) and reagent grade n-propanol (Matheson, Coleman and Bell) were used to vary the density of the carrier solution. Double distilled water containing 0.1% of a low foaming, low alkalinity detergent (FL-70, Fisher Scientific Co.) was used as the carrier for most of the studies.

The centrifuge temperature was set at 2 °C. Sample introduction occurred before the initiation of either flow or rotation. A microsyringe was used to inject sample into the injection port under these static conditions. The syringe needle was then withdrawn slowly and the carrier pump and the centrifuge were turned on at the desired settings. The carrier pump was turned off (stop-flow) after 0.5 ml of carrier had been supplied to the channel to allow relaxation to the steady state within the column itself. After the interval sufficient for relaxation to occur, flow was resumed and the SFFF process was thereby initiated.

Retention ratio R was measured as the volume of a nonretained peak divided by the elution volume of the sample peak. Both volumes were corrected for dead volume, 0.2 ml, the latter being measured as the retention volume in a flow system by-passing the column. Plate height was measured in terms of the width at half height and calculated in the standard manner.

RESULTS AND DISCUSSION

The utility of columns with a rectangular cross-section having a large breadth/width ratio is best illustrated by comparing the significant retention in such columns (documented below) with the lack of retention in columns of round and square cross-sections. A column of length 200 cm and a circular cross-section with a diameter of 0.1 cm provided no retention for beads from 907 to 4808 Å in diameter using field strengths up to 1432g. A 500-cm long tube with a square cross-section of side 0.1 cm likewise failed to provide retention under similar circumstances. This tube did provide minor retention—several orders of magnitude below the predicted value—with yeast particles from 1 to 5 μm in diameter and 1.13 g/ml in density. However both tubes failed to provide retention anywhere near the expected magnitude, a result that is consistent with the existence of major secondary flow disturbances in such columns.

Retention and plate height parameters for the column of rectangular cross-section were measured for a great variety of conditions to test the equations developed in the theory section.

Retention. The retention ratio, R, was measured as the void volume of the column divided by the retention volume of the peak. The column void volume was calculated from the column dimensions to be 7.38 ml. Acetone and phenol both exhibited a retention volume of 7.4 ml with rotation; subtraction of the end volume (0.2 ml) suggests a column volume of 7.2 ml, in good agreement with 7.38 ml. The latter was rounded to 7.4 ml and used for the calculation of R values. All measured retention volumes were corrected by subtracting out the dead volume of the ancillary flow path, 0.2 ml.

Retention increases with increasing spin velocity and field strength, as expected. Figure 2 shows the trend of decreasing retention parameter, R (increasing retention), with increasing field strength, G, for different particle diameters. The trend of increasing retention with increasing particle size is also apparent.

To test the theoretical validity of these results, we must extract the dimensionless retention parameter λ from each R value using Equation 9. Equations 7 and 8 predict that a plot of λ vs. $1/G$ will produce a straight line through the origin. The slope can be calculated in terms of size and density parameters. Figure 3 shows the experimental data plotted in this fashion, and the corresponding theoretical prediction. The agreement of theory and experiment is seen to be excellent.

Equation 8 predicts that λ will vary as the inverse cube of particle diameter, d. This is verified in Figure 4, in which a plot of λ vs. $1/d^3$ at a constant rotational velocity yields a straight line passing through the origin. The straight line of Figure 4 has been obtained from Equation 8, and once again shows excellent agreement with experimental values.

Retention is affected by variations in solvent density as well as by changes in spin velocity and particle size. Equations 7 and 8 show that λ should be inversely proportional to the density difference, $\Delta\rho$, between sample and solvent. This conclusion was tested using various aqueous solutions. The experimental data and theoretical lines are compared in Figure 5. The agreement is, again, most satisfactory.

Plate Height. Plate height values at different flow velocities were measured for two different sizes of polystyrene

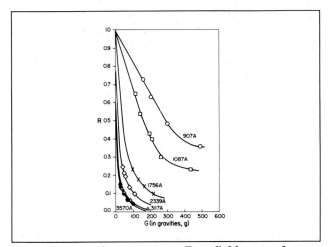

Figure 2. Retention parameter R vs. field strength measured in gravities for various polystyrene beads. Bead diameters are indicated in angstroms

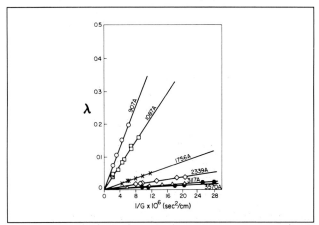

Figure 3. Variation of retention parameter, λ, with reciprocal field strength, $1/G$ for various particle diameters. The straight lines are obtained theoretically

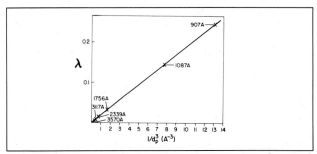

Figure 4. Plot of λ vs. $1/$(particle diameter)^3 at $G = 127.6g$ and 6 ml/hr flow rate. The straight line is predicted from theory

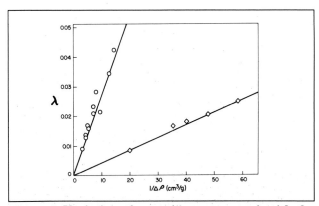

Figure 5. Variation of retention parameter λ with the solute–solvent density difference $\Delta\rho$, in aqueous solutions of n-propanol (top) and sucrose (bottom)

beads. Samples of 2339-Å beads were rotated at 1080 rpm ($G = 104g$), producing a retention of $R = 0.117$ and $\lambda = 0.0203$. Beads of 3117-Å diameter were run at 715 rpm ($46g$) and were retained at the level $R = 0.098$ and $\lambda = 0.0168$. Plate height curves for the two are shown in Figure 6. These tend to form a straight line cutting the H axis well above the origin.

Also shown in Figure 6 are the respective theoretical plots for these two samples. These were obtained from Equation 14, with χ calculated from exact nonequilibrium expressions by Y. H. Yoon of this laboratory. The error introduced by ignoring the longitudinal diffusion term is negligible over the

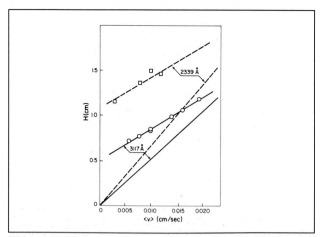

Figure 6. Plate height *vs.* mean flow velocity, ⟨v⟩, for polystyrene beads of diameter 2339 Å (top) and 3117 Å (bottom). The solid lines are obtained theoretically

present experimental range. The diffusion coefficients were obtained from Stoke's law for friction coefficient, f, and Einstein's diffusion equation, $D = k T/f$.

The deviations between the experimental and the theoretical plate height curves are considerable. It is, therefore, necessary to look carefully at various perturbations to see what role, if any, they play in this departure.

Relaxation Effects. The precautions described above (stop-flow injection) should eliminate any significant relaxation effects stemming from the injection process. Any other disturbance in geometry or flow is potentially capable of causing similar disturbances. However, it is difficult to imagine how disturbances major enough to create the observed plate height discrepancies could occur. The general agreement of theory and experiment insofar as retention is concerned support this conclusion. Furthermore, relaxation effects could cause the greatest plate height distortion at high flow velocities, a trend opposite to that observed.

End Effects. End effects, including dead volume, frequently augment chromatographic plate height. Along with a detector and inlet and outlet tubing similar to those in chromatographic systems, the special rotor seal has its own set of unknown characteristics. The entire dead-volume train was evaluated by substituting a short column (1.8 cm) into the system, measuring the plate height, and isolating its contribution by the subtraction method of Giddings and Seager (*16*). Results for the 3117-Å beads are shown in Table I. It is apparent from this table that end effects constitute only a minor part of the observed plate height.

Polydispersity. The contribution of sample heterogeneity to plate height can be estimated by the method outlined in the theory section. For the cases at hand, retention is high and λ is small, making Equation 23 applicable to good approximation. Quite generally in the high retention range, as shown by Equation 23, plate height is independent of retention temperature, and all other parameters of the SFFF system except column length. We can therefore calculate, for sample beads with a given dispersion in size, a heterogeneity contribution per unit length, H_p/L, that will be applicable to any SFFF sys-

tem working at high retention. In Table II, we show these values for the polystyrene beads used in this study. We also show the contribution, H_p, for our particular column. Particle size dispersion values are those reported by the manufacturer, Dow Chemical Co.

Table II shows that the heterogeneity effect is negligible in the present study for all but the smallest beads. However, the effect would be quite generally noticeable in a high efficiency system in that a number of the H_p contributions are in the vicinity of 1 mm—rather large in terms of ultimate plate-height potential. Heterogeneity of the magnitude shown here could obviously be measured as an independent parameter in a high efficiency SFFF system.

Concentration Effects. Nonlinear concentration effects might also be responsible for extra peak spreading. To investigate this possibility, sample content was varied by utilizing a range of concentrations of 3117-Å beads in a 5-μl injection. Figure 7 shows the resultant effect on both retention parameter λ and plate height H. Clearly there is a major concentration effect whose onset in this particular experiment begins for samples containing well under 0.1 mg of polystyrene. While the retention parameters reported earlier are not much disturbed, the influence on plate height is very substantial and appears to explain the plate height discrepancies noted earlier.

The foregoing conclusion is supported in a qualitative way by the trend in the discrepancies between theory and experiment shown in Figure 6. The discrepancies tend to disappear

Table I. Contribution of End Effects to the Plate Height of 3117-Å Beads

⟨v⟩, cm/sec	H obsd, cm	H corr, cm	% of H contributed by end effects
0.0145	0.98	0.95	3.3
0.0103	0.85	0.81	3.4
0.0062	0.69	0.68	2.4

Table II. The Polydispersity Contribution to Plate Height, H_p, and to Plate Height per Unit of Column Length, H_p/L, under Conditions of High Retention According to Equation 23, $H_p = 9 L (\sigma_d/d)^2$

Particle dimensions, $d \pm \sigma_d$ (in Å)	H_p/L	H_p (present column, $L = 45.7$ cm) (in cm)
907 ± 57	0.036	1.6
1087 ± 27	0.0056	0.25
1756 ± 23	0.0015	0.070
2339 ± 26	0.0011	0.051
3117 ± 22	0.0004	0.020
3570 ± 56	0.0022	0.101
4808 ± 18	0.0001	0.006

at high flow velocities. At high flow velocities, of course, the "theoretical" peak is relatively broad—wide enough to contain the sample without undue concentration and distortion.

A quantitative view of this phenomenon can be gained by using Equation 31 to calculate c_{00}^L, the maximum concentration within the zone predicted by theory to exist at the column exit. This quantity is plotted against mean solvent flow velocity, (v), in Figure 8. Along with it is shown a curve of the departures of plate height from their theoretical values. Certainly the trend in the two curves is similar, suggesting some connection. We note that the calculated c_{00}^L values are below the maximum possible concentration of about 630 mg/ml by a factor of about 500. (The value 630 mg/ml comes from the assumption that the beads are packed as closely as possible in a random structure, thereby filling about 60% of the local volume element with material of density 1050 mg/ml.) Thus, the physical crowding is not excessive, but forces between particles might be sufficient to disturb their normal equilibrium distribution.

The following processes would be expected to occur if significant repulsion existed between the particles. Following injection, the particles sediment toward the wall but fail to reach the normal exponential distribution because of repulsion. The particles which cannot adequately sediment are held at abnormally high altitudes and are thus swept downstream rapidly, once flow commences, until adequate dilution occurs

beneath for normal sedimentation. The high velocity contrasts with the low velocity of particles next to the wall, and is responsible for widening the zone along the flow axis. This excessive spreading is accompanied by a slight displacement forward in the zone's center of gravity, a slightly earlier elution, and thus a slightly higher value for measured R and λ values. While this process would occur with the greatest intensity near the column entrance, it would continue to occur along the path of migration up to the point that dilution is able to restore the normal distribution. The effects of the disturbance would, of course, increase with increase in sample size.

These hypothetical processes would explain why the effect on plate height is much greater than that on retention, why the consequences are greatest at low flow velocities, and why both effects are magnified by further increases in sample size.

The results obtained here show that nonlinear disturbances can occur with very small samples (under 0.1 mg). It is probable that the effect is most marked for highly retained peaks and high efficiency columns. However, more work is needed to clarify these factors. The availability of sensitive detectors for this work is obviously desirable.

Particle Fractionation. The differential retention of particles of different size, as exemplified in Figure 2, makes particle fractionation feasible. This is illustrated in Figure 9, which shows the elution pattern and separation of polystyrene beads of different diameters. This figure also demonstrates the ease of controlling retention and separation by variations in field strength. A change from 1140 rpm in Figure 9a to 1520 rpm in Figure 9b increases the sedimentation field strength by a factor of 1.78 and thus increases considerably the retention volume of the peaks. This, in turn, increases the resolution volume of the peaks. This result confirms the potential utility of a programmed field SFFF system; such a system will be reported in a subsequent paper.

POTENTIAL OF SFFF

It was mentioned earlier that the present system is a prototype system with characteristics far removed from those expected in an optimal system. Most importantly, the rotational velocity and the field strength must be increased to gain any significant advances. The degree to which performance might be enhanced is shown below.

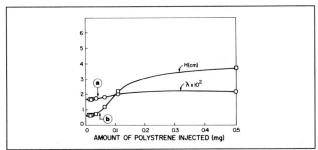

Figure 7. Plate height and retention *vs.* mg of injected polystyrene

Bead diameter = 3117Å; field strength = 45.8g; flow rate = 6 ml/hr. Points *a* and *b* are the concentrations used for the retention and plate height studies, respectively

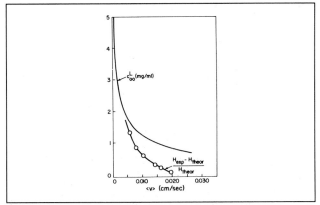

Figure 8. Maximum concentration in the exiting peak and the fractional departure of plate height from theory plotted as a function of solvent flow velocity

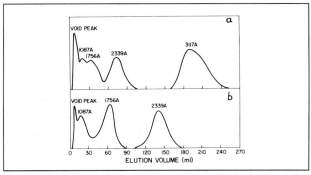

Figure 9. Separation of polystyrene fractions at (*a*) 1140 rpm and (*b*) 1520 rpm

Under conditions of high retention, Equation 13 can be substituted into Equation 14 to yield the potential plate height

$$H = 24\lambda^3 w^2 \langle v \rangle / D \qquad (33)$$

With the aid of Equation 10 this expression becomes

$$H = 4\lambda^2 w^2 v / D \qquad (34)$$

where the zone velocity in the column, $R\langle v \rangle$ has been written simply as v. If now Equation 8 is used to eliminate λ^2, we have

$$H = \frac{144(kT)^2 v}{\pi^2 d^6 G^2 (\Delta\rho)^2 D} \qquad (35)$$

If diffusion coefficient D is replaced by the Stokes-Einstein value, $D = k\,T/3\pi\eta d$, where η is the coefficient of viscosity, we get

$$H = \frac{432}{\pi} \frac{kT\eta v}{d^5 G^2 (\Delta\rho)^2} \qquad (36)$$

If H is written as Cv, then the nonequilibrium coefficient C not only reflects the level of H achievable, but, more importantly, C can be shown to equal the minimum time in which a theoretical plate can be generated. We have

$$C = \frac{432}{\pi} \frac{kT\eta v}{d^5 G^2 (\Delta\rho)^2} \qquad (37)$$

This equation shows that the efficacy of SFFF can be improved by manipulating a number of variables, including viscosity, temperature, and density. The method will, in theory, work much better with large particles than small, as reflected in the fifth power dependence on particle diameter d. However there are practical limits to this gain which will become apparent as d approaches either layer thickness l or surface-roughness dimensions in magnitude.

Equation 37 shows that C for a given particle is inversely proportional to the square of the sedimentation field strength, G. The dependence on rotational velocity is therefore inverse fourth power. In the present study the maximum G was about 500g. If ultracentrifuges, with field strengths up to 300 times greater than this, were adapted to SFFF, C values could in theory be reduced by $(300)^2 = 90,000$. While such gains would not be totally applicable to particles in the size range used here because of the previously mentioned restrictions on size, the above factor would be applicable to much smaller particles and to macromolecules, thus making their separation convenient also.

Particle size analysis is significant in many fields of environmental control and industrial operation. The present method is promising in such analyses by virtue of its predictable dependence on simple mass and density parameters and its potential for further improvements in fractionating power.

REFERENCES

(1) J. C. Giddings, *Separ. Sci.*, **1**, 123 (1966).
(2) J. C. Giddings, *J. Chem. Phys.*, **49**, 1 (1968).
(3) M. E. Hovingh, G. H. Thompson, and J. C. Giddings, *Anal. Chem.*, **42**, 195 (1970).
(4) E. Grushka, K. D. Caldwell, M. N. Myers, and J. C. Giddings, *Separ. Purification Methods*, **2**, 129 (1973).
(5) J. C. Giddings, *J. Chem. Educ.*, **50**, 667 (1973).
(6) J. C. Giddings and K. Dahlgren, *Separ. Sci.*, **6**, 345 (1971).
(7) J. C. Giddings, *Separ. Sci.*, **8**, 567 (1973).
(8) H. C. Berg and E. M. Purcell, *Proc. Nat. Acad. Sci. USA*, **58**, 862 (1967).
(9) H. C. Berg, E. M. Purcell, and W. W. Stewart, *Proc. Nat. Acad. Sci. USA*, **58**, 1286 (1967).
(10) H. C. Berg and E. M. Purcell, *Proc. Nat. Acad. Sci. USA*, **58**, 1821 (1967).
(11) T. Svedberg, "The Ultracentrifuge," T. Svedberg and K. O. Pederson, Eds., Clarendon Press, Oxford, 1940.
(12) J. W. Williams, "Ultracentrifugation of Macromolecules," Academic Press, New York, N.Y., 1972.
(13) W. R. Dean, *Phil. Mag.*, **V**, 673 (1928).
(14) D. J. McConalogue and R. S. Srivastava, *Proc. Roy. Soc., Ser. A*, **303**, 37 (1968).
(15) G. S. Benton and D. Boyer, *J. Fluid Mech.*, **26**, 69 (1966).
(16) J. C. Giddings and S. L. Seager, *Ind. Eng. Chem., Fundam.*, **1**, 277 (1962).

Received for review March 21, 1974. Accepted July 12, 1974. This investigation was supported by Public Health Service Research Grant GM 10861-17 from the National Institutes of Health.

Reprinted from *Anal. Chem.* **1974**, *46*, 1917–24.

To many analytical chemists, kinetic aspects of chemical processes are viewed as problems to be avoided whenever possible rather than as potentially useful features to be exploited. This article, like many others from the authors' laboratory, demonstrates how kinetic aspects of chemical processes can be used to develop innovative approaches to quantitative determinations that can offer significant advantages relative to more traditional equilibrium-based methods.

Harry L. Pardue
Purdue University

Novel Approach to Reaction-Rate Based Determinations by Use of Transient Redox Effects

V. V. S. Eswara Dutt and Horacio A. Mottola
Department of Chemistry, Oklahoma State University, Stillwater, Okla. 74074

Industrial quality control and pollution studies, inevitably, encounter situations where a large number of samples of similar nature have to be processed for the determination of a single species whose level of concentration is variable. Obviously, in such situations, the use of precalculated concentrations of reagents for each sample increases the operating cost and time. The use of the same reaction mixture for repetitive determinations in aliquots of the same or different samples can obviate this problem with the additional convenience that multiple reagent handling is eliminated.

A novel approach to fast, continuous kinetic-based determinations of a variety of chemical species using a flow-through cell system and transient oxidation–reduction signals is present here. All necessary reagents contained in a single reservoir are continuously circulated, at constant flow, through the cell into which an aliquot of the sample containing the species to be determined is quickly injected. The sought-for species participates in a very rapid redox reaction which dramatically perturbs the concentration of the monitored species. A subsequent but slower reaction either regenerates the monitored species to its original concentration level or brings about a situation in which the original signal level is reinstated. This sequence of signal perturbations results in a recorded peak whose height is proportional to the amount of sought-for species in the sample. Because of the high level of concentrations of some of the chemicals involved in the reservoir solution and the nature of the systems studied up to now, signal monitoring seems to be confined to absorbance of radiant energy, although some other specialized cases may lead to a situation amenable to a similar treatment by use of other monitoring procedures.

In summary, the development of reaction rate (kinetic)

methods based on the characteristics outlined above requires rapid, initial "indicator reaction" resulting in a drastic, almost instantaneous change in signal level, products generated by the subsequent reactions (other than the monitored species) absorbing at wavelength(s) different from those of the monitored species (or the reagent mixture as an entity), and restitution of the initial signal level by a subsequent reaction(s). Fast "indicator reactions" also minimize errors due to changes in flow rates. It should be noted that suitable adjustment of catalyst (if involved) and/or reagent concentrations, as well as other parameters affecting rates, can provide reaction rates of the desired values without sacrificing sensitivity.

Recently we came across a series of reactions which have these characteristics and show promise for future developments along these lines. The oxidation of tris(1,10-phenanthroline)iron(II), ferroin, by chromium(VI) in presence of oxalic acid as a promoting activator (1), is one of them. The oxidation of ferroin is normally slow and rather minute concentrations of oxalic acid markedly accelerate the rate. The use of high concentrations of oxalic acid, on the other hand, considerably increases the rate of oxidation of ferroin by mass action, but after a few moments the ferroin is regenerated (1). This is because a single species, *viz.*, oxalic acid acts first as an accelerator for ferroin oxidation and then is a reducing agent of ferriin when the concentration of this species has built up a level which results in the predominance of the ferroin regeneration step. The reactions involved may be represented as follows:

$$\text{Cr(VI)} + \text{ferroin} \xrightarrow{\text{oxalic acid}} \text{Cr(III)} + \text{ferriin} + \text{oxalic acid}$$

Experiments have been conducted to see whether this reagent regeneration cycle could be used for continuous determination of Cr(VI). A 250-ml aqueous reaction mixture containing $8.0 \times 10^{-5}M$ ferroin and $0.40M$ oxalic acid was constantly circulated through an absorption cell by means of a peristaltic pump. The outlet from the cell was fed continuously back to the reservoir. The reaction mixture was constantly stirred both in the reservoir and in the cell with the help of magnetic and air driven stirrers (2) respectively. Figure 1 shows the schematic of this setup. A typical signal profile is shown in Figure 2; the peak height is proportional to chromium(VI) concentration. As many as 50 samples containing 0.90 ppm of chromium(VI) were manually injected in a time period of about 30 minutes and the chromium(VI) con-

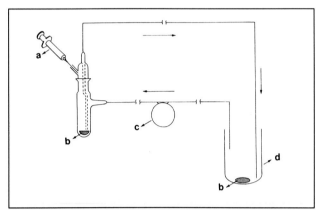

Figure 1. Flow-through cell and flow-through loop for repetitive and continuous injection determinations

(a) Hypodermic syringe and sample injection port, (b) Magnetic stirring bars, (c) Peristaltic pump (Masterflex with SRC Model 7020 speed controller and 7014 pump head), (d) Reagent solution reservoir. (The flow-through cell is 13-mm o.d. and 70 mm high, equipped with $ 12/80 for easy disassembly and cleaning. A flow rate of 15 ml/minute was used in the determinations illustrated in this note)

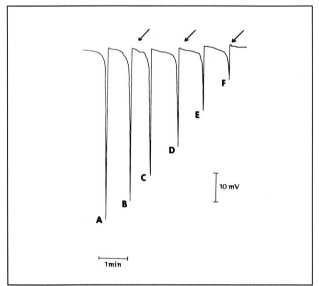

Figure 2. Signal profiles in the determination of chromium(VI)

Experimental conditions are as described in the text. Concentration of chromium(VI) in flow-through cell: (A) 1.38 µg/ml, (B) 1.15 µg/ml, (C) 0.92 µg/ml, (D) 0.69 µg/ml, (E) 0.46 µg/ml, (F) 0.23 µg/ml. Typical moments of sample injection indicated by arrows

centration was evaluated with 4.9₇% standard deviation. Better precision, of course, can be achieved by mechanically controlled sample injection.

In the COD determination of water samples, where the final step necessitates the determination of unreacted chromium(VI), the adaptation of the repetitive method described here can offer convenience. The same indicator reaction could be utilized for the repetitive determination of a variety of reducing species. Consider, for instance, a mixture containing sulfuric acid $(0.25M)$, large excess of chromium(VI), $(1.0 \times 10^{-3}M)$, ferroin $(2.0 \times 10^{-4}M)$ and a low concentration of oxalic acid $(4.0 \times 10^{-3}M$, large enough to promote the oxidation of ferroin but not sufficient to react with ferriin), functioning as the reservoir solution. In this situation, ferriin reacts with any reducing agent like ascorbic acid, uric acid, or hydroquinone to yield ferroin, which subsequently gets oxidized by the excess chromium(VI) present in the system (Figure 3). The transiently formed ferroin concentration would be, then, directly proportional to the amount of reducing agent injected.

Although this reagent-regeneration type of chemical system is not often encountered, these observations seem to open up interesting possibilities for the fast determination of a variety of chemical species. If a reagent does not contribute to absorbance at the wavelength chosen for monitoring, it would, for instance, be introduced in a large excess and the transient signals of even irreversible chemical reactions could be analytically used for species determination. Observations on some reactions belonging to this category are briefly discussed below.

A transient blue chromophore can be seen in the oxidation of sodium diphenylamine sulfonate by vanadium(V) also in the presence of oxalic acid. Injection of cerium(IV) or manganese(VII) into a large concentration of diphenylamine sulfonate produces a similar transient signal. In this case, however, it is not yet clear if the disappearance of color is due to regeneration, destruction, or simply to the formation of a new chemical species. In any event, vanadium(V), cerium(IV), and maganese(VII) can be determined at the microgram level by repetitive injection and signal monitoring at 550 nm. Oxidation of brucine by chromium(VI) is markedly accelerated in presence of oxalic acid and, once again, the colored bruciquinone can be only transiently seen in presence of high concentrations of oxalic acid. This can be used for the continuous determination of either chromium or brucine by signal monitoring at 530 nm. Signals proportional to brucine concentration have been recorded by injecting 2 to 15 ppm of brucine into a reaction mixture $0.25M$ in H_2SO_4, $0.10M$ in oxalic acid, and $5.50 \times 10^{-4}M$ in Cr(VI).

Metal phthalocyanines, an important class of industrial pigments, have been considered inferior redox indicators because of the very unstable colored free radical intermediates produced upon oxidation with cerium(IV) (3). With the set-up described above, continuous quantitative determination of metal phthalocyanines is possible. Injection of 0.5 to 5.0 ppm of cobalt phthalocyanine into $0.001M$ Ce(IV)–$0.1M$ H_2SO_4 solutions gave transient signals proportional to concentra-

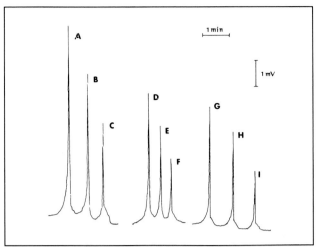

Figure 3. Signal profiles in the determination of vanadium(V), manganese(VII), and cerium(IV) using the diphenylamine sulfonate indicator system

Conditions for curves A, B, and C: Diphenylamine sulfonate 0.025%, perchloric acid 1.5M, oxalic acid 0.25M, total reservoir volume 200 ml, concentration of vanadium(V) in flow-through cell: (A) 0.51 µg/ml, (B) 0.34 µg/ml, (C) 0.17 µg/ml. Conditions for curves D, E, F, G, H, and I: diphenylamine sulfonate 0.01%, sulfuric acid 0.1M, total reservoir volume 200 ml, concentration of manganese(VII) in flow-through cell: (D) 0.21 µg/ml, (E) 0.14 µg/ml, (F) 0.07 µg/ml; concentration of cerium(IV) in flow-through cell: (G) 2.79 µg/ml, (H) 1.86 µg/ml, (I) 0.93 µg/ml

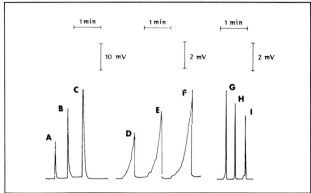

Figure 4. Signal profiles in the determination of hydroquinone, isoniazid, and chloropromazine hydrochloride

Conditions for curves A, B, and C: cerium(IV) $8.0 \times 10^{-4}M$ in 0.1M sulfuric acid (monitored at 525 nm), total reservoir volume 100 ml, concentration of chloropromazine hydrochloride in flow-through cell: (A) 5 µg/ml, (B) 10 µg/ml, (C) 15 µg/ml. Conditions for curves D, E, and F: vanadium(V) $2.0 \times 10^{-3}M$, phosphoric acid 1.5M, osmic acid 0.002% (monitored at 425 nm), total reservoir volume 100 ml. Concentration of isoniazid in flow-through cell: (D) 1 µg/ml, (E) 2 µg/ml, (F) 3 µg/ml. Conditions for curves G, H, and I: chromium(VI) $3.33 \times 10^{-4}M$, sulfuric acid 0.10M, oxalic acid 0.01M, ferroin $1.00 \times 10^{-4}M$ (monitored at 510 nm) total reservoir volume 100 ml, concentration of hydroquinone in flow-through cell: (G) 2.9 µg/ml, (H) 2.2 µg/ml, (I) 1.6 µg/ml

tion. Common tranquilizers, viz., chloropromazine hydrochloride and promethazine hydrochloride, similarly can be determined by use of a strongly oxidizing mixture and by monitoring the absorbance of the transient semi-quinone formed.

The most extensively used anti-tuberculosis drug "isoniazid" (isonicotinic acid hydrazide) can be assayed through a transient signal by utilizing the catalytic effect of osmium on the internal oxidation–reduction of vanadium(V)—"isoniazid complex" (4). A reaction mixture comprising osmic acid, phosphoric acid, and an excess of vanadium(V) acts as a reservoir, and injection of isoniazid to this results in the transient appearance of vanadium(V)–isoniazid complex which absorbs at 425 nm. This is because, in the presence of osmium, the complex is highly unstable and rapidly decomposes to give vanadium(IV), nicotinic acid, and nitrogen. This procedure seems to be well suited for the quick assay of isoniazid content in pharmaceutical preparations.

Use of a mixture of arsenic(III) and iodide as a reaction reservoir permits fast and continuous determination of cerium(IV) and halates (such as bromate and periodate) by monitoring the transient generation and reduction of iodine (coupled with or without starch). The reactions are:

$$As(III) + I_2 \rightarrow As(V) + I^-$$
$$I^- + Ce(IV) \rightarrow Ce(III) + I_2$$
$$(\text{Reservoir mixture: } As(III) + I^-)$$

Likewise, the bromate and periodate oxidation of organic dyes like α-naphthoflavone and p-ethoxychrysoidine, cata-

lyzed by vanadium(V), can be applied for the repetitive determination of low concentrations of halates.

Some signal profiles obtained in most of the above itemized cases are shown in Figures 3 and 4. It should be mentioned that in all the above cases the determination times are in the order of seconds, and effects arising from the decrease of reagent concentration during continuous analysis could always be made negligible by employing high initial concentrations.

On the basis of the observations reported above, it is believed that this new approach to non-equilibrium determinations can be extended to several other systems, and studies along these lines are currently under consideration.

LITERATURE CITED

(1) V.V.S. Eswara Dutt and H. A. Mottola, Anal. Chem., **46**, 1090 (1974).
(2) H. Hall, B. E. Simpson, and H. A. Mottola, Anal. Biochem., **45**, 453 (1972).
(3) J. N. Brazier and W. I. Stephen, Anal. Chim. Acta., **33**, 625 (1965)
(4) P. V. K. Rao and G. B. B. Rao, Analyst (London), **96**, 712 (1971).

Received for review August 15, 1974. Accepted November 8, 1974. This work is supported by Grant GP-38822X, from the National Science Foundation.

Reprinted from Anal. Chem. **1975**, 47, 357–59.

This watershed publication instantly revolutionized the determination of inorganic analytes, especially anions, in terms of speed, sensitivity, and convenience. Rapid commercialization led to extensive applications and an explosive growth of the literature. Measuring small inorganic ions, such as chloride and sulfate, at the low-ppm level was a major challenge in the 1970s, whereas today these species are routinely measured at the part-per-trillion level. The major, ultimate impact was a resurgence of interest in chromatographic analysis of all types of ions in diverse areas. This is evidenced by the fact that the present worldwide market for ion chromatographic equipment and supplies exceeds $125 million.

Purnendu K. Dasgupta
Texas Tech University

Novel Ion Exchange Chromatographic Method Using Conductimetric Detection

Hamish Small
Central Research, The Dow Chemical Company, Midland, MI 48640

Timothy S. Stevens
Michigan Division Analytical Laboratories, The Dow Chemical Company, Midland, MI 48640

William C. Bauman
Inorganic Process Research, Texas Division, The Dow Chemical Company, Freeport, TX 77541

Ion exchange resins have a well-known ability to provide excellent separation of ions, but the automated analysis of the eluted species is often frustrated by the presence of the background electrolyte used for elution. By using a novel combination of resins, we have succeeded in neutralizing or suppressing this background without significantly affecting the species being analyzed which in turn permits the use of a conductivity cell as a universal and very sensitive monitor of all ionic species either cationic or anionic. Using this technique, automated analytical schemes have been devised for Li^+, Na^+, K^+, Rb^+, Cs^+, NH_4^+, Ca^{2+}, Mg^{2+}, F^-, Cl^-, Br^-, I^-, NO_3^-, NO_2^-, SO_4^{2-}, SO_3^{2-}, PO_4^{3-} and many amines, quaternary ammonium compounds, and organic acids. Elution time can take as little as 1.0 min/ion and is typically 3 min/ion. Ions have been determined in a diversity of backgrounds, e.g., waste streams, various local surface waters, blood serum, urine, and fruit juices.

The demand for the determination of ionic species in a variety of aqueous environments is increasing rapidly and, as a result, there is an expanding need for automated or semiautomated analysis of chemical plant streams, environmentally important waters such as waste streams, rivers, and lakes, and fluids of biological interest such as blood, urine, etc. There are many examples where there is a continual need for routine analysis of common species such as Li^+, Na^+, K^+, NH_4^+, Ca^{2+}, Mg^{2+}, F^-, Cl^-, Br^-, I^-, SO_4^{2-}, NO_2^-,

NO_3^-, PO_4^{3-}, etc. Ion exchange resins have a well-known ability to provide excellent separations of ionic species and there are a number of instances where ion exchange chromatography has been successfully applied (1).

In recent years, however, liquid chromatography has moved in the direction of high speed separations and continuous effluent monitoring by detector–analyzer systems which can yield almost instant readout of analytical data. Consequently, in light of this present day practice, the usefulness of a chromatographic separation is often measured by the extent to which it can be coupled to such continuous detectors. If the ions eluting from an ion exchange column have some property which distinguishes them from the background electrolyte, e.g., absorption in the UV or visible range, then the solution to the analytical problem is fairly straightforward and many modern high speed chromatographic procedures have been developed using ion exchange resins in conjunction with spectrophotometric detectors. However, the analysis of certain ionic species is frustrated by the presence of the background electrolyte in that available detectors are not able to detect the species of interest against this background, and it has been stated (2) that this limitation in detectors is one of the main factors retarding a more widespread penetration of ion exchange into modern chromatographic methodology. The detector problem is exemplified by the case of conductimetric detection which has often been proposed and occasionally applied with very limited success in ion exchange chromatography. It would be desirable to employ some form of conductimetric detection as a means of monitoring ionic species in a column effluent since conductivity is a universal property of ionic species in solution and since conductance shows a simple dependence on species concentration. However, the conductivity from the species of interest is generally "swamped out" by that from the much more abundant eluting electrolyte. We have solved this detection problem by using a combination of resins which strips out or neutralizes the ions of the background electrolyte leaving only the species of interest as the major conducting species in the effluent. This has enabled us to successfully apply a conductivity cell and meter as the detector system.

The Principle of the Method. The technique employs the following train of columns, etc. (A) An eluant reservoir, (B) a pump, (C) a sample injection device, (D) the *separating column* wherein the species are resolved by conventional elution chromatography followed by the *eluant stripper column* (referred to henceforth simply as the *stripper column* or *stripper*) wherein, as the name implies, the eluant coming from the separating column is stripped or neutralized. Thus, only the species of interest leave the bottom of the stripper in a background of deionized water where they are monitored by the conductivity cell/meter/recorder (integrator) combination.

Some general principles for choosing useful eluant–separation–stripper column combinations will be outlined later but an understanding of the technique may best be gained at this stage by giving specific details of how the method is applied. Consider, for example, the analysis of a sample containing Li^+, Na^+, and K^+. The eluant, in this case dilute HCl, is pumped to the two columns in tandem (Figure 1) which contain a cation exchanger in the separating column and a strong base resin, OH^- form, in the stripper column. If a sample containing Li^+, Na^+, and K^+ is injected at the head of the first column, the ions will be resolved in the separating bed and will exit at various times from the bottom of this column in a *background of HCl eluant*. On entering the stripper column two important reactions take place: HCl is removed by the strong base resin

$$HCl + Resin\ OH^- \rightarrow Resin\ Cl^- + H_2O \qquad (1)$$

The alkali metal chlorides are converted to their hydroxides

$$M^+Cl^- + Resin\ OH^- \rightarrow M^+OH^- + Resin\ Cl^- \qquad (2)$$

which pass unretarded through the stripper column and into the conductivity cell where they are monitored and quantified in a background of de-ionized water by either measuring the height of, or area under, the conductivity peak.

An analogous scheme for anion analysis can be envisaged where sodium hydroxide is the eluant, an anion exchanger is in the separating column, and a strong acid cation exchanger in the H^+ form is used as stripper.

An important way in which this method differs from most, if not all chromatographic methods, is in the need to regenerate the stripper. Too frequent a need for this step would clearly be a drawback, so an important feature of this technique is the means which have been devised to render this regeneration step as unobtrusive as possible. The problem of stripper regeneration can be expressed in the form of a question-

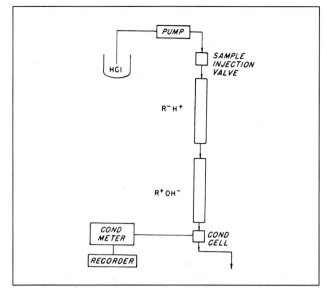

Figure 1. System for cation analysis by conductimetric chromatography

tion, namely, how many samples can be analyzed by the system before stripper regeneration becomes necessary? The answer can in turn be approximately represented in the form of a simple equation, viz.,

$$N = \frac{V_B C_B}{V_A C_A K_{y^\pm}{}^{x^\pm}} \qquad (3)$$

where N = number of samples injected during the stripper's lifetime (assuming maximum utilization of the available time), V_A = volume of separating bed, V_B = volume of stripper bed, C_A = specific capacity of the separating bed, C_B = specific capacity of the stripper bed, $K_{y^\pm}{}^{x^\pm}$ = selectivity coefficient (relative to the eluting ion $y^\pm$) of the ion $x^\pm$, which, in a series to be analyzed, has the greatest affinity for the separating resin.

To obtain a high chromatographic efficiency it is necessary to keep V_B/V_A as low as possible; otherwise the resolution obtained in the separation bed will be offset by mixing in the relatively massive void volume of the stripper bed. A value of V_B/V_A close to unity is excellent but a value of 10 or less would be acceptable. In order, therefore, that N be as large as possible it is necessary that the quantity $C_B/C_A K_{y^\pm}{}^{x^\pm}$ be kept, within certain limits, as large as possible. This can be achieved by 1) Maintaining C_B as large as possible, i.e., by using conventional resins of a high degree of crosslinking. 2) Maintaining C_A as small as possible. This has been achieved by using specially prepared resins of very low capacity. However, a lower limit on the capacity of the separating resin is set by the need to avoid overloading of this column by the sample injected. 3) Maintaining $K_{y^\pm}{}^{x^\pm}$ as small as possible. The options in the choice of eluting ion $y^\pm$ are limited by the further requirement that this ion be amenable to removal or neutralization in a stripping reaction. Nor should the affinity of the eluting ion $y^\pm$ for the separating resin be too high ($K_{y^\pm}{}^{x^\pm}$ too small) or the species of analytical interest will elute too rapidly with lack of resolution.

Closely tied to point 3 is the requirement that the ions of interest not undergo any neutralization or removal reactions in the stripper column. Selecting eluting ions or resins to satisfy one of these criteria is a simple matter but the choice of those which fit all three is limited. Nevertheless a number of schemes have been proposed and are summarized in Table I.

Hardware. The pump used was Milton Roy miniPump with a maximum pumping speed of 160 or 460 ml/hr. The columns were obtained from Chromatronix Inc. (now Laboratory Data Control, a division of Milton Roy Company), Berkeley, CA—the sizes used will be identified in the individual examples quoted later. A Chromatronix conductivity cell was used as detector in conjunction with a conductivity meter designed and built in the Physical Research Laboratory of The Dow Chemical Company. More recently we have used commercially available conductivity meters. The output from the conductivity meter, which is proportional to the conductivity of the sample in the cell, was expressed on a strip chart recorder to provide chromatograms.

Samples were injected to the columns by means of a Chromatronix sample injection valve and a pressure gauge T-ed off ahead of this valve served as a depulser.

Resins and Eluants. The discussion of the methods for preparing the special resins will be taken up at appropriate places later in the paper. Eluant solutions were prepared from deionized water and reagent grade chemicals.

CATION ANALYSIS

The Separating Resin. A resin of very low cation exchange capacity was prepared by surface sulfonation of a styrene divinylbenzene (S/DVB) copolymer (2% DVB) by a method previously described by one of the authors (3). It involves briefly heating (minutes) a quantity of S/DVB copolymer (2% DVB) with an excess of hot (~100 °C) concentrated sulfuric acid, which leads to formation of a thin surface shell of sulfonic acid groups.

The capacity of a typical separating resin is around 0.02 mequiv/gram of starting copolymer. Apart from being a resin of low capacity, the pellicular nature of the sulfonated material would be expected to have favorable mass transfer characteristics due to the proximity of all of the active sites to the eluant resin interface (4). The dimensions of the separating bed will be given in the individual examples.

The Stripper Resin. The stripper resin was Dowex 1 X8 (OH⁻) 200–400 mesh. The dimensions of the stripper bed will be given in the individual examples.

The Eluant. The eluant in all the examples to be described was aqueous HCl, in the range of 0.01 to 0.02N, except for some of the amine separations where pyridine hydrochloride and aniline hydrochloride solutions were used.

Separations. *Alkali Metal Ions.* Details of the columns, resins, eluants, etc., used in the various examples cited, are summarized in Table II. Figure 2 shows the separation of the alkali metal ions. Linear flow rate through the separating column was increased by using microbore columns and complete separations of sodium from potassium were attainable within as little as three minutes. Figures 3–5 show examples of the application of the technique to Na^+, K^+, and NH_4^+ determination. Figure 3 shows the chromatogram of human urine—the peak between Na^+ and K^+ was identified as NH_4^+. Figure 4 shows the elution of Na^+ and K^+ in dog blood serum. In Figure 4A, the K^+ peak is barely perceptible above the base line but shows as a substantial peak in Figure 4B as a result of increasing the detector sensitivity after the Na^+ peak had passed through. This example illustrates the ability in this case to determine a small amount of one ion, K^+, in a relatively much higher background concentration of another ion, Na^+. The concentrations of the Na^+ and K^+ in the serum were 4050 and 165 ppm, respectively. The chromatograms of orange juice and grape juice are illustrated in Figure 5.

Organic Amines and Quaternary Ammonium Compounds. Several organic amines and quaternary ammonium compounds were successfully detected and/or resolved using dilute hydrochloric acid as the eluant as explained in Table I. In the stripper, the protonated amine is converted to the free amine and the quaternary salt to quaternary ammonium hydroxide. Table III lists the elution data obtained. Note that elution times vary with the amount of sample injected. The effect is minor

Table I. Schemes for Conductimetric Chromatography

Cation analysis

Separating column	Eluant	Stripper column	Stripping reaction
Resin–H⁺	HCl	Resin–OH⁻	Resin–OH⁻ + HCl → Resin Cl⁻ + H₂O
Resin–Ag⁺	AgNO₃	Resin–Cl⁻	Resin Cl⁻ + AgNO₃ → Resin NO₃⁻ + AgCl↓
Resin–Cu²⁺	Cu(NO₃)₂	Resin–amine	Resin amine + Cu(NO₃)₂ → Resin amine · Cu(NO₃)₂
Resin–anilinium	Aniline hydrochloride	Resin–OH⁻	Resin OH⁻ + aniline HCl → aniline + Resin Cl⁻ + H₂O
Resin–Ag⁺	AgNO₃—HNO₃	1st Resin–Cl⁻	Resin–Cl⁻ + AgNO₃ → Resin NO₃⁻ + AgCl↓
		2nd Resin–OH⁻	Resin–OH⁻ + HNO₃ → Resin NO₃⁻ + H₂O

Anion analysis

Separating column	Eluant	Stripper column	Stripping reaction
Resin–OH⁻	NaOH	Resin–H⁺	Resin H⁺ + NaOH → Resin Na⁺ + H₂O
Resin–phenate	Na–phenate	Resin–H⁺	Resin H⁺ + PhO⁻Na⁺ → Resin Na⁺ + PhOH

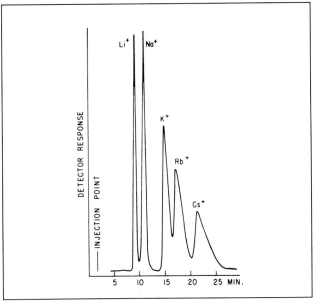

Figure 2. Separation of alkali metal ions

at low to moderate concentrations but becomes significant at higher concentrations. Peak width at half height is listed to permit calculations of resolution and HETP. As an example, a resolution of mono-, di-, and trimethyl amine is shown in Figure 6.

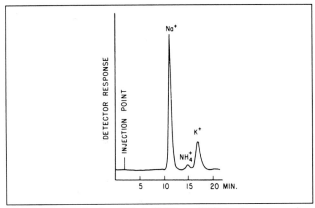

Figure 3. Analysis of Na⁺, NH₄⁺, and K⁺ in human urine

Table II. Elution Conditions Used in the Examples Illustrated in Various Figures

Fig.	Separating column	Stripping column	Eluant	Flow rate, ml/hr	Sample size, ml	Sample composition
2	9 mm × 250 mm SSS/DVB 180–325 mesh Sp. cap 0.016 mequiv/g	9 mm × 270 mm Dowex 1 X8 OH⁻ 200–400 mesh	0.01N HCl	160	0.1	0.01N in each of Li, Na, K, Rb, and Cs chlorides
3	Same as in Fig. 2	Same	0.01N HCl		0.1	Urine diluted 10-fold
4	Same as in Fig. 2	Same	0.01N HCl		0.1	Serum diluted 10-fold
5	Same as in Fig. 2	Same	0.01N HCl		0.1	Undiluted juices
6	9 mm × 250 mm SSS/DVB 180–325 mesh Sp. cap 0.024 mequiv/g	9 × 250 mm Dowex 1 X8 OH⁻ 200–400 mesh	0.01N HCl	460	0.1	8 ppm monomethylamine 8 ppm dimethylamine 20 ppm trimethylamine
7	2.8 mm × 300 mm SSS/DVB 180–325 mesh Sp. cap 0.024 mequiv/g	1st column 9 mm × 250 mm Dowex 1 X8 Cl⁻ 200–400 mesh 2nd column 9 mm × 250 mm Dowex 1 X8 OH⁻ 200–400 mesh	0.002N AgNO₃ 0.0004N HNO₃	230	0.1	20 ppm tetraethyl ammonium bromide 60 ppm tetra-n-butyl ammonium bromide
8	9 mm × 250 mm SSS/DVB 180–325 mesh Sp. cap 0.024 mequiv/g	9 mm × 250 mm Dowex 1 X8 OH⁻ 200–400 mesh	0.001M aniline hydrochloride 0.001M HCl	460	0.1	14 ppm monoethylamine 20 ppm diethylamine 40 ppm triethylamine
9	2.8 mm × 500 mm SSS/DVB 180–325 mesh Sp. cap 0.024 mequiv/g	9 mm × 300 mm Dowex 1 X8 Cl⁻ 200–400 mesh	0.05N AgNO₃	92	0.1	Untreated Saginaw river water
13A	2.8 mm × 300 mm SA Dowex 2 resin SA Dowex 2 resin	2.8 mm × 300 mm Dowex 50W X8 H⁺ 200–400 mesh	0.015F sodium phenate	60	0.01	Untreated Midland city water
13B	SA Dowex 2 resin	200–400 mesh	0.015F sodium phenate	60	0.01	Untreated Lake Huron water
14	2.8 mm × 300 mm SA Dowex 2 resin	9 mm × 250 mm Dowex 50W X8 H⁺ 200–400 mesh	0.035F NaOH 0.015F sodium phenate	64	0.01	0.01N in each of sodium mono-, di-, and tri-chloroacetate
15	Same as Fig. 14	2.8 mm × 250 mm Dowex 50W X8 H⁺ 200–400 mesh	0.015F sodium phenate	64	0.01	0.01N in each of chloride, bromide, and iodide

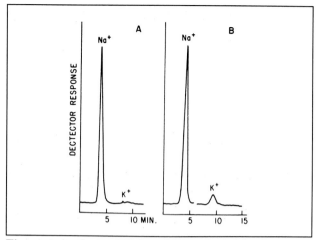

Figure 4. Analysis of Na⁺ and K⁺ in dog's blood serum

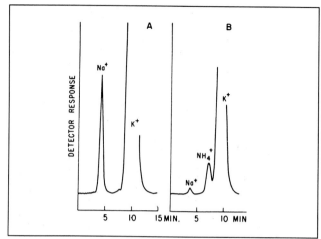

Figure 5. Analysis of (A) orange juice and (B) grape juice for Na⁺, K⁺, and NH_4^+

Table III. Elution Data for Amines[a]

Sample	ppm Concentration	Peak max. elution time, min.	Peak width at half height, min
Monomethyl amine	8	6.5	0.6
Monomethyl amine	80	6.3	0.7
Monomethyl amine	800	5.2	1.2
Dimethyl amine	8	8.2	0.9
Dimethyl amine	80	7.7	1.0
Dimethyl amine	800	6.3	1.4
Trimethyl amine	8	10.3	1.1
Trimethyl amine	80	9.9	1.3
Trimethyl amine	400	8.7	1.1
Tetramethyl ammonium Br	10	12.0	1.4
Tetramethyl ammonium Br	100	12.3	1.6
Tetramethyl ammonium Br	1000	9.8	1.8
Monoethyl amine	14	7.2	0.8
Monoethyl amine	140	6.8	0.8
Diethyl amine	20	10.8	1.3
Diethyl amine	200	10.0	1.3
Triethyl amine	40	15.5	1.8
Triethyl amine	400	14.0	2.2
Tetraethyl ammonium Br	100	27.5	3.0
Tetraethyl ammonium Br	1000	22.0	4.4
n-Butyl amine	20	14.2	1.8
n-Butyl amine	200	12.7	1.8
Cyclohexyl amine	200	23.6	3.8
Cyclohexyl amine	1000	19.5	3.8
Tri-n-butyl amine		>30 min.	
Tetra-n-butyl ammonium Br		>30 min.	
Monoethanol amine	20	5.3	0.7
Diethanol amine	20	5.5	0.8
Triethanol amine	200	5.9	1.2
Monoisopropanol amine	20	5.6	0.7
Diisopropanol amine	40	6.2	1.0
Triisopropanol amine	400	6.6	1.7
Ammonia	1	5.1	0.5
Ammonia	10	5.1	0.5
Ammonia	100	4.9	0.7

[a] Separating column: 9 mm × 250 mm SS/DVB; 0.024 mequiv/g; 180–325 mesh. Stripper column: 9 mm × 250 mm Dowex 1 X8 OH⁻, 200–400 mesh. Eluant: 0.01N HCl. Flow rate: 460 ml/hr.

Some of the amines listed in Table III took more than 15 minutes to elute so eluants with a higher affinity for the surface sulfonated resin were used to reduce the elution time. These eluants were silver nitrate–nitric acid, a mixed eluant, and a double stripper, see Table I, and aniline hydrochloride–hydrochloric acid mixed eluant. The presence of the acid is needed to protonate the amines to cations. Figure 7 shows the resolution of tetraethylammonium bromide with the silver nitrate–nitric acid eluant and Figure 8 shows a resolution of mono-, di-, and triethyl amine using the aniline hydrochloride–hydrochloric acid eluant. Note that both systems decreased the time needed for the determinations. The detection limit of conductimetric chromatography for the lower molecular weight amines is normally 0.1–1 ppm.

Divalent Metal Ions. A scheme such as the one used in the alkali metal separation has a drawback when applied to the case of Ca^{2+}, Mg^{2+} determination. The problem derives from the high affinity that these divalent ions have for the separating resin. Consequently, the selectivity factor becomes dominant (see Equation 3), and the large amounts of acid required to elute these species off the resin greatly shortens the lifetime of the stripper.

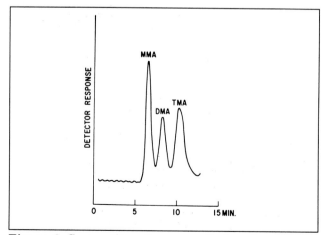

Figure 6. Separation of monomethyl amine (MMA), dimethyl amine (DMA), and trimethyl amine (TMA)

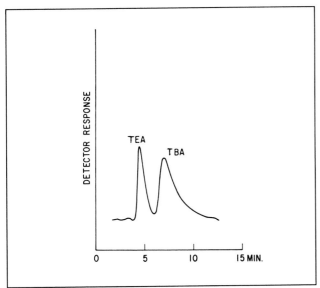

Figure 7. Separation of tetraethyl ammonium (TEA) and tetra-*n*-butyl ammonium (TBA) ions

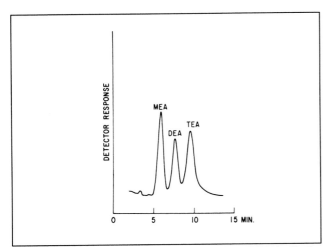

Figure 8. Separation of monoethylamine (MEA), diethylamine (DEA), and triethylamine (TEA) using aniline hydrochloride as eluant

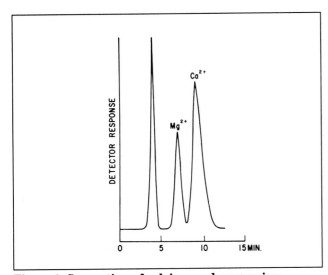

Figure 9. Separation of calcium and magnesium

To get around this problem, we therefore sought an eluting species which would have the following characteristics: 1) Have a significantly higher affinity for the sulfonated resin than H^+. 2) Be easily and completely removed by the stripper by a reaction which would not remove the species being determined, viz., Ca^{2+} and Mg^{2+}.

The scheme devised employs Ag^+ as the eluting ion which has roughly ten times the affinity for a sulfonic acid resin that H^+ does. After passing from the separating column, the Ag^+ is removed by precipitating it as AgCl on the Cl^- form of an anion exchanger. The spent stripper is regenerated by a solution containing ammonium hydroxide ($1N$) and ammonium chloride ($1N$). A typical chromatogram of a sample which contained Ca^{2+} and Mg^{2+} is shown in Figure 9—the first peak is due to the alkali metals, Na^+ and K^+, which were also present in the sample. Recently a *p*-phenylene diamine/HCl eluant with a strong base resin stripper has been found to be a superior system to the silver system.

ANION ANALYSIS

The Separating Resin. Anion exchangers of very low (surface) capacity have been used in the separating bed. At the inception of this work, it was known that resins of this type were available commercially, but since none was readily at hand an alternate route to such a resin was devised. It has been known for some time that cation and anion exchangers have a marked tendency to clump together—a manifestation of the strong electrostatic interaction between polycationic and polyanionic materials in general. This clumping property is exploited in the following manner to produce an effective anion exchanger of very low capacity.

A strong base anion exchanger (Dowex 1, Dowex 2, IRA 900) is thoroughly ground in a rod mill and the larger particles are removed by sedimentation. A very dilute suspension of the fine particles is then passed through a column packed with a surface sulfonated S/DVB resin where the fine anion exchange particles are agglomerated with the surface of the SSS/DVB. Eventually the surface of the SSS/DVB becomes saturated with the fine resin particles and they break through from the column, at which time, after a brief water rinse, the column is considered ready for use.

It has been established by electron micrographs that the dry particle size of a useful ground material is around 0.5–2 micron. Furthermore, by careful size fractionation, it is possible to prepare columns of a variety of capacities and resolving powers. For convenience in referring to these surface agglomerated resins, we will use the simple designation SA followed by the name of the anion exchange resin used, e.g., SA Dowex 2.

The Stripper Resin. Dowex 50W X8 (H^+) was used as stripper and was regenerated by $0.25N$ H_2SO_4 when exhausted.

The Eluant. In devising a scheme for anion analysis, an obvious choice for eluting species is the hydroxide ion since it is so conveniently neutralized by a strong acid resin in the stripper bed. On the other hand, on the basis of its selectivity in anion exchange reactions (*5–7*), the choice is not such a desirable one—within a large series of anions, the hydroxide ion is one of the least tightly held, i.e., $K_{OH^-}{}^{X^-}$ is in most cases large. Furthermore, with respect to other ions, it is less tightly held on a Dowex 1 than on Dowex 2 and it is for this reason that in this work a Dowex 2 type resin has been employed in the separating column. As a result of this unfavorable selectivity factor, fairly high concentrations of OH⁻ have to be used to elute ions of even moderate affinity, which in turn shortens the lifetime of the stripper. The low displacement potential of the hydroxide ion also leads to extensive tailing of the elution curves of the more tightly bound ions which has the undesirable effect of reducing the sensitivity for detecting the species in the effluent. It was, therefore, clearly desirable to find a suitable alternative to OH⁻ as an eluting ion and a species that seemed a likely candidate was the phenate ion. Not only did it have a more favorable selectivity coefficient on Dowex 2, 0.14 for phenate vs. 1.5 for OH⁻, but an acid form resin as stripper would convert it to phenol which being a very weak acid, would be only feebly dissociated and contribute little to the conductivity of the effluent from the stripper. Accordingly, the phenate ion was extensively investigated as a displacing ion.

The Phenate System. Tables IV and V summarize the results of an extensive series of elutions of a wide variety of anions wherein the concentration of OH⁻ and PhO⁻ and the size of the stripper was varied.

In general, the order of elution of ions is in accord with previous work (5–7). However, there are a number of exceptions. 1) Using the large stripper bed (9 × 250 mm) F⁻ elutes at the same time or later than Cl⁻ (Table IV) which is not consistent with their selectivities for anion exchange on Dowex 2 (6). 2) Likewise, formate, acetate, and monochloroacetate elute later than would be expected on the basis of anion exchange selectivities alone. 3) Phosphate elutes after sulfate under some conditions (Table IV) and before it under other conditions (Table V).

The reason for some of these inconsistencies (1 and 2) becomes apparent when one makes a more detailed study of the function of the stripper bed for it is this bed that is the cause of these elution reversals.

Consider Figure 10 which illustrates a stripper bed in the process of being exhausted and lists the various ionic species existing in the exhausted sodium form zone and in the unexhausted hydrogen form zone.

Table IV. Elution of Anions by Sodium Hydroxide/Sodium Phenate Mixtures[a]

Anion	Eluant: 0.045F OH⁻/0.005F PhO⁻	Eluant: 0.035F OH⁻/0.015F PhO⁻
	Elution vol ml	
F⁻	6.72	6.51
Cl⁻	6.72	6.19
Br⁻	9.65	7.47
I⁻	V. large	18.0
IO_3^-	5.55	5.81
NO_3^-	9.87	7.47
NO_2^-	7.36	6.29
SO_3^{2-}	8.05	6.72
SO_4^{2-}	10.2	7.15
PO_4^{3-}	16.5 (?)	8.48
Formate	7.25	5.14
Acetate	8.27	8.59
Cl acetate	6.67	6.67
Cl_3 acetate	V. large	13.6
Oxalate	11.3	7.15

[a] Separating column: 2.8 mm × 300 mm "SA Dowex 2 Resin." Stripper column: 9 mm × 250 mm Dowex 50W X16 (H⁺) 200–400 mesh. Flow rate: 60 ml/hr.

Table V. Elution of Anions by Sodium Phenate[a]

Anion	0.005F PhO⁻	0.01F PhO⁻	0.015F PhO⁻
F⁻	1.92	1.8	1.68
Cl⁻	3.48	2.5	2.2
Br⁻	7.8	4.62	3.46
I⁻			16.3
NO_3^-	7.8	4.62	3.52
NO_2^-	4.02	2.9	2.6
IO_3^-	1.88	1.62	1.68
BrO_3^-			1.94
SO_4^{2-}	>34.2	11.6	5.76
SO_3^{2-}	21.0		3.88
CO_3^{2-}	13.5	5.0	3.78
CrO_4^{2-}		>25.2	21.8
PO_4^{3-}			4.72
Formate	2.30	2.06	
Acetate	2.34	2.22	1.84
Propionate	2.64		
Cl acetate	2.74	2.22	
Cl_2 acetate	5.34	3.58	
Cl_3 acetate			10.1
Glycolate			1.84
Oxalate		11.6	5.76
Maleate		7.46	4.2
Fumarate		10.4	5.56
Succinate			3.66
Malonate			3.78
Itaconate		7.56	
Benzoate	10.9	7.18	6.3
Ascorbate			2.1/5.88
Citrate			42.4

[a] Separating column: 2.8 mm × 300 mm "SA Dowex 2 Resin." Stripper column: 2.8 mm × 300 mm Dowex 50W X8 (H⁺) 200–400 mesh.

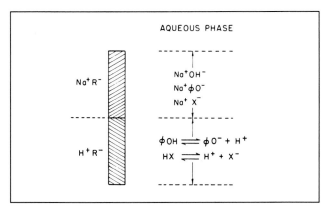

Figure 10. The species in the aqueous phase in the phenate stripper column

The elution volume (V_E) through this stripper bed of species containing the ion X^-, the ion of analytical interest, is given by the expression

$$V_E = V_v + K_1 \chi_{Na} V_B + K_2 (1 - \chi_{Na}) V_B \qquad (4)$$

The first term, V_v, is simply the retention volume attributable to displacement of the void volume of the stripper bed.

The second term, $K_1 \chi_{Na} V_B$, describes the retention of the species X^- by the sodium form of the stripper where K_1 is the distribution coefficient of X^- between the resin and mobile phase, χ_{Na} is the fraction of the bed that has been exhausted, and V_B is the volume of the stripper. Due to Donnan Exclusion (8), K_1 is very small and term 2 in Equation 4 is therefore negligible.

The third term, $K_2 (1 - \chi_{Na}) V_B$, describes the retention of X^- and related species by the unexhausted H^+ form resin. Since the X^- ion is converted to its corresponding acid at the Na^+R^-/H^+R^- boundary, the elution behavior of X type species through the H^+R^- zone will be very dependent on the strength of this acid. If the acid is strong, then the predominant species in this region of the bed will be X^- which will be effectively excluded from H^+R^-, that is K_2 will be very small, and term 3 will be negligible. If, on the other hand, the acid is weak, a fraction of X will exist as the undissociated HX species which, being uncharged, is not subject to ion exclusion forces and can enter the hydrogen form resin quite readily. In other words K_2 is no longer of very small magnitude and term 3 can become so significant that it is the dominant factor controlling the elution order through the complete train of columns and is the cause of the elution reversals already noted.

Not only can term 3 be of significant magnitude in certain cases but it also varies in magnitude as X_{Na} varies from zero (fresh stripper) to unity (exhausted stripper). This gives rise to the undesirable result that the elution volume of the species of interest depends on the degree of exhaustion of the stripper column. Since one has no control over the value of K_2 or χ_{Na} the effect of the third term can be reduced and the drifting of peak position minimized by keeping V_B as small as possible. A parallel case exists for chromatography of cations which form weak bases and *it is therefore recommended that, in the chromatography of species which convert to partially dissociated forms in the stripper bed, the volume of this bed be kept as small as*

practicable. On the basis of the arguments already presented, the elution of species which form highly dissociated derivatives in the stripper bed is sensitive only to the first term in Equation 4 and the size of the stripper bed is therefore determined solely by the considerations expressed in Equation 3.

The shifting elution position of phosphate (anomaly 3 above) is for a reason quite different from those just discussed. It is believed that the conditions in the separating column dictate its elution position rather than those in the stripper column since phosphoric acid is a strong acid and should be subject only to term 1 in Equation 4. It is well known from ion exchange studies that ion affinities are strongly dependent on the valence of the exchanging ions with higher valence ions being the more tightly held, all other factors being equal. In the case of the phosphate ion the species involved in exchange in the separating column (pH 10 to 11) is not a single species but rather a mixture of HPO_4^{2-} and PO_4^{3-}. Consequently, conditions which favor the formation of the trivalent PO_4^{3-} at the expense of the divalent HPO_4^{2-} will promote the retention of phosphate in the separating column and an excess of sodium hydroxide above that necessary to neutralize the phenol (Table IV) is such a condition.

Separations. A number of separations and analyses of anions are illustrated in Figures 11 to 15.

Figure 11 shows the comparison between the hydroxide and phenate systems in the elution of Cl^- and SO_4^{2-}. Calibration plots were constructed for these two ions and are shown in Figure 12 which illustrates the close linear dependence between peak height and the amounts of ion present over a wide range of concentration.

Figure 13A illustrates the chromatogram of a local municipal water. Three of the peaks were attributable to Cl^-, CO_3^{2-} and SO_4^{2-}, and F^- was suspected as being the origin of the first peak to elute, particularly since it did not appear in the chromatogram of water from Lake Huron (Figure 13B) which

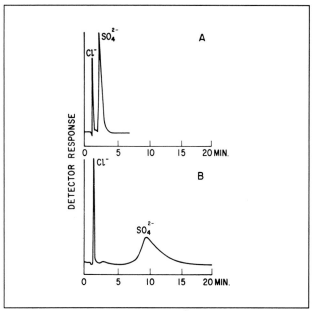

Figure 11. Elution of Cl^- and SO_4^{2-} by (A) sodium phenate, (B) NaOH

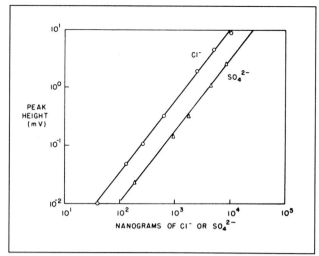

Figure 12. Calibration plots for Cl⁻ and SO₄²⁻

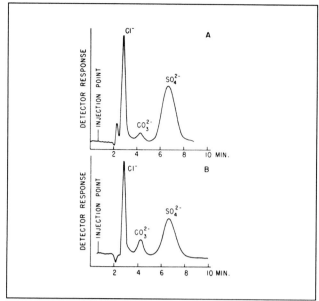

Figure 13. Chromatograms of (A) Midland City water, (B) raw Lake Huron water

is the source of the municipal water. Confirmation that the first peak was very probably due to F⁻ was obtained from chromatograms of raw Huron water spiked with known amounts of fluoride. It is worth noting that, in these experiments, a distinct peak due to fluoride was obtained for only 3.8 nanograms of added F⁻ which attests to the sensitivity of the technique. Figures 14 and 15 illustrate two other separations obtained with the anion analysis system.

Table VI shows results of analyses of a local river water at four different sampling locations made by the subject method and at least one accepted analytical method. The average overall variance was 7.6%. This level of agreement was ob-

tained with samples of river water generally believed to contain many components, in addition to the ones determined, none of which apparently affected the results under the conditions used.

CONCLUSIONS

A practical ion exchange chromatographic method for anions and cations has been developed which uses a conductiv-

Table VI. Comparison of Analysis by Conductimetric Chromatography (CC) and Other Methods

Ion determined	Method	Concn, ppm			
		1[a]	2[a]	3[a]	4[a]
Na⁺	CC	22	49	24	45
Na⁺	AA[b]	23	48	24	45
K⁺	CC	1.4	4.0	1.8	2.3
K⁺	AA	2.2	3.7	2.5	2.6
Ca²⁺	CC	43	74	44	52
Ca²⁺	AA	47	74	49	56
Ca²⁺	EDTA titration	46	74	47	53
Mg²⁺	CC	13	21	14	15
Mg²⁺	AA	12	19	12	13
Mg²⁺	EDTA titration	14	20	14	16
Cl⁻	CC	46	122	57	91
Cl⁻	AgNO₃ titration	54	117	55	100
Cl⁻	Neutron activation	50	118	56	99
SO₄²⁻	CC	28	44	28	32
SO₄²⁻	Turbidimetry	30	49	28	32

[a] Sampling locations on the Saginaw River.
[b] Atomic absorption.

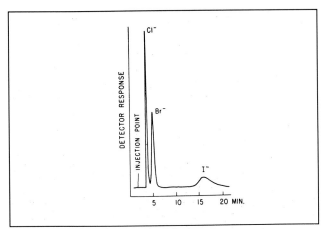

Figure 14. Separation of halide ions

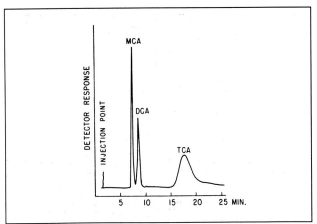

Figure 15. Separation of monochloroacetate (MCA), dichloroacetate (DCA), and trichloroacetate (TCA)

ity cell as detector. This is made possible by a unique combination of resins which separates the ions of interest and strips or neutralizes the eluant from the background.

Analytical schemes have been devised for a large variety of organic and inorganic cations and anions which can be determined with good accuracy and precision. The method has a large dynamic range and is capable of determining small amounts of one species in a high background of other ions.

The method is particularly attractive for anions in light of the special and varied techniques usually required for their analysis. With this technique, a great diversity of anionic species may be quantified by a simple universal detector, the conductivity cell.

Note. The methods described in this publication are the subject of pending patents.

LITERATURE CITED

(1) W. Reiman III and H. F. Walton, "Ion Exchange in Analytical Chemistry," Pergamon Press, Oxford, 1970.
(2) J. E. Salmon, Third IUPAC Analytical Chemistry Conference, Budapest, 1970.
(3) H. Small, *J. Inorg. Nucl. Chem.*, **18**, 232 (1961).
(4) J. J. Kirkland, "Modern Practice of Liquid Chromatography," Wiley Interscience, New York, 1971.
(5) S. Peterson, *Ann. N.Y. Acad. Sci.*, **57**, 144 (1953).
(6) R. M. Wheaton and W. C. Bauman, *Ind. Eng. Chem.*, **43**, 1088 (1951).
(7) R. Kunin and R. J. Myers, "Ion Exchange Resins," Wiley, New York, p 154.
(8) F. Helfferich, "Ion Exchange," McGraw Hill, New York, 1962.

Received for review December 5, 1974. Accepted February 24, 1975.

Reprinted from *Anal. Chem.* **1975**, *47*, 1801–09.

Electrochemists had long been aware of the effects of adsorption, oxide film formation, and other surface effects on the electrochemical behavior of an electrode. However, this article introduced the concept of deliberate and controlled modification of an electrode surface by covalent attachment. Thus, this work, along with studies by Kuwana, Miller, and others, led to the enormous growth of chemically modified electrodes, in which the electrode surface properties are designed to exhibit interesting and useful behavior (e.g., catalysts, sensors, and electrical devices). Fundamental studies of modified electrodes (e.g., those with polymer layers) have also provided a better understanding of charge transfer and charge transport processes in thin films.

Allen J. Bard
University of Texas—Austin

Chemically Modified Tin Oxide Electrode

P. R. Moses, L. Wier, and R. W. Murray
Kenan Laboratories of Chemistry, University of North Carolina, Chapel Hill, N.C. 27514

Surface synthetic procedures are described whereby, via silane chemistry, amine, pyridyl, and ethylenediamine ligands can be attached to SnO₂ electrodes. The surface reactions and confirmatory chemical tests involving amine protonation and metal coordination are followed by X-ray photoelectron spectroscopy (ESCA). The modified electrodes retain electrochemical activity toward solution reactants.

Chemical modifications transforming heterogeneous, unpredictive surfaces into chemically predictive ones is an area of research gradually percolating into diverse fields. Separation scientists have for several years chemically modified solid supports to improve chromatographic column performance. Such "bonded phase" work has depended on organosilane reactions with the surfaces of silica or alumina particles (1, 2). In another separations application, glass wool surfaces have been derivatized with dithiocarbamate ligands, scavenging trace metals for X-ray photoelectron spectroscopy (ESCA) measurement (3). Acid–base dye indicators attached (4) to silica surfaces create "solid indicators". Homogeneous catalysts such as rhodium diphenylphosphine complexes have been attached to silica, again using organosilane chemistry, to provide heterogeneous hydroformylation catalysts (5, 6).

Electrochemistry is a field where chemically predictive surfaces are at a premium. As yet no stable covalently bonded electrode surfaces have been described. An important step in this direction was taken by Lane and Hubbard (7), who described the strong chemisorption of electroactive allyl compounds on Pt electrodes.

We report here surface synthetic techniques by which several ligands (amine, pyridine, and ethylenediamine) can be co-valently bonded to the surfaces of tin oxide electrodes. Preparation of these surface ligand sites is a first step toward surface-bound metal complex redox centers, among several applications. The surface bonding depends on reactions of organosilane reagents with hydroxyl groups on the tin oxide surfaces. Reacted surfaces were examined by electron spectroscopy as one means of demonstrating successful surface synthesis. The electrochemical viability of the derivatized tin oxide electrodes was demonstrated by electrochemical reactions of model electrochemical couples (ferrocyanide, *o*-tolidine).

EXPERIMENTAL

The tin oxide electrodes were obtained as antimony-doped, transparent films on glass from PPG Industries, Pittsburgh, Pa. Four-point probe measurements indicated a film resistivity of ca. 5 ohms square⁻¹. Interference patterns were used to estimate film thickness at 6×10^{-5} cm. Specimens for surface modification and ESCA or electrochemical examination were prepared from $12 \times 12 \times 1/8$ inch stock by either cutting the glass into 1/2-inch squares under a flowing stream of water using a diamond saw or by epoxying a 4×4 inch plate onto a plate of uncoated glass and drilling 1/4-inch disks (for ESCA) with a 5/16-inch diamond drill. The epoxy is removed by pyrolysis at 450 °C. Freshly cut specimens were either extracted with heptane overnight to remove surface grease and then heated at 450 °C in air for several hours, or treated with hot concentrated HCl for several hours followed by copious washing with distilled water and then alcohol. Surface acidification does not seem to be an essential prerequisite for reactivity of tin oxide surfaces toward silanes, as no gross differences in reactivity were discerned for the two pretreatment procedures. Minor differences undoubtedly exist.

The organosilanes are commercially available, from Petrarch Systems, PCR Chemical Company, or Silar Chemical Company. They were used without further purification, employing serum cap and syringe techniques to protect the reagents and reacting solutions from moisture. Reactions of trichlorosilanes and triethoxysilanes were selected to emphasize high surface yields. Specific reagents include aminopropyltriethoxysilane, 3-(2-aminoethylamino)propyl-trimethoxysilane (Dow Corning Z-6020 Silane), 3-dichloropropyl-trichlorosilane (Dow Corning Z-6010 Silane), and β-trichlorosilyl-2-ethylpyridine.

Silanization reactions were carried out in dry deaerated benzene or xylene. Electrode specimens were reacted under 50 ml of a ca. 10% solution of refluxing silane, under nitrogen, for several hours with trichlorosilanes to several days with triethoxysilanes. Minimum reaction times actually required were not investigated. After reaction, the solution was decanted and the specimens were washed under nitrogen with several portions of fresh solvent. With the ethylenediamine reagent, washing was with water and then alcohol.

All organosilanes are liquids except the pyridine reagent, which is solid and less reactive than most. After refluxing overnight with this reagent, electrode specimens were especially carefully washed with hot solvent and then alcohol to remove the last traces of unreacted silane from the electrode surface.

Electron spectroscopy proved invaluable in following the surface synthesis. ESCA spectra were obtained with a DuPont Model 650B spectrometer with Mg anode. Under optimum conditions a 1/4-inch gold disk yields a $4f_{5/2}$ peak intensity of 400,000 counts/sec with FWHM of 1.2 eV. The pumping system consists of a cryogenic forepump with titanium ion main pump; typical operating vacuum is 1×10^{-7} Torr. At this pressure, and using specimens necessarily exposed to laboratory atmosphere, a contamination C 1s peak of ca. 30,000 counts/sec appears on all samples. Our experience in this and other ESCA-electrochemical studies shows that alterations occur in the contamination carbon film during spectrometer X-ray exposure, and an adherent, insulating film often forms which adversely affects subsequent electrochemical use. Accordingly, electrochemical data were always obtained prior to ESCA examination, or on separate specimens. No examples of electrochemical destruction of the organosilicon layer were encountered except where extensive scans of potential beyond the background limits were involved.

As the tin oxide electrode films were deposited on an insulating substrate, ESCA charging shifts of several eV were common. The C 1s contamination peak was employed as a reference peak and was assigned a value of 285.0 eV (8, 9). All binding energies reported are referenced to this value.

Electrochemical experiments on tin oxide were carried out in miniature Lucite cells of design similar to that employed by Kuwana and associates (10), using a peripheral copper ring for electrical contact and an O-ring seal. The geometrical area of the electrode exposed to the cylindrical solution cavity is determined by the sealing O-ring; this area is 0.031 cm² for one cell and 0.079 cm² for another. The auxiliary electrode was a Pt wire coil; potentials are referenced to an SCE of design after Adams (11). Electrochemical instrumentation was conventional. No iR compensation was employed. The cyclic voltammetric experiment is used here as the principal electrode-characterizing technique.

RESULTS AND DISCUSSION

Schematically the reaction of an organosilane with a surface hydroxyl group is representable as

$$-\underset{\text{surface}}{\text{M}}\!\!-\!\!OH + \underset{\text{silane}}{XSiYZR} \rightarrow -\underset{\text{surface}}{\text{M}}\!\!-\!\!OSiYZR + HX \quad (1)$$

In the case M equals Si (e.g., silica surface), the Si–O–Si bond is known to be very stable to acids and bases as well as thermally stable to as high as 300 °C (1). Where M equals Sn, the modified surface chemical stability does not equal that of silica, but is nonetheless quite good. Treatment of tin oxide electrodes derivatized with, for example, aminopropyltriethoxysilane (designated $SnO_2/PrNH_2$ electrodes) with room temperature $1M$ mineral acid is without effect. Hot concentrated HCl, however, within 2–3 hours has hydrolyzed approximately 50% of the surface groups. More immediate hydrolysis is observed by soaking in $0.1M$ NaOH.

It is known from chromatographic (1) and other studies that the order of reactivity of halosilanes with silica is $X_3SiR > X_2SiR_2 > XSiR_3$. This relationship was also observed with SnO_2 surfaces in a series of reactions where $X \equiv Cl$ and $R \equiv$ methyl. The Si 2p ESCA band was most intense for SnO_2 treated with Cl_3SiCH_3, and least for $ClSi(CH_3)_3$. The $ClSi(CH_3)_3$, although less reactive, did produce Si 2p bands well above background silicon.

Organosilanes with more than one reactive group have the potential of binding to more than one surface site. A trichlorosilane does not, however, necessarily bind to three surface sites. Boucher et al. (5), for instance, demonstrated during the immobilization of diphenylphosphine groups on silica that the trichlorosilane reagent claimed slightly less than two sites per silane. The fate of the remaining silane reactive group is of some importance. The formation of linear siloxane polymers bound to silica surfaces at only a few sites has been claimed by Aue and Hastings (12, 13) in connection with chromatographic bonded phases. Polymer formation can occur if a dichloro or trichlorosilane forms one >Sn–O–Si< surface link, and then a second Si–Cl bond becomes hydrolyzed by water, forming >Sn–O–Si(OH)<. If this occurs in the presence of unreacted solution chlorosilane, a polymer chain can be initiated at the Si(OH) site. Organosilane polymerization is thought to be more severe for smaller, more reactive silanes, such as trichloromethylsilane (14). It can be avoided altogether by the use of monochlorosilanes.

We employed X_3SiR silanes to emphasize high surface coverages at this phase of our surface modification research, and also because most suitable ligand-bearing commercial organosilanes are of this category. We have attempted to minimize the possibilities for polymer formation by utilizing dry solvents, by carefully and thoroughly washing excess silane from the tin oxide specimens with fresh aprotic solvent before exposing the specimens to moisture or other hydrolyzing substances, and by ESCA checks of the Si 2p and Sn 3d bands of at least one SnO_2 electrode disk from each synthetic batch. Occasionally a poor reaction is spotted in which high Si 2p intensities and usually low Sn 3d intensities indicate extensive polymer formation. Suitable functioning of a tin oxide specimen as an electrode serves as an additional criterion of at least minimal polymer formation, and electrochemical data were taken for each chemically modified surface described below.

Unreacted SnO_2 Electrodes. Unreacted SnO_2 specimens were characterized by ESCA preparatory to study of chemically modified ones. A survey ESCA spectrum of unreacted SnO_2 is shown in Figure 1, and data are summarized in Table I. The spectrum contains strong bands for O 1s at 532 eV, oxygen Auger at 750 eV, Sn $3p_{3/2}$ at 715 eV, Sn $3d_{3/2}$ at 497 eV, Sn $3d_{5/2}$ at 487 eV, and C 1s at 285 eV (reference peak). A small N 1s peak of about 300 counts/sec is present on unreacted electrodes which have not been first heated to ca. 450 °C for several hours. The binding energy for this peak, 400.5 eV, suggests it is a reduced form (e.g., not nitrate), either adsorbed dinitrogen or atmospheric amine impurities (*15*).

Unreacted SnO_2 specimens also exhibit a minute Si 2p band. This band is reduced only about 10% by masking the cut edge of the SnO_2–silica disk. It is reduced 10–20% by thorough washing of the SnO_2 surface with a 0.1M NaOH solution. Our initial interpretation of the residual signal was in terms of pinholes which penetrate the SnO_2 film to expose the underlying silica. Pinholes in SnO_2–silica obtained from Corning have been observed in scanning electron micrographs (*16*). Using a comparison to the Si 2p band intensity (14,200 counts/sec) on a clean soft glass disk, the pinhole interpretation leads to ~3% of the total SnO_2 area being penetrated. We subsequently have determined that argon sputtering of ~30 Å of SnO_2 surface completely eliminates the Si 2p signal on unreacted SnO_2 disks. This result indicates that the source of the Si signal is more likely a surface species deposited during or after the film preparation. The pinhole interpretation

would require sputtering-induced topological changes which are probably unlikely.

As an incidental illustration of the detection of adsorption on SnO_2 by ESCA, we soaked several unreacted SnO_2 specimens in $Cu(NO_3)_2$ solutions using either water or acetone as solvent, and following with extensive rinsing. A fairly intense copper spectrum with bands at Cu $2p_{1/2}$ 954.6 eV and Cu $2P_{3/2}$ 934.7 eV resulted. Other metal cations, such as Pb^{2+} and Fe^{3+}, also strongly adsorb on SnO_2 (*17*). A spectrum of crystalline $Cu(NO_3)_2$ exhibits the Cu 2p bands at 957.0 and 937.0 eV plus the multiplet satellites associated with the

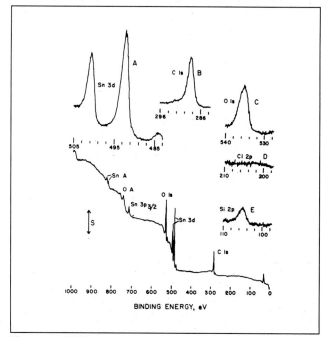

Figure 1. ESCA spectra of unreacted SnO_2 electrode

Lower curve, survey scan, S (sensitivity) = 40,960 counts/sec, 4 scans averaged; Curve A, Sn $3d_{5/2}$, $3d_{3/2}$, doublet at S = 81,920 counts/sec, one scan; Curve B, C 1s at S = 20,480 counts/sec, one scan; Curve C, O 1s at S = 20,480 counts/sec, one scan; Curve D, Cl 2p at S = 10,240 counts/sec, 8 scans averaged: Curve E, unresolved Si $2P_{3/2}$, $2p_{1/2}$ doublet, S = 10,240 counts/sec, 8 scans averaged

Table 1. Summary ESCA Data for SnO_2 Electrodes[a]

Electrode	Sn $3d_{5/2}$	Si 2p[b]	N 1s	NH^+ 1s	Cl $2p_{3/2}$[c]	d, Å[d]
Unreacted	487.7 ± 0.3	102.4 ± 0.2	400.5 ± 0.2		199.3	
	(114 ± 33)	(0.51 ± 0.23)	(0.96 ± 0.26)		(0.31)	
SnO_2/PrCl	487.7 ± 0.2	102.3 ± 0.2			200.2	9.1
	(50.1 ± 24)	(3.6 ± 0.4)			(25 ± 0.3)	
SnO_2/PrNH$_2$	487.5	102.4	400.3 ± 0.1	401.9 ± 0.1		5.1
	(72 ± 40)	3.9	(Sh)	(4.4 ± 2)		
SnO_2/en	487.4 ± 0.2	102.2 ± 0.2	399.3 ± 0.3	400.6 ± 0.2		10.6
	(43.5 ± 39)	(3.3 ± 2.0)	(2.8 ± 0.9)	(3.0 ± 0.4)		
SnO_2/Py	487.3 ± 0.2	102.4 ± 0.2	399.4 ± 0.4	401.8 ± 0.3		~1
	(105.0 ± 40)	(5.0 ± 1.7)	(3.7 ± 1.3)			

[a] Binding energies in eV; intensities (kilocounts sec^{-1}).
[b] Center of unresolved Si $2p_{1/2}$, Si $2p_{3/2}$ doublet.
[c] Cl $2p_{1/2}$ is a partially resolved shoulder on this band.
[d] Calculated from Equation 2 using λ = 11 Å.

Cu(II) state (18). These satellites are very weak in the SnO₂-adsorbed copper spectrum, and the 2p binding energies of the adsorbed copper are more typical of the Cu(I) or Cu(0) states than Cu(II) (18). While the apparent reduction of the copper might occur concurrently with its adsorption onto SnO₂, it is more probable that the adsorbed Cu(II) is thermally reduced to Cu(I) during the ESCA X-ray irradiation. Thermal reduction of CuO to Cu₂O has been observed (19) and we have seen prominent Cu(I) bands in spectra of small amounts of Cu(NO₃)₂ deposited on metal surfaces.

Cyclic voltammetry experiments on blank aqueous solutions (Figure 2) and solutions of model electrochemical reactants (Figure 3) gave results on unreacted PPG SnO₂ generally similar to literature reports (20, 21). Blank voltammograms do not depend particularly on solution pH except for the background potential limits. The cathodic and anodic limits (as defined at an arbitrary 5-μA level) move in the expected directions with pH change, the cathodic limit moving 60 mV/pH unit, the anodic limit moving at a lower pace. Cyclic voltammetric data for $Fe(CN)_6^{4-}$ in Table II show diffusion control and reversibility; resistance effects increase ΔE_p slightly over the reversible value at the higher sweep rates.

SnO₂/PrCl Electrodes. Tin oxide electrodes were reacted with 3-chloropropyltrichlorosilane as an example of an "inert" grouping. The 3-chloropropyltrichlorosilane reagent is quite reactive, fuming vigorously in moist air. ESCA data summarized in Table I show silicon (unresolved Si 2P₃/₂ and Si 2p₁/₂) and chlorine (Cl 2p₃/₂) bands appear after reaction, and the Sn 3d doublet diminishes in intensity as a consequence of the overlaying molecular film. The binding energy for Sn 3d

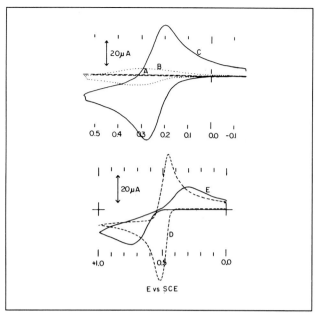

Figure 3. Cyclic voltammograms in 0.1M KCl, 0.5M glycine, pH 2.4, 100 mV/sec

Curve A, blank solution, unreacted SnO₂; Curve B, blank solution, SnO₂/en; Curve C, 2mM $Fe(CN)_6^{4-}$, SnO₂/en. Cyclic voltammograms in 0.1M KCl, pH 2.3, 2mM o-tolidine, 50 mV/sec. Curve D, unreacted SnO₂; Curve E, SnO₂/en; blank solution scans indistinguishable from zero on this current scale

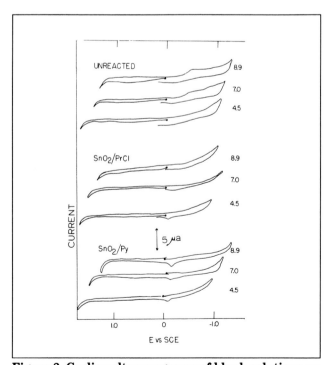

Figure 2. Cyclic voltammograms of blank solutions on unreacted SnO₂, SnO₂/PrCl, and SnO₂/Py electrodes

Numbers beside curves are pH values of phosphate buffers. 50 mV/sec. Dots represent sweep origin

Table II. Cyclic Voltammetric Data[a] for SnO₂/Py and SnO₂/PrCl

1m M ferrocyanide, 0.1M KCl, 0.5 M glycine, pH 2.4

Unreacted SnO₂

V, mV/sec	E_p^a	E_p^c	ΔE_p	$i_p^a/ac\,V^{1/2}$	i_p^c/i_p^a
5	0.268	0.214	0.054	10.7×10^2	0.91
10	0.270	0.212	0.058	10.6×10^2	1.00
50	0.279	0.205	0.074	11.1×10^2	1.00
100	0.289	0.197	0.092	11.1×10^2	1.00

2m M ferrocyanide, 0.1 M KCl, 0.5 M glycine, pH 2.4

SnO₂/Py

5	0.270	0.216	0.054	9.6×10^2	0.99
10	0.274	0.214	0.060	8.8×10^2	1.00
50	0.282	0.205	0.077	9.9×10^2	0.99
100	0.293	0.197	0.096	10.4×10^2	0.97

1m M ferrocyanide, 0.1 M KCl, 0.5 M glycine, pH 2.4

SnO₂/PrCl

10	0.294	0.176	0.118	7.1×10^2	1.00
20	0.290	0.176	0.114	7.5×10^2	0.80
40	0.294	0.167	0.127	7.1×10^2	0.95
100	0.305	0.167	0.138	6.1×10^2	0.93

[a] Current constant $i_p/ac\,V^{1/2}$ in A sec$^{1/2}$ cm/volt$^{1/2}$ mole. E vs. SCE, Electrode area = 0.031 cm².

electrons is insufficiently altered by the –Sn–O–Si– bonds to be resolved from lattice tin; we see no change in the Sn 3d binding energy for this or any other modified surface. Likewise, Si 2p binding energies are unchanged throughout the series of modified surfaces.

The intensity of the Sn 3d bands varies substantially from specimen to specimen (Table I). The average diminution of the Sn 3d band (from unreacted SnO_2) can be used to estimate the surface coverage achieved by the silanization reaction, using the relation

$$I_{Sn}/I_{Sn, \text{ unreacted}} = \exp[-d/\lambda] \quad (2)$$

where d is the silane film depth and λ is the escape depth ($\mathring{A}$) of Sn 3d photoelectrons through the surface film. (No electron escape angle correction is needed since $\theta = 90°$.) Taking $\lambda \sim 11 \mathring{A}$ for 487 eV Sn 3d electrons, which have 767 eV kinetic energy (22), $d \sim 9 \mathring{A}$ for the SnO_2/PrCl electrode film. Assuming unity density, this depth translates to 6×10^{-10} mole/cm^2, roughly monolayer coverage. The assumptions involved in this estimate are crude, particularly that of λ. Escape depth of photoelectrons through organics may well be several times larger than those in metals (23, 24), and so d values given in Table I may be low.

Blank solution cyclic voltammograms (Figure 2) at SnO_2/PrCl electrodes are not particularly distinguished from those on unreacted electrodes except for generally smaller charging currents, as would be expected from replacement of surface –SnOH groups with the neutral silane moiety. Cyclic voltammograms of $Fe(CN)_6^{4-}$ are likewise well-formed and undistinguished, except that ΔE_p (Table II) indicates lowered reversibility on the SnO_2/PrCl electrode. The derivatized surface is completely accessible to electrochemical reactant in a diffusional sense, but the effective (microscopic) electrode area must nonetheless be lowered by the attachment of the silane groups. The smaller microscopic area means higher current density, increased charge transfer rate limitations, and mild irreversibility.

SnO_2/Py Electrode. ESCA data for electrodes treated with β-trichlorosilyl-2-ethylpyridine are summarized in Table I. Bands for Si 2p and N 1s appear at intensities roughly the same as for other silanization reactions. Sn 3d intensity decreases as before, but to a significantly smaller extent. The meaning of the latter is not clear; it may reflect a surface structural difference which leads to a different λ for the Sn 3d photoelectrons.

When an electrode fresh from the acidic silanization reaction is rinsed thoroughly with hot benzene and its N 1s band observed, the N 1s appears as a doublet of approximately equal peaks. Rinsing with 0.1M HCl largely eliminates the lower binding energy peak; rinsing with base or distilled water does the reverse. These effects, illustrated in Figure 4, clearly are due to protonation of the pyridyl nitrogen. Electrodes treated with 3-aminopropyltriethoxysilane (SnO_2/PrNH$_2$) exhibit similar behavior on acid–base treatment (Table I). The shift in binding energy, 1.5 eV, between N 1s and NH$^+$ 1s agrees with earlier observations of protonation effects (23, 25). The absolute N 1s binding energies for the

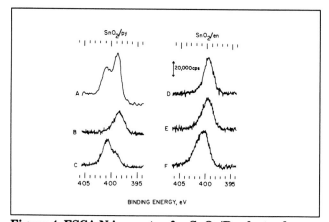

Figure 4. ESCA N is spectra for SnO_2/Py electrodes

Curve A, freshly prepared electrode washed with benzene, 32 scans summed; Curve B, washed with H$_2$O, 8 scans summed; Curve C, washed with 0.1M I, 8 scans summed. N 1s spectra for SnO_2/en electrodes, 15 scans summed: Curve D, rinsed In 0.1M NaOH; Curve E, rinsed in pH 6.9 buffer; Curve F, rinsed In 0.1M HCl

surface-bound pyridines and amines are roughly 1.5 eV higher than those reported for free bases (23).

The reactivity of SnO_2/PY as a ligand surface was further demonstrated by soaking the electrode in a dilute, neutral $Cr(NO_3)_2$ solution obtained by passing degassed 0.1M $Cr(NO_3)_3$ through a Jones reductor. Oxygen was passed through the solution for several minutes to reoxidize the Cr^{2+}. This procedure was repeated three more times, each reaction being followed by a copious (water) rinsing of the SnO_2/Py electrodes and removal of an electrode specimen for ESCA examination. After each of the first three steps, the Cr 2p band (575 eV) increased, with no increase recorded on the fourth reaction cycle. Concurrently, the N 1s band broadened toward higher binding energy, indicating overlap of two closely spaced peaks. No Cr 2p bands were observed on similarly treated unreacted SnO_2 or on SnO_2/Py electrodes soaked in unreduced $Cr(NO_3)_3$ solution.

Cyclic voltammograms of blank solutions (Figure 2) and of $Fe(CN)_6^{4-}$ (Table II) on SnO_2/Py show only minor differences from unreacted SnO_2 electrodes.

SnO_2/en Electrodes. A comparison of ESCA data for unreacted SnO_2 and that for SnO_2 reacted with 3-(2-aminoethylamino)propyltrimethoxysilane shows diminution of the Sn 3d band and appearance of Si 2p and N 1s bands. Protonation studies gave results analogous to those above. Exposure of a given specimen (or a series of SnO_2/en specimens) to continually higher pH solutions enhances the lower binding energy N 1s band at the expense of the higher (Figure 4). The total shift observed with pH was ~1.5 eV. We are unable at present to discern whether both amine nitrogens become protonated.

Electrochemical experiments on SnO_2/en in buffered blanks at various pH illustrate the double layer-modifying result of derivatization with a neutral base. In a pH 9.2 buffer, the level of charging current, which reflects the double layer capacitance, for unreacted SnO_2 and SnO_2/en electrodes is indistinguishable. When the pH is lowered, on the other hand, to pH 4, where the amine becomes protonated, the capacitance current for the SnO_2/en electrode is significantly enhanced

(Figure 5). The effect is that of an electrode positive of its point of zero charge, on which a cation is strongly adsorbed.

The charge given the SnO_2/en electrode surface by protonation is also apparently reflected in electrochemical behavior of a cationic reactant, o-tolidine. Figure 3 and Table III compare cyclic voltammograms for this reactant on unreacted SnO_2 and on SnO_2/en. It is apparent that the o-tolidine reaction on the latter is much less reversible than on unreacted SnO_2 surfaces. The cyclic voltammetric behavior of the anionic $Fe(CN)_6^{4-}$ reactant, on the other hand, is very nearly the same on unreacted SnO_2 and on SnO_2/en. An adsorption wave, traced to the glycine buffer component and peculiar to the SnO_2/en electrode, is probably the source of the lower quality i_p^c/i_p^a data for $Fe(CN)_6^{4-}$ on that electrode.

General Comments. The results demonstrate that SnO_2 surfaces exhibit a reactivity toward organosilanes quite analogous to that of silica surfaces. Further, in the absence of double layer complications, silanization of the SnO_2 surface does not adversely affect its usefulness as an electrode. We anticipate that organosilanes will provide a versatile route to many varieties of modified SnO_2.

As we noted in the introduction, preparation of surface sites is only the first step in attaining electrochemically useful chemically modified electrodes. We will report on further attachments of redox centers in a future communication. A variety of factors connected with the $-Sn-O-Si-$ surface linkages are also evident objects of continued study. With respect to the structure of the bonded layer, the questions of chain formation and the number of $Sn-O-Si$ bonds formed per silicon center will require synthesis and investigations of $XSiR_3$ and X_2SiR_2 reagents. The role of SnO_2 surface water in the silanization process has probably been oversimplified here, and metal ion adsorption on SnO_2 as opposed to coordination by attached surface ligands requires discriminating tests. Finally, the response of the double layer parameters to chemical events like amine protonation invites probing of the chemical events through this medium. The pK_b of a surface amine is potential dependent (7) and on a positively charged electrode is probably, for instance, substantially larger than the "ordinary," unattached amine.

LITERATURE CITED

(1) E. Grushka, Ed., "Bonded Stationary Phases in Chromatography," Ann Arbor Science Pub., Ann Arbor, Mich., 1974.
(2) H. H. Weetall, *Science*, **166**, 615 (1969).
(3) D. M. Hercules, L. E. Cox, S. Onisick, G. D. Nichols, and J. C. Carver, *Anal. Chem.*, **45**, 1973 (1973).
(4) G. Bruce Harper, *Anal. Chem.*, **47**, 348 (1975).
(5) A. A. Oswald, L. L. Murrel, and L. J. Boucher, Abstracts Div. Petroleum Chem., 168th National Meeting of the American Chemical Society, Los Angeles, Calif., 1974.
(6) K. G. Allum, R. D. Hancock, I. V. Howell, S. McKenzie, R. C. Pitkethly, and P. J. Robinson, *J. Organomet. Chem.*, **87**, 203 (1975).
(7) R. F. Lane and A. T. Hubbard, *J. Phys. Chem.*, **77**, 1401, 1411 (1973).
(8) G. Johansson, J. Hedman, A. Berndtsson, M. Klasson, and R. Nilsson, *J. Electron Spectrosc. Relat. Phenom.*, **2**, 295 (1973).

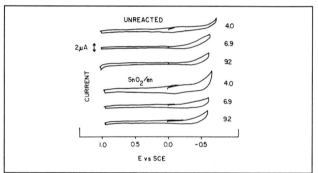

Figure 5. Cyclic voltammograms of blank solutions on unreacted SnO_2 and SnO_2/en electrodes

Numbers beside curves are pH values of buffers (4.0, acid phthalate; 6.9, phosphate; 9.2, borax)

Table III. Cyclic Voltammetric Data[a] for SnO_2/en Electrodes

2mM ferrocyanide, 0.1m M KCl, 0.5 M glycine, pH 2.4

	Unreacted SnO_2					SnO_2/en				
V, mV/sec	E_p^a	E_p^c	ΔE_p	i_p^a/ac $V^{1/2}$	i_p^c/i_p^a	E_p^a	E_p^c	ΔE_p	i_p^a/ac $V^{1/2}$	i_p^c/i_p^a
50	0.266	0.205	0.061	9.49×10^2	0.97	0.271	0.205	0.066	8.87×10^2	0.81
100	0.268	0.202	0.066	9.36×10^2	0.99	0.276	0.200	0.076	8.91×10^2	0.86
300	0.278	0.185	0.093	9.73×10^2	0.95	0.291	0.185	0.106	9.11×10^2	0.83
500	0.288	0.180	0.108	9.66×10^2	0.92	0.300	0.175	0.125	9.13×10^2	0.82

2mM o-tolidine, 0.1 M KCl, pH 2.3

	Unreacted SnO_2					SnO_2/en				
50	0.517	0.460	0.057	1.53×10^3	0.91	0.628	0.360	0.268	8.83×10^2	0.70
100	0.526	0.449	0.077	1.52×10^3	0.82	0.657	0.342	0.315	8.36×10^2	0.63
300	0.549	0.432	0.117	1.44×10^3	0.79	0.690	0.332	0.358	7.62×10^2	0.57
500	0.581	0.405	0.176	1.34×10^3	0.76	0.717	0.320	0.397	7.33×10^2	0.49

[a] Current constant i_p/ac $V^{1/2}$ in A $sec^{1/2}$ $cm/volt^{1/2}$ mole. Electrode area = 0.079 cm^2. E vs. SCE.

(9) J. C. Carver, R. C. Gray, and D. M. Hercules, *J. Am. Chem. Soc.*, **96**, 6851 (1974).

(10) T. Kuwana and N. Winograd, "Spectroelectrochemistry," in "Electroanalytical Chemistry," Volume 7, A. J. Bard, Ed., Dekker, 1974.

(11) R. N. Adams, "Electrochemistry at Solid Electrodes," Dekker, New York, N.Y., 1969.

(12) W. A. Aue and C. R. Hastings, *J. Chromatogr.*, **42**, 319 (1969).

(13) W. A. Aue, C. R. Hastings, J. M. Augl, M. K. Norr, and J. V. Larson, *J. Chromatogr.*, **56**, 295 (1971).

(14) I. V. Borisenko, A. V. Kiselev, R. S. Petrova, V. K. Chuikina, and K. D. Shcherbakova, *Russ. J. Phys. Chem.*, **39**, 1436 (1965).

(15) J. B. Sorrell and R. Rowan, *Anal. Chem.*, **42**, 1712 (1970).

(16) T. Kuwana, Ohio State University, private communication, 1975.

(17) H. A. Laitinen, University of Florida, private communication, 1975.

(18) P. E. Larson, *J. Electron Spectrosc. Rel. Phenom.*, **4**, 213 (1974).

(19) A. Rosencwaig and G. K. Wertheim, *J. Electron Spectrosc. Rel. Phenom.*, **1**, 493 (1973).

(20) J. W. Strojek and T. Kuwana, *J. Electroanal. Chem.*, **16**, 471 (1968).

(21) D. Elliot, D. L. Zellmer, and H. A. Laitinen, *J. Electrochem. Soc.*, **117**, 1343 (1970).

(22) P. W. Palmberg, *Anal. Chem.*, **45**, 549A (1973).

(23) K. Siegbahn et al., "ESCA, Atomic, Molecular, and Solid State Structure Studied by Means of Electron Spectroscopy," Almquist and Wiksells, Uppsala, 1967.

(24) B. L. Henke, *J. Phys. (Paris)*, **C4**, 115 (1971).

(25) L. E. Cox, J. J. Jack, and D. M. Hercules, *J. Am. Chem. Soc.*, **94**, 6575 (1972).

Received for review May 14, 1975. Accepted July 3, 1975. This research has been facilitated by the U.N.C. Materials Research Center under Defense Advanced Research Projects Agency Grant DAHC-157369, and by National Science Foundation Grant GP-38633X.

Reprinted from *Anal. Chem.* **1975**, *47*, 1882–86.

The great advances in technology experienced in the past 30 years have brought about sophisticated systems whose performance may be severely affected by small amounts of foreign materials. The need to identify such materials, along with the development of FT-IR spectroscopy in the 1970s, pointed to the potential for ultramicro IR spectroscopic analysis. The realization of this potential is described in this article, which led to the modern IR microscope. This instrument, combined with the high-sensitivity mercury cadmium telluride detector, has made IR microspectroscopy a standard technique in today's analytical chemistry laboratory.

J. E. Katon
Miami University

Fourier Transform Infrared Analysis below the One-Nanogram Level

R. Cournoyer,* J. C. Shearer, and D. H. Anderson
Industrial Laboratory, Eastman Kodak Company, Rochester, New York 14650

The interfacing of microscopy with Fourier Transform infrared (FTIR) spectroscopy is a useful combination allowing samples of less than 1 ng to be identified. The sample, usually supported by a thin sodium chloride plate, is centered in an aperture 50 to 200 μm in diameter. The sample mount is oriented in an 8X beam condenser in an FTIR spectrometer where multiscan signal averaging techniques produce a spectrum with the desired signal-to-noise ratio. One or two hours of total analysis time is generally required. Polymers and other solids as well as oils and various liquids have been identified. The small amounts of material require that the entire sample preparation be done under a microscope.

The characterization of samples too small to be visible to the naked eye is restricted to the domain of the microscope. In many cases characterizations can be made accurately and quickly with the microscope alone, but some microsamples do not yield to such analysis or do so only with great difficulty. These samples require the interfacing of modern instrumental techniques with classical microscopic analysis. The interfacing of microscopy with Fourier Transform infrared (FTIR) spectroscopy is a useful combination allowing preparations of less than 1 ng of sample to be identified. The sample, usually supported by a thin sodium chloride plate, is centered in an aperture 50 to 200 μm in diameter. The sample mount is oriented in an 8X beam condenser in an FTIR spectrometer where multiscan signal averaging techniques produce a spectrum with the desired signal-to-noise ratio. One or two hours of total analysis time is generally required. Polymers and other solids as well as oils and various liquids have been identified using this approach.

EXPERIMENTAL

Instrumentation. A stereo and compound microscope as well as microtomes, hot stages, or any other apparatus appropriate for the particular samples at hand are required. A fine pointed probe (dissecting needle), forceps, and other paraphernalia suitable for micromanipulation are also necessary. A Digilab FTS-14 Fourier Transform spectrometer equipped with the standard nichrome wire source and TGS detector, and a Perkin-Elmer 8X reflecting beam condenser were used to produce the spectra.

Sodium Chloride Plates. Thin sodium chloride plates (200–500 μm thick) are prepared by cleaving rock salt used in standard infrared work. The salt crystals are first cut into rectangles (approximately 1/4 × 1/2 cm) with a clean, single edged stainless steel blade. This rectangle is transferred with forceps to a clean microscope slide and placed under a stereo microscope. The rectangle is stood on edge and cleaved into two plates of equal thickness with a clean, unused single edged blade. The process is repeated until plates of the desired thickness are obtained. Plates without both faces freshly cleaved are discarded to avoid possible contamination. A plate is selected that is flat and has flawless domains, and is trimmed to 1–2 × 3–4 mm. Some infrared salt plates will not cleave satisfactorily and much aggravation can be avoided by searching out plates that are easily cleaved. A single (38 × 19 × 4 mm) infrared rock salt crystal will provide scores of the desired thin plates. Cleanliness throughout the preparation is imperative because of the small amount of sample involved in the analysis.

Aperture Disk. Perkin-Elmer micropellet disks with 250 or 500 μm apertures are the bases for sample mounts. Apertures, 50 to 200 μm, are fashioned under the microscope by punching holes of appropriate diameter in 1-mil brass or stainless steel shim stock with tapered watchmaker's reamers. The apertures should be flat and have clean edges. Some polishing is necessary with an Arkansas stone. The aperture, centered on the micropellet disk, is attached with glue or adhesive tape. The aperture disk surface and aperture should be inspected under the microscope for cleanliness prior to use.

Sample Preparation. The technique used for sample preparation is determined by the nature of the sample, the available equipment, and the manipulative skills and preferences of the microscopist. The sample should be prepared so that it will approximately fill the aperture, and so that it is only several micrometers thick. The thin salt plates are used to support the sample in an aperture of appropriate diameter. The smaller samples require the use of the smaller apertures. The sample may be too thin or, as more often the case, too thick. A general approach has been to make the sample as thin as possible. One approach for solids has been to place the sample on a microscope slide and roll it under a fine pointed probe tip; another approach is to press the sample between two clean microscope slides or a slide and a coverslip, then use a new, solvent washed razor blade to peel the material from the glass surface. Another technique which is suitable for solids and liquids is to place the sample between two freshly cleaved thin salt plates and apply gentle pressure with a probe. The plates must be flat and without irregularities; otherwise the pressure applied will crack the plates. When the two salt plates are separated, most or all of the sample generally remains on one of the plates. If the sample will not stay flat because of elasticity or surface tension, the plates need not be separated. Troublesome samples sometimes require the use of a small vise to hold the salt plates together.

Liquid sample handling is sometimes facilitated by transferring the liquid to a flexible film. The film is then arched with the liquid at the high point, and the sample transferred by touching the salt plate to the droplet. These are only the most generally applicable sample preparation techniques used in our laboratory. Other techniques exist and additional ones are continually being developed as different samples require. Small amounts of volatile materials may be handled by adapting cryogenic techniques familiar to many microscopists and other workers involved with low temperature sample manipulation.

Sample Mounting. The sample side of the salt plate is placed against the aperture disk surface when a single salt plate is being used. With this orientation, the sample sits in the aperture rather than some distance above it. The thin salt plates with the samples attached are secured to the aperture disks with a small amount of soft wax. The wax is applied to the end of the salt plate surfaces with a fine pointed probe. An aperture disk containing the sample is then placed in the FTIR spectrometer, and aligned by adjusting the disk position until the detector signal is maximized. FTIR spectra are accumulated until a spectrum with usable signal to noise ratio results. Higher gains and longer scanning times are required as the sample or aperture size decreases. We have found 50 μm to be the smallest usable aperture for our instrument. A Globar source and/or Mercury–Cadmium–Telluride detector will en-

able the use of somewhat smaller apertures; however, diffraction effects will prevent the use of apertures less than 30 μm.

RESULTS AND DISCUSSION

The sensitivity of the FTIR spectrometer has enabled infrared identification of increasingly smaller amounts of material (1–3). The exploitation of this sensitivity requires new approaches for sample presentation to the instrument. Investigations carried out in this laboratory indicate that miniaturization of the familiar KBr pellet would not give suitable preparations for samples at the nanogram level because of dilution and loss of sample, or impurities in the KBr or solvents. Even when these limitations were overcome, atmospheric contamination would often swamp out the sample signal. A single atmospheric "dust speck" is frequently 10 to 100 times the weight of the samples being considered. It was decided for these reasons that the samples would have to be examined neat. Sodium chloride was chosen as the most suitable support for the sample because of its infrared transmission and ease of cleavage. The use of microscopic techniques throughout the sample preparation has two advantages; the analyst is able to work with materials not visible to the naked eye, and microscopic physical and chemical separations enable relatively pure samples to be used, which greatly increases the probability of successful identification by infrared spectroscopy.

Figure 1 is an infrared curve of 6 ng of triphenyl phosphate. This curve was used to identify an isolated 6-ng sample found in a manufacturing environment. Total analysis time was approximately 2 h, of which 80% involved sample preparation.

Figure 2 is the spectrum of 3.4 ng of cellulose acetate film base in a 100-μm aperture, and that in Figure 3 is of 0.9 ng of the same sample in a 50-μm aperture. Both of these spectra are unsmoothed and were recorded after 2000 scans. The increased noise in Figure 3 shows a reduction in sample size and aperture. The spectrum in Figure 3 is identifiable as cel-

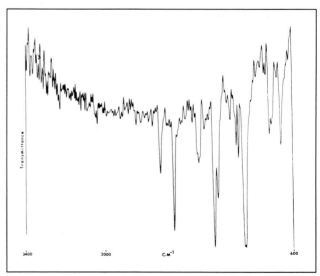

Figure 1. Spectrum of 6 ng of triphenyl phosphate in a 100-μm aperture scanned 1000 times at 8 cm⁻¹ resolution

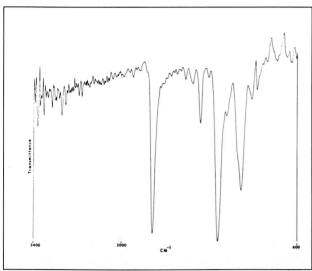

Figure 2. Spectrum of 3.4 ng of cellulose acetate with 10% triphenyl phosphate in a 100-μm aperture scanned 2000 times at 8 cm⁻¹ resolution

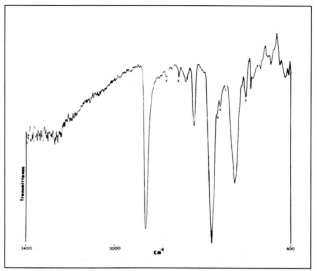

Figure 4. Spectrum of 0.9 ng of cellulose acetate with 10% triphenyl phosphate in a 50-μm aperture scanned 96 000 times at 8 cm⁻¹ resolution

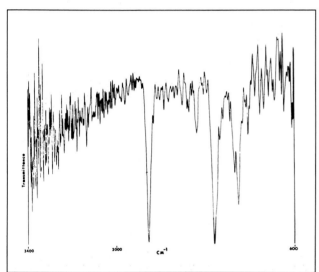

Figure 3. Spectrum of 0.9 ng of cellulose acetate with 10% triphenyl phosphate in a 50-μm aperture scanned 2000 times at 8 cm⁻¹ resolution

lulose acetate even though less than 1 h of spectrometer time and less than 1 ng of sample were used. Figure 4 shows the improvement obtained by prolonged scanning (~40 h) of the sample described in Figure 3. The noise has been reduced to the extent that absorptions due to 90 pg of triphenyl phosphate, which is present at approximately 10% in the film base, are recognizable. These absorptions have been marked with x's and may be compared with those in Figure 1.

Microscopic sample manipulation combined with the sensitivity of FTIR spectroscopy allows us to identify previously unidentifiably small amounts of material. Work is under way to construct a stronger beam condenser to reduce data collection time and to improve sensitivity. The combination of this condenser with refinements in sample preparation should result in at least an additional 10-fold reduction in minimum identifiable sample size. The obtainable spectra are of sufficiently high quality that spectral subtractions will be feasible. Work is now under way to demonstrate the practicality and utility of this technique. Successful infrared spectroscopic analysis of materials at the nanogram to picogram level indicates that this technique, with all its advantages, can be considered as an ultra micro method.

ACKNOWLEDGMENT

We thank Rex Wooton for his assistance in producing the spectra in Figures 1 through 4.

LITERATURE CITED

(1) P. R. Griffiths and F. Block, Paper 329, 23rd Pittsburgh Conference on Analytical Chemistry and Applied Spectroscopy, Cleveland, Ohio, 1972.
(2) S.S.T. King, *J. Agric. Food Chem.*, **21**, 526 (1973).
(3) D. H. Anderson and T. E. Wilson, *Anal. Chem.*, **47**, 2482 (1975).

Received for review July 18, 1977. Accepted September 26, 1977.

Reprinted from *Anal. Chem.* **1977**, *49*, 2275–77.

Packed Microcapillary Columns in High Performance Liquid Chromatography

Takao Tsuda[1] and Milos Novotny*
Department of Chemistry, Indiana University, Bloomington, Indiana 47401

New high-pressure LC (packed microcapillary) columns are described which consist of small adsorbent particles drawn inside glass capillaries of 50–200 μm internal diameters. A typical ratio of the column inner diameter to particle size is 2, and a significant plate-height reduction is achieved with decreasing particle size. With typical flow rates of several μL/min through such columns, modified injection and detection techniques are necessary. The columns of different inner diameters and particle sizes were evaluated through the reduced plate height vs. velocity plots. The effect of column coiling diameter on chromatographic performance was also studied. Whereas sample capacity of packed microcapillary columns is low, typical column efficiencies are significantly higher than those obtained with the hitherto available LC columns.

The basic types of columns available in analytical gas chromatography (GC) are conventional packed columns, open tubular (capillary) columns, micropacked columns (with typical internal diameters around 1 mm), and packed capillary columns. These columns differ widely in terms of separation efficiency and sample capacity. Selection of a column type in liquid chromatography (LC) has been more restricted because of viscosities and solute diffusivities in the liquid phase which are orders of magnitude different from the values in the gas phase.

The most efficient columns presently used in LC are those packed with totally porous small particles (with particle size down to several micrometers). The other column types, as known in GC, have not been sufficiently explored. This is pri-

marily due to the fact that the mass transfer is a diffusion-controlled process. Thus, capillary LC is unlikely to give desired efficiencies under the conditions of laminar flow.

Packed capillary columns studied extensively in GC by Halasz and Heine (1) and Landault and Guiochon (2) have rather unique characteristics. Their technology is different from other column types in that the adsorptive material is drawn inside the glass capillaries, in a way that is somewhat similar to the very common method for preparation of glass capillary helices (3). However, according to Halasz and Heine (1), the most important feature of such columns is the ratio of particle size to the internal column diameter. Whereas packed capillaries possess usual values of this ratio between 0.2 and 0.5, the conventional packed columns are typically well below 0.1. Thus, packed capillaries have certain geometrical characteristics of their own which are reflected in their analytical performance. Considering a rather loose packing of these columns used in GC, their efficiencies are quite high. Halasz and Heine (1) attribute this to the mobile-phase mixing effect that aids the radial diffusion and stress the importance of high column permeability.

When comparing the theoretical separating power of GC and LC with the actual situation, Giddings (4, 5) and Golay (6) note that, unlike in GC where a reasonable approach has been made to its theoretical limit through the capillary column, there is a discrepancy in LC of many orders of magnitude.

Giddings (4) derived a simple relationship between the limiting number of theoretical plates, $N_{\lim}$, and certain parameters involved in the chromatographic separation (time of analysis disregarded):

$$N_{\lim} = \frac{d_p^2 \Delta p}{4 \psi \gamma \eta D_M} \tag{1}$$

where ψ and γ are geometrical constants, d_p is the particle

---

[1] On leave from Department of Applied Chemistry, Faculty of Engineering, Nagoya University, Nagoya, Japan.

size, Δp is pressure gradient, η is viscosity, and D_M is solute diffusivity in the mobile phase. Thus, for the columns of the same geometrical characteristics used in GC and LC it applies that

$$\frac{N_{\text{lim}}\,(\text{GC})}{N_{\text{lim}}\,(\text{LC})} = \frac{\eta_L D_L}{\eta_G D_G} \approx \frac{1}{1000} \qquad (2)$$

When substituting typical values into Equation 1, N_{lim} for LC can be as much as 10^8 theoretical plates.

It also appears from the above considerations that the pressure gradient Δp (and, subsequently, inlet pressure) is the most important variable in achieving higher column efficiencies. In modern LC using columns packed with micrometer-size particles, pressures up to several hundred atmospheres are typically employed. Although substantial reductions of the plate height are achieved while decreasing the particle size, there are some practical limits to this procedure. As pointed out by Halasz et al. (7), there are difficulties in uniform packing of very small particles, as well as the problem with the evolved heat of friction.

The present investigation explores a version of packed capillary columns for high-performance LC. However, since both particle diameters and internal radii of such columns are typically an order of magnitude lower than those used in GC, we will refer to them as "packed microcapillaries" throughout the text. Special technology has been developed to achieve uniform packing of these columns. As demonstrated by the presented results, capacities of packed microcapillaries are significantly lower than those of typical LC analytical columns. However, column efficiencies are significantly higher. Since the injector and detector connections (split sampling and the use of a make-up liquid flow) are crucial to the column performance, this chromatographic technique is strongly reminiscent of the use of capillary columns in GC.

In the present work, several column types with different particle size and the column internal diameter were investigated. Van Deemter plots obtained with the packed microcapillaries are presented and their shapes are discussed. Since the plate height, H, of such columns appears to be strongly affected by hydrodynamic factors, a possible role of column coiling was briefly investigated. Evaluations were made for columns of identical internal diameter and particle size, but different column coil radii.

EXPERIMENTAL AND RESULTS

Preparation of Packed Microcapillaries. A commercial glass drawing machine (Hupe and Busch, Groetzingen, West Germany) was used for the preparation of columns. The drawing apparatus was placed in the vertical position while the packing material was introduced uniformly into the drawn microcapillary by means of gravity. The packing uniformity of the original glass column (typically, 5.5-mm o.d. and 0.25-mm i.d.) is very important; packing under the vibrator action aids this uniformity. The inner and outer column diameters can be varied through selection of the initial dimensions of glass tubing and drawing conditions. Because of

its relatively low melting temperature, soft glass is a preferred material. Reproducible preparation of columns with internal diameters below 100 μm is relatively easy. A typical wall thickness of our columns is 250–300 μm.

The uniformity of packing inside the microcapillaries will strongly affect column efficiency. Thus, selection of a uniform particle size fraction is initially desirable. Whereas it is technically feasible preparing packed microcapillaries from a variety of materials that do not undergo undesirable irreversible changes during the drawing process, such columns require some variations of technology. The present fundamental investigations of packed microcapillaries were limited to the use of acidic and basic alumina packing materials. These were 30 μm and 10 μm activated alumina (LiChrosorb, AloxT, from E. Merck Reagents, Darmstadt, West Germany), 100 μm activated alumina (from M. Woelm, Eschwege, West Germany), as well as the corresponding particle size fractions obtained from the coarse column chromatography materials after grinding, mechanical sieving, and sedimentation sizing process.

Although various columns with different particle sizes and diameters were prepared and their analytical properties are described below, we presently consider columns with internal diameters between 50 to 80 μm, packed with particles around 30 μm, to be a good compromise between analytical performance and minimum difficulties of operating conditions (sample size, available inlet pressure, and sample introduction). The columns have a uniform appearance along their entire length. A microphotograph of a typical section of the alumina-packed microcapillary is shown in Figure 1. Technological importance of the ratio of capillary inner diameter to particle size must be stressed. Although the columns of larger diameters packed with small particles (e.g., 140-μm i.d. and 30-μm particle size) have uniform appearance, their structure collapses under the operating conditions, and clogging occurs.

Whereas the radius of most studied columns was 6 cm, the

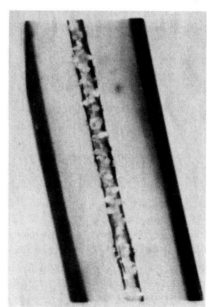

Figure 1. Microphotograph of a section of alumina-packed microcapillary. Column inner diameter, 75 μm; average particle size, 30 μm

drawing apparatus had to be provided with home-made coiling tubes for the preparation of columns coiled into greater or smaller radii.

Analytical System. The high-pressure syringe pump and UV-detector of the Varian Model 4100 liquid chromatograph were used throughout the measurements. Since the problems of dead volume are considerably more severe with these column types than with the conventional LC columns, both injection and detection techniques were modified. In order to overcome these problems, we resorted to the commonly practiced techniques of capillary GC (sample splitting and the use of additional liquid flow at the column outlet). Figure 2 shows the schematic diagram of the system used, whereas Figure 3 details the parts of injecter/splitter and the make-up solvent assembly.

Typical split ratios around 1:1000 resulted in the use of maximum detector sensitivity. However, the detector response is decreased by the use of an extra liquid flow to overcome the detector cell dead volume. Whereas such flow was around 20–50 µL/min, typical flow rates through 70-µm i.d. packed microcapillaries are less than 10 µL/min. As judged from reproducibility of retention times, flow appears constant among individual runs, once the inlet pressure is set to a desired value.

It should be emphasized that this work primarily addresses itself to fundamental column performance investigations. Thus, better sample introduction techniques and detectors with small internal volumes and greater sensitivities are needed for future work.

Column Efficiency Studies. Three different ratios of particle size and inner diameter were used in preparation of packed microcapillaries: 100/200, 30/70, and 10/50 µm. Van Deemter plots were measured for standard UV-absorbing solutes, introduced onto the columns in the amounts close to 10^{-8} g. As expected, values of the height equivalent to a theoretical plate, H, were reduced with decreasing particle size, dp. Figure 4 demonstrates differences in H vs. v (average linear velocity of the mobile phase) curves for a 100-µm compared to a 30-µm particle column. Whereas the minima of these H vs. v plots apparently lie at considerably lower velocity values than measurable, the plate heights at the lowest velocity values of Figure 4 roughly correspond to those theoretically predicted (7):

$$H_{\min} = A + 2\sqrt{BC} \approx 3.6\, d_{\mathrm{p}}$$

Shapes of H vs. v curves are important for both understanding the physical phenomena occurring in the columns as well as their analytical utility. Obviously, the measured curves fail to fit the usually proposed equations of liquid chromatography (7, 19).

Whereas the initial nonlinearity of the measured curves is not unusual for some LC columns and can be explained by the coupling theory of Giddings (8), there appears no simple explanation for the further rise of H with flow rate. It should be stressed that this general curve shape (rise, a plateau, and a second steeper rise) is highly reproducible with sequentially prepared columns of the same particle size. As indicated below, the curve shape holds also for other d_{p} values, although

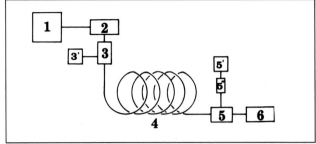

Figure 2. Microcapillary liquid chromatograph: 1, high-pressure syringe pump; 2, injection point; 3, splitter; 3′, metering valve to adjust split ratio; 4, packed microcapillary column; 5, make-up liquid mixing piece; 5′, liquid reservoir; 5″, flow restricting valve; 6, detector

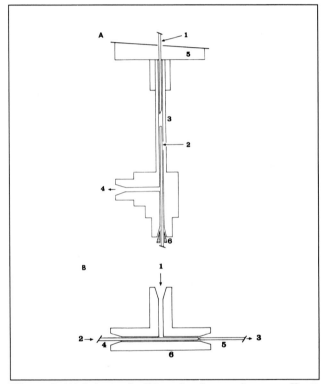

Figure 3. (A) Injector/splitter: 1, microsyringe needle; 2, column inlet; 3, 1.5-mm i.d. tube; 4, direction of discard flow; 5, injector flange; 6, polyimide ferrule. (B) Make-up liquid mixing piece: 1, entering make-up liquid; 2, column effluent; 3, liquid entering detector cell; 4, column outlet; 5, detector connecting capillary; 6, mixing tee

particle size or packing geometry appears to determine the second onset of the curves. It is also unlikely that mechanical changes of the column bed due to high pressure and/or flow shear are implicated; repeatedly measured curves gave identical shapes.

Plots of reduced values (8) are generally considered as good indication of the column's chromatographic quality across a range of particle sizes. Figure 5 shows the reduced plots for three columns of different particle size. The plots were obtained using a nonretained (or only slightly retained) solute, benzene, in order to dissociate mass-transfer processes from

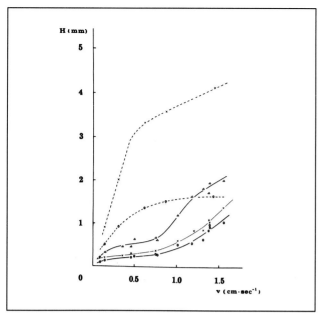

Figure 4. Plate height vs. velocity curves for standard solutes run on 100-μm (broken line) and 30-μm (solid line) particle size columns. O = benzene; x = methyl benzoate (retention relative to benzene, 1.34 for 100-μm column and 1.14 for 30-μm column); Δ = quinoline (retention relative to benzene, 1.71). To adjust approximately the same relative retention on both columns, the mobile phase was hexane with 0.03% and 0.05% methanol, respectively, for a 30-μm and 100-μm column. Column lengths were 14 m and 40 m for 30-μm and 100-μm columns, respectively

those of hydrodynamic nature. The initial h values of the smallest-particle column (10 μm) are somewhat higher than those of other two columns. This is most likely a result of our present insufficient packing technology at this particle size and less homogeneous column bed. Figure 5 further demonstrates the effect of particle size on band broadening under the dynamic conditions. The shapes of curves obtained with retained solutes on the columns with different particles are nearly identical (Figure 6). With the current column technology, less than 10% column-to-column variation in H at a given velocity and retention time was observed.

If hydrodynamic factors are involved in the zone broadening phenomena, it is of some interest to investigate whether the column coil radius can influence the plate height as well. This situation received some attention previously (9–11) with different column types. The results obtained with the columns of identical inner diameter and particle size, but different column coil radii, are shown in Figure 7.

Although analytical applications of packed microcapillaries will be a matter of future research, two important questions are of immediate interest: first, what are the typical column efficiencies to be expected and, second, what sample sizes can be tolerated.

A chromatogram of a standard mixture obtained on a "typical" packed microcapillary is shown in Figure 8. Whereas the total number of theoretical plates obtained on this run is not identical with the so-called effective plates (it is estimated that some 70% of the column space is occupied by the mobile

phase), there is every indication of achieving column efficiencies approximately an order of magnitude higher than those currently available in practice of modern LC.

Sample capacity of a 10-μm particle column was investigated using benzene, methyl benzoate, and dimethyl phthalate introduced through the splitter in varying amounts (Figure 9). Obviously, the plate height increases with sample size. Microcapillaries of somewhat wider diameters and greater lengths will be more tolerant to slightly larger samples. Nevertheless, a need for sensitive detection is clearly suggested for these columns.

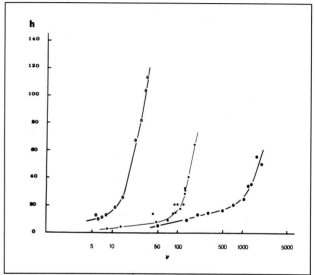

Figure 5. Reduced plate height vs. reduced velocity curves for benzene measured on O = 10-μm particle size column (length, 3.6 m), ■ = 30 μm (length, 14 m), and ● = 100 μm (length, 40 m). Mobile phase: same as Figure 4

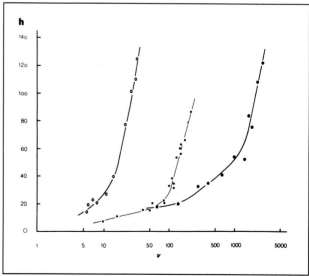

Figure 6. Reduced plate height vs. reduced velocity curves for retained solutes on same columns as indicated in Figure 5, under the same conditions. Solutes of similar relative retention were used: pyridine (10-μm column); quinoline (30-μm column); and methyl benzoate (100-μm column)

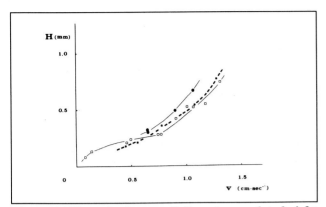

Figure 7. Effect of column coil diameter on plate height vs. velocity curves. Solute: benzene; all columns were 12 m × 75 μm i.d., packed with 30-μm alumina. Mobile phase: hexane with 0.05% methanol. Column coiling diameters: ○, 30 cm; □, 12 cm; and ■, 5.7 cm

DISCUSSION

The novel LC columns described in this paper may have utilization in three analytically important directions: (1) availability of greater resolving power; (2) the overall miniaturization of LC systems; and (3) increased compatibility of LC with mass spectrometry or other ancillary techniques that require low mobile-phase flow rates. The obtained results are discussed in view of these trends below.

Small-particle column technology has drastically improved the state of modern LC since several years ago. *H* values lower than 20 μm have been reported (7) for short columns packed with 4-μm particles. Whereas difficult-to-resolve pairs of solutes (e.g., various isomers) can frequently be separated through mobile-phase selectivity adjustments in LC, a similar degree of resolution in GC may necessitate a "brute force" approach, i.e., a great number of theoretical plates. However, it is the overall efficiency of a given chromatographic column which is instrumental in separating complex mixtures. Alternatively, it is often of advantage to combine high plate numbers with selectivity. Just as with GC where the plate height values of capillary and best packed columns are roughly comparable, the column length that can be practically utilized for a given separation problem becomes crucially important.

Since the inlet pressure remains a single most important parameter in gaining higher efficiencies (5), longer columns are likely to be more emphasized in future LC separations. Present pressure limitations of the chromatographic equipment are less severe than is commonly believed. On the other hand, the approach of decreasing particle size with a simultaneous increase of column length in a usual way (slurry-packed columns) has its limitations. Namely, these include the evolved heat of friction, difficulties in packing extremely small particles, and joining shorter column sections together. The packed microcapillary columns described in this paper may provide an alternative solution. Their column technology is relatively easy.

The described columns are quite unique, both structurally and analytically. Whereas they are a microversion of packed capillaries used previously in GC, their analytical characteristics differ significantly from the irregularly packed columns

investigated by Halasz and Walkling (12). These authors observed a turbulence-caused decrease of *H*. Their plate height values typically ranged from 2 to 10 mm.

Van Deemter plots constructed for various solutes and measured on the packed microcapillaries of varying internal diameter and particle size show quite unusual behavior. Whereas *H* values as low as 100 μm were achieved at very low flow rates and the curves remained flat over an analytically useful range of flow velocities, a sudden increase of *H* is observed at higher velocities. Interestingly enough, this increase demonstrates itself in the areas of velocities roughly comparable to those of decreasing *H* of the irregularly packed columns of Halasz and Walkling (12). Thus, some secondary flow effects are suggested. However, their interpretation may be a difficult matter.

Reduced plate heights are comparatively low for microcapillaries filled with different particles. Furthermore, such columns can be prepared with good reproducibility. The preparation of columns with 10-μm particles packed into 50-μm capillaries is somewhat exceptional, since they yield consistently less satis-

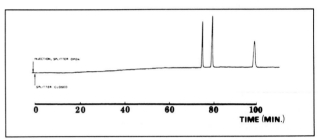

Figure 8. Chromatogram of the standard mixture on a 29 m × 75 μm i.d. (30-μm particle size) alumina-packed microcapillary column. Mobile phase: *n*-hexane with 0.05% methanol, flow linear velocity, 0.65 cm/s. Solutes: benzene, methyl benzoate, and quinoline (in the order of elution); number of theoretical plates for quinoline = 85 000; inlet pressure, 1500 psi; column permeability, 6.1×10^{-8}

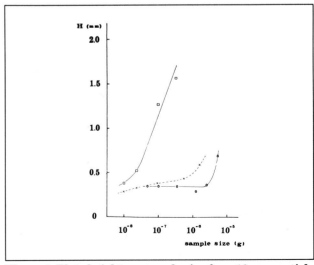

Figure 9. Plate height vs. sample size for a 10-μm particle size (3.6 m long) microcapillary. Mobile phase: hexane with 0.05% methanol, flow linear velocity, 0.4 cm/s. Solutes: ○, benzene; ×, methyl benzoate; □, dimethyl phthalate; UV-detector sensitivity, 0.02–0.16 OD, full scale

factory results. The most likely reason for this is insufficient uniformity of 10-μm packing in the original (0.25-mm i.d.) glass tube. Indeed, particles of this size are known to be packed with difficulties in a dry state. In addition, the higher ratio of column diameter to particle size may also play some role.

Although the packing technology for small particles may still be improved in the future, we presently consider columns with 70-μm i.d., packed with 30-μm particles, to be a good compromise. Without excessive inlet pressures, good efficiencies are obtained. Adequate sample capacity is yet another consideration.

Since the packed microcapillaries of great lengths are coiled into helices, it has been of some interest finding whether there is some effect of column coiling on their performance. Long ago, such a possibility was considered in chromatography by Giddings (9, 10). A simple relation was derived (10) for the plate height contribution due to coiling:

$$H = \frac{7 v r^4}{12 R^2 \gamma D_M} \qquad (3)$$

where v is flow velocity, r is column radius, R is coil radius, γ is the obstruction factor of the B term of the van Deemter equation, and D_M is the solute diffusion coefficient in the mobile phase. This contribution to the plate height primarily arises from the different velocities of the molecules encountered in different distances from the column wall. Whereas in gas chromatography, this contribution can be significant perhaps only in preparative-scale work, greater caution must be exercised in high-pressure liquid chromatography because of the considerably smaller values of D_M. The effect of column shape in LC was experimentally explored by Barth et al. (11). Significant H increases were observed with columns of conventional diameters, but no particular changes were observed for a 0.76-mm i.d. tube.

Even though the microcapillary radius is very small to amount to an appreciable increase of H through the coil effect (9, 10) or the limitations by flow of the lateral mass transfer (13), some secondary flow phenomena could occur in the complex structure of packed microcapillaries. Because of the complexity of hydrodynamic conditions inside the column, Reynolds number would probably be a useless criterion. Whereas mass transfer is increased because of the secondary flow in open tubular columns (14), coiling may have only a minor effect on the performance of packed microcapillaries. Certainly, Figure 7 seems to indicate that. The differences in measurements with the columns of different coil radii are indeed very small and within the reproducibility range of column technology.

Miniaturization of equipment without sacrificing its resolving power is an important trend in analytical separation methods. As recently shown by Ishii (15), a miniature LC equipment is feasible that employs detector cells of very small volume. The packed microcapillaries described here naturally lend themselves to miniaturization, with typical flows of a few microliters per minute. Admittedly, the instrumental conditions used in this work are less than ideal and the necessary instrumentation has yet to be designed. In particular, smaller detection cells are needed.

In order to utilize more effectively the separating power of this technique, longer columns and correspondingly higher pressures are needed. The columns with diameters and lengths as described in this work seldom required inlet pressures over 200 atm. However, with the typical liquid flow volumes of these columns, there should be no major technological problems to utilize pressures around 1000 atm. Chromatography carried out under the extreme pressure conditions was demonstrated for both dense gases (16) and liquids (17).

Possibilities of combining LC with MS have now been under focus for several years. It appears that some technical problems of a successful coupling of these methods arise from excessive column liquid flows. Again, the described columns (or any other small-diameter columns) can potentially reduce these problems.

Whereas the present study is confined to one particular type of chromatography (the liquid–solid separation principle) and the alumina adsorbent, other column materials are worth exploring. Thus, in situ modifications of alumina with either salts (18) or liquids are clearly feasible, as is the preparation of bonded phases with this material (19). As described by Halasz and Heine (1), particles of graphitized carbon black can also be drawn inside the glass capillaries if special precautions are made. Obviously, a choice of sorption materials is limited to those of sufficient thermal stability. Finally, if the very useful siliceous materials are to be used, special techniques must be developed to regenerate the thermally lost surface hydrated structures. The surface silanol groups are needed for the preparation of chemically bonded stationary phases of different selectivities.

LITERATURE CITED

(1) I. Halasz and E. Heine, Adv. Chromatogr., 4, 207 (1967).
(2) C. Landault and G. Guiochon, in "Gas Chromatography 1964," A. Goldup, Ed., British Petroleum Institute, London, 1965, p 121.
(3) D. H. Desty, J. N. Haresnape, and B.H.F. Whyman, Anal. Chem., 32, 302 (1960).
(4) J. C. Giddings, Anal. Chem., 36, 1890 (1964).
(5) J. C. Giddings, Ref. 2, p 3.
(6) M.J.E. Golay, Chromatographia, 6, 242 (1973).
(7) I. Halasz, R. Endele, and J. Asshauer, J. Chromatogr., 112, 37 (1975).
(8) J. C. Giddings, "Dynamics of Chromatography," Marcel Dekker, New York, N.Y., 1965.
(9) J. C. Giddings, J. Chromatogr., 3, 520 (1960).
(10) J. C. Giddings, J. Chromatogr., 16, 444 (1964).
(11) H. Barth, E. Dallmeier, and B. L. Karger, Anal. Chem., 44, 1726 (1972).
(12) I. Halasz and P. Walkling, J. Chromatogr. Sci., 7, 129 (1969).
(13) J. C. Giddings, "Dynamics of Chromatography," Marcel Dekker, New York. N.Y., 1965, p 52.
(14) R. Tijssen, Chromatographia, 3, 525 (1970).
(15) D. Ishii, K. Mochizuki, and Y. Mochida, Abstracts of the 1977 Pittsburgh Conference on Analytical Chemistry and Applied Spectroscopy, Cleveland, Ohio, No. 385.
(16) L. McLaren, M. N. Myers, and J. C. Giddings, Science, 159, 197 (1968).
(17) B. A. Bidlingmeyer, R. P. Hooker, C. H. Lochmüller, and L. B. Rogers, Sep. Sci., 4, 439 (1969).
(18) C. G. Scott and C.S.G. Phillips, Ref. 2, p 226.
(19) J. H. Knox and A. Pryde, J. Chromatogr., 112, 171 (1975).

Received for review July 20, 1977. Accepted November 7, 1977.

Reprinted from Anal. Chem. 1978, 50, 271–75.

This pioneering article on laser desorption reported the production of protonated and cationized molecules from bioorganics with little fragmentation. This surprising result was thought to arise from rapid heating, allowing the desorption of intact species to occur before decomposition. This article set the stage for further developments that have revolutionized the ability to study large biomolecules by MS. For example, matrix-assisted laser desorption ionization has become a valuable tool for the analysis of proteins with masses of tens, and even hundreds, of thousands of daltons.

Michelle Buchanan
Oak Ridge National Laboratory

Laser Desorption–Mass Spectrometry of Polar Nonvolatile Bio-Organic Molecules

M. A. Posthumus[1], P. G. Kistemaker,* and H. L. C. Meuzelaar
FOM-Institute for Atomic and Molecular Physics, Kruislaan 407, 1098 SJ Amsterdam, The Netherlands

M. C. Ten Noever de Brauw
Central Institute for Nutrition and Food Research CIVO-TNO, Zeist, The Netherlands

Polar, nonvolatile organic molecules were analyzed with a laser induced desorption technique. In this technique a thin layer of sample coated on a metal surface is exposed to a submicrosecond laser pulse, producing cationized species of the *intact* molecules and some fragments. These ions are mass analyzed by a magnetic sector type mass spectrometer equipped with a simultaneous electro-optical ion detection system or in high resolution experiments by a double focusing mass spectrometer with photoplate detection. For oligosaccharides, notorious for their nonvolatility and thermal lability, the ions due to cationized original molecules and building blocks are the most prominent. Despite the intense laser pulse, there is little elimination of water from the molecule, if any. With this method, we have successfully analyzed, for instance, digitonin, a digitalis pentaglycoside with a mass of 1228 amu, and underivatized adenylyl-(3′,5′)-cytidine (ApC, M = 572 amu).

In mass spectrometry, the study of "nonvolatile" polar and thermally labile compounds gives rise to many problems. These can be partially circumvented by chemical derivatization of the polar groups, thereby producing molecules with higher vapor pressures. However, the large increase in molecular weight, proportional to the number of polar groups present, constitutes a basic limitation to this approach. As shown for a number of oligopeptides, spectra of involatile substances can be obtained in some cases by direct exposure of the sample to the ion plasma of a chemical ionization source (1). Also field desorption can be used for a wide range of non-

volatile bioorganic molecules as recently reviewed by Schulten (2). This method is perhaps most widely used now for mass spectrometric analysis of nonvolatiles although routine application is still hampered by a moderate reproducibility due to many influencing factors like emitter quality, emitter temperature, and sample impurities (3).

A different approach is to attempt reproducible flash pyrolysis in vacuum to obtain volatile fragments, still bearing structural information on the original compound (4–6). This method has been applied successfully in the classification of highly complex biological material.

Recently a new class of desorption techniques has been introduced. In the last year, a number of authors reported the desorption of intact molecules of highly involatile substances upon the impact of energetic particles on the sample material coated on a metal substrate. Ions, neutrals as well as photons, have been used as bombarding particles in quite different energy domains. Macfarlane et al. (7) and Krueger (8) used MeV energy fission products of californium-252 to induce the desorption of protonated and deprotonated molecular particles. Primary ions in the key energy range were used by Benninghoven, with results comparable to the ^{252}Cf–plasma desorption technique data (9). Moreover, already in 1968, Vastola and co-workers observed the production of ions during the irradiation of organic samples with a high intensity ruby-laser pulse (10).

In this article we report new results of laser induced desorption mass spectrometry. It is shown that this technique is not confined to organic salts (11), but can also be applied successfully to the analysis of an extensive range of polar, nonvolatile organic substances as often encountered in biochemical or clinical practice. Ionization occurs mostly by attachment of an alkali cation to the desorbed molecules. To demonstrate the possibilities of laser ionization we report,

[1] Present address: Laboratory of Organic Chemistry, Agricultural University, Wageningen, Netherlands.

among other things, on the analysis of underivatized oligosaccharides, cardiac glycosides, and nucleotides.

EXPERIMENTAL

Simultaneous Ion Detection. The short duration of the ion burst upon the laser pulse precludes the use of a scanning type of mass spectrometer. Therefore time-of-flight mass spectrometers were generally used in experiments thus far reported in the literature (10, 11). In this research, however, we used a home built magnetic sector mass spectrometer (see Figure 1), in which part of the spectrum is projected on a chevron CEMA detector (channeltron electron multiplier array, Galileo Optics Corp., Sturbridge, Mass.). The secondary electron output of the CEMA is proximity focused on a phosphor screen, coated on a fiber optic window, which serves also as a vacuum feedthrough for the resulting optical line spectrum. This image is detected and digitized with a vidicon camera, coupled to a 500-channel OMA (optical multichannel analyzer, SSR Instrument Co., Santa Monica, Calif.). The characteristics of this simultaneous ion detector are described elsewhere in full detail (12). To obtain optimal benefit of this detector system, the mass spectrometer is equipped with an ion optic "zoom" lens, consisting of an electrostatic quadrupole lens between the magnet and the detector and a magnetic quadrupole lens at the source side of the sector magnet. With the electrostatic lens, it is possible to change the mass dispersion, while the magnetic lens can move and rotate the focal plane to the desired position producing a sharp line spectrum on the detector plane (13). The ion source is normally operated at an accelerating voltage of 1500 V. With the maximum field strength of our sector magnet (9 kG, radius 15 cm) this results, for example, in a simultaneously covered mass range of m/e 530–635 with a resolution of 300. Sacrificing ion source efficiency, the mass range could be extended to about m/e 1450 by lowering the accelerating voltage to 650 V. By adjusting the quadrupoles the mass ratio of 1.2 can be extended to 1.6 with some loss of resolution. This relatively small mass range is

due to the limited dimensions of the flight tube and channelplate. Optimum design of these dimensions may allow a mass ratio of 4:1 (14).

During the laser desorption process, ionized particles are produced. Therefore, the filament of the electron impact source was turned off during the measurements. Only for taking calibration spectra of perfluorotributylamine and tri(perfluoroheptyl)-s-triazine (PCR Inc., Gainesville, Fla.), the normal EI mode was used.

The laser used to irradiate the sample was a TEA-CO_2 laser (GenTec, Dalton, Quebec, Canada), with a pulse energy of 0.1 J released in 0.15 μs. The laser pulse was focused by a germanium lens (f = 128 mm) on the sample material, resulting in a power density of about 1 MW/cm^2. By translation of the lens, the focal spot, and thus the power density, could be varied.

Sample Handling. The samples were dissolved in analytical grade methanol or in distilled water (0.1–1 mg/mL). Of these solutions, 5-μL drops were applied to the stainless steel probe, 6 mm in diameter. Subsequent evaporation of the solvent generally results in an invisible surface layer of the compound. For the methanolic solutions, electrospraying (15) proved to be an excellent coating procedure, resulting in a homogeneous surface layer. The probe was then positioned in the ion source via the direct insertion system.

High Resolution Measurements. High resolution measurements were performed with a SM 1 double focusing mass spectrometer (Varian MAT, Bremen, G.F.R.) with photoplate detection. The mass spectrometer was equipped with a Neodymium-glass laser (Korad K-MV, Union Carbide, Santa Monica, Calif.) mounted on a modified ion source. The output of this laser is 1 J in a 100-μs pulse, and the spot size is about 0.1 mm. In this case the samples were compressed to small pellets, about 2-mm diam., in a probe tip on which the laser could be focused via a f = 63 mm objective. The Ilford Q2 photoplates were evaluated with a Leitz comparator connected to a Varian SS100 computer system. Several laser pulses were needed to obtain sufficient blackening on the plate.

RESULTS AND DISCUSSION

Oligosaccharides. To start our investigations on laser ionization, we chose the disaccharide sucrose as a model compound. Contrary to glucose, which can still be evaporated without extensive thermal fragmentation (16), sucrose is practically nonvolatile, as recently confirmed by Li$^+$ ion attachment ionization experiments in high electrical fields (17). So raising the temperature of a sucrose sample in a direct probe tip by several hundred degrees is of no use to generate intact sucrose molecules in the gas phase. Also heating of 1 μg sucrose in vacuum within 0.2 s to 510 °C by means of the Curie-point method (18) results in extensive thermal fragmentation to products with low masses (19). Also laser heating of sucrose with relatively long exposure times results in complete thermal decomposition (20). However, the heating rate rather than absolute temperature appears to govern the competition between evaporation and degradation processes (21).

Further, the use of a mild ionization method based on ion attachment, e.g., through protonation or cationization, is essen-

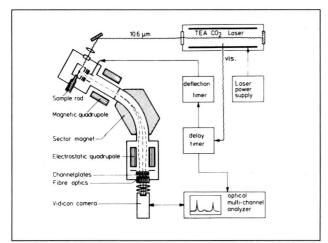

Figure 1. Block diagram of the experimental set-up. The delay circuit and time window controller have a resolution of 0. 1 μs and are designed for time resolved experiments. The delay circuit is connected with the laser cavity by an optical guide line and triggered by the visible light produced in the predischarge

tial in these cases because of the instability of the molecular radical cation of compounds like saccharides generated upon electron impact. This prerequisite, now, is fulfilled under our experimental conditions as, upon laser heating, alkali metal ions are generated in high density.

In our low resolution experiments, we used therefore submicrosecond laser pulses. Hitting the probe surface, coated with 2 µg sucrose on 28 mm², with a slightly defocused beam (spot diameter ca. 0.3 mm) generates an intense signal at m/e 365 and m/e 381, as shown in Figure 2a. These peaks are obviously due to alkali cation attachment in the laser-plasma to intact desorbed sucrose molecules. Not *any* trace of expulsion of water or methanol, as found for example in field desorption using activated emitters at BAT conditions (2), could be detected. The first peaks next to these quasimolecular ion peaks are found at m/e 185 and 203 (fragments + Na⁺) and their potassium analogues at m/e 201 and 219, respectively. These fragments must be due to cleavage at the glycosidic oxygen accompanied by hydrogen transfer. The assignment of the above-mentioned peaks as cationized species has been confirmed by exact mass measurements of the corresponding peaks found with the high resolution mass spectrometer.

Figure 2b gives the spectrum of sucrose obtained with the Nd-glass laser and the double focusing instrument. Here the potassium ion complexes are more abundant, but except for the occurrence of a peak corresponding to the fragment (180 − 2H₂O + K⁺), the results are essentially the same as compared to the CO₂ laser desorption spectrum. In the low mass range of both spectra, peaks corresponding with Na⁺, K⁺, H₂ONa⁺ and H₂OK⁺ can be observed.

In spite of the fact that the lasers in the two experiments emit at totally different wavelengths, the resulting mass spectra are highly similar, indicating that the desorption process

is not strongly influenced by the wavelength. Moreover, the desorption of large organic molecules by the Nd-glass laser, emitting at a wavelength of 1.06 µm where most organic molecules show only weak absorption bands, points to the importance of fast heating of the substrate rather than to direct heating (viz., vibrational excitation) of the organic molecules.

Obviously the laser desorption spectrum gives information on the molecular weight and the mass of the saccharide units. Because the two constituents of sucrose, viz., fructose and glucose, are isobaric, no further conclusions can be drawn in this case without labeling experiments.

It should be noted that the occurrence of the alkalis in the spectrum of sucrose, as shown in Figure 2, must be due to impurities in the probe surface (stainless steel, 18% Cr, 8% Ni) and in the sucrose sample. A second laser pulse on the same spot still produces a spectrum, although with much lower intensity. However, the abundancy of the potassium ion species is then considerably reduced with respect to the sodium species. Doping the sample with an alkali halogenide simplifies the spectrum since the corresponding cationized species then dominate completely. This is also true for doping with lithium salts, but the expected increase in ion production, as found by Giessmann in lithium ion attachment experiments in high electrical fields (17), is not observed under our conditions. All spectra discussed in this article are measured without addition of alkali halogenide salts to the sample.

Besides sucrose some other underivatized oligosaccharides have also been studied. The results, summarized in Table I, are essentially the same as for sucrose, except for a slight elimination of water, observed in some cases. The molecular weight can be determined without any difficulty and the mass of the glycosidic units (in these cases only 180 amu) can be derived from the spectrum. Neither formation of sugar cluster ions of the type [nM + C]⁺, as found in field desorption mass spectrometry of saccharides (22, 23) nor formation of doubly charged ions [M + 2C]²⁺ is observed in our experiments.

Glycosides. The usefulness of the laser induced desorption–cationization method in the analysis of oligosaccharides made it very attractive to investigate glycosides also, a very important class of compounds in pharmacology. We confined our attention to some *strophantus* and *digitalis* glycosides (Serva Feinbiochimia, Heidelberg, and E. Merck, Darmstadt, G.F.R.). Upon laser irradiation, the cationized species of the intact molecule is formed in high yield for all these compounds. Furthermore, many fragments due to expulsion of one or more glycosidic units (in the higher mass range), are observed as well as the glycosidic units themselves and possible fragments of these units and of the aglycone (in the lower mass range). This is illustrated for digoxin in Figure 3. As for the oligosaccharides, there is no loss of water and of other small molecules from the original molecule itself. If the structure of digoxin (C₄₁H₆₄O₁₄, M = 780 amu) is represented as M = HOSOSOSOR (in which HOSOH stands for the glycosidic unit digitoxose and ROH for the aglycon digoxigenine), besides the cationized molecule M (m/e 803) also the cationized fragments OSOSOR (m/e 673), S:2OR (m/e 507), ROH (m/e 414), SO (m/e 153), HOSOH (m/e 171), SOSO (m/e 283),

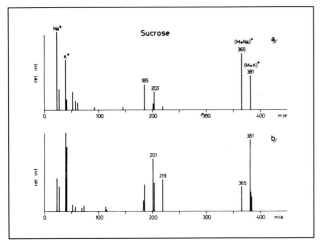

Figure 2. Laser desorption mass spectrum of sucrose (C₁₂H₂₂O₁₁, M = 342 amu). The upper spectrum (a) has been obtained with the CO₂ laser and channelplate detector. Isotopic peaks have been omitted in this and all other figures. Figure 2b shows the spectrum obtained with the double focusing instrument equipped with photoplate detection. In this case, potassium metal ion complexes are more abundant probably due to contamination of the probe by KCl (note the small KCl · K⁺ peaks at m/e 113 and 115

Table I. Intensity of Cationized Molecules and Building Block Fragment Peaks in the Laser Desorption Spectra of Underivatized Oligosaccharides[a]

	M, amu	M^b	M − 18	M − 162	M − 180	M − 324	M − 342	M − 486	M − 504	M − 522
Glucose	180	100	0	—	—	—	—	—	—	—
Sucrose	342	100	0	32	24	—	—	—	—	—
Gentiobiose[c]	342	100	6	15	6	—	—	—	—	—
Raffinose	504	98	15	100	25	28	26	—	—	—
Stachyose[d]	766	81	4	100	30	64	34	33	92	68

[a] Note: Generalizing the formula of the oligosaccharides as $HO(SO)_nH$, the (M − 18) species has the structure $(SO)_n$, (M − 162) represents $HO(SO)_{n-1}H$, etc.
[b] M stands for the cationized species.
[c] In field desorption MS, the protonated molecular ion gives rise to a peak of only 30%, the (M + 1 − 18) peak being 100% (22).
[d] For stachyose in field desorption instead of the molecular ion, a (M + 18 + 1) species has been found (22).

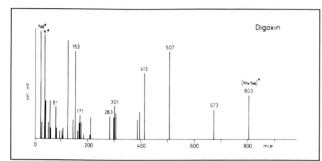

Figure 3. Laser desorption mass spectrum of digoxin ($C_{41}H_{64}O_{14}$, M = 780 amu)

HOSOSOH (m/e 301), and SOSOSO (m/e 413) can be recognized. The off-scale peaks at lower mass range are due to K^+ (m/e 39) and $NaCl \cdot Na^+$ (m/e 81 and 83), which are present as impurities. As also observed for other investigated glycosides, there are on the one hand cationized $M-(SO)_n$ and in some cases $M-HO(SO)_nH$ fragments, and on the other hand $(SO)_n$ and $HO(SO)_nH$ cationized species. All underivatized glycosides show a relatively intense molecule peak. Still difficult to understand in this connection, however, is the formation of the fragment S:2OR (m/e 507); generated by loss of $C_{12}H_{24}O_8 \cdot 2(HOSOH)$, from digoxin.

A clear exception to the above-mentioned characteristics of LD-spectra of glycosides is formed by the monoglycoside ouabain (M = 584 amu). For this substance only the quasi-molecular ion peak at m/e 607 and the cationized aglycon ouabagenin at m/e 461 are found, but not any trace of ions due to the rhamnose residue could be detected. These results suggest that fragmentation occurs after the sodium attachment to the molecules, since in case of cleavage of the glycosidic bond prior to cationization, the rhamnose moiety should have been found.

Contrary to the oligosaccharides investigated (vide supra), the building blocks in most glycosides do not all have the same elemental composition, but show more variety, for example: digitoxose $C_6H_{12}O_4$ (M = 148), xylose $C_5H_{10}O_5$ (M = 150), and rhamnose $C_6H_{12}O_5$ (M = 164). Therefore more differences in the spectra could be observed although diffentiation between isobaric constituents is also not possible in this case. Figure 4 displays a typical spectrum trace of the pentaglycoside digitonin $C_{56}H_{92}O_{29}$, directly obtained from the memory of the optical multichannel analyzer, and showing the quasi-molecular ion $(M + Na^+)$ at m/e 1251, accompanied by lithium and potassium analogues. Further some other peaks can be observed, the most prominent ones probably due to $(M-HOS_1OH)$ and $(M-S_2O)$ fragments, in which S_1 corresponds with the terminal xylose residue and S_2 with the terminal glucose residue. One has to keep in mind that, because of the relatively weak sector magnet (9 kG, 15-cm radius) in our apparatus, the cationized molecule of this compound at m/e 1251 could be detected only by reducing the accelerating voltage to a value as low as 650 V. This results in a lower ion source efficiency and in a reduced resolving power of 150 under static conditions.

Note that the peaks in the laser ionization spectrum appear to be broader than those of the reference compound, taken with electron impact ionization at 70 eV. There are two effects contributing to this peak broadening: (i) Because of the low resolution, isotope peaks are not separated from the main peak. In the case of digitonin the (M + 1) and the (M + 2) isotope peaks should have an intensity of 64% and 26% respectively, of the main peak. The m/e 1166 peak of the reference compound corresponds with a composition $C_{24}F_{44}N_3$, for which the isotopic peak intensity can be calculated to be only 26% and 3%, respectively. (ii) Under laser irradiation conditions, ions can be formed with higher kinetic energy than upon electron impact ionization. In a single focusing instrument, this gives rise to peak broadening, especially at reduced acceleration voltage as used for this spectrum.

Nucleotides. One of the most interesting classes of nonvolatile bioorganic compounds, where ionization techniques for intact molecules and fragments would be of great value, are nucleotides. Because of their extremely low volatility, derivatization is an absolute necessity for EI-mass spectrometric investigation of these compounds (24). Using FD-MS, Schulten et al. were able to analyze unprotected mononucleotides (25) and some dinucleoside (26, 27). However, there are still many pitfalls in the FD-MS of nucleotides (3).

The nature of these compounds is quite different as compared to the saccharides and glycosides and the building

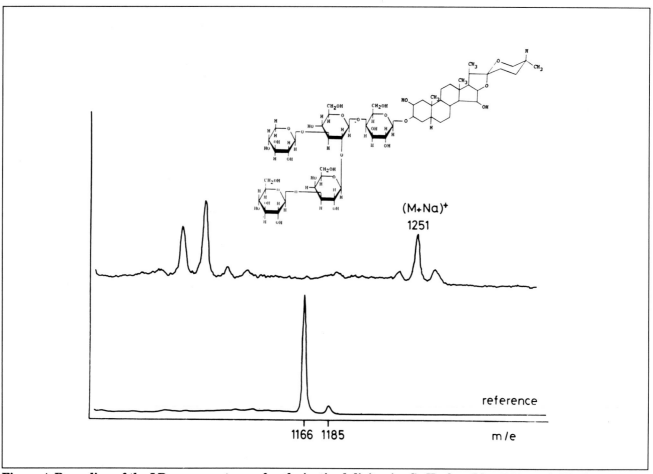

Figure 4. Recording of the LD mass spectrum of underivatized digitonin, $C_{56}H_{92}O_{29}$, (M = 1228 amu). The lower trace gives the spectrum of a reference compound which has been taken with the ion source in the electron impact mode (70 eV). Note: the relatively low resolution and signal intensity are mainly due to the low accelerating voltage (see text)

blocks are not isobaric, thus more structural information should be obtainable from fragments, if present.

As expected, nucleotides do not present special problems, since they can still be analyzed by CI mass spectrometry (28). In the spectra of adenosine (Figure 5a) and guanosine (not shown), the quasimolecular ions form the highest peaks and also strong fragment peaks corresponding with the intact base are present, but peaks due to the ribose moiety are quite low or absent. Besides cationized species, protonated ones also can be observed for these compounds.

In the spectra of the underivatized mononucleotides, both the free acid as well as the disodium salt yield cationized and protonated molecular ions in high abundance as can be seen in Figure 5b and c for adenosine 5′-monophosphoric acid and the disodium salt, respectively (AMP, AMP-Na$_2$, obtained from Serva, Heidelberg, G.F.R.). But there is also extensive fragmentation, generating the nucleoside (m/e 290), the base (m/e 136, 158, and 174) corresponding with (B + H)⁺, (B + Na)⁺, and (B + K)⁺, respectively, and the phosphoric acid residues, for example HPO$_3$H⁺ (m/e 81) in AMP, NaPO$_2$Na⁺ (m/e 109), NaPO$_3$Na⁺ (m/e 125), and Na$_2$HPO$_4$Na⁺ (m/e 165) in AMP-Na$_2$. As found for nucleosides, the contribution of the sugar moiety to the LD spectrum is only small (m/e 137 and 155). In the spectrum of AMP also, a peak corresponding to

M + 1–H$_3$PO$_4$ (m/e 250) and the product of expulsion of phosphoric acid via the 3′,5′-cyclic AMP intermediate can be observed (m/e 232 and 254), which product is also found in FD spectra (26) and in CI, using activated emitters as direct probe, as recently reported by Hunt et al. (29). Also, small contributions due to cyclic phosphate esters of the ribose moiety can be recognized at m/e 217 and 235.

Preliminary experiments involving dinucleoside phosphates gave results that are essentially the same as those for the mononucleotides, as can be seen in Figure 6. This spectrum shows the laser induced desorption mass spectrum of underivatized adenylyl-(3′,5′)-cytidine (ApC, M = 572 amu). Also for this compound the identification of the molecular weight is unequivocal because of the presence of (M + H)⁺, (M + Na)⁺, and (M + K)⁺ ions at m/e 573, 595, and 611, respectively.

Furthermore, the cationized peaks corresponding to the bases (m/e 150 and 174), cytidine (m/e 282), cytidinephosphoric acid (m/e 362), the loss of adenine from ApC (m/e 476), the loss of adenine and H$_2$O from ApC (m/e 458), and metaphosphoric acid (m/e 119) can also be observed.

Amino Acids and Oligopeptides. Much attention has been paid already to the mass spectrometry of underivatized compounds of this group. Most of the common α-amino acids can still be evaporated from a direct probe tip and subse-

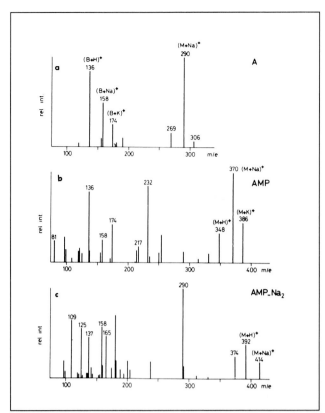

Figure 5. LD-mass spectra of adenosine (A), adenosine-5'-monophosphoric acid (AMP), and adenosine-5'-monophosphate disodium salt (AMP-Na$_2$)

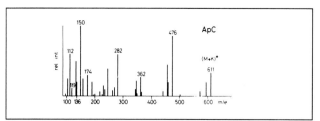

Figure 6. LD mass spectrum of underivatized adenylyl-(3',5')-cytidine (ApC, M = 572 amu)

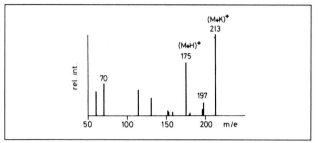

Figure 7. LD mass spectrum of arginine (M = 174 amu)

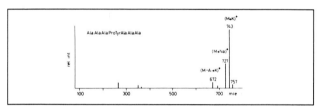

Figure 8. LD mass spectrum of the octapeptide Ala$_3$ProTyrAla$_3$ (M = 704 amu). The peak at m/e 757 is likely due to an impurity (M = 718 amu)

quently analyzed by EI or CI, but arginine, for example, fails to give molecular ions under these conditions (30, 31). However, with FD (32) or special CI probe techniques like evaporation from Teflon (33) or from an activated FD emitter probe (29), spectra of arginine have been obtained with a protonated molecular ion and intense fragment ion peaks corresponding to the loss of ammonia and the guanidino group.

Recently it has been reported that a special FD technique, in which electrolytic solutions were coated onto untreated wire emitters, yields quasi-molecular ions for arginine without fragmentation (34).

Also laser induced desorption yields merely unfragmented arginine ions as is demonstrated by the high relative intensities of the peaks at m/e 175, 197, and 213 (M + H$^+$, M + Na$^+$, and M + K$^+$, respectively) in Figure 7. Very small peaks due to the loss of ammonia are seen at m/e 158, 180, and 196, but peaks due to the loss of water are not observed. The peak at m/e 70 is the base peak in CI spectra (29, 31) and can be explained by expulsion of HCOOH and HNC(NH$_2$)$_2$ from the protonated molecular ion (31).

Unfortunately the tendency of little fragmentation of amino acids as found for arginine (Figure 7), cystine, and methionine (not shown) extends to polypeptides. The few oligopeptides that we have investigated so far yield intense cationized molecular ion peaks, for the pentapeptide AlaProTyrAlaAla (M = 491 amu) m/e 514 (100% rel. int.) and 530 (16% rel. int.) are the only peaks observed, except for the usual alkali cation peaks themselves. Also for the octapeptide AlaAlaAlaPro-

TyrAlaAlaAla (M = 704 amu), the cationized molecular ions at m/e 727 and m/e 743 form the base peaks and some fragmentation occurs; see Figure 8. Besides the M–A$_1$ fragment at m/e 672, no sequential information is obtained. In these cases the use of collisional activated dissociation of the molecular ion peaks would be necessary (vide infra).

Miscellaneous Compounds. Finally, to explore the potentials of laser induced ionization, we further analyzed steroid conjugates (glucuronides, sulfates and phosphates, free acids, as well as salts), antibiotics (penicillin derivatives), chlorophyll, and sodium acetate.

All these compounds yield cationized molecular ions in high abundance and therefore not all spectra will be shown here. However, two compounds invite some remarks. In the spectrum of estriol-3-phosphate disodium salt (Steraloids, Pawling, N.Y.) shown in Figure 9, many intense peaks can be observed. Nearly all of these can be assigned to inorganic ions, viz. sodium chloride (NaCl · Na$^+$ at m/e 81 and 83) and polyphosphate ions, which are probably due to impurities (buffer residue?) in the sample. Laser desorption of Na$_2$HPO$_4$ results in a spectrum that contains all these phosphate peaks, though with different relative intensities. These peaks are of interest because of their complex nature and their supposed involatility, and are listed in Table II.

It is interesting to note that they resemble the inorganic cluster ions found in pyrolysis-field desorption (Py-FD) mass

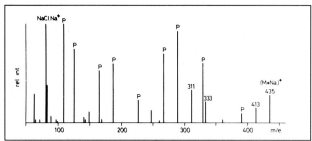

Figure 9. LD mass spectrum of a commercial estriol-3-phosphate disodium salt preparation (M = 412 amu). Peaks marked with letter P are assigned as corresponding with polyphosphate ions (cf. Table II)

Table II. Inorganic Phosphate Ions Found in the Laser Desorption Spectra of Estriol-3-phosphate-Na$_2$ Salt (see Figure 9) and of Na$_2$HPO$_4$

m/e	Elemental composition	m/e	Elemental composition
109[a]	NaPO$_2 \cdot$ Na$^+$	227	Na$_2$P$_2$O$_6 \cdot$ Na$^+$
125	NaPo$_3 \cdot$ Na$^+$	267	Na$_3$HP$_2$O$_7 \cdot$ Na$^+$
143[b]	NaH$_2$PO$_4 \cdot$ Na$^+$	289	Na$_4$P$_2$O$_7 \cdot$ Na$^+$
165	Na$_2$HPO$_4 \cdot$ Na$^+$	329	Na$_3$P$_3$O$_9 \cdot$ Na$^+$
187[a]	Na$_3$PO$_4 \cdot$ Na$^+$	391	Na$_5$P$_3$O$_{10} \cdot$ Na$^+$

[a] Also found for Na$_3$PO$_4$ (other ions not searched for).
[b] Very small intensity.

spectrometry of DNA (35), besides that the ions found in LD contain in a few cases only a hydrogen atom. In the Py-FD experiments these clusters are the result of initial thermal breakdown of the organic sample, but in the case of LD of estriol-3-phosphate-Na$_2$, the phosphate ions are more likely to originate from inorganic impurities as they are not found for other organic phosphates (e.g., nucleotides, vide supra). Besides these phosphate ions and the molecular ion peaks, fragments at m/e 311 and 333 can also be seen. These peaks, also present in the LD-spectrum of estriol-3-glucuronide, correspond with estriol and probably the estriol sodium derivative, respectively.

Real cluster ions have been found in laser ionization of organic compounds only in the case of sodium acetate thus far. The spectrum is shown in Figure 10, where intense peaks corresponding to ions of the type $(n\,M + Na)^+$ (M stands for CH$_3$COONa) can be observed. Obviously the desorption process for this relatively simple ionic substance proceeds in a different way as compared to substances such as sucrose, which could be due to the stability of the sodium acetate salt lattice. It is interesting to compare this result with essentially similar results obtained by field desorption mass spectrometry (36) and completely different results obtained by spark source mass spectrometry (37), where fragments dominate the spectrum of sodium acetate and no clusters are observed. The greater complexity of spark source spectra of organic compounds was also found in the analysis of amino acids and their salts, where lines are seen at nearly every mass position up to about m/e 80 (38).

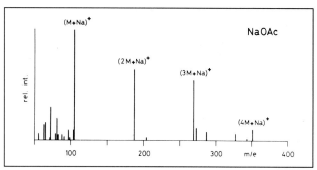

Figure 10. LD mass spectrum of sodium acetate (M = 82 amu)

On the Ionization Process. The mechanism of the generation of cationized molecular ions is not fully understood yet. However, the fast heating in vacuum under nonequilibrium conditions appears to promote an uptake of sufficient translational energy to cause desorption before relaxation into unstable vibrational modes leads to decomposition (21). Further, the lag of vibrational temperature is a well known phenomenon in flash processes.

At the same time, either thermoionization of alkali atoms from the metal surface (unavoidably present as impurity) or thermal mobilization of alkali metal ions in the sample occurs. Experiments by Holland et al. on heated emitters without a strong external field (39) suggest that field-independent chemical ionization may occur in a thin semi-fluid layer, in which the temperature is a determining factor. These considerations favor desorption of the molecules as cationized species.

On the other hand in a laser plasma, induced on a solid target, a high concentration of alkali ions is always present (40). Although ionization potentials of alkali atoms from metals (work functions) are lower than gas phase ionization potentials, the presence of a metal surface seems not to be necessary since in the high resolution experiments the laser pulses were directed on bulk organic material. The energy of the alkali ions appears to be rather low. From peak broadening of the m/e 23 and 39 peaks, it could be concluded that this energy is well below 10 eV. Thus the conditions in our experiments are much milder than those described in the literature on laser interaction with solids (41), where 200 eV alkali metal ions are measured at a laser power density estimated to be 10^7 W/cm^2.

The alkali ions have a high affinity to hydroxylic and other polar functions (42), thus enabling the generation of cationized species (43). It is likely that in the complexes formed the charge remains localized on the alkali moiety, because of their low ionization potential (ca. 5 eV) as compared to the ionization potential of organic molecules (8–10 eV). This is favorable for the stability of the ion (44) as is the even electron character of the complex.

Whether ionization takes place in the semi-fluid layer or in the relatively dense plasma plume, in both cases the laser generates a high density of alkali metal ions, thus facilitating efficient ionization of the sample molecules. Time resolved experiments (see caption, Figure 1) and detailed studies on negative ions are planned to get more insight into the ionization process. We have measured already relatively intense (M–H)$^-$

peaks for amino acids, peptides, and nucleotides. More detailed results will be published in a separate article.

Also the mechanism of the fragmentation reactions, thermal fragmentation prior to cationization or decomposition of the cationized molecular ion complex, is not clear yet. The almost total absence of elimination of water and other small stable molecules, being the dominating reactions in flash pyrolysis (5, 45), and the absence of the rhamnose peak in the LD spectrum of ouabain is most noteworthy and indicates the fragmentation of the molecule after the cationization.

CONCLUSIONS AND FUTURE DEVELOPMENTS

Upon laser pulse irradiation, distinctly nonvolatile substances (nucleotide-Na_2 salt) or compounds liable to decompose upon heating (sucrose) generate cationized molecular ions in high abundance. Furthermore, a number of other ions are produced, which appear to represent building blocks and other essential fragments, thereby giving valuable structural information. But in some cases (peptides), only little fragmentation is observed. In order to obtain more structural information on stable quasimolecular ions we are now building a tandem mass spectrometer for collisional activated (CA)-MS. By application of simultaneous ion detection this instrument will be capable of recording CA spectra of cationized molecular ions generated by short lasting phenomena as LD or FD (46).

The results obtained up to now strongly indicate that laser induced desorption–cationization has promising potentials for the analysis of nonvolatile and thermally labile compounds in mass spectrometry.

ACKNOWLEDGMENT

Thanks are due to M.M.J. Lens for taking part of the spectra and to F. L. Monterie for expert technical support. Furthermore the authors are indebted to J. Kistemaker for his continuous stimulation of this work.

LITERATURE CITED

(1) M. A. Baldwin and F. W. McLafferty, *Org. Mass Spectrom*, **7**, 1353 (1973).
(2) H.-R. Schulten, in "Methods of Biochemical Analysis," Vol. 24, D. Glick, Ed., Interscience-Wiley, New York, N.Y., 1977, pp 313–448.
(3) H. Budzikiewicz and M. Linscheid, *Biomed. Mass Spectrom.*, **4** 103 (1977).
(4) P. P. Schmid and W. Simon, *Anal. Chim. Acta*, **89**, 1 (1977).
(5) M. A. Posthumus, N.M.M. Nibbering, A.J.H. Boerboom, and H.-R. Schulten, *Biomed. Mass Spectrom.*, **1**, 352 (1974).
(6) P. G. Kistemaker, A.J.H. Boerboom, and H.L.C. Meuzelaar, *Dyn. Mass Spectrom.*, **4**, 139 (1975).
(7) R. D. Macfarlane and D. F. Torgerson, *Int. J. Mass Spectrom. Ion Phys.*, **21**, 81 (1976).
(8) O. Becker, N. Fürstenau, W. Knippelberg, and F. R. Krueger, *Org. Mass Spectrom.*, **12**, 461 (1977).
(9) A. Benninghoven and W. Sichtermann, *Anal. Chem.*, in press.
(10) F. J. Vastola and A. J. Pirone, *Adv. Mass Spectrom.*, **4**, 107 (1968).
(11) R. O. Mumma and F. J. Vastola, *Org. Mass Spectrom.*, **6**, 1373 (1972).
(12) H. H. Tuithof, A.J.H. Boerboom, and H.L.C. Meuzelaar, *Int. J. Mass Spectrom. Ion Phys.*, **17**, 299 (1975).
(13) H. H. Tuithof and A.J.H. Boerboom, *Int. J. Mass Spectrom. Ion Phys.*, **20**, 107 (1976).
(14) H. H. Tuithof, Ph.D. Thesis, Delft (1977).
(15) E. Bruninx and G. Rudstam, *Nucl. Instrum. Methods*, **13**, 131 (1961).
(16) H. D. Beckey, "Field Ionization Mass Spectrometry," Pergamon Press, Oxford, 1971, p 307.
(17) U. Giessmann and F. W. Röllgen, *Org. Mass Spectrom.*, **11**, 1094 (1976).
(18) M. A. Posthumus, A.J.H. Boerboom, and H.L.C. Meuzelaar, *Adv. Mass Spectrom.*, **6**, 397 (1974).
(19) M. A. Posthumus, Amsterdam, unpublished results, 1977.
(20) W. K. Joy and B. G. Reuben, *Dyn. Mass Spectrom.*, **1**, 183 (1970).
(21) R. D. Macfarlane and D. F. Torgerson, *Science*, **191**, 920 (1976).
(22) J. Moor and E. S. Waight, *Org. Mass Spectrom.*, **9**, 903 (1974).
(23) J.-C. Prome and G. Puzo, *Org. Mass Spectrom.*, **12**, 28 (1977).
(24) A. M. Lawson, R. N. Stillwell, M. M. Tacker, K. Tsyboyama, and J. A. McCloskey, *J. Am. Chem. Soc.*, **93**, 1014 (1971).
(25) H.-R. Schulten and H. D. Beckey, *Org. Mass Spectrom.*, **7**, 861 (1973).
(26) H.-R. Schulten and H. M. Schiebel, *Fresenius' Z. Anal. Chem.*, **280**, 139 (1976).
(27) H.-R. Schulten and H. M. Schiebel, *Nucl. Acids Res.*, **3**, 2027 (1976).
(28) M. S. Wilson, I. Dzidic, and J. A. McCloskey, *Biochim. Biophys. Acta*, **240**, 623 (1971).
(29) D. F. Hunt, J. Shabanowitz, F. K. Botz, and D. A. Brent, *Anal. Chem.*, **49**, 1160 (1977).
(30) G. Junk and H. Svec, *J. Am. Chem. Soc.*, **85**, 839 (1963).
(31) P. A. Leclercq and D. M. Desiderio, *Org. Mass Spectrom.*, **7**, 515 (1973).
(32) H. U. Winkler and H. D. Beckey, *Org. Mass Spectrom.*, **6**, 655 (1972).
(33) R. J. Beuhler, E. Flanigan, L. J. Greene, and L. Friedman, *J. Am. Chem. Soc.*, **96**, 3990 (1970).
(34) H. J. Heinen, U. Giessmann, and F. W. Röllgen, *Org. Mass Spectrom.*, **12**, 710 (1077).
(35) H.-R. Schulten, H. D. Beckey, A.J.H. Boerboom, and H.L.C. Meuzelaar, *Anal. Chem.*, **45**, 2358 (1973).
(36) H.-R. Schulten and F. W. Röllgen, *Org. Mass Spectrom.*, **10**, 649 (1975).
(37) R. P. Buck and J. R. Hass, *Anal. Chem.*, **45**, 2208 (1973).
(38) W. L. Baun and O. W. Fischer, *Anal. Chem.*, **34**, 294 (1962).
(39) J. F. Holland, B. Soltmann, and C. C. Sweeley, *Biomed. Mass Spectrom.*, **3**, 340 (1976).
(40) J. F. Ready, "Effects of High-Power Laser Radiation," Academic Press, New York, N.Y., 1971, p 149.
(41) E. Bernal, J. F. Ready, and F. J. Allen, *J. Appl. Phys.*, **45**, 2980 (1974).
(42) I. Dzidic and P. Kebarle, *J. Phys. Chem.*, **74**, 1466 (1970).
(43) B. E. Knox. *Dyn. Mass Spectrom.*, **2**, 61 (1971).
(44) F. W. Röllgen and H.-R. Schulten, *Z. Naturforsch. A*, **30**, 1685 (1975).
(45) M. A. Posthumus and N.M.M. Nibbering, *Org. Mass Spectrom.*, **12**, 334 (1977).
(46) H. H. Tuithof, A.J.H. Boerboom, P. G. Kistemaker, and H.L.C. Meuzelaar, *Adv. Mass Spectrom.*, **7**, in press (1977).

Received for review December 20, 1977. Accepted March 6, 1978. Part of this work has been presented at the 4th International Symposium on Mass Spectrometry in Biochemistry and Medicine, held June 20–22, 1977, in Riva del Garda, Italy, and at the International Mass Spectrometry Symposium Natural Products, 28 August–2 September, 1977, in Rehovot, Israel. This work is part of a research project sponsored by the Organization for Fundamental Research on Matter (F.O.M.) and the Ministry of Health and Environmental Hygiene in The Netherlands. One of the authors (M.A.P.) acknowledges the financial support of the University of Amsterdam.

Reprinted from *Anal. Chem.* **1978**, *50*, 985–91.

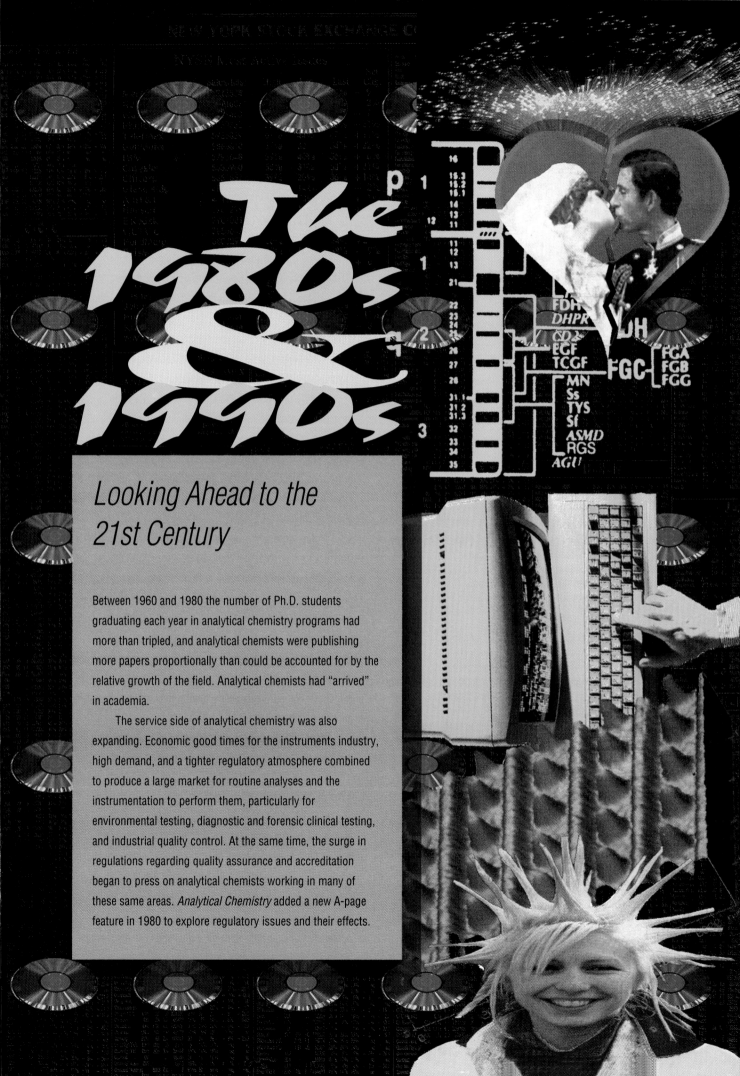

The 1980s & 1990s

Looking Ahead to the 21st Century

Between 1960 and 1980 the number of Ph.D. students graduating each year in analytical chemistry programs had more than tripled, and analytical chemists were publishing more papers proportionally than could be accounted for by the relative growth of the field. Analytical chemists had "arrived" in academia.

The service side of analytical chemistry was also expanding. Economic good times for the instruments industry, high demand, and a tighter regulatory atmosphere combined to produce a large market for routine analyses and the instrumentation to perform them, particularly for environmental testing, diagnostic and forensic clinical testing, and industrial quality control. At the same time, the surge in regulations regarding quality assurance and accreditation began to press on analytical chemists working in many of these same areas. *Analytical Chemistry* added a new A-page feature in 1980 to explore regulatory issues and their effects.

1980
Andrei Sakharov, Soviet Nobel physicist and dissident, is stripped of honors by the USSR and exiled in Gorky
Paul Berg, Walter Gilbert, and Frederick Sanger win the Nobel Prize in Chemistry for work in DNA sequencing
Mount St. Helens erupts
Voyager 1 space probe sends back images of Saturn; six new moons are discovered

1981
Walter Cronkite retires
Acquired immune deficiency syndrome (AIDS) is identified
IBM introduces the personal computer
Hans Krebs dies
Nicholas Nickelby and *Cats* play on Broadway and in London

1982
The 100-year anniversary of Charles Darwin's death
Recombinant human insulin is synthesized
Thelonius Monk and Artur Rubinstein die
The first artificial heart is implanted
Tracy Kidder wins the Pulitzer Prize for *The Soul of a New Machine*

1983
R. Buckminster Fuller dies
Barbara McClintock wins the Nobel Prize in Physiology for the discovery of "jumping genes"
Compact discs are invented
The length of the meter is redefined
Vincent Price narrates Michael Jackson's "Thriller"

1984
Indira Gandhi is assassinated by her guards
U.S. and French researchers identify the AIDS virus
The Human Genome Initiative is proposed
Union Carbide chemical leak kills 2500 in Bhopal, India
Alec Jeffreys invents variable-number terminal repeat genetic fingerprinting
James Joyce's *Ulysses* is republished with 5000 corrections

1985
Herbert Hauptman and Jerome Karle win the Nobel Prize in Chemistry for mathematical analysis of molecular structure
Marc Chagall, Orson Welles, and Dante Giacometti die
A large hole in the ozone layer is discovered over Antarctica
The U.S. becomes the world's largest debtor nation with a national deficit of $130 billion

Inevitably, as analytical methods became faster, more sensitive, and more powerful, demand for them rose. Instruments that had been used almost exclusively in research settings—some of which were still generally "homemade" systems—were adapted for routine analysis with the development of automated, modular, process-scale, remote, and even portable versions.

The type of information chemists look for has also changed dramatically in the past 14 years. In his Special Award Address at the 1980 Pittsburgh Conference, *Analytical Chemistry* Editor Herbert Laitinen named several trends for the future of the field that have held true in the 1980s and 1990s. Laitinen predicted a strong emphasis on speciation, spatial distribution of analytes, direct measurement of complex mixtures with matrix effects, surface chemistry, time resolution, and combined or "hyphenated" techniques.

In addition, the analytes and specimens that can be studied have proliferated. Among these are environmental analytes such as trace and ultratrace metals, dioxins, and polycyclic aromatic hydrocarbons; clinical analytes such as serum enzymes, hormones, drug metabolites, genetic mutations, and the AIDS virus; and polymers, semiconductors, caustic brines, foods, and pharmaceuticals.

With a recession brewing in the late 1980s, federal organizations such as the National Science Foundation began to foster the concept of technology transfer and to encourage partnerships between academic researchers and industry. As a reflection of this shift, in 1988 the National Bureau of Standards was renamed the National Institute of Standards and Technology. Specialized centers such as the Center for Process Analytical Chemistry and large collaborative projects such as the Human Genome Initiative were funded to address some of the new analytical problems.

What brought about these changes in analytical chemistry, and how has the field advanced in the past 14 years?

The instrument revolution

The introduction of powerful microcomputers and affordable PCs during the early 1980s was dubbed "the computer revolution." PCs and system computers were soon used as laboratory information management systems. As the decade progressed, PCs were used to perform advanced calculations such as Fourier transforms, multivariate statistical analyses, and peak deconvolution algorithms. Commercial software made these functions widely accessible to researchers.

Optical fibers, developed for the telecommunications

industry, were quickly incorporated into probes for spectroscopy, microscopy, and electrochemistry. Silicon chip technology was used for charge-coupled and charge-injection device (CCD and CID) 2D array detectors or "cameras." By 1988 CCD cameras were being used for high-resolution X-ray tomography, 2D and 3D imaging for crystallography and fluorescence microscopy, and time-resolved spectrometry.

Tunable dye lasers and acoustooptic tunable filters increased the accuracy and speed of wavelength selection in spectrometers and enhanced their capacity for simultaneous determinations.

Micromachining techniques, also borrowed from silicon and semiconductor technology, were used to create miniaturized and portable instruments. Samples have dropped from the milliliter range to nanoliter volumes, and revamped methods now detect pico-, femto-, and attomole levels of analytes.

Izaak Maurits Kolthoff

Methods of the new era

Separations. In 1981 James Jorgenson and co-workers developed capillary zone electrophoresis. Shigeru Terabe and co-workers described micellar electrokinetic separation in open-tubular capillaries, and Barry Karger and co-workers added polyacrylamide gel to the capillaries to separate large oligonucleotides. Long DNA and RNA strands were also separated by pulsed 2D slab gel electrophoresis.

By the end of the decade chiral phases for HPLC had been introduced by William Pirkle and others. The pharmaceutical industry has been quick to adopt the new chiral separation strategies in order to lower toxicity and improve the effectiveness of such chiral drugs as ibuprofen. In the environmental sector, solid-phase extraction and supercritical fluid extraction were used to improve the recovery of semivolatile and volatile organic compounds from complex matrixes. Although SFC was developed in the 1960s, it wasn't until the 1980s that Steven Hawthorne and others began to test the effects of methanol, acetonitrile, and other modifiers on the solvating properties of supercritical CO_2.

Mass spectrometry. MS methods changed so rapidly in the 1980s that textbooks could hardly keep up. Hyphenated techniques, MS/MS and MS^n, and soft ionization methods enabled researchers to perform peptide and DNA sequencing,

ionization of larger and more fragile proteins and protein complexes, structural analyses of neutral compounds, and the identification of unstable reaction intermediates in the gas phase. Makers of ion trap and FT-ion cyclotron resonance mass spectrometers took advantage of the refinements in radio frequency mass selection methods developed by Alan Marshall and others. Fast atom bombardment, glow discharge, and time-of-flight instruments were also introduced. Soft ionization methods included atmospheric pressure, thermospray, plasma desorption, and electrospray ionization.

These and continuous-flow methods became successful interfaces for both LC/MS and CE/MS. Laser desorption, enhanced by Franz Hillenkamp and Michael Karas' addition of a light-absorbing matrix, has been used to desorb large proteins gently so that they remain intact during ionization. MALDI has also been used to interface MS with slab gel electrophoresis, and ICP has been combined with quadrupole MS for speciation.

If there is anyone whose work can be said to encompass analytical chemistry as a whole, it may be Izaak Maurits ("Piet") Kolthoff, who began his professional career in the U.S. in 1927 and published more than 900 articles before his death in 1993. Though primarily an electrochemist, his interests and those of his students—and his students' students—branched out to include almost all the fields of analytical science.

Electrochemistry. Janet Osteryoung introduced a square-wave voltammetric detector for LC in 1980. Microelectrodes developed by Mark Wightman and others were combined with antibodies and enzymes as biosensors. They were also combined with scanning electron microscopy by Allen Bard, Royce Engstrom, and others. Ion-sensitive electrodes continued to be developed for clinical applications such as the measurement of free magnesium in blood.

Spectroscopy. Near-IR methods were developed for analyzing wet tissues. Atomic and electronic methods such as Auger electron spectroscopy, EELS, and Zeeman-corrected graphite furnace AAS were introduced in the mid-1980s, and increased synchrotron radiation power enabled the use of high-sensitivity X-ray methods such as Laue XRD, XPS, XANES, and EXAFS. Numerous double-quantum and imaging methods were developed for NMR. As of 1993, the fabrication of scanning

1986
Rita Levi-Montalcini and Stanley Cohen win the Nobel
  Prize in Physiology for research on growth factors
A nuclear accident occurs at Chernobyl
The space shuttle *Challenger* explodes in flight
Simone de Beauvoir dies
Les Misérables and *Phantom of the Opera* are hits
  on Broadway

1987
Mathias Rust (FRG) lands his Cessna airplane in Red
  Square
The world population tops 5 billion
The reputation of *Dr. Zhivago* author Boris Pasternak is
  posthumously restored in the Soviet Union
25th anniversary of Telstar communications satellite;
  3.6 billion phone calls have been logged

1988
Stephen Hawking's book on cosmology theory, *A Brief
  History of Time*, is published
Crack cocaine abuse grows
Earthquake in Armenia
Mikhail Gorbachev, elected President of the USSR,
  declares "glasnost" policy

1989
F. W. de Klerk is elected President of South Africa
T. R. Cech and S. Altman win the Nobel Prize in
  Chemistry for RNA analysis
Stanley Pons and Martin Fleischmann announce "cold
  fusion" discovery
Exxon *Valdez* spills 11 million gallons of oil in Alaskan
  waters
The Berlin Wall comes down; reunification of East and
  West Germany begins

1990
Nelson Mandela is freed
Iraq invades Kuwait
Lech Walesa is elected President of Poland
Jim Henson, inventor of "The Muppets" for *Sesame
  Street*, dies
Gene therapy is administered to a human for the first
  time

1991
The USSR is dissolved into separate states
War rocks the former Yugoslavia
The Hubble Space Telescope is launched out of focus

1992
Bill Clinton is elected President of United States
U.N. and U.S. troops attempt to distribute food and
  medical supplies in Bosnia and Somalia

1993
The World Trade Center is bombed
Israel and the PLO sign a peace agreement

tunneling and atomic force microscopy probes has brought 3D imaging capabilities down to the single-atom scale.

Spectroscopy lost one of its brightest stars in 1986. Tomas Hirschfeld, who published more than 200 papers and patents in his varied career, contributed to the development of near-IR and Raman spectroscopic methods both for in-lab and industrial process analyses, and he was a strong proponent of hyphenated techniques such as GC-IR. His accomplishments were honored at Pittcon '87 in a special memorial symposium, and Technicon Instruments introduced a series of commemorative graduate student awards in his name.

Biotechnology. DNA amplification methods such as the polymerase chain reaction were developed by Nobel laureate Kary Mullis and others in the mid- to late 1980s and permit the rapid detection of as few as five original copies of target DNA from a sample. These methods, along with automated DNA sequencing techniques, have led to forensic identification tests and clinical assays for genetic disorders. Immunoassays and enzyme-based assays have been used to determine picomolar quantities of clinical and environmental analytes.

Chemometrics. James Callis, Bruce Kowalski, Sarah Rutan, and others have demonstrated the power of partial least-squares and other multivariate statistical techniques to pull specific information out of poorly resolved data in complex systems.

Marking our progress

All this expansion in analytical chemistry has been reflected in the progress of the Pittsburgh Conference over the years. Pittcon moved from Cleveland to Atlantic City in 1980, where it enjoyed larger crowds every year, especially at the technical exposition. From 845 presentations in 1980, the conference has grown to more than 1800 presentations and 2900 exhibitor booths in 1993!

Analytical Chemistry has also kept pace with this progress. With George H. Morrison as Editor beginning in 1980, the Journal underwent two design changes, one in 1980 and the other in 1987, when it was decided to publish two issues each month. Five Associate Editors were appointed for the first time in 1985. In 1991 Royce W. Murray succeeded Morrison as Editor, and the peer review process for technical manuscripts was transferred to Murray's office at the University of North Carolina. As of 1994 the Journal has been redesigned for the 21st century and includes new features such as a survey of literature in the field and software reviews. We look forward to discovering what lies ahead in *Analytical Chemistry*.

DEBORAH NOBLE

In this classic article, theory and experiments are combined to give a rational and clear explanation of the electrochemical response at microelectrodes—an emerging field in the 1980s. In the practical sense, this work demonstrated the utility of microelectrodes for voltammetric studies. Furthermore, this article represents a more profound change that has taken and is taking place in the field of electroanalytical chemistry. The surface of the electrode is now recognized as a significant factor in electrochemical experiments—a factor that can be modified, if desired.

James Anderson
DuPont Merck

Faradaic Electrochemistry at Microvoltammetric Electrodes

M. A. Dayton, J. C. Brown, K. J. Stutts, and R. M. Wightman*
Department of Chemistry, Indiana University, Bloomington, Indiana 47405

Carbon fibers have been used to fabricate voltammetric electrodes with an active area of approximately 5×10^{-7} cm^2. The response of these electrodes has been evaluated in aqueous solutions containing K$_3$Fe(CN)$_6$ for cyclic voltammetry, differential pulse voltammetry, and chronoamperometry. Because the diameter of these electrodes is smaller than the distance for molecular diffusion on the time scale of these experiments, the current is essentially time independent. The electrodes are shown to be useful for analytical determination of concentration over 3 orders of magnitude. The properties of the electrochemical response for the microelectrodes permit direct residual current correction. In addition, the contribution of homogeneous chemical reactions to the faradaic current is diminished.

Recently a report appeared in this journal describing the use of carbon fibers (8-μm diameter) as voltammetric electrodes (*1*). The carbon fibers were found to give satisfactory responses for the oxidation of catecholamines. We have also investigated the response of single carbon fibers; however, our electrodes differ from those previously described since only the diameter of the carbon fiber determines the electrode area. For an electrode of this geometry, *the diameter of the electrode is much smaller than the diffusion distance* for the time scale of a typical electrochemical experiment. Because of this, the electrochemical response of these electrodes is greatly different from conventional voltammetric electrodes. Nevertheless, the response of these electrodes will be shown to be predicted by existing electrochemical theory.

Consideration of the documented morphology and manufacture of carbon fibers (*2*) suggests that the surface of the fiber perpendicular to the fiber axis is ideal for voltammetric applications. Carbon fibers are formed by the high temperature pyrolysis of polymeric materials such as polyacrylonitrile or pitch, while plastically stretching the polymeric carbon material. The structure of these fibers appears to be similar to that of glassy carbon; that is, sheets of graphite are arranged in microfibrils which form a pattern similar to intertwined ribbons. In carbon fibers, the order of the carbon layers depends on the mode of manufacture. The high-modulus fibers have a high degree of alignment of the carbon–carbon bond parallel to the fiber axis which results in a "tree ring" morphology (*3*). Thus, in contrast to glassy carbon, the surface of a high-modulus carbon fiber perpendicular to the fiber axis has a high degree of edge orientation of the graphitic microfibrils, while the surface parallel to the fiber axis has a high degree of basal plane character. Faradaic electron transfer has been shown to occur preferentially at the edge orientation of oriented graphite (*4, 5*). In addition, faradaic electrochemistry may be impaired on the axial surface of carbon fibers since they are frequently coated with a sizing material such as polyvinylacetate.

Electrochemical measurements with carbon fiber electrodes have been diverse in scope but limited in number. Jennings and Bailey have investigated their use as working electrodes for coulometric titrations (*6*). Dietz and Peover have used chronocoulometry to determine the degree of porosity of carbon fiber electrodes (*7*). Das Gupta and Fleet have taken advantage of the large surface area of bundles of carbon fibers to construct a high yield electrochemical reactor (*8*). Microvoltammetric electrodes constructed with 1 mm of the carbon fiber exposed have also been used for in vivo measurements of the concentration of easily oxidized neurotransmit-

Table I. Properties and Sources of Various Carbon Fibers

manufacturer	type	diameter, μm	resistance kΩ/cm	% carbon[a]
Stackpole Fiber Co.	Panex 30	10.64 ± 2.59	3.97 ± 0.27	—
Union Carbide Corp.	Thornell P-55 (Grade VSB-32)	10.24 ± 0.91	0.949 ± 0.119	99
	Thornell 300 (Grade WYP 30 1/0)	7.00 ± 0.34	5.28 ± 0.446	92
Courtald, Ltd.	Grafil	7.19 ± 0.33	4.40 ± 0.872	—
Hercules, Inc.	Magnamite (Type HTS2)	7.03 ± 0.31	3.68 ± 0.815	94
	Magnamite (Type HMS)	7.55 ± 0.31	2.09 ± 0.123	99+
	Magnamite (Type HMPVA)	7.38 ± 0.86	2.12 ± 0.281	99.5+
AERE Harwell	H28G	5.64 ± 0.16	2.85 ± 0.304	—

[a] Manufacturer's specification.

ters (9). For this latter application, electrodes of extremely small diameter are required so that damage of neuronal tissue is minimal. In all of these cases, the electrode area has been predominantly defined by the axial surface of the carbon fiber.

In this paper we compare the electrochemical response of carbon fiber electrodes ($A = 5 \times 10^{-7}$ cm^2) to conventionally sized carbon paste electrodes ($A = 2 \times 10^{-2}$ cm^2). Existing electrochemical theory predicts the predominantly steady-state response we observed for the microvoltammetric electrodes. As will be shown, this steady-state response permits double-layer correction during data acquisition via double-potential step techniques. In addition, our data demonstrate that a decrease in the electrode radius in effect decreases the time scale for which homogeneous reactions become apparent. Both of these factors improve the accuracy of the voltammetric measurement.

EXPERIMENTAL

Electrodes. Carbon paste was prepared by thoroughly mixing carbon powder (UCP 1M, Ultra Carbon, Bay City, Mich.) with mineral oil (25% mineral oil by weight). The electrode was constructed by force fitting 1.5-mm i.d. Teflon tubing over a piece of 1.56-mm diameter copper wire in a manner which left a small well into which the carbon paste was packed.

The carbon fibers were obtained from various sources. The sources and some of the physical properties are described in Table I. To determine the resistance, a carbon fiber was placed between two mercury droplets that were a known distance apart. The droplets were then connected to an ohm meter. The diameter of the fibers was determined with an optical microscope. The microelectrodes were constructed by aspirating a carbon fiber into a 0.5-mm i.d. capillary tube (GCF-100-4, A-M Systems, Inc., Toledo, Ohio) which contains a glass microfilament. The capillary tube containing the carbon fiber was then pulled with a vertical pipet puller (Model 700C, David Kopf Instruments, Tujunga, Calif.). The heat and

solenoid controls on the vertical pipet puller were adjusted so that the glass formed an almost intact seal around the carbon fiber which extends from the electrode tip. This end was then dipped in an epoxy which is nonviscous at 70 °C (Epon 828, Shell Chemical Co., Houston, Tex., with 14% by weight metaphenylenediamine Cl, Miller Stephenson, Chicago, Ill.). The epoxy filled the glass tip to a height of about 0.5 mm by capillary action, promoted by the glass microfilament. The electrodes were then cured at 100 °C for 2 h. The tip of the electrode with the protruding carbon fiber was then carefully cut with a scalpel so that the fiber tip is virtually flush with the glass. In other words, the active area of the electrode is defined by the diameter of the carbon fiber. Even though the surface is rough, we assume the geometry of the electrode is best represented by a disk. A scanning electron micrograph of the electrode tip showing the glass, epoxy, and fiber is shown in Figure 1. Electrical contact from the carbon fiber to a stainless steel wire was made by filling the pipet with Hg. During

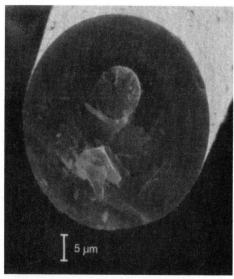

Figure 1. Scanning electron micrograph of the tip of a carbon fiber microelectrode

each step of construction, the electrodes were visually examined with a microscope. Those electrodes which did not fill with epoxy or were not cut to provide a flat tip were discarded. To prevent convective effects, the electrode tips were sheathed with a 1.0-mm i.d. glass tube.

Chemicals. Potassium ferricyanide solutions were prepared freshly for each experiment in doubly distilled water with 1.0 M potassium chloride supporting electrolyte (pH adjusted to 4.4 with HCl). Dopamine solutions were freshly prepared in a pH 7.4 phosphate–citrate buffer.

Apparatus. Potentiostatic control was maintained with a polarographic analyzer (Princeton Applied Research Corporation, Model 174A, Princeton, N.J.). Wave forms for cyclic voltammetry and differential pulse voltammetry were generated by the polarographic analyzer. The wave form generation and data acquisition for chronoamperometry were accomplished with a locally constructed microcomputer based on an Intel 8080 microprocessor. The digital-to-analog converter of the computer system was interfaced to the electrochemical cell through the multipin connector on the rear of the polarographic analyzer. Currents from the polarographic analyzer were amplified by 10 with a locally constructed amplifier and then sent to an analog-to-digital convertor. The accuracy of the computer-applied voltage is ±5 mV and the accuracy of the digitized data is 0.04% for a full scale measurement. With this system, the maximum data acquisition rate is 10 kHz. The digitized current measurements were transferred to a Texas Instruments 980 computer for data analysis via a 9600 baud communication multiplexor. Unless otherwise noted, chronoamperometric data were corrected for background contributions by subtraction of current obtained from a potential step in the supporting electrolyte/solvent system. Chronoamperometric data at times less than 50 ms were discarded in the data analysis. All chronoamperometric data presented from the microelectrodes are the result of ensemble averaging at least ten runs. A large amount of 60 Hz noise is apparent in a single run. This is not surprising since the current measured in these experiments is less than 1 nA.

The electrochemical cell was a 30-mL glass bottle with a plastic top that was drilled with holes for the electrodes and nitrogen purging. A saturated calomel reference electrode and platinum wire auxiliary electrode were employed. The cell was mounted on a board which was suspended by rubber tubing from a metal frame. This arrangement is a remarkably simple method to dampen bench top vibration and, thus, minimize this source of convection. The cell and damping platform were all located in a faraday cage. Experiments were temperature controlled at 22 ± 1 °C.

RESULTS AND DISCUSSION

Properties of Carbon Fibers. The diameter, resistance, and composition of the carbon fibers investigated in this work are summarized in Table I. The Magnamite and Panex fibers are manufactured from polyacrylonitrile, while the Thornell P-55 is manufactured from pitch. The Magnamite HMS and HMPVA differ only in that the latter is coated with polyvinyl acetate sizing. The currents measured in this application are

in the nanoampere range, and, therefore, the resistance of the fiber has a negligible effect on the electrochemical response.

The response of the microelectrodes is affected by their construction. For example, if the epoxy does not seal the electrode tip properly, leakage of solution into the micropipet results in a large resistive component in addition to the faradaic current. Because of this, we have had more success with Magnamite HMPVA than HMS, even though they differ only in sizing material. In our preliminary investigation of the electrochemistry of 4 mM $Fe(CN)_6^{3-}$, we found electrodes containing Panex 30 or Thornell 300 carbon fibers to give the most reproducible results, and the remainder of this work has been done with these fibers. Panex 30 exhibits higher residual current than the other carbon fibers. Therefore, it is unsatisfactory for determination of dilute solutions.

Voltammetry. Cyclic and differential pulse voltammograms of potassium ferricyanide obtained with the carbon paste electrode described in the Experimental section are shown in Figure 2, A and B. The shape of the cyclic voltammogram is qualitatively similar to that predicted by Nicholson for a quasi-reversible electrochemical system ($\Delta E_p = 207$ mV) (10). As expected, the differential pulse voltammogram has a lower amplitude than predicted for a reversible system (11). For this electrode, the peak potential is at 0.238 V and the peak has a width at half-height of 0.163 V. The results of cyclic voltammetry and differential pulse voltammetry under identical conditions are given for a microelectrode in Figure 2, C and D. In direct contrast to the shape of the conventional cyclic voltammogram, the cyclic voltammogram obtained with the microelectrode is sigmoidal in shape. In addition, during the reverse potential scan, no wave is present for the oxidation of ferrocyanide. However, the differential pulse voltammogram obtained with the microelectrode is similar to that

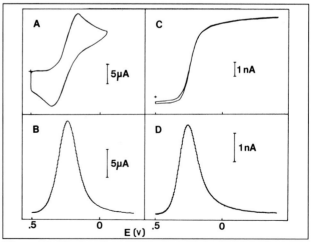

Figure 2. Electrochemical reduction of 4.00 mM potassium ferricyanide in 1.0 M KCl. Conditions: cyclic voltammetry, 100 mV/s scan rate; differential pulse voltammetry, 10 mV/s scan rate, 0.5 s between pulses, 100 mV pulse. (A) Carbon paste electrode ($r = 7.87 \times 10^{-2}$ cm), cyclic voltammetry. (B) Carbon paste, differential pulse voltammetry. (C) Carbon fiber electrode (Thornell 300, $r = 5.1 \times 10^{-4}$ cm), cyclic voltammetry. (D) Carbon fiber, differential pulse voltammetry

obtained with the carbon paste electrode. The peak potential is at 0.250 V and the width at half-height is 0.165 V. However, the differential current per unit area is about 10 times larger than that obtained at the carbon paste electrode.

As will be shown, the data in Figure 2 are consistent with the statement made in the introduction concerning the relative sizes of the electrode diameter and the diffusion layer. Indeed, these data are similar to those in the original paper where the term voltammetry was introduced (12). However, we have also considered other possible effects. To minimize convection, which would also contribute to the sigmoidal shape of the cyclic voltammogram, a glass tube (i.d. = 1.0 mm) shielded the microvoltammetric electrode tip. As noted in the Experimental section, all electrochemical experiments were conducted on a platform which effectively damped vibration. Sigmoidal cyclic voltammograms for the oxidation of ferrocyanide indicate the shape of the current–voltage curves is not an artifact caused by unusual electron transfer at carbon fiber electrodes. In addition, when a length of the carbon fiber protrudes from the glass capillary such that the electrode area is greatly increased, oxidative current is seen on the reverse scan following the reduction of ferricyanide (also note that this is the behavior seen in ref. 1). As would be expected, a bundle of carbon fibers epoxied together and sealed in a glass tube, such that the electrochemically active surface of the electrode is determined by approximately the diameter of 100 carbon fibers, exhibits a more conventional electrochemical response.

Chronoamperometry. To experimentally confirm our contention that the unconventional response of our microvoltammetric electrodes is a function of the relative size of the diffusion layer to the electrode diameter, we have used the technique of chronoamperometry. In cyclic voltammetry, correction terms for steady-state diffusion to microelectrodes are available for spherical (13) but not for planar electrodes. However, several investigators have considered this problem for chronoamperometry. The equation for the current (I) obtained in chronoamperometry in quiescent solutions under conditions where the potential step is of sufficient magnitude to cause the surface concentration of electroactive species to be zero is given by

$$I = \frac{nFAD^{1/2}C}{\pi^{1/2}t^{1/2}} + arnFDC \qquad (1)$$

where A is the electrode area, r is the electrode radius, a is a coefficient which depends on the electrode geometry, and the other terms have their usual electrochemical meanings. Thus, a linear regression of the current vs. $t^{-1/2}$ should give a straight line with a slope which can be used to determine the area and with an intercept which gives the steady-state term. Values of the coefficient a are given in Table II for various electrode geometries. The hemispherical steady-state coefficient has not been reported in the electrochemical literature. The value in Table II is taken from derivations of the same problem by workers in the area of diffusion kinetics (15, 21). Several values of the coefficient for a disk electrode have been reported because different approximations have been made to solve the

Table II. Values of Coefficient *a* for Various Electrode Geometries

electrode geometry	coefficient *a*	ref.
sphere	4π	14
hemisphere	2π	15
disk	4	16, 17
	3.4	18
	3.75[a]	19
	3.13[a]	20

[a] Experimentally determined values.

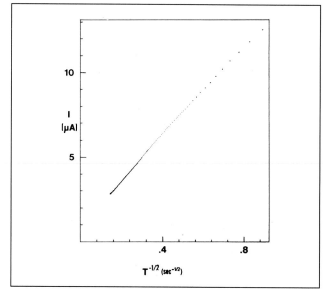

Figure 3. Background corrected chronoamperometric current for the reduction of 4.00 mM potassium ferricyanide in 1.0 M KCl with a carbon paste electrode. $E_{step} = -0.20$ V, 50-s duration

differential equations. Furthermore, convection hampers experimental determination of the time independent portion of Equation 1 at electrodes of conventional size. We believe that the best value of the coefficient a is 4.0 for a disk electrode since this same result has been obtained by workers considering diffusion-controlled rate constants (22). It should be noted that the derivations of the coefficients for the disk electrode that are given in Table II assume that the disk was located in an infinite plane. Since the thickness of the glass insulating sheet around our carbon fiber is the same size as the carbon fiber for the microelectrodes described here, an experimentally determined value larger than 4.0 can be expected.

The chronoamperometry experiments have been done in 1 M KCl with potassium ferricyanide since the diffusion coefficient is known for this system (23). The results of a linear regression analysis for the data obtained at a carbon paste electrode (following residual current subtraction) show an excellent fit to Equation 1 for times as long as 50 s (Figure 3). For chronoamperometric experiments lasting 10 s, the average correlation coefficient for a linear regression analysis of the current vs. $t^{-1/2}$ data was 0.9999 (short-time data were disregarded because of charging current and surface roughness contributions), the average radius determined from the slope was 0.0859 cm (compared to a geometric radius of 0.0787 cm), and the value of the coefficient a determined from the intercept was 3.82 with an SEM of 0.62 for 12 different determinations. A similar regression analysis from chronoamperometric data at a microelectrode is given in Figure 4. For 14 different microelectrodes constructed from Panex 30 or Thornell 300 carbon fibers run at 10-s step-times, the correlation coefficients for the linear regression varied from 0.955 to 0.999. The decreased value of the correlation coefficient can be attributed to the increased noise with the very low current measured at the microelectrodes. The radius determined from the slope varied from 3.90×10^{-4} to 6.97×10^{-4} cm, which is slightly larger than that determined geometrically. This presumably arises since the fibers are not cut perfectly perpendicular to their axis resulting in an increased area. In addition, surface roughness may play a role in the apparent enlarged electrochemical area. The value of the coefficient a determined by using the electrochemically determined radius is 5.69 with an SEM of 0.48. The linearity of the regression analysis, the value of the coefficient for the steady-state term, and the value of the experimentally obtained radii are all in excellent agreement with those expected and, thus, Equation 1 apparently describes the processes occurring at our microelectrode.

Utilization in Analysis. An understanding of the diffusion equation (Equation 1) has several implications in the application of microvoltammetric electrodes to chemical analysis. Equation 1 predicts that the current (rather than $It^{1/2}$ product) should vary in a linear fashion with changes in concentration. The logarithmic calibration curve given in Figure 5A shows that the average current during the last 10% of a 1.4-s voltage pulse, when those data have been corrected for background contributions, shows a linear response to the ferricyanide concentration from 10^{-5} to 10^{-2} M (correlation coefficient = 0.9963, slope = 1.044).

Of particular interest is double-potential step chronoamperometry with microelectrodes using step times of equal duration. When the potential is returned to its initial value following reduction of ferricyanide, very little current is observed for the last 10% of the second step as is shown in a semilogarithmic plot (Figure 5B). The current in this figure is the difference of the residual and faradaic currents, and it is seen that this difference is virtually zero. This phenomenon occurs because the species generated by the electrolysis diffuse away from the electrode and are not present to be reelectrolyzed during the backstep. Therefore, under conditions of semiinfinite linear diffusion, the current during the second step in double-potential step chronoamperometry can be used to correct for residual current at microelectrodes, while this is not the case with conventional electrodes (24).

Perhaps even more intriguing is the ability of these electrodes to mask the presence of consecutive electron transfer reactions following the initial transfer steps. Reaction mechanisms which consist of an electrochemical electron transfer followed by a chemical homogeneous reaction which generates a species which is also electroactive at the applied potential are well known (25). With a microelectrode, the reaction has to occur extremely rapidly to be detected; otherwise, the newly formed material diffuses away from the electrode. This is

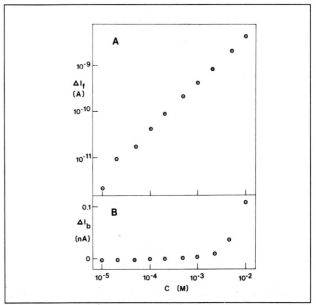

Figure 5. (A) Logarithmic plot of average background corrected chronoamperometric current (last 10%) for the reduction of potassium ferricyanide in 1.0 M KCl with a carbon fiber electrode (Thornell 300). $E_{step} = -0.4$ V, 1.4-s step duration. (B) Semilogarithmic plot of average background corrected current (last 10%) for a step back to the initial potential (0.50 V) subsequent to the reduction of potassium ferricyanide in 1.0 M KCl

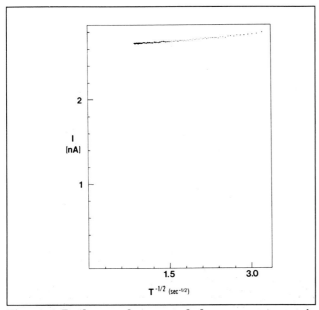

Figure 4. Background corrected chronoamperometric current for the reduction of 4.00 mM potassium ferricyanide in 1.0 M KCl with a carbon fiber electrode (Thornell 300). $E_{step} = -0.40$ V, 10-s duration

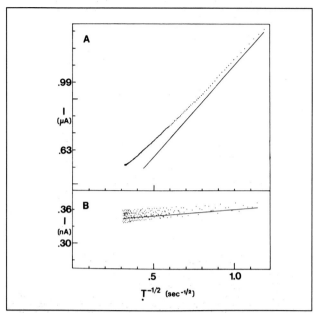

Figure 6. Backstep corrected chronoamperometric current for the oxidation of 0.10 mM dopamine (pH 7.4). E_{step} = 0.50 V, 10-s step duration. Solid line: current calculated from Equation 1. Points: experimental measurement. (A) Carbon paste electrode ($r = 7.87 \times 10^{-2}$ cm). (B) Carbon fiber electrode (Panex 30)

demonstrated in Figure 6, A and B, where the respective chronoamperometric responses of a carbon paste electrode and a microelectrode are presented for the oxidation of dopamine. As can be seen, the current increases over that predicted by Equation 1 with the macroelectrode indicating the presence of a subsequent electrochemical reaction while the current for the microelectrode shows no such contribution. Dopamine is known to cyclize to form an easily oxidized product under these conditions (26). This observation has particular import for in vivo analysis of catecholamines because ascorbic acid interferes via a similar reaction (26). Studies of this reaction at microelectrodes are in progress.

CONCLUSIONS

Microvoltammetric electrodes provide quite different information than conventionally sized electrodes. Microelectrodes constructed from carbon fibers exhibit properties predicted by the Cottrell equation, modified to account for the steady-state contributions from diffusion. The electrodes appear to have utility in chemical analysis. They are linear in response to concentration; they provide a method for immediate residual current correction; and they avoid the complication of competing chemical and electrochemical reactions. It should also be noted that these electrodes are virtually nondestructive. For a potential step of 1-s duration in a 10^{-5} M solution, only 10^{-16} mol are electrolyzed. Furthermore, these electrodes give an almost steady-state current response at a fixed potential

which suggests that they can be used under conditions where variations in potential are disadvantageous. All of these properties suggest that microvoltammetric electrodes are ideal for applications where small size is required such as the in vivo determination of biogenic amines.

ACKNOWLEDGMENT

Discussions with A. Szabo are gratefully acknowledged.

LITERATURE CITED

(1) Ponchon, J.-L.; Cespuglio, R.; Gonon, F.; Jouvet, M.; Pujol, J.-F. *Anal. Chem.* **1979**, *51*, 1483–1486.
(2) Jenkins, G. M.; Kawamura, K. "Polymeric Carbons-Carbon Fibre, Glass and Char"; Cambridge University Press: New York, 1976.
(3) Stewart, M.; Feughelman, M.; Gillin, L. M. *Nature (London)* **1972**, *274*, 235.
(4) Morcos, I.; Yeager, E. *Electrochim. Acta* **1970**, *15*, 953–975.
(5) Wightman, R. M.; Paik, E. C.; Borman, S.; Dayton, M. A. *Anal. Chem.* **1978**, *50*, 1410–1414.
(6) Jennings, V. A.; Bailey, T. D. *Anal. Chim. Acta* **1976**, *84*, 61–65.
(7) Deitz, R.; Peover, M. E. *J. Mater. Sci.* **1971**, *6*, 1441–1446.
(8) Fleet, B.; Das Gupta, S. *Nature (London)* **1976**, *263*, 122–123.
(9) Gonon, F.; Cespuglio, R.; Ponchon, J.-L.; Buda, M.; Jouvet, M.; Adams, R. N.; Pujol, J.-F. *C. R. Hebd. Seances Acad. Sci., Ser. D* **1978**, *286*, 1203–1206.
(10) Nicholson, R. S. *Anal. Chem.* **1965**, *37*, 1351–1355.
(11) Keller, H.E.; Osteryoung, R. A. *Anal. Chem.* **1971**, *43*, 342–348.
(12) Laitinen, H.A.; Kolthoff, I.M. *J. Phys. Chem.* **1941**, *45*, 1062–1079.
(13) Nicholson, R. S.; Shain, I. *Anal. Chem.* **1964**, *36*, 706–723.
(14) MacGillavry, D.; Rideal, E. K. *Recl. Trav. Chem. Pays Bas* **1937**, *56*, 1013–1021.
(15) Alberty, R. A.; Hammes, G. G. *J. Phys. Chem.* **1958**, *62*, 154–159.
(16) Soos, Z. G.; Lingane, P. J. *J. Phys. Chem.* **1964**, *68*, 3821–3828.
(17) Saito, Y. *Rev. Polarogr.* **1968**, *15*, 178–186.
(18) Flanagan, J. B.; Marcoux, L. *J. Phys. Chem.* **1973**, *77*, 1051–1055.
(19) Lingane, P. J. *Anal. Chem.* **1964**, *36*, 1723–1726.
(20) Ito, C. R.; Asukura, S.; Nobe, K. *J. Electrochem. Soc.* **1972**, *119*, 698–701.
(21) Hill, T. L. *Proc. Natl. Acad. Sci. (U.S.A.)* **1975**, *72*, 4918–4922.
(22) Szabo, A. *Proc. Natl. Acad. Sci. (U.S.A.)* **1978**, *75*, 2108–2111.
(23) Von Stackelburg, M.; Pilgram, M.; Toome, V. *Z. Electrochem.* **1953**, *57*, 342–350.
(24) Schwarz, W. M.; Shain, I. *J. Phys. Chem.* **1965**, *69*, 30–36.
(25) Hanafey, M. K.; Scott, R. L.; Ridgway, T. H.; Reilley, C. N. *Anal. Chem.* **1978**, *50*, 116–137.
(26) Tse, D. C. S.; McCreery, R. L.; Adams, R. N. *J. Med. Chem.* **1976**, *19*, 37–40.

Received for review November 16, 1979. Accepted January 28, 1980. This research was supported by the National Science Foundation (Grant No. BNS 77-28254). M.A.D. is a combined Medical-Ph.D. candidate, Indiana University. R.M.W. is the recipient of a Research Career Development Award from the National Institutes of Health (Grant No. 1 K04 NS 00356-01).

Reprinted from *Anal. Chem.* **1980**, *52*, 946–50.

This paper is seminal to the development of thermospray ionization. Earlier work by this group demonstrated that ions could be produced by heating the liquid jet nozzle with a laser beam. Here Vestal and co-workers describe the use of oxy-hydrogen flames to vaporize the LC effluent. Although this technique later proved impractical, it led to the use of cartridge heaters and eventually resistive heaters, which is still incorporated today in thermospray interfaces.

Jack D. Henion
Cornell University

Liquid Chromatograph–Mass Spectrometer for Analysis of Nonvolatile Samples

C. R. Blakley, J. J. Carmody, and M. L. Vestal*
Department of Chemistry, University of Houston, Houston, Texas 77004

A liquid chromatograph–mass spectrometer system has been developed for application to analyses of molecules of extremely low volatility. This LC-MS system uses oxy-hydrogen flames to rapidly vaporize the total LC effluent and molecular and particle beam techniques to efficiently transfer the sample to the ionization source of the mass spectrometer. The instrument is comparable in cost, complexity, and performance to a combined gas chromatograph–mass spectrometer (GC-MS) but extends the capabilities of combined chromatograph–mass spectrometry to a broad range of compounds not previously accessible.

The combination of high pressure liquid chromatography with mass spectrometry has been recognized for some time as possessing enormous potential for analyses of polar, nonvolatile, or thermally unstable compounds not amenable to GC-MS. It is equally well known that the problems which must be solved to achieve a practical LC-MS combination are much more difficult than those encountered in the development of GC-MS. The approaches to LC-MS interfacing which have been developed were reviewed recently by Arpino and Guiochon (1).

In our original approach to the problem of LC-MS coupling, we employed laser heating to rapidly vaporize both the solvent and the sample and molecular beam techniques to transport and ionize the sample with minimal contact with solid surfaces (2). The rationale for our approach is based in part on the work of Friedman and co-workers (3) in which it was shown that quite nonvolatile samples can be vaporized intact by employing rapid heating and by vaporizing the sample from weakly interacting surfaces such as Teflon.

In our initial work, we found that we were unable to achieve stable vaporization of a liquid jet with the laser beam intersecting the liquid jet in free space; however, if the laser beam were shifted so that the liquid jet nozzle was heated by the laser beam, stable vaporization could be achieved. Furthermore, even though the tip of the nozzle was often operated at red heat, there was little indication that pyrolysis of the sample was occurring during the vaporization process. We also found in this earlier work that the pumping system in our original system was overdesigned. In particular, the large diffusion pump (4200 L/s) used to evacuate the chamber between the nozzle and skimmer was unnecessary and could be replaced by its mechanical backing pump (17 L/s).

Using the results of our earlier studies, we have developed the completely redesigned system described in the present paper. This new instrument uses oxy-hydrogen flames to rapidly vaporize the total LC effluent. The pumping system has been drastically simplified so that the overall cost and complexity of the new instrument is comparable to that of a combined GC-MS. This new system has been applied to a variety of relatively nonvolatile biological molecules, and it appears to provide a useful and practical LC-MS system with a wide range of potential applications.

DESCRIPTION OF THE INSTRUMENT

A schematic diagram of the apparatus is shown in Figure 1. The effluent from the LC enters the vaporizer through a stainless steel capillary tube (typically 0.015-cm i.d.) which is threaded about 0.3 cm into the end of a copper cylinder 0.8 cm o.d. × 1 cm long. The copper cylinder is heated to bright red heat (ca. 1000 °C) by four small oxy-hydrogen flames positioned as indicated in Figure 1. As the LC effluent approaches the end of the stainless capillary, it is very rapidly heated and partially vaporized producing a jet of vapor and aerosol. This

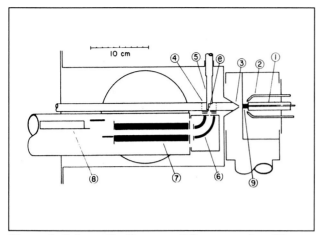

Figure 1. Schematic diagram of the apparatus: (1) capillary tube carrying liquid effluent from LC; (2) oxyhydrogen torch; (3) skimmer; (4) ion source; (5) heated probe; (6) electrostatic deflector; (7) quadrupole mass filter; (8) electron multiplier; (9) copper cylinder. The electron beam enters the ion source perpendicular to the plane of the figure at the point (e)

jet is further heated as it passes through a 0.075-cm diameter stainless steel lined channel through the copper; it then undergoes an adiabatic expansion, and a portion passes through the skimmer to the ion source where the beam impinges on a nickel plated copper probe which is electrically heated to ca. 250 °C.

All of the work described in the present paper employed an ion source configured for chemical ionization work; however, electron impact ionization can be employed with the same basic instrument with decreased skimmer aperture and "open" ion source. For the present work, the skimmer aperture was 0.075 cm and it was located about 0.4 cm downstream from the vaporizer. The apertures in the ion source for electron beam entrance and ion exit were each 0.05 cm in diameter. The source is equipped with a differentially pumped electron gun of a simple design described previously (4). This allows the rhenium filament to be removed from the hostile environment adjacent to the ion source and has increased filament life drastically. Ions produced by the source are accelerated to 10–20 eV and deflected into the quadrupole mass analyzer by a 2.5-cm radius, 90° cylindrical condensor. This condensor acts as a low resolution energy analyzer but its main function is to prevent high energy neutrals or photons produced in the ion source from being transmitted to the vicinity of the electron multiplier and producing ionization of the residual gas. This measure has proven effective in removing almost all of the "mass independent" background observed in our earlier work with the more conventional axial source and quadrupole alignment. The electron multiplier is mounted off-axis with deflectors for directing the ion beam to the first dynode of the multiplier. The final deflector is attached to a guard ring which is part of the structure of this particular electron multiplier. For the detection of positive ions, this final deflector and guard ring are connected to first dynode potential of ca. −3 kV; for the detection of negative ions, the deflector and guard ring are biased at ca. +2 kV to serve as a conversion

dynode in the manner developed by Stafford (5) and described by Hunt and Crow (6).

Several stages of differential pumping are required to maintain sufficiently low pressure in the mass spectrometer and ion optics while injecting and vaporizing liquid flows up to 1 mL/min. In this instrument, the outside of the vaporizer in the vicinity of the torches is open to the atmosphere. The region between the vaporizer and the skimmer is maintained at about 1 Torr by a 17 L/s mechanical pump. The ion source housing and ion optics are maintained at about 1×10^{-4} Torr by a 1200 L/s diffusion pump. The quadrupole is pumped to ca. 1×10^{-5} Torr by a 285 L/s diffusion pump, and the electron gun housing is maintained below 10^{-5} Torr by a 150 L/s diffusion pump. In operation, the ion source pressure is typically in the 1- to 2-Torr range. The ion source is connected by 1-cm i.d. tubing to a small mechanical pump; flow through this line can be regulated, if required, by a throttling valve. Usually, the best sensitivity has been found with this valve fully open. Under this condition 50–70% of the total flow of vapor into the ion source exits through this auxiliary pump and the remainder through the electron entrance and ion exit apertures to the 1200 L/s pump.

The detailed configuration of the apparatus described above has been developed through a fairly extensive series of experiments aimed at maximizing the sensitivity of the technique while minimizing pyrolysis of nonvolatile, thermally labile samples separated by reversed phase LC using water/methanol as the mobile phase at flow rates between 0.2 and 1.0 mL/min. At first glance, it may be somewhat surprising that this can be accomplished by passing the sample through copper heated nearly to its melting point! The reasons that vaporization without pyrolysis can be accomplished in this manner appear to be that the sample spends a very short time in this high temperature region and that during this time it is protected from overheating by the solvent. The liquid entering the hot region is heated from ambient to the vaporization point in a few milliseconds and the vapor is expelled from the hot region in a millisecond or less after it is formed. In our studies we have found conditions under which nearly complete vaporization of liquid inputs up to at least 2 mL/min can be accomplished on this time scale; however, these conditions do not appear to correspond to optimum performance of the instrument. Under the operating conditions described above, it appears that about 95% of the liquid input is vaporized and the remainder is in the form of a highly collimated particle or aerosol beam which is accelerated to approximately sonic velocity by the rapid expansion of the vapor in the vaporizer. We estimate that these particles may typically have a mass of 10^{-7} g and a velocity of ca. 10^{5} cm/s. Nonvolatile samples appear to be carried preferentially by the particles rather than the vapor. Since the particles have a very large axial momentum compared to the momentum of individual vapor molecules, they are transmitted with high probability through the skimmer to the ion source. As a result, high transmission efficiencies can be achieved for nonvolatile samples with removal of a significant fraction of the solvent. When the particle strikes the heated probe, it is wholly or partially vaporized and chemical ionization of the sample is produced in more or less the conventional manner, with ions produced from the solvent serving as the reagent ions.

EXPERIMENTAL

All of the results presented in this paper were obtained using the mass spectrometer system described above coupled to a Perkin-Elmer Model 601 Liquid Chromatograph and a Finnigan-Incos Model 2300 data system. Samples were injected into the liquid effluent stream from the liquid chromatograph using a Rheodyne Model 7120 injection valve with a 20-µL sample loop. Liquid phases used were dilute aqueous formic acid (0.01 to 0.2 M) or ammonium formate buffer (0.2 M, pH 5) at flow rates in the range from 0.3 to 1.0 mL/min. The total LC effluent was vaporized continuously in the mass spectrometer interface and the resulting vapor served as the reagent gas for the chemical ionization source.

RESULTS AND DISCUSSION

During the course of the development of the LC-MS system, spectra have been obtained on several hundred biologically important molecules; however, particular attention has been given to amino acids, peptides, nucleosides, and nucleotides. A few examples of recent results are presented in Figures 2 through 6, where both positive and negative CI mass spectra are presented for a few representative examples. These spectra were obtained under typical LC operating conditions but without a column installed. Results on two amino acids are presented in Figures 2 and 3. The relatively volatile amino acid, phenylalanine, presents no particular difficulty for conventional mass spectrometry, and spectra similar to those shown in Figure 2 may be obtained by conventional CI techniques. Arginine, on the other hand, is considered a difficult compound for conventional EI or CI mass spectrometry and ions characteristic of the intact molecule are often not observed. In the present work the spectrum of arginine is qualitatively similar to that obtained for phenylalanine in that intense ions indicative of the molecular weight (M + H+ or M − H−) are observed along with a few structurally significant fragments. These results are typical of the 20 amino acids which have been investigated.

Several dipeptides have been studied; examples of the results are shown in Figure 4. The dipeptides appear similar to the amino acids in that they also generally yield intense ions characteristic of the molecular weight; however, the fragmentation patterns are considerably more complex. Insufficient data are presently available to determine the utility of the spectra for determination of the structure of the dipeptides.

In most of our earlier work we used the elements of nucleic acids—bases, nucleosides, and nucleotides—as test compounds which provided a convenient series of biologically important molecules of increasing difficulty. The purine and

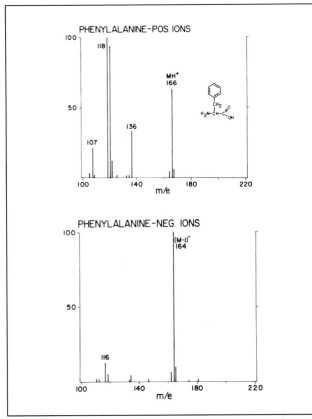

Figure 2. Chemical ionization mass spectra of phenylalanine above 100 amu obtained using the new LC-MS system. Sample was injected as 20 µL of 0.01 M formic acid solution (1 mg/mL) and carried to the mass spectrometer interface in 0.01 M formic acid flowing at 0.3 mL/min. (a) Positive ion and (b) negative ion spectra

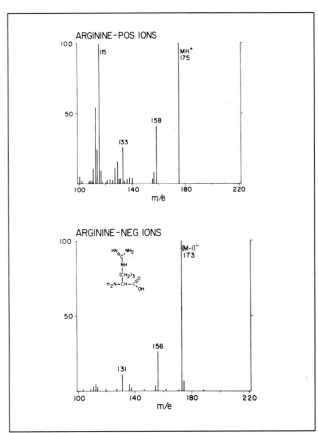

Figure 3. Chemical ionization mass spectra of arginine above 100 amu obtained using the new LC-MS system; conditions were the same as given in Figure 2

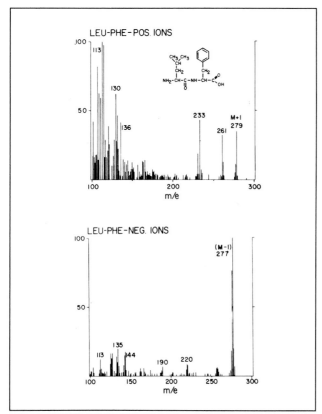

Figure 4. Chemical ionization mass spectra of the dipeptide, leucylphenylalanine, above 100 amu obtained using the new LC-MS system for an injection of 1 µg of sample; flow rate was 0.5 mL/min and the other conditions were as given in Figure 2

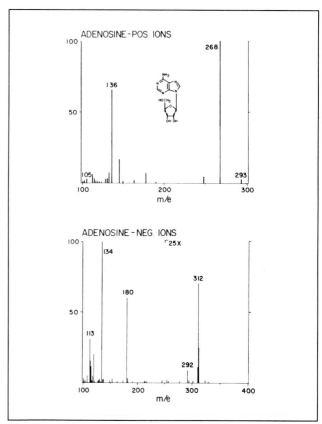

Figure 5. Chemical ionization mass spectra of adenosine above 100 amu obtained using the new LC-MS system; mobile phase was 0.2 M formic acid at 0.5 mL/min and otherwise the conditions were as given in Figure 2

pyrimidine bases present no particular problem; we have generally obtained spectra in which the protonated molecular ion is the base peak in the positive ion CI spectrum. The nucleosides are substantially more difficult. While the intensities of the fragment ions are relatively stable and easily reproduced, the intensity of the MH$^+$ ion in positive CI is very sensitive to the operating conditions. A spectrum corresponding to perhaps the "best" positive CI result on adenosine is given in Figure 5a, and a fairly typical negative ion CI result is shown in Figure 5b. The positive CI result is in good agreement with the conventional CI results of Wilson and McCloskey (7). For all of the nucleosides studied the (b + 2H)$^+$ and b$^-$ ions are always observed as very intense fragments in the positive and negative ion CI, respectively, and are usually the base peaks in the spectra. In positive ion CI the MH$^+$ ion is nearly always observed, but its relative intensity varies from being the base peak as shown in Figure 5a down to a few percent of the base peak depending on the nucleoside, the size of the sample, and details of the operating conditions which are not yet fully understood. Using formic acid or formate buffers, the formate adduct ion (mass 312 in Figure 5b) is usually observed as the highest mass ion identifiable in the spectrum. With pure water the (M − 1)$^-$ ion is observed instead.

The nucleotides represent the most difficult molecules which we have studied. Positive and negative CI spectra for the 5′-AMP are shown in Figure 6. To the best of our knowl-

edge, protonated molecular ions have not been observed in positive ion CI spectra of the nucleotides by conventional techniques; however, protonated AMP has been observed using a moving belt LC-MS interface (8). In our work, the intensities of the ions characteristic of molecular weight are very sensitive to the operating conditions. Spectra showing pseudo-molecular ion intensities comparable to or greater than those shown in Figure 6 have been obtained frequently for several nucleotides; on the other hand, under superficially similar operating conditions, the pseudo-molecular ions are sometimes not observed. Some further improvements in our techniques for operating the instrument and particularly in controlling and monitoring the vaporizer will probably be required before application of the combined LC-MS to molecules as nonvolatile and thermally labile as the nucleotides can be considered routine. Nevertheless, the present results do show that vaporization and ionization of intact nucleotide molecules are feasible.

EFFICIENCY AND SENSITIVITY

An important consideration in combined chromatography–mass spectrometry is the efficiency with which sample eluting from the chromatograph is transferred to the ion source of the mass spectrometer. At the same time, it is usually necessary to allow only a small fraction of the solvent or carrier to be transferred. In our present apparatus about 3–5% of the sol-

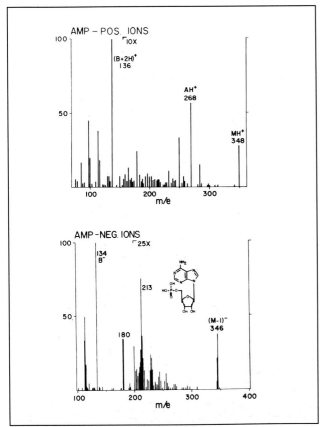

Figure 6. Chemical ionization mass spectra of the nucleotide 5′-adenosine monophosphate above 100 amu obtained using the new LC-MS system under the same conditions given in Figure 5

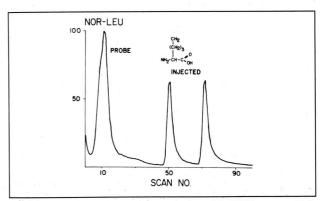

Figure 7. Mass chromatogram for the MH⁺, mass 132, from three 900-ng samples of norleucine. The peak on the left corresponds to placing the sample directly into the ion source using the direct insertion probe and the two later peaks correspond to duplicate injections using the LC interface

vent vapor is transferred to the ion source when the liquid flow is 0.5 mL/min of aqueous medium. The transfer efficiency for nonvolatile solutes is much higher. The results of a direct measurement of transfer efficiency are shown in Figure 7. In this experiment, the normal solvent vaporization conditions were employed with 0.2 M ammonium formate buffer (pH 5) at a flow rate of 0.5 mL/min. A sample of norleucine was placed on the direct insertion probe and the mass spectrum was scanned repetitively from 70 to 300 amu as the probe was heated at a rate of 50 °C/min to 250 °C. Immediately following this experiment, an identical sample of norleucine was injected using the LC injection valve. This was followed by a second injection of the same sample of norleucine. The mass chromatogram for the MH⁺ ion, mass 132, for these three samples of norleucine is shown in Figure 7. The areas of the peaks corresponding to the injected samples are 49 and 52%, respectively, of the peak area for the probe sample. In all cases, the MH⁺ ion was the base peak in the spectrum above mass 70, and it represented more than 50% of the increase in the total ion current due to the sample. The experimental parameters were constant for these experiments but, in the first case, the sample was inserted directly into the chemical ionization source and, in the others, the sample entered via the LC effluent and the vaporizer. Thus, we conclude that approximately 50% of the norleucine sample was transferred to the ion source of the mass spectrometer by the

LC-MS interface. Similar experiments have been performed a number of times using several relatively nonvolatile amino acids and nucleosides with rather similar results. The less volatile samples, for example, adenosine, tend to give rather higher measured transfer efficiencies, but this may be because they do not vaporize cleanly from the direct insertion probe.

An example of the response of the mass spectrometer as a function of sample size is shown in Figure 8, where the integral of the MH⁺ intensity and the integral of the increase in total ion intensity are plotted as functions of nanograms of norleucine injected. The dashed lines on this log–log plot are drawn with unit slope corresponding to the relation expected if the response is directly proportional to sample size. As can be seen from Figure 8, the total ion current response is proportional to sample size at the high end but the MH⁺ response increases faster than linearly with sample size above ca. 500 ng. For an injection of 3.4 μg, the MH⁺ intensity is nearly 90% of the total ionization attributable to the sample, while below 340 ng, MH⁺ is about 20% of total ionization. For an injection

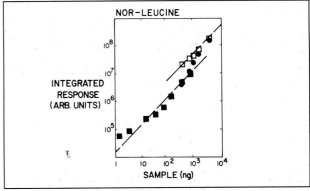

Figure 8. Log–log plot of the integral of the MH⁺ intensity, mass 132, as a function of nanograms of norleucine injected (■). The points corresponding to the MH⁺ intensity at larger sample sizes (●) and to the total ion current (□) were obtained with the multiplier gain reduced by a factor of 64. The raw data for the larger samples were multiplied by this gain factor in preparing the plot

of ca. 1 µg, we estimate that the maximum sample concentration in the ion source (expressed as mole fraction) is about 5×10^{-4}. At this concentration, collisions between sample ions and sample molecules occur frequently. If the fragment ions react with the parent molecules to produce protonated parent ions, the apparent degree of fragmentation would decrease with increasing sample size in the manner shown in Figure 8. Similar effects have been observed for a number of the molecules studied.

The smallest amount of sample detectable in this experiment was 1.4 ng which gave a response about twice the blank response. This result corresponds to a minimum detectable input rate of about 60 pg/s. The sensitivity at mass 132 is limited by the presence of a background peak which is at least 100 times the minimum detectable ion signal. At higher masses where the background ionization is substantially lower, somewhat lower detection limits may be attained.

The present system is applicable to a variety of LC separations including those requiring aqueous buffers and gradient elution. Our best results have been obtained with ammonium formate buffers, but alkali salts may also be used. Phosphate buffers are troublesome in that they cause buildup of phosphate deposit in the interface and may eventually lead to plugging of the inlet capillary. An example of a reversed phase separation of bases and nucleosides using gradient elution from ammonium formate buffer to methanol is shown in Figure 9. In this particular example, the $(b + 2H)^+$ ions for all of the nucleosides are prominent in the mass chromatograms, but MH^+ ions were detected only for adenosine and 1-methyladenosine.

A substantial amount of work remains to establish the full utility of the new LC-MS system. A feature of the present system which may prove somewhat troublesome in practice is that the LC solvent vapor also serves as the CI reagent gas. This dual role of the same substance may cause difficulty in the simultaneous optimization of both the LC separation and the mass spectrometric detection. Otherwise, the present system appears generally applicable to a wide range of LC separation techniques presently employed. For very large molecules which are nonvolatile or thermally labile, some pyrolysis will undoubtedly occur during vaporization; however, the mass spectrometer may still prove to be a useful detector even in these cases. These problems will be addressed in the evaluation and applications phase of this work which is now under way.

CONCLUSION

While we are just beginning to apply the new LC-MS system to real analyses, it appears that we have met all of the design goals which we set for ourselves at the beginning of this development effort (2). In particular, up to 1 mL/min of aqueous LC effluent can be vaporized without degrading the LC performance and at least 50% of the sample present can be transferred to the ion source of the mass spectrometer along with, at most, 5% of the solvent vapor. In its present configuration, the instrument can normally be operated for periods of at least several weeks without shutdown for cleaning or replacement of components. The sample transfer efficiencies and detection limits appear to be comparable to those currently obtained in GC-MS, but the LC-MS system sub-

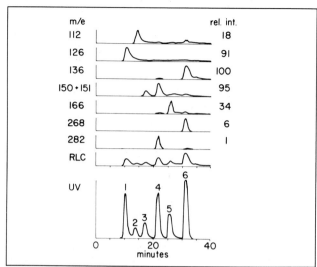

Figure 9. LC-MS analysis of a mixture of nucleosides and bases using gradient elution from ammonium formate buffer (0.2 M, pH 5) to methanol on 25 cm × 4.6 mm Partisil 10-ODS/2 column at flow rate of 0.5 mL/min. Linear gradient started 12 min after injection, completed to pure methanol at 32 min. The lower trace is from UV detector (254 nm, 0.2 AUFS); the second trace from the bottom is the reconstructed liquid chromatogram (RLC) obtained by summing all ions above 100 amu. Individual mass chromatograms correspond to m/e 112, $(b + 2H)^+$ cytidine; 126, MH^+ methylcytosine; 136, $(b + 2H)^+$ adenosine; 150 and 151, $(b + 2H)^+$ methyladenosine and methylinosine, respectively; 166, $(b + 2H)^+$ methylguanosine; 268, MH^+ adenosine; and 282, MH^+ methyladenosine. Components in the mixture were (1) 5-methylcytosine, (2) cytidine, (3) 7-methylinosine, (4) 1-methyladenosine, (5) 7-methylguanosine, (6) adenosine

stantially expands the range of samples for which combined chromatography–mass spectrometry is applicable.

ACKNOWLEDGMENT

The authors thank L. M. Marks for preparing the figures for publication, and are particularly grateful to J. A. McCloskey for providing samples of nucleosides and nucleotides.

LITERATURE CITED

(1) P. J. Arpino and G. Guiochon, *Anal. Chem.*, **51**, 683A (1979).
(2) C. R. Blakley, M. J. McAdams, and M. L. Vestal, *J. Chromatogr.*, **158**, 261 (1978).
(3) R. J. Buehler, E. Flanigan, L. J. Green, and L. Friedman, *J. Am. Chem. Soc.*, **96**, 3990 (1974).
(4) M. L. Vestal, C. R. Blakley, P. W. Ryan, and J. H. Futrell, *Rev. Sci. Instrum.*, **47**, 15 (1976).
(5) G. C. Stafford, Jr., patent pending.
(6) D. F. Hunt and F. W. Crow, *Anal. Chem.*, **50**, 1781 (1978).
(7) M. S. Wilson and J. A. McCloskey, *J. Am. Chem. Soc.*, **97**, 3436 (1975).
(8) W. H. McFadden, Finnigan Instruments, 845 W. Maude Ave., Sunnyvale, Calif., private communication.

Received for review March 21, 1980. Accepted May 20, 1980. This work was supported by the Institute of General Medical Sciences (NIH) under Grant GM 24031.

Reprinted from *Anal. Chem.* **1980**, *52*, 1636–41.

This article will take you back almost 15 years to the time when "hyphenation" was de rigueur. In this case the hyphenated technique is liquid chromatography–electrochemistry, and the authors show us some of their best work—chromatographic detection with square-wave voltammetry—emphasizing its inherent sensitivity and as much selectivity as one can ever achieve with voltammetry.

Dennis Evans
University of Delaware

Rapid Scan Square Wave Voltammetric Detector for High-Performance Liquid Chromatography

Robert Samuelsson
Department of Analytical Chemistry, University of Umeå, S-90187 Umeå, Sweden
John O'Dea and Janet Osteryoung*
Department of Chemistry, State University of New York at Buffalo, Buffalo, New York 14214

Sir: Voltammetric detectors for high-pressure liquid chromatography (HPLC) are widely used because of their good sensitivity for many electroactive compounds (1). In the most commonly employed constant potential mode, this technique has only modest selectivity. Although electroanalytical techniques are inherently unselective, the selectivity should be improved dramatically by using a scanning technique which provides potential as well as time resolution. As pointed out by Johnson (2), despite pessimistic attitudes toward pulse techniques in this application (1), square wave voltammetry (3,4) should provide the full resolution obtainable with respect to potential without degradation of sensitivity. In principle square voltammetry should have the following attributes: (1) discrimination against charging currents and against "background" currents because of the current measurement scheme, (2) little electrode fouling because of the small amount of material converted, and (3) good potential resolution on a time scale short in comparison with that required for chromatographic resolution because high scan rates can be employed. We describe here the use of rapid scan square wave voltammetry for HPLC detection. The model compounds chosen are N-nitrosodiethanolamine (NDELA) and N-nitrosoproline (NPRO), and the chromatographic separation is based on work reported elsewhere (5, 6). The advantages and possibilities of the technique are discussed, and the selectivity and sensitivity of the detector are compared with others.

The chromatographic system, which is described in detail elsewhere (5), employed a detector based on the PARC 310 polarographic detector (Princeton Applied Research, Princeton, NJ).

The excitation wave form was generated by D/A converters controlled by Digital Equipment Corp. PDP 8/e computer equipped with 32K words of main memory, floating point processor, cathode ray tube display, 12 bit D/As, programmable real time clock, 12 bit A/D, and 2.5M byte hard disk for mass storage. This system supports Digital Equipment Corp. OS/8 operating system and Fortran IV, with assembly language subroutine calls to handle the clock and converters. The real time clock was used to initiate A/D conversions immediately before the next potential step of the wave form was applied (3, 4). The individual forward and reverse currents or their difference was displayed in real time for the benefit of the experimenter and was also stored on the disk for analysis at a later time.

The potentiostat, which was fast settling and immune to voltage saturation, was homemade.

The theory for square wave voltammetry for irreversible systems is discussed by O'Dea and Osteryoung (7) and by O'Dea (8), and experimental studies on such systems have been reported by O'Dea (8). NPRO exhibits one 4e⁻ totally irreversible reduction wave in acidic solution (9). Figure 1, which shows a square wave polarogram of NPRO, displays the expected behavior. Sensitivity studies indicated that concentrations of the order 10^{-7} M are easily determined.

Buchanan and Bacon have reported the use of square wave polarography at constant potential for detection in ion exchange chromatography (10). They measured the square wave current at four different potentials applied in sequence to the detector cell.

In our case the potential was scanned over a potential range of 500 mV every 2 s. A total number of 102 forward and reverse current samples were collected and stored. The step height and frequency were 10 mV and 100 Hz, respectively. A predetermined number of scans were recorded and stored for each chromatogram. The data were analyzed with a program

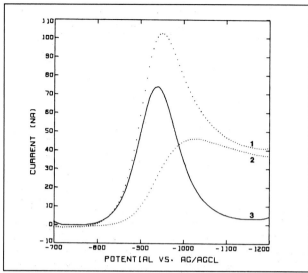

Figure 1. Square wave polarogram of NPRO in 1% phosphate buffer (pH 3.5): (1) forward current; (2) reverse current; (3) net current. Conditions: [NPRO] = 6.03×10^{-6} M, step height = 5 mV, square wave amplitude = 25 mV, square wave frequency = 10 Hz. Background currents have been subtracted.

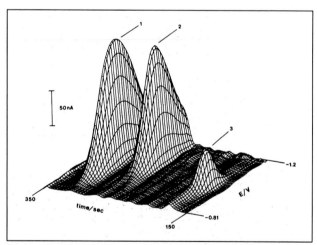

Figure 2. Three-dimensional chromatopolarogram of NDELA and NPRO: (1) NPRO; (2) NDELA; (3) unknown impurity. Conditions: [NPRO] = 1.34×10^{-5} M, [NDELA] = 1.2×10^{-5} M, step height = 10 mV, square wave amplitude = 25 mV, square wave frequency = 100 Hz, mobile phase = 1% phosphate buffer (pH 3.5). Background currents have been subtracted.

that was able to subtract the background, filter the noise, and find the peak position and the peak height for each N-nitrosamine.

Figure 2 shows a three-dimensional chromatopolarogram of NDELA and NPRO. The nitrosamines are completely resolved in time (5) and the peak potentials are −1.07 and −1.04 V for NDELA and NPRO, respectively. A peak is also observed at t_0 which corresponds to an unknown impurity in the standard solution. A calibration plot for NPRO over the range 1–15 mM was linear with intercept −3.4 nA, slope 15.2 nA/µM, and pooled standard deviation 3.23 nA, which gives an estimate of the detection limit of 0.6 µM at the 95% confidence level. The values of retention time and peak potential and their standard deviations for six concentrations in the same range were 266 ± 2 and 308 ± 3 s and −1.07 ± 0.01 and −1.04 ± 0.01 V for NDELA and NPRO, respectively. Thus, peak potential provides a reliable additional parameter for compound identification. In addition, the resolving power of the chromatopolarographic system is increased because separation in both time and potential can be made.

As reported earlier (5), detection is limited by a high noise level caused by pulsations in flow rate. By improvement of the pumping system, a decrease in the detection limit by 1 order of magnitude may be possible. Optimization of the square wave parameters and the geometry of the flow cell should give even better performance. Work on this and various applications to organic and inorganic systems is proceeding in our laboratory.

LITERATURE CITED

(1) Kissinger, Peter T. *Anal. Chem.* **1977**, *49*, 447A–456A.
(2) Johnson, Dennis C. *Anal. Chem.* **1980**, *52*, 131R–188R.
(3) Christie, J. H.; Turner, John A.; Osteryoung, R. A. *Anal. Chem.* **1977**, *49*, 1899–1903.
(4) Turner, John A.; Christie, J. H.; Vukovic M.; Osteryoung, R. A. *Anal. Chem.* **1977**, *49*, 1904–1908.
(5) Samuelsson, Robert; Osteryoung, Janet. *Anal. Chim. Acta*, in press.
(6) Iwaoka, W.; Tannenbaum, S. R. *J. Chromatogr.* **1976**, *124*, 105–110.
(7) O'Dea, John; Osteryoung, R. A., unpublished work.
(8) O'Dea, John. Ph.D. Dissertation, Colorado State University, Ft. Collins, CO, 1979.
(9) Hasebe, Kiyoshi; Osteryoung, Janet. *Anal. Chem.* **1975**, *47*, 2412–2418.
(10) Buchanan, E. B., Jr.; Bacon, J. R. *Anal. Chem.* **1967**, *39*, 615–620.

Received for review June 24, 1980. Accepted August 7, 1980. This work was supported by the National Science Foundation under Grant No. CHE-7917543. R.S. wishes to thank the Swedish Work Environment Fund for financial support.

Reprinted from *Anal. Chem.* **1980**, *52*, 2215–16.

The early development of ICPMS occurred in the laboratories of Alan Gray (initially at ARL, later at the University of Surrey), at the Iowa State University–Ames laboratory, and at Sciex. Two of the most important papers appeared in *Anal. Chem.* The first (below) resulted from a collaboration between Gray and Houk, Fassel, and co-workers at Ames. The second, by Douglas and French (*Anal. Chem.* **1981**, *52*, 37–41), described the differentially pumped sampling interface that became standard for all ICP instruments. In rereading these classic papers, one notes that many observations remain true and that the authors' expectations about ICPMS development were remarkably accurate.

James W. McLaren
National Research Council Canada

Inductively Coupled Argon Plasma as an Ion Source for Mass Spectrometric Determination of Trace Elements

Robert S. Houk, Velmer A. Fassel,* Gerald D. Flesch, and Harry J. Svec
Ames Laboratory–USDOE and Department of Chemistry, Iowa State University, Ames, Iowa 50011

Alan L. Gray
Department of Chemistry, University of Surrey, Guildford, Surrey, England GU2 5XH

Charles E. Taylor
Southeast Environmental Research Laboratory–USEPA, Athens, Georgia 30601

Solution aerosols are injected into an inductively coupled argon plasma (ICP) to generate a relatively high number density of positive ions derived from elemental constituents. A small fraction of these ions is extracted through a sampling orifice into a differentially pumped vacuum system housing an ion lens and quadrupole mass spectrometer. The positive ion mass spectrum obtained during nebulization of a typical solvent (1% HNO_3 in H_2O) consists mainly of ArH^+, Ar^+, H_3O^+, H_2O^+, NO^+, O_2^+, HO^+, Ar_2^+, Ar_2H^+, and Ar^{2+}. The mass spectra of the trace elements studied consist principally of singly charged monatomic (M^+) or oxide (MO^+) ions in the correct relative isotopic abundances. Analytical calibration curves obtained in an integration mode show a working range covering nearly 4 orders of magnitude with detection limits of 0.002-0.06 μg/mL for those elements studied. This approach offers a direct means of performing trace elemental and isotopic determinations on solutions by mass spectrometry.

Despite the demonstrated utility of mass spectrometry for the analysis of a wide variety of gaseous or solid samples, this technique is scarcely used for the routine determination of elemental constituents in aqueous solutions. Commonly used ion sources are not suitable for the rapid, direct examination of aqueous samples because extensive sample preparation procedures are required (*1, 2*). Thus,

the sample is evaporated onto a filament for thermal ionization or incorporated into an electrode for spark ionization before the sample-containing substrate is physically mounted in the vacuum system. The associated time requirement for these operations renders the routine analysis of large numbers of solutions impractical.

Elemental constituents in solution samples are commonly determined by atomic absorption or emission spectrometry. In these techniques solution aerosols are injected directly into a variety of high-temperature atomization cells at atmospheric pressure for vaporization, atomization, and excitation. These flames and plasmas often provide significant populations of positive ions, which can be extracted through an appropriate sampling orifice into a vacuum system for mass analysis and detection (*3–20*). Ions derived from elemental constituents of injected solution aerosols should also be extractable by a similar approach. Thus the analytical capabilities of mass spectrometry can, in principle, be combined with the convenience and efficiency of solution introduction into an appropriate plasma ion source.

A.L.G. has previously evaluated a system for trace element determinations based on the introduction of solution aerosols into a dc capillary arc plasma (CAP) (*21*). A small fraction of plasma gas along with its ions was extracted from the CAP through a pinhole-like sampling orifice into a differentially pumped vacuum system containing an electrostatic ion lens, quadrupole mass analyzer, and electron multiplier. Background mass spectra obtained from the CAP had few peaks above 50 amu, and thus facilitated use of a low-resolution mass analyzer. Analyte elements were detected essentially as singly charged, monatomic, positive ions, i.e., the simplest possible mass spectrum. Detection limits of 0.000 02–0.1 µg/mL were obtained; those elements with ionization energies below 9 eV had the best powers of detection (22–24). The relative abundances of the various isotopes of Sr and Pb were determined with relative precisions of ±0.5% in dissolved mineral samples (25, 26). These results indicated the feasibility of obtaining elemental mass spectra from analytes in solution with a plasma ion source. However, matrix and interelement interferences were severe (26).

Although both the CAP and the inductively coupled plasma (ICP) were originally developed for trace element determinations by atomic emission spectrometry, the ICP has found much wider application. Most of the characteristics of the ICP that have vaulted it to supremacy as an excitation source for atomic emission spectrometry are also highly desirable in an ion source for mass spectrometry (27–29). In particular, a high number density of trace element ions is implied by the common use of emission lines from excited ions for the determination of trace elements by atomic emission spectrometry. For example, cadmium, despite its relatively high ionization energy (8.99 eV), is often determined by using an ion line (30). Also, the ICP as an excitation source is remarkably free from such interferences as (a) incomplete solute vaporization and atomization and (b) ionization suppression or enhancement caused by changes in the solution concentration of easily ionized concomitant elements, e.g., Na (31–34). The objective of the present work is to present results that demonstrate the feasibility of inductively coupled plasma-mass spectrometry (ICP-MS) for the determination of elemental concentrations and isotopic abundance ratios in solutions.

APPARATUS AND PROCEDURES

The ICP-MS apparatus used in the present work is shown schematically in Figure 1. The components and operating conditions are listed in Table I. The apparatus has been described in greater detail elsewhere (35).

Inductively Coupled Plasma. The ICP was generated in a horizontal torch fitted with an extended outer tube as shown in Figure 1. The tube extension merely elongated the ICP relative to its dimensions in torches of conventional length. As viewed from its end, the extended ICP had the usual toroidal appearance. Thus, the injected aerosol particles remained localized in the central or axial channel of the ICP, where vaporization, atomization, and ionization of analyte species occurred as in conventional ICPs (27–29). The torch was enclosed in a grounded, copper-lined shielding box.

Plasma Sampling Interface. The function of the interface was to extract a small fraction of plasma gas, along with its ions, into the vacuum system. The extraction was performed in two steps with the skimmer and sampler shown in Figure 1. The axial channel region of the ICP flowed through the central hole of the water-cooled, stainless steel skimmer, forming a well-defined plume. Analyte species derived from the sample aerosol streamed through the skimmer hole with

the plume, while the outer portions of the vortex of the ICP were deflected outside the skimmer. The plume, still near atmospheric pressure, next impinged on the sampler, which consisted of a water-cooled copper cone mounted on the vacuum system. Plume particles (atoms, ions, and electrons) were extracted through a 50 µm diameter orifice drilled through the center of a molybdenum disk. The disk was mounted in the tip of the sampler behind a retaining copper lip as shown in Figure 2. The copper lip held the disk firmly in position, served as a vacuum seal, and provided thermal contact between the disk and the cooled sampler cone.

The stainless steel skimmer glowed orange hot (~1000 K) when immersed in the ICP. The tip of the sampler glowed red hot (~800 K) when thrust inside the skimmer. These elevated temperatures greatly inhibited the condensation of analyte-derived solids on either the skimmer or sampler tip. When such condensation became extensive, ion sampling was unstable, i.e., solid deposits plugged the sampling orifice or the ICP arced sporadically to the skimmer and sampler. Maximum count rates for analyte ions were obtained when the sampler tip was thrust inside the skimmer about 2 mm behind the skimmer tip.

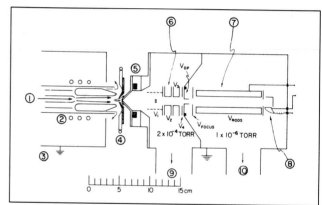

Figure 1. Schematic diagram of ICP, ion sampling interface, and vacuum system: (1) analyte aerosol from nebulizer; (2) ICP torch and load coil; (3) shielding box; (4) skimmer with plasma plume shown streaming through central hole; (5) sampler cone with extraction orifice (detailed diagram in Figure 2); (6) electrostatic ion lens assembly; (7) quadrupole mass analyzer; (8) channeltron electron multiplier; (9) pumping port to slide valve and diffusion pump (first pumping stage); (10) pumping port to slide valve, liquid nitrogen baffle, and diffusion pump (second pumping stage).

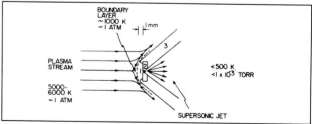

Figure 2. Cross-sectional diagram of sampler tip: (1) sampling orifice (50 µm diameter); (2) molybdenum disk containing orifice; (3) copper cone with spun copper seal to retain molybdenum disk.

Table I. Instrumental Facilities

Component description and manufacturer	Operating conditions	Component description and manufacturer	Operating conditions
Plasma generator: Type HFP-2500D with impedance matching network Plasma-Therm, Inc. Kresson, NJ	Forward power 1000 W, reflected power < 10 W, 27.12 MHz	Ion lens voltage supply: Model 275-L25 Extranuclear Laboratires, Inc. Pittsburgh, PA	Operated in atmospheric pressure ionization mode (electron impact ionizer off)
Plasma torch: all quartz Ames Laboratory design and construction (29) with outer tube extended 50 mm above tip of aerosol tube	Argon flow rates: plasma flow 12 L/min, aerosol carrier flow 1 L/min, auxiliary flow used only during ignition	Quadrupole mass spectrometer: Model 100C Uthe Technologies, Inc. (UTI) Sunnyvale, CA	Minor modifications described in text
Ultrasonic nebulizer: Model UNS-1 Plasma Therm, Inc. Kresson, NJ similar to Ames Laboratory design (36) modified Margoshes-Veillon desolvation system (37)	Sample introduction rate 2.5 mL/min by peristaltic pump, transducer power ~ 50 W, transducer and condenser ice water cooled	Detector: Channeltron electron multiplier Model 4717 Galileo Electro-Optics Corp. Sturbridge, MA Supplied by UTI	Cathode bias −4 kV, gain ≈ 10^6, anode electrically isolated from channel
Orifice disk: molybdenum disk Agar Aids Stansted, Essex, England	2 mm o.d., 0.5 mm nominal thickness, orifice length ≈ orifice diameter, 50 μm orifice diameter	Pulse counting system: Model 1121 preamplifier-discriminator Model 1109 counter EG&G Princeton Applied Research Princeton, NJ	Single discriminator mode, threshold 3 mV
Vacuum system: welded stainless steel assembly, differentially pumped Ames Laboratory construction	First stage pressure: 1×10^{-3} torr (air, 1 atm, 25 °C), 4×10^{-4} torr (ICP sampling); second stage pressure: 1×10^{-6} torr (ICP sampling)	Data acquisition (scanning mode): active low pass filter-amplifier Model 1020 Spectrum Scientific Corp. Newark, NJ	Spectrum recorded on X-Y recorder: Y axis, filtered and amplified dc voltage (proportional to count rate from counter); X axis, dc voltage (0 to +10 V) from mass spectrometer controller (proportional to transmitted mass)
Ion lens elements: stainless steel, based on Model 275-N2 Extranuclear Laboratories, Inc. Pittsburgh, PA	Voltage values: $V_1 = -200 V$, $V_2 = -80$, $V_3 = -95$, $V_4 = -60$, $V_{DP} = -60$, $V_{FOCUS} = -18$, $V_{RODS} = -11$	Data acquisition and handling (integration mode): Counter interfaced to teletype for paper tape, hard copy record	Mass spectrometer manually peaked on mass of interest, count period 10 s, 5–10 count periods recorded and averaged at each mass and for each solution

A typical sampler operated in a stable fashion for nebulization of dilute (<150 μg/mL) analyte solutions for 8–10 h before sampling conditions deteriorated due to gradual condensation of solid on the tip of the sampler. The sampler was readily cleaned by immersing it in an ultrasonically agitated water bath for a few minutes. An individual sampler remained useful for a total of 50–100 h. During this time the disk gradually became pitted and discolored, and the orifice developed an irregular cross section.

A Teflon gasket and nylon bolts were used to retain the cooling flange and to isolate it electrically from the vacuum system. The skimmer and sampler were each grounded through separate inductive–capacitive filters (36). This grounding scheme reduced RF interference in the ion gauges, counting electronics, and recording equipment.

Vacuum System. The sampler cone and orifice assembly were mounted on a two-stage, differentially pumped vacuum

system of welded stainless steel construction. The first stage was evacuated by an oil diffusion pump (1600 L s⁻¹, Lexington Vacuum Division, Varian Associates, Lexington, MD). The electrostatic ion lens was mounted in the first stage. As shown in Table I, the first-stage pressure was sufficiently low for ion collection and beam formation but too high for mass spectrometer operation. A second stage of differential pumping was therefore required. The ions were directed through a 3 mm diameter × 8 mm long aperture into the second stage, which housed the quadrupole mass spectrometer. The second stage was pumped by a second 1600 L s⁻¹ oil diffusion pump equipped with a liquid nitrogen cooled baffle. Both pumping stations were provided with slide valves to permit rapid venting for sampler installation or modification of internal components.

Electrostatic Ion Lens System. An ion lens system was used to collect positive ions from the supersonic jet of sampled gas while neutral particles were pumped away. The ions were then focused and transmitted to the mass analyzer. As shown in Figure 1, the lens system consisted of a set of coaxial, sequential cylinders, each biased at a particular dc voltage. Maximum ion signals were obtained at the voltages specified in Table I (35). The shapes, width, resolution, and symmetry of the ion peaks were unaffected by the voltage settings on the ion optical elements. The cylindrical section of the first element was made of no. 16 mesh screen to provide fast pumping of neutral species from the ion collection and collimation region. A 4.6 mm diameter solid metal disk was positioned in the center of the first element. This disk acted as an optical baffle, i.e., it blocked the line of sight from the ICP through the sampling orifice, lens system, and quadrupole axis, and thus helped to prevent optical radiation from the ICP from reaching the electron multiplier.

Mass Analyzer. The quadrupole mass analyzer (originally supplied as a residual gas analyzer) was modified as follows. First, the filaments, grid, and reflector of the electron impact ionizer were removed; the focus plate was retained as the quadrupole entrance aperture. The latter was aligned visually with the center of the lens system by shimming under the rod mounting bracket. Second, the rods were biased below ground by connecting separate dc supplies into the dc rod driver circuit. The mass analyzer had a mass range of 1–300 amu with resolution sufficient to resolve adjacent masses unless one peak was much more intense than the adjacent one. Because the transmission of the mass analyzer dropped significantly as the transmitted mass increased, the observation of relatively low analyte masses was emphasized in this feasibility study.

Electron Multiplier and Pulse Counting Electronics. The Channeltron electron multiplier detector as supplied with the mass analyzer was operated in the pulse counting mode. Although this multiplier had a much lower gain than those designed specifically for pulse counting, it still performed adequately for the following reasons. First, there was no evidence of loss of gain or pulse overlap at count rates up to at least 5×10^4 counts/s. The multiplier therefore had a linear dynamic range of at least 5×10^4. At -4 kV the threshold setting on the pulse counting equipment could be set over a broad range (0.2–30 mV) without attenuating the observed count rate. The pulses were conducted from the multiplier anode to a preamplifier–discriminator–counter system. The counting threshold was set just above the height of RF noise pulses from the ICP.

Mass Spectra, Analytical Calibration Curves, and Detection Limits. The reference blank solution and the matrix for the reference solutions used for calibration consisted of 1% (volume) nitric acid, prepared by diluting doubly distilled, concentrated nitric acid with deionized water. The reference calibration solutions were prepared by appropriate dilution of stock solutions. The stock solutions were prepared by dissolving pure metals or reagent grade salts in dilute nitric acid.

Mass spectra were acquired in the scanning mode as described in Table I. Individual points for analytical calibration curves were obtained in the integration mode. The average to-

tal count for the reference blank solution at the mass of interest was evaluated first, followed by the average total count for each reference calibration solution, in ascending order of concentration. The average total count for the reference blank was then subtracted from the average total count for each reference standard solution before plotting. The detection limit was calculated as the analyte concentration required to give an average net count equal to twice the standard deviation observed at the mass of interest for the blank solution, $2\sigma_b$.

RESULTS AND DISCUSSION

Boundary Layer Formation. As the flowing plasma plume approached the sampler, the plume gas was deflected around the blunt sampler tip. As shown in Figure 2, an aerodynamically stagnant layer of gas formed between the flowing plume and the sampler tip (10, 15, 16, 38–40). Formation of a space–charge sheath or electrical double layer in contact with the sampler was also probable (40–44). This composite boundary layer was in thermal contact with the relatively cool sampler. Thus, the temperatures in the boundary layer were intermediate between the plume and sampler temperatures. The boundary layer extended across the sampler tip and was visibly unbroken by the gas flow drawn into the sampling orifice. Ion extraction into the vacuum system therefore occurred only after transport through the boundary layer, which would take 1–2 ms and involve up to $\sim 10^6$ collisions (35).

Such collisions in a medium temperature environment probably facilitated ion–electron recombination, ion neutralization at the sampler walls, charge exchange, ion–neutral attachment, nucleation and condensation of solid deposits, or other reactions (10, 16, 38). The metal surface of the orifice disk may have catalyzed some of these reactions occurring in the boundary layer or just inside the channel-like orifice (15). Also, collisions leading to clustering, ion–electron recombination, or charge exchange occurred in the supersonically expanding jet of extracted gas (10, 11). The effects of these reactions are described below.

Mass Spectra of Reference Solutions. The mass spectrum of the major positive ions from the ICP plume observed during nebulization of a reference blank solution is shown in Figure 3. The two most intense peaks corresponded to Ar$^+$ (40 amu) and ArH$^+$ (41 amu). A comparable peak for H$^+$ (1 amu)

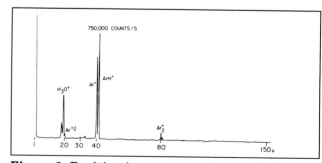

Figure 3. Positive ion mass spectrum of reference blank solution (1% HNO₃ in deionized distilled water). Vertical scale is linear with count rate; base peak count rate is indicated. The background ranged from 30 to 100 counts/s.

was evident, with its low mass edge obscured by "zero blast," i.e., ions anomalously transmitted through the quadrupole field region at the beginning of a scan because the low applied potentials led to very weak fields within the rod structure (2).

The major ions observed in the mass spectrum of the reference blank solution are identified in Table II. All of the major ions have been observed previously by other investigators in the mass spectra of flames and plasmas (5, 17, 18, 20, 22, 23) or as analogous cluster species formed during supersonic jet expansions. Of the major ions only Ar^+ has been identified in analytical ICPs by optical spectrometry, and even that identification is tentative or disputed (30, 45). The existence of an intense peak due to ArH^+, along with the observation of other cluster ions such as H_3O^+ (19 amu) and Ar_2^+ (80 amu), indicated that some clustering reactions occurred during ion extraction. Some minor ions (≤1000 counts/s) were observed at times at 2, 45–48, 50, 54–59, 68–70, 73, and 76 amu. Many of these were also formed during the extraction process, e.g., O_2^+•H_2O at 50 amu. Despite the opportunities for complicating reactions described above, the mass spectrum of the reference blank solution had usefully clear mass regions from 2 to 13 amu, from 21 to 29 amu, and from 42 amu up.

Because there was no ion source inside the vacuum system, the residual gas was not ionized. Ions derived from pump oil were not observed. Thus, a major background contribution from ionization of residual gas in conventional ion sources was not observed with the plasma ion source (22, 23).

The count rate obtained for the reference blank spectrum at those masses free of major or minor ions was 30–100 counts/s, well above the dark current count rate characteristic of the electron multiplier (≤1 count/s). This background count rate was the same at all masses and was independent of the ion lens voltages and mass spectrometer operating conditions. Apparently, this background was caused by vacuum UV photons striking the electron multiplier. These photons probably were radiated directly from the ICP and also from the decay of metastable argon atoms within the vacuum system. Although the direct line-of-sight from the orifice through the quadrupole field region was blocked by a disklike baffle (Figure 1) and the multiplier was offset from the quadrupole axis, numerous photons still struck the multiplier. The background count rate (photons + minor ions, if present) at each mass of interest of the reference blank solution was reproducible during a 5–10-h period, and integration data for reference standards were adequately corrected by subtraction of the reference blank spectrum.

The recorded peaks of monatomic, singly charged positive ions from solutions of Mn, Cu, Rb, and Ag are shown superimposed on the reference blank spectrum in Figure 4. The metal ion spectra are plotted on the same mass and count rate scales as the reference blank spectrum but are displaced vertically by a change in the recorder zero. As shown in the figure, the accepted relative abundances of the isotopes of Cu, Rb, and Ag were observed. The mass spectrum of Cd is shown in Figure 5; again, the count rates for the various isotopes corresponded to the accepted relative isotopic abundances. The peaks were symmetrical and nearly triangular, as expected for quadrupole mass analysis of ions having a low kinetic energy spread. The least abundant Cd isotope ($^{108}Cd^+$, 0.88%) was clearly detected. For the elements shown in Figures 4 and 5, and for most of the elements studied, only monatomic, singly charged ions (M^+) were observed. Several elements were detected as a distribution of M^+ and MO^+ ions, e.g., Ti, As, and Y. The only doubly charged analyte ions observed were Ba^{2+} and Sr^{2+}, i.e., from the two elements with the lowest second ionization energies. Also, no Cu^+ or Mo^+ ions were ob-

Table II. Major Ions Observed in Mass Spectrum of Reference Blank Solution

mass	ion(s)[a]	rel count rate[b]	
		ArH^+ = 100	$^{55}Mn^+$ = 100
16	O^+	0.4	6
17	HO^+, NH_3^+	1	15
18	H_2O^+, NH_4^+	12	180
19	H_3O^+	40	600
20	Ar^{2+}	4	60
30	NO^+	4	60
32	O_2^+	2	30
33	$(O_2^+)\cdot H$	4	60
36	$^{36}Ar^+$	0.8	12
37	$^{36}ArH^+$	1	15
	$(H_3O^+)\cdot H_2O$		
40	$^{40}Ar^+$	75	1125
41	$^{40}ArH^+$	100	1500
	$(Na^+)\cdot H_2O$		
80	Ar_2^+	8	120
81	$(ArH^+)\cdot Ar$	3	45

[a] Possible ions at same mass number are listed in decreasing order of likelihood or probable intensity.
[b] $ArH^+ \approx 750\,000$ counts/s; $^{55}Mn^+ \approx 50\,000$ counts/s at 50 μg/mL solution concentration.

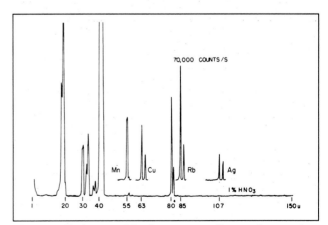

Figure 4. Reference blank spectrum (bottom); superimposed spectra are from 50 μg/mL solutions of the indicated element in 1% HNO_3. Vertical scale sensitivity is 10 times that of Figure 3.

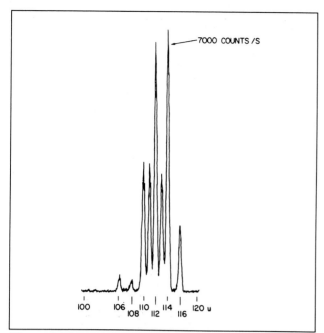

Figure 5. Mass spectrum of Cd at 50 μg/mL in 1% HNO₃.

served from the orifice assembly. Thus, the mass spectra obtained were remarkably simple, which facilitated use of a low-resolution mass analyzer.

Isotopic Abundance Determinations. The utility of the ICP-MS approach for the direct determination of isotopic abundances of elemental constituents in solutions is illustrated further by the data shown in Table III. The agreement with the accepted values of the relative abundances of $^{63}Cu^+$ and $^{65}Cu^+$ was within the estimated uncertainty in the determined values. The absolute standard deviation of the count was approximately 5 times greater than the square root of the average count, which indicated that the uncertainty in the count was significantly greater than the uncertainty expected from counting statistics. This increased uncertainty was undoubtedly due to instability of some instrumental parameter; instability in nebulizer efficiency and ion extraction efficiency through the boundary layer were likely culprits. Thus the precision of the isotopic ratio determinations in Table III is expected to improve with continued development of the ICP-MS technique.

These isotope ratio determinations were performed directly

Table III. Relative Isotopic Abundance Determination of Naturally Occuring Copper Isotopes, 2.5 μg/mL Cu in 1% HNO₃

isotope	N^a	σ	$N^{1/2}$	% abundance determined	accepted
$^{63}Cu^+$	19786	740	143	69.9 ± 1.1	69.1
$^{65}Cu^+$	8505	546	98	30.1 ± 1.1	30.9

$^a N$ = background subtracted average count, obtained in integration mode as described in Apparatus and Procedures section. Uncertainties are indicated at 95% confidence level for 15 determinations, counting time 10 s for each determination.

on a trace level of copper in solution. Also, the total time for the determination of both isotopes was 5 min, including the time required for sample interchange, nebulizer equilibration, and adjustment of the mass transmitted by the mass analyzer. Thus, 100 isotopic ratio determinations could easily be performed in a single day, indicating the potential of the ICP-MS approach for rapid isotopic abundance determinations of trace levels of elements in large numbers of solutions.

Analytical Calibration Curves and Detection Limits. The analytical calibration curves shown in Figure 6, which were obtained in the integration mode, show a useful working range of 3 to 4 orders of magnitude. These data were obtained from reference solutions containing only one element.

When the plasma plume was first moved into contact with the sampler, the count rates of all the ions increased rapidly. After about 1 h this rate of increase tapered off so that the calibration data could be obtained. The Co and Mn curves in Figure 6 show replicate determinations at the 0.02 μg/mL level. For Co and Mn the point labeled by the arrow was determined first. The unlabeled points at 0.02 μg/mL were determined after the calibration data at higher concentrations were obtained, i.e., after about 30 min. This small positive deviation was not caused by memory; instead it reflected the general tendency of the count rates of all the ions to increase slowly with time (∼ 20% every hour) in the absence of orifice plugging. This gradual increase was not accompanied by any discernible increase in orifice diameter. Some subtle phenomena related to the plasma sampling process apparently caused the number of extracted ions to increase with time. This increase should not affect isotope ratio measurements provided they are performed by rapid, repetitive scanning or peak switching techniques. For determination of elemental concentrations, normalization of the ion count rate to an internal standard ion or to a total beam monitor signal should provide internal compensation for the increasing ion signals.

The detection limits obtained for selected elements are listed in Table IV. The detection limits for the major isotopes

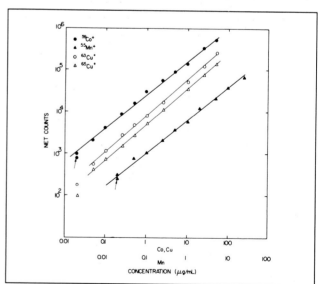

Figure 6. Analytical calibration curves obtained in integration mode. Read bottom scale for Mn curve.

Table IV. Detection Limits Obtained

element	ion detected	% abundance	detection limit ($2\sigma_b$) μg/mL	ppma[a]
Mg	$^{24}Mg^+$	78.6	0.006	0.004
Cr	$^{52}Cr^+$	83.8	0.002	0.0007
	$^{53}Cr^+$	9.6	0.01	0.003
Mn	$^{55}Mn^+$	100	0.003	0.001
Co	$^{59}Co^+$	100	0.006	0.002
Cu	$^{63}Cu^+$	69.1	0.009	0.002
	$^{65}Cu^+$	30.9	0.02	0.005
Rb	$^{85}Rb^+$	72.2	0.008	0.002
	$^{87}Rb^+$	27.8	0.02	0.004
As	$^{75}AsO^+$	100 (^{75}As)	0.06	0.01
Y	$^{89}YO^+$	100 (^{89}Y)	0.04	0.008

[a] ppma = DL(μg/mL) × (18/(atomic weight)).

of Cr, Cu, and Rb were lower than those for the corresponding minor isotopes by factors approximately equal to the relative isotopic abundances. Because the reference blank spectrum had minor peaks (<1000 counts/s) at 54, 66, and 59 amu, the standard deviation of the reference blank count rate increased in the order 52 amu (photons) < 55 amu (photons + ions from peak edges at 54 and 56 amu) < 59 amu (photons + minor ions). Thus the detection limits were degraded in the same order, i.e., $^{52}Cr^+ < {}^{55}Mn^+ < {}^{59}Co^+$, although the net counts at these masses were similar for equimolar solutions of these three elements. It is clearly desirable to reduce both the number and count rates of minor ions in the reference blank spectrum and the count rate of the photon background.

The detection limits listed in Table IV were obtained with a sampler that provided more AsO$^+$ and YO$^+$ than As$^+$ and Y$^+$ ions; hence the oxide ions were used for the determinations of the detection limits of these two elements. Useful analytical calibration curves for As and Y were obtained up to at least 50 μg/mL for either the metal or metal oxide ion. Although formation of metal oxide ions was not desirable, these ions were still useful analytically. Also, metal oxide ion formation is not expected to cause serious mass spectral interferences, because most of the elements are not detected as oxide ions, and the masses and isotopic distribution of MO$^+$ ions are predictable for those elements with a high probability of oxide formation.

The ICP-MS detection limits in Table IV were poorer, i.e., larger numbers, than the CAP-MS detection limits by factors of 10–100. This discrepancy is largely accounted for by the longer integration time (30 s), larger orifice diameter (75 μm), and much lower background levels and background standard deviations obtained in the CAP-MS study. Indeed, the standard deviation of the background in the CAP-MS study was so low (~ 1 count/s) that the conventional definition of detection limit was considered inappropriate (22, 23). The CAP-MS detection limits were superior to those characteristic of all other techniques for trace elemental analysis. Unfortunately,

these powers of detection were of little analytical value because of severe interelement interference effects (26).

The ICP-MS detection limits in Table IV represent potentially useful powers of detection, which can undoubtedly be enhanced by experimental improvements in ion extraction and transmission efficiency and optical baffling of the electron multiplier. Furthermore, the ICP-MS approach should be less susceptible to interelement interference effects than CAP-MS because of the superior sample injection, vaporization, and atomization capabilities and higher electron number density of the ICP (27–29, 32). This expectation is partially demonstrated below.

Ionization Type Interelement Effects. As mentioned above, the ICP as an atomization–excitation source for atomic emission spectrometry is relatively free from ionization interferences caused by easily ionizable elements in solution. For example, the intensities of certain emission lines of excited ions of Ca, Cr, and Cd show little dependence on Na concentration from 0 to 7000 μg/mL (32, 33). In the present work, the count rates of mass resolved analyte ions were suppressed to a somewhat greater extent in the presence of Na as shown in Figure 7. Each point plotted in this figure was normalized to a reference count rate for the same solution concentration of the analyte ion in the absence of Na. Because solute concentrations ≥150 μg/mL gradually clogged the orifice with condensed solid, the reference count rate gradually decreased and was therefore determined repeatedly. The arbitrary nature of this correction procedure, coupled with a significant long-term drift in aerosol intensity produced by the ultrasonic transducer, led to the scatter of the points shown in Figure 7. These plots exhibited a shape similar to those observed for atomic emission spectrometry by Larson et al. (32, 33). Because of the ultrasonic nebulizer used in the present work, 1000 μg/mL Na corresponded roughly to 10 000 μg/mL Na in Larson's work, which was performed with a pneumatic nebulizer (36).

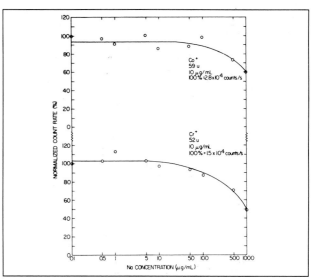

Figure 7. Effect of Na concomitant concentration on count rates of $^{52}Cr^+$ and $^{59}Co^+$, integration mode, 10 s count time, ~5 count periods averaged for each point.

The magnitude of suppression of analyte ionization shown in Figure 7 is approximately twice as large as that observed by atomic emission spectrometry by Larson et al. (32) if the latter data are adjusted for the approximately tenfold difference in nebulization efficiency. This difference may be rationalized as follows. In the present work, ion extraction occurs through an unbroken boundary layer that is somewhat cooler than the unperturbed plasma. Ion–electron recombination or electron loss to the sampler wall may occur at a significant rate in this layer, leading to an effective electron number density (n_e) in the vicinity of the orifice that is less than that prevalent in the unperturbed plasma. For this lower value of n_e, the "extra" electrons contributed by the ionization of Na should be more significant, causing a proportionately greater increase in the total n_e near the orifice. Thus, the greater suppression of analyte ionization observed in the present work may be a characteristic of the boundary layer rather than the unperturbed plasma. A strict comparison of the magnitudes of the ionization suppressions observed in the present work with those observed from the ICP by atomic emission spectrometry is therefore not necessarily valid.

To place the ionization type interelement effect measured in the present work into perspective, we note that the degree of suppression of analyte ionization was far less severe than that observed for the capillary arc plasma–mass spectrometric approach (26) or for flames and other plasmas that have been used as atomization sources in atomic emission or absorption spectrometry (31, 33). Furthermore, the >100 µg/mL Na range, where ionization suppression was significant, represented the analytical equivalent of determining the Co and Cr content of NaCl. Thus, analytical calibrations established for the determination of Co and Cr in a deionized water matrix would have yielded analytical results only ~ 12% lower if the sample calibrations were used for the analysis of a NaCl sample prepared as a solution of approximately 100 µg/mL Na. Even only approximate matching of the total concentration of easily ionizable elements in reference calibration solutions and samples would essentially eliminate analytical bias caused by ionization type interferences, including samples in which these elements (e.g., Na or K) represent varying major fractions of the total metal content.

Solid Deposition in the Sampling Orifice. Solid condensation in or near the orifice remains an operational problem. As mentioned above, normalization of analyte ion count rates either to an internal standard ion or to a beam monitor signal should correct for the gradual decrease in extraction efficiency of analyte ions caused by progressive solid condensation expected from solutions such as hard water. However, progressive deposition of sample material does restrict the useful life of orifices exposed to solutions whose total solute concentrations are above approximately 150 µg/mL. Thus, biological fluids such as urine or blood serum would require a dilution factor of several hundred before analyses of such solutions could be performed for more than about 1 h. The consequent deterioration in powers of detection for analyte elements may not be acceptable for various applications.

Sample deposition in the orifice is primarily caused by the formation of involatile metal compounds in the relatively cool, stagnant gas of the boundary layer shown in Figure 2. Potential solutions to the troublesome deposition of sample material in the orifice undoubtedly lie in the geometry and operating conditions of the sampler tip, which govern boundary layer formation. For example, solid deposition should be less significant in orifices of diameter >50 µm, which would also extract more ions into the vacuum system. A more streamlined, conical orifice assembly should deflect the plasma stream smoothly around the cone tip, instead of allowing a stagnant layer of gas to build up outside the extraction orifice (Figure 2) (10, 11, 15, 16, 38, 40). Such refinements are expected to relax the compromise between powers of detection, dilution factors, and orifice lifetimes, thus facilitating application of the ICP-MS approach to elemental and isotopic determinations in samples of total solute content greater than 150 µg/mL.

ACKNOWLEDGMENT

The contributions of Tom Johnson and Garry Well of the Ames Laboratory machine shop are gratefully acknowledged.

LITERATURE CITED

(1) Ahearn, A. J., Ed. "Trace Analysis by Mass Spectrometry"; Academic Press: New York, 1972; Chapter 1.
(2) Dawson, P. H., Ed. "Quadrupole Mass Spectrometry and Its Applications"; Elsevier: New York, 1976; p 323.
(3) Drawin, H. W. In "Plasma Diagnostics"; Lochte-Holtgreven, W., Ed.; Wiley: New York, 1968; Chapter 13.
(4) Fristrom, R. M. Int. J. Mass Spectrom. Ion Phys. 1975, 16, 15–32.
(5) Goodings, J. M.; Bohme, D. K.; Ng, C. W. Combust. Flame 1979, 36, 27–43.
(6) Hasted, J. B. Int. J. Mass Spectrom. Ion Phys. 1975, 16, 3–14.
(7) Hastie, J. W. Int. J. Mass Spectrom. Ion Phys. 1976, 16, 89–100.
(8) Hayhurst, A. N.; Mitchell, F. R. G.; Telford, N. R. Int. J. Mass Spectrom. Ion Phys. 1971, 7, 177–187.
(9) Hayhurst, A. N.; Telford, N. R. Combust. Flame 1977, 28, 67–80.
(10) Hayhurst, A. N.; Kittelson, D. B.; Telford, N. R. Combust. Flame 1977, 28, 123–135, 137–143.
(11) Burdett, N. A.; Hayhurst, A. N. Chem. Phys. Lett. 1977, 48, 95–99; Combust. Flame 1979, 34, 119–134.
(12) Horning, E. C.; Horning, M. G.; Carroll, D. L.; Dzidic, J.; Stillwell, R. N. Anal. Chem. 1973, 45, 936–943; 1975, 47, 1308–1312, 2369–2373: 1976, 48, 1763–1768; J. Chromatogr. 1974, 99, 13–21.
(13) Knewstubb, P. F. "Mass Spectrometry and Ion–Molecule Reactions"; Cambridge University: London, 1969; Chapter 2.3.
(14) Milne, T. A.; Greene, F. T. Adv. Chem. Ser. 1968, No. 72, Chapter 5.
(15) Morley, C. Vacuum 1974, 24, 581–584.
(16) Pertel, R. Int. J. Mass Spectrom. Ion Phys. 1975, 16, 39–52.
(17) Prokopenko, S. M. J.; Laframboise, J. G.; Goodings, J. M. J. Phys. D 1972, 5, 2152–2160; 1974, 7, 355–362, 563–568; 1975, 8, 135–140.
(18) Rowe, B. Int. J. Mass Spectrom. Ion Phys. 1975, 16, 209–223.
(19) Siegel, M. W.; Fite, W. L. J. Phys. Chem. 1976, 80, 2871–2881.
(20) Vasile, M. J.; Smolinsky, G. Int. J. Mass Spectrom. Ion Phys. 1973, 12, 133–146; 1975, 18, 179–192; 1976, 21, 263–277.
(21) Jones, J. L.; Dahlquist, R. L.; Hoyt, R. E. Appl. Spectrosc. 1971, 25, 628–635.
(22) Gray, A. L. Proc. Soc. Anal. Chem. 1974, 11, 182–183; Anal. Chem. 1975, 47, 600–601; Analyst (London) 1975, 100, 289–299.
(23) Gray, A. L. In "Dynamic Mass Spectrometry"; Price, D., Todd, J. F. J., Eds.; Heyden: London, 1975; Vol. 4, Chapter 10.
(24) Applied Research Laboratories, Ltd., British Patent 1 261 596, 1969; U.S. Patent 3 944 826, 1976.

(25) Anderson, F. J.; Gray, A. L. *Proc. Anal. Div. Chem. Soc.* **1976**, *13*, 284–287.

(26) Gray, A. L. In "Dynamic Mass Spectrometry"; Price, D., Todd, J. F. J., Eds.; Heyden: London, 1978; Vol. 5, Chapter 8.

(27) Barnes, R. M. *CRC Crit. Rev. Anal. Chem.* **1978**, *7*, 203–296.

(28) Fassel, V. A. *Science* **1978**, *202*, 183–191; *Anal. Chem.* **1979**, *51*, 1290–1308A; *Pure Appl. Chem.* **1977**, *49*, 1533–1545.

(29) Fassel, V. A.; Kniseley, R. N. *Anal. Chem.* **1974**, *46*, 1110A–1120A, 1155A–1164A.

(30) Winge, R. K.; Peterson, V. J.; Fassel, V. A. *Appl. Spectrosc.* **1979**, *33*, 206–219.

(31) Rubeska, I.; Rains, T. C. In "Flame Emission and Atomic Absorption Spectrometry"; Dean, J. A., Rains, T. C., Eds.; Marcel Dekker: New York, 1969; Vol. 1, Chapters 11 and 12.

(32) Larson, G. F.; Fassel, V. A.; Scott, R. H.; Kniseley, R. N. *Anal. Chem.* **1975**, *47*, 238–243.

(33) Larson, G. F.; Fassel, V. A. *Anal. Chem.* **1976**, *48*, 1161–1166.

(34) Kalnicky, D. J.; Fassel, V. A.; Kniseley, R. N. *Appl. Spectrosc.* **1977**, *31*, 137–150.

(35) Houk, R. S. Ph.D. Dissertation, Iowa State University, Ames, Iowa, 1980; Report IS-T-989; U.S. Department of Energy, Washington, DC, 1980.

(36) Olson, K. W.; Haas, W. J., Jr.; Fassel, V. A. *Anal. Chem.* **1977**, *49*, 632–637.

(37) Veillon, C.; Margoshes, M. *Spectrochim. Acta, Part B* **1968**, *23B*, 553–555.

(38) Biordi, J. C.; Lazzara, C. P.; Papp, J. F. *Combust. Flame* **1974**, *23*, 73–82.

(39) Reed, T. B. *J. Appl. Phys.* **1963**, *34*, 2266–2269.

(40) Clements, R. M.; Smy, P. R. *Combust. Flame* **1977**, *29*, 33–41.

(41) Boyd, R. L. F. *Proc. Phys. Soc. London, Sect. B* **1051**, *64*, 795–804.

(42) Oliver, B. M.; Clements, R. M. *J. Phys. D* **1975**, *8*, 914–921.

(43) Smy, P. R. *Adv. Phys.* **1976**, *25*, 517–553.

(44) Böhme, D. K.; Goodings, J. M. *J. Appl. Phys.* **1966**, *37*, 362–366, 4261–4268.

(45) Robin, J. *Analusis* **1978**, *6*, 89–97; *ICP Inf. Newsl.* **1979**, *4*, 495–509.

Received for review November 12, 1979. Resubmitted June 19, 1980. Accepted August 19, 1980. Presented in part at the Federation of Analytical Chemistry and Spectroscopy Societies 6th Annual Meeting, Philadelphia, PA, September 1979, and at the Pittsburgh Conference on Analytical Chemistry and Applied Spectroscopy, Atlantic City, NJ, March 1980. This work was supported by the U.S. Environmental Protection Agency and was performed at the Ames Laboratory, U.S. Department of Energy, Contract No. W-7405-Eng-82, under Interagency Agreement EPA-IAG-D-X0147-1.

Reprinted from *Anal. Chem.* **1980**, *52*, 2283–89.

This widely referenced article was the first practical demonstration of zone electrophoresis in open-tubular glass capillaries. It marked the first time that scientists were able to put together the package of a 75-μm-i.d. capillary, a "homemade" on-column detector, and an applied voltage of up to 30 kV to detect amino acids, dipeptides, and amines. Although an induction period of several years followed publication of this work, by the mid- to late 1980s the area was expanding. Capillary electrophoresis is now one of the major growth areas of separation science.

Barry L. Karger
Northeastern University

Zone Electrophoresis in Open-Tubular Glass Capillaries

James W. Jorgenson* and Krynn DeArman Lukacs
Department of Chemistry, University of North Carolina, Chapel Hill, North Carolina 27514

A system for performing zone electrophoresis in open-tubular glass capillaries of 75 μm inside diameter and with applied voltages up to 30 kV is described. The small inside diameter of these capillaries allows efficient dissipation of the heat generated by the application of such high voltages. However, the small inside diameter also necessitates the use of a sensitive on-column fluorescence detector to record the separation of solute zones. With this system, separation efficiency is proportional to the applied voltage, with efficiencies in excess of 400 000 theoretical plates demonstrated. Strong electroosmotic flow in the capillary allows both positive and negative ions of a variety of sizes to be analyzed in a single run with relatively short analysis times. High-efficiency separations of fluorescent derivatives of amino acids, dipeptides, and amines as well as separation of a human urine sample were obtained with analysis times of 10–30 min.

Several important causes of zone broadening may be identified when considering separation efficiency in zone electrophoresis. Molecular diffusion will certainly cause zone broadening, although its effects are generally negligible. More serious difficulties often arise from convection currents in the electrophoretic medium. These are usually minimized through the use of gels, paper, or other stabilizers. However, this approach may introduce additional zone-broadening problems such as adsorptive interactions between the solutes and stabilizer and "eddy migration" in the channels created by some stabilizers (1). Mikkers, Everaerts, and Verheggen (2) sought to solve these convection problems through the use of the "wall effect" by performing zone electrophoresis in narrow-bore Teflon tubes. This approach ap-

peared to solve the problem of convection in a simple way, avoiding the difficulties associated with stabilizers. They found that the concentration of sample ions must be kept well below the concentration of carrier electrolyte in order to achieve symmetric peak shapes. When the sample concentration is too high, the sample alters the conductivity of the medium in its own vicinity, resulting in a distorted electric field gradient and an asymmetric peak shape. If zone electrophoresis is performed in narrow-bore tubes using low concentrations of sample relative to carrier electrolyte, conditions arise where molecular diffusion, originally negligible, may become the predominant cause of zone broadening. The difficulty with this approach is in finding any suitable detection system capable of detecting minute quantities of solutes in small capillary tubes. In this study, zone electrophoresis was attempted in glass capillary tubes. Detection of solute zones was accomplished with an "on-column" fluorescence detector which detects fluorescent solutes while they are still in the glass capillary tube.

THEORY

Consider an electrophoresis system consisting of a tube filled with a buffering medium across which a voltage is applied. Charged species introduced at one end of the tube migrate under the influence of the electric field to the far end of the tube. If a suitable detection device is placed at the far end of the tube, the passage of each solute zone may be recorded, yielding an electropherogram.

The migration velocity of a particular species is given by

$$v = \mu E = \mu V/L \tag{1}$$

where v is the velocity, μ the electrophoretic mobility, E the electric field gradient, V the total applied voltage, and L the length of the tube. The time, t, required for a zone to migrate the entire length of the tube is

$$t = L/v = L^2/\mu V \tag{2}$$

If molecular diffusion alone is responsible for zone broadening, the spatial variance, σ_L^2, of the zone after a time, t, is given by the Einstein equation

$$\sigma_L^2 = 2Dt \tag{3}$$

where D is the molecular diffusion coefficient of the solute in the zone. Substituting the expression for time from eq 2 into this expression yields

$$\sigma_L^2 = 2DL^2/\mu V \tag{4}$$

The concept of separation efficiency expressed in terms of theoretical plates may be borrowed from chromatography as suggested by Giddings (3). The number of theoretical plates, N, is defined as

$$N = L^2/\sigma_L^2 \tag{5}$$

Substituting eq 4 into this expression results in

$$N = \mu V/2D \tag{6}$$

One may note several interesting aspects of this simple result. N is directly proportional to the applied voltage, which suggests the use of the highest voltages possible for high separation efficiency. Somewhat surprisingly, N is independent of tube length and analysis time. Finally, N is proportional to the ratio of the mobility to the diffusion coefficient, factors more or less intrinsic to the solute species and not easily manipulated to improve efficiency. Thus, the most direct route to improved separation efficiency seems to be increasing the voltage applied to the separation medium. Since eq 2 predicts that the analysis time is proportional to the square of the tube length and inversely proportional to the applied voltage, it appears that high voltages applied to short tubes would generate the greatest number of theoretical plates in the shortest length of time.

The principal difficulty with this approach lies in the limited ability to dissipate heat generated in the electrophoretic process. Heat is generated uniformly throughout the medium but is only removed at the inner surface and ends of the tube. Once thermal equilibrium is established, there will be a parabolic temperature gradient across the tube (1, 4, 5). Under extreme circumstances the temperature in the center of the tube will become high enough for the solvent to boil, leading to total breakdown of the electrophoretic process. However, before this effect is observed the undesirable consequences of a radial temperature gradient will be felt in the form of zone broadening. Electrophoretic mobility will increase as the temperature of the medium is increased, at a rate of approximately 2% $°C^{-1}$ (1). Solutes in the warmer center of the tube will migrate faster while those at the wall will migrate more slowly, resulting in zone broadening. The most effective way to minimize this effect is to reduce the tube radius. This approach should have two beneficial results. First, heat dissipa-

tion will be more efficient. According to Wieme (1), the temperature difference from the center to the wall of the tube is proportional to the square of the radius, so reducing the radius should reduce temperature differences markedly. There is a second beneficial effect to be expected from a reduction in tube radius. A temperature gradient is only undesirable to the extent that a solute molecule spends a larger than average fraction of its time in a particular portion of the radius of the tube. The radial position of individual solute molecules is constantly changed by diffusion. In a tube of reduced radius a solute molecule will diffuse back and forth across the tube radius more often and thus be less likely to spend an abnormally large fraction of time in any one particular portion of the radius. By effectively randomizing or averaging the solute's radial occupancy of the tube the solute's migration velocity will also be averaged, and for a collection of solute molecules, deviations from the average will be small. Thus a reduction in tube radius not only should reduce radial temperature differences but should also diminish the impact of any temperature differences that remain. These two effects argue strongly for the use of small tube diameters.

EXPERIMENTAL SECTION

Apparatus. Straight lengths of glass tube (80–100 cm long; 75 μm i.d., 550 μm o.d.) were drawn from Corning Type 7740 Pyrex glass on a glass drawing machine (Shimadzu GDM-1B, Kyoto, Japan). These tube dimensions provided sufficient cooling efficiency to minimize difficulties associated with the aforementioned thermal effects. Larger diameter tubes inevitably led to problems with heat dissipation resulting in poor separation efficiency. The inside surface of these tubes was not modified in any way prior to filling with the electrophoresis buffer medium. A regulated high-voltage dc power supply (Megavolt Model RDC-30-10, Hackensack, NJ) delivering from 0 to +30 kV was used to drive the electrophoretic process. The operator was protected from accidental contact with the high voltage through an interlock system. The high voltage end of the system was enclosed in a Plexiglass box which automatically cut off the high voltage when opened. Detection was carried out by using a homemade "on-column" fluorescence detector. The detector used a high-pressure mercury arc lamp as the source of ultraviolet light, glass filters for isolation of excitation and emission wavelengths, and a photomultiplier tube detector.

Chemicals. Dansyl amino acids and fluorescamine were obtained from Sigma Chemical Co. (St. Louis, MO). Dipeptides were provided by C. Horvath of Yale University. Alkylamines were obtained from RFR Corp. (Hope, RI).

Procedure. Tubes were filled with 0.05 M pH 7 phosphate buffer. Filling was accomplished by dipping one end of the tube in the buffer solution and allowing capillary action to draw the buffer into the tube. The filling process could be accelerated by allowing the liquid to flow "downhill." In this way, complete filling of a tube required 15 min. After the tubes were filled, both ends of the tube were dipped in beakers containing the buffer medium. The end at which samples were introduced was connected via a graphite electrode to the positive high voltage supply. The detector end was connected via a graphite electrode to ground. Samples were applied to the tube by removing the beaker containing the buffer and replacing it with one containing the sample. High voltage was applied for a few seconds and then turned off. The beaker containing the buffer was replaced, the high voltage applied, and electrophoresis allowed to proceed.

RESULTS AND DISCUSSION

It is evident from the various electropherograms shown that high separation efficiencies may be achieved within short analysis times by performing zone electrophoresis in glass capillaries. Figure 1 shows the separation of several amino acids as their fluorescent dansyl derivatives. The separation of basic, neutral, and acidic amino acids is completed in 25 min. Figure 2 shows the separation of several dipeptides as fluorescamine derivatives. Here, separation is based primarily on size rather than charge. Figure 3 shows the separation of normal propyl, butyl, and hexylamines as their fluorescamine derivatives; separation is based purely on size. The three amine derivatives are well resolved despite their very minor differences in overall size. Finally, Figure 4 is the separation of a human urine sample in which the compounds containing primary amine groups have been derivatized with fluorescamine. In this case, labeling was carried out by adding 10 mg of fluorescamine dissolved in 2 mL of tetrahydrofuran to 20 mL of diluted urine (diluted 10-fold with buffer). Electrophoresis reveals a complex mixture of amines of unknown identity and demonstrates the ability of this technique to deal with "real" samples. In all these separations the applied potential was 30 kV, the limit of the power supply. The current was approximately 0.1 mA.

It is important to note that at pH 7 all these substances bear a net negative charge and yet migrate toward the negative electrode. This apparent contradiction is the result of a strong electroosmotic flow occurring in the capillary. The magnitude of this flow is such that even small, triply charged anions are carried toward the negative electrode. Thus, the order of appearance of solutes in the electropherograms is cations, neutrals, and finally anions. The electroosmotic flow proves to be quite convenient for two reasons. First, without this flow only cations or anions could be analyzed in a single run, as sample ions of the "wrong" charge would not enter or migrate through the capillary. Second, without the flow, very large and/or weakly charged ions would require great lengths of time to travel the length of the tube. With the electroosmotic flow, ions of a variety of size and charge may be analyzed in a single run.

Electroosmotic flow will somewhat modify the equations describing the separation efficiency. Fortunately, the electroosmotic flow profile approximates a "plug" shape (6, 7) and thus the flow profile itself leads to minimal zone broadening. The velocity of electroosmotic flow may be given as

$$v_{osm} = \mu_{osm} E = \mu_{osm} \frac{V}{L} \qquad (7)$$

where μ_{osm} is a coefficient similar to the electrophoretic mo-

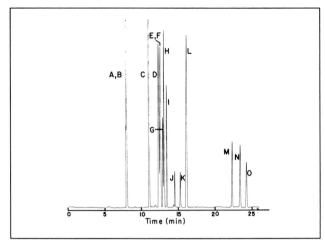

Figure 1. Zone electrophoretic separation of dansyl amino acids: A = unknown impurity, B = ε-labeled lysine, C = dilabeled lysine, D = asparagine, E = isoleucine, F = methionine, G = serine, H = alanine, I = glycine, J and K = unknown impurities, L = dilabeled cystine, M = glutamic acid, N = aspartic acid, O = cysteic acid. The concentration of each derivative is approximately 5×10^{-4} M, dissolved in operating buffer.

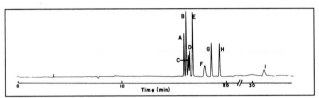

Figure 2. Zone electrophoretic separation of fluorescamine derivatives of dipeptides: A = phenylalanylleucine, B = phenylalanylvaline, C = valylleucine, D = glycyltyrosine, E = phenylalanylalanine, F = glycylproline, G = glycylalanine, H = glycylglycine, I = glycylaspartic acid. The concentration of each derivative is approximately 50 μg mL^{-1}, dissolved in operating buffer.

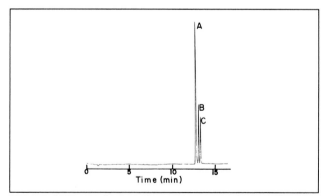

Figure 3. Zone electrophoretic separation of fluorescamine derivatives of amines: A = n-hexylamine, B = n-butylamine, C = n-propylamine.

bility, relating the electroosmotic velocity to the electric field gradient. The net migration velocity of a substance is then given by

$$v = \mu \frac{V}{L} + \mu_{osm} \frac{V}{L} = (\mu + \mu_{osm}) \frac{V}{L} \qquad (8)$$

The signs as well as the magnitudes of μ and μ_{osm} will be important, as the signs indicate the relative directions of the flow and the electrophoretic migration. The time it takes an ion to migrate the entire length of the tube is

$$t = \frac{L^2}{(\mu + \mu_{osm})V} \tag{9}$$

and the resulting spatial variance is

$$\sigma_L^2 = \frac{2DL^2}{(\mu + \mu_{osm})V} \tag{10}$$

The resulting separation efficiency is

$$N = \frac{(\mu + \mu_{osm})V}{2D} \tag{11}$$

quite similar to the original expression for separation efficiency, eq 6. This new equation still predicts that the number of theoretical plates is proportional to the applied voltage. This prediction was tested by using the fluorescamine derivative of hexylamine as a solute. The number of theoretical plates was computed from peak profiles by using the formula

$$N = 5.54 \left(\frac{t}{w}\right)^2 \tag{12}$$

where w is the full peak width at the half-maximum points. In Figure 5 N is plotted vs. the applied voltage. The plot is essentially a straight line with two notable features. First, the data at low voltages extrapolate to a nonzero intercept, the meaning of which is not clear. The peaks themselves appeared symmetric, suggesting that there was no serious overloading problem. However, overloading is a possible explanation of the nonzero intercept. Second, the plot shows a negative deviation from linearity at high voltages. This is interpreted as a result of the increased temperatures and temperature gradients that are expected at high voltages and currents.

In the presence of electroosmosis, eq 9 predicts an inverse relationship between applied voltage and analysis time. In Figure 6 reciprocal time is plotted vs. applied voltage giving the expected linear relationship. The positive deviation at high voltages is again indicative of higher temperatures, leading to increased mobilities and shorter analysis times.

Equation 11 suggests a misleading approach to improved separation efficiency. This is to promote very large values of μ_{osm}, electroosmotic flow, in the same direction as the electrophoretic mobility. Giddings (3) derived an expression for resolution in electrophoresis as

$$R_s = \frac{N^{1/2}}{4} \frac{\Delta v}{\bar{v}} \tag{13}$$

where R_s is the resolution and $\Delta v / \bar{v}$ is the relative velocity difference of the two zones being separated. This ratio is equal to

$$\frac{\Delta v}{\bar{v}} = \frac{\mu_1 - \mu_2}{\bar{\mu}} \tag{14}$$

where μ_1 and μ_2 are the mobilities of the two zones, and $\bar{\mu}$ is

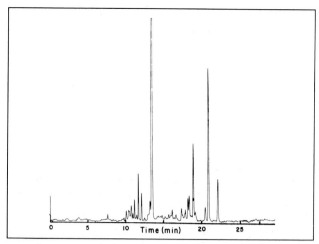

Figure 4. Zone electrophoretic separation of human urine derivatized with fluorescamine.

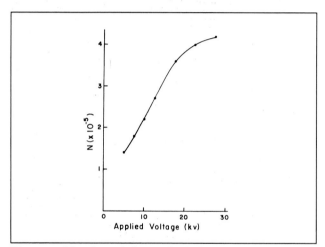

Figure 5. Number of theoretical plates as a function of applied voltage.

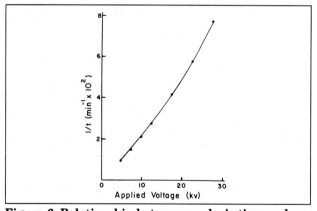

Figure 6. Relationship between analysis time and applied voltage.

their average mobility. However, in the presence of electroosmosis this becomes

$$\frac{\Delta v}{\bar{v}} = \frac{\mu_1 - \mu_2}{\bar{\mu} + \mu_{osm}} \tag{15}$$

It is readily apparent that a large value of μ_{osm} will decrease the relative velocity difference of the two zones. By substituting the expressions for the relative velocity difference (eq 15) and number of theoretical plates (eq 11) into the expression for resolution (eq 13) we obtain

$$R_s = \frac{1}{4} \left[\frac{(\bar{\mu} + \mu_{osm})V}{2D} \right]^{1/2} \left[\frac{\mu_1 - \mu_2}{\bar{\mu} + \mu_{osm}} \right] \quad (16)$$

and by rearranging

$$R_s = 0.177(\mu_1 - \mu_2) \left[\frac{V}{D(\bar{\mu} + \mu_{osm})} \right]^{1/2} \quad (17)$$

Now it is clear that a large component of electroosmotic flow in the same direction as the electrophoretic migration will decrease the actual resolution of two zones. In fact, it may be seen that the best resolution will be obtained when the electroosmotic flow just balances the electrophoretic migration or

$$\mu_{osm} = -\bar{\mu} \quad (18)$$

at which point substances with extremely small differences in mobility may be resolved. This resolution will be obtained, however, at a large expense in time, as may be seen by referring to eq 9 and imagining μ and μ_{osm} being nearly equal but opposite. In Figure 7 the effect of electroosmosis on resolution and analysis time is illustrated. The upper portion of the figure shows the separation of some neutral amino acids as dansyl derivatives. This separation was performed in a glass capillary as previously described. The lower portion of the figure shows the same separation, but this time in a capillary in which the inner surface was treated with trimethylchlorosilane in order to reduce electroosmotic flow. The improvement in resolution is obvious, especially in peaks C and D. However, this is at the expense of a large increase in analysis time.

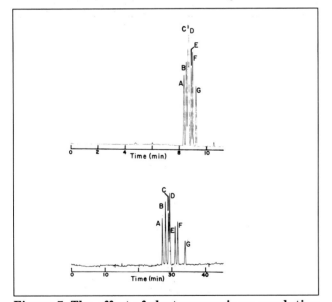

Figure 7. The effect of electroosmosis on resolution and analysis time of some dansyl amino acids: (upper) untreated capillary, (lower) capillary pretreated with 10% trimethylchlorosilane in dichloromethane for 20 min; A = asparagine, B = isoleucine, C = threonine, D = methionine, E = serine, F = alanine, G = glycine.

The sample introduction technique described here was used because of the ease with which it could be carried out, as well as the fact that it introduces minimal zone broadening. This approach also eliminates any need for leak-free connections that would be required if introduction were accomplished with hydrostatic pressure. However, sample introduction with applied voltage will introduce substances based on their electrophoretic mobilities, which may complicate the matter of quantitative analysis. A detailed comparison of electric and hydrostatic sample introduction techniques with respect to zone broadening and quantitation is needed. If electric introduction proves viable with respect to the requirements of quantitation, the simplicity with which it could be automated would be an attractive advantage.

Extending the technique to a wide range of solutes will require development of alternative modes of detection. Conductometric, UV absorption, refractive index, and thermometric detectors (2, 8) have been described in conjunction with electrophoresis in larger bore tubes. Their application to detection of low concentrations of solutes in micron-sized tubes will be difficult, but any developments along these lines will be of great utility. Detectors of higher sensitivity will allow the use of even smaller diameter tubes, permitting the application of higher electric field gradients. This would open the way for the use of higher voltages and/or shorter tubes, with the result of even higher separation efficiencies and shorter analysis times.

Extension of the technique to separation of proteins and other macromolecules and particles is also of interest. Surfaces more inert and nonadsorptive than untreated glass will probably be necessary. Alternative detection modes will also be an advantage here. If these difficulties can be overcome, the possibilities for high-resolution separations of macromolecules will be quite promising.

The prospects for yet higher separation efficiencies hinge directly on the use of higher voltages in an apparently straightforward manner. Significant improvements in separation efficiency over the present work will require potentials in excess of 100 kV. These higher voltages may present certain practical difficulties in operation and safety, but these problems are probably resolvable. With higher voltages, the simple assumption that molecular diffusion is the dominant cause of zone broadening may also break down. Achievement of higher efficiencies will place stricter limitations on sample overloading, thermal gradients, adsorption, and the electroosmotic flow profile. Solving these difficulties in order to realize the benefits of higher voltages will require further refinement of the technique.

ACKNOWLEDGMENT

The authors gratefully acknowledge the kind gift of dipeptides from Csaba Horvath.

LITERATURE CITED

(1) Wieme, R. J. In "Chromatography: A Laboratory Handbook of Chromatographic and Electrophoretic Methods," 3rd ed., Heftmann, E., Ed.; Van Nostrand Reinhold: New York, 1975; Chapter 10.

(2) Mikkers, F. E. P.; Everaerts, F. M.; Verheggen, Th. P. E. M. *J. Chromatogr.* **1979**, *169*, 11–20.

(3) Giddings, J. C. *Sep. Sci.* **1969**, *4*, 181–189.

(4) Hinckley, J. O. N. *J. Chromatogr.* **1975**, *109*, 209–217.

(5) Brown, J. F.; Hinckley, J. O. N. *J. Chromatogr.* **1975**, *109*, 218–224.

(6) Pretorius, V.; Hopkins, B. J.; Schieke, J. D. *J. Chromatogr.* **1974**, *99*, 23–30.

(7) Rice, C. L.; Whitehead, R. *J. Phys. Chem.* **1965**, *69*, 4017–4024.

(8) Bier, M. In "An Introduction to Separation Science," Karger, B. L., Snyder, L. R., Horvath, C., Eds.; Wiley: New York, 1973; Chapter 17.

Received for review January 16, 1981. Accepted April 24, 1981. Support for this work was provided by the donors of Petroleum Research Fund, administered by the American Chemical Society, and the University Research Council of the University of North Carolina.

Reprinted from *Anal. Chem.* **1981**, *53*, 1298–1302.

This paper characterizes a time of rapid change in ion mobility spectrometry, one symbolized by use of the term plasma chromatography in the title and ion mobility spectrometer in the text. Since its inception in the late 1960s, the plasma chromatograph (now universally called the ion mobility spectrometer) remained unchanged until the early 1980s, when reports of novel sample introduction methods, ionization sources, and modes of operation began to appear. Today, nonradioactive-based ionization sources such as the one described here continue to be of interest as ion mobility spectrometry is applied to real-time analysis of environmental, industrial, and military samples.

Herbert Hill
Washington State University

Plasma Chromatography with Laser-Produced Ions

David M. Lubman* and Mel N. Kronick
Quanta-Ray, Inc., 1250 Charleston Road, Mountain View, California 94043

Laser multiphoton ionization (MPI) is used to produce ions in an ion mobility spectrometer. The ions created are then separated by gaseous electrophoresis, i.e., according to their mobility in a drift gas under the influence of an applied electric field. The MPI process allows direct ionization of organic compounds with production of only one peak which is either the molecular ion or MH^+. The problem of multiple peaks occurring due to the ion-molecule process in plasma chromatography is thus significantly reduced. This technique can provide great sensitivity, i.e., at least down to 1 ppb in the case of benzene. In addition, the laser wavelength can provide an additional means of discrimination of molecules in an ion mobility spectrometer.

We introduce a unique method for producing ions for plasma chromatography (PC). This technique involves using laser multiphoton ionization to ionize molecules directly under atmospheric conditions in a commercial ion mobility spectrometer.

There are many reviews of the theory of plasma chromatography (*1–15*). In conventional plasma chromatography, ions are initially produced in a carrier gas by a ^{63}Ni β-decay source. These ions initiate a sequence of ion–molecule reactions that eventually yield a few molecular ions which then undergo ion–molecule reactions with the trace compound to form the ions measured in a drift tube. In the drift region the ions are moved through a column of gas by an electric field and separated according to their mobilities. Nitrogen is usually used as the drift gas to prevent further ion–molecule reactions. Eventually the ions diffuse to a detector where either positive or negative ions are detected. The output of the detector is in the form of current as a function of time. An ion mobility spectrum is thus obtained which fingerprints molecules in a manner similar to gas chromatography. Since the drift time of the ions is in the millisecond regime, the separation and detection can be performed in real time. In addition, N_2 can serve as a universal column thus alleviating the problem of column choice inherent in gas chromatography.

The initial creation of ions through the use of the ion–molecule reaction technique often produces data difficult to interpret. The ion–molecule reaction often creates several different ion–molecule combinations with the trace compounds. Plasma chromatography with laser-produced ions (PCLI) minimizes the problem of nonspecific ionization. The ionization source for PCLI in our experiments is ultraviolet laser light from a Nd:YAG pumped pulsed laser system. The laser radiation allows direct ionization of molecules, thus circumventing limitations of the ion–molecule method. Laser photoionization provides extremely efficient ionization of large molecules with only the molecular ion appearing when the laser operates at sufficiently low intensity (*16–22*). The ionization process can be accomplished directly in air (*23, 24*).

The laser ionization process occurs when ionization follows the absorption of several photons by a molecule in the presence of an intense visible or UV light source, i.e., it is a multiphoton process (Figure 1). When the laser source is tuned to an allowed n-photon transition, the process is greatly enhanced. This is referred to as resonance enhanced multiphoton ionization, i.e., REMPI. In the case of REMPI, ionization occurs via a real intermediate state. Since the density of states above the lowest energy state populated is usually quite high, subsequent absorptions are resonant or nearly resonant. Excitation from the intermediate state to the ionization continuum is quite rapid and may involve the absorption of as

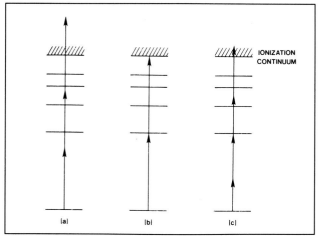

Figure 1. Energy level diagram showing MPI transitions: (a) nonresonant MPI, (b) one-photon resonant REMPI, (c) two-photon resonant REMPI.

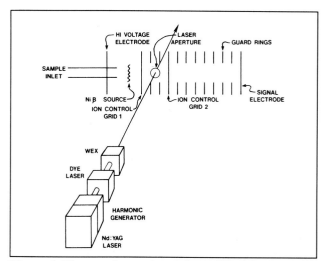

Figure 2. Plasma chromatograph modified for use with laser ionization.

many as six additional photons (25). The transition to the lowest intermediate real state is generally the rate-limiting step. When the laser is not tuned to a real state, the probability for multiphoton ionization (MPI) is nearly negligible.

Most large aromatic and aliphatic compounds have two salient features: strong absorptions in the UV and low ionization potentials, typically between 7 and 13 eV. The UV absorption spectra of these molecules is rather broad at room temperature due to the large number of rovibronic states populated. Most larger molecules can thus be ionized with the fourth harmonic of the Nd:YAG laser (266 nm) with the absorption of one photon to resonance and, subsequently, one to three more photons to ionization. The process is thus extremely efficient.

Since REMPI is based upon the resonance absorption of molecules, there are several other advantages which we gain from this technique. Laser radiation at wavelengths greater than 200 nm will not ionize air at the power levels required to ionize organic compounds. Molecules such as CO_2, O_2, and N_2 from the environment will thus not interfere with the data. In addition, by using the laser ionization technique the unique absorptions can be used to spectrally separate the different compounds in the environment. Klimcak and Wessel have demonstrated the spectral separation of two isomers, anthracene and phenanthrene, at atmospheric pressure in a proportional counter using a tunable dye laser (23). They in fact suggested in the same reference that time-resolved ion mobility measurements might actually provide additional discrimination in those cases where molecules have similar one-photon absorptions or where ionization potentials are nearly equivalent.

EXPERIMENTAL SECTION

The experimental setup is shown in Figure 2. The ion mobility spectrometer is a modified version of a commercial plasma chromatograph obtained from PCP, Inc., West Palm Beach, FL. The present setup can be operated as a conventional plasma chromatograph with a Ni-β source or a laser

chromatograph by changing the bias circuit and introducing a laser beam through the cell window. When used as a conventional plasma chromatograph, ion control grid 1 must allow ions to pass. The ion–molecule reactions occur between the β source and grid 2. The ions are injected into the drift region by pulsing grid 2. The drift distance from grid 2 to the ion collector is 8.0 cm. The injection pulse can vary from 0.05 to 0.5 ms. A smaller pulse provides greater resolution but also less sensitivity. Typically, a pulse length of 0.2 ms was used to monitor the presence of compounds using PC. In the case of PCLI, grid 1 should be biased to block the ionization from the radioactive source. Grid 1 then provides the bias for ions created by MPI in the interaction region. The drift distance from the center line of the laser beam to the ion collector is 12.5 cm. Suprasil quartz windows (2.54 cm diameter) were used in order to transmit UV laser radiation. The apparatus was operated at above 200 °C at all times in order to keep the apparatus free from contamination. The quartz inlet was operated at 275 °C. A countercurrent flow of dry N_2 was filtered with a molecular sieve trap before it entered the PC in order to remove water vapor or other contaminants.

The laser source consisted of a Quanta-Ray DCR-1A Nd:YAG laser used alone at its fourth harmonic (266 nm) or used at its second harmonic (532 nm) to pump various dyes in a Quanta-Ray PDL-1 dye laser. In order to produce tunable UV light, the output from the dye laser was frequency doubled in a phase-matched KD*P crystal. This was performed for the various dyes using the Quanta-Ray WEX-1 wavelength extender device which can produce scannable UV radiation over the frequency doubled range of each dye. The dye laser wavelength was scanned by use of a stepping-motor controlled by a Quanta-Ray CDM-1 control display module.

The signal from the ion mobility spectrometer was digitized into 800 points over a 40-ms interval using a DEC ADV-11 A/D interface to our DEC PDP-11 computer. The digitized signal was then signal averaged over 500 laser pulses. The resulting ion mobility spectrum was then displayed on an x-y recorder. The MPI spectra as a function of wavelength for a particular compound were taken by a PAR Model 160 boxcar averager with the integration window placed over the peak of interest in the ion mobility spectrum.

The samples used in the various experiments were prepared in several ways. For our initial studies, benzene and toluene samples were prepared for us commercially as mixtures in concentrations of 1.0 ppm and 2.0 ppm in nitrogen respectively (Scott Specialty Gases, San Bernardino, CA). When the sample gas is flowed very slowly, i.e., 1–5 cm^3/min, then the counterflow of N$_2$ gas at approximately 600 cm^3/min dilutes the sample gas in the interaction region approximately according to the ratio of flow rates (26). This method does not provide an accurate concentration but does provide an estimate of the concentration, probably within 50%, that is sufficient for characterizing the physical properties of this technique.

For more accurate measurement of the limits of sensitivity of PCLI the exponential dilution flask method was used (27). The apparatus consists of a 250-mL flask containing a magnetic stirrer. The magnetic stirrer used had extension fins to increase the mixing between the benzene and N$_2$ gas. The benzene was diluted by injecting 1 mL of benzene in a 100-mL volumetric flask and then diluting with methanol. One milliliter of this solution was further diluted in a 500-mL volumetric flask. When 1 μL of this solution was injected into the 250-mL flask a concentration of approximately 32 ppb was obtained. The flask was heated to 150 °C and kept at a constant temperature using a temperature proportional controller. All gas sample inlet lines were kept at above 200 °C in order to prevent benzene from absorbing on the surfaces. The N$_2$ carrier flow rate was typically approximately 100 cm^3/min. In order to check for complete mixing, we used a stop-flow method. A sample was examined at a given time. The flow was stopped and mixing allowed to continue. A sample was then taken at a later time. The signal was the same for each of these samples thus indicating that mixing was complete. The dilution flask method was repeated ten times and the results were averaged.

In the case of the solid samples, azulene and naphthalene, the diffusion tube method was used (28). Diffusion tubes (Vici Metronics, Santa Clara, CA) with apertures of either 0.5 or 0.2 cm and a length of 7.62 cm were used to create samples in concentrations on the order of 10 ppm. Provided the carrier N$_2$ gas was flowing slowly the sample was further diluted by the drift gas to concentrations on the order of 10–20 ppb. The diffusion tubes were used at room temperature.

RESULTS AND DISCUSSION

We have found that resonant multiphoton ionization can be used to ionize softly various hydrocarbons at atmospheric pressure in an ion mobility spectrometer. The MPI process directly ionizes the compound under study but does not ionize background gases. The ion mobility spectrum subsequently allows identification of each component. Figure 3 illustrates laser MPI mobility spectra of benzene and toluene produced by 266-nm laser light. Only one mobility peak is obtained from the ionization of each compound. We identify this peak

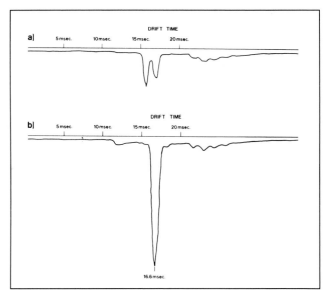

Figure 3. (a) Laser-induced mobility spectrum of benzene and toluene. The peak at 15.4 ms corresponds to benzene and that at 16.6 ms, to toluene. The conditions under which this spectrum was recorded are as follows: λ = 266 nm, T = 200 °C, input laser energy = 2 mJ, laser beam diameter = 6 mm, electric field = 171 V/cm, benzene concentration = 30 ppb, toluene concentration = background. (b) Laser induced toluene mobility spectrum at a concentration of 15 ppb. The amplifier sensitivity and gain settings are the same as in Figure 3a.

as either the molecular ion or MH$^+$ ions based upon data obtained from β source ionization as well as the work of Griffin et al. (11). In Figure 3 other unidentified peaks are present at longer mobility time even with no sample present. These are due to background contamination present in the device. If the beam was tightly focused a spectrum with several peaks was obtained as in the case of toluene (Figure 4). Under vacuum conditions we know that fragmentation of the molecules occurs at high power densities as identified by a mass spectrometer. It thus is probable that fragmentation is occurring in the interaction region of this device. There were no ions present in the negative ion mobility spectrum, although there was a large peak due to electrons produced as a result of the MPI process.

The laser ionization technique appears to be very efficient. We have been able to detect easily mobility spectra for toluene and benzene at concentrations below 10 ppb in nitrogen at 266 nm with no more than approximately 2 mJ of input laser energy per pulse. We used the rate equations derived by Reilly and Kompa (29) to estimate the number of ions which could in theory be produced at a given input intensity. According to these authors the number of molecular benzene ions produced by the 2-photon ionization process is given by

$$C_6H_6^+ = \left(\frac{\sigma_1 \sigma_2 I^2 A_{1g}}{\gamma_+ - \gamma_-} \right) \left(\frac{e^{\gamma_+ T}}{\gamma_+} - \frac{e^{\gamma_- T}}{\gamma_-} + \frac{1}{\gamma_-} - \frac{1}{\gamma_+} \right) \quad (1)$$

where I is the laser intensity in photons cm^{-2} s^{-1}, τ is the intermediate state lifetime, A_{1g} is the ground state concentration, T is the duration of the laser pulse, σ_i are cross sections

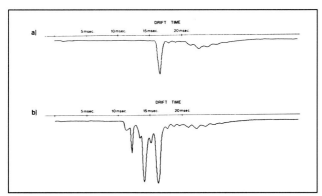

Figure 4. (a) Laser induced mobility spectrum of toluene, parent peak: λ = 266 nm, T = 210 °C, laser input energy = 2 mJ, laser beam diameter = 6 mm, toluene concentration = background, electric field = 171 V/cm. (b) Laser induced mobility spectrum of toluene, fragmentation pattern: λ = 266 nm, T = 210 °C, laser input energy = 8 mJ, laser beam focused to a point using a 10-cm focal length lens, electric field = 171 V/cm, toluene concentration = background. The amplifier sensitivity and gain settings are the same as in Figure 4a.

for the respective stimulated transitions

$$\gamma \pm = \frac{-\alpha_1 \pm \sqrt{\alpha_1{}^2 - 4\alpha_2}}{2} \qquad (2)$$

and

$$\alpha_1 = (\sigma_1 + \sigma_2 + \sigma_3)I + 1/\tau \qquad (3)$$

$$\alpha_2 = \sigma_1 I(\sigma_2 I + 1/\tau) \qquad (4)$$

We have performed our experiments with the fourth harmonic of the Nd:YAG laser (266 nm). The beam was an annulus with a 6 mm outer diameter and 2 mm inner diameter. The laser input intensity was approximately 2 mJ. We calculate $I \simeq 1.08 \times 10^{24}$ photons cm^{-2} s^{-1}. Literature data for benzene (30, 31) indicate that, at 266 nm, $\sigma_1 = 5 \times 10^{-20}$ cm^2. The value of σ_2 (266 nm) was taken to be σ_2 (249 nm) = 3.4×10^{-17} cm^2 (29). This approximation can be made since, for absorption into a largely structureless continuum, the cross section for one-photon ionization of the intermediate excited states of benzene should be independent of wavelength above threshold (32). The value of τ_{266} is taken to be 100 ns (33). Using these values we obtain a value of $C_6H_6{}^+ = (2.8 \times 10^{-3}) A_{1g}$. At 220 °C, 10 ppb is equivalent to 1.36×10^{11} molecules/cm^3 of neutral benzene molecules. If the interaction region is 0.28 cm^3, then the laser can interact with 3.8×10^{10} molecules. If, as is usually the case, $A_{1g} \simeq$ number of neutrals $\simeq 3.8 \times 10^{10}$ molecules (34), then the above calculation would predict that approximately 1×10^8 ions are produced per laser pulse. This is true since typically a small fraction of the neutral molecules (<10%) is excited from A_{1g} to the first excited state, and only a fraction of these will ionize due to competing processes. At 10 ppb benzene we obtain a signal of 1.5×10^{-10} A. The pulse width is approximately 650 μs fwhm so that the total charge produced is 9.7×10^{-14} C. This is equivalent to 6.1×10^5 ions

actually detected. Therefore $(6.1 \times 10^5)/(1 \times 10^8) \sim 0.6\%$ is a lower bound estimate of the efficiency of detection of the PCLI technique after ionization. Of course the efficiency of ionization itself can be increased by increasing the laser power. In fact, the signal has been found to increase linearly with laser power at 266 nm for toluene and benzene. However, if too many ions are created, space charge effects begin to appear and thus resolution decreases as the signal width starts to broaden.

In order to determine the experimental limits of detection of PCLI, we used the exponential decay flask method. Benzene was chosen for study since absorption of benzene on the cell walls was less of a problem than in the case of toluene, naphthalene, or azulene. In this method the gas concentration, C, decays with time according to the relationship (27)

$$C = C_e \exp\left(-\frac{U}{V}t\right)$$

where V is the volume of the vessel, U the gas flow rate, C_e the initial gas concentration, and t the time elapsed after introducing the test gas. If the signal amplitude from the detector is plotted against elapsed time, a perfect detector would indicate a log–linear decay of gas concentration with a slope of $1/\tau = U/V$. The response can be followed down to the background noise level of the device and any departure from linearity can be detected immediately. We were able to detect at least 1 ppb of benzene in nitrogen by using this technique. The lower limit of detection was determined by residual benzene from the cell walls and not by signal-to-noise limitations. The value of τ was typically on the order of 3.5 min.

As mentioned above, space charge effects can influence resolution and sensitivity. Provided we consider the space charge effect produced mainly by widely separated ions, then the criterion for negligible space charge distortion of an applied field E_0 is then

$$n \ll \frac{E_0}{4\pi e L}$$

where L is the relevant dimension of the apparatus (1). In our experimental setup $L = 12.4$ cm and $E_0 = 171$ V/cm. In order to obtain the correct units V/cm must be divided by 300 to be expressed in statvolts/cm. The unit "statvolts/cm" is equivalent to statcoulombs/cm^2 and $e = 4.8 \times 10^{-10}$ statcoulombs. Thus, signal space charge should occur at ion densities on the order of 10^7 ions/cm^3. We have estimated previously that a maximum of 1×10^8 ions of benzene are produced per laser pulse at 266 nm with an energy of 2 mJ in a volume of 0.3 cm^3 when the benzene concentration is 10 ppb. The number of ions produced, hence the amount of signal, depends upon the size of the laser beam. The peak signal amplitude would be expected to increase in proportion to the volume intersected by the laser provided there are no space charge effects. This was found to be the case under these conditions, so that in fact we are probably not creating the maximum theoretical number of ions that could be produced. At higher concentrations such as 1 ppm, space charge effects could be readily observed and the mobility peaks became very broad.

The dimension of the laser beam along the drift direction is analogous to the grid pulse length in the conventional PC case. As the laser beam size or grid pulse length increases, the signal width increases. The time width of the peak is found to increase approximately as the diameter of the laser beam. At a beam diameter of 1 mm we were able to obtain a fwhm peak width of approximately 350 µs. We were never able to reach the 250 µs width obtained with the 200 µs gate of the PC pulsing grid. The longer pulse width obtained in PCLI appears to result from spreading of the ion packet due to the longer drift distance. Provided there is no space charge or ion–molecule reactions with the drift gas, the width of the ion pulse in space should be determined by diffusion with a width proportional to the square root of the drift time, t_d (4). The pulse width in space $\sigma_{1/2}$ is related to the width in time by the relation $\sigma_{1/2} = v_d \Delta t$ (4), where v_d is the drift velocity. It can be shown that $\Delta t = C(t_d / V_{app}^{1/2})$ (4). V_{app} is the effective bias voltage applied to the ions. Thus if we plot Δt vs. $t_d / V_{app}^{1/2}$ a straight line should be obtained if the process is diffusion controlled. This appears to be true within experimental limits. Thus, at a given input voltage to the first electrode, V_{in}, we can directly compare the experimental peak widths from PCLI and PC. Of course, V_{app} is different in the two cases because of the different drift distances. Because of the voltage gradient across the guard rings, V_{app} for PCLI is $V_{in} \times 12.5/16$ and for PC, $V_{app} = V_{in} \times 8/16$. At an input voltage of +2750 V we obtain a peak width ratio of PCLI/PC of 350 µs/250 µs = 1.4. A comparison of $t_d / V_{app}^{1/2}$ for PCLI/PC should provide the same ratio. This holds within the experimental limits of error. Thus, the wider peak width in PCLI appears to be due to the longer path length in our particular experimental system.

One can increase the magnitude of the signal without losing resolution by increasing the dimension of the beam perpendicular to the drift direction. By use of a variable length 2-mm slot it was found experimentally that the signal increased as the slot length was increased, but the peak width was not affected. By expansion of the laser beam in this fashion, the signal can be increased without increasing the power density, thus avoiding space charge effects and simultaneously increasing sensitivity.

One especially valuable feature of PCLI is that the signal can be controlled by the laser input energy per pulse. It might be expected in a 2-photon process that a 2-photon power dependence would be obtained (35). However, the observed power dependence for toluene and benzene was 1. This indicates that a rate-determining step is present in the process. The power dependence of the signal in PCLI can be used not only to increase the signal for low concentrations but to decrease the signal where the concentration is very high. Thus, the laser intensity can be used to extend the practical upper limit of detectability of this technique. In the case of laser chromatography illustrated in Figure 5, the signal may saturate at one particular value of the laser input energy, but saturation can be avoided by decreasing the input energy in order to limit the total number of ions produced.

One characteristic of conventional plasma chromatography is that in the presence of a strong base such as NH_3 the sen-

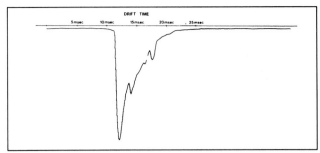

Figure 5. Saturated benzene spectrum at a concentration of 1 ppm in N_2: T = 200 °C, input laser energy = 5 mJ, laser beam diameter = 6 mm, electric field = 171 V/cm. The large peak at 12 ms is probably NH_4^+ which is produced in ion–molecule reactions with benzene ions.

sitivity of the technique can be greatly reduced. A strong base can effectively steal the positive charge from weaker bases in ion–molecule reactions. Because of the presence of NH_3 in environmental samples this effect cannot be underestimated in real analysis. Laser ionization can help minimize this effect. Ammonia is not directly ionized in a 2-photon process at 266 nm. In the case of PCLI, once ions are created by the laser, there is only 15–20 ms in which they can react with background contamination in the drift region. In Figure 3b note the broad peak present in the PCLI case at 11.5 ms. This peak corresponds to the drift time for NH_3 as identified with the β ionization source (26). We believe that the broad width is due to the continuous reaction of ions with NH_3 as the ions migrate toward the collector. In the case of the conventional β source plasma chromatograph a large NH_3 signal is obtained under the same conditions. In fact, it is large enough to prevent efficient detection of the benzene peak. This is probably because the sample ions are exposed for a longer period of time to contaminants in the reactor region than in PCLI and also because many other ions are formed which can transfer charge to NH_3.

A potentially powerful and unique feature of the laser ionization technique is the selectivity afforded by a tunable laser source. In the case of benzene and toluene distinct spectral features allow an added dimension in the discrimination of these two compounds. Figure 6 demonstrates ionization spectra taken in our ion mobility spectrometer at 210 °C of toluene and benzene. The data were taken using a boxcar averager with the integrating window placed over the peak of interest in the plasma chromatogram. Note the very sharp spectral feature in the benzene spectrum at 266.69 nm obtained even at atmospheric pressure. Thus, we could conceivably obtain a three-dimensional plot with ion mobility as one axis, wavelength as the second axis, and intensity as the third axis.

Wavelength selectivity has been shown in previous work to be an important technique in distinguishing isomers such as azulene and naphthalene (17). We therefore measured multiphoton ionization spectra in the ion mobility spectrometer taken in the region between 278 and 282.5 nm. This region was chosen because the spectra of azulene and naphthalene as taken in a UV spectrophotometer at T = 50 °C led us to believe that there should be distinguishing features between

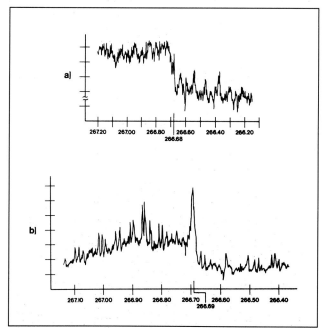

Figure 6. (a) Ionization spectrum of toluene as a function of wavelength in an ion mobility spectrometer. The concentration is approximately 10 ppb: T = 210 °C, input laser energy = 1 mJ. (b) Ionization spectrum of benzene as a function of wavelength in an ion mobility spectrometer. The concentration is approximately 15 ppb, T = 210 °C, and input laser energy = 1 mJ.

these compounds in this region. The power was carefully monitored and kept at a constant value by adjusting the Q-switch delay on the Nd:YAG laser. In fact at T = 200 °C in an ion mobility spectrometer the ionization spectra taken using a boxcar integrator (gate width, 80 μs) were shown to be featureless in both cases.

In the work of Lubman, Naaman, and Zare (17) a multi-color scheme using the second (λ = 533 nm) and third harmonics (λ = 355 nm) of the Nd:YAG laser aligned in time and space was used to ionize azulene and allow it to be discriminated from naphthalene. This experiment was originally performed under vacuum at a pressure of approximately 2×10^{-5} torr. The same experiment was attempted in order to ionize azulene under atmospheric pressure at 220 °C in our plasma chromatograph. No signal was obtained. The ionization under vacuum probably proceeds through an intermediate Rydberg state. Rydberg states are very sensitive to quenching by high pressure because of their large effective radii (36). It can be seen in these examples that the high temperature and pressure of a plasma chromatograph may reduce the potential wavelength selectivity of the MPI based on detailed spectral features.

Azulene and naphthalene could not successfully be discriminated on the basis of their mobilities. An attempt was made to use the ionization potentials of azulene (7.7 eV) and naphthalene (8.1 eV) in order to discriminate the pair. As shown by Lubman and Kronick (37) azulene should ionize at wavelengths longer than 306 nm but naphthalene should not in a 2-photon process. At wavelengths longer than 306 nm neither compound produced a signal within the range of con-

centration and power limits used. At 293.77 nm, the room temperature vapor pressure from a diffusion tube, both compounds produced a signal. However, the azulene signal was very weak. The naphthalene signal was approximately 15–20 times stronger than the azulene signal when both were compared at a concentration of 30 ppb in N_2. At 280 nm both compounds produced a signal. However, the naphthalene signal was still at least 5–10 times the azulene signal. The azulene signal was large enough in this region for a power dependence to be performed. The power dependence was found to be 2. Naphthalene at 280 nm had a power dependence of 1. At 266 nm the signals for both compounds were comparable under the same conditions and the power dependences were both equal to one. Thus, the two compounds can be distinguished at 280 nm based upon their different power dependences. At 280 nm apparently some process prevents saturation, thus allowing the true rate dependence to appear at the expense of the signal magnitude. Figure 7 shows ion mobility spectra of naphthalene produced with laser radiation at λ = 266 nm and λ = 280 nm. Note the presence of toluene and benzene at 266 nm but not at 280 nm. Thus the wavelength can be used to select components and simplify spectra in a mixture so long as the high temperature and pressure do not significantly interfere with the MPI process. This appears to be true for the benzene/toluene mixture but not for azulene/naphthalene.

A report describing the use of an ion mobility spectrometer as a capillary gas chromatography detector has recently appeared in the literature (38). The selectivity of such a detector might be greatly enhanced by the use of laser multiphoton ionization as described in this paper.

ACKNOWLEDGMENT

We wish to thank Martin Cohen and Roger Wernlund of PCP, Inc., West Palm Beach, FL, for construction of the

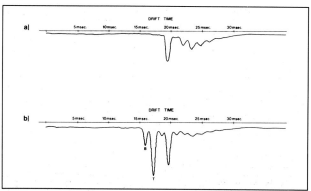

Figure 7. (a) Laser induced mobility spectrum of naphthalene at λ = 280 nm: C = 20 ppb naphthalene, T = 210 °C, laser input energy = 1.5 mJ, electric field = 171 V/cm. No toluene background is present. The small peaks at longer drift times than the naphthalene peak are due to background present in the machine. (b) Laser induced mobility spectrum of naphthalene at λ = 266 nm: C = 20 ppb naphthalene, T = 210 °C, laser input energy = 2 mJ, electric field = 171 V/cm, T = toluene background, B = benzene background. The amplifier sensitivity and gain settings are the same as in Figure 7a.

present apparatus and for helpful suggestions. W. Earl Bell of Quanta-Ray also helped provide useful ideas and advice during the formative stages of this project.

LITERATURE CITED

(1) McDaniel, E. W.; Mason, E. A. "The Mobility and Diffusion of Ions in Gases"; Wiley: New York, 1973.
(2) McDaniel, E. W. "Collision Phenomena in Ionized Gases"; Wiley: New York, 1964; pp 457–461.
(3) Loeb, L. B. "The Kinetic Theory of Gases"; Dover Publications, Inc.: New York, 1927; Chapter 11.
(4) Revercomb, H. E.; Mason, E. A. *Anal. Chem.* **1975**, *47*, 970.
(5) Karasek, F. W. *Anal. Chem.* **1974**, *46*, 710A.
(6) Karasek, F. W.; Kim, S. H.; Hill, H. H., Jr. *Anal. Chem.* **1976**, *48*, 1133.
(7) Karasek, F. W.; Kane, D. M. *Anal. Chem.* **1974**, *46*, 780.
(8) Karasek, F. W.; Cohen, M. J.; Carroll, D. I. *J. Chromatogr. Sci.* **1971**, *9*, 391.
(9) Karasek, F. W.; Denney, D. W. *J. Chromatogr.* **1974**, *93*, 141.
(10) Hagen, D. F. *Anal. Chem.* **1979**, *51*, 870.
(11) Griffen, G. W.; Dzidic, I.; Carroll, D. I.; Stillwell, R. N.; Horning, E. C. *Anal. Chem.* **1973**, *45*, 1204.
(12) Karasek, F. W. *Int. J. Environ. Anal. Chem.* **1972**, *2*, 157.
(13) Spangler, G. E.; Collins, C. I. *Anal. Chem.* **1975**, *3*, 393.
(14) Carr, A. W. *Anal. Chem.* **1979**, *6*, 705.
(15) Cohen, M. J.; Karasek, F. W. *J. Chromatogr. Sci.* **1970**, *8*, 330.
(16) Dietz, T. G.; Duncan, M. A.; Liverman, M. G.; Smalley, R. E. *Chem. Phys. Lett.* **1980**, *70*, 246.
(17) Lubman, D. M.; Naaman, R.; Zare, R. N. *J. Chem. Phys.* **1980**, *72*, 3034.
(18) Seaver, M.; Hudgens, J. W.; DeCorpo, J. J. *Int. J. Mass Spectrom. Ion. Phys.* **1980**, *34*, 159.
(19) Zandee, L.; Bernstein, R. B. *J. Chem. Phys.* **1979**, *70*, 2574.
(20) Zandee, L.; Bernstein, R. B. *J. Chem. Phys.* **1979**, *71*, 1359.
(21) Lichtin, D. A.; Datta-Ghosh, S.; Newton, K. R.; Bernstein, R. B. *Chem. Phys. Lett.* **1980**, *75*, 214.
(22) Boesl, U.; Neusser, H. J.; Schlag, E. W. *J. Chem. Phys.* **1980**, *72*, 4327.
(23) Klimcak, C.; Wessel, *J. Anal. Chem.* **1980**, *52*, 123.
(24) Brophy, J.; Rettner, C. T. *Opt. Lett.* **1979**, *4*, 337.
(25) Boesl, U.; Neusser, H. J.; Schlag, E. W., private communication.
(26) Cohen, Martin, PCP, Inc., private communication.
(27) Lovelock, J. E. *Anal. Chem.* **1961**, *33*, 162.
(28) Altshuller, A. P.; Cohen, J. R. *Anal. Chem.* **1960**, *32*, 802.
(29) Reilly, J. P.; Kompa, K. L. *J. Chem. Phys.* **1980**, *73*, 5468.
(30) Potts, W. J., Jr. *J. Chem. Phys.* **1955**, *23*, 73.
(31) Callomon, J. H.; Dunn, T. M.; Mills, I. M. *Philos. Trans. R. Soc. London, Ser. A* **1966**, *259*, 499.
(32) Rettner, C. T.; Brophy, J. H. *Chem. Phys.* **1981**, *56*, 53.
(33) Spears, K. G.; Rice, S. A. *J. Chem. Phys.* **1971**, *55*, 5561.
(34) Parker, D. H.; El-Sayed, M. A. *Chem. Phys.* **1979**, *42*, 379.
(35) Johnson, P. M. *Acc. Chem. Res.* **1980**, *13*, 20.
(36) Robin, M. "Higher Exited States of Polyatomic Molecules"; Academic: New York, 1974; Vol. I.
(37) Lubman, D. M.; Kronick, M. N. *Anal. Chem.* **1982**, *54*, 660.
(38) Balm, M. A.; Hill, H. H., Jr. *Anal. Chem.* **1982**, *54*, 38.

Received for review January 28, 1982. Accepted May 3, 1982. This work received financial support from the U.S. Army Research Office, Contract No. DAAG 29-81-C-0023.

Reprinted from *Anal. Chem.* **1982**, *54*, 1546–51.

This is a classic paper from one of the pioneering groups in MS/MS. The structure of "China White" had been determined the previous year on the basis of a combination of MS, IR, and NMR spectra (requiring µg to mg quantities). Although the traditional EI mass spectrum provided limited structural information, the addition of a second MS stage permitted unambiguous structural identification directly from the MS/MS data. The use of a homemade three-sector MS/MS instrument (with its limited resolution) was a sign of the times; today's commercial tandem mass spectrometers (triple quadrupoles, multisectors, hybrid sector/quadrupole, and even ion traps) have revolutionized MS/MS, bringing the power of the technique to the masses.

Richard Yost
University of Florida

Structural Information from Tandem Mass Spectrometry for China White and Related Fentanyl Derivatives

Michael T. Cheng, Gary H. Kruppa, and Fred W. McLafferty*
Department of Chemistry, Cornell University, Ithaca, New York 14853

Donald A. Cooper
Special Testing Laboratory, U.S. Drug Enforcement Administration, McLean, Virginia 22102

The potential of tandem mass spectrometry utilizing collisionally activated dissociation (CAD) for molecular structure determination is illustrated with α-methylfentanyl ("China White"), whose complex structure required several methods for its original elucidation. CAD spectra of fragment ions in its electron ionization (EI) mass spectrum provide information on dissociation pathways and fragment structures; the latter information comes from both interpretation and matching against reference spectra. Although the EI spectrum shows no odd-electron and few primary fragment ions, ion types most useful for such CAD studies, CAD data from the even-electron, secondary fragment ions gave sufficient structural information.

Elucidation of the molecular structure of China White, an illicit narcotic implicated in drug overdose deaths, attracted unusually wide publicity (*1–3*). Structure **1** was assigned on the basis of mass, infrared, and nuclear magnetic resonance spectra; synthesis corroborated this identification, eliminating **3** as a candidate. The unusual narcotic activity of this compound meant that available samples contained very small concentrations, so that it would have been especially advantageous if the complete identification could have been done with a method requiring only a submicrogram sample, such as mass spectrometry. However, spectral interpretation was difficult for this polyfunctional compound; the electron ionization (EI) mass spectrum (Figure 1) "was totally unfamiliar" (*1*). From this spectrum the Self-Training Interpretive and Retrieval System (STIRS) (*4, 5*) correctly identified the β-phenylethylamine moiety and the cyclic amine but gave little indication of the *N*-phenylpropionamide portion of the molecule. It thus appeared that these compounds might provide an interesting test of the additional structural information available from tandem mass spectrometry (MS/MS) (*6–10*).

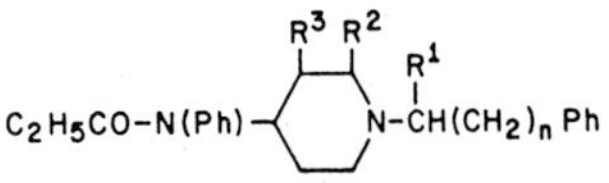

1: $R^1 = CH_3$, $R^2 = R^3 = H$, $n = 1$
2: $R^1 = R^3 = H$, $R^2 = CH_3$, $n = 1$
3: $R^1 = R^2 = H$, $R^3 = CH_3$, $n = 1$
4: $R^1 = R^2 = R^3 = H$, $n = 1$
5: $R^1 = R^2 = R^3 = H$, $n = 0$

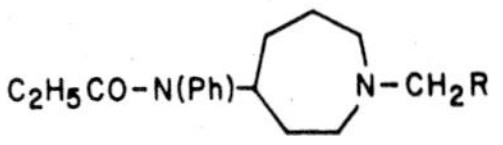

6: R = cyclopropyl
7: R = Ph
8: $R = CH_2Ph$
9: $PhNHCH(CH_3)$-$CH{=}CH^2$
10: $PhNHCH_2CH{=}CHCH_3$
11: $PhN(CH_3)CH_2$-$CH{=}CH_2$
12: $PhCH_2N[CH(CH_3)$-$CH{=}CH_2]_2$

In MS/MS the first mass spectrometer (MS-I) is operated as a conventional instrument, forming ions from the sample by methods such as electron and chemical ionization (EI and CI). Ions of a specific mass separated by MS-I are then fragmented further, usually by metastable ion (MI) or collisionally activated dissociation (CAD), with separation of the resulting product ions in MS-II. This secondary mass spectrum is indicative of the structure of its precursor fragment ion; the same general fragmentation rules of EI or CI mass spectra

 2855-8/94/0337$08.00/0

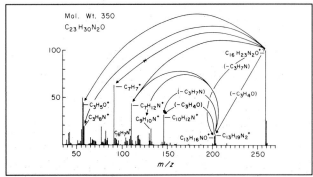

Figure 1. The electron-ionization mass spectrum of China White, compound 1.

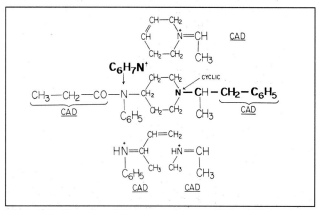

Figure 2. Structural information on China White from the normal EI mass spectrum (MS-I) in bold, plus additional information from CAD (MS-II).

pertain. The CAD mass spectrum produced by multikilovolt ion collisions is particularly valuable. The relative abundances of these fragment ions, excluding those formed by the lowest energy reactions (peaks observed in the MI spectrum), are independent of the precursor ion internal energy and thus are quantitatively characteristic of the ion's structure (6–10). In normal EI and CI mass spectra the peak masses provide valuable information as to the mass of the molecule and some of its fragments; by exact measurement of these masses the corresponding elemental compositions can be determined. Early studies of CAD mass spectra (8) pointed out the potential utility for molecular structure determination of the additional information concerning the structures of these fragment ions available from CAD mass spectra, but little has been done to apply this technique to larger molecules for which much more information is required for structural definition. A separate study (11) describes the applicability of MS/MS to structure determination of oxygenated steroids, for which the CAD spectra of odd-electron fragment ions were found to be the most useful. However, the important peaks of Figure 1 represent even-electron (EE) ions; another objective of this study was to assess the value of CAD spectra of EE fragment ions for substructural assignment.

EXPERIMENTAL SECTION

CAD and MI mass spectra were measured on an MS/MS instrument utilizing a double-focusing Hitachi RMH-2 as MS-I, a He molecular beam to produce CAD, and an electrostatic analyzer as MS-II (12), using an ion source temperature of 130 °C, ion accelerating potential of 9.8 kV, and a collision gas pressure giving a precursor transmittance of 25%. Each CAD spectrum is a computer average of at least 20 scans. Peaks are often poorly resolved, so that the accuracy of abundance measurements of neighboring peaks is sometimes compromised. EI mass spectra (Figure 1 and Table II) were measured on a Finnigan quadrupole GC/MS instrument.

Compound **9** was synthesized from 3-chloro-1-butene and aniline with purification by silica column chromatography and **12** from 3-chloro-1-butene and benzylamine with purification by distillation. PhNHCH(CH₃)CH₂OH (**14**) was synthesized from propylene oxide and aniline under acidic condition and isolated by distillation. PhNHCH₂CH(OH)CH₃ (**15**) was synthesized from lithioanilide and propylene oxide and puri-

fied by column chromatography. Proton NMR and EI-MS indicated the compounds to be pure. Compounds **6–8** were supplied by Thomas N. Riley (13). Other compounds were purchased from Aldrich Chemical Co.

RESULTS

The CI mass spectrum and exact mass measurements of China White (**1**) indicate the molecular ion composition $C_{23}H_{30}N_2O$, m/z 350 (1). This corresponds to a "rings-plus-double-bonds" value of 10 (14); fragment ion elemental compositions (Figure 1) indicate that major parts of this involve a hydrocarbon ($C_7H_7^+$) and a one-nitrogen ($C_6H_7N^+$) fragment. STIRS indicates molecular weight 350 (4) and the substructures (5) methyl (99% reliability), benzyl (96%), $C_6H_5CH_2CN$ (12/15 best matching compounds, match factor 11.2), and saturated N-containing three- to six-membered ring (9/15, MF 11.0 and 11.1). This information is shown in bold face in Figure 2.

The small size (~700 spectra) of the current reference file of CAD mass spectra (15) was a serious disadvantage for this study, and so the following CAD reference spectra were catalogued. The sources of major fragment ions indicated by CAD are shown by the arrows of Figure 1.

m/z **57, C₃H₅O⁺.** CAD spectra of six isomers have been measured (16); these are easily distinguishable, as each gives a separate base peak. For example, that from the propionyl ion corresponds to the loss of CO to form $C_2H_5^+$. The CAD spectrum of the propionyl ion is identical with those of the $C_3H_5O^+$ ions obtained from **1–4**.

m/z **58, C₃H₈N⁺.** Five of these isomeric ions can be distinguished by using their CAD mass spectra (17). The spectrum of $C_3H_8N^+$ from **1** corresponds to the reference ion $CH_3N^+H=CHCH_3$.

m/z **91, C₇H₇⁺.** CAD spectra of six isomeric $C_7H_7^+$ ions have been described (18, 19). The CAD spectrum of $C_7H_7^+$ ions from **1** matches that for the benzyl reference ion.

m/z **132, C₉H₁₀N⁺.** CAD spectra of $C_9H_{10}N^+$ ions from **9**, PhNHCH₂CH=CH₂ (**13**), **14**, and **15** were measured as references. Spectra from **1–3** were the same and matched the spectrum of these ions from **9** and **13**.

Table I. CAD Mass Spectra of $C_{10}H_{12}N$ Ions, m/z 146

compound	ion structure	51	63–6[a]	77	91	92	94	103	104	105	115	117	130–1[a]	144
1	a	54	28	100	44	13	20	52	37	30	(23)[b]	(50)	(115)	(230)
2	b	36	18	100	34	25	15	29	55	28	(13)	(35)	(135)	(230)
3[c]	c	54	19	100	54	8	5	65	54	24	(41)	(61)	(195)	(171)
5	e, a	58	42	100	(850)	(26)	< 0.2	36	48	(52)	36	125	(80)	(200)
6	?	44	17	100	33	6	23	38	74	29	21	42	60	(71)
7	b, e	22	10	100	85	20	10	15	55	27	20	44	(39)	(100)
8	?	17	9	100	17	17	7	33	(66)	(44)	17	37	(43)	(90)
9[c]	a	56	23	100	49	13	13	51	37	28	(44)	(94)	(240)	(125)
10[c]	b	31	22	100	31	23	10	25	46	20	(32)	(84)	(320)	(190)
11[c]	d	34	36	100	25	20	< 0.1	49	17	76[d]	(14)	(9)	(190)	(150)
12[c]	e	31	40	49	(440)	(185)	< 0.1	10	34	12	37	35	32	100

[a] Incompletely resolved peaks; values are average abundances for the indicated mass range. [b] Peaks in parentheses also formed by metastable ion decompositon. [c] Used as a reference spectrum of the ion structure indicated. [d] Additional peaks at m/z 106–107 (40%).

m/z 146, $C_{10}H_{12}N^+$. CAD spectra of these ions from **1–3** and **5–12** are shown in Table I.

m/z 202, $C_{13}H_{16}NO^+$. After separation of a small amount of $C_{14}H_{20}N^+$ by high-resolution MS, the CAD spectrum of these ions from **1** has prominent peaks at m/z 146*, 109, 77, 118, 187*, 130–131, and 91 (in order of decreasing abundance; peaks with asterisk in MI spectrum).

m/z 203, $C_{13}H_{19}N_2^+$. The CAD mass spectrum of these ions from **1** has prominent peaks at m/z 146*, 111, 119, 132, 201, 93, and 57–58.

m/z 259, $C_{16}H_{23}N_2O^+$. The CAD spectrum of these ions from **1** has prominent peaks at m/z 202–203*, 110–111, 57–59, 146, 154, 132, 118–120, 216, 77, 81–86, 92–97, 68–71, 159, 160, 105, 30, and 43.

DISCUSSION

For information on the whole molecule it is advantageous to examine complementary pairs of ions, whose sum of masses or elemental compositions equals that of the molecule (20). Fortunately, two significant primary ions, $C_7H_7^+$ (m/z 91) and $C_{16}H_{23}N_2O^+$ (m/z 259), comprise such a complementary pair; the CAD spectrum of the latter does not indicate that it is a precursor to the former. (The only two other peaks of significance which could be higher-mass members of a complementary pair are $C_{13}H_{19}N_2$ and $C_{13}H_{16}NO$; there are no pairing ions for these.) For the $C_7H_7^+$ ion, the CAD spectrum provides persuasive evidence for the benzyl substructure. Although $C_7H_7^+$ ions can undergo facile rearrangement, this usually produces a significant proportion of the more stable tropylium ion. In the variety of benzylic molecules studied, a high benzyl/tropylium ratio was only observed for compounds such as $PhCH_2I$ in which the benzylic bond is relatively weak (18, 19, 21).

The complementary $C_{16}H_{23}N_2O^+$ fragment has so many possible isomers that choosing the correct reference ion with-

out other information was unlikely, and a further subdivision into complementary pair(s) of ions was sought. However, because $C_{16}H_{23}N_2O^+$ is an even-electron (EE) ion, a complementary pair must consist of an EE ion plus an odd-electron ion. For example, complementary ions for the EE $C_{13}H_{16}NO^+$ and $C_{13}H_{19}N_2^+$ ions (m/z 202 and 203) would be the OE ions $C_3H_7N^{+\bullet}$ and $C_3H_4O^{+\bullet}$, respectively; these are not observed, as expected from the "even-electron" rule (22). However, the EI mass spectrum has prominent EE ions $C_3H_8N^+$ and $C_3H_5O^+$ which could represent the same substructures; a common decomposition of EE$^+$ ions involves loss of a stable molecule through rearrangement of a hydrogen from that portion of the ion to the product EE$^+$ ion (22, 23). The CAD spectra of these ions define their substructures as CH_3NH^+ = $CHCH_3$ and $C_2H_5CO^+$ and suggest that the molecules CH_3N = $CHCH_3$ and CH_3CH = C = O were lost from exterior skeletal portions of the $C_{16}H_{23}N_2O$ fragment to form the m/z 202 and 203 ions. Because $C_2H_5CO^+$ ions are not abundant products of hydrogen rearrangements (23), it is likely that this substructure is also on the exterior of the molecule.

Supporting these assignments are the CAD spectra of the $C_{13}H_{16}NO^+$ and $C_{13}H_{19}N_2^+$ ions, whose numbers of possible isomers are still large for direct CAD structural identification. For both of their CAD spectra the largest peak is $C_{10}H_{12}N^+$ (m/z 146), again due to the losses of C_3H_4O and C_3H_7N. Thus the $C_{10}H_{12}N^+$ ion should represent the last unidentified piece of **1**. Structural information can be deduced from its CAD spectrum. The base peak at m/z 77 and the aromatic ion series are strong evidence for the presence of a phenyl ring. Also $-CH_2-$ is probably not adjacent to the phenyl ring, as this should cause [m/z 91] > [m/z 77].

Appropriate reference CAD spectra (Table I) were measured of isomeric $C_{10}H_{12}N^+$ ions which should have the structures PhN^+H = $C(CH_3)CH$ = CH_2 (a), PhN^+H = $CHCH$ = $CHCH_3$ (b), PhN^+H = $CHC(CH_3)$ = CH_2 (c), $PhN^+(CH_3)$ =

Table II. Even-Electron Ions from Rearrangements

compound	R¹	R²	R³	ion product, m/z							
				132	146	160	188	202	216	203	259
4[a]	H	H	H	25[b,c]	100[d,e]	5[f]	4[g,h]	15[i,j]	1[f]	2	3
1	CH₃	H	H	16[b,c]	32[d,e]	1[f]	<1[g,h]	8[i,j]	2[f]	17	100
2	H	CH₃	H	37[c]	13[b]	33[d]	<1[h]	48[g,j]	12[i]	25	100
3	H	H	CH₃	39[c]	8[b]	100[d,e]	<1[h]	3[g]	28[i,j]	43	77

[a] m/z 189, 43%; 245, 93%. [b] Can be formed by mechanism A as $PhN^+H = CHCR^3 = CHR^2$. [c] By A' as $PhN^+H = CHCH = CH_2$. [d] By B as $PhN^+H = C(CH_3)CR^3 = CHR^2$. [e] By B' as $PhN^+H = C(CH_2R^3)CH = CH_2$. [f] Formed by none of these mechanisms. [g] By A as $C_2H_5CON^+(Ph) = CHCR^3 = CHR^2$. [h] By A' as $C_2H_5CON^+(Ph) = CHCH = CH_2$. [i] By B as $C_2H_5CON^+(Ph)C(CH_3)CR^3 = CHR^2$. [j] By B' as $C_2H_5CON^+(Ph) = C(CH_2R^3)CH = CH_2$.

$CHCH = CH_2$ (d) [could also contain $PhN^+(=CH_2)CH_2CH=CH_2$], and $PhCH_2N^+H = CHCH = CH_2$ (e), based on their formation from **9, 10, 3** (vide infra), **11**, and **12**, respectively. These spectra make it possible to identify the $C_{10}H_{12}N^+$ ions from 1 as (a) (Figure 2). The combination of (a) with $CH_3N = CHCH_3$ should indicate the structure of the $C_{13}H_{19}N_2$ piece of the molecule. In addition to the phenyl ring, and the presumed double bond to the quaternary nitrogen atom, this piece must contain one ring or double bond; STIRS indicates the former. Possible structures for $C_{13}H_{19}N_2^+$ were tested by seeing which is the most rational for the formation of (a) and $CH_3N^+H = CHCH_3$; the best candidate and postulated mechanisms are shown in Scheme I, which is discussed below. A key indicator that the methyl group of (a) arises from a ring CH_2 group in the $C_{13}H_{19}N_2^+$ ions is the structure of m/z 132 ($C_9H_{10}N$) in the EI spectrum of **1**; its CAD spectrum matches that of the reference ion $PhNH^+ = CHCH = CH_2$, a homologue of (a). (A more straightforward way to identify the piperidine ring of **1** would be to show that the CAD spectrum of its m/z 110 ions matched that of (f) Scheme I.) Consistent with the STIRS (4, 5) prediction of β-phenylethylamine, the $PhCH_2$– group should be located on the α-C of the cyclic N (Figure 2). A logical place for the C_2H_5CO– group is on the other nitrogen, as propionamide ions commonly lose $CH_3CH = C = O$ (23). Thus the evidence leads directly to the China White structure. Although the data are much less consistent with other possible structures, this opinion may be biased by our prior knowledge of the correct structure.

Mechanisms of Piperidine Ion Decompositions. The uniqueness of the assignment of structure **1** is reinforced by the mechanistic rationalization (Scheme I) of major peaks in the EI mass spectra of compounds **1–4** and **8** (Table II). The CAD spectra indicate that a major portion of these secondary ions is formed through the $(M - \cdot CH_2Ph)^+$ peak.

Besides the methyl ketene elimination discussed above, for these EE ion decompositions two pathways (A and B) are proposed which are applied to both cyclic EE ions, $(M - \cdot CH_2Ph)^+$ and $(M - \cdot CH_2Ph - C_3H_4O)^+$. Pathway A (illustrated in Scheme I only for m/z 203) could be triggered by hydrogen rearrangement to the charge site (23), after which olefin elimination would give the resonance-stabilized immonium ion. For pathway B inductive cleav-

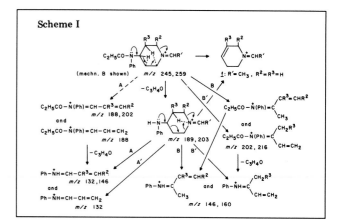

Scheme I

age at the charged ring nitrogen followed by two H rearrangements would yield the homologous immonium ion incorporating an additional ring carbon as a methyl group. In general (Table II), pathway B is favored over A, in particular for those ions which still contain the propionyl group. This would be expected on enthalpic grounds; both the ion and neutral products of B have an extra methyl group to provide stabilization, while the extra C_2H_4 formed by path A has a slightly positive heat of formation. For the 2-methyl compound (**2**), however, the CAD spectrum of the m/z 146 formed by EI matches that of $PhN^+H = CHCH = CHCH_3$ (b), so that these ions arise mainly by pathway A, not B'. The initial C–N bond cleavage in path A could be enhanced by the presence of the 2-methyl to stabilize the charge transferred to the 2-carbon.

The analogous compounds **5–8** could conceivably decompose by similar mechanisms following ionization, but the CAD spectra of their $C_{10}H_{12}N^+$ ions (Table I) are substantially different than that of isomer (a) (path B). Those ions originating from **5** appear to be largely $PhCH_2N^+H = CHCH = CH_2$ (e), and so contain the terminal $PhCH_2N$ group of this compound; this isomer is formed to a lesser extent in the analogous compound **7**. This interference should not be a problem with seven-membered ring compounds **6** and **8**, but the CAD spectra of their m/z 146 ions indicate that these contain isomers in addition to (a–e). Although the large m/z 77 peaks in these CAD spectra indicate that the nitrogen is adjacent to the phenyl, we have not been able to prepare a reference ion whose

CAD spectrum alone, or combined with (a–e), is consistent with these. If **6** and **8** had been unknowns, however, these *m/z* 146 CAD spectra would at least have indicated the absence of the piperidinyl substructure.

CONCLUSIONS

The utility of mass spectrometry for molecular structure determination of larger molecules is enhanced by substructure identification using CAD spectra of fragment ions. Expansion of the data base of CAD reference spectra (*15*) should make such substructural information much more readily usable for submicrogram samples for which methods such as nuclear magnetic resonance do not have sufficient sensitivity.

ACKNOWLEDGMENT

We are grateful to M. P. Barbalas and F. Turecek for valuable advice and to T. N. Riley for samples of **6–8.**

LITERATURE CITED

(1) Kram, T. C.; Cooper, D. A.; Allen, A. C. *Anal. Chem.* **1981**, *53*, 1379A.
(2) Stinson, S. *Chem. Eng. News* **1981**, Jan 19, 72.
(3) Wilford, J. N. *New York Times* **1980**, Dec 30, C1.
(4) Mun, I. K.; Venkataraghavan, R.; McLafferty, F. W. *Anal. Chem.* **1981**, *53*, 179–182.
(5) Haraki, K. S.; Venkataraghavan, R.; McLafferty, F. W. *Anal. Chem.* **1981**, *53*, 386–392.
(6) McLafferty, F. W. *Acc. Chem. Res.* **1980**, *13*, 33–39.
(7) McLafferty, F. W. *Science* **1981**, *214*, 280–287.
(8) McLafferty, F. W.; Kornfeld, R.; Levsen, K.; Haddon, W. F.; Sakai, I.; Bente, P. F., III; Tsai, S.-C.; Schuddemage, H. D. R. *J. Am. Chem. Soc.* **1973**, *95*, 3886.
(9) Levsen, K. "Fundamental Aspects of Organic Mass Spectrometry"; Verlag Chemie: Weinheim, 1978.
(10) McLafferty, F. W. *Phil. Trans. R. Soc. London, Ser. A* **1979**, *293*, 93–102.
(11) Cheng, M. T.; Barbalas, M. P.; Pegues, R. F.; McLafferty, F. W. *J. Am. Chem. Soc.*, in press.
(12) McLafferty, F. W.; Todd, P. J.; McGilvery, D. C.; Baldwin, M. A. *J. Am. Chem. Soc.* **1980**, *102*, 3360–3363.
(13) Finney, Z. G.; Riley, T. N. *J. Med. Chem.* **1980**, *23*, 895.
(14) McLafferty, F. W. "Interpretation of Mass Spectra," 3rd ed.; University Science Books: Mill Valley, CA, 1980; p 23.
(15) McLafferty, F. W.; Hirota, A.; Barbalas, M. P. *Org. Mass Spectrom.* **1980**, *15*, 327–328.
(16) McLafferty, F. W.; Dymerski, P. P., unpublished results.
(17) Levsen, K.; Wipf, H.-K.; McLafferty, F. W. *J. Am. Chem. Soc.* **1974**, *96*, 139.
(18) McLafferty, F. W.; Winkler, J. *J. Am. Chem. Soc.* **1974**, *96*, 5182.
(19) Bockhoff, F. M.; McLafferty, F. W. *Org. Mass Spectrom.* **1979**, *14*, 181–184.
(20) Turecek, F.; Cheng, M. T.; Proctor, C. J.; McLafferty, F. W., submitted for publication in *Org. Mass Spectrom.*
(21) McLafferty, F. W.; Bockhoff, F. M. *J. Am. Chem. Soc.* **1979**, *101*, 1783–1786.
(22) Karni, M.; Mandelbaum, A. *Org. Mass Spectrom.* **1980**, *15*, 53.
(23) Reference 14, Sections 8.7–8.10, in particular pp 162–164.

Received for review April 6, 1982. Accepted August 10, 1982. The Army Research Office, Durham, and the National Institutes of Health provided financial support.

Reprinted from *Anal. Chem.* **1982**, *54*, 2204–07.

In this work, Small and Miller converted the standard absorption detector to a novel universal detector for ion chromatography (IC), a development that immediately made IC accessible to all chromatographic laboratories. The concept of displacement quickly led to similar detection schemes in conventional LC and in capillary electrophoresis, attracting theoretical studies to explain the unusual response factors and unique applications requiring sensitive detection.

Edward S. Yeung
Iowa State University

Indirect Photometric Chromatography

Hamish Small* and Theodore E. Miller, Jr.
1712 Building, M. E. Pruitt Research Center, The Dow Chemical Company, Midland, Michigan 48640

Indirect photometric chromatography is a sensitive single-column ion analysis method developed from the concept that photometers may be used to detect transparent ionic species. The use of light-absorbing eluent ions in an ion-exchange mode enables sample ions to appear as "troughs" in the base line absorbance as transparent sample ions substitute for the light-absorbing displacing ions. The elution times of these troughs vary with the ion injected and their depths (or areas) are proportional to the amount of sample injected. Notable advantages of the new technique are its single column simplicity, its applicability to a wide range of ionic species, and an inherently greater sensitivity than single-column conductometric approaches.

I on determination by liquid chromatography is often frustrated not by separation problems but by detection problems. An example is the problem of determining the many important inorganic ions that are not light-absorbing. Whereas the separation of such transparent ions may be conveniently effected on ion exchange resin columns, their detection and measurement by conventional photometric means are thwarted since they are optically indistinguishable from the transparent eluents commonly prescribed.

The technique known as ion chromatography (*1, 2*) was developed to circumvent the detection problem posed by transparent sample ions. In just 6 years it has become a widely practiced and popular method addressing problems in a great variety of areas (*2*). Ion chromatography (IC) usually comprises a two-column arrangement followed by a conductance detector where the first column serves to separate the ions of interest while the second column, the suppressor, serves to lower the conductance of the eluent while usually increasing the conductance of the eluted sample ions. The suppressor column in IC becomes exhausted in the course of normal usage and must be periodically regenerated or replaced—usually regenerated. Whereas this regeneration step has been automated in commercial instruments so that it is relatively unobtrusive or is made continuous as in the recently developed hollow fiber suppressor (*3*), it would nevertheless be desirable and advantageous for the following reasons to develop single column (suppressorless) methods for the many nonchromophoric ions:

(1) Decreased complexity of the instrumentation should yield a concomitant increase in reliability. This is a very important factor in penetrating the process control area with chromatographic methods where the demands for unattended and relatively maintenance-free operation have high priority.

(2) Reduced dead volume as a result of eliminating the suppressor will yield faster analysis and somewhat improved resolution.

Suppressorless single-column conductometric methods of ion analysis have been described in earlier literature (*4–7*). The limitations of these approaches have been elaborated by Pohl and Johnson (*8*) who point to the problem inherent in attempting to determine accurately the oftentimes small changes in eluent conductivity that accompany the replacement of eluent ions by sample ions—both of which are conducting. They argue the sensitivity advantage to be gained by adding a suppressor that effectively "removes" the conductance of the eluent.

We would like to report a single column approach that solves the monitoring problem in a different manner while retaining much of the sensitivity of the original ion chromatography method. This new technique is derived from a comprehensive development of the concept that photometers may be used to monitor the many "transparent" ionic species commonly thought not to be amenable to this type of detection. Although mentioned earlier by Laurent and Bourdon (9), lack of sensitivity was expected to limit application of the new idea. The independent discovery and development of the method as reported here demonstrate its abundant versatility, particularly in optimizing sensitivity.

From the standpoint of sensitivity we will also discuss how this novel means of detection overcomes the problems that are intrinsic to suppressorless conductometric monitoring.

A feature of this new photometric approach is the use of light-absorbing (usually UV absorbing) eluents, made so by including in the eluent light-absorbing ions of the same charge as the ions to be separated.

These light-absorbing ions have a dual role: (1) of selectively displacing the sample ions from the chromatographic column and (2) of revealing the sample ions in the effluent.

The appearance of sample ions in the effluent is signaled by "dips" or "troughs" in the base line absorbance of the effluent as the transparent sample ions substitute for the light-absorbing displacing ions.

In applying this new approach to the very sensitive detection and determination of ions, it is essential to understand the interplay of such factors as concentration of eluent, concentration of sample, capacity of the ion exchanger, and the optical properties of the eluent. The definition of these critical relationships occupies a large part of this contribution.

We have chosen the name indirect photometric chromatography (IPC) to describe this rapid, sensitive, broad-scope exploitation of ion exchange and photometric monitoring.

PRINCIPLE OF THE METHOD

Consider an ion exchange column—for illustrative purposes specifically an anion exchanger—which has been pumped and equilibrated with an electrolyte denoted Na^+E^- so that the sites in the exchanger are occupied exclusively by eluent ions E^-. A concentration monitor capable of sensing all ionic species and placed at the outlet of such an operating column would reveal a steady level of Na^+ and E^- if the feed concentration of the eluent is maintained constant (Figure 1A). If an injection is made of sample electrolyte, denoted Na^+S^-, then the sample anion, S^-, will generally be retarded by the stationary phase and will exit at a characteristic elution volume determined by such factors as the capacity of the exchanger, the concentration of the solution, and the affinity of the stationary phase for S^- relative to E^-. A suitable monitor at the column exit would indicate the concentration of S^- to rise and fall in a familiar fashion as it leaves the column (Figure 1B). Conventional ion exchange LC art has been concentrated on devising suitable detectors for *directly* monitoring the magnitude (height or area) of these sample peaks. Generally ignored has been the fact that accompanying the appearance of S^- there must be a concerted and equivalent change in E^- since, by the principles of electroneutrality and equivalence of exchange, the total equivalent concentration of anions (S^- and E^-) must remain fixed since the concentration of sodium coions is fixed. It therefore follows that the concentration of S^- in the effluent could be indirectly monitored by continuously monitoring the level of eluent ion E^-.

On the basis of this argument it follows how this somewhat latent feature of the ion exchange mode may be usefully tapped in the case of problematical sample ions. Thus, if sample ions are inconveniently lacking in a particular property, for example, optical absorbance, one may exploit this deficiency in the sample species by deliberately choosing an eluent ion that is light absorbing and monitoring the "troughs" generated in the base line absorbance as transparent sample ions elute.

The development of this combination of ion exchange and indirect photometric monitoring is the main concern of this contribution. An example may serve to illustrate how the method works.

A column containing an anion exchanger was equilibrated with a dilute (10^{-3} M) solution of sodium phthalate until the effluent absorbance was stable as indicated by the UV photometer monitoring the column effluent. When a sample containing chloride, nitrite, bromide, nitrate, and sulfate was injected, the chromatogram of Figure 2 was obtained. By making separate injections of the individual anions in the mixture, we established the identities of the troughs in phthalate absorption. The off-scale positive deflection is due to the ion exchange displacement of phthalate by the injected sample anions as a whole. Since the total equivalent concentration of the sample exceeded

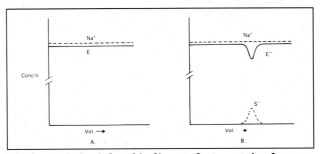

Figure 1. Principle of indirect photometric chromatography.

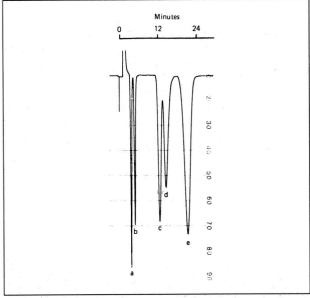

Figure 2. Separation and indirect photometric detection of several "transparent" sample ions: (a) chloride, (b) nitrite, (c) bromide, (d) nitrate, (e) sulfate.

that of the eluent, the void disturbance was positive. When the total concentration of the sample is less than that of the eluent, the disturbance is negative.

An elution order was established experimentally for a large variety of anions using phthalate as displacing ion. This is recorded in Table I which may be used as a rough guide in predicting the feasibility of certain separations. But it must be used judiciously for, as will be evident later, elution order depends upon a variety of experimental conditions. The feasibility of determining ions by this indirect approach brings a number of questions to mind, notably: How does one choose an eluent species from the enormous number that qualify through being ionic and having appropriate spectral properties? What are appropriate spectral properties? What is the sensitivity of the method and how is it related to the various elution conditions? These questions were addressed in a systematic manner—the results and recommendations follow.

CONSIDERATIONS IN THE CHOICE OF ELUTING IONS AND CONDITIONS

In IPC, sample ions are revealed and quantified by the decrements they produce in eluent concentration. Since the displacing species is usually in much greater abundance than sample species—a feature of elution chromatography-these decrements would ordinarily represent rather small fractional changes in eluent level. Thus, the successful application of IPC is directly related to how precisely we can measure these fractional differences (the signal) in the presence of the random fluctuations (the noise) of the base line response. To this end it is critical to the understanding of IPC to appreciate how signal to noise ratio and, in turn, sensitivity are related to an important variable in the system, namely, the concentration of the eluent.

(A) Concentration of the Eluent. Let us consider the case of elution of a sample ion through an anion exchanger operating in the IPC mode. It is assumed that the conditions have been chosen so that the eluted sample is adequately remote from the void disturbance. Typically this would result in a Gaussian-shaped change in sample ion concentration, but for simplicity of treatment we will assume that it emerges as a square wave pulse with maximum concentration, C_S. This pulse of sample will cause a concomitant and identically shaped pulse change in the eluent level as indicated in Figure 3A. The signal to be measured, ΔS, is the difference between the signals due to the eluent at base line concentration, C_E, and when sample elutes, $C_E - C_S$.

This may be expressed as follows:

$$\text{SIGNAL} = \Delta S \propto C_S A_S + (C_E - C_S)A_E - C_E A_E = C_S(A_S - A_E) \quad (1)$$

where A_S and A_E denote the absorptivities of the sample and eluent ions, respectively. Equation 1 assumes that the signal response is directly proportional to species concentration.

Figure 3A depicts an ideal, that is, a noiseless detection sit-

Table I. Elution Volume of Anions[a]

ion	V_E, mL	ion	V_E, mL
void	2.0	fluoride	3.0
acetate	3.2	glycolate	3.4
azide	55	iodate	3.8
bromate	14	maleate	78
bromide	69	malonate	62
carbonate	6.5	nitrate	84
chlorate	86	nitrite	24.1
chloride	17.1	o-phosphate	7.5
monochloracetate	8.0	propionate	3.3
dichloracetate	27.5	succinate	47
trichloracetate	[b]	sulfate	114
citrate	[b]	sulfite	86
cyanate	6.5		

[a] Eluent: 10^{-3} M sodium phthalate, 10^{-3} M boric acid, pH 9. Column: 4×250 mm, SAR-20-0-6. [b] Very large.

uation. Reality is represented in Figure 3B, which depicts the noise within which the signal must be detected, that is, the uncertainty in measuring the concentration of eluent C_E. At a given base line absorbance, noise represents a fixed (random) fluctuation, represented by N. The signal as a fraction of the base line absorbance is given by the expression

$$\frac{C_S(A_S - A_E)}{C_E A_E} \quad (2)$$

from which it follows that

$$\frac{\text{signal}}{\text{noise}} \propto \frac{C_S(A_S - A_E)}{N C_E A_E} \quad (3)$$

For transparent ions A_S is zero so, neglecting signs which are not significant for our purposes,

$$\frac{\text{signal}}{\text{noise}} \propto \frac{C_S}{N C_E} \quad (4)$$

This simple expression incorporates the conclusion that sensitivity improves, the lower the concentration of eluent employed.

There are other important considerations, however, that impose lower limits to the concentration of eluent preferred. Perhaps most significant among them are in the penalties of overlong run times and loss of sensitivity due to band spreading that result from using an eluent that is too dilute. Ideally, the run time should be no longer than the time necessary to adequately resolve the troughs.

When using LC in an ion exchange mode, there is a further very fundamental consideration to be kept in mind when choosing the concentration of the eluent, that is, the saturation level in the eluting capacity of the eluent. How close one

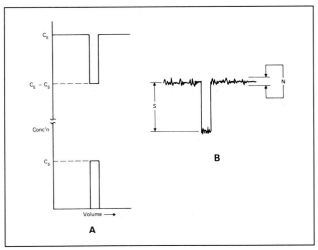

Figure 3.

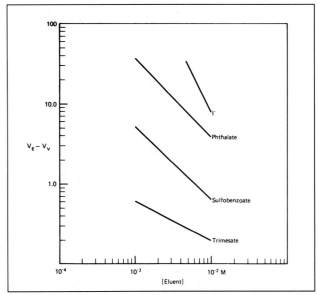

Figure 4. Elution volumes of sulfate ion using four different eluents. Resin: 2.8 × 250 mm, SAR-40-0.6.

operates to this level, or to what extent one exceeds it, exercises strong control over the sharpness of the eluted peaks and hence resolution. Basically the source of the limitation derives from the inability of eluent to displace sample ion at a higher concentration than that of the eluent.

(B) The Relative Affinity of E⁻ and S⁻. As well as its concentration, the displacing power of the eluent ion with respect to the sample ions is an extremely important factor in the practice of IPC. Different ions vary widely in their displacing power. In an attempt to reduce these eluent options to a reasonable number, we examined a large variety of candidate eluents having a wide range of ion exchange affinities and from these limited our recommendations, somewhat arbitrarily, to just a few. For anion separations, in addition to o-phthalate, the 1,2-sulfobenzoate, 1,3,5-benzenetricarboxylate (trimesate), and iodide ions are useful displacing species. Trimesate and sulfobenzoate are generally more potent displacing ions than phthalate while iodide is less so. This is illustrated in Figure 4 which is a plot of the elution volume of sulfate ion on a surface agglomerated anion exchanger for the four different eluent species as a function of eluent concentration. Within the concentration range of eluent depicted, the eluting power of the displacing ion is seen to follow the expected trend of polyvalent ions being more potent displacing species than monovalent ions. This order of potency is not always obeyed for, as will be seen shortly, it depends on the charge on the eluent and sample ions and on the concentration of the eluent.

(C) The Effect of the Charge of E^{x-} and S^{y-} on Elution Rate and Elution Order. The elutability of several sample ions by the four candidate eluents was measured, and the results are illustrated in Figure 5. The quantity $V_E - V_v$ is the corrected elution volume of the ion, that is, the effluent volume between sample injection and trough elution less the void volume of the column. A number of features of the data are noteworthy: (1) the linear dependence of log $(V_E - V_v)$ on log (concentration of eluent); (2) the differing slopes of the log $(V_E - V_v)$ vs. log C plots (there are in fact five different slopes in the 15 plots of Figure 5); (3) the cross-over in certain elution orders.

The elution can thus be described in the convenient mathematical form

$$\log(V_E - V_v) \propto -m \log [E^{x-}]$$

where $[E^{x-}]$ is the molarity of the eluent.

A relationship of this form, namely

$$\log (V_E - V_v) = \text{constant} - y/x \log [E^{x-}] \qquad (5)$$

may be developed from basic ion exchange theory (10).

This expression, besides accounting for the experimentally observed linearity in the log $(V_E - V_v)$ vs. log [eluent] plots, also defines the slope as the ratio y/x of the charge of the sample ion to that of the eluent ion. Consequently, it is of interest to examine how the experimentally observed slopes agree with the values predicted by the expression. Accordingly, the slopes of the plots in Figure 5 were measured and are compared with the values predicted by eq 5. The comparison is provided in Table II and with one exception—trimesate eluent, bromide sample ion—the agreement between theoretical and observed is good to excellent.

This has an important practical implication. It suggests that from a knowledge of the charges of the sample and eluent ions and a single determination of V_E at a single concentration of eluent, it is possible to predict V_E for that particular species at other eluent concentrations. Furthermore, the dependence of slope on ion charge ratio explains the observed cross-overs in elution order. Knowing the origin of this effect in turn affords us another means of controlling resolution and avoiding the condition of nonresolvability that is represented by a cross-over point.

PHOTOMETRIC FACTORS

The precise determination of eluent absorbance (concentration) is a very important part of IPC, so principles that apply to conventional spectrophotometric measure-

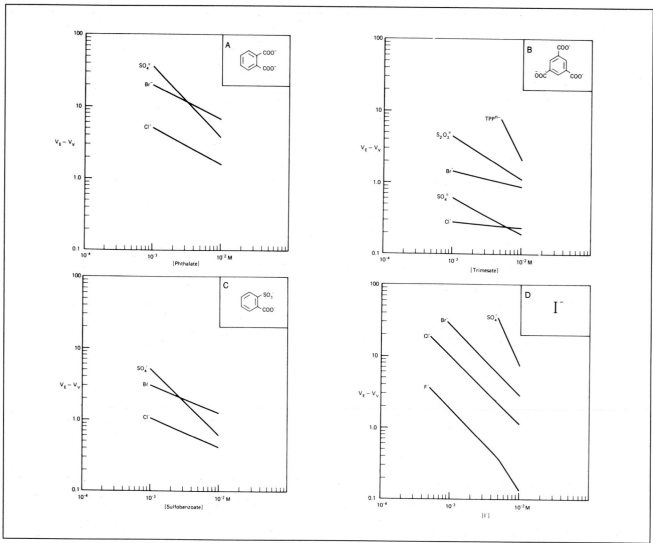

Figure 5. Elution volumes of sample ions with: (A) phthalate; (B) trimesate; (C) sulfobenzoate; (D) iodide. Resin: 2.8 × 250 mm, SAR-40-0.6.

ments also apply to IPC. It is known from classical spectrophotometry that the most accurate measurements are obtained when the optical density is 0.43 (*11*). Actually this value is not critical and the error of measurement varies little within a range of optical density from about 0.2 to 0.8. For this reason it is important to the accuracy of IPC to monitor the eluent under conditions where its absorbance falls within this range.

Since the concentration of eluent to be used will generally be dictated by such other considerations as column capacity and eluent ion affinity and since cell path length is fixed, by what means may optical absorbance be adjusted? The molar absorptivity of a given eluent ion generally exhibits so great a dependence upon wavelength that the desired eluent absorbance is obtained merely via selection of detector wavelength. A photometric detector with multiwavelength monitoring capabilities is therefore a very useful adjunct to IPC although under certain conditions fixed wavelength devices have operated quite effectively. Eluent concentrations as dilute as 10^{-6} M and as concentrated as 1 M have conformed to the

"optimum absorbance" requirement of IPC through appropriate choice of detection wavelength.

APPLICATIONS

Experimental Section. The apparatus used in IPC is conventional, usually consisting of an eluent reservoir, a pump, a sample injection valve, an ion exchange column, a flow-through photometric detector, and a recorder. A trough integrator is optional.

Any of a large variety of LC pumps is suitable—in this work the Laboratory Data Control (LDC) Constametric I, the LDC minipump, and the Altex Model 110A were used. Photometers found to be useful were two fixed-wavelength types, the LDC Model 1203 and the Altex 153, and two multiwavelength instruments, the Perkin-Elmer LC-75 and the Varian UV-50. Note that the photometric detector responds to small fractional reductions in an elevated base line as solute ion zones traverse the optical cell. This feature calls for techniques to suppress the base line level without reducing or otherwise obscuring the small troughs. We have achieved this in two ways. In dual beam instruments with reference cells we have operated with the reference cell containing either air or eluent. Operating with air in the reference cell we found that some of the commercially available photometers did not have enough zero adjustment to null out the high absorbance of some eluents, and we found it necessary, therefore, to develop additional base line suppression circuitry to handle this problem. Operating with eluent in the reference cell avoids the problem.

In a few cases we found careful temperature control of the column to be necessary. In one eluent system—the trimesate at pH 10—we found an exceptional sensitivity to temperature change in the column. For example, it would manifest itself in marked biphasic base line disturbances on even slight warming of a short section of column as might be brought about by briefly gripping the column between finger and thumb. Enclosing the column in a thermostating bath maintained at 37 ± 0.1 °C eliminated this problem. As a rule, however, we found such close temperature control to be unnecessary.

The eluent in IPC is only slightly changed on passing through the column and may be recycled to the eluent reservoir with negligible penalty. This ability to recycle eluent is desirable in unattended process control applications.

Stainless steel columns (4.1 × 250 mm) or glass columns (2.8 × 250 mm) were used to contain the ion exchangers, and sample was introduced via a Rheodyne Model 7010 injector.

The column packings are one of the most important features of IPC and a variety of materials are effective. We have made wide use of ion exchangers originally developed for ion chromatography (1, 12–14). Especially useful are the surface agglomerated pellicular anion exchangers which are prepared by depositing a monoparticulate layer of submicron anion exchanging spheres onto a much larger (20–50 μm) substrate bead whose surface is anionic. In the original IC work (1), surface sulfonated styrene divinylbenzene copolymer particles were much used as substrate spheres but, in the present research, conventional strong acid cation exchangers such as Dowex 50 have been used exclusively for this purpose. The very small particle anion exchangers of uniform particle size distribution were prepared by quaternizing emulsion copolymers of vinylbenzyl chloride and divinylbenzene (14). Surface agglomeration was carried out by adding a quantity of the substrate cation exchange resin to a suspension of the colloidal anion exchanger. The anion exchange capacity of these surface agglomerated separating resins is controlled by the size of the colloidal particles and by the size of the substrate, being greater the smaller the substrate size and decreasing as the colloidal resin size decreases. Surface agglomerated resins are described in the text by a code that indicates the sizes of the substrate and of the colloidal anion exchanger. For example, the designation SAR-20-0.6 denotes a surface agglomerated resin prepared by coating a 20 μm diameter cation exchanger with a 0.6 μm diameter colloidal anion exchanger.

Ion exchangers found to be useful in IC for cation analysis are also useful in IPC applications. In this regard, surface sulfonated styrene divinylbenzene copolymer spheres (12) have been used to develop a number of IPC analysis schemes. Besides these low capacity ion exchangers, high specific capacity materials have been successfully applied to IPC. They are especially useful in applications where direct injection of concentrated samples is desirable and column overloading becomes an important consideration.

A number of commercial anion and cation exchangers have been used with success.

The following applications have been chosen to exemplify the scope, selectivity, and sensitivity of IPC. Details on columns, packings, eluents, etc., are provided in Table III.

Anion Separations. Figure 6 represents the separation of a mixture of five anions. Noteworthy is the trough produced by carbonate ion. This ability to determine anions of high pK acids gives IPC an advantage over ion chromatography which by its nature is very insensitive to such species.

Figures 7 and 8 show the determination of nitrite in chloride and of traces of sulfate in a high background of salt. Both illustrate the excellent selectivity and sensitivity of which IPC is capable.

As discussed earlier, sensitivity may be improved by decreasing eluent concentration. An example of this is provided in Figure 9. The fairly potent sulfobenzoate ion was chosen as the displacing species so that relatively low concentrations would suffice to elute the sample ion, in this case sulfate, with reasonable speed. An appropriately low capacity column was also used. The trough due to sulfate and showing good signal to noise resulted from a 100-μL injection of 10^{-6} M sodium sulfate. The detectability of sulfate under these conditions is approximately 1 ng, attesting to the high sensitivity attainable by the technique.

IPC has exceptional capabilities for handling ionic species with high affinity for anion exchangers. Noteworthy is the rapid elution of the polyphosphate species (Figure 10), an ion that is normally very difficult to displace.

Cation Separations. A number of cation separation schemes were developed by using the IPC approach. Figure 11 illustrates a rapid separation of sodium, ammonium, and potassium. This separation is noteworthy in that it was obtained on a very small column (2.8 × 20 mm) of high specific capacity cation exchanger. The eluent was copper sulfate (0.01 M), copper being the UV-absorbing displacing ion. A column and eluent such as this have been used to determine small amounts of sodium and potassium in concentrated (20%) cal-

Table II. Comparison of y/x (Theoretical) with y/x (Observed)

E^{x-}	S^{y-}	y/x	y/x observed
I^-	Br^-	1	1.00
I^-	SO_4^{2-}	2	2.00
trimesate^{3-}	Br^-	1/3	0.21
trimesate^{3-}	$SO_4^{2-}, S_2O_3^{2-}$	2/3	0.60
trimesate^{3-}	TPP^{y-}	?	1.9
phthalate^{2-}	Br^-	1/2	0.47
	SO_4^{2-}	1	0.98

cium chloride solution. Only moderate dilution of the sample is required in view of the high capacity of the resin employed. After the monovalent ions had been eluted, the resin was flushed briefly with a concentrated (1 M) solution of copper nitrate in order to displace divalent ions which would otherwise have appeared at a much later time and interfered with subsequent chromatograms. We found it convenient to introduce this purge solution by way of another sample injection valve equipped with a large (0.5 mL) loop. A separation of sodium, potassium, calcium, and magnesium (Figure 12) was achieved by using a split column technique wherein two columns of equal length but containing resins of different specific capacities were connected in series and appropriately switched.

Joint Anion and Cation Determination. We have demonstrated that indirect photometric chromatography may be extended to simultaneous joint anion and cation analysis by combining chromophoric anion and cation mobile phase ions with suitable ion exchange columns. For extension of IPC to

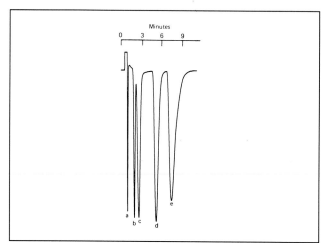

Figure 6. IPC of several anions: (a) carbonate, 1.8 µg; (b) chloride, 1.4 µg; (c) phosphate, 3.8 µg; (d) azide, 5.0 µg; (e) nitrate, 10 µg.

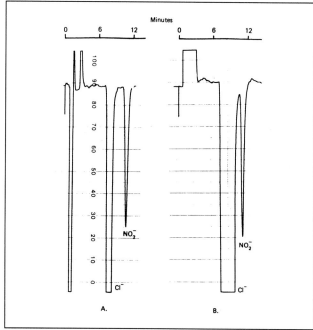

Figure 7. Determination of nitrite in chloride: (A) 58.5 ppm chloride, 4.6 ppm nitrite; (B) 585 ppm chloride, 4.6 ppm nitrite.

combined analysis of anions and cations in a single chromatograph, a special eluent must be chosen. In accordance with the principles already outlined, the eluent for analysis of, say, only anions in a sample has a twofold function: to displace anion bands individually from the column and to render them detectable as transparencies in contrast to eluent anion UV absorbance. Joint analysis of anions and cations with a single eluent and UV detection, then, requires mobile phase anion and cation *both* with UV absorbance and appropriate sample elution power. A necessary further consideration is that mobile phase anion and cation each contribute approximately equally to the absorbance at base line since we have seen already how sensitivity to eluting sample ions relates to mobile phase absorbance.

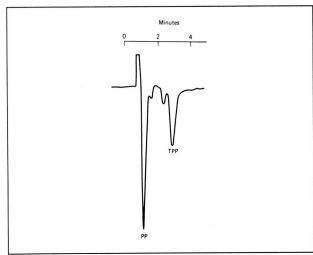

Figure 10. Separation of pyro- (PP) and tripoly- (TPP) phosphates.

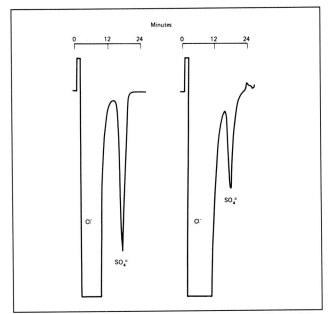

Figure 8. Determination of sulfate in 1% sodium chloride: (A) 100 ppm sulfate; (B) 10 ppm sulfate.

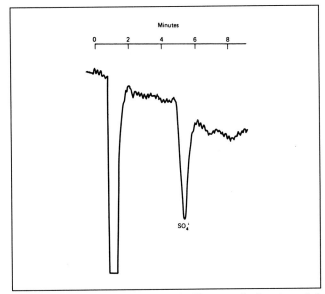

Figure 9. Sensitivity of IPC. Sulfate peak due to a 0.1 mL injection of 10^{-6} M sodium sulfate.

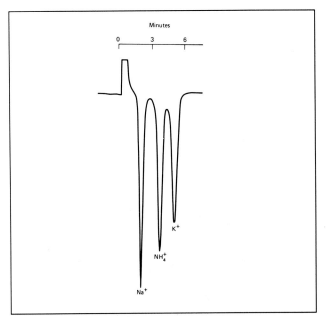

Figure 11. IPC of cations.

Taking the various factors into account, we found copper nitrate to be a suitable eluent. As determined by a Cary 15 spectrophotometer at 241 nm, ϵ_{Cu} = 37.5 L equiv $^{-1}$ cm^{-1} and ϵ_{NO_3} = 77.5 L equiv $^{-1}$ cm^{-1}. It follows that a 5×10^{-3} M $Cu(NO_3)_2$ solution would produce a base line absorbance of

$$A_{TOT} = A_{Cu} + A_{NO_3} =$$

$$(\epsilon_{Cu} + \epsilon_{NO_3}) \times \text{normality} \times \text{pathlength}$$

and so

$$A_{TOT} = 1.15$$

in a detector with a 1-cm pathlength. This is an acceptable base line absorbance value and thus a 5×10^{-3} M copper nitrate eluent was selected along with a detection wavelength of 241 nm.

The separating columns chosen were commercially available strong anion and cation exchange columns arranged in series. The cation exchanger was 4.6 × 250 mm Partisil 10-SCX from Whatman containing 10-μm microparticulate packing with siloxane-bonded sulfonic acid exchange groups. The anion exchanger was a 4.6 × 250 mm Partisil 10-SAX with 10-μm microparticulate packing and siloxane-bonded quaternary anion exchange sites.

The chromatogram obtained from an injection of a synthetic mixture comprising 0.2 M NaF, 0.2 M RbCl, and 0.1 M MnCl₂ is shown in Figure 13. The first three troughs are the deficiencies in the copper absorbance caused by the emergence of the three sample cations, the nitrate absorbance (concentration) remaining constant within this region. The last two troughs are the deficiencies in nitrate absorbance due to sample anions while the copper absorbance remains constant.

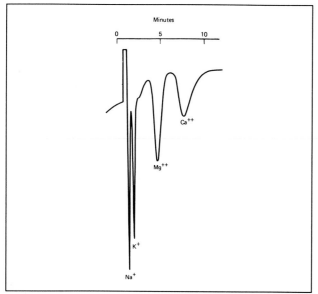

Figure 12. Separation of mono- and divalent cations by a split column technique.

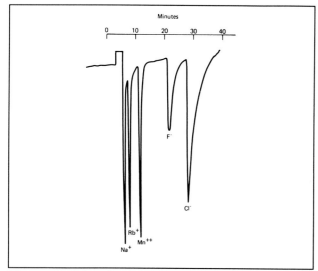

Figure 13. Joint determination of anions and cations by IPC.

Calibration. Calibration runs for the three ions—sulfate, nitrate, and phosphate—yielded curves that indicate a convenient linear dependence of trough depth on the amount of ion injected.

There is an interesting aspect to calibration in the IPC mode in that for many ions the area of the trough is not dependent on the ion injected but only on its amount. This is a natural result of the method of monitoring since each equivalent of sample ion displaces the same amount of monitoring ion from the mobile phase irrespective of the sample ion. To demonstrate this, separate injections of accurate amounts of nitrate, sulfate, and phosphate were eluted by sodium phthalate (pH 8) and the areas of the troughs measured. The results are shown in Table IV.

The area of trough per equivalent of ion is indeed approximately independent of the ion injected—for these three ions. On the basis of this observation we expect anions to adhere to this rule. Anions of acids with medium to high pKs should give responses determined by their valence at the ambient pH of the eluent. Phosphate, for example, exists predominantly as the HPO_4^{2-} species at pH 8 so that 1 mol of phosphate injected would be expected to displace 2 equiv of monitor ion. The data of Table IV support this expectation.

CONCLUSIONS

This work has demonstrated that photometers may be used as chromatographic detectors for accurate, sensitive determination of transparent ionic species commonly considered photometrically undetectable. The technique, indirect photometric chromatography as we have called it, shows considerable potential in the area of inorganic and organic ion analysis.

In IPC, detection of the eluting sample is accomplished by monitoring a change in property of the column effluent as the eluent ion is displaced by the sample ions. Clearly, sensitivity in such a case is impaired when the eluent and the sample ions possess this property to a comparable extent. Such is the case with suppressorless conductometric monitoring—both sample ion and eluent ion are conducting and the closer the values of their equivalent conductances the poorer the sensitivity for that sample ion. In IPC on the other hand one chooses a monitoring wavelength where the displacing ion is absorbing and the sample ion is not, hence satisfying a condi-

Table IV. Calibration Data for Nitrate, Sulfate, and Phosphate

injection	area of trough (arbitrary units)	area of trough/ (mequiv/L) of ion injected
5 × 10⁻³ M sodium nitrate	117.5	23.5
2.5 × 10⁻³ M sodium sulfate	111.0	22.2
1.67 × 10⁻³ M sodium orthophosphate	80.4	24.1

tion for obtaining maximum sensitivity. This we claim places IPC in a superior position among suppressorless ion chromatographic methods.

Indirect photometric chromatography is a promising new approach to a number of ion analysis problems.

ACKNOWLEDGMENT

The authors have appreciated the experimental assistance of D. F. Scheddel during the course of this work.

LITERATURE CITED

(1) Small, H.; Stevens, T. S.; Bauman, W. C. *Anal. Chem.* **1975**, 47, 1801–1809.
(2) Small, H. In "Applications of Ion Chromatography in Trace Analysis"; Lawrence, J. F., Ed.; Academic Press: New York, 1981.
(3) Stevens, T. S.; Davis, J. C.; Small, H. *Anal. Chem.* **1981**, *53*, 1488–1492.
(4) Gjerde, D. T.; Fritz, J. S. *J. Chromatogr.* **1979**, 176, 199–206.
(5) Gjerde, D. T.; Fritz, J. S.; Schmuckler, G. *J. Chromatogr.* **1979**, *186*, 509–519.
(6) Gjerde, D. T.; Schmuckler, G.; Fritz, J. S. *J. Chromatogr.* **1980**, *187*, 35–45.
(7) Fritz, J. S.; Gjerde, D. T.; Becker, R. M. *Anal. Chem.* **1980**, *52*, 1519–1522.
(8) Pohl, C. A.; Johnson, E. L. *J. Chromatogr. Sci.* **1980**, *18*, 442–452.
(9) Laurent, A.; Bourdon, R. *Ann. Pharm. Fr.* **1978**, *36* (9–10), 453–460.
(10) Ringbom, A. "Complexation in Analytical Chemistry"; Interscience: New York, 1963; pp 198–199.
(11) Kolthoff, I. M.; Sandell, E. B. "Textbook of Quantitative Inorganic Analysis"; Macmillan: New York, 1949; pp 662–668.
(12) Small, H.; Stevens, T. S. U.S. Patent 4 101 460, July 18, 1978.
(13) Stevens, T. S.; Small, H. *J. Liq. Chromatogr.* **1978**, *1* (2), 123–132.
(14) Small, H.; Solc, J. Proceedings of an International Conference on "The Theory and Practice of Ion Exchange"; Streat, M., Ed.; The Society of Chemical Industry: London, 1976; pp 32-1 to 32-10.

Received for review September 16, 1981. Accepted November 6, 1981. The methods described in this publication are the subject of pending patents which are licensed to Dionex Corporation for commercial use.

Reprinted from *Anal. Chem.* **1982**, *54*, 462–69.

Many of us who were working in open-tubular capillary chromatography in the early 1980s were startled to read the results in this paper. Jorgenson had demonstrated the very high efficiency of separation that is possible with capillary electrophoresis in 1981, but the technique was limited to the separation of charged species. Here, Terabe applies a chromatographic separation mechanism—the use of surfactants above their critical micellar concentration—to the newly established capillary electrokinetic format. His innovation made it possible to separate a wide variety of neutral species by CE.

Michael Sepaniak
University of Tennessee, Knoxville

Electrokinetic Separations with Micellar Solutions and Open-Tubular Capillaries

Shigeru Terabe,* **Koji Otsuka, Kunimichi Ichikawa, Akihiro Tsuchiya, and Teiichi Ando**
Department of Industrial Chemistry, Faculty of Engineering, Kyoto University, Sakyo-ku, Kyoto 606, Japan

Sir: The applicability of the solubilization by micelles to chromatography as a distribution process has briefly been discussed by Nakagawa (1). The point of his discussion may be summarized as follows: Micelles of an ionic surfactant can migrate in an aqueous solution by electrophoresis. When a solubilizate is added into a micellar solution, some portion of the solubilizate may be solubilized into the micelle. Thus the solubilization by micelles can constitute a mechanism of retention in chromatography. The distribution ratio of a solubilizate will increase with an increase of the micellar concentration but will be constant regardless of the concentration of the solubilizate. If a solubilizate is soluble in an aqueous solution to a certain extent, the distribution equilibrium is considered to be established very rapidly because, e.g., the stay time of benzene in the micelle of sodium dodecyl sulfate has been estimated to be less than 10^{-4} s (2).

The electrokinetic separation method described in this paper may be classified as a type of liquid–liquid partition chromatography requiring no solid support to hold the stationary liquid phase, although micelles are considered to be a pseudophase. It should be noted that this technique is distinctly different from the reversed-phase liquid chromatography with micellar mobile phase (3): In the latter, the separation is based on distribution processes among three phases, stationary bonded phase, micelle, and water, and micelles migrate with water as an aqueous pseudophase.

This paper presents the results of some preliminary studies on electrokinetic separation with micellar solution in open-tubular capillaries, in which use was made of the technique of free zone electrophoresis in open-tubular capillaries (4, 5).

EXPERIMENTAL SECTION

Apparatus. Electrokinetic separation was performed in microbore vitreous silica tubing, 650 or 900 mm long, 0.05 mm i.d. (Scientific Glass Engineering Inc.), with a Model HSR-24P regulated high-voltage dc power supply (Matsusada Precision Devices, Otsu, Japan) delivering +3 to +25 kV. Each end of the capillary tube was dipped in a small glass beaker containing a surfactant solution covered with a silicone-rubber stopper having two small-bore holes, one for a platinum electrode and the other for the capillary tube. The electric current was monitored between the negative electrode and the negative terminal of the power supply with an ammeter throughout the operation. Detection was carried out by on-column measurement of UV absorption through a slit of 0.05 mm × 0.75 mm, the long axis of which was placed parallel to the column axis at a position 150 mm from the negative end of the tube. The polymer coating of the vitreous silica tubing was partly burned out at the detection point of the tube to make an on-column UV cell. A Jasco UVIDEC-100-II spectrophotometric detector (Tokyo, Japan) was used with minor modification to obtain a higher amplifier gain and a shorter response time than the conventional one.

Reagent. Sodium dodecyl sulfate (SDS) of protein-research grade purchased from Nakarai Chemicals (Kyoto, Japan) was used as it was received. Water was purified with a Milli-Q system. Other reagents were of analytical-reagent grade and were used without further purification. Borate–phosphate buffer solution, pH 7.0, was prepared by mixing a 0.025 M sodium tetraborate solution and a 0.05 M sodium dihydrogen phosphate solution in an appropriate ratio to indicate pH 7.0. An SDS solution was prepared by dissolving 1 mmol of SDS in 20 mL of the borate–phosphate buffer solution followed by filtration of the solution through a membrane filter of 0.5-μm pore size.

Procedure. A capillary tube was filled with an SDS solution by use of a microsyringe and 1.5 mL each of the same

2855-8/94/0352$08.00/0 © 1994 American Chemical Society *Milestones in Analytical Chemistry*

SDS solution was introduced in two beakers placed at the same level. For the sample injection, the positive end of the tube was moved into a vessel containing a sample solution and the level of the sample solution was raised about 4 cm higher than that of the SDS solution to allow the sample solution to flow downward into the capillary tube. After 5 to 90 s, depending on the desired amount of injection, the end of the tube was returned to the beaker and a high voltage was applied.

RESULTS AND DISCUSSION

The chromatogram shown in Figure 1 illustrates the high resolution obtained by electrokinetic separation with micellar solution. The applied voltage between both ends of 900 mm tube was ca. 25 kV. The separation performed in the 750 mm portion from the injection end to the detector cell was recorded by the on-column detection technique. Fourteen phenol derivatives injected as a water solution were completely resolved within 19 min. The injected amount of each phenol was estimated to be 0.7–1 ng and the total injection volume about 12 nL. Theoretical plate numbers calculated from the chromatogram were 210 000 for phenol, 260 000 to 350 000 for cresols and chlorophenols, and 300 000 to 400 000 for xylenols and p-ethylphenol, corresponding to plate height equivalent to a theoretical plate of 1.9–3.6 μm. Lower plate numbers observed for the peaks at shorter retention times may be attributable to the adverse effect of the large sample volume.

It has been reported that the electroosmotic flow is much stronger than the electrophoretic migration of an ion in the case of electrophoresis in open-tubular glass capillaries (5). Similar results were observed in this study: The SDS solution as a whole was carried from the positive electrode to the negative one, and negatively charged micelles of SDS also migrated toward the negative electrode as opposed to the electrophoretic attraction. This means that every sample injected at the positive end of the tube can be detected at the negative side of the tube. When a cationic surfactant such as cetyltrimethylammonium bromide was employed instead of an anionic one, the situation was reversed and hence the inversion of polarities of electrodes was needed.

The volume flow by electroosmosis in a narrow cylindrical capillary increases linearly with the applied electric field and also with the current (6). When the electrokinetic radius κα is larger than 50, where κ is the reciprocal of the Debye length and α is the radius of the capillary tube, the velocity profile has been calculated to be flat in the range $0 \leq r \leq 0.9\alpha$ by Rice and Whitehead (6), where r is the point distance from the axis. The value κα is estimated much larger than 50 under the conditions employed in this study. Therefore, the plug-shape flow of electroosmosis is one of the reasons for the high efficiency attained in this study. The linear relationship between the electroosmotic migration velocity v_{EO} and the current was always recognized. However, the plot of the velocity v_{EO} vs. the total applied voltage showed a positive deviation from linearity at higher voltages, although the actual strength of the applied field in the tube was not measured.

The retention parameters in electrokinetic separation are different from those in the conventional elution chromatogra-

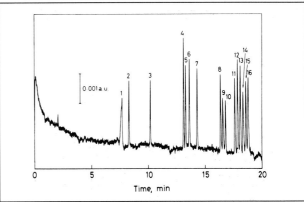

Figure 1. Electrokinetic separation of phenols with an SDS solution: (1) water, (2) acetylacetone, (3) phenol, (4) o-cresol, (5) m-cresol, (6) p-cresol, (7) o-chlorophenol, (8) m-chlorophenol, (9) p-chlorophenol, (10) 2,6-xylenol, (11) 2,3-xylenol, (12) 2,5-xylenol, (13) 3,4-xylenol, (14) 3,5-xylenol, (15) 2,4-xylenol, (16) p-ethylphenol; micellar solution, 1 mmol of SDS in 20 mL of borate–phosphate buffer, pH 7.0; current, 28 μA; detection wavelength, 270 nm; temperature, ca. 25 °C.

phy, because the retention time of any sample, if it is electrically neutral, should fall between the retention times of an insolubilized solute and a micelle itself in this method. Two assumptions are made for simplicity of the discussions below. One is that solute molecules are electrically neutral under the separation conditions. The other is that the electroosmotic velocity is larger than the electrophoretic velocity of a micelle and that their migrating directions are opposite.

A solute which is not solubilized by micelles at all should migrate with the same velocity as the electroosmotic flow v_{EO} and be eluted first at the retention time t_0. On the other hand, a solute which is completely solubilized with micelles should migrate with the same velocity as that of a micelle v_{MC} and be eluted last at the retention time t_{MC}. The velocity v_{MC} is the difference between v_{EO} and the electrophoretic velocity of a micelle v_{EP}, or $v_{MC} = v_{EO} - v_{EP}$. The retention time of an ordinary sample should depend on the capacity factor k', which is given by the ratio of the total moles of the solute in the micelle n_{MC} to those in the aqueous phase n_W, or $k' = n_{MC}/n_W$. The retention time t_R should appear in the range $t_0 \leq t_R \leq t_{MC}$. The R value, the fraction of the solute in the aqueous phase, is given by

$$R = \frac{v_S - v_{MC}}{v_{EO} - v_{MC}} \quad (1)$$

where v_S is the migration velocity of the solute. Now, the R value can be related to k' by (7)

$$R = \frac{1}{1 + k'} \quad (2)$$

Inserting the relationship, $v_{EO} = L/t_0$, $v_{MC} = L/t_{MC}$, and $v_S = L/t_R$, where L is the tube length from the injection end to the detector cell, into eq 1, followed by combination with eq 2 gives

$$k' = \frac{t_R - t_0}{t_0(1 - (t_R/t_{MC}))} \qquad (3)$$

The term $(1 - (t_R/t_{MC}))$ comes from the retention behavior characteristic of electrokinetic separations. When t_{MC} becomes infinite, eq 3 is equivalent to the well-known equation for conventional chromatography.

The chromatogram shown in Figure 2 clearly reflects the situation described above. Methanol was chosen as an insolubilized solute to measure t_0 and Sudan III to determine t_{MC}. The capacity factors of the solutes in Figure 2 are 0, 0.49, 1.28, 2.27, 3.08, and infinity in the order of elution. A capacity factor of infinity means the solute will not be eluted by traditional chromatography and it also means that the solute is totally associated with the micelle in this case. The linear decrease of k' with the increase of the current or v_{EO} was observed although slopes of the plots k' vs. v_{EO} were different among solutes. The reason for this dependence remains to be clarified but can probably be found in the increase of the solution temperature by Joule heating with increasing applied voltage and/or the possible change in the physical property of a micelle by the strong external electric field.

Electrokinetic separations with micellar solutions in open-tubular capillaries have been proved to be a high-resolution chromatographic method. It is limited to an analytical application because of a small sample size at present, but the zone electrophoretic technique with some kinds of stabilizing media widely utilized in the field of electrophoresis may be employed for preparative purposes in electrokinetic separations. The use of a surfactant solution in an aqueous organic solvent will expand the applicability of this method to water-insoluble compounds. Electrokinetic separations with micellar solutions would be useful for studying chemistry of micelles as well as for analytical purposes. Further extensive investigations are being continued to develop the possibilities of this technique.

ACKNOWLEDGMENT

We thank T. Nakagawa who has proposed the application of solubilization by micelles to chromatography and H. Jizomoto for their helpful suggestions and discussions.

LITERATURE CITED

(1) Nakagawa, T. *Newsl., Div. Colloid Surf. Chem., Chem. Soc. Jpn.* **1981**, *6*, No. 3, 1.

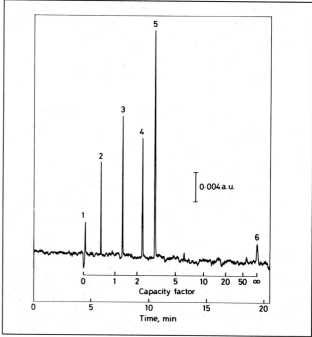

Figure 2. Chromatogram by electrokinetic separation indicating the total range of elution: (1) methanol, (2) phenol, (3) *p*-cresol, (4) 2,6-xylenol, (5) *p*-ethylphenol, (6) Sudan III; total tube length, 650 mm; tube length from the injection end to the detector cell, 500 mm; total applied voltage, ca. 20 kV; current, 33 μA; detection wavelength, 220 nm. Other conditions are the same as in Figure 1.

(2) Nakagawa, T.; Tori, K. *Kolloid Z. Z. Polym.* **1964**, *194*, 143–147.
(3) Armstrong, D. W.; Nome, F. *Anal. Chem.* **1981**, *53*, 1662–1666.
(4) Mikkers, F. E. P.; Everaerts, F. M.; Verheggen, Th. P. E. M. *J. Chromatogr.* **1979**, *169*, 11–20.
(5) Jorgenson, J. W.; Lukacs, K. D. *Anal. Chem.* **1981**, *53*, 1298–1302.
(6) Rice, C. L.; Whitehead, R. *J. Phys. Chem.* **1965**, *69*, 4017–4024.
(7) Karger, B. L.; Snyder, L. R.; Horvath, C. "An Introduction to Separation Science"; Wiley: New York, 1973; Chapter 5.

Received for review July 8, 1983. Accepted September 19, 1983.

Reprinted from *Anal. Chem.* **1984**, *56*, 111–13.

This article was the first in a series that demonstrated the capabilities of electrospray ionization for mass spectrometric characterization of proteins, nucleotides, and other large biopolymers. This "soft" ionization method extended the range of polymers that could be ionized intact to those with molecular masses of 55–100 kDa. In this paper, Fenn and his colleagues explored the advantages of ESI as an interface between HPLC and MS—a study that presaged its development as the interface of choice for this hyphenated technique. Fenn and his colleagues have subsequently led the development of mechanistic models for the ionization and desorption of molecules from solution to the gas phase.

Catherine Fenselau
University of Maryland Baltimore County

Electrospray Interface for Liquid Chromatographs and Mass Spectrometers

Craig M. Whitehouse, Robert N. Dreyer,[1] Masamichi Yamashita,[2] and John B. Fenn*
Department of Chemical Engineering, Yale University, New Haven, Connecticut 06520-2159

Electrospraying LC effluent into a dry bath gas creates a dispersion of charged droplets which rapidly evaporate. As the droplets grow smaller the increase in surface charge density and the decrease in radius of curvature result in electric fields strong enough to desorb solute ions. Part of the resulting dispersion of ions in bath gas passes through a small orifice or channel into an evacuated region to form a supersonic free jet. The core of this jet passes through a conical skimmer orifice and transports the ions to the inlet of a mass analyzer. The reported results suggest that from the standpoints of flexibility, convenience, sensitivity, cleanliness, and ease of maintenance, this ESPI source may comprise an effective and practical LC–MS interface.

The successful union of gas chromatography with mass spectrometry spawned a revolution in chemical analysis. An important impediment to the marriage was a basic incompatibility between vital features of each partner's mode of operation. The lifeblood of a gas chromatograph is the flow of carrier gas in which the species of analytical concern are usually present in only trace amounts. To the mass spectrometer, which is happy only under high vacuum, the prospective flood of carrier gas diluent was anathema. Clearly needed to overcome this impediment was a means of removing and discarding a large fraction of the carrier gas before the stream to be analyzed entered the mass spectrometer. One of the early and most successful devices for accomplish-

ing this removal was the so-called jet separator first described by Ryhage and still widely used in one form or another (1). This device derived from the research of E. W. Becker and his colleagues who were exploring the use of supersonic free jets expanding into vacuum for molecular beam sources as had been proposed by Kantrowitz and Grey (2). Surprisingly large species separation effects were observed when the free jet gas comprised a mixture of heavy and light molecules (3, 4). Originally attributed to pressure diffusion during the expansion, this separation was subsequently shown to stem from preferential inertial penetration of heavier species into the stagnation zone behind the bow shock wave on the sampling probe immersed in the supersonic flow (5, 6).

More recently the rapid and highly successful development of high-performance liquid chromatography (HPLC) has reached the stage where its practitioners have for some time also been contemplating the advantages of mass spectrometric detection. As was the case in GC–MS there is an impediment to the union of LC with MS that is similar but even more formidable. To be successful an LC–MS interface must not only divert a large fraction of the mobile phase from the mass spectrometer inlet but it must also make possible the transformation of nonvolatile and fragile species from solutes in a liquid to ions in a vacuum ready for mass analysis. To accomplish the latter step has been an exceedingly refractory problem to which until recently there have been no really satisfactory solutions even when the sample is static and there are no harsh constraints on the time available for preparation.

In the last few years there has emerged a new breed of ion sources that to a much greater extent than their predecessors seem at once to be effective, easy to use, and compatible with both partners of the LC–MS union. Similar in some sense to the field ionization (FI) and field desorption (FD) sources in-

[1] Present address: Department of Pharmacology, Yale University.
[2] Present address: Institute of Space and Astronautical Science, Tokyo 153, Japan.

troduced by Beckey and his collaborators, the newcomers apparently share a common mechanism: field ion desorption from liquids (FIDL) (7). Their membership includes the electrohydrodynamic (EHD) source of Evans and co-workers, the atmospheric pressure ion evaporation (APIE) source of Iribarne and Thomson, and the thermospray (TS) source of Vestal and his colleagues (8–18). Along with other techniques for the ionization of nonvolatile molecules, these FIDL sources have recently been reviewed by Vestal (19).

In this paper we report some experience and results with another variation on the FIDL theme, an electrospray ion (ESPI) source that has its roots in some experiments performed 15 years ago by Malcolm Dole and his collaborators. They attempted to produce beams of charged macromolecules by electrospraying a solution of polystyrene molecules into a bath gas to form a dispersion of macroions that was expanded through an orifice as a supersonic free jet into vacuum (20–22). Those authors believed that as the droplets evaporated, the increasing coulomb repulsive forces came to exceed the surface tension so that the droplet underwent fission. Successive fissions ultimately gave rise to droplets containing a single solute molecule that retained the droplet charge as the remaining solvent evaporated. In the case of smaller solute ions with which we have been working, it appears that the high field formed at the surface of an evaporating charged droplet desorbs solute ions into the bath gas in much the same way as in APIE and TS (23).

ESPI is operationally different from TS and APIE in the way it produces charged droplets. The latter two techniques disperse liquid into droplets by hydrodynamic forces. Statistical fluctuations in the distribution of solute anions and cations result in a net positive or negative charge on each droplet. In TS the sample liquid is passed through a small bore tube whose walls are hot enough to vaporize most of the solvent. The consequent expansion accelerates the flow and atomizes the remaining liquid so that a dispersion of approximately equal numbers of positively and negatively charged droplets in solvent vapor issues from the tube as a supersonic jet into the first stage of the vacuum system. In APIE the liquid sample is nebulized by a jet of air in the vicinity of a polarizing electrode at high voltage. The resulting field not only determines the sign of the droplet charge but also greatly enhances its magnitude. Whereas TS and APIE produce charging by atomization, ESPI produces atomization by charging. The sample liquid is injected into the bath gas through a metal hypodermic tube at a potential of several kilovolts relative to the surrounding chamber walls. Charge is thus deposited on the surface of the emerging liquid and produces coulomb repulsion forces sufficient to overcome surface tension so that the liquid is dispersed in a fine spray. In all three techniques evaporation of solvent from the droplets increases the surface charge density and decreases the radius of surface curvature. The resulting increase in electric field strength finally reaches levels high enough to desorb ions into the ambient gas. Also common to APIE, TS, and ESPI is the use of a supersonic free jet to transport the ions from relatively high pressure in the desorption region into the vacuum of the mass spectrometer chamber. In sum, it would appear that history has once again succumbed to repetition, if not bigamy. The same supersonic free jet that aided the marriage of MS and GC now promises to abet the union of MS with LC!

Experimental Section.

Figure 1 shows the essential features of our most recent version of an ESPI source. Sample solution at flow rates typically between 5 and 20 µL/min enter the electrospray chamber through a stainless steel hypodermic needle. Representative values of applied voltage are in parentheses after each of the following components: needle (ground), cylindrical electrode (−3500), metalized inlet and exit ends of the glass capillary that passes ion-bearing gas into the first stage of the vacuum system (−4500 and +40, respectively), the skimmer aperture between the first and second vacuum stages (−20), the ion lens in front of the quadrupole (−100). To produce negative ions, voltages of similar magnitude but opposite sign are applied. We have not yet operated this new apparatus in the negative ion mode but in an earlier apparatus we obtained equivalent results with both polarities (24).

A first glance the indicated potential differences of 4540 V between the inlet and exit ends of the glass capillary may seem startling. We have found that with carrier gas at a pressure of about 1 atm, the ion mobility is so low that the flow can lift the ions out of the potential well at the capillary inlet. Consequently, only a few elements inside the apparatus need be at high voltage. All the external parts are at ground potential and pose no hazard to an operator. Indeed, we have been able to "pump" ions through a potential difference of at least 15 kV. Thus we can readily provide the ion energies necessary for injection into a magnetic sector analyzer. This possibility is not so readily realized with some of the other sources. The capillary, with a bore of 0.2 × 60 mm, for given inlet conditions passes just about the same flux of both gas and ions as the thin plate orifice, 0.1 mm in diameter, that it replaced.

The high field at the needle tip charges the surface of the emerging liquid which, as mentioned above, becomes dispersed by coulomb forces into a fine spray of charged droplets. Driven by the electric field the droplets migrate to the inlet of the glass capillary through a countercurrent stream of bath gas (nitrogen) at a pressure that is slightly above 1 atm, typically 1000 torr, a temperature that we have varied from 50 °C

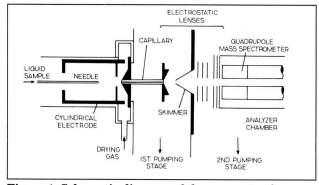

Figure 1. Schematic diagram of the apparatus for mass spectrometry with an electrospray ion source.

to 80 °C or more, and a flow rate typically in the range of 150 cm³/s. The performance of the source is not at all sensitive to these variables. The droplets rapidly evaporate and the solvent vapor along with any other uncharged material is swept away by the flow of bath gas. Desorbed ions arriving in the vicinity of the capillary inlet are entrained in dry bath gas and transported into the first vacuum chamber where they emerge in the supersonic jet of carrier gas leaving the capillary. Pressure in this first vacuum chamber is maintained at 5×10^{-4} or less by an oil diffusion pump with an effective speed of about 1000 L/s. A portion of the free jet flow passes through the 2-mm aperture of the skimmer into a second vacuum chamber containing a quadrupole mass spectrometer (VG Micromass 1212) whose output is monitored by a UV chart recorder. Pressure in the quadrupole chamber is maintained at about 10^{-6} torr.

When the field at the needle tip exceeds a critical value determined by temperature, pressure, and composition of bath gas, configuration of electrodes, and solution composition, electrical breakdown occurs and results in a sustained corona discharge. Typically, this discharge starts at voltages between 4 and 6 kV. Under these "high voltage" conditions the spectra are markedly different and are characterized by substantial attenuation of the parent solute peaks along with the appearance of other peaks corresponding to species found in discharges as a result of ion molecule reactions. All of the spectra shown in this report were obtained at voltages below the breakdown value. Thus, essentially all of the peaks correspond to solute ions present in the original solution, sometimes in aggregation with one or more un-ionized solvent or solute molecules. We have not yet studied them in detail, but it seems quite likely that the differences between spectra obtained at high and low voltage will contain information useful in identifying the structure of the unfragmented parent ions.

RESULTS AND DISCUSSION

The ESPI results in our previous papers (23, 24) were obtained with an apparatus in which the upper limit for the mass range of the quadrupole analyzer was between 300 and 400 daltons. They showed that, for a wide variety of species, the primary peaks in both positive and negative modes comprised solute ions in aggregation with one or more nonionic solvent or solute molecules. In the new apparatus described in the previous section the quadrupole analyzer can resolve masses up to 1500 daltons. Thus, one principal objective of the present study was to determine whether ESPI would work as effectively in the higher mass range that our new apparatus can probe. However, for a first test of the new system we tried some of the same species we had used in the earlier studies. Figure 2a shows the spectrum obtained with a solution of adenosine in a 50–50 methanol–water solvent at a concentration 0.1 μg/mL. The liquid flow rate was 6 μL/min during the 5 min it took to obtain this spectrum. The principal peak is at a mass of 268 daltons and corresponds to a protonated adenosine molecule. Also labeled is the peak at 300 daltons which we think corresponds to the same species solvated with one molecule of methanol. The fairly pronounced peak

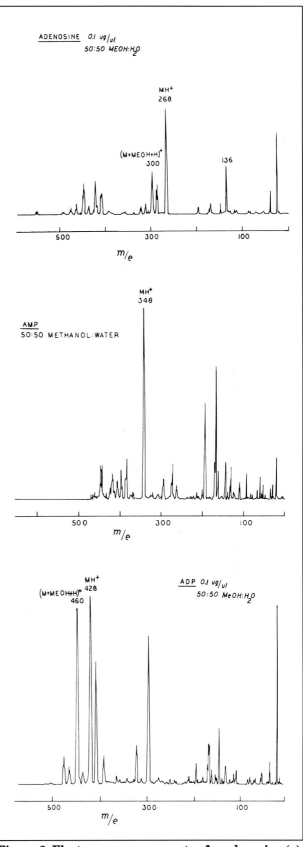

Figure 2. Electrospray mass spectra for adenosine (a), adenosine monophosphate (b), and adenosine diphosphate (c). All were present at concentrations of 0.1 μg/μL in 50–50 methanol–water. The liquid flow rate was 6 μL/min. The full mass scan took 5 min.

labeled 136 could be due to a doubly charged parent ion peak but is more likely due to a combination of the mass 133 and 134 fragments into which adenosine can readily split. We emphasize that all of the peak assignments in this report must be regarded as tentative. The quadrupole is still not working well, so that with these relatively dilute sample solutions the actual masses may be up to three units in error in the low mass range and ten or more at the high masses. We believe that all of the other peaks correspond to small cations, e.g., Na^+ or H^+, with one or more molecules of solvation, as was the case in the very similar spectrum obtained in the old system. Figure 2b,c shows spectra obtained respectively for the mono- and diphosphates of adenosine. The peak corresponding to the protonated parent molecule is identified in each case. It is interesting that a strong peak corresponding to the parent together with one methanol molecule of solvation appears in the AD case but not the AMP case. We have not tried to make definite assignments for the other peaks because of the uncertainty in mass mentioned above. However, the similarity between these results and those obtained in the original system, where we were able to make assignments with more confidence, persuade us that no appreciable fragmentation of the parent species occurs.

Figure 3 shows the spectrum obtained with a solution comprising 0.1 µg/µL of the peptide cyclosporin A, an antilymphocytic agent, in an 85–15 acetonitrile–water solution. The primary peak is the singly protonated parent molecule. The width of the peak at the base clearly reveals the inadequate resolution of our quadrupole, so we make no attempt to interpret the small shoulders on either side of the peak base. The smallest peak of the triplet near mass 600 is probably due to a doubly protonated parent molecule. Our confidence in this assignment stems from experiments with high concentrations at which the improved resolution makes it quite clear that the apparent mass difference in isotopic peaks is half that to be expected for singly charged ions. The two slightly larger peaks seem likely to stem from the presence respectively of one and two water molecules of solvation. Relative to the singly

charged ion the higher polarity of the doubly charged ion may enhance its affinity for water enough to compensate for the relatively hydrophobic exterior of the parent molecule.

Figure 4 shows the spectrum obtained with the antibiotic gramicidin S at a concentration of 0.01 µg/µL in 50–50 methanol–water. It is interesting that the predominant peak is due to the doubly charged ion, identified from isotopic mass intervals when experiments were carried out with higher concentrations of analyte as mentioned above. In this case, the presence of double charge does not give rise to the solvation that seemed to occur with cyclosporin A. It is appropriate to remark on sensitivity. The spectrum shown in Figure 4 required 5 min for the complete mass scan because at higher scan rates the mass markers do not appear in the chart output of the recorder that is our only present means of collecting data. The sample liquid flow rate was 6 µL/min so that 0.3 µg of analyte passed through the system during the scan. We estimate that the fraction of the total scan devoted to the primary peak was roughly 15/1500 or 0.01. Thus, the amount of analyte in the peak is about 3 ng when we use an oscilloscope to monitor the output signal, we can carry out a complete scan in 1 s and obtain relative peak heights exactly the same as those for the slow scan. We conclude that we would thus need only 1/300 of 3 ng, or 10 pg, of analyte to obtain a peak equivalent to that in Figure 4. The apparent signal/noise in the recorder spectrum is at least 200 and probably more. If we assume that S/N of 2 represents detectability, we conclude that we can detect as little as 100 fg of this material. This argument assumes that S/N is not significantly greater in the fast scan, a not unreasonable assumption for the dc mode in which we operate. Just because the peak that we discern con-

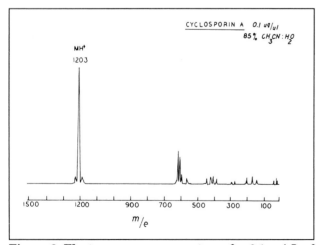

Figure 3. Electrospray mass spectrum for 0.1 µg/µL of cyclosporin A in an 85-15 acetonitrile-water solution. Liquid flow rate was 6 µL/min. The mass scan took 5 min.

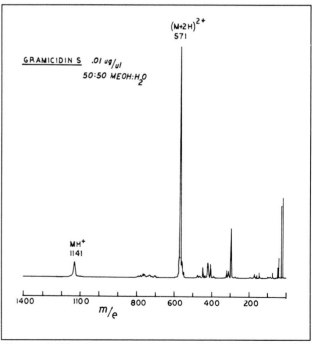

Figure 4. Electrospray mass spectrum for 0.01 µg/µL gramicidin S in 50-50 methanol-water. Note that the predominant peak is the doubly charged ion. Flow rate was 6 µL/min. Mass scan took 5 min.

tains only 100 fg of analyte does not mean that we can detect this amount of material in any arbitrary sample introduced at random into the ion source. Detectability in any unconstrained sense will always have a marked dependence on scan rate, resolution, concentration, and injection time. These factors will all be different from laboratory to laboratory and experiment to experiment. A perhaps more meaningful basis upon which to compare instrument sensitivities, especially in the case when sample is introduced continuously, albeit for short periods, rather than in batches, is the minimum steady state flow rate of analyte that will give rise to a discernible mass peak. On this basis the experiments that led to Figure 5 provide a useful perspective. For the lowest concentration of gramicidin S, 5×10^{-4} µg/mL, we could obtain a discernible peak with S/N of order 2 at a flow rate of 1.0 µL/min with a single mass scan at a rate of 20 amu/s. In other words we were detecting an analyte flow rate of 8 pg, or roughly 7 fmol, per second. If enough sample is available to permit multiple scans and signal averaging, the minimum detectable flow rate could be substantially less. These numbers are not ultimate in any sense. We are still on the steep part of the learning curve for this technique and have every reason to expect substantial improvements in sensitivity.

We have obtained entirely similar results with the peptides bleomycin and "Substance P" that have molecular weights respectively of about 1375 and 1348. The actual predominant peak with the sample of commercial bleomycin that we used (Blenoxane) was the complex with copper having a mass 1440. For both of these substances, as well as for gramicidin S, the doubly charged ion peaks predominate. This propensity toward multiple charging is provocative. If we can determine when, why, and how it occurs, we may be able to promote it and thus to extend substantially the effective mass range, not only of the ESPI source but of available mass analyzers as well.

It is important for quantitative analysis that the mass spectrometer signal have a near-linear dependence upon the concentration of analyte. In Figure 5 is an example that shows such a near-linear dependence over 4 orders of magnitude in the concentration of gramicidin S in methanol–water. Moreover, the data for gramicidin S in Figure 6 demonstrate that the signal does not depend strongly on the liquid flow rate. These features of ESPI behavior are reassuring because they mean that peak heights (areas) obtained with LC effluents will not be distorted by an ESPI interface.

Of course, the key concern in this forum is how well the ESPI source will stabilize the LC–MS union. Before definitive conclusions can be reached, we must actually couple a chromatograph with a mass spectrometer by means of an ESPI source. Thanks to Whatman, Ltd., we have in fact obtained a microbore column with a 1 mm i.d. and a length of 250 mm and, except for appropriate injection valves, are ready to make the acid test. Indeed, in some preliminary experiments, we have injected a sample of cyclosporin A into the column and passed the effluent into the source. We obtained a spectrum essentially identical with the one shown in Figure 3. Unfortunately, the injected sample was 20 µL, so large that we overloaded the column with the result that the "elution" time or peak width was 52 min for 2 µg of analyte! With an oscillating mass scan of 8 daltons centered at mass 1203, we were able to determine the shape of the leading and trailing edges of the peak. They were very abrupt and sharp, indicating minimal tailing. Thus, the ESPI source has virtually no dead volume.

It is clear that until we have acquired more experience with an actual LC–ESPI–MS combination we will be in no position to make any meaningful judgment on the extent to which ESPI can be considered a solution to the interface problem. However, our experience to date seems to provide grounds for optimism. We close with some observations on practical operational features of ESPI that seem pertinent.

1. The system is not prone to fouling. The countercurrent flow of bath gas sweeps away all the solvent vapor and other uncharged material. In effect, only ions are carried into the vacuum system by the free jet of bath gas. We have run pretty steadily for periods as long as a month without any decrease in performance. Even then, there was no appreciable fouling

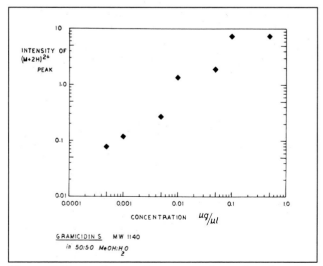

Figure 5. Dependence of mass spectrometer signal on analyte concentration for the case of gramicidin S in methanol–water solution.

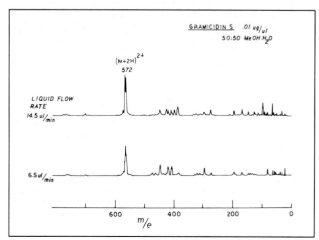

Figure 6. Independence of mass spectrometer signal on sample flow rate for the case of gramicidin S in methanol–water solution.

in the vacuum system. The major problem was a slight accumulation of salt deposit at the end of the hypodermic injection tube and on the side walls of the electrospray chamber. Both of these elements are in the high pressure section of the system and thus readily accessible for cleaning.

2. Thus far we have found no evidence to suggest that sensitivity depends upon the molecular weight of the analyte beyond what one would expect for decreases in quadrupole transmission and multiplier response for ions of increasing mass. In other words, the ionization efficiency seems to be relatively constant over the mass range. Of course, we have as yet tried relatively few materials, but our confidence in this conclusion is increasing.

3. The system seems inherently stable. Small fluctuations in liquid flow rate do not cause ripples in the signal output. Modest changes in temperature and flow rate of bath gas do not markedly affect the signal. Consequently, these variables do not need to be controlled all that carefully. The variable of most immediate and direct influence on performance is the voltage applied to the liquid injection needle, so that a reasonably well stabilized power supply is needed. Optimal voltage, bath gas flow rate, and temperature settings are identical for all compounds run thus far. Consequently, adjustments need not be made to optimize sensitivity from one compound to the next.

4. The ability to use viscous drag forces during bath gas flow through the capillary to "pump" ions to any voltage that would be required for injection into a mass magnetic sector mass analyzer is very important. It means that all the advantages of such instruments in the analysis of very heavy ions will be accessible to LC–MS analysis.

In sum, though our experience is thus far limited, results with ESPI along with those obtained by Vestal with TS and by Iribarne and Thomson with APIE give more reasons for hope than despair in the quest for a practical LC–MS interface.

ACKNOWLEDGMENT

We gratefully acknowledge the interest, support, and encouragement of VG Analytics who provided the quadrupole mass spectrometer and other reenforcement. We also record our gratitude for the cooperation and support that our colleagues at Yale have so generously provided. In particular, the yeoman efforts of Chin-Kai Meng and Ted Grabowski have been invaluable.

LITERATURE CITED

(1) Ryhage, R. *Anal. Chem.* **1964**, *51*, 359.
(2) Kantrowitz, A.; Grey, J. *Rev. Sci. Instrum.* **1951**, *22*, 328.
(3) Becker, E. W.; Bier, K.; Burghoff, H. Z. *Naturforsch., Teil A* **1955**, *10*, 565.
(4) Becker, E. W.; Bier, K.; Ehrfeld, W.; Schubert, K.; Schutte, R.; Seidel, D. In "Nuclear Energy Maturity"; Zaleski, P., Ed.; Pergamon: Oxford, 1975; p. 172.
(5) Reis, V. H.; Fenn, J. B. *J. Chem. Phys.* **1963**, *39*, 3240.
(6) Sherman, F. S. *Phys. Fluids* **1965**, *8*, 773.
(7) Beckey, H. D. "Principles of Field Ionization and Field Desorption Mass Spectrometry"; Pergamon: New York, 1977.
(8) Stimpson, B. P.; Simons, D. S.; Evans, C. A., Jr. *J. Phys. Chem.* **1978**, *82*, 660.
(9) Evans, C. A. Jr.; Hendricks, C. D. *Rev. Sci. Instrum.* **1972**, *43*, 1527.
(10) Simons, D. S.; Colby, B. N.; Evans, C. A., Jr. *Int. J. Mass Spectrom. Ion. Phys.* **1974**, *15*, 291.
(11) Stimpson, B. P.; Evans, C. A., Jr. *Biomed. Mass Spectrom.* **1978**, *5*, 52.
(12) Iribarne, J. B.; Thomson, B. A. *J. Chem. Phys.* **1976**, *64*, 2287.
(13) Thomson, B. A.; Iribarne, J. B. *J. Chem. Phys.* **1979**, *71*, 4451.
(14) Iribarne, J. B.; Dziedzic, P. J.; Thomson, B. A. *Int. J. Mass Spectrom. Ion. Phys.* **1983**, *50*, 331.
(15) Blakley, C. R.; Carmody, J. J.; Vestal, M. L. *J. Am. Chem. Soc.* **1980**, *102*, 5931.
(16) Blakley, C. R.; Carmody, J. J.; Vestal, M. L. *Clin. Chem. (Winston-Salem, N.C.)* **1980**, *26*, 1467.
(17) Blakley, C. R.; Carmody, J. J.; Vestal, M. L. *Anal. Chem.* **1980**, *52*, 1636.
(18) Blakley, C. R.; Vestal, M. L. *Anal. Chem.* **1983**, *55*, 750.
(19) Vestal, M. L. *Mass Spectrom. Rev.* **1983**, *2*, 447.
(20) Dole, M.; Mack, L. L.; Hines, R. L.; Mobley, R. C.; Ferguson, L. D.; Alice, M. B. *J. Chem. Phys.* **1968**, *49*, 2240.
(21) Mack, L. L.; Kralik, P.; Rheude, A.; Dole, M. *J. Chem. Phys.* **1970**, *52*, 4977.
(22) Dole, M.; Cox, H. L., Jr.; Gieniec, J. *Adv. Chem. Ser.* **1973**, No. *125*, 73.
(23) Yamashita, M.; Fenn, J. B. *J. Phys. Chem.* **1984**, *88*, 4451.
(24) Yamashita, M.; Fenn, J. B. *J. Phys. Chem.* **1984**, *88*, 4671.

Received for review October 8, 1984. Accepted December 6, 1984. This research has been sponsored in part by the National Science Foundation (Grant ENG-7910843), the U.S. Department of Energy (Grant ET-78-G-01-3246), and the National Cancer Institute (Grant C288-52/04). C.M.W. acknowledges with appreciation a fellowship award from the Exxon Corp. M.Y. acknowledges support from the Institute of Space and Aeronautical Sciences that made his participation possible.

Reprinted from *Anal. Chem.* **1985**, *57*, 675–79.

The continuous-flow sample probe developed by Caprioli, Fan, and Cottrell was a milestone on the road to practical soft ionization mass spectrometric techniques for analyzing polar biomolecules. Many industrial and research labs have incorporated cf-FAB as a standard method of sample introduction, either in combination with high-resolution chromatographic separations or as a convenience for rapid, automated sample introduction. Caprioli et al. demonstrated that cf-FAB offered higher sensitivity than conventional FAB, principally by lowering chemical background and offering a time-resolved signal.

Sanford P. Markey

National Institute of Mental Health, NIH

Continuous-Flow Sample Probe for Fast Atom Bombardment Mass Spectrometry

Richard M. Caprioli* and Terry Fan
The Analytical Chemistry Center and Department of Biochemistry and Molecular Biology, University of Texas Medical School, Houston, Texas 77030

John S. Cottrell
Kratos Analytical Instruments, Urmston, Manchester, United Kingdom

The design and performance of a sample probe that allows a continuous flow of solution to be introduced into a fast atom bombardment (FAB) ion source are described. Samples can be injected into a solvent flow that contains water/glycerol (8:2) and dilute buffers. Samples containing 13.5 ng of peptides injected in 0.5-μL portions show peaks in the total ion chromatogram emerging over 30 s, corresponding to a volume of 2.5 μL. Ion intensities recorded with varying sample amounts show a linear relationship from 0.7 to 200 ng. Quantitatively, calculations of peak areas from replicate injections show standard deviations of approximately ±10% of the mean. With regard to sensitivity, the peptide substance P (mol wt 1347) at 0.3 ng gave a signal-to-noise ratio of 5:1. Comparison of background chemical noise between the continuous-flow probe and the standard FAB probe (using an 80% glycerol matrix) showed a significant improvement in signal-to-chemical noise using the flow probe. High mass performance is demonstrated by showing the resolved molecular ion regions of injected samples of oxidized bovine insulin B chain (mol wt 3493) and intact bovine insulin (mol wt 5730). Conditions required for the stable operation of the probe are discussed.

Samples are usually introduced into the ionization chamber of a fast atom bombardment (FAB) mass spectrometer through the use of a direct insertion probe. The sample is first dissolved in glycerol or some other suitable viscous matrix, and then several microliters of the solution are placed on the probe tip. Exposure of this sample to a beam of energetic xenon atoms inside the mass spectrometer source causes surface layers of molecules to be sputtered and the resulting ions are subsequently analyzed (1). Although this type of sample introduction is simple and easy to use, it has several shortcomings. First, it does not easily lend itself to following dynamic processes especially where rapid changes in reactants or products are expected. Since each sample becomes an isolated analysis, the number of samples taken is a matter of speed and/or endurance both in the sampling process and also in the subsequent analyses. Second, comparisons of ion intensities from sample to sample are difficult and the results uncertain without the use of internal standards. Third, substantial amounts of glycerol (or other matrix material) are required, usually 80–95%, so that the liquid droplet can survive introduction into the vacuum system. This precludes direct sampling of reactions that must proceed in a substantially aqueous environment such as, for example, enzyme reactions.

Several investigators have reported work involving the continuous introduction of liquid samples into a FAB source of a mass spectrometer, although these were mostly aimed at on-line HPLC applications. One approach involves use of a moving belt onto which is deposited fractions of the HPLC eluant (2, 3). The belt is then continuously cycled into the source of the mass spectrometer where the sample spots are bombarded. More recently, Ito et al. (4) reported on the use of a capillary inlet device for the direct connection of a microbore HPLC column to a FAB ionization source. These workers demonstrated the separation and analysis of bile acids using a mobile phase of glycerol/acetonitrile/water (10:27:63) at a flow rate of 0.5 µL/min. A stainless-steel mesh frit was used at the terminus of the capillary in the ion source to disperse the mobile phase and concentrate the solute and glycerol. Other approaches to continuously flowing aqueous solutions into mass spectrometers, although not involving FAB ionization sources, include thermospray (5, 6) and direct liquid injection (DLI) methods (7, 8).

We have constructed and tested a sample introduction probe for use with mass spectrometers equipped with fast atom bombardment sources that permits a continuous flow of solution to be brought directly into the source. The solution can be essentially aqueous, containing as little as 10% glycerol, and is brought into the source at a flow rate of about 5 µL/min. Samples may be injected into this flow of solvent or included in the solvent, depending on the application. Dilute buffers, acids, and salts can be used in the aqueous solvent. Total ion current chromatograms obtained with injected samples in the nanogram range produce sharp peaks with little tailing and no significant memory effects. Further, since the spectrometer can operate at full accelerating potential, the probe permits the full sensitivity of the instrument to be utilized, an important factor when making measurements at high mass.

EXPERIMENTAL SECTION

Design and Operation of the Continuous-Flow FAB Probe. The probe, shown in Figure 1, consists of a hollow shaft that is capped with an angled tip through which a 0.3-mm hole is drilled. A 0.075-mm-i.d. (0.26 mm o.d.) × 1-m fused silica capillary (SGE, Victoria, Australia) is passed through the shaft and allowed to protrude no more than 0.2 mm beyond the tip. A vacuum seal consisting of a vespel ferrule and compression nut seals the probe shaft to the capillary at the base of the probe. A pump-out port is provided in the shaft to evacuate the hollow shaft during the rough pumping of the probe in the insertion lock. The capillary is connected to a Rheodyne injection valve: a Model 7520 for injection of submicroliter amounts or a Model 7010 for injection of 5 µL and larger amounts. The continuous-flow solvent is provided by a suitable pump (an Isco Model µLC 500 syringe pump or Waters Model 590 pump) normally operated at a flow of 5 µL/min.

A Kratos MS50RF high-resolution mass spectrometer was used to provide the data presented in this work. The instrument was not specially modified except that the FAB source

block was fitted with a heater and thermocouple sensor that are controlled by existing circuits on the operator's console. The source block is generally maintained at 40 °C. Since the source fitted with the continuous flow probe normally operates at a relatively high pressure (about 5×10^{-4} torr), extreme care was taken to electrically ground the instrument properly and protect delicate components from the effects of source arcs. The mass spectrometer was operated at full sensitivity (8-keV accelerating potential). The FAB gun is the Ion Tech Model B11NF saddle field source and is operated at 7 keV at a current of 50 µA with xenon gas. Data were taken by use of the Kratos DS90 software with the DG S/280 computer system. Mass spectra were generally obtained by using a wide-range multichannel analyzer program with a magnet scan and the spectrum calibrated with CsI. Voltage scans (varying both the accelerating and electric sector voltages) were used to provide narrow mass scan ranges for the protonated molecular ion regions of bovine insulin and oxidized bovine insulin B chain.

Reagents. All chemical supplies were obtained from Sigma Chemical Co. (St. Louis, MO) except where noted. Substance P, des-(Gln)[6]-substance P, angiotensin II, and angiotensin III were obtained from Vega Chemical Co. (Phoenix, AZ) and maltoheptaose from Boehringer Mannheim (Indianapolis, IN).

RESULTS

Injection of Samples into a Solvent Flow. In order to test the performance of the probe as a microsampling device, 0.5-µL samples of several peptides were injected into the device. The continuously flowing carrier solution consisted of a mixture of water and glycerol (8:2), containing 0.3% trifluoracetic acid, which was fed into the mass spectrometer at a rate of 5 µL/min. The acid produces maximal sensitivity for a number of compounds, especially peptides. The samples were dissolved in this same solution. Figure 2A shows the total ion chromatogram resulting from the single injection of 13.5 ng of the peptide substance P (mol wt 1347). The peak emerged over a time of approximately 30 s (seven scans) and in a total volume of about 2.5 µL. Figure 2B shows the FAB mass spectrum at the peak maximum in the chromatogram, and Figure 2C shows the mass spectrum just after the peak (around 4 min in the chromatogram). Only residual traces of the peptide $(M + H)^+$ ion can be seen in Figure 2C; the $(M + H)^+$ ion intensity is less than 0.5% of that of the mass spectrum taken from the chromatogram peak maximum. Thus, at these concentrations, the memory effect of the probe is quite small. Similar results were obtained for a variety of polypeptides injected in this manner, including ribonuclease S peptide, oxidized insulin B chain, angiotensin II and III, and several smaller peptides.

Injections of much larger sample volumes such as 5–10 µL can be used where it is desirable to maintain signal currents for much longer periods of time. For example, injection of 5 µL of substance P, containing 27 ng/µL of solution, gave a broad flat-topped tailing peak in the total ion chromatogram and produced intense $(M + H)^+$ ion currents for about 5 min.

Linearity of Response. In order to determine the linearity of response of the probe, solutions containing 0.7, 1.4, 6.8,

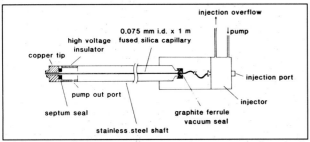

Figure 1. Schematic diagram of the continuous-flow probe for fast atom bombardment mass spectrometers.

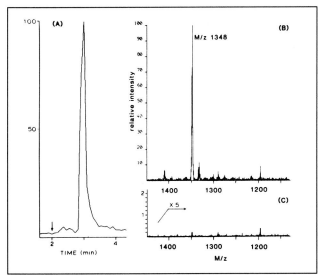

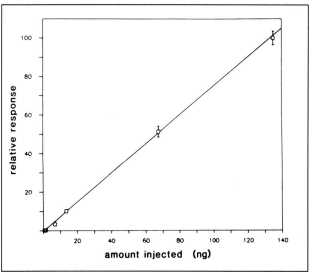

Figure 2. (A) Single injection of 13.5 ng of substance P (mol wt 1347) in 0.5 μL of water/glycerol (8:2) containing 0.3% TFA. The arrow denotes the point of injection of the sample. (B) FAB mass spectrum at chromatogram peak maximum. (C) Background after the sample peak at about 4 min in the chromatogram. The resolution of the instrument was 1500 (the compressed mass plot in the figure gives an apparent lower resolution).

Figure 3. Linearity of response for the range of 0.7–135 ng of the peptide substance P (mol wt 1347) injected on the probe. The response was calculated from the area of the selected ion chromatograms of the (M + H)+ ions from triplicate injections. The average deviation from the mean is shown as error bars; for response values below 20 ng, this deviation was too small to be shown in the figure.

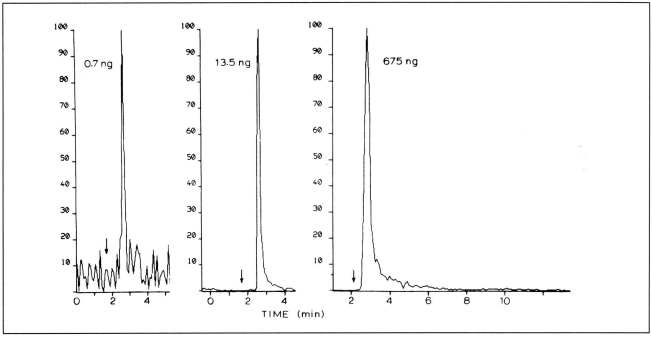

Figure 4. Selected ion chromatograms of the molecular ion region (m/z 1345–1351) from injections of various amounts of substance P. The arrows denote the points of injection of the sample.

13.5, 67.5, 135, and 675 ng of the peptide substance P/0.5 μL of water/glycerol (8:2, containing 0.3% TFA) were injected into the probe. By use of measurements of the areas of selected ion chromatograms of the (M + H)+ ions, the response was found to be linear in the range of approximately 0.7–140 ng as shown in Figure 3. Above approximately 200 ng, a positive deviation from linearity was observed as a result of overloading the probe sur-

face. In this case, several minutes were required for the solvent to sweep the samples outside the area on the tip from which ions can be focused into the mass spectrometer. Figure 4 compares the selected ion chromatogram peak profiles of the (M + H)+ ion region for injections of 0.7, 13.5, and 657 ng of the peptide.

Reproducibility. Figure 5 shows the selected ion chromatograms of the (M + H)+ ions for injections of 10 replicate

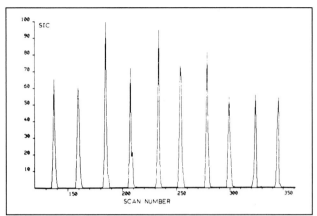

Figure 5. Reproducibility of multiple injections of 13.5 ng of substance P on the continuous-flow probe. The selected ion chromatogram of the _m/z_ 1345–1351 region is shown.

samples of the peptide substance P, each containing 13.5 ng (injected in 0.5 μL). Measurements of peak heights show a standard deviation of ±42% of the mean with a range of 86–156% of the mean. However, _areas_ calculated for each of these peaks showed a standard deviation of ±10.5% from the mean with a range of 86–125% of the mean. The variation in peaks widths indicates that flow conditions probably change somewhat throughout the run and give rise to slightly different peak profiles. Some of the variability can be attributed to irreproducibility involved in the injection of submicroliter volumes of sample.

Sensitivity. The sensitivity of the mass spectrometer for samples introduced via the probe was tested with several peptides. The peptide substance P gave a _S/N_ level of approximately 5:1 at 340 pg (252 fmol) injected onto the probe. For most samples, improvement in signal to background chemical noise is seen using the continuous-flow probe relative to the standard FAB probe where samples are dissolved in 80–90% glycerol. Figure 6 shows the mass spectrum of a mixture of peptides taken with both the standard probe and the continuous-flow probe. The compounds employed were angiotensin II, angiotensin III, and des-(Gln)⁶-substance P. Approximately 20 nmol of each peptide was used in each sample. In the case of the flow probe, the sample was injected in 0.5 μL of solution (which becomes diluted to about 2.5 μL on exit from the capillary), while for the standard probe the sample was dispersed in a total volume of 2 μL (containing 80% glycerol and 0.3% TFA on the tip). Generally, the two spectra are quite similar, although there appears to be considerably greater background chemical noise in the case of the standard probe. A portion of each spectrum is detailed in Figure 7, showing quite clearly the increase in the signal-to-chemical noise obtained from the flow probe (Figure 7B) relative to the standard probe (Figure 7A). Of great significance in this regard is the fact that the background "matrix" ions, producing a peak at every mass, are greatly reduced in intensity while those related to sample emerging from the capillary remain intense. For example, the ion at _m/z_ 1282 in Figure 7 corresponds to (M + ⁶³Cu)⁺ for des-(Gln)⁶-substance P, and the ion at _m/z_

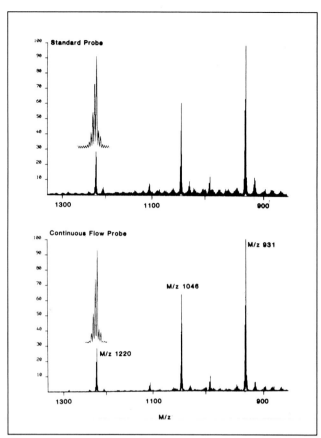

Figure 6. Comparison of mass spectra taken from (top) the standard FAB probe and (bottom) the continuous-flow probe from a sample containing angiotensin III (mol wt 930), angiotensin II (mol wt 1045), and des-(Gln)⁶-substance P (mol wt 1219). Experimental details are given in the text. The peak profiles above the _m/z_ 1220 ion show the actual resolution of the scan (the compressed plot gives an apparent lower resolution). Approximately 20 nmol of each peptide was used for each sample.

1234 corresponds to (M + ⁶³Cu₃)⁺ for angiotensin II. Similarly, in the lower mass region of Figure 7, the ion at _m/z_ 1108 corresponds to (M + ⁶³Cu)⁺ for angiotensin II. Results of calculations of isotope patterns from elemental compositions fit well for these ions. Of course, these ions are of relatively low intensity (less than 5%) in comparison to the (M + H)⁺ ions for the peptides and arise from bombardment of the copper tip and sample, as described earlier (9). These data presented in the figure show an advantage of the continuous-flow probe in reducing background chemical noise, thereby improving the observation of sample-related ions. This reduced background effect is presumed to be due to the ejection of ions from the capillary tip and surrounding surface from a solution containing much less glycerol than that obtained from the standard probe. The effect is seen with other compounds as well: for example, the FAB mass spectrum obtained from an injection of a sample of the oligosaccharide maltoheptaose (mol wt 1152) gave similar results.

High-Mass Performance. The mass spectra of bovine insulin (mol wt 5730) and oxidized bovine insulin B chain (mol

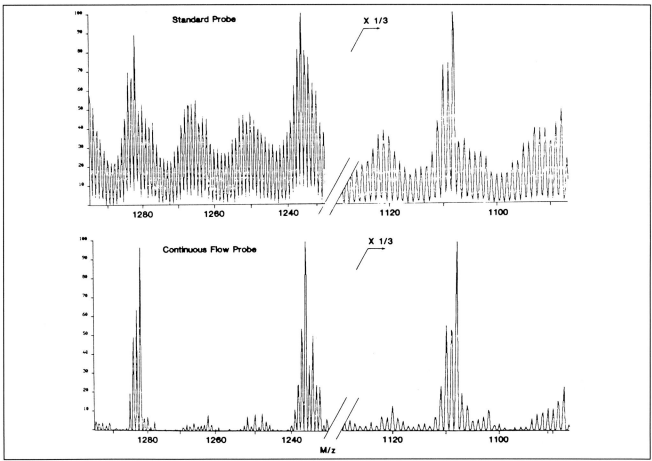

Figure 7. Expanded background regions from the mass spectra shown in Figure 6.

wt 3493) were obtained to determine the performance of the flow probe with high-mass samples. Figure 8A shows the narrow-range voltage scan for the (M + H)$^+$ ion of oxidized insulin B chain. The sample solution containing 1.5 µg/µL of peptide (8:1:1 aqueous 50 mM Tris buffer, pH 7.2/glycerol/thioglycerol) in 0.5 µL was injected onto the flow probe using a continuous flow of an aqueous solution containing 20% glycerol and 0.2% TFA. The spectrum shown in the figure is the sum of 60 scans with the instrument resolution set at about 3000.

Figure 8B shows the (M + H)$^+$ ion region for bovine insulin from 0.5 µL injected onto the flow probe of a sample containing 2 µg/µL of protein in 8:1:1 water/glycerol/thioglycerol with 0.2% TFA. The continuous-flow solution had this same composition. The spectrum is the sum of 20 scans taken using a voltage scan with an instrument resolution of about 5000. This spectrum shows a signal-to-chemical noise response of approximately 10:1. A spectrum taken under the same conditions using the standard FAB probe where the sample is dissolved in a thioglycerol matrix gave a signal-to-chemical noise response of approximately 4:1.

Operation and Stability. The stable operation of the probe can be defined simply as the condition where a constant ion current is obtained from ions sputtered from the tip under the flow condition described. Such stable operation can be achieved in a mass spectrometer when the rate of evaporation of the solvent from the probe surface and the pumping speed of the source of the mass spectrometer are in balance. The latter is generally not easily controlled, and so it is more practical to control the rate of evaporation. In the system described, this may be altered within limits by changing either the flow rate of liquid to the tip, the total vapor pressure of the solution, or the amount of heat applied to the tip. Too slow a flow and insufficient heat will lead to freezing within the capillary tip and will give rise to unsteady ion currents, as described by Arpino and co-workers (10) from their calculations of flow dynamics through capillaries. On the other hand, too high a flow rate will lead to spattering of the liquid surface, again giving unsteady ion currents. The pressure in the source is quite high, approximately 5×10^{-4} torr for the system described here. Nevertheless, under stable operating conditions, this pressure varies surprisingly little. With respect to the use of salts and buffers, relatively dilute buffers (e.g., 5 mM) seem not to present a problem. On the other hand, use of concentrated nonvolatile buffers can eventually lead to high-voltage breakdown if they are used for extended periods of time. When the conditions for stability are achieved, the probe tip contains a wet viscous surface but no definable droplet. Usually, this condition requires 15–30 min to develop on initial insertion of the probe.

It is speculated that when the solution emerges from the capillary tip after the probe is stabilized, ions are produced and analyzed from this highly aqueous area of the tip. This zone is pushed outward by oncoming solution and, as the zone

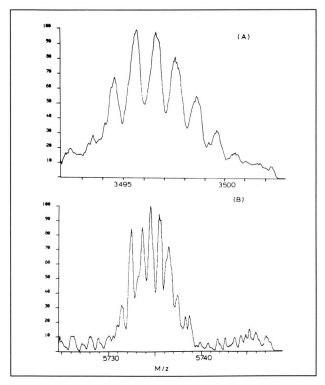

Figure 8. Mass spectra of the molecular ion regions of high molecular weight samples injected onto the continuous-flow probe: (A) 1.5 µg/µL of oxidized bovine insulin B chain (mol wt 3493), the injected volume was 0.5 µL in water/glycerol/thioglycerol (8:1:1), the spectrum represents 60 signal-averaged voltage scans at a resolution of 3000; (B) bovine insulin, the injected volume was 0.5 µL of a solution containing 2 µg/µL in water/glycerol/thioglycerol (8:1:1), the spectrum represents 20 signal-averaged voltage scans at a resolution of 5000.

migrates toward the edge, water and glycerol are evaporated until there remains only a very viscous matrix at the edge. Fortunately, ions seem not to be collected from this outer region. The buildup of viscous material with time is remarkably small; the probe is often operated 3–4 h without being removed from the instrument with as many as 50–60 injections of samples containing nanomole amounts of compounds during this time without significant deterioration of performance.

DISCUSSION

The continuous-flow FAB probe can be used effectively for several types of applications. It provides an easy and fast method for the analysis of compounds in aqueous solution. For dilute solutions of sample, 0.5-µL injections may be made every 2 min without showing a significant memory effect. Small volumes of solutions containing only water as solvent may also be used when the continuous-flow buffer contains some glycerol. Also, solutions may be sampled from outside the mass spectrometer on a continuous basis using a suitable pump.

A combination of several factors is responsible for the performance of the instrument equipped with this continuous-flow FAB probe. These include the maintenance of a relatively high solvent flow rate, the unobstructed flow of the sample

into and out of the region on the probe tip from which ions are collected, and the dynamic balance of the input solvent flow and its evaporation. Ito et al. (4) emphasize in their report that the function of the frit in their device is to vaporize the volatile portion of the mobile phase, leaving the solute and glycerol matrix on the frit surface to be bombarded by the atom beam. In addition, they report that flow rates of about 0.5 µL/min are required in order to maintain the production of stable ion currents. We have found that, with the continuous-flow FAB probe described in the current paper, conditions or physical restrictions that impede solvent flow tend to result in greater memory effects and accumulation of glycerol on the probe tip, which increases chemical background noise and, generally, give less satisfactory performance.

One of the major disadvantages of the use of the continuous-flow probe involves the high pressures under which the ion source must operate. For magnetic instruments that operate at high accelerating voltages, this may lead to high-voltage breakdown especially when the probe is unstable. Also for this reason, high salt concentrations are to be avoided in samples or solvent. Since the stable operation of the probe is a dynamic one, any change in one of the major parameters can lead to unstable operation, such as drastically changing flow rates as the result of clogging of the capillary with undissolved material or precipitation or crystallization of a sample in the capillary.

The sharp response of the mass spectrometer equipped with the probe with samples contained in relatively small volumes suggests that it may be of utility for other applications such as combined HPLC/mass spectrometry. Although this was demonstrated by Ito et al. (4), the higher flow rates used in the present work may provide much greater general utility for such an interface. Although flow rates of about 5 µL/min can be used with microbore HPLC, the device could also be of use for large-bore applications where stream splitting is acceptable.

LITERATURE CITED

(1) Barber, M; Bordoli, R. S.; Sedgwick, R. D.; Tyler, A. N. *J. Chem. Soc., Chem. Commun.* **1981**, 325–327.
(2) Stroh, J. G.; Cook, J. C.; Milberg, R. M.; Brayton, L.; Kihara, T.; Huang, Z.; Rhinehart, K. L. *Anal. Chem.* **1985**, *57*, 985–991.
(3) Dobberstein, P.; Karte, E.; Meyerhoff, G.; Pesch, R. *Int. J. Mass Spectrom. Ion Phys.* **1983**, *46*, 185–188.
(4) Ito, Y.; Takeuchi, T.; Ishi, D.; Goto, M. *J. Chromatogr.* **1985**, *346*, 161–166.
(5) Blakley, C. R.; Carmody, J. J.; Vestal, M. L. *Anal. Chem.* **1980**, *52*, 1636–1641.
(6) Pilosov, D.; Kim, H. Y.; Dykes, D. F.; Vestal, M. L. *Anal. Chem.* **1984**, *56*, 1236–1239.
(7) Arpino, P. J.; Bounine, V. P.; Dedieu, M.; Guiochon, G. *J. Chromatogr.* **1983**, *271*, 43–48.
(8) Covey, T.; Henion, J. D. *Anal. Chem.* **1983**, *55*, 2275–2279.
(9) Martin, S. A.; Costello, C. E.; Biemann, K. *Anal. Chem.* **1982** *54*, 2362–2368.
(10) Arpino, P. J.; Krien, P.; Vajta, S.; Devant, G. *J. Chromatogr.* **1981**, *203*, 117–130.

Received for review March 24, 1986. Accepted July 25, 1986. Support for this work by NSF Grant PCM-8404230 and NIH Grant RR-01720 is gratefully acknowledged.

Reprinted from *Anal. Chem.* **1986**, *58*, 2949–54.

The preparation and preservation of uncontaminated electrode surfaces are elusive goals that are as old as the science of electrochemistry. Many treatments for glassy carbon electrodes had already been published; however, the in situ laser activation procedure introduced by McCreery represented a novel and appealing innovation. This article drew renewed attention to how crucial electrode surface properties can be in determining the rate of electron transfer by demonstrating large improvements with the new technique.

Fred Anson
California Institute of Technology

In Situ Laser Activation of Glassy Carbon Electrodes

Melanie Poon and Richard L. McCreery*
Department of Chemistry, The Ohio State University, Columbus, Ohio 43210

Laser pulses of short duration (10 ns) and high intensity (20 MW cm^{-2}) can increase the rate of heterogeneous electron transfer at a glassy carbon electrode by 1–3 orders of magnitude. The laser pulse may be delivered in situ, directly in the solution of interest, repeatedly if desired. The heterogeneous electron transfer rate constant, $k°$, for the ferri-/ferrocyanide redox system increases from 0.004 to 0.20 cm s^{-1} with laser activation, resulting in the highest $k°$ yet observed for this system on glassy carbon. Laser activation results in minor morphological changes to the surface, as observed by scanning electron microscopy, mainly removal of an apparent layer of carbon microparticles. The technique holds promise as a means to repeatedly activate glassy carbon electrodes in situ, thus circumventing the need for renewal or reactivation by polishing or other ex situ treatments.

The wide use of the dropping mercury electrode (DME) stems from its renewable surface, a reproducible electrochemical response for each drop, and measurements that are devoid of any electrode history effects. Solid electrodes have been studied extensively because they provide a wider potential range than mercury, have better mechanical properties, and can act in a catalytic role for reactions of importance to energy conversion, electrosynthesis, and electroanalysis. However, it has long been recognized that solid electrode behavior, unlike that of the DME, is highly dependent on history and that performance may be drastically altered by pretreatment procedures or processes occurring in the solution of interest (1–7). These alterations take the form of changes in heterogeneous electron transfer rate, increases in capacitance or surface Faradaic reactions, or in severe cases, total deactivation of the electrode. They often result in unstable and irreproducible analytical performance or complete destruction of electrocatalytic behavior.

Glassy carbon (GC) has been studied extensively in recent years because of its wide potential range, chemical inertness, relatively low cost, and low porosity. Mechanical polishing (8–13) and chemical (14, 15), electrochemical (16–26), thermal (27, 28), and rf plasma (29, 30) pretreatments have been used on carbon electrodes preceding an electrochemical experiment. The treatments were used to produce a surface as free as possible from contamination and to activate the electrode toward electron transfer. Hu, Karweik, and Kuwana (13) have compiled a variety of treatments for glassy carbon and noted that the heterogeneous electron transfer rate constant for the ferri-/ferrocyanide redox system can vary from 1.5×10^{-4} to 0.14 cm s^{-1} depending on pretreatment procedure. These workers reported the highest rate constant yet observed for this system on GC, 0.14 cm s^{-1}, and noted that this value is comparable to that for a platinum electrode.

While they can have large beneficial effects on electrode performance, none of the pretreatment procedures mentioned so far can be carried out quickly or repeatedly in the solution of interest, in a fashion analogous to the DME. In most cases, the electrode behavior degrades with time due to adsorption of impurities from the solution or chemical changes to the electrode surface, and the solid electrode becomes unsuitable for quantitative measurements. We recently reported a new, in situ method for cleaning and activating platinum and glassy carbon electrode surfaces using a short laser pulse delivered to the electrode directly in solution (31). The method is rapid and repeatable and was able to remove surface films and accelerate electron transfer by several orders of magnitude.

The present work was undertaken to provide a more quantitative and detailed understanding of in situ laser activation of glassy carbon electrodes. The overall goal of the work is a rapid in situ technique for producing GC surfaces with reproducibly high electron transfer rates.

EXPERIMENTAL SECTION

Electrochemical measurements were performed with a Bioanalytical Systems CV-1B potentiostat, a BAS Ag/AgCl (3 M NaCl) reference electrode, and Pt wire auxiliary electrode. For cyclic voltammetry above 0.2 V s^{-1}, a custom potentiostat driven by a function generator was used, with data recorded by a Tektronix 7854 digital oscilloscope. Heterogeneous electron transfer rate constants were determined from the anodic/cathodic peak separation using the method of Nicholson (32) with $\alpha = 0.5$. Apparent electrode capacitance was measured by the method introduced by Fagan, Hu, and Kuwana (28), where chronocoulograms are measured for each of a series of potential steps ranging in height from 50 to 150 mV. A plot of charge vs. potential step size yields the apparent capacitance.

The electrochemical cell shown in Figure 1 was constructed of Teflon, with a Pyrex window to allow laser light to impinge on the electrode after passage through the solution. The reference and auxiliary electrodes were placed without special attention to geometry, except at scan rates above 0.2 V s^{-1}, where a Luggin capillary was used between the reference and working electrodes. Unless noted otherwise, the working electrode consisted of a Tokai GC-20 or GC-20s disk pressed against a Teflon washer, which was exposed to the solution. The electrode area was defined by the hole in the washer, and equaled 0.012 cm^2 as measured by chronoamperometry. The entire exposed area was illuminated by the laser. In several experiments, a commercial (Bioanalytical Systems) GC electrode consisting of GC-30s carbon press fitted in Kel-F was employed. Most of the results reported here were obtained with the Tokai disk electrode design, but only minor differences were observed with the commercial electrode after activation.

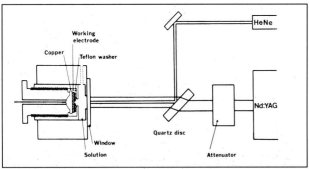

Figure 1. Experimental apparatus and optical arrangement. The He–Ne and Nd:YAG laser beams were colinear, and the Nd:YAG power was adjusted with the variable attenuator. The working electrode was defined by the Teflon washer pressed against the working electrode disk. Reference and auxiliary electrodes are not shown; the cell body was made of Teflon.

Two different polishing procedures were employed. The procedure referred to as "conventional" consisted of polishing with 600-grit SiC paper followed by 1.0-, 0.3-, and 0.05-μm alumina in a slurry with Nanopure water on a Texmet (Buehler) polishing cloth. A more rigorous procedure was that of Hu, Karweik, and Kuwana (13), involving alumina on a glass plate with no cloth. In both procedures, the final electrode was thoroughly rinsed and sonicated. As noted below, the polishing procedure had no observable effect on the electrode performance after laser activation.

The laser was a Quantel 580-10 Nd:YAG laser capable of producing a 10-ns, 300-mJ pulse at 1064 nm. Experience with a different laser (Quanta-Ray DCR-2) indicated that beam quality is very important for this experiment, and erratic results were obtained if significant power density variations occurred across the electrode area. The laser beam was 6 mm in diameter, but only the center 1 mm reached the electrode. The power density across the 1-mm irradiated electrode area varied by approximately ±20%, as measured by a linear photodiode array after attenuation. A conventional thermal power meter (Scientech Model 38-0101) was used to measure the laser power passing through the Teflon washer, which defined the working electrode. The power density was calculated by assuming the laser output was a square pulse of 10 ns duration. Power densities quoted here have an absolute accuracy of about ±20%, with some variation being unavoidable due to beam shape changes with laser tuning and alignment from experiment to experiment. The laser wavelength was 1064 nm for all experiments, and the power level was adjusted with a Newport Research Co. 935-5 beam attenuator. The optics were arranged as shown in Figure 1, such that the Nd:YAG beam was collinear with a low-power He–Ne beam, which was used to aim the activating pulse onto the electrode.

Despite reports on electrode filming with the ferri-/ferrocyanide redox system (33), it was used as a test system for comparing electron transfer rate constants because of the large volume of data available in the literature. Ascorbic acid, phenol, potassium ferrocyanide, o-chlorophenol, catechol, hydroquinone, Na$_2$SO$_4$, H$_2$SO$_4$, HClO$_4$, and KCl were reagent grade and used as received. Dopamine hydrochloride, NADH (98% grade III), 3,4-dihydroxyphenylacetic acid (DOPAC), and 3,4-dihydroxybenzylamine (DHBA) were obtained from Sigma Chemical Co. (St. Louis, MO). Unless noted otherwise, all solutions were prepared fresh daily using either NANOpure II water (Sybron Barnstead, Boston, MA) or water that had been triply distilled from dilute alkaline permanganate. Nitrogen purified by a Pall activated carbon filter (Pall Trinity Micro Corp., Cortland, NY) and an oxygen trap was used to degas solutions.

A Tencor Alpha Step profilometer based on vertical deflection of a 5-μm-diameter needle pulled across the surface was used to measure surface roughness. The vertical resolution was 100Å, but the tip size resulted in integration of the measurement over a 5-μm region. Scanning electron microscopy was carried out with an ISI SX-30 microscope operated at 5 kV with a practical resolution of 100Å.

RESULTS

The effect of laser activation on the cyclic voltammogram of the ferri-/ferrocyanide, $Fe(CN)_6^{3-/4-}$, redox couple is shown in Figure 2. After the dashed curve was taken, three 10-ns, 22 MW cm^{-2} laser pulses were delivered in rapid succession to the GC electrode; then the solid curve was obtained about 30 s after the last laser pulse. There is an obvious improvement in heterogeneous rate constant, resulting in a significant decrease in ΔE_p. Measurement of the heterogeneous electron transfer rate constant, $k°$, before and after activation indicated an increase from 0.004 ± 0.002 (N = 7) to 0.15 ± 0.01 (N = 7) cm s^{-1}, or a factor of 37. Quantitative effects of laser activation on $k°$ for $Fe(CN)_6^{3-/4-}$ are shown in Figure 3 as a function of laser power density. Little or no change in $k°$ is observed at powers below 9 MW cm^{-2}. Over the range from 12 to 24 MW cm^{-2}, $k°$ increases from 0.002 to a plateau at 0.20 ± 0.05 cm s^{-1}, which remains flat at least up to 60 MW cm^{-2}. As shown in Figure 4, $k°$ decreases slowly with time after the laser pulse when the electrode is allowed to stand in Nanopure electrolyte, with a faster decay rate observed in less pure double-distilled water. Figure 5 demonstrates that the activation process could be repeated by delivering single laser pulses in situ after deactivation with standing in double-distilled water.

The effect of laser pulses on the voltammetry of ascorbic acid on GC is shown in Figure 6. The conventional polishing procedure yields a voltammetric oxidation peak at 0.6 V vs. Ag/AgCl with considerable variation from run to run. Laser activation at 20 MW cm^{-2} results in partial activation, with a peak shift to 0.450 V. An increase in power density to 28 MW cm^{-2} results in full activation to a peak potential of 0.295 V, and further increases in power did not yield further peak potential shifts. The present results differ from those in our previous communication on laser activation (31) because the entire electrode is irradiated rather than small spots, yielding

a fully activated surface and a voltammogram comparable to that obtained from careful polishing or vacuum heat treatment. As estimated previously (31), the 300-mV shift in the ascorbic acid peak potential corresponds to an increase in $k°$ of greater than a factor of 10^3. The effect of activation on the voltammetry of a mixture of dopamine and ascorbic acid is shown in Figure 7, with complete resolution of the overlapped peaks being observed after the laser pulse. After activation, both ascorbic acid and dopamine oxidize at nearly their thermodynamic potentials, whereas before activation, both systems exhibit slow electron transfer kinetics. The lifetime of activation for the ascorbic acid and dopamine systems is comparable to that observed for the $k°$ of $Fe(CN)_6^{3-/4-}$ noted in Figure 4. It was possible to repeatedly activate an electrode

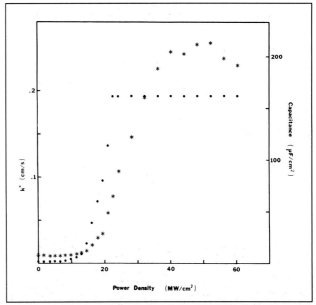

Figure 3. Observed $k°$ (points) for $Fe(CN)_6^{3-/4-}$ (1.0 mM in 1.0 M KCl) and capacitance (asterisks) as functions of laser power density.

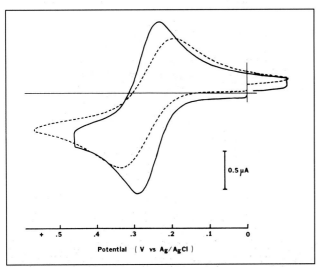

Figure 2. Effect of three 22 MW cm^{-2}, 10-ns laser pulses on the voltammogram of 1.0 mM ferrocyanide in 1 M KCl, scan rate = 0.1 V s^{-1}. Dashed curve is after conventional polishing with a polishing cloth. Solid curve is after laser activation.

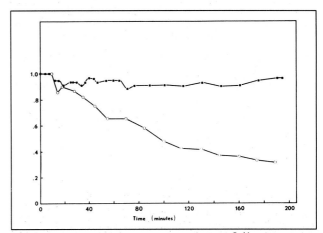

Figure 4. Time course of $k°$ for $Fe(CN)_6^{3-/4-}$ after laser activation. Values are normalized to their magnitudes immediately after the pulse; conditions are same as in Figure 3. Squares are $k°$ in Nanopure water; points are $k°$ in double-distilled water.

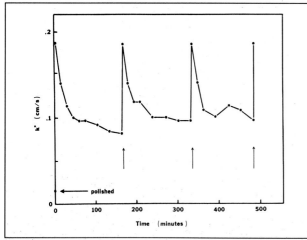

Figure 5. $k°$ vs. time for $Fe(CN)_6^{3-/4-}$ in 1.0 M KCl, in double-distilled water. Conventional polishing led to the first point, at 0.01 cm s^{-1}; then an increase to 0.19 cm s^{-1} occurred after three laser pulses at 22 MW cm^{-2}. Single laser pulses were applied at 180, 340, and 490 min.

Table I. Effect of Laser Activation on the Voltammetry of Several Redox Systems[a]

system	before[b]			after[c]		
	$E_{p,a}$	$E_{p,c}$	ΔE_p	$E_{p,a}$	$E_{p,c}$	ΔE_p
ascorbic acid	0.250			0.000		
NADH	0.480			0.335		
dopamine	0.260	0.135	0.125	0.200	0.168	0.032
O$_2$ 1st wave		−0.725			−0.450	
2nd wave		−1.625			−1.475	
DOPAC	0.255	0.095	0.160	0.177	0.150	0.027
DHBA	0.355	0.120	0.235	0.230	0.200	0.030
catechol	0.290	0.132	0.158	0.206	0.181	0.025
ferrocyanide	0.325	0.225	0.100	0.300	0.243	0.057
hydroquinone	0.230	0.000	0.230	0.113	0.065	0.048

[a] 1 mM in 1 M KCl, 0.1 M phosphate, pH 7.0, scan rate = 0.1 V s^{-1}. [b] Conventional polishing with polishing cloth. [c] After three 20 MW cm^{-2} laser pulses in situ.

toward ascorbic acid oxidation over a period of 3 days, without removal or polishing of the electrode. Table I presents the results of voltammetry on a variety of compounds before and after laser activation. In all cases, the voltammetry is improved with laser activation, and in some cases the improvement is dramatic.

Activation was observed for either the $Fe(CN)_6^{3-/4-}$ or ascorbic acid redox system if the laser pulse occurred in air before the solution was added. Activation also occurred if the laser pulse was applied in blank electrolyte, which was then replaced with the ascorbic acid solution. These observations eliminate the possibility of any photolytic reactions of solution species contributing to the observed results, although such processes would be unexpected with 1064-nm light. It also in-

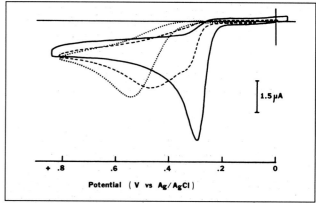

Figure 6. Effect of in situ laser irradiation on the voltammetry of 1.0 mM ascorbic acid in 0.1 M H$_2$SO$_4$. Dotted curve was obtained on conventionally polished GC, dashed curve after three 20 MW cm^{-2} pulses, solid curve after three 28 MW cm^{-2}.

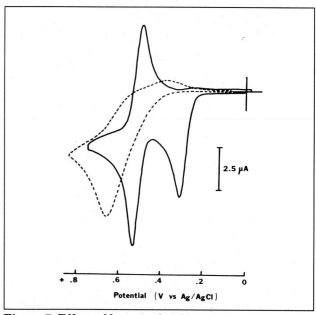

Figure 7. Effect of laser activation on the voltammetry of a mixture of 1.0 mM dopamine and 1.0 mM ascorbic acid in 0.1 M H$_2$SO$_4$. Dashed curve is after conventional polishing; solid line is after three 25 MW cm^{-2} laser pulses.

dicates that liquid water is not essential for the activation process. In some cases, the electrode lost some of its activity during solution transfer or other handling, so in all experiments reported here, the laser pulse occurred in situ directly in the solution of interest.

The effects of laser activation on the morphology and elemental composition of the GC surface were assessed with optical and scanning electron microscopy (SEM), scanning Auger spectrometry (SAM), and a profilometer. Profilometer traces for an electrode polished with 1.0-μm alumina are reproducible and easily distinguished from those of an electrode polished with 0.05-μm alumina, as shown in Figure 8. There is no apparent difference between a nonirradiated, polished surface and an adjacent area that had received three 40 MW

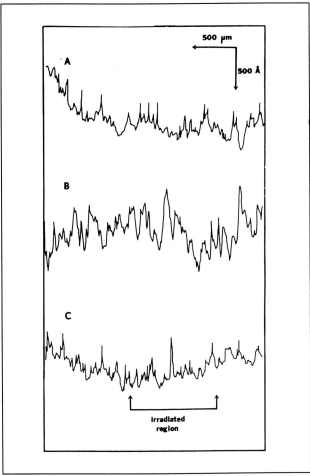

Figure 8. Profilometer traces of GC surfaces following polishing or laser activation. Upper trace was taken after polishing with 0.05-μm alumina, middle trace after 1.0-μm alumina. Lower trace was taken on an electrode polished with 0.05-μm alumina, then irradiated by three 40 MW cm^{-2} laser pulses in the indicated region.

cm^{-2} laser pulses. These results indicate that a laser power density which is sufficient to activate the surface does not cause ablation or other gross changes in the GC surface on the scale of a few hundred angstroms. Scanning electron micrographs of polished and irradiated surfaces are shown in Figure 9. Micrograph A is an ordinary polished surface, showing the small pits common to glassy carbon, but no polishing scratches. The lack of apparent scratches has been attributed to a surface layer of carbon microparticles generated during polishing (34). After a 22 MW cm^{-2} laser pulse (Figure 9B), surface scratches appear, but no other changes are observable. At 40 MW cm^{-2}, surface scratches are more apparent and randomly oriented rings appear, presumably caused by diffraction from dust on the cell window.

SAM analysis of the GC surface was carried out after transferring the electrode through air from the cell to the SAM vacuum chamber. After laser irradiation (24 MW cm^{-2}) in 1 M KCl, the oxygen/carbon ratio was slightly lower for irradiated vs. nonirradiated regions of the same GC surface. For six electrodes the O/C ratio in the irradiated region was 62% (±17%) of its value in the nonirradiated region (average

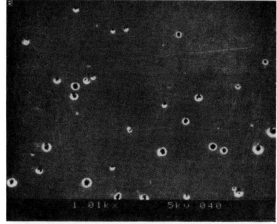

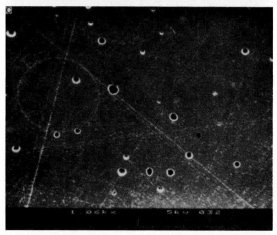

Figure 9. Scanning electron micrographs of polished and laser-activated GC surfaces: (A, top) conventionally polished surface, (B, middle) after activation with three 20 MW cm^{-2} laser pulses in solution, (C, bottom) after three 40 MW cm^{-2} pulses.

O/C ratio decreased from 0.07 to 0.04 upon irradiation). The approximately 40% decrease in the O/C ratio implies removal of oxygen functional groups even when the irradiation occurs in water. No incorporation of potassium or chlorine was observed with the SAM at these power levels. Laser irradiation of a GC electrode in a high-vacuum environment yielded a decrease in the O/C ratio of 43% with a single pulse, as measured by ESCA. Further pulses and exposure to the ultrahigh-vacuum environment lowered the surface oxygen to an immeasurably low level, implying that the laser is able to remove layers from the GC surface.

Laser activation at 20 MW cm^{-2} caused the voltammetric background current to increase by approximately a factor of 2. No new couples were observed with cyclic voltammetry, such as those attributed to surface quinone species (11, 12, 35–37). The apparent capacitance measured by chronocoulometry is shown in Figure 3. A plot of capacitive charge vs. potential step amplitude was linear from −50 to 100 mV, with a slope proportional to the apparent capacitance. For an electrode polished conventionally (with a polishing cloth), the slope of such a plot corresponds to 10 ± 2.7 μF cm^{-2}, and this value increases to 23 ± 6.2 μF cm^{-2} after laser activation at 21 MW cm^{-2}. Using the same method for capacitance measurement, Kuwana and co-workers obtained 70 μF cm^{-2} for a freshly polished (no polishing cloth) activated electrode and a value of 10 μF cm^{-2} for a vacuum heat treated GC electrode (28). Note that irradiation with a power density of 21 MW cm^{-2} produces an increase of a factor of 37 in $k°$ for $Fe(CN)_6^{3-/4-}$, but only a factor of 2.3 in apparent capacitance.

DISCUSSION

The rate constant of 0.20 cm s^{-1} observed for $Fe(CN)_6^{3-/4-}$ on laser-activated GC is comparable to the best values obtained with other pretreatment procedures (0.12–0.14 cm s^{-1}) (13) and is in the range of the highest values observed for $Fe(CN)_6^{3-/4-}$ on a platinum electrode (0.20–0.24 cm s^{-1}) (5, 38). The plateau in the $k°$ vs. power density curve (Figure 3) implies that the surface condition no longer affects the rate constant and the observed rate is the upper limit for $Fe(CN)_6^{3-/4-}$ on either platinum or glassy carbon electrodes. The excellent performance of the laser-treated electrodes was obtained with a 10-ns pulse applied in situ, repeatedly if desired, and therefore circumvents the requirement for removal and often extensive polishing. Laser activation provides many of the features of the DME, particularly in situ renewability and reproducibility, but also retains the wide potential range of GC compared to mercury. In addition, solid electrode surfaces can exhibit electrocatalytic behavior provided they can be properly cleaned. The in situ laser activation technique circumvents the problem of slow electrode deactivation by adsorbed solution species.

The improvement in the voltammograms of ascorbic acid with laser activation is also comparable to that observed for other pretreatment procedures (27, 28), with the added feature of rapid in situ repeatability. The decrease in peak potentials for the oxidation of DOPAC, DHBA, ascorbic acid, and $Fe(CN)_6^{3-/4-}$ with 10-ns in situ laser pulses is comparable to or better than a heat treatment procedure involving several hours at reduced pressure (27). The ability to activate a variety of electron transfer reactions to exhibit near-Nernstian behavior may be of significant analytical value, particularly since a deactivated analytical sensor may be reactivated quickly without removal from the cell.

The extent of laser activation was largely independent of the polishing procedure. In many cases we purposely used a brief conventional polishing protocol employing a polishing cloth that is known to produce a low activity surface. Upon laser activation, this surface yielded $k°$ values comparable to the rigorous polishing procedure of Hu, Karweik, and Kuwana (13). Laser activation produced comparable $k°$ values, whether or not the rigorous polishing procedure was used, and was equally effective on commercial electrodes with Kel-F bodies. We chose to use a polishing cloth for most experiments presented here because the capacitance of the activated electrode was lower and the effects of laser activation were more obvious. Regardless of which preparation procedure we used, an active surface with $k°$ for $Fe(CN)_6^{3-/4-} > 0.15$ cm s^{-1} could be produced repeatedly in situ.

The profilometer and SEM results indicate that no gross change has occurred to the GC surface at laser powers sufficient to activate the surface. The only change observable at 20 MW cm^{-2} or below was the apparent removal of the surface microparticle film proposed by Kazee, Weisshaar, and Kuwana (34). It is clear that not only should powers above 20 MW cm^{-2} be avoided to minimize increases in capacitance and background current but also that 20 MW cm^{-2} is sufficient to activate the reactions studied here. The decrease in the surface O/C ratio revealed by Auger spectrometry implies that desorption or ablation of at least some surface layers occurs, but on a scale too small to be observed by SEM or the profilometer.

The increase in capacitance caused by laser activation could be caused by increases in microscopic surface area or by an increase in the capacitance per unit of microscopic area or both. The capacitance values observed here are intermediate between those for vacuum heat treatment (28) and highly polished surfaces (13), and the laser pulse can increase the capacitance by up to a factor of 24 at high power densities. Since there is no rigorous method to measure the microscopic surface area on carbon, it cannot be stated unequivocally that increases in capacitance imply proportional increases in microscopic area. However, it can be concluded rigorously that the electron transfer rate enhancements resulting from laser activation are not caused solely by increases in microscopic surface area. As noted earlier, an increase in capacitance by a factor of 2.3 is accompanied by an increase in $k°$ for $Fe(CN)_6^{3-/4-}$ by a factor of 37. Carefully polished surfaces (13) with more than 3 times higher capacitance (70 μF cm^{-2} vs. 23 μF cm^{-2}) have slightly lower rate constants than laser-activated GC. Furthermore, the apparent $k°$ for ascorbic acid increased by a much larger factor than the capacitance upon laser activation. The optimum power density range of 18–24 MW cm^{-2} was chosen as a compromise between capacitance and rate enhancement. At this power density, the $k°$ was not at its maximum, but the power density was not high enough to cause large increases in capacitance.

The areas of laser desorption mass spectrometry (39) and laser processing of materials have stimulated a large body of research on laser effects on surfaces (40, 41). However, there are few studies of laser effects carried out in solution, and those are limited to examinations of corrosion (42) and accelerated electroplating (43). Consequently, it is difficult to estimate the temperature achieved during a laser pulse or to predict what processes will occur on the surface.

Although other processes cannot yet be ruled out, it is possible to explain all of the observations made here, as well as many others that have been reported, by a simple desorption mechanism. As proposed by others (13, 28), it is possible that the active sites on the GC surface are deactivated by the adsorption of trace materials encountered during preparation or during exposure to the solution of interest. In many preparation procedures, an active electrode is not produced because sufficient adsorbates are present in the polishing materials. If an active surface is achieved, whether by heat treatment, proper polishing, or laser activation, it is deactivated on a time scale of tens of minutes (or more quickly) by adsorption of pollutants from the solution. The laser activation process may be merely an effective means to remove such impurities, probably by a desorption process. Redox mediation and proton transfer have been proposed to explain accelerated electron transfer on carbon surfaces (13, 21, 30), and these mechanisms may be important for certain reactions. However, they do not appear to be involved in the accelerated oxidation of ascorbic acid or $Fe(CN)_6^{3-/4-}$ observed with laser activation or vacuum heat treatment. The changes in capacitance or possible changes in microscopic area with preparation procedure appear to be secondary to the desorption process, since they are not quantitatively correlated with electrode activity. Although the observed rate constant will depend upon the microscopic surface area, it is possible to observe high $k°$ values (≥ 0.14 cm s^{-1}) for $Fe(CN)_6^{3-/4-}$ on electrodes with a wide range of apparent capacitance ($10-70$ μF cm^{-2}) and presumably wide range of microscopic surface area.

ACKNOWLEDGMENT

We thank Daniel Fagan for carrying out the ESCA measurements. We also thank T. Kuwana, I.-F. Hu, R. Mark Wightman, and Royce Engstrom for useful discussions and for providing preprints of their papers on carbon activation.

Registry No. C, 7440-44-0; $Fe(CN)_6^{3-}$, 13408-62-3; $Fe(CN)_6^{4-}$, 13408-63-4; NaDH, 58-68-4; O_2, 7782-44-7; DOPAC, 102-32-9; DHBA, 37491-68-2; ascorbic acid, 50-81-7; catechol, 120-80-9; hydroquinone, 123-31-9; dopamine, 51-61-6.

LITERATURE CITED

(1) Adams, R. N. *Electrochemistry at Solid Electrodes*; Marcel Dekker: New York, 1969.
(2) Bishop, E.; Hitchcock, P. H. *Analyst (London)* **1973**, *98*, 475.
(3) Kinoshita, K. In *Modern Aspects of Electrochemistry*; Bockris, J. O'M., Conway, B. E., White, R. E., Eds.; Plenum: New York, 1982; p 557, and references therein.
(4) Amatore, C.; Saveant, J. M.; Tessier, D. *J. Electroanal. Chem.* **1983**, *146*, 37.
(5) Goldstein, E. L.; Van de Mark, M. R. *Electrochim. Acta* **1982**, *27*, 1079.
(6) Gilman, S. *Lectroanal.* **1967**, *2*, 111.
(7) Conway, B. E.; et al. *Anal. Chem.* **1973**, *45*, 1331.
(8) Rusling, J. F. *Anal. Chem.* **1984**, *56*, 578.
(9) Kamau, G. N.; Willis, W. S.; Rusling, J. F. *Anal. Chem.* **1985**, *57*, 545.
(10) Thornton, D. C.; Corby, K. T.; Spendel, V. A.; Jordan, J.; Robbat, A.; Rutstrom, D. J.; Gross, M.; Ritzler, G. *Anal. Chem.* **1985**, *57*, 150.
(11) Laser, D.; Ariel, M. *J. Electroanal. Chem.* **1974**, *52*, 291.
(12) Gunsingham, H.; Fleet, B. *Analyst (London)* **1982**, *107*, 896.
(13) Hu, I. F.; Karweik, D. H.; Kuwana, T. *J. Electroanal. Chem.* **1985**, *188*, 59.
(14) Plock, C. E. *J. Electroanal. Chem.* **1969**, *22*, 185.
(15) Taylor, R. J.; Humffray, A. A. *J. Electroanal. Chem.* **1973**, *42*, 347.
(16) Engstrom, R. C. *Anal. Chem.* **1982**, *54*, 2310.
(17) Engstrom, R. C.; Strasser, V. A. *Anal. Chem.* **1984**, *56*, 136.
(18) Blaedel, W. J.; Jenkins, R. A. *Anal. Chem.* **1974**, *46*, 1952.
(19) Moiroux, J.; Elving, P. J. *Anal. Chem.* **1978**, *50*, 1056.
(20) Wightman, R. M.; Paik, E. C.; Borman, S.; Dayton, M. A. *Anal. Chem.* **1978**, *50*, 1410.
(21) Cabaniss, G. E.; Diamantis, A. A.; Murphy, W. R., Jr.; Linton, R. W.; Meyer, T. J. *J. Am. Chem. Soc.* **1985**, *107*, 1845.
(22) Wang, J.; Hutchins, L. D. *Anal. Chim. Acta* **1985**, *167*, 325.
(23) Wang, J. *Anal. Chem.* **1981**, *53*, 2280.
(24) Gonon, F. G.; Fombarlet, C. M.; Buda, M. J.; Pujol, J. F. *Anal. Chem.* **1981**, *53*, 1386.
(25) Rice, M. E.; Galus, Z.; Adams, R. N. *J. Electroanal. Chem.* **1983**, *143*, 89.
(26) Falat, L.; Cheng, H. Y. *J. Electroanal. Chem.* **1983**, *157*, 393.
(27) Stutts, K. J.; Kovach, P. M.; Kuhr, W. G.; Wightman, R. M. *Anal. Chem.* **1983**, *55*, 1632.
(28) Fagan, D. T.; Hu, I. F.; Kuwana, T. *Anal. Chem.* **1985**, *57*, 2759.
(29) Miller, C. W.; Karweik, D. H.; Kuwana, T. *Anal. Chem.* **1981**, *53*, 2319.
(30) Evans, J.; Kuwana, T. *Anal. Chem.* **1979**, *51*, 358.
(31) Hershenhart, E.; McCreery, R. L.; Knight, R. D. *Anal. Chem.* **1984**, *56*, 2257.
(32) Nicholson, R. S. *Anal. Chem.* **1965**, *37*, 1351.
(33) Kawaik, J.; Jedral, T.; Galus, Z. *J. Electroanal. Chem.* **1983**, *145*, 163.
(34) Kazee, B.; Weisshaar, D. E.; Kuwana, T. *Anal. Chem.* **1985**, *57*, 2736.
(35) Murray, R. W. *Electroanal. Chem.* **1984**, 13.
(36) Panzer, R. E.; Elving, P. J. *Electrochim. Acta* **1975**, *20*, 635.
(37) Blurton, K. F. *Electrochim. Acta* **1973**, *18*, 869.
(38) Daum, P. H.; Enke, C. G. *Anal. Chem.* **1969**, *41*, 653.
(39) Harden, D. J.; Fan, T. P.; Blakley, C. R.; Vestal, M. L. *Anal. Chem.* **1984**, *56*, 2.
(40) Duley, W. W. *Laser Processing and Analysis of Materials*; Plenum: New York, 1983; p 89.
(41) Ready, J. F. *Effects of High Power Laser Radiation*; Academic Press: New York, 1971.
(42) Ulrich, R. K.; Alkire, R. C. *J. Electrochem. Soc.* **1981**, *128*, 1169.
(43) Kuiken, H. K.; Mikkers, F. E. P.; Wierenger, P. E. *J. Electrochem. Soc.* **1983**, *130*, 554.

Received for review March 26, 1986. Accepted July 1, 1986. This work was funded by the OSU Materials Research Laboratory and by the NSF Division of Chemical Analysis.

Reprinted from *Anal. Chem.* **1986**, *58*, 2745-50.

This article represents one of the first studies to examine the impact of microenvironment on the optical transduction properties of a fiber-optic chemical sensor. Hieftje and co-workers present a novel class of fiber-optic sensors for iodide based on the dynamic fluorescence quenching of an immobilized indicator dye. The quality of the analytical performance is correlated with the different physical and chemical environments associated with three unique dye immobilization schemes.

Mark Arnold
University of Iowa, Iowa City

Characterization and Comparison of Three Fiber-Optic Sensors for Iodide Determination Based on Dynamic Fluorescense Quenching of Rhodamine 6G

Wayde A. Wyatt,[1] Frank V. Bright,[2] and Gary M. Hieftje*
Department of Chemistry, Indiana University, Bloomington, Indiana 47405

Three different support media are evaluated as substrates for iodide ion sensing optrodes. The optrodes are based on the dynamic quenching of adsorbed rhodamine 6G by iodide ion. Results indicate that the local environment of the rhodamine 6G molecules dictates strongly the sensitivity of a given optrode system. For example, the new sensor's sensitivity clearly parallels the fluorophore's accessibility by iodide ion.

As the field of fiber-optic sensors grows, interest in the development of indirect optrodes is increasing. The ability to detect chemical species whose spectroscopic properties cannot be directly probed (e.g., metal and halide ions) in a remote, continuous manner is highly desirable. To date, several indirect optrodes have been developed for metal ions (1–3), but progress in the halide ion sensor area has not been as extensive. Several industrial processes, from plating baths to bromine-producing process streams, would benefit greatly from the implementation of halide-ion optrodes. In these applications, the ability to monitor halide-ion concentrations continuously could provide quantitative information that would result in tighter control and increased productivity.

Because the spectroscopic properties of iodide ion cannot be directly probed with wavelengths compatible with optical fibers, indirect approaches must be adopted. One such method uses chemical conversion of iodide to iodine which is then

(1) Current address: 14607-4 Grenadine Dr., Tampa, FL 33613.
(2) Current address: Department of Chemistry, State University of New York at Buffalo, Buffalo, NY 14214.

monitored by absorption at 540 nm (4). A flow-injection analysis (FIA) detector has been developed which incorporates this method; however, an optrode has not yet been developed which exploits it.

Dynamic fluorescence quenching by halide ions is well understood (5–11) and can be adapted for use in an optrode system. Recently, Wolfbeis et al. reported an optrode for halide-ion sensing based on fluorescence quenching of acridinium and quinolinium indicators (7). With this system, iodide-ion concentrations could be detected down to 0.15 mM. The optrode's precision was 1% relative standard deviation in the 0.01 to 0.1 M range.

As a continuation of earlier work which characterized fluorophore-immobilization effects (12), the present study describes the development of an iodide optrode based on fluorescence quenching of immobilized rhodamine 6G, also a well-studied quenching system (10, 11). Three optrodes have been constructed by use of Teflon tape, XAD-4 resin beads, and crushed XAD-4 resin as solid supports for immobilization. These optrodes are compared in terms of several analytical figures of merit in an attempt to elucidate critical factors in optrode design.

EXPERIMENTAL SECTION

Optrode Design and Preparation. All optrodes utilize the same optical design. A 2-m single optical fiber (Valtec Corp. 240-μm step index or Galileo 400 μm) guides the excitation beam to the fiber's distal end; six additional fibers then surround the first fiber, collect the resultant fluorescence, and carry it to the detection system. The solid support used for immobilization is attached to the sealed end of a glass tube (2.5 cm length) which fits over the bundle's distal end and serves as a spacer. This tube is positioned so the cone of exciting radiation will exactly fill the area of its inner diameter at its far end (Figure 1). Advantages of the glass-tube spacer are (1) the laser power is distributed over a greater area and thereby provides a larger sampling surface and reduced photodegradation effects and (2) the reagent-phase component of the optrode can be constructed separately and interchanged easily without disturbing the optical-fiber component. We found that placing the rhodamine 6G treated resins or Teflon directly at the distal end of the fiber (no glass tube spacer) leads to irreversible photolization of the rhodamine 6G. Importantly, photodegradation is achieved with as little as 5 mW of laser power. However, the glass-tube spacer eliminated these effects.

XAD-4 Resin Optrodes. In order to attach the resin to the glass-tube spacer, a microscope-slide coverslip was fixed to the tube's end with Super Glue (Super Glue Corp.) and its edges were sanded off. Epoxy (Epotek 320) was then spread in a uniform layer over the sealed end of the glass tube and four 40-mesh XAD-4 polystyrene beads (Rohm and Haas Co.) were placed symmetrically in the center of the converslip. After the epoxy cured overnight, the immobilization was effected by immersing for a few seconds the attached resin beads in a 1.0 mM ethanolic solution of rhodamine 6G. The resin was then withdrawn from the solution, washed with deionized water to remove excess fluorophore, and allowed to equilibrate in deionized water for approximately 1 h. The crushed-resin optrode was prepared in exactly the same manner except that the XAD-4 resin was first crushed into a fine powder (approximately 25 μm particles) and then attached in a uniform layer by using the epoxy.

Teflon Optrode. In this design, Teflon tape was placed over one end of the glass tube and held in place with heat-shrink tubing. One drop of ethanolic 1.0 mM rhodamine 6G was placed on the Teflon surface and allowed to stand until the tape became transparent, at which time it was washed with deionized water and allowed to equilibrate in deionized water for approximately 1 h.

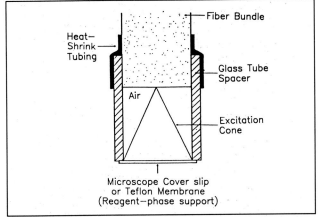

Figure 1. Fiber-bundle optrode design with glass-tube spacer.

Iodide-Optrode Instrumental Design. The instrument used for iodide concentration measurements is similar to that described elsewhere (12). The 514.5-nm line of a continuous wave (CW) argon-ion laser (Spectra Physics, Inc., Model 171), operating at a plasma current of 30 A, passes through a chopper and is then focused onto the input fiber. Radiant power measured at the fiber's distal end is approximately 3.0 mW. The subsequent fluorescence of the immobilized fluorophore is collected by the fiber bundle and is imaged into a double monochromator (Spex, Inc., Model 1680 Spectramate). The spectral band-pass for all measurements is 9 nm. The fluorescence is detected by a photomultiplier tube (PMT) (RCA R928) operated at a biasing voltage of −900 V dc. The output of the PMT is connected to a lock-in amplifier (EG&G Princeton Applied Research Model 5101 with a time constant of 0.1 s) which is referenced to the chopper frequency. The lock-in amplifier signal is sent to a computer (Digital Equipment Corp., MINC 11-23) where it is ratioed to the laser power as registered by the laser's integral power meter. Response curves are obtained by placing the optrode in a beaker of magnetically stirred deionized water to which iodide (from KI, Mallinckrodt, Inc.) is added by buret. After each addition of I⁻, exactly 2 min elapse before the fluorescence intensity is sampled. Data collection is accomplished by recording 1000 points at 30 Hz, three times in succession. The result is the average of these three data sets. Calculations of average signal, noise on the signal, and percent drift (<2%) in signal are performed on the MINC computer by means of a BASIC program. Importantly, the noise level and the signal-to-noise (S/N) ratio depend on the quencher concentration (cf. Figure 4).

Determination of Fluorescence Lifetimes with a Sampling Oscilloscope. The instrument used for the determination of fluorescence lifetimes is similar to that described elsewhere (12). The 514.5-nm line of a mode-locked argon-ion laser (Spectra Physics, Inc., Model 171 laser, Model 342 mode locker, and Model 452 mode-locker drive) is mechanically chopped and focused into one end of the optical fiber described in the previous section. The mode-locker frequency is between 40.9900 and 40.9940 MHz, the laser plasma current is 35 A, and the radiant power measured at the fiber's distal end is 9.0 mW. The optical fiber delivers excitation radiation to the sample holder and a bundle of six fibers collects the resulting fluorescence and sends it to a fast photomultiplier tube (RCA 31024) operated at a biasing voltage of −3500 V dc. A longpass filter which cuts off below 520 nm is placed in a sliding mount and interposed between the return fiber bundle and the photomultiplier tube (PMT). The filter is placed into or out of the path of the radiation depending on whether a fluorescence decay curve or instrument-response curve is being collected, respectively. The output from the PMT is directed to a sampling oscilloscope (Tektronix, Inc., Model 7844 mainframe, Model S4 sampling head, and Model 7S11 sampling unit) which is triggered by the synchronous output of the mode-locker driver. The output from the oscilloscope is connected to a lock-in amplifier (EG&G Princeton Applied Research Model 5101, with 0.1-s time constant), which is referenced to the mechanical chopper frequency. The lock-in amplifier reduces additive noise introduced after the chopper (13). The output from the lock-in amplifier is sampled by a computer (Digital Equipment Corp., MINC 11-23) which collects 128 points across a 12-ns time window. One hundred twenty-eight points were chosen for convenience in the event that a fast-Fourier digital filter would need to be employed.

Operation of the Time-Resolved Fluorescence Instrument. Depending on the required signal-to-noise ratio, between 200 and 500 data points are collected and averaged at each time increment on a fluorescence decay curve. An instrument-response curve is measured before and after each fluorescence decay curve by collecting and averaging 200 data points of scattered radiation with the longpass filter and fluorescent sample (resin or Teflon) displaced from the light path. The lifetime of the sample is extracted by use of a convolute-and-compare algorithm (14–16) on a VAX 11-780 computer.

RESULTS AND DISCUSSION

The purpose of this work was to compare the effect of alternative immobilization methods on the performance of an actual optrode system. Previous work showed that 1.0 mM rhodamine 6G had a longer fluorescence lifetime when immobilized on XAD-4 (5.14 ns) than on XAD-2 resin (4.58 ns) (12). In addition, very similar values were found for the crushed XAD-4 (5.16 ns) and XAD-2 (4.53 ns) resin beads, indicating little difference in the local rhodamine 6G environments. A longer lifetime is desired in dynamic fluorescence quenching because it increases the probability of excited-state quenching and thereby provides a more sensitive measure of quencher that is present (8). In order to determine if porosity (XAD-4 has a 400-nm average pore diameter and a surface area of 725 m^2/g) of the resin was a limiting factor in optrode response, both crushed XAD-4 resin and the original resin beads were employed. Teflon tape was also used because of its structural difference from the XAD resins and its prospect as an immobilization means in preliminary experiments. It is believed that the rhodamine 6G molecules are "layered" on the Teflon tape, a situation that is in marked contrast to that encountered with the resins. As a result, response might be

quite different. Rhodamine 6G (1.0 mM) immobilized on Teflon tape was found to have a fluorescence lifetime of 2.04 ± 0.05 ns.

Stern–Volmer Plots. Iodide-response curves were obtained for each optrode and plotted (Figure 2) in terms of the Stern–Volmer equation which, if it is strictly obeyed, describes collisional quenching of fluorescence (17)

$$F_0/F = 1 + K_D[I^-] \qquad (1)$$

In this equation, F_0 is the fluorescence intensity in the absence of quencher and F is the fluorescence intensity with $[I^-]$ quencher present. Plotting $[(F_0/F) - 1]$ vs. $[I^-]$ should yield a straight line with zero intercept and a slope equal to K_D, the quenching constant, which is a measure of the fluorophore's sensitivity to the quencher. From these plots the Teflon optrode is much more sensitive than either resin-based sensor and thus provides greater resolution in iodide concentration.

If a linear relationship is assumed, a least-squares fit of the data provides Stern–Volmer quenching constants (95% confidence intervals used) for the resins beads, crushed resin, and Teflon optrodes of 8.82 ± 0.96, 11.61 ± 1.43, and 28.7 ± 0.87 M^{-1}, respectively. Thus, even though the rhodamine 6G immobilized on Teflon has a much shorter lifetime than the other optrodes, it provides a stronger response. This trend might be attributed to accessibility of the fluorophore molecules when they are immobilized. In the resin-bead optrode, a large fraction of fluorophore molecules might reside in the bead's pores and thus require diffusion of I^- into the pores (if the I^- is not size-excluded) before quenching can take place. This effect would be reduced in the crushed resin because its porous structure is virtually destroyed; the Teflon optrode would provide the most accessible environment for quenching, since the rhodamine 6G molecules are surface-adsorbed.

Deviation from the linear Stern–Volmer relation is observed in Figure 2 at high iodide concentrations. In addition, the plots do not go through zero, indicating further that the Stern–Volmer model is not followed. A similar trend has been noted in quenching experiments when a fraction of the fluorophore molecules are unavailable for quenching, such as in iodide quenching of tryptophan residues (18). A modified Stern–Volmer equation based on this hypothesis has been derived and can be applied to the immobilized rhodamine 6G/iodide system under study here (18). For the modified relation

$$F_0/(F_0 - F) = 1/f_a K[I^-] + 1/f_a \qquad (2a)$$

where

$$f_a = F_{0a}/(F_{0a} + F_{0b}) \qquad (2b)$$

In these equations, F is again the total fluorescence intensity with quencher present and F_0 is that in the absence of quencher. F_{0a} is the fluorescence due to accessible fluorophore with no quencher present and F_{0b} is that for the fraction of inaccessible fluorophore. Plots of $F_0/(F_0 - F)$ vs. $1/[I^-]$ for each optrode appear in Figure 3. If one assumes a linear relationship, the reciprocal of the y intercept of these plots gives the percentage of fluorophore which is available to quenching.

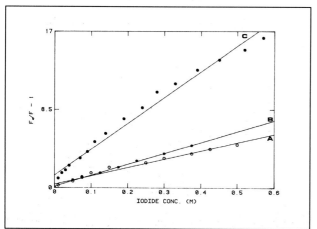

Figure 2. Stern–Volmer plots for (A) XAD-4 resin-bead optrode, (B) XAD-4 crushed-resin optrode, and (C) Teflon optrode response to iodide.

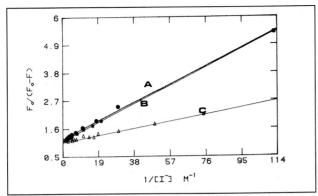

Figure 3. Modified Stern–Volmer plot (see eq 2a) for (A) XAD-4 resin-bead optrode, (B) crushed-resin optrode, and (C) Teflon optrode.

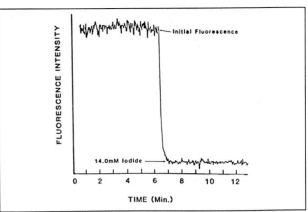

Figure 4. Response curve for the Teflon optrode after the addition of 14.0 mM iodide.

These values are 93% for Teflon, 89% for crushed resin, and 83% for resin beads. This information qualitatively parallels data derived from the previous plot (Figure 2) which presumably indicated the accessibility of the rhodamine 6G molecules to iodide. The Teflon optrode appears not to offer 100% of the fluorophore available for quenching, which supports the theory of "layered" molecules on the Teflon surface. The percentage available is lowest for the resin-bead optrode and again is believed to be due to rhodamine 6G molecules being trapped in the resin's pores and being inaccessible by I⁻.

Of course, in Figure 3, the data do not exactly follow the predicted linear relationship, and the true behavior might be much more complicated than even the modified Stern–Volmer relationship assumes. It is possible that there is a population of fluorophore molecules at an intermediate stage of sequestration. Several workers have described negative deviations from the Stern–Volmer relation at high halide-ion concentrations (6, 7, 9, 19) and have circumvented the problem simply by adding an exponent to the equation (19)

$$(F_0/F - 1)^m = k[I^-] \qquad (3)$$

Calculations for detection limit and dynamic range have been obtained by curve fitting the data to this exponential form of the Stern–Volmer equation (Table I).

Figures of Merit. Figures of merit obtained for each optrode are compiled in Table II. The detection limits were calculated as the concentration of I⁻ that would produce a signal (taken from the curve-fit exponential form of the Stern–Volmer equation) equal to three times the noise on the initial fluorescence signal. As expected (because of its greater Stern–Volmer quenching constant), the Teflon optrode yielded the lowest detection limit, 0.18 mM, which is comparable to the value of 0.15 mM reported by Wolfbeis et al. (7). The dynamic range of the three kinds of optrode were similar at about 4 orders of magnitude and were limited at the upper end by the highest concentration at I⁻ attainable in aqueous solution.

Response times for the optrodes again parallel the ease with which I⁻ ions can come into contact with immobilized fluorophore molecules. Figure 4 shows a typical response to I⁻ addition for the Teflon optrode. The Teflon optrode gave response times that were slightly faster than those reported by

Table I. Parameters for the Curve-Fit Stern–Volmer Relation (Equation 3)

$$(F_0/F - 1)^m = k[I^-]$$

	m	k	r^2
Teflon	1.34	76.5	0.997
resin, crushed	1.15	14.5	0.998
resin, beads	1.51	20.0	0.998

Table II. Figures of Merit Summary

	solid support for immobilization		
	Teflon	crushed resin	resin beads
detection limit, mM	0.18	1.10	0.30
dynamic range	~ 4 orders of magnitude		
response time[a] (quench)	~ 25 s	~ 7 min	~ 20 min
response time[b] (unquench)	~ 35 s	~ 10 min	~ 35 min
reversibility,[c] %	86	60	67
precision[d] (% RSD)	3	5	11

[a] Measured after placing optrode in 0.01 M I⁻. [b] Measured after placing optrode in distilled-deionized water. [c] Percent of original signal recovered after removing optrode from 0.05 M I⁻ and placing in distilled-deionized water. [d] Determined using a 10 mM solution of I⁻ (10 replicates).

Wolfbeis (~25 s vs. ~40 s) (7). As expected, the resin-bead optrode has the slowest response, as I⁻ ions must diffuse into the resin's pores.

The precision and reversibility to each optrode suggest that some iodide might remain trapped in the immobilization matrix. This effect would be most pronounced for the resin-bead optrode because of its porous nature and would be least significant for the fluorophore adsorbed on Teflon. It was found by repeating the reversibility experiments several times that the recovered signal indeed reached a fixed level. This indicates that iodide was somehow trapped in the support. However, the crushed-resin optrode deviates from this

Table III. Iodide Optrode Sensitivity[a] (mM) to 1.0 M Solutions of Various Anions and Cations

ion	Teflon	crushed resin
bromide	10.0	0.79
chloride	3.30	0.41
sulfate	—	—
nitrate	1.31	0.34
perchlorate	—	—
cyanide	2.04	0.61
phosphate	—	—
bromate	0.73	0.84
isothiocyanate	0.45	0.83
Zn(II)	—	—
Fe(II)	—	—
Fe(III)	—	—
Mn(II)	—	—
Cu(II)	—	—

[a] Stated in terms of the iodide concentration that the interfering signal would represent. — indicates no measurable effect.

trend in its reversibility, a fact that suggests another possibility. Because the fluorophore molecules are not chemically bound to the various solid supports, bleeding of the rhodamine 6G is observed. This bleeding is greatest for the crushed resin because the resin's ability to "secure" molecules is altered upon crushing. Thus, it would appear that both residual I⁻ trapped in the optrodes and bleeding contribute to the observed trends in reversibility and precision. It should be noted that the bleeding effect is quite slow and might be overcome by measuring fluorescence lifetimes instead of total fluorescence intensity (8). Although the lifetime of the fluorophore molecules has been shown to be concentration-dependent when they are resin-immobilized, there might be a concentration range where this trend "levels off." No work has yet been performed to analyze the Teflon immobilization method for the same effect.

Interferences from Other Ions. Each optrode was tested for its sensitivity to several potentially interfering anions and cations (Table III). After each optrode was placed in a 1.0 M solution of a selected ion, the fluorescence intensity was measured and converted into the amount of I⁻ it would represent on the basis of the curve-fit Stern–Volmer equation (eq 3). Results show noticeable interference from bromide, chloride, nitrate, cyanide, bromate, and isothiocyanate. This result is similar to data obtained by Wolfbeis (7), who also reported interferences from bromide, chloride, sulfite, isothiocyanate, cyanide, and cyanate, but not with sulfate, phosphate, perchlorate, or nitrate. Results are shown also for several positively charged metal ions; however, none caused any significant effect on the immobilized rhodamine 6G fluorescence. This experiment is intended only to indicate roughly the degree of interference that other ions would cause. In reality,

the interference will not be additive; instead, other ions might compete with the iodide as collisional quenchers and thereby produce a nonlinear interference. Also, because iodide is a much more effective quencher than other ions, the interference effect might be negligible at high I⁻ concentrations, especially if interferent concentrations are below the 1.0 M level used in this study.

Fluorophore molecules immobilized on the surface of Teflon tape produced better figures of merit than immobilization on resin even though the resin-bound rhodamine 6G has a fluorescence lifetime which is twice as long. Thus, accessibility has proven to be a more important factor in optrode performance than the fluorescence lifetime of an immobilized fluorophore. Resin immobilization does provide an effective solid support but, for optimal optrode performance, surface adsorption or chemical bonding directly to the fiber's surface should be the method of choice.

Registry No. Rhodamine 6G, 989-38-8; XAD-4, 37380-42-0; Teflon, 9002-84-0; iodide, 20461-54-5.

LITERATURE CITED

(1) Saari, S. A.; Seitz, W. R. *Anal. Chem.* **1983**, *55*, 667–670.
(2) Saari, S. A.; Seitz, W. R. *Analyst (London)* **1984**, *109*, 655–657.
(3) Zhujun, Z.; Seitz, W. R. *Anal. Chim. Acta* **1985**, *171*, 251–258.
(4) Fuwa, K.; Fujiwara, K. *Anal. Chem.* **1985**, *57*, 1012–1016.
(5) Wolfbeis, O. S.; Urbano, E. Z. *Fresenius' Z. Anal. Chem.* **1983**, *314*, 577–586.
(6) Wolfbeis, O. S.; Urbano, E. Z. *Anal. Chem.* **1983**, *55*, 1904–1906.
(7) Urbano, E.; Offenbacher, H.; Wolfbeis, O. S. *Anal. Chem.* **1984**, *56*, 427–429.
(8) Hieftje, G. M.; Haugen, G. R. *Anal. Chim. Acta* **1981**, *123*, 255–261.
(9) Stroughton, R. W.; Rollefson, G. K. *J. Am. Chem. Soc.* **1939**, *61*, 2634–2638.
(10) Merkelo, H.; Hartman, S. R.; Mar, T. *Science* **1969**, *164*, 301–302.
(11) Harris, J. M.; Lytle, F. E. *Rev. Sci. Instrum.* **1977**, *48*, 1469–1476.
(12) Wyatt, W. A.; Poirier, G. E.; Bright, F. V.; Hieftje, G. M. *Anal. Chem.* **1987**, *59*, 572–576.
(13) Vickers, G. H.; Miller, R. M.; Hieftje, G. M. *Anal. Chim. Acta* **1987**, *192*, 145–153.
(14) Ware, W. R. *Transient Luminescence Measurements*; Marcel Dekker: New York, 1971; Vol. 1A, Chapter 5.
(15) Ramsey, J. M. Ph.D. Dissertation, Indiana University, 1979.
(16) Demas, J. N. *Excited State Lifetime Measurements*; Academic: New York, 1983; pp 128–129.
(17) Stern, O.; Volmer, M. *Phys. Z.* **1919**, *20*, 183.
(18) Lakowicz, J. R. *Principles of Fluorescence Spectroscopy*; Plenum: New York, 1983; pp 260–281.
(19) Peterson, J. I.; Fitzgerald, R. V.; Buckhold, D. K. *Anal. Chem.* **1984**, *56*, 62–67.

Received for review March 2, 1987. Accepted June 1, 1987. Supported in part by the Office of Naval Research, The Upjohn Co., and the National Science Foundation through Grant CHE 83-20053. We also thank the National Science Foundation for its support of the VAX 11-780 used in this work (Grants CHE 83-09446 and CHE 84-05851). In addition, we also thank Dean Gerachi and Galileo Fiber Optics for providing the fiber-optic materials.

Reprinted from *Anal. Chem.* **1987**, *59*, 2272–76.

The rising popularity of MS in biological research has its roots in Klaus Biemann's pioneering efforts at MIT, which go back more than 30 years. The characterization of polypeptide and protein sequences, particularly those with post-translational modifications, is one of the most important applications of MS. Until 1987, however, MS methods were unable to differentiate leucine from isoleucine. In this article, Biemann's group and his collaborators describe a tandem MS method, based on high-energy collisionally induced dissociation, for distinguishing between these amino acid residues and polypeptide and protein sequences.

Richard D. Smith
Pacific Northwest Laboratory

Novel Fragmentation Process of Peptides by Collision-Induced Decomposition in a Tandem Mass Spectrometer: Differentiation of Leucine and Isoleucine

Richard S. Johnson, Stephen A. Martin,[1] and Klaus Biemann*
Department of Chemistry, Massachusetts Institute of Technology, Cambridge, Massachusetts 02139

John T. Stults and J. Throck Watson
Department of Biochemistry, Michigan State University, East Lansing, Michigan 48824

The mass spectra produced upon collision-induced decomposition of the protonated molecules of peptides often exhibit peaks that correspond to ions that are formed by cleavage of the –N–CR– bond along the peptide chain followed by cleavage of the β,γ bond if R has the general structure $-C_\beta-C_\gamma-R'$. Ions produced in this manner are assigned the notation w_n and are helpful in the characterization of the amino acid at that position. Most important is the differentiation of the amino acids leucine and isoleucine, which is generally difficult or impossible by mass spectrometry. Aromatic amino acids do not undergo this fragmentation because it would involve cleavage of a C–Ar bond; neither does alanine, which would involve loss of a hydrogen radical; nor does glycine, which lacks a β,γ bond. The nature of this fragmentation process is demonstrated by exact mass measurements and precursor–product ion studies.

T he collision-induced decomposition (CID) mass spectra of peptides are very well suited for the determination of their amino acid sequence. In contrast to the normal fast atom bombardment (FAB) spectra (1), they exhibit extensive series of the same ion type. The most frequently occurring are ions of type a_n and b_n, generally observed if a basic amino acid is located at or near the N-terminus or if there is no basic amino acid, and y_n, if one is lo-

---

[1] On leave of absence from the Department of Cell and Molecular Pharmacology and Experimental Therapeutics, Medical University of South Carolina, Charleston, SC.

Table I. Fragments Produced from Protonated Linear Peptides (I)

cated at the C-terminus. Often all three ion series are present, with one or two of them dominating (2). This is in contrast to normal FAB mass spectra, which rarely exhibit continuous series of ions of one type and are difficult to interpret in the low-mass region, which is obliterated by the contributions of the matrix. One way to improve the signal-to-background ratio is to use a very high concentration of pure peptide in the matrix, a situation that rarely prevails when working with peptides of unknown sequence derived by proteolytic cleavage of a protein of unknown structure.

The generalized structures of fragments produced upon FAB from peptides of the typical structure I are shown in Ta-

Table II. Exact Mass Measurements

peptide[a]	no. in Table IV	path A		path B or C		measd mass
		elem comp of w_n ion	calcd mass	elem comp of w_n ion	calcd mass	
RPPGFSPFR[b]	13	$C_{22}H_{31}N_7O_5$	473.2387	$C_{23}H_{33}N_6O_5$	473.2513	473.2510
LQQ*I*GALK	17			$C_{21}H_{38}N_5O_6$	456.2822	456.2841
LQQIGALK	17	$C_{25}H_{45}N_7O_7$	555.3381	$C_{26}H_{47}N_6O_7$	555.3506	555.3510

[a]Amino acids in single-letter code; for three-letter code see Table III. [b]w_n ion at italized amino acid.

ble I. Detailed fragmentation processes leading to these ions have been suggested previously and have been summarized in a recent review (3). The nomenclature used here for the various ion types is a variation of that proposed by Roepstorff and Fohlman (4).

In the course of our work on peptide sequencing with high-performance tandem mass spectrometry, which permits the selection of the ^{12}C species of a $(M + H)^+$ isotopic cluster in MS-1 and the determination of its CID spectrum in MS-2 with an accuracy of better than ±0.3 u (5), we have often observed a peak which occurred 54 mass units higher than that corresponding to an ion of type y. This mass increment implied the retention of a remnant of the next N-terminal amino acid and would nominally correspond to the loss of the R group plus two hydrogens from the next y ion (y_n) as shown in path A.

These fragment ions were assigned the notation w_n (6). However, as CID spectral data of a large number of peptides of known sequence (7) as well as those derived from proteins of unknown sequence accumulated, it became clear that this fragmentation process is much more dependent on the nature of the N-terminal amino acid of the fragment ion than is the case with the other sequence-specific ions (a_n, b_n, c_n, x_n, y_n, z_n), which involve only the cleavage of a bond along the peptide backbone, sometimes accompanied by the rearrangement of a hydrogen atom (Table I). Most notable was the absence of w ions when aromatic amino acids, glycine, or alanine were involved or those that carry a substituent on the β carbon, such as valine, isoleucine, and threonine. For the latter three, ions were observed at higher mass corresponding to the retention of one of the substituents at the β carbon. This observation suggested that the formation of the w_n ions involves the cleavage of the β,γ bond because such a fragmentation would be difficult for aromatic amino acids where it would require cleavage of a CH_2–Ar bond. It also would cause retention of the substituents at the β carbon, but not at the N-terminal nitrogen atom. Furthermore, path A was ruled out on the basis of exact mass measurements in the normal FAB mass spectrum of two peptides that exhibited w ions of sufficient intensities (Table II).

EXPERIMENTAL SECTION

The CID mass spectra were determined either with a two-sector magnetic deflection mass spectrometer (the JEOL HX110 at MSU) operated in the linked-scan mode and using the collision cell in the first field-free region (after the ion source) or in a four-sector tandem mass spectrometer (the JEOL HX110/HX110 at MIT) with the collision cell in the third field-free region (5).

Two-Sector Linked Scans. A 1-μL aliquot of the peptide solution (1–3 nmol/μL) was mixed with 1 μL of matrix (either glycerol or 5:1 dithiothreitol/dithioerythritol) on the FAB probe tip. The precursor ions were formed by fast atom bombardment with a 6-keV Xe^0 beam in a JEOL HX110 mass spectrometer (EB geometry) operated at 10-kV accelerating voltage, resolving power = 3000. Helium was admitted to the collision cell (first field-free region) to the pressure required to attenuate the precursor ion to 30% of its original abundance. The JEOL DA5000 data system generated the linked scan at constant B/E ratio. The raw data profiles were acquired (30 s/decade) with the accumulation software, summing 10–15 spectra/sample.

Two of the peptides used (Table III, no. 10 and 12) were commercially available. Other peptides were from trypsin digests of proteins (Table III: no. 8, rubisco from spinach; no. 11 and 17, thioltransferase from rat liver; no. 23, H_{20} from tobacco) for which the sequences are known.

Four-Sector Tandem Mass Spectra. To a solution of 1–2 nmol/μL in 30% aqueous acetic acid was added an equal volume of glycerol and 1 μL of this mixture was placed on the FAB target, which was then inserted into the ion source of the JEOL HX110/HX110 spectrometer. Ionization was achieved by bombardment with a Xe^0 beam of 6 keV kinetic energy. The accelerating voltage in MS-1 was 10 keV. The ^{12}C species of the $(M + H)^+$ isotopic ion cluster was selected by MS-1 and this precursor ion beam was transmitted through the collision cell into MS-2. Helium was admitted into the cell at such a pressure (approximately 10^{-3} Torr) that the precursor ion was attenuated to about 30% of its original abundance (as measured at detector 3, after the collector slit of MS-2). The CID mass spectrum was recorded by the JEOL DA5000 data system which also generated the B/E linked scan of MS-2. The spectrum shown in Figure 2 is from a single scan recorded in 1.9 min and represents the raw data profile.

To record the product ion spectra (Figure 1) of fragment ions produced in MS-1, the latter was set to transmit these

Table III. Partial Listing of Peptides Used in This Study with the Amino Acids That Give Rise to w_n Ions Underlined

1. Tyr-Gly-I̲l̲e̲-Arg
2. Gly-G̲l̲n̲-Leu-Lys
3. L̲e̲u̲-T̲h̲r̲-V̲a̲l̲-Ala-Lys
4. <Glu-L̲e̲u̲-Tyr-G̲l̲u̲-A̲s̲n̲-Lys (Neurotensin 1–6)
5. Tyr-Gly-Gly-Phe-L̲e̲u̲-Lys
6. V̲a̲l̲-Gly-Ala-L̲e̲u̲-S̲e̲r̲-Lys
7. Trp-C̲y̲s̲[a]-Gly-P̲r̲o̲-Cys[a]-Lys
8. Ac-P̲r̲o̲-Gln-T̲h̲r̲-Glu-Thr-Lys[b]
9. V̲a̲l̲-His-L̲e̲u̲-T̲h̲r̲-Pro-V̲a̲l̲-Glu-Lys
10. V̲a̲l̲-His-L̲e̲u̲-T̲h̲r̲-P̲r̲o̲-Val-Glu-Lys[b]
11. Ac-G̲l̲n̲-Ala-Ala-Phe-V̲a̲l̲-A̲s̲n̲-Ser-Lys[b]
12. Tyr-Gly-Gly-Phe-M̲e̲t̲-Arg-Gly-Leu[b]
13. Arg-P̲r̲o̲-P̲r̲o̲-Gly-Phe-S̲e̲r̲-Pro-Phe-Arg (Bradykinin)
14. Arg-Pro-Pro-Gly-Phe-Ser-Pro-Phe
15. Pro-Pro-Gly-Phe-Ser-Pro-Phe-Arg
16. Asn-Gly-G̲l̲u̲-Val-Ala-Ala-Thr-Lys
17. L̲e̲u̲-G̲l̲n̲-G̲l̲n̲-I̲l̲e̲-Gly-Ala-Leu-Lys[b]
18. N-Formyl-nLeu[c]-Leu-Phe-nLeu[c]-Tyr-Lys
19. Gly-I̲l̲e̲-P̲r̲o̲-T̲h̲r̲-Leu-Leu-Leu-Phe-Lys
20. Thr-T̲h̲r̲-L̲e̲u̲-S̲e̲r̲-G̲l̲n̲-T̲h̲r̲-Leu-Glu-Lys
21. Ser-I̲l̲e̲-P̲r̲o̲-T̲h̲r̲-Leu-M̲e̲t̲-Leu-Phe-Lys
22. Pro-His-P̲r̲o̲-Phe-His-Phe-Phe-Val-Tyr-Lys
23. Ser-H̲y̲p̲-H̲y̲p̲-H̲y̲p̲-H̲y̲p̲-Thr-Hyp-Val-Tyr-Lys[b]
24. V̲a̲l̲-Asp-M̲e̲t̲-V̲a̲l̲-V̲a̲l̲-Gly-Ala-V̲a̲l̲-Pro-Lys
25. Pro-Lys-Pro-Gln-Gln-Phe-Phe-Gly-Leu-Met-NH₂
26. Arg-P̲r̲o̲-Lys-Pro-Gln-Gln-Phe-Phe-Gly-Leu-Met-NH₂ (Substance P)
27. Arg-P̲r̲o̲-Lys-Pro-Gln-Gln-Phe-Phe-Gly-Leu-Met
28. L̲e̲u̲-A̲s̲n̲-I̲l̲e̲-Asp-G̲l̲n̲-A̲s̲n̲-P̲r̲o̲-Gly-Thr-Ala-Pro-Lys
29. Ser-Ala-Ala-Ala-G̲l̲n̲-V̲a̲l̲-T̲h̲r̲-Asp-S̲e̲r̲-Thr-Phe-Lys
30. Tyr-Gly-Gly-Phe-Met-Thr-Ser-Glu-Lys-Ser-Gln-Thr-Pro-Leu-Val-Thr
31. Ala-Asp-Ser-Gly-G̲l̲u̲-Gly-A̲s̲p̲-Phe-L̲e̲u̲-Ala-G̲l̲u̲-Gly-Gly-Gly-Val-Arg (Fibrinopeptide A)
32. <Glu-Gly-Pro-Trp-Leu-Glu-Glu-Glu-Glu-Glu-Ala-Tyr-Gly-Trp-Met-Asp-Phe-NH₂
33. N-Formyl-Val-Gly-Ala-Leu-Ala-Val-Val-Val-Trp-Leu-Trp-Leu-Trp-Leu-Trp-NHCH₂CH₂OH (Val-1 gramicidin A)

[a]Carbamidomethylated cysteine. [b]B/E scan with EB two-sector instrument. [c]nLeu=norleucine.

---

fragment ions (one after the other) into the collision cell. In each case, the product ion spectra were recorded in the same manner as outlined above for $(M + H)^+$ ions.

The peptide samples were either commercially available (Table III: no. 4, 5, 9, 13–15, 18, 22, 25–28, and 30–33) or generated by proteolytic digestion of thioredoxins from *E. coli* and from *Anabaena*, small proteins of known structure (9, 10) (Table III: no. 1–3, 6, 7, 16, 19–21, 24, and 29).

Mass Measurements. Exact mass measurements for ions in the normal FAB spectrum were made with a JEOL HX110 mass spectrometer at a resolving power of 10 000 in the peak-

matching mode. Glycerol cluster ions were used as reference masses.

RESULTS AND DISCUSSION

Two pathways (B or C) could account for the formation of such a fragment ion from either a z_n or a $z_n + 1$ ion as outlined in the case of N-terminal leucine in Scheme I.

Scheme I

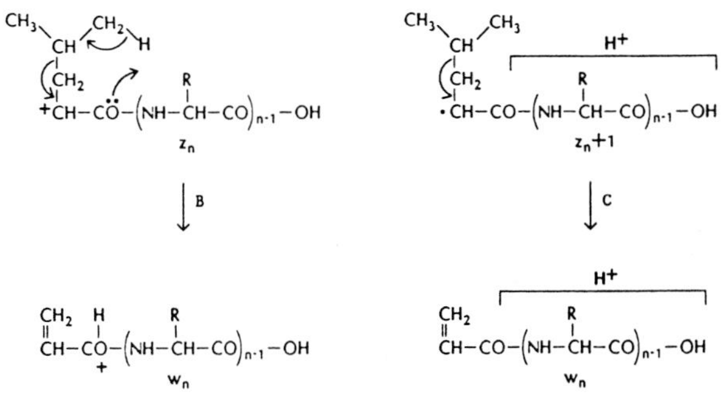

Of these, path B is intuitively less likely because it would require the elimination of CH_2 or O in the case of valine or threonine, respectively. The decision in favor of path C was finally made by the determination of the CID spectrum of a z_n and a $z_n + 1$ ion that was reasonably abundant in the normal FAB mass spectrum of fibrinopeptide A (sequence no. 31 in Table III), $M_r = 1535.69$. When m/z 741.4 (z_8) was selected by MS-1 of a four-sector instrument, no signal at m/z 699.3 was observed (Figure 1a) but the CID spectrum of m/z 742.4 ($z_8 + 1$) (Figure 1b) exhibited a strong signal at m/z 699.1, proving the validity of path C.

A particularly illustrative example is shown in Figure 2, which represents the CID spectrum of a tryptic peptide (Gly-Ile-Pro-Thr-Leu-Leu-Leu-Phe-Lys, no. 19 in Table III) derived from the red-ox protein thioredoxin isolated from *E. coli* (8). Although it is somewhat atypical, because it exhibits a contiguous series of w_n ions (from w_3 to w_8) it serves to illustrate the information provided by this fragmentation process. The important regions of the spectrum are expanded in Figure 3a–g, which incidentally also demonstrates the resolution of MS-2 of the JEOL HX110/HX110 four-sector mass spectrometer used in this work. Table IV lists the experimentally determined values for the w_3–w_8 ions as well as those calculated based on path C.

Figure 3a demonstrates the absence of a w_2 ion (at m/z 201.1) which would require the cleavage of a CH_2–C_6H_5 bond (path C). Figures 3b–d show the w_3, w_4, and w_5 ions due to the elimination of the isopropyl radical from the side chain of leucine. Figure 3e illustrates the multiplicity of the w_6 ions at m/z 701.6 for the cleavage of the C_β–OH bond (w_{6a}) and m/z 703.6 (w_{6b}) for the cleavage of the C_β–CH_3 bond (in those cases where the β carbon is substituted by two groups of different mass, such as threonine and isoleucine, two w_n ions are usually observed and they are labeled w_{na} and w_{nb} by increasing mass). There is also present a peak at m/z 688.6 which nominally corresponds to the elimination of C_2H_5OH from y_6. The peak labeled w_{6L} (m/z

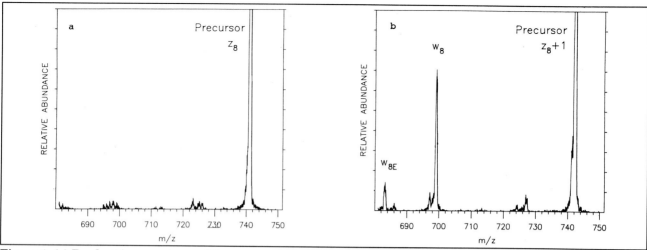

Figure 1. (a) Product ion spectrum of the precursor ion m/z 741.39 (z_8) in the FAB spectrum of fibrinopeptide A, M_r 1535.69 (no. 31 in Table III). (b) Product ion spectrum of the precursor ion m/z 742.39 ($z_8 + 1$) in the same FAB spectrum.

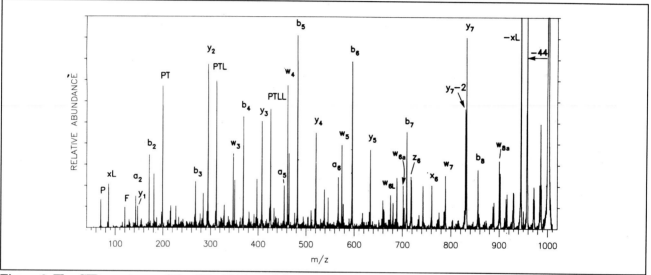

Figure 2. The CID spectrum of the $(M + H)^+$ ion of the peptide Gly-Ile-Pro-Thr-Leu-Leu-Leu-Phe-Lys (no. 19 in Table III). To avoid cluttering, some of the peaks ranging from 10 to 40% full scale are not labeled. Most of them are x_n ions or are due to the loss of 18 u (H_2O) or 28 u (CO) from the internal fragments PT, PTL, and PTLL or from some of the b_n ions (see also Figure 3b–d).

Table IV. List of w_n Ions in Figure 2

	m/z found	m/z calcd		m/z found	m/z calcd
w_3	348.2	348.18	w_{6b}	703.5	703.43
w_4	461.3	461.26	w_7	788.8	788.48
w_5	574.4	574.35	w_{8a}	899.8	899.55
w_{6a}	701.5	701.45	w_{8b}	913.8	913.56

found = 675.5, calcd = 675.46) is due to the $z_6 + 1$ ion that eliminates the side chain from the next C-terminal amino acid (Leu-5), possibly through formation of a ring rather than a double bond (the side chain lost is that of the amino acid indicated as the single-letter subscript).

We have made the observation that this "secondary" w_n-type cleavage, while not frequently observed, mainly occurs when the N-terminal amino acid of the $z_n + 1$ ion is of the type that cannot undergo cleavage of the β,γ bond, such as with aromatic amino acids, and when the next amino acid is one that

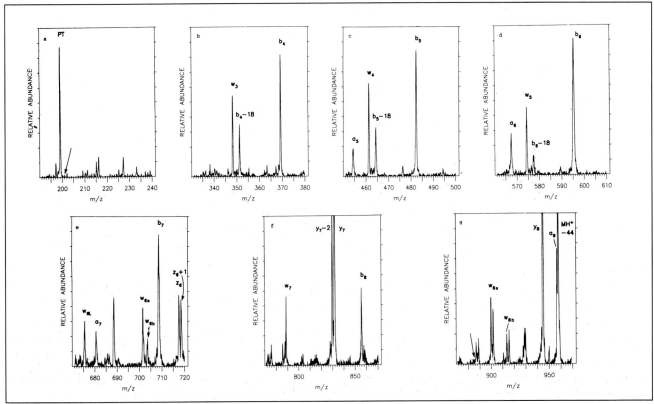

Figure 3. Expanded regions of the spectrum shown in Figure 2: (a) arrow points to the absence of the w_2 ion (m/z 201) that would have to be produced by cleavage of the C_6H_5–CH_2 bond of Phe in position 8; (b)–(d) cover the regions in which peaks due to ions w_3–w_5 appear; (e) the complex set of peaks due to the multiplicity of w_n ions produced by threonine (see text); (f) the region around w_7 and y_7 (see text); (g) the region that permits identification of the amino acid in position 2 as isoleucine rather than leucine (see text). In some of these figures the y axis has been somewhat expanded for clarity. The relative abundances can be gleaned from Figure 2.

readily cleaves at the β,γ bond, such as leucine or glutamic acid (see also Figure 1b).

Figure 3f is an expanded view of the region where the w_7 ion is observed, which corresponds to the loss of the "side chain" of proline. For this cyclic amino acid, homolytic cleavage of the N–C bond does not lead to fragmentation, i.e., does not lead to a normal $z_n + 1$ ion but still permits further cleavage of the β,γ bond to form the w_n ion.

This segment of the spectrum also clearly shows a $y_n - 2$ ion, which corresponds to a C-terminal immonium ion (rather than the most abundant ammonium ion, y_n). This process is particularly noticeable if the N-terminal amino acid of this fragment is proline, as had been previously noted by Williams et al. (*11*).

Figure 3g expands the region around the w_8 ion group, which involves the cleavage at isoleucine. In agreement with path C, the w_{8a} ion is found at m/z 899.8 (calcd 899.53) and there is no significant signal at m/z 885.5 (arrow in Figure 3g), which should be observed if the amino acid at this posi-

tion were leucine. In addition, there is also a peak (w_{8b}) at m/z 913.9 resulting from the elimination of the β-methyl group.

CONCLUSIONS

The above represents a detailed discussion of the characteristics of w_n ions and the conclusions that can be drawn from their mass, if present, or from their absence. Table III is a partial list of the large number of CID spectra we have accumulated. A significant w_n ion is observed for the underlined amino acids. To be representative, a number of peptides are included that do not exhibit any w_n ions although their CID spectra exhibit other abundant fragment ions, chiefly of the type a_n, b_n, and y_n. One can draw a number of conclusions about the structure of the peptide from the presence or absence of this ion type. The most important of these is the fact that the presence of a basic amino acid at or near the C-terminus seems to be required for the formation of reasonably abundant w_n ions. In this respect there is a certain analogy to other C-terminal ions, particularly y_n. An interesting example is the pair of peptides no. 26 and 27 (in Table III) which indicates that a basic amino acid must be part of the w_n ion to trigger its formation, because only Pro-2 but neither Pro-4 nor Gln-5, Glu-6, or Leu-10 give rise to a w_n ion. Presumably the proton shown above the brackets in Scheme I, path C is attached to the basic side chain; the process thus involves "remote site" fragmentation, first proposed by Gross et al. (12). Table III also supports the statement made at the outset that aromatic amino acids, glycine, or alanine do not exhibit this fragmentation.

The most significant aspect of the fragmentation process discussed here is the possibility of distinguishing leucine from isoleucine in peptides that give rise to C-terminal ions. This complements the similar significance of the a_n ion types, a_n-42 for Leu and a_n-28 for Ile, observed for peptides that favor retention of the charge on N-terminal fragments (6). Earlier efforts to differentiate leucine from isoleucine by mass spectrometry had met with only limited success (3). Of secondary importance is the confirmatory aspect of the presence or absence of a w_n ion corresponding to amino acids at the N-terminal positions of these ions, indicating the presence or absence of amino acids that possess a readily cleaved β,γ bond. Finally the m/z value of w_n ions allows the confirmation of the position of amino acids substituted at the β carbon, such as threonine or valine. A note of caution should be added, however, because the fragmentation of a β,γ bond of a side chain of an amino acid next to that representing the N-terminus of the fragment may occur.

ACKNOWLEDGMENT

The authors are indebted to H. A. Scoble (MIT) for his useful comments.

Registry No. 1, 110143-90-3; 2, 110143-91-4; 3, 110143-92-5; 4, 87620-09-5; 5, 83404-43-7; 6, 110143-93-6; 7, 110143-94-7; 8, 110143-95-8; 9, 62526-81-2; 11, 110143-96-9; 12, 80501-44-6; 13, 58-82-2; 14, 15958-92-6; 15, 16875-11-9; 16, 110143-97-0; 17, 110143-98-1; 18, 71901-21-8; 19, 110143-99-2; 20, 110144-00-8; 21, 110144-01-9; 22, 75645-19-1; 23, 110144-02-0; 24, 110144-03-1; 25, 53749-61-4; 26, 33507-63-0; 27, 71977-09-8; 28, 110144-04-2; 29, 110144-05-3; 30, 59004-96-5; 31, 25422-31-5; 32, 10047-33-3; 33, 4419-81-2; leucine, 61-90-5; isoleucine 73-32-5.

LITERATURE CITED

(1) Barber, M.; Bordoli, R. S.; Sedgwick, R. D.; Tyler, A. N. *J. Chem. Soc., Chem. Commun.* **1981**, 325–327.
(2) Martin, S. A.; Biemann, K. *Int. J. Mass Spectrom. Ion Processes* **1987**, *78*, 213–228.
(3) Martin, S. A.; Biemann, K. *Mass Spectrom. Rev.* **1987**, *6*, 1–76.
(4) Roepstorff, P.; Fohlman, J. *Biomed. Mass Spectrom.* **1984**, *11*, 601.
(5) Sato, K.; Asada, T.; Ishihara, M.; Kunihiro, F.; Kammei, Y.; Kubota, E.; Costello, C. E.; Martin, S. A.; Scoble, H. A.; Biemann, K. *Anal. Chem.* **1987**, *59*, 1652–1659.
(6) Martin, S. A.; Biemann, K. Presented at the 34th Annual Conference on Mass Spectrometry and Allied Topics, Cincinnati, OH, 1986.
(7) Martin, S. A., unpublished results.
(8) Biemann, K.; Martin, S. A.; Scoble, H. A.; Johnson, R. S.; Papayannopoulos, I. A.; Biller, J. E.; Costello, C. E. In *Mass Spectrometry in the Analysis of Large Molecules*; McNeal, C. J., Ed.; Wiley: Sussex, England, 1986; pp 131–149.
(9) Hoog, J.-O.; Von Bahr-Lindstrom, H.; Josephson, S.; Wallace, B. J.; Kushner, S. R.; Jornvall, H.; Holmgren, A. *Biosci. Rep.* **1984**, *4*, 917–923.
(10) Gleason, F. K.; Whittaker, M. M.; Holmgren, A.; Jornvall, H. *J. Biol. Chem.* **1985**, *260*, 9567–9573.
(11) Williams, D. H.; Bradley, C. V.; Santikarn, S.; Bojesen, G. *Biochem. J.* **1981**, *201*, 105–117.
(12) Jensen, N. J.; Tomer, K. B.; Gross, M. L. *J. Am. Chem. Soc.* **1985**, *107*, 1863–1868.

Received for review May 14, 1987. Accepted July 20, 1987. This work was supported by research grants from the National Institutes of Health to K. Biemann (RR00317 and GM05472) and to J. T. Watson (RR00480).

Reprinted from *Anal. Chem.* **1987**, *59*, 2621–25.

This article describes a method, now known as matrix-assisted laser desorption/ionization (MALDI), for converting protein molecules in a solid into naked intact ions in the gas phase—a remarkable physical transition. By embedding proteins in a matrix of small organic molecules that strongly absorbed UV laser light, the authors made the conditions for the physical transition essentially independent of the protein's specific physical properties. The new method produced large quantities of isolated protein ions whose molecular masses were readily determined by time-of-flight MS. Karas and Hillenkamp's marvelous development has since been refined and has become a method of choice for characterizing proteins and other macromolecules.

Brian T. Chait
Rockefeller University

Laser Desorption Ionization of Proteins with Molecular Masses Exceeding 10 000 Daltons

Michael Karas* and Franz Hillenkamp
Institute of Medical Physics, University of Münster, D-4400 Münster, Federal Republic of Germany

Sir: Until recently the desorption of ions of bioorganic compounds in the mass range above 10 000 daltons seemed to be exclusively the domain of plasma desorption mass spectrometry (PDMS) (1–4). In 1987 Tanaka et. al. (5) reported laser desorption of protein molecular ions up to a mass of 34 000 daltons. Oligomers of lysozyme containing up to seven monomeric units have also been observed by this group, using a pulsed N_2 laser and a matrix of a metal powder, finely dispersed in glycerol. Fast atom bombardment or liquid secondary ion mass spectrometry (SIMS) data on compounds above 10 000 molecular weight show weak signal intensities and poor signal-to-noise (S/N) ratios. The only exception is the results reported on the analysis of small proteins in the range of 10–24 000 daltons obtained with a 30-keV cesium ion source in a conventional double-focusing mass spectrometer (6), but sample amounts in the 10-μg range were necessary in that case.

In the following, the first results on ultraviolet laser desorption (UVLD) of bioorganic compounds in the mass range above 10 000 daltons will be reported. Strong molecular ion signals were registered by use of an organic matrix with strong absorption at the wavelength used for controlled energy deposition and soft desorption (7).

EXPERIMENTAL SECTION

A reflector-type time-of-flight (TOF) mass spectrometer equipped with a Q-switched quadrupled Nd-Yag-Laser (pulse length 10 ns, wavelength 266 nm) was used for ion generation and analysis. The laser was focused to a spot between 10 and 50 μm in diameter. Samples were used as supplied commercially and dissolved in water at a concentration of about 10^{-5} M. One microliter of sample solution (10^{-11} mol) was mixed with 1 μL of 10^{-3} M aqueous solution of nicotinic acid serving as the absorbing matrix. The mixture was dripped onto a metallic substrate and air-dried. The sample then covered an area of ~5 mm^2 on the substrate. Individual spectra obtained with single laser shots of about 10^8 W/cm^2 irradiance and an area of about 10^{-3} mm^2 already showed detectable signal intensities. Sum spectra as shown in Figures 1–4 are obtained by irradiating the same spot 20–50 times. Provided suitable sample preparation techniques are developed, a much smaller amount of sample would suffice to obtain the same results. A time of 2–5 min, needed to acccumulate 20–50 single spectra, was limited by the laser repetition frequency of 1 Hz and by the PC data processing time. The LAMMA 1000 instrument used was designed for microprobe analysis of inorganic and organic compounds in the low mass range. It operates at an ion kinetic energy of 3 keV and the postacceleration potential was limited to a maximum of 9 kV. The ion velocity of lysozyme (molecular weight 14 306), e.g., at the conversion electrode, amounts to a value of only 1.1×10^4 m/s and is even lower for the higher mass compounds. Note that a minimum velocity of ca. 1.7×10^4 m/s was determined for insulin ion detection in PDMS (2). Signals were registered by a Biomation transient recorder with an aquisition memory of only 2048 channels resulting in a very low time resolution in the high mass range (for details see results). Currently this also limits the accuracy of mass calibration, which is done with the sodium and matrix signals in the low mass range. Thus all results have to be understood as a documentation of the general feasibility of the technique for the generation of high molecular mass ions.

RESULTS AND DISCUSSION

Spectra of four proteins are reported here: lysozyme (from chicken egg white, molecular weight 14 306), β-lactoglobulin

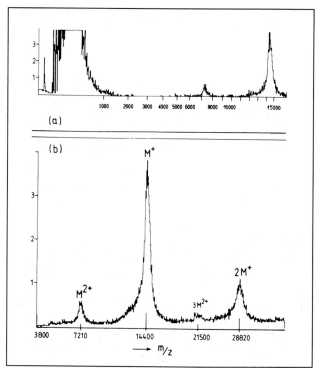

Figure 1. Matrix-UVLD spectrum of lysozyme (from chicken egg): (a) total registered mass range, 25 accumulated individual spectra; (b) mass range of molecular ion region, 50 accumulated individual spectra.

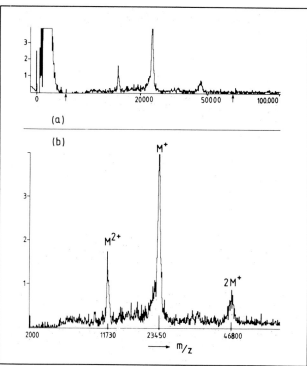

Figure 3. Matrix-UVLD spectrum of porcine trypsin, 50 accumulated individual spectra: (a) mass range 0–100 000 daltons; (b) mass range of molecular ion region.

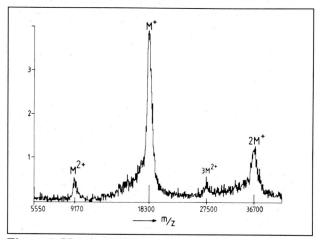

Figure 2. Matrix-UVLD spectrum of β-lactoglobulin A, mass range of molecular ion region, 50 accumulated individual spectra.

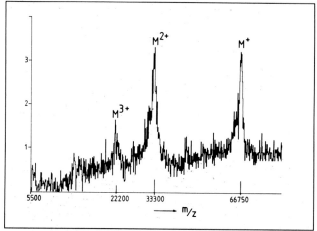

Figure 4. Matrix-UVLD spectrum of bovine albumin, mass range of molecular ion region, 100 accumulated individual spectra.

A (from bovine milk, molecular weight 18 277), porcine trypsin (molecular weight 23 463), and albumin (bovine, approximate molecular weight 67 000) (all molecular weights are averages). Comparable results have been obtained for a variety of other proteins.

Figure 1 shows the laser desorption (LD) spectrum of lysozyme. Without time delay the mass range at an analog to digital time resolution of 200 ns/channel extends from 0 to ~16 000 daltons. For higher masses a time delay or decreased time resolution had to be used. In the low mass range up to 1000 daltons, strong signals of nicotinic acid matrix ions and

possibly also analyte fragment ion signals are observed. Besides the dominating peak of the molecular ion, a dimer and doubly charged molecular ion are clearly detected as well as a $3M^{2+}$ signal. The same holds for β-lactoglobulin A (Figure 2). The geometrically determined mass centroids of the peaks are indicated at the mass axis of the spectrum. For porcine trypsin (Figure 3) a time per channel of 500 ns was used. The upper trace in this figure shows the total registered mass range of 0–100 000 daltons. Again the background signal of the matrix below 1000 daltons is seen. The lower trace shows a section of the upper one as indicated by the arrows. Again molecular ions are found as M^+, $2M^+$, and M^{2+}. Figure 4 shows the LD-spectrum of bovine albumin with a molecular weight of about 67 000. Molecular ion signals are still obvious though the signal-to-noise ratio has declined compared to 20 000 daltons range. The absolute flight time for the bovine albumin molecular ion already amounts to 0.8 ms. For bovine albumin, 100 single spectra were accumulated. These results clearly document the ability of LD to produce ions in this mass range so far inaccessible even by PDMS.

It is also worth noting that in all spectra multiply charged ions are observed. Multiple charges have so far never been reported for laser desorption ionization. They are quite common in PDMS of high mass proteins and presumably reflect the potential of these large molecules to remain stable even with several charge centers rather than a peculiarity of the excitation mechanism.

The width (full width at half maximum) of the molecular ion signal (600 daltons in the case of trypsin) is in all cases much larger than that expected from the mass resolution of the instrument of approximately 600 as measured in the low mass range. A resolution of the molecular ion signal that is expected to show both protonated and cationized species is not possible, but this is at least partly due to the low time resolution of the transient recorder (one channel of 200 ns is equivalent to 15 daltons at 14 000 daltons, and one channel of 500 ns to 80 daltons at 67 000 daltons). The large width may also originate from so far unknown effects in the ion–electron conversion at the low ion velocity. An improved detector system is expected to yield much better results. This expectation is supported by the fact that the transition from the commonly used 6 to 9 kV postacceleration resulted in a 10-fold improvement in signal-to-noise ratio.

Another unique feature of the matrix-UVLD technique as compared to PDMS or liquid-SIMS ion generation in the high mass region is a very low (chemical) noise level. Besides the molecular ion signals, ions are only registered in the mass range below about 1000 daltons. The matrix ion intensities are only about 10–50 times higher than the sample ion signals.

CONCLUSION

The results reported demonstrate the ability of matrix-UVLD to generate a large number of intact molecular ions as well as dimers and doubly charged molecular ions of proteins in the mass range above 10 000 daltons, stable at least up to times of about a millisecond. Singly charged molecular ions were in all cases the base peak of the analyte signal; no fragment ions were observed in the mass range above 1000 daltons. Additionally, multimers of the molecular ions and doubly charged molecular ions were detected improving the molecular ion detection and increasing the molecular weight determination accuracy. Also, a remarkable sensitivity is demonstrated. A detection limit in the subnanogram range for total sample mass needed for a sum spectrum appears to be realistic. All these features, no doubt, can still be optimized as discussed above. Though the general applicability of UVLD still needs to be shown by the successful desorption of a larger variety of different compounds, matrix-UVLD promises to be able to extend the accessible range for mass spectrometry of nonvolatile bioorganic compounds considerably with the added advantages of low sample consumption, ease of preparation, and short measurement time.

Registry No. Lysozyme, 9001-63-2; trypsin, 9002-07-7.

LITERATURE CITED

(1) MacFarlane, R. D.; Hill, J. C.; Jacobs, D. L.; Phelps, R. G. In *Mass Spectrometry in the Analysis of Large Molecules*; Wiley: Chichester, 1986; pp 1–12.
(2) Sundquist, B.; Hedin, A.; Hakansson, P. I.; Kamensky, M.; Salehpour, M.; Säwe, G. *Int. J. Mass Spectrom. Ion Processes* **1985**, *65*, 69–89.
(3) Kamensky, I.; Craig, A. G. *Anal. Instrum. (N.Y.)* **1987**, *16*, 71–91.
(4) Chait, B. T.; Field, F. H. *Int. J. Mass Spectrom. Ion Processes* **1985**, *65*, 169–180.
(5) Tanaka, K.; Ido, Y.; Akita, S.; Yoshida, Y.; Yoshida, T. Presented at the Second Japan–China Joint Symposium on Mass Spectrometry (abstract), Takarazuka Hotel, Osaka, Japan; Sept 15–18, 1987.
(6) Barber, M.; Green, B. N. *Rapid Commun. Mass Spectrom.* **1987**, *1*, 80–85.
(7) Karas, M.; Bachmann, D.; Bahr, U.; Hillenkamp, F. *Int. J. Mass Spectrom. Ion Processes* **1987**, *78*, 53–68.

Received for review May 16, 1988. Accepted July 5, 1988. This work was supported by the Deutsche Forschungsgemeinschaft under Grant No. Hi 285/2-5 and by a grant from the Ministerium für Wissenschaft und Forschung des Landes Nordrhein-Westfalen.

Reprinted from *Anal. Chem.* **1988**, *60*, 2299–2301.

In this paper, Bard and co-workers established scanning electrochemical microscopy as a unique and important tool for spatially resolved in situ studies of surface reactivity. In particular, they introduced the powerful "feedback mode," which made scanning electrochemical microscopy applicable to high-resolution studies of conducting and insulating surfaces alike and permitted quantitative assessment of microscopically local reactivity. The work of several other groups depended on the foundations laid in this and other Bard papers.

Royce Engstrom
University of South Dakota

Scanning Electrochemical Microscopy: Introduction and Principles

Allen J. Bard,* Fu-Ren F. Fan, Juhyoun Kwak, and Ovadia Lev
Department of Chemistry, University of Texas, Austin, Texas 78712

The technique of scanning electrochemical microscopy (SECM) is described. In this technique the electrolysis current that flows as an ultramicroelectrode tip (diameter ca. 10 μm) immersed in a solution is moved above a substrate surface is used to characterize processes and structural features of the substrate. Modes of operation considered include collection modes, where products electrogenerated at the substrate are detected at the tip (held at constant potential or operated in the cyclic voltammetric mode), and feedback modes, where the effect of substrate on the tip current is monitored. The feedback mode can be used with both conductive and insulating substrates and is less sensitive to electrical coupling between substrate and tip. An alternating current generation/collection mode is also described. Experimental results for the different modes of operation and proposed extensions of the SECM technique are presented.

INTRODUCTION

Scanning electrochemical microscopy (SECM) is a technique in which the current that flows through a very small electrode tip (generally an ultramicroelectrode with a tip diameter of 10 μm or less) near a conductive, semiconductive, or insulating substrate immersed in solution is used to characterize processes and structural features at the substrate as the tip is moved near the surface. The tip can be moved normal to the surface (the z direction) to probe the diffusion layer, or the tip can be scanned at constant z across the surface (the x and y directions). The tip and substrate are part of an electrochemical cell that usually also contains other (e.g., auxiliary and reference) electrodes. The device for carrying out such studies involves means of moving the tip with a resolution down to the Å region, for example, by means of piezoelectric elements or stepping motors driving differential springs, and is called a scanning electrochemical microscope. The abbreviation SECM is used interchangeably for both the technique and the instrument. The principle of an SECM operating in the generation/collection mode is illustrated in Figure 1B. As discussed below, other modes of operation are also used.

While the SECM resembles the scanning tunneling microscope (STM) (1) (Figure 1A) in its use of a tip to scan over a substrate surface and in the methods of moving the tip, there are fundamental differences in the principles of operation and range of applications. Since the STM depends upon the flow of a tunneling current between tip and substrate, the distance between them is of the order of 1 nm or less and surface topographic x–y resolution of only this size scale is usual (typical surface scan areas in STM are 30 nm by 30 nm). Moreover, even for STM studies in solution, the tunneling current is a nonfaradaic one (i.e., no chemical changes in solution components or substrate surface species occur because of the current flow), so that the tip current cannot be related directly to the substrate potential by consideration of the redox potentials of appropriate half-reactions. In SECM the current is carried by redox processes at tip and substrate and is controlled by electron transfer kinetics at the interfaces and mass transfer processes in solution, so that measurements at large spacings, e.g., the range of 1 nm to 10 μm, can be made. This should allow probing of the diffuse and diffusion layers near the substrate through z scans of about 1 nm to 10 μm, as well as lower resolution surface x-y scans. In this case the tip current (i_T) should be related to the potential of the tip (E_T) and substrate (E_S) by the usual electrochemical considerations (2).

2855-8/94/0388$08.00/0
© 1994 American Chemical Society *Milestones in Analytical Chemistry*

In an alternative approach the tip can be a reference electrode (e.g., a metal wire quasireference electrode or a drawn-out glass capillary connected to a true reference electrode via a salt bridge) and the potential at the tip monitored in the z direction or x–y plane. This type of scanning potentiometry has already been employed at much lower resolution, for example, to probe corroding substrates (3–5). The use of an ultramicroelectrode (UME) to probe the diffusion layer concentrations by z-direction movement over a range of about 2–150 μm by detecting substrate-generated species at the UME has been described recently (6).

We describe here several different types of experiments with the SECM. In principle SECM can provide surface topography information, potential distributions, and analytical data. Since electrochemical reactions occur in SECM, micro-

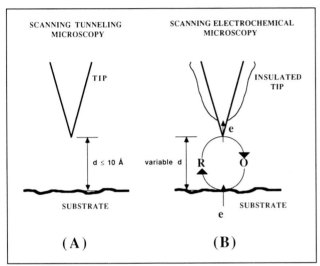

Figure 1. Comparison of scanning tunneling microscopy (STM) (A) and scanning electrochemical microscopy (SECM) (B). d is the spacing between tip and substrate.

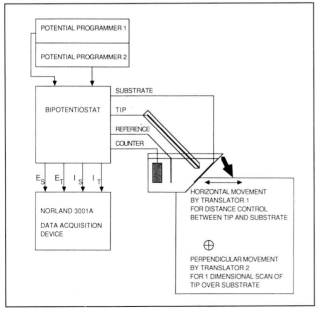

Figure 2. Block diagram of apparatus in dc experiment.

fabrication can also be carried out with this apparatus. For example, etching and deposition of metals and semiconductors and synthesis (e.g., electropolymerization) are possible. Accounts of such applications have appeared (7, 8) and this area will not be discussed here.

EXPERIMENTAL SECTION

Apparatus. A block diagram of the apparatus used for tip movement and control, which generally follows previous STM designs (1), is shown in Figure 2. In this apparatus the working electrode tip was held on a tripod (x', y', and z') piezoelectric scanner. The substrate, held at a 45° angle to the tip axis, was carried by a movable x–y stage which was moved independently by two piezoelectric translators (Burleigh Instruments, Fishers, NY, Type IW-502-2), which were controlled by a programmable controller (Type CE-2000-2A00). These permitted movements in the z direction of 0.81 μm/step and in the x direction of 1.22 μm/step. Higher resolution is obtained by acquisition of data during the step (from one step position to another). The tip potential (E_T) and substrate potential (E_S) were controlled vs a reference electrode, either a silver quasi-reference electrode (AgQRE) or a saturated calomel electrode (SCE) with a home-built bipotentiostat of conventional design (9) controlled with a Model 175 programmer (Princeton Applied Research, Princeton, NJ). The potentials, E_T and E_S, and the tip and substrate currents, i_T and i_S, were recorded with a four-channel data acquisition device (Digital Oscilloscope, Model 3001, Norland Corp., Fort Atkinson, WI). The data were either analyzed on the oscilloscope, employing the build-in functions, or transferred to a microcomputer (Apple Macintosh II or IBM XT) for analysis and plotting of the data.

The ac experiments were carried out with the apparatus shown in Figure 3. The phase angle between i_T and i_S and the relative magnitudes of the currents were obtained with a lock-in amplifier (Model 5206, Princeton Applied Research), which also supplied an ac potential, i.e., operating in the internal reference mode, to the substrate via the potentiostat. As shown in Figure 3, the tip potential was not controlled by the potentiostat but rather by a high-capacity SCE reference electrode in series with a battery power supply. This controlling circuit worked well as long as the tip current was very small (usually on the order of nanoamperes) and the solution resistance was low.

Tip Preparation. Several different tips were used in these experiments. Construction of the Pt disk-in-glass ultramicroelectrodes (with 12.5- and 5-μm radii) followed previous practice (10, 11) and is described in more detail elsewhere (12, 13).

Pt wire of the desired radius was washed with 30% HNO_3 and dried overnight. This was placed in a 10 cm long, 1 mm i. d. Pyrex tube sealed at one end. The open end of the tube was connected to a vacuum line and heated with a helix heating coil for 1 h to desorb any impurities on the Pt. One end of the wire was then sealed in the glass by increasing the heating coil temperature. The sealed end was polished with sandpaper until the wire cross section was exposed, and then successively with 6-, 1-, and 0.25-μm diamond paste (Buehler, Ltd., Lake Bluff, IL). Electrical connection to the unsealed end of the Pt wire was made with silver paint to a Cu wire. The glass-wall surrounding the Pt disk was conically sharpened by emery paper (Grit 600, Buehler, Ltd.) and 6-μm diamond paste, with frequent checking by optical microscope until the diameter of the flat glass section surrounding the Pt disk was less than 100 μm. This decreased the possibility of contact between glass and substrate because of any slight deviations in the axial alignment of the tip as the tip was moved close to the substrate. Carbon tips (10, 11) were used in the ac experiments. These were prepared from 7 μm diameter C fibers that were burned slowly in a heating coil to decrease the diameter to 1–3 μm. These were sealed in predrawn glass capillaries in a gas flame.

RESULTS AND DISCUSSION

Resolution and Tip Shape. The ultimate resolution of the SECM depends primarily upon the tip size and shape. However, the solution resistance and mass and charge transfer process rates that affect the current density distribution are also important factors. Approximate expressions for the steady-state current that flows between substrate and tip as a function of z spacing for a conical and spherical tip in the generation/collection mode have been presented (14). The steady-state current mode would be of most interest for topographic information during x–y scans. The approximate expressions for the steady-state current in the generation/collection mode, for example, in the configuration shown in Figure 1B for different tip geometries above a much larger (assumed infinite) planar substrate are given below. Typical current (i)–distance curves are shown in Figure 4. For a planar tip, in the absence of edge effects, the familiar plane/plane thin layer expression, eq 1, applies (15)

$$i = nFADC^*/d \qquad (1)$$

where d is the spacing between substrate and tip and C^* is the sum of the concentrations of oxidant, C_O, and of reductant, C_R. D is the diffusion coefficient (assuming $D_O = D_R = D$, in which D_O and D_R are the diffusion coefficients of oxidant (O) and reductant (R), respectively) and A is the electrode area. An actual planar disk electrode will show behavior more like a hemispherical electrode, especially when $d \gg a$ (where a is the disk radius) (16, 17). Under these conditions, or when the tip actually has a spherical or hemispherical shape, the approximate expression for i is (14)

$$i = 2\pi nFDC^*r \ln[1 + (r/d)] \qquad (2)$$

where r is the radius of the sphere. For a cone-shaped tip, the approximate current is given by (14)

$$i = 2\pi nFDC^*a^*(1 + a^2)^{1/2}[1 - \gamma' \ln(1 + 1/\gamma')] \qquad (3)$$

where $\alpha = a^*/H$, where a^* is the cone base radius and H is the height, and $\gamma' = d/H$. Note that in this latter case (Figure 4), the current becomes independent of tip-to-substrate spacing at very small d (as $\gamma \to 0$) (14). Thus a conical or tapered cylin-

drical tip would be less useful than an inlaid disk or hemisphere for probing the topography of a substrate. Actual tips may have more complex shapes as are sometimes observed in ultramicroelectrodes (10, 11).

Collection Experiments. In the collection mode the tip is used as the detector only. The collection experiments can be performed in several operation modes. In the transient generation/constant potential collection (TG/CPC) mode, the potential of the substrate is stepped to a value, E_S, where an electrochemical reaction, for example, the oxidation R → O + ne, occurs. The tip is held at a potential, E_T, where reduction of O back to R takes place. The tip current, i_T, is monitored as a function of time at constant z. The TG/CPC mode is analogous to transient experiments at the rotating ring–disk electrode (RRDE), where a product produced at the disk is swept convectively to the collector ring. With the TG/CPC mode it should be possible to calculate the distance between the tip and substrate from the time for the onset of the collection current, i.e., the time required for substrate-generated O to transit the gap and reach the tip. This approach has also been used to determine the concentration profile of oxidized species, $C_O(z,t)$, near a substrate for $2 < z < 75$ μm at $0 < t < 4$ s by neglecting any distortion of the diffusion layer caused by the tip (6). A problem with this type of experiment is that the large current generated at the substrate by the potential step, which is composed of both nonfaradaic (charging) and faradaic components, is coupled via the interelectrode capacitance and resistance to the tip and produces a transient tip current that

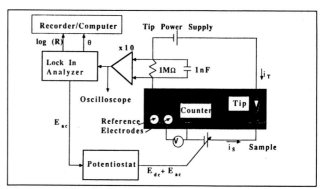

Figure 3. Block diagram of apparatus in ac experiments.

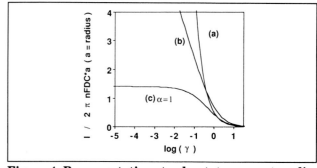

Figure 4. Representative steady-state current vs distance for generation/collection experiments for a planar substrate and different shaped tips: (a) planar disk; (b) spherical; (c) conical.

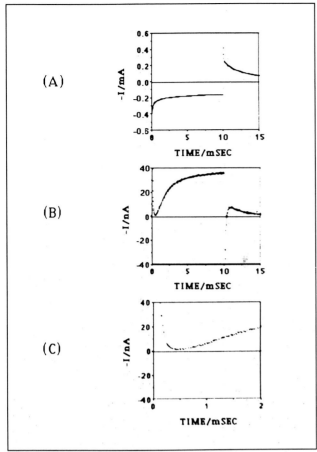

(A)

(B)

(C)

Figure 5. Substrate current (i_S) (A) and tip current (i_T) (B, C) for a 2.5 mm radius Pt disk substrate stepped from 0 to 1.0 V vs AgQRE with detection at a 5 μm radius Pt disk tip held at 0 V vs AgQRE at a distance, d, of about 2.5 μm from substrate. The solution was 2.5 mM ferrocene and 25 mM TBABF$_4$ in MeCN.

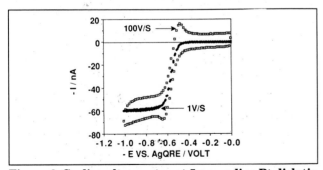

Figure 6. Cyclic voltammetry at 5 μm radius Pt disk tip above a 2.5 mm radius Pt disk substrate held at $E_S = 0.0$ V vs AgQRE with interelectrode separation, d, of 2.5 μm. Solution contained 2.5 mM ferrocene and 25 mM TBABF$_4$ in MeCN.

interferes with the measurement of the faradaic i_T, especially at smaller t and z. This transient, in the direction of the substrate current, is clearly seen in the experiment in Figure 5 and makes it difficult to estimate the time for the onset of the collection current (i.e., the diffusional transit time between substrate and tip). The magnitude of this transient increased with the size of the substrate (i.e., with i_S) and would be of

greater importance as the transit time decreased, i.e., as z decreased.

Alternatively, the tip potential can be scanned while the substrate potential is changed. For example, a linear potential sweep can be applied to the tip to observe products generated at the substrate in the generation/cyclic voltammetric collection (G/CVC) mode. The observed tip current depends upon the scan rate (v), d, and E_S. Typical experimental curves for a 5 μm radius Pt tip over a 2.5 mm radius Pt disk substrate in a solution of 2.5 mM ferrocene (Fc) in MeCN, 25 mM tetrabutyl-ammonium fluoroborate (TBABF$_4$) are shown in Figures 6 and 7. When d is large (e.g., 5 mm), the substrate has no effect upon the tip current, and typical ultramicroelectrode behavior (10, 11) is observed. At slow scan rates (e.g., ≤1 V/s), the anodic current for Fc oxidation attains a steady-state value given by (10, 11)

$$i = 4nFDC^*a \tag{4}$$

where a is the radius of the ultramicrodisk and C^* is the bulk concentration of Fc and no cathodic current appears. At fast scan rates (e.g., ≥100 V/s) a cyclic voltammogram closer to that expected of semiinfinite linear diffusion is obtained. When the tip is moved close to the substrate (e.g., ca. 2.5 μm), as in Figures 6 and 7, the tip current depends upon the nature of the substrate and its potential. For example, when the substrate is held at 0 V (Figure 6), Fc$^+$ generated at the tip is reduced to Fc and the anodic current is increased, compared to that when d is large. This feedback effect, discussed in more detail later, also results in a greater similarity of the anodic currents at large and small v. The cathodic current at larger v is reduced, since Fc$^+$ generated at the tip is consumed at the substrate. When the substrate is held at 1.0 V and d is small (Figure 7), the anodic current is greatly decreased, since there is little unoxidized Fc in the region of the tip. The cathodic current is increased, since the tip is now acting in a collection mode, with the substrate acting as a generator of Fc$^+$. This CV mode of operation would be especially useful in identifying species or in elucidating the kinetics of unstable species dissolved from the substrate.

ac Mode. Another collection mode (called the ac generation/collection mode) involves applying an alternating potential or current variation to the substrate (optionally superimposed on a controlled dc substrate potential, E_S) and determining the ac component of the tip current at a given E_T

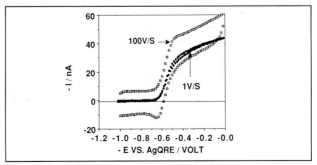

Figure 7. As in Figure 6, except $E_S = 1.0$ V vs AgQRE.

Table I: Phase Angle (θ) and Relative Amplitude (R) Slopes for an ac Collection/Generation Experiment[a]

frequency (f), Hz	forward scan[b]				reverse scan[c]	
	$\partial\theta/\partial z$,[d] deg μm^{-1}	$\partial \log R/\partial z$,[e] μm^{-1}	$(\partial\theta/\partial z)/f^{1/2}$	$(\partial \log R/\partial z)$, $f^{1/2}$	$\partial\theta/\partial z$,[d] deg μm^{-1}	$\partial \log R/\partial z$,[e] μm^{-1}
0.75	3.0	0.0245	3.46	0.028	3.2	0.0245
1	3.64	0.029	3.64	0.029		
1.5	4.25	0.037	3.47	0.030	4.25	0.037
2	5.1	0.043	3.61	0.030	5.8	0.041
3	6.4	0.059	3.69	0.034	6.1	0.059
4	7.3	0.067	3.65	0.034	7.3	0.070
5	9.4	0.081	4.2	0.036	9.5	0.085

[a] Pt electrode substrate in 5 mM $K_4Fe(CN)_6$, 5mM $K_3Fe(CN)_6$, 1 M KCl, modulated with 0.1 mA at frequency f with 6 μm C tip at –0.3 V vs SCE. [b] Tip moving toward substrate. [c] Tip moving away from substrate. [d] Slope of θ vs z curve for z of 0 to 3.7 μm. [e] Slope of $\log R$ vs z curve.

with a lock-in amplifier referenced to substrate signal. The principles of this technique and some preliminary experiments are included here. When an alternating current, $i = I \sin \omega t$, is applied to the conductive substrate electrode immersed in a solution containing both forms of a redox couple (O/R), the electrolysis that occurs induces a sinusoidal variation of the concentrations of O and R at the electrode surface. This type of electrolysis has been considered originally in connection with electrochemical studies by Warburg and Kruger and is treated, for example, by Vetter (18), where further details of the mathematical development can be found. The concentration variations in the vicinity of the substrate electrode surface are in the form of an exponentially damped sinusoid of the form (j = O, R):

$$C_j(z,t) = C_j + A^* \exp(-z/z_0) \sin[\omega t - z/z_0 - \pi/4] \quad (5)$$

where $z_0 = (2D/\omega)^{1/2}$, C_j is equilibrium concentration of j (j = O, R, both with a diffusion coefficient, D) and $A^* = I/nF(D\omega)^{1/2}$. The tip, held at a suitable potential, will sense the local concentrations in the vicinity of the tip through variations in i_T. Thus from eq 5, i_T will vary with z, the tip distance from the substrate electrode, as an exponentially decaying sinusoidal function about a steady-state value. If i_T is measured with respect to the current applied to the substrate, i_S, e.g., with a lock-in amplifier, the following characteristics are expected: (1) The measured peak-to-peak tip current should decay exponentially with tip-to-sample distance, z. (2) The phase angle between the substrate and tip current, $\theta = |z/z_0 + \pi/4|$, should be proportional to z and should approach 45° as $z \to 0$. (3) $\partial\theta/\partial z = (\omega/2D)^{1/2}$ and thus $\partial\theta/\partial z$ varies as $f^{1/2}$, where f is the frequency of the applied ac ($\omega = 2\pi f$). (4) The amplitude of the sinusoidal concentration wave, $R = A^* \exp(-1 z/z_0)$ so that $|\partial \ln R/\partial z| = 1/z_0 = (\omega/2D)^{1/2}$. Since the tip current is proportional to concentration, $\partial \log i_T/\partial z$ should also vary as $f^{1/2}$.

Experiments were carried out to determine $\partial \log R/\partial z$ and $\partial\theta/\partial z$ as functions of f. These involved a 6 μm diameter C tip above a Pt substrate immersed in a solution containing 5 mM each $K_4Fe(CN)_6$ and $K_3Fe(CN)_6$ in 1 M KCl. The applied ac peak substrate current was 0.1 mA (producing a variation of

±30 mV around the rest potential) and E_T was –0.3 V vs SCE. The modulation frequency, f, was set at a given value between 0.75 and 5 Hz, and θ and R recorded from the lock-in amplifier outputs as the tip was advanced toward the substrate at a rate of 0.1 μm/s and then retracted. Recordings of θ vs z and $\log R$ vs z were linear over the z–region investigated, 0–3.7 μm. The slopes of these lines are listed in Table I. As predicted, $\partial\theta/\partial z$ and $\partial \log R/\partial z$ are proportional to $f^{1/2}$, with values of $(\partial\theta/\partial z)f^{1/2} = 3.59$ (±0.10) deg μm^{-1} $Hz^{-1/2}$ and $(\partial \log R/\partial z)f^{-1/2} = 0.031$ (±0.003) μm^{-1} $Hz^{-1/2}$ for f or 0.75–4 Hz. These values are close to those predicted from the relations for these values [i.e., $(1/\omega^{1/2}z_0)$ or $(2D)^{-1/2}$], which in the same units are 3.64 deg μm^{-1} $Hz^{-1/2}$ and 0.028 μm^{-1} $Hz^{-1/2}$, respectively. At higher frequencies, the results deviated from these values, even though the values of θ and R were reproducible for the tip advancing toward and retracting from the substrate, and θ vs z and $\log R$ vs z plots became nonlinear. Deviations were particularly pronounced in the longer distance region (≥ 3 μm). These deviations can probably be attributed to some capacitive coupling between tip and substrate. Other factors that may be important in this mode of operation include feedback from the tip to substrate (not taken into account in eq 5) and the disturbance of the concentration profiles near the substrate surface caused by the tip. This latter effect will be especially important when the tip is scanned over the substrate. The nature of the tip, e.g., the quality of the seal between metal or carbon and the glass insulating sheath, is also an important factor in tip/substrate coupling, since poor sealing increases the tip capacitance.

These results suggest that the ac generation/collection mode, with measurements of θ or R, can be used in scans of substrate surfaces. High resolution, based on the slopes of the θ vs z and $\log R$ vs z curves found here, will require application of higher modulation frequencies and smaller tips. A preliminary experiment involving a scan with the 6-μm C tip across a grid of 7-μm Au wires spaced about 24 μm apart showed small oscillations in θ and $\log R$ at the expected ca. 25 μm spacing, but the resolution was poor. As suggested by a reviewer, application of ac approaches that decrease double layer capacitance contributions, e.g., second-harmonic methods, may be useful.

Feedback Experiments. In the feedback mode a potential is applied to the moving tip which is used as both the source and the detector; the tip current (i_T) is a function of the nature of the substrate and the distance between substrate and tip. When the substrate is conductive or semiconductive, the oxidized form, O, formed at the tip can be reduced at the substrate producing R, which diffuses back to the tip. This causes i_T to be higher than it is when the tip is far ($d \rightarrow \infty$) from the substrate. The magnitude of this "feedback" component of the current is a function of the distance between the generator tip and the substrate (the smaller d, the larger the feedback current). This type of feedback effect is unique to closely spaced electrodes in quiescent solutions, e.g., it is not found with the RRDE where hydrodynamic flow from disk to ring prevents any detectable disk feedback current, and has been addressed experimentally and theoretically at microelectrode arrays (*18, 19*). One can think of this as a form of "electrochemical radar," where the wave of O generated at the tip interacts at the substrate to "reflect back" a wave of R that is detected at the tip. As opposed to actual radar, which is basically a transient technique, this feedback mode can also be used at steady state. This mode has the advantage over the transient collection mode that the coupling transient is absent, since only measurements at the tip are made, permitting the examination of very small distances and short times. Moreover, the feedback mode is also useful when the substrate cannot be conveniently held at an externally applied potential (e.g., when it is a small isolated conductive zone) or when application of a potential might cause undesirable reactions or decomposition of the substrate (e.g., for Pt sputtered on mica (*21*), where hydrogen evolution can cause the Pt to peel off). In

the feedback mode the substrate itself need not necessarily be connected to an external potential source (e.g., a potentiostat), because most of the conductive substrate is located away from the tip reaction and is bathed in a solution containing mainly R, maintaining it at a potential negative to the formal potential, $E^{\circ\prime}$, of the O/R couple. In this case, the localized reduction current at the portion of the substrate under the tip can be driven by the oxidation of R at those portions of substrate away from the tip region. The feedback mode can also be applied to an insulating substrate, as discussed in the next section.

We will describe several experiments in the feedback mode and give a more quantitative treatment of the feedback current at the tip. Consider the current at an ultramicroelectrode ($a = 5$ μm) immersed in a solution of 2.5 mM Fc in 25 mM TBABF$_4$, MeCN over a 2.5-mm Pt disk substrate (Figure 8). When the tip is far (> 100 μm) from the substrate, i_T is not affected by the presence of substrate; i_T ($d \rightarrow \infty$) is given by eq 4 (*10, 11*). When the tip is close to the substrate, i.e., as $d \rightarrow 0$, the tip current can be approximated by eq 1 or with $A = \pi a^2$ by

$$i_T(d) = nFDC^*\pi a^2 / d \qquad (6)$$

Normalizing $i_T(d)$ with respect to $i_T(d \rightarrow \infty)$ yields

$$i_T(d) / i_T(d \rightarrow \infty) = (\pi / 4) (a / d) \quad (d \rightarrow 0) \qquad (7a)$$

$$i_T(d) / i_T(d \rightarrow \infty) = 1 \quad (d \rightarrow \infty) \qquad (7b)$$

As shown in Figure 8, the theoretical computer simulation curve (*22*) fits the experimental results quite well. The curve exhibits an expected straight line (eq 7a) at small d and $i_T(d)/4nFDC^*a$ deviates to 1 (eq 7b) at large d. With this calibration curve the distance between substrate and tip can be found from the measured tip current and known tip diameter. Thus, i_T, which varies as $1/d$, is a measure of distance, d, as compared to the situation in STM, where the tunneling current varies as exp(−const d) (*1*).

The feedback mode can also be used with a nonconductive substrate. In this case, i_T is perturbed because the substrate interferes with the essentially hemispherical diffusion field to the tip, causing a decrease in i_T as the tip-to-substrate distance is decreased. When the tip gets close to the insulating substrate, diffusion to the tip is hindered (i.e., a portion of the hemispherical diffusion zone is blocked). To demonstrate this effect, preliminary experiments were carried out with a 50 μm diameter glass fiber on glass microscope cover slip substrate (Figure 9). When the tip was above (ca. 60 μm) the cover slip, the current was slightly smaller than $i_T(d \rightarrow \infty)$ (17.3 nA). As the tip was scanned over the 50-μm fiber, the current decreased because the fiber interfered with diffusion to the tip. This decrease in current was smaller when the tip was scanned at a larger distance above the fiber (Figure 9B). Note that the width (in time) of the current decrease region corresponds approximately to that of the fiber at this scan rate (dx/dt) and that the symmetrical shape of the i_T vs t (or x) is that

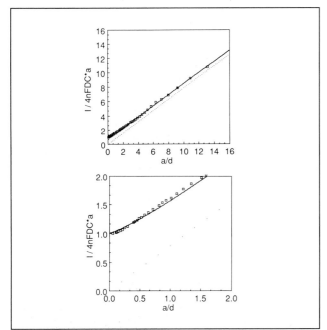

Figure 8. Normalized current vs a/d, where a = tip radius (5 μm). Dotted line with zero intercept is plot of thin layer eq 1. Solid curve is the computer-simulated theoretical curve. Experimental data are shown by open squares.

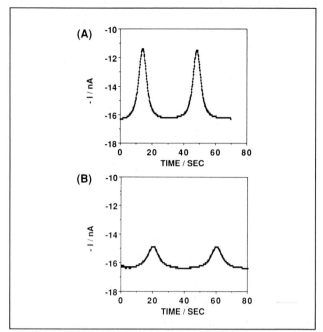

Figure 9. (A) Tip current for 5 μm radius Pt disk held at $d < 60$ μm with a 50 μm diameter glass fiber and scanned at 6 μm/s. $E_T = 1.0$ V vs AgQRE. Solution as in Figure 6. Scan direction reversed at 30 s. (B) $d > 68$ μm, scan direction reversed at 40 s.

expected for the circular cross section of the glass fiber. Moveover, the i_T vs t trace, when the tip scan direction is reversed, is essentially the mirror image of the forward scan at both values of d. The ability to examine an insulating substrate is unique to this mode of operation and to the SECM as compared to the STM. Insulators can be examined by the atomic forces microscope based on different principles (23).

Thus in the feedback mode, the difference in the conductive nature of the substrate determines whether $i_T > i_T(d \to \infty)$ (conductor) or $i_T < i_T(d \to \infty)$ (insulator). To demonstrate this effect on a substrate with both types of regions, the tip was scanned above a structure made by folding a 40 μm thick Pt foil, insulating between the folds with epoxy cement, and cementing this between two glass microscope slides (Figure 10B). The surface of this structure was polished flat with sandpaper (grit 600) and 5-μm diamond paste. When the tip was scanned ($d < 10$ μm) above the glass region, I_T was smaller than the $i_T(d \to \infty)$ of 17.3 nA (Figure 10A). When the tip was above the conductive Pt region, $i_T > i_T$ ($d \to \infty$), with the epoxy region represented by a current dip. Again, the pattern produced a mirror image when the scan direction was reversed.

This SECM feedback approach is potentially a powerful one for obtaining a two-dimensional (x–y) scan across a surface with both conductive and insulating regions. Scans with the substrate held at different potentials may also be informative in distinguishing different conductive regions with different heterogeneous electron transfer rates for the substrate reaction (O $\to$ R). Under these conditions the choice of the redox couple (O/R) will be important. The couple should also be selected with regard to the chemical nature of the substrate, i.e., couples with very positive $E°$ values should be avoided with

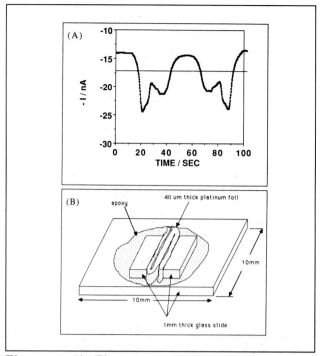

Figure 10. (A) Tip current for 5 μm radius Pt disk scanned at 6 μm/s about 10 μm above a Pt/glass structure shown in (B). Conditions as in Figure 6. Scan direction reversed at 60 s.

oxidizable substrates, such as copper, and couples that might cause gas evolution, e.g., $MV^{2+/+}$ (MV^{2+} = methyl viologen), should not be used with Pt substrates, where H_2 formation could occur. This would be especially important with substrates such as Pt on mica, which are unstable in the hydrogen evolution region. Although the resolution seen here is low, because the tip diameter is rather large, the principles demonstrated should apply with smaller tips. As with the STM, the tip shape is important in obtaining good resolution.

CONCLUSIONS

The results given here suggest that SECM can be employed to investigate surfaces immersed in liquids with good resolution. Although the x–y resolution demonstrated here is only of the order of micrometers, improvements in tip preparation and control should allow higher resolution to be obtained. The preparation of flat, small, disk-shaped electrodes surrounded by a thin insulating layer is probably the most important factor in improving the resolution. Another problem that must be considered as the tip is scanned above a substrate is the perturbation of the diffusion layer by the tip structure itself. Clearly, the smaller the tip, the less the diffusion layer perturbation will be. The feedback mode seems to be especially useful for several reasons. It can be applied with both conductive and nonconductive substrates and will probably be less sensitive to the tip perturbation effects mentioned above, since the diffusion layer of interest in the measurement originates at the tip rather than at the substrate. It should also be less sensitive to interelectrode and external coupling effects and to charging current perturbation.

Although the results discussed here emphasized only electrochemical measurements in SECM, other modes of operation should be possible. For example, electrogenerated chemiluminescence (ECL) (24) or inverse photoemission spectroscopic (IPS) (25, 26) modes of investigating conductive substrates would involve detection of light emitted as the tip is scanned across the surface with tip and substrate potentials adjusted to generate suitable intermediates. Similarly, an electroluminescent (EL) method might be of interest in the investigations of semiconductors (27, 28). Note that in these methods the resolution attainable is a function of the tip size and distance from substrate rather than the wavelength of the emitted radiation. Alternatively, the effect of irradiating the substrate (e.g., a semiconductor) with suitable radiation (larger than the band gap) and recording the tip photocurrent as a function of tip position might be useful in the mapping of a surface in photoelectrochemical (PEC) experiment. This would be analogous to the technique previously described where a laser beam was rastered over a semiconductor surface to obtain information about surface states and structures (29). Note that this PEC mode of operation has already been employed to etch thin lines in a GaAs substrate immersed in solution (8).

LITERATURE CITED

(1) Binnig, G.; Rohrer, H. *Helv. Phys. Acta* **1982**, *55*, 726–735.
(2) Bard, A. J.; Faulkner, L. R. *Electrochemical Methods, Fundamentals, and Applications*; Wiley: New York, 1980; Chapter 1.
(3) Isaacs, H. S.; Kendig, M. W. *Corrosion* **1980**, *36*, 269–274.
(4) Isaacs, H. S.; Vyas, B. In *Electrochemical Corrosion Testing*, ASTM STP 727; Mansfeld, F., Bertocci, U., Eds.; American Society for Testing and Materials: Philadelphia, PA, 1981; pp 3–33.
(5) Rosenfeld, I. L.; Danilov, I. S. *Corros. Sci.* **1967**, *7*, 129–142.
(6) Engstrom, R. C.; Meany, T.; Tople, R.; Wightman, R. M. *Anal. Chem.* **1987**, *59*, 2005–2010.
(7) Lin, C. W.; Fan, F.-R. F.; Bard, A. J. *J. Electrochem. Soc.* **1987**, *134*, 1038–1039.
(8) Craston, D. H.; Lin, S. W.; Bard, A. J. *J. Electrochem. Soc.* **1988**, *135*, 785–786.
(9) Bard, A. J.; Faulkner, L. R. *Electrochemical Methods, Fundamentals, and Applications*; Wiley: New York, 1980; p 567.
(10) Wightman, R. M.; Wipf, D. O. In *Electroanalytical Chemistry*; Bard, A. J., Ed.; Marcel Dekker: New York, 1988; Vol 15.
(11) Fleischmann, M.; Pons, S.; Rolison, D.; Schmidt, P. P. *Ultramicroelectrodes*; Datatech Systems: Morgantown, NC, 1987.
(12) Garcia, E.; Kwak, J.; Bard, A. J. *Inorg. Chem.*, in press.
(13) Kwak, J. Ph.D. Dissertation, University of Texas at Austin, 1989.
(14) Davis, J. M.; Fan, F.-R. F.; Bard, A. J. *J. Electroanal. Chem.* **1987**, *238*, 9–31.
(15) Hubbard, A. T.; Anson, F. C. In *Electroanalytical Chemistry*; Bard, A. J., Ed.; Marcel Dekker: New York, 1970; Vol. 4, p 129.
(16) Albery, W. J.; Hitchman, M. L. *Ring-Disc Electrodes*; Clarendon Press: Oxford, 1971.
(17) Bruckenstein, S.; Miller, B. *Acc. Chem. Res.* **1977**, *10*, 54–61.
(18) Vetter, K. J. *Electrochemical Kinetics*; Academic Press: New York, 1967; pp 200–213.
(19) Bard, A. J.; Crayston, J. A.; Kittlesen, G. P.; Varco Shea, T.; Wrighton, M. S. *Anal. Chem.* **1986**, *58*, 2321–2331.
(20) Varco Shea, T.; Bard, A. J. *Anal. Chem.* **1987**, *59*, 2101–2111.
(21) Liu, H.-Y.; Fan, F.-R. F.; Bard, A. J. *J. Electrochem. Soc.* **1985**, *132*, 2666–2668.
(22) Kwak, J.; Bard, A. J. Manuscript in preparation.
(23) Binnig, G.; Quate, C. F.; Gerber, C. *Phys. Rev. Lett.* **1986**, *56*, 930–933.
(24) Bard, A. J.; Faulkner, L. R. *In Electroanalytical Chemistry*; Bard, A. J., Ed.; Marcel Dekker: New York, 1977; Vol. 10, Chapter 1.
(25) MacIntyre, R.; Sass, J. K. *J. Electroanal. Chem.* **1985**, *196*, 199–202.
(26) Ouyang, J.; Bard, A. J. *J. Phys. Chem.* **1987**, *91*, 4058–4062.
(27) Beckmann, K. H.; Memming, R. *J. Electrochem. Soc.* **1969**, 116, 368–373.
(28) Fan, F.-R. F.; Leempoel, P.; Bard, A. J. *J. Electrochem. Soc.* **1983**, *130*, 1866–1875.
(29) Folmer, J. C. W.; Turner, J. A.; Parkinson, B. A. *Inorg. Chem.* **1985**, *24*, 4028–4030.

Received for review September 12, 1988. Accepted October 17, 1988. The support of this research by the Robert A. Welch Foundation and the Texas Advanced Technology Research Program is gratefully acknowledged. We thank the U.S.–Israel Education Foundation for providing a Fulbright Fellowship to O. Lev.

Reprinted from *Anal. Chem.* **1989**, *61*, 132–38.

Although the term "microanalytical chemistry" has been used for many years, few analytical chemists have successfully confronted the challenge of trace analysis on a sample with a total volume of a few nanoliters ("nanoanalytical chemistry"?). In this article, Kennedy and Jorgenson accomplish this goal on a sample of interdisciplinary interest: the different neurotransmitter types and concentrations present in a single neuron. Their work not only demonstrates the great power of the separation and detection techniques developed in Jorgenson's laboratory during the previous decade, it also shows that neurons can contain more than one neurotransmitter, a principle not previously demonstrated at this fundamental level.

R. Mark Wightman
University of North Carolina–Chapel Hill

Quantitative Analysis of Individual Neurons by Open Tubular Liquid Chromatography with Voltammetric Detection

Robert T. Kennedy and James W. Jorgenson*
Department of Chemistry, University of North Carolina at Chapel Hill, Chapel Hill, North Carolina 27599-3290

The ability to analyze individual cells is often important in biology because of the heterogeneity of tissue; this is especially true in the area of neurobiology. A method is described for the determination of trace levels of organic compounds in individual cells by open tubular liquid chromatography with voltammetric detection. In the method, a cell is isolated, an internal standard is added, the cell is homogenized and centrifuged, and the supernatant is injected directly onto the chromatography column. Since data are collected in both the electrochemical and chromatographic domains, the resolution of the method is better than that obtained by using amperometric detection. The combination of voltammetry and chromatography also aids in the identification of compounds. By use of this method three different neurons, D2, E4, and F1, from the land snail *Helix aspersa* are analyzed. The data show that the cells give certain unique and repeatable chemical profiles. Dopamine, serotonin, tyrosine, and tryptophan were identified and quantified in two of the cells at the femtomole level. In the third cell, only the two amino acids were observed and measured. The quantitative data indicate that the method is at least as reliable as other methods that have been applied to single cells and considerably more sensitive. The combination of qualitative and quantitative information allows for the chemical mapping of cells.

INTRODUCTION

In a number of areas of biology it is useful to know the chemical contents of individual cells and how cells differ chemically. Such knowledge is valuable in determining a cell's function. This is especially true in the area of neurobiology where cell heterogeneity is well known. An area of active research concerns the role of individual neurons in memory and learning and how biochemical plasticity may be involved in these processes (1). Another, older question concerns the use of neurotransmitters and individual neurons. Dale's hypothesis, as expounded by Eccles, states that individual neurons contain and release only one neurotransmitter at all of its branches and terminals. This concept, as stated, appears to be too limited in light of modern evidence but is still a subject of interest (2). It is apparent that in both of these areas it would be valuable to be able to analyze cells individually.

Because of the interest in the chemistry of individual neurons, a number of analytical methods have been developed for them. These methods include micro thin-layer chromatography (TLC) (3), gas chromatography/mass spectrometry (GC/MS) (4), high-performance liquid chromatography (HPLC) with amperometric detection (5), and enzymatic radiolabeling (6). All of these methods provide good information; however they have important limitations. Micro TLC is not sensitive enough to actually analyze individual cells and therefore requires the pooling of four to six giant cells for analysis. Also the method requires an extensive sample preparation, including the derivatizing of compounds for detection. GC/MS of individual cells is limited, because of a lack of sensitivity, to using the mass spectrometer in the selected ion monitoring mode; therefore a limited number of compounds can be determined at one time. This limitation is important because it means that the compounds to be determined must be chosen before the experiment. Also this method often requires sample derivatization in order to improve volatility of analytes. HPLC with amperometric detection is also limited by sensitivity and can only be used for cells that contain high levels of analyte. Enzymatic radiolabeling is sensitive; however it requires that the analyte be selected before the experiment, and the specificity of the method is limited by cross-reactivities.

We have recently begun to develop methods for the analysis of individual cells by using open tubular liquid chromatography (OTLC) with voltammetric detection (7). The separation is performed inside of a capillary tube which typically has an internal diameter of 15–20 μm. The stationary phase is bound to the inner wall of the capillary rather than to particles packed inside the column as in conventional liquid chromatography. With voltammetric detection, the potential on the detector electrode is scanned so that voltammograms can be obtained on compounds as they elute from the column.

A number of properties of this method appear to make it a good choice for the analysis of single cells. First of all, OTLC is a high-resolution separation method. Columns with inner diameters of 15 μm, such as those used in this report, equal or exceed the resolving power of modern HPLC columns. Another important aspect of the columns is that, because of their small dimensions, they require small samples. A 15 μm i.d. column that is 2 m long has a total volume of 390 nL and a typical injection volume of 5 nL. Finally, the detectors that have been developed for OTLC have good mass sensitivity. The voltammetric detector used in these experiments has a limit of detection of 0.1 fmol for hydroquinone. This same detector, used in an amperometric mode, has a detection limit of 1 amol. In both cases the detection limit is well below the detection limit of any other method that has been applied to individual cells. Because of this combination of properties, OTLC allows for the simultaneous determination of a variety of compounds at trace levels in samples of nanoliter volumes.

The original method which we developed using OTLC had some drawbacks with respect to quantification and therefore was not as reliable as desired (7). First of all, there was no way to account for variables associated with the sample preparation and injection. In the method, the cell was homogenized and centrifuged and the supernatant removed and injected onto the chromatography column. Unfortunately it was not possible to remove all of the supernatant or to know how much of the supernatant was removed. Analyte could also be lost due to adsorption to glassware that the sample came in contact with. These problems resulted in artificially low values for compounds and an extra source of variability. Quantification was also limited because reponse of the detector electrode could change from run to run because of adsorption to the electrode surface and other forms of electrode fouling.

In this report we describe a new method, also based on open tubular liquid chromatography with voltammetric detection, for analyzing individual cells which takes into account these problems and apply it to the analysis of three, identified neurons from the land snail *Helix aspersa*. Variables in the sample preparation are monitored by using an internal standard. The quantification by the detector is improved by using an electrochemical cleaning procedure and by using daily calibrations. The data obtained indicate that cells have unique and repeatable chromatograms that allow for the chemical mapping of cells. The data also show that the putative neurotransmitters dopamine and 5-hydroxytryptamine (serotonin) coexist in two of the neurons.

EXPERIMENTAL SECTION

Sample Preparation. The cells used in this study were three giant neurons (diameters 120–140 μm) from the land snail *Helix aspersa*. Specifically, the neurons were those labeled F1 (right parietal ganglion), E4 (visceral ganglion), and D2 (left parietal ganglion), according to the map of Kerkut et al. (8). These neurons were used because their positions were well established in the maps and so could be readily identified from preparation to preparation.

To perform the analysis, the *Helix* brain was dissected out as previously described and mounted on a microscope slide under Ringer solution (9). The cells were isolated by using microdissection techniques (10). Only the cell body, not the axon, was used for the analysis. Once isolated, the cells were rinsed in Ringer solution and then transferred, with approximately 5–10 nL of Ringer, to a 500-nL microvial by using a pipet which had an inner diameter of approximately 120 μm.

Next 1–2 nL of internal standard solution, which was 0.1 mM 3,4-dihydroxybenzylamine (DHBA) dissolved in the mobile phase with 2 mM ascorbic acid, was added with a pneumatic microsyringe. The ascorbic acid was added to prevent the oxidation of analytes. The syringe used in this step is similar in design to other syringes that were developed to perform intracellular injections (11, 12). The syringe was calibrated by injecting aqueous solutions into mineral oil and determining the volume dispensed from the size of the resultant spherical droplet. The size of the droplet was determined by measuring its diameter by use of a microscope fitted with a reticle. The details of the syringe and its use as a dispenser of internal standard have been described previously (13).

Once the internal standard was added, the cell was homogenized inside the microvial with a miniature glass rod. The resultant solution was then centrifuged at 3000*g* for 10 min and the supernatant removed and injected directly into the column by using the microinjector which has been described before (7). All of the sample preparation and injections were done with the aid of a Wolfe Selectra II stereomicroscope, a Narishiga MM-33 micromanipulator, and an Oriel micropositioner.

Chromatography. The chromatography columns were borosilicate glass capillaries with inner diameters of 15–19 μm and lengths of 220–250 cm and were fabricated in the lab (14). Dimethyloctadecylsilane was chemically bound to the inner wall of the capillary to serve as the stationary phase. The mobile phase was 0.1 M citrate buffer adjusted to pH 3.1 with sodium hydroxide. The mobile phase also contained 0.21 mM dimethyloctylamine and 0.62–0.95 mM sodium octyl sulfate. The octyl sulfate was varied as needed to keep the chromatographic retention of the biogenic amines constant. This was necessary due to changes in capacity factors for the analytes as the columns aged.

Detection. The detector used in these experiments was a carbon fiber microelectrode which has been described before (15–18). The fiber had a diameter of 9 μm and a length of 1 mm and was inserted into the outlet end of the column by using micropositioners. The potential on the electrode was ramped from 0.0 to +1.3 V versus Ag/AgCl at 1.0 V/s. After a ramp was completed, the potential was stepped back to 0.0 V and held for 1.3 s before initiating the next voltage ramp. Current measurements were made only during the forward scan and the data were collected at 100 points per second. The effect of charging current was reduced by using a background subtraction technique. The applied voltage, data acquisition, and data manipulation were all done with a microcomputer.

Chemicals. All standards, reagents, and mobile phase constituents were used as received from Sigma Chemical Co. except for the dimethyloctylamine, which was from Aldrich Chemical Co.

RESULTS AND DISCUSSION

An example of the data that is obtained by using this method is shown in surface plot form in Figure 1. In the plot, each line parallel to the time axis corresponds to a chromatogram taken at the voltage indicated on the potential axis. Each line parallel to the potential axis represents a voltammogram taken at the time indicated. All of the voltammograms were subtracted from a scan obtained prior to the elution of any compounds. The chromatovoltammogram is shown to begin just after the elution of the first compound because there was a large unretained peak due to the added ascorbic acid which interferes with the other peaks. When the plot is studied, it is important to realize that it is from an individual cell and that each of the larger peaks represents hundreds of femtomoles and the smaller peaks represent just a few femtomoles injected onto the column. Another important aspect of these data is the added resolution that is obtained as a result of combining voltammetry with chromatography. A number of peaks that would not have been resolved chromatographically were resolved in the electrochemical domain. The voltammetry also adds qualitative information to the data which improves the reliability of the identification of compounds.

Unknown compounds can be identified by matching their

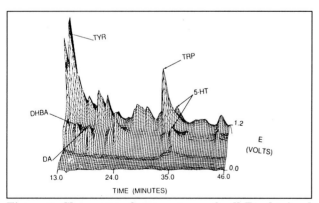

Figure 1. Chromatovoltammogram of cell F1 obtained by using the method described in the text; the column had a 19-μm i.d. The peak labels are as follows: DA is dopamine, DHBA is 3,4-dihydroxybenzylamine, 5-HT is serotonin, TYR is tyrosine, and TRP is tryptophan. The voltammogram for serotonin has two peaks due to the two functionalities which are oxidizable at the electrode, namely the phenol group and the indole nitrogen.

chromatographic retention times and voltammetric peaks with standards. With this method, four compounds, tyrosine, tryptophan, dopamine, and serotonin, were identified in the cell F1 as shown. The same four compounds were found in the cell E4, but only the two amino acids were present in D2. The fact that serotonin and dopamine were not detected in cell D2 indicates that they are present at less than 0.14 and 0.15 fmol respectively, based on the detection limit for those compounds. Other compounds that were tested for, but not observed, include 3,4-dihydroxyphenylalanine (DOPA), 3,4-dihydroxyphenylacetic acid (DOPAC), 5-hydroxytryptophan (5-HTP), and 5-hydroxyindolacetic acid (HIAA). The detection limits for these compounds set their upper limits in the cells as follows: DOPA, <1.5 fmol; DOPAC, <0.54 fmol; 5-HTP, <0.11 fmol; and HIAA, <0.36 fmol. In all cases the detection limits were calculated by determining the amount injected which gave a response near the detection limit, that is, a signal to root mean square noise ratio (S/N) less than 10, and then extrapolating to the amount that would give a response of three times the S/N. These particular compounds were chosen for possible identification because they are common precursors and metabolites of the neurotransmitters that had been found in the neurons.

Quantification. Once the compounds were identified, it was possible to determine their amounts in the cell bodies. Quantification can be done a number of ways because of the nature of the data obtained by using this method. Compounds can be measured according to the area or height of chromatographic peaks at a given voltage, area or height of volammetric peaks at the chromatographic maximum, or the volume of the chromatovoltammetric peaks. Quantification of identified compounds for this work was done by measuring the height of the voltammetric wave of the compound of interest at its chromatographic peak. This method was chosen because it was the least susceptible to interference from closely retained compounds.

Electrode Calibration. The reliability of this method of quantitating compounds was tested initially by calibrating the electrode for the four identified compounds and the internal standard for five different amounts injected. The data for the test calibration are summarized in Table I. The data are treated in a log–log format because the amounts injected cover 3 orders of magnitude. The slopes of the log–log data show that there is curvature to the calibration lines. This curvature is possibly due to the effect of the iR drop in the detector cell and other nonideal behavior such as adsorption to the electrode; however the correlation coefficients of the logarithmically transformed data indicate that the curvature does not diminish the reliability of the calibrations.

For actual measurements the electrode was calibrated daily with respect to amount injected by running standards of the identified compounds at four different concentrations. The calibration curves encompassed the amounts found in the cells. All of the daily calibration plots, used in log–log format, had linear correlation coefficients greater than 0.99 for all compounds measured. The daily calibrations helped to account for the day-to-day variations in the electrode response

Table I. Calibration Data for Compounds Which Were Quantitated[a]

	DA	DHBA	TYR	5-HT	TRP
slope	0.678	0.702	0.795	0.786	0.649
intercept	9.181	9.551	10.603	10.370	9.012
linear corr coeff	0.9994	0.9960	0.9996	0.9959	0.9937
min injected amt, fmol	1.35	3.41	1.41	3.18	3.11
max injected amt, pmol	1.68	1.71	1.05	1.59	1.55

[a]The compound abbreviations are the same as in Figure 1.

Table II. Measured Amounts of Identified Compounds in *Helix aspersa* Neurons[a]

	measured amount, fmol			
cell	TYR	DA	TRP	5-HT
D2	340 ± 98	N.D.	163 ± 20	N.D.[b]
E4	546 ± 422	6.2 ± 1.9	59.1 ± 22.2	30.6 ± 13.9
F1	486 ± 144	70.7 ± 6.2	89.4 ± 24.2	42.8 ± 27.6

[a]The compound abbreviations are the same as in Figure 1. The values in the table are the average of five runs with the standard deviation given to the side. The conditions of analysis and method of quantification are given in the text. The cells had an average volume of 1.2 nL as determined from the diameter and assuming that the cell was a sphere. [b]One run of cell D2 contained 2.4 fmol of serotonin, but the other runs had no detectable serotonin. This occurrence was presumably due to a contamination.

Table III. Measured Amounts of Identified Compounds in *Helix aspersa* Neurons Uncorrected for Recovery of the Internal Standard[a]

	measured amount, fmol			
cell	TYR	DA	TRP	5-HT
D2	92.0 ± 16.6	N.D.	45.5 ± 17.7	N.D.
E4	249 ± 213	3.0 ± 1.1	26.21 ± 7.5	15.7 ± 9.8
F1	99.1 ± 41.6p	12.2 ± 4.2	18.2 ± 7.0	8.5 ± 5.3

[a]Table is in the same format as Table II.

due to changes in flow rates and electrode conditions. For a period of 21 days, the slope and intercept of the daily calibration lines had a relative standard deviation of around 10% for all of the compounds indicating that the day-to-day variability of the electrode response is significant enough to make the daily calibrations necessary. Within a given day, however, the electrode response was more reproducible. A series of five injections, the injection volume containing 100 fmol of each of the four compounds identified in the cells, had the following relative standard deviations of voltammetric wave height: 2.2% for TYR, 3.1% for DA, 2.0% for TRP, and 3.9% for serotonin. The electrodes were cleaned electrochemically, using a previously described procedure (17), before each run to ensure that they were in the same condition for each run. If the electrochemical cleaning was not used, the reproducibility was considerably worse and calibration of the electrode was impossible. The irreproducibility of the electrode response without the electrochemical cleaning is presumably due to fouling of the electrode surface by one or more of the analytes or their oxidation products.

Quantitative Results. The identified compounds were quantified in the cells as described above and the measured levels with the standard deviations are shown in Table II. The values shown have been corrected for losses during sample preparation by dividing them by the ratio of DHBA detected to DHBA added. This ratio represents the fraction of the DHBA recovered during the sampling, and its value ranged from 0.2 to 0.5. The importance of using the internal standard can be evaluated by comparing the results in Table II with those in Table III, which show the uncorrected values, that is, the numbers that would have been obtained had the internal standard not been used. The tables show that, if the internal standard had not been used, considerably lower averages would have been obtained due to low recoveries, indicating that in terms of accuracy the internal standard is extremely important. Comparing the relative standard deviations, however, shows that the internal standard is not as important in terms of precision. In most cases the relative standard deviations is slightly lower with the internal standard than without it, although there are a few exceptions. This result implies that the major source of variability in the observed values is not the recovery.

The accuracy of the values obtained is difficult to evaluate because there has been little quantitative work done on these particular cells; however, they can be compared to what has been obtained with other methods on other cells. The concentration of the neurotransmitters, dopamine and serotonin, is comparable in these cells to what has been measured for other neurotransmitters in other cells (3–5, 19, 20). For example, we found 2.6×10^{-5} M for serotonin in cell E4 and 3.6×10^{-5} M in cell F1, based on a typical cell volume of 1.2 nL, compared to 3.8×10^{-4} M for serotonin in a neuron from the cerebral ganglion of *Helix pomatia* as measured by Osborne (3) and 1.1×10^{-5} M for serotonin in a neuron from *Aplysia* measured by Brownstein et al. (20). The lowest level measured for one of the putative neurotransmitter compounds in this report was for dopamine in cell E4, where the concentration was roughly 5.2×10^{-6} M. Neurotransmitter levels as low as this are not unprecedented; octopamine has been measured at 2.5×10^{-6} M and lower in certain *Aplysia californica* neurons (19). The significance of the wide range of concentrations of neurotransmitter compounds in single neurons is not well understood. The levels of the amino acids are even more difficult to evaluate since no quantitative methods have previously been applied for their measurement at the single cell level.

The variance of the values, as indicated by the standard deviations and relative standard deviations, obtained by this method are higher in several cases than what is normally encountered in biological analysis. The source of variation may be experimental error or natural biological variation. The evidence that has been gathered so far indicates that biological variation may be the most important contributor. The reproducibility of the standards runs and the results from the comparison of the data with and without the internal standard suggest that the variability cannot be attributed to changes in the electrode response or the amount of recovery from the sample vial. Possible experimental sources of error that have not been accounted for include reproducibility of the homogenization procedure and contamination of the sample from an exogenous source. This latter possibility, however, does not seem as likely to be important given that the cells were rinsed before they were analyzed. It seems more likely that the source of the variability is the cells themselves. Slight differences in the diameter of a cell result in large differences in the volume of the cell which will presumably affect the amount of the compounds being measured. Differences in the condition of the snails and the individual neurons, brought about by external factors such as food consumption, at the time of the analysis may also affect the measured amounts and cause variance. The view that biological variance is the primary source of the observed variability is supported by the fact that analysis of a variety of neurons by a variety of techniques have given similar relative standard deviations to those obtained here. As an example, the relative standard deviation of serotonin in leech neurons measured by GC/MS was 60% (4), similar to the values of 45% in cell E4 and 64% in cell F1 obtained here. Another example is the measured relative standard deviation of dopamine in a different leech neuron, measured by HPLC with electrochemical detection, was 17.8% (5), again similar to the value of 8.8% in cell F1 that was obtained here. Experiments are currently being planned to determine more completely the sources of variation. The comparisons with other methods indicate that the quantitative data obtained by the method as described in this report is at least as reliable as what is obtained by other methods. Furthermore, these data were obtained simultaneously on a variety of compounds at significantly lower levels than possible by other methods.

Qualitative Results: Coexistence of Neuroactive Substances. As indicated before, dopamine and serotonin were found to coexist in neurons F1 and E4. This result is interesting because it demonstrates the coexistence of two classical neurotransmitters in individual neurons. Since these data were obtained on neurons that were isolated by hand, there is the possibility that the coincidence of both compounds was due to contamination from surrounding tissue. This explanation does not seem likely however, since both compounds appeared consistently in all runs of the cells. If the compounds were from an exogenous source, a greater degree of variability would be expected. Also, the fact that cell D2 did not show any dopamine or serotonin indicates that the levels measured in the other cells were not due to an artifact of the analysis procedure, such as contamination from glassware or the Ringer solution that the brain is bathed in.

Although the coexistence of dopamine and serotonin is un-usual, there is other evidence for it in these cells in the literature. Kerkut and Walker, using the Falck-Hillarp staining method to study the *Helix* brain, reported that certain cells gave results that indicated that they contained both serotonin and dopamine (21). These cells were not identified, but they were found in the same region as F1 and E4, that is, at the border between the right parietal and visceral ganglia. Unfortunately, this result was not conclusive because, in the staining method, a high concentration of serotonin gives a result similar to dopamine. Kerkut and Walker also observed, using the same method, that these same cells could uptake both DOPA and 5-HTP and convert them to dopamine and serotonin, respectively (21). Emson and Fonnum developed a method that allowed the enzymatic activity of individual neurons to be measured and used it to study *Helix* neurons (22). They reported that cells which were identified as containing monoamines from histochemical fluorescence, contained aromatic L-amino acid decarboxylase. The aromatic L-amino acid decarboxylase had approximately equal activity with DOPA, to form dopamine, as with 5-HTP, to form serotonin. These data would indicate it is at least possible for the cells to contain both neurotransmitters and that, if either of the precursors were in extracellular space, these neurons could take them up and convert them to the corresponding neurotransmitter.

It is also interesting to note that Emson and Fonnum observed that none of the cells in the left parietal ganglion contained aromatic L-amino acid decarboxylase. This correlates well with the results in Table II for cell D2, which is from the left parietal ganglion, that indicate the absence of both of the monoamines.

Cell Mapping. The method used to analyze the cells in this work also allows for individual cells to be compared on a qualitative basis. Figure 2 is a series of single potential chromatograms from cell E4 of three different snails obtained on the same column in a 3-day period. The chromatograms shown represent a "slice" taken out of the chromatovoltammetric data and were constructed by using the computer. It is interesting that, in addition to the identified compounds which are observed in the chromatogram of each of the specimens, a number of unknown peaks appear reproducibly in each case. The relative abundance of each of the peaks appears to fluctuate somewhat, perhaps indicating differences in the state of the neuron at the time of the analysis, but in general the relative size of the peaks remains constant. The different current levels are indicative of different levels of the recovery of the cellular content as determined from the internal standard. Thus the top chromatogram was from a relatively low recovery run and the bottom two were from higher recoveries. This is also reflected in the fact that a number of the lower level compounds detected in the bottom two chromatograms are lost in the noise in the top chromatogram. A similar consistency of results is observed for the D2 cells shown in Figure 3.

It is also interesting to make intercellular comparisons based on the chromatograms in Figures 2 and 3. The cells had similar amounts of amino acids, but D2 contained no detectable levels of dopamine or serotonin as E4 did, as mentioned

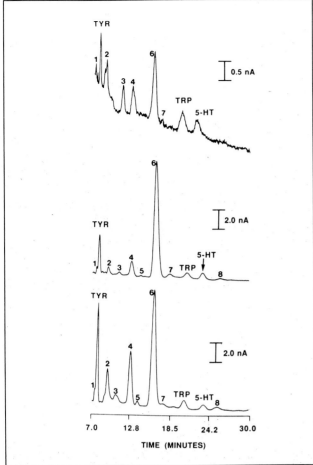

Figure 2. Single potential chromatograms at +1.00 V from cell E4. The abbreviations are the same as in Figure 1. The data were obtained by using a 15-μm i.d. column. The numbered peaks indicate compounds of unknown identity but appear reproducibly in the chromatograms and are presumably due to the same compound in each case. The dopamine and DHBA peaks do not show up in this chromatogram because they are oxidized at a lower potential.

before. The single potential chromatograms also reflect that there are other important differences in the chemical makeup of the cells as well. These chromatograms indicate that the method can distinguish between cells based on their chemical contents and that specific cells give reproducible chromatograms.

A significant difference between the cells is the presence of the large peak labeled "6" in the chromatogram of cell E4 and its absence in both D2 and F1. Its unique occurrence and high concentration suggest some significant role in the cell's function. This compound has a voltammetric peak of +0.98 V vs Ag/AgCl indicating that it is not a catechol or hydroxyindole. An attempt was made to identify peak "6" based on matching its retention time and voltammetric peak, as had been done for the other compounds, with several derivatives of the aromatic amino acids. No match could be found. This problem highlights the need for the development of improved instrumentation for OTLC. It would be advantageous to have detectors, such as mass spectrometry, for OTLC which could give

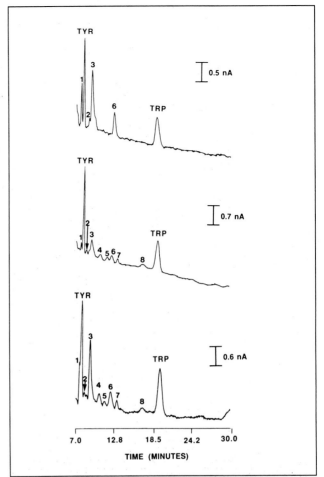

Figure 3. Single potential chromatograms from cell D2. See Figure 2 legend for description of data. Peak numbers do not correspond to peaks with the same number in Figure 2.

more qualitative information. This is a direction we are currently working in (23).

CONCLUSIONS

In spite of this limitation, this study has shown that the method is of value in the analysis of single cells and that chemical information from single cells is of use in the study of nervous tissue. This study has shown that two neurons consistently contain both serotonin and dopamine in femtomole levels while a third neuron contains neither. The data also indicate that the amino acids tyrosine and tryptophan are present in similar levels in all three neurons. The chromatograms from each cell type are characteristic of that cell type, indicating a unique chemical makeup. The presence of a high level of an unidentified compound in cell E4 was found. These conclusions were possible only because individual cells were used in the analysis and because of the ability to analyze simultaneously for several compounds with high sensitivity. If whole brain or brain sections had been used, much of the information would have been lost.

The type of information obtained by using this method suggests a number of applications as the method is refined. They include pharmacological studies on single cells, studies on the chemical basis of single cell function, and studies on the chemical basis of cellular differentiation. In principle, the method can be scaled to work with smaller cells, or less concentrated compounds in giant cells. This will require improvements in OTLC instrumentation, such as the development of smaller columns, and more sensitive detectors, such as laser-induced fluorescence. Future work will also be required to determine the major source of the variance observed in the quantitative data.

LITERATURE CITED

(1) Black, I. B.; Adler, J. E.; Dreyfus, C. F.; Friedman, W. F.; La-Gamma, E. F.; Roach, A. H. *Science* **1987**, *236*, 1263–1268.
(2) *Coexistence of Neuroactive Substances in Neurons*; Chan-Palay, V., Palay, S., Eds.; Wiley: New York, 1984.
(3) Osborne, N. N. *Nature* **1977**, *270*, 622–623.
(4) McAdoo, D. J. In *Biochemistry of Characterized Neurons*; Osborne, N. N., Ed.; Pergaman: New York, 1978; pp 19–45.
(5) Lent, C. M.; Meuller, R. L.; Haycock, D. A. *J. Neurochem.* **1983**, *41*, 481–490.
(6) McCaman, R. M.; Weinrich, D.; Borys, H. *J. Neurochem.* **1973**, *21*, 473–476.
(7) Kennedy, R. T.; St. Claire, R. L.; White, J. G.; Jorgenson, J. W. *Mikrochim. Acta* **1987**, *II*, 37–46.
(8) Kerkut, G. A.; Lambert, J. D. C.; Gayton, D.; Walker, R. J. *Comp. Biochem. Physiol. A* **1975**, *50A*, 1–25.
(9) Walker, R. J. *Exp. Physiol. Biochem.* **1974**, *1*, 331–345.
(10) Osborne, N. N. *Microchemical Analysis of Nervous Tissue*; Pergaman: New York, 1974.
(11) Corson, D. W.; Fein, A. *Biophys. J.* **1983**, *44*, 299–304.
(12) McCaman, R. E.; McKenna, D. G.; Ono, J. K. *Brain Res.* **1977**, *136*, 141–147.
(13) Kennedy, R. T.; Jorgenson, J. W. *Anal. Chem.* **1988**, *60*, 1521–1524.
(14) St. Claire, R. L., III. Ph.D. Thesis, University of North Carolina at Chapel Hill, 1986.
(15) Knecht, L. A.; Guthrie, E. J.; Jorgenson, J. W. *Anal. Chem.* **1984**, *56*, 479–482.
(16) St. Claire, R. L.; Jorgenson, J. W. *J. Chromatogr. Sci.* **1985**, *23*, 186–191.
(17) White, J. G.; St. Claire, R. L.; Jorgenson, J. W. *Anal. Chem.* **1986**, *58*, 293–298.
(18) White, J. G.; Jorgenson, J. W. *Anal. Chem.* **1986**, *58*, 2992–2995.
(19) Saavedra, J. M.; Brownstein, M. J.; Carpenter, D. O.; Axelrod, J. *Science* **1974**, *185*, 364–365.
(20) Brownstein, M. J.; Saavedra, J. M.; Axelrod, J.; Zeman, G. H.; Carpenter, D. O. *Proc. Natl. Acad. Sci. U.S.A.* **1974**, *71*, 4662–4665.
(21) Kerkut, G. A.; Sedden, G. B.; Walker, R. J. *Comp. Biochem. Physiol.* **1967**, *23*, 159–162.
(22) Emson, P. C.; Fonnum, F. *J. Neurochem.* **1974**, *22*, 1079–1088.
(23) de Wit, J. S. M.; Parker, C. E.; Tomer, K. B.; Jorgenson, J. W. *Anal. Chem.* **1987**, *59*, 2400–2404.

Received for review July 12, 1988. Accepted November 23, 1988. This work was supported by the donors to the Petroleum Research Fund, administered by the American Chemical Society, and the University Research Council of the University of North Carolina. R.T.K. received support from a North Carolina Governor's Board of Science and Technology Fellowship and from an American Chemical Society Analytical Fellowship sponsored by the Society for Analytical Chemistry of Pittsburgh.

Reprinted from *Anal. Chem.* **1989**, *61*, 436–41.

When we started the Center for Process Analytical Chemistry at the University of Washington, we polled industrial analytical chemists in order to find challenging problems. One of the most difficult was pH measurement in hot caustic brine—any direct probe would disintegrate. Callis used the partial least-squares technique of multivariate calibration to determine NaOH and NaCl concentrations in caustic brine by near-IR spectroscopy even though NaCl is spectrally transparent in the near-IR region.

Bruce Kowalski
Center for Process Analytical Chemistry, University of Washington

Measurement of Caustic and Caustic Brine Solutions by Spectroscopic Detection of the Hydroxide Ion in the Near-Infrared Region, 700–1150 nm

M. Kathleen Phelan, Clyde H. Barlow,[1] Jeffrey J. Kelly,[1] Thomas M. Jinguji, and James B. Callis*
Center for Process Analytical Chemistry, Department of Chemistry, BG-10, University of Washington, Seattle, Washington 98195

We have explored the feasibility of caustic measurement by direct detection of hydroxide ion using optical absorption spectroscopy in the near-infrared wavelength range 700–1150 nm. Unfortunately, the spectral features of the hydroxide ion are obscured by strong bulk water absorptions whose intensities and peak shapes are dependent upon temperature and the presence of electrolytes. Nevertheless, with the aid of difference spectra, second-derivative techniques, and multivariate spectral reconstruction, we have obtained a clear indication of the spectrum of the hydroxide ion. Its features include (a) a sharp absorption band at 965 nm that arises from the second overtone of the OH stretching motion localized on the hydroxide ion, (b) a broad absorption band centered at 1100 nm that arises from the binding of two water molecules to the hydroxide ion, and (c) a second sharp absorption band that is attributed to a combination stretch–bend transition arising from the concerted motion of the hydrated ion. Using this knowledge as a guide, we have developed multivariate analysis methods for determining hydroxide concentration of caustic brines in the range 0.01–5.0 M that are successful even in the presence of a large variable excess of NaCl. Such methods are suitable for implementation as process monitoring tools.

INTRODUCTION

Measurement of hydroxide concentration in caustic and caustic brine solutions is an important industrial problem. In the past, a number of techniques have been employed for caustic analysis (1). Among these methods, pH measurement by glass electrode is most common. However, at high pH, conventional glass electrodes become unstable and begin to suffer from interferences caused by the presence of other cations, most notably sodium (2). As a result, other techniques have been developed for caustic determination, such as index of refraction, conductivity, on-line titration, and flow injection analysis. All of the above methods have the drawback that they require physical invasion of the process with some sort of probe or sampling device. In a review on the current status of process analysis, the potential of remote, noninvasive methodologies was described (3). One particularly promising approach to noninvasive analysis involves the use of short-wavelength near-infrared (SW-NIR) spectroscopy in the wavelength region 700–1150 nm. This technique has the following advantages: (a) Measurements can be made remotely through quartz windows, using fiber optics to guide the light to and from the window; (b) path lengths can be long (many centimeters); (c) scanning fiber-optic spectrophotometers are available that are rugged, small, and relatively inexpensive (4); (d) signal-to-noise ratios are very high; and (e) good quantitative results are obtained on highly scattering samples. The major disadvantage of SW-NIR spectroscopy is low spectral resolution, which results in severe overlap of absorption spectra. This disadvantage is overcome with the aid of multivariate calibration methods (5), which usually work well in this spectral region due to the excellent signal-to-noise ratio.

Unfortunately, it is not straightforward to develop a spectroscopic method for determining hydroxide ion concentration because the major spectral features of this ion are expected to overlap the broad bands of water. To make matters worse, the exact details of the water absorption spectrum in the SW-NIR region remain a matter of some controversy (6, 7). In addition, the water spectrum is a strong function of temperature and is affected by the presence of dissolved ions (8).

Various types of qualitative studies have been carried out

[1]Permanent address: Department of Chemistry, The Evergreen State College, Olympia, WA 98505.

Table I. Composition of Mixtures Analyzed

	set 1	set 2	set 3	set 4
salt				
NaOH, M	0–5.027	0–5.015	0–0.465	0–1.011
NaBr, M				0–1.031
NaCl, M			2.33–3.72	0–1.037
temp, °C	20.8 ± 0.6	25.5 ± 0.2	20 ± 1	26 ± 1
no. of samples	20	11	21	20
reference	air	water	air	water

on ions in aqueous solutions (*9–11*). Most research relevant to the present work has been performed in the mid-infrared region and has shown that the symmetric OH vibration of aqueous alkali–metal hydroxide occurs around 3610 cm^{-1}, indicating strong hydrogen bonding from the hydroxide oxygen to a water hydrogen (*12, 13*). At the same time, as the hydroxide concentration rises, an intense continuous absorption grows in the 3000–2000-cm^{-1} region at the expense of the OH stretching vibration (3600–3000 cm^{-1}) of the bulk water (*14*).

Information provided by structural and ionic studies can provide a basis for developing quantitative methods. Hirschfeld (*15*) has used the effects of NaCl on the water bands in the long-wavelength region of the near-infrared (1200–2400 nm)—in particular, the effect of NaCl on the 1450- and 1900-nm water bands—to develop a quantitative method for salinity. Of greatest relevance here is the recent excellent work of Watson and Baughman (*16*), who explored the use of the near-IR spectral region from 1100 to 2500 nm for determining caustic concentration. In this paper, we have extended the spectroscopic approach of Watson and Baughman to a spectral region better suited to in situ measurement, used multivariate calibration methods to improve quantitation, and developed a rationale for the functional basis of the method.

EXPERIMENTAL SECTION

Materials. Sodium hydroxide, sodium chloride, and sodium bromide were all of reagent grade. Water was distilled in house.

Spectroscopy. Absorption spectra were measured on a Pacific Scientific 6250 scanning near-infrared spectrophotometer equipped with a holographic grating and silicon detector for the spectral region 680–1235 nm. Subsequent manipulations of data sets were limited, however, to 700–1150 nm since excessive stray light interferences were noted above 1150 nm and severe noise increases were observed below 700 nm. Samples were held in a quartz sample cuvette that had a 1.00-cm path length. The spectra are referenced to an air or water blank as noted in Table I. All spectra are the result of an averaging of 50 scans that took approximately 30 s.

Data Analysis. Stagewise linear regression (SLR) (*17*), as supplied by Pacific Scientific Co. (Gardner/Neotec Instrument Division, Stage-Wise Regression Analysis, 1987), was performed to search for correlations between various features in the spectra and the independently measured constituent values, namely, molarity of the solutions. Further analysis was carried out by using the method of partial least squares (PLS), (Veltkamp, D.; Kowalski, B. R.; Center for Process Analytical Chemistry, BG-10, University of Washington, Seattle, WA 98195; PLS 2-Block Modeling Version 1.0 (IBM), 1986). The selection of the optimal number of latent variables was done by using predictive residual sum of squares plots as provided in the PLS package. The standard error of prediction is from a cross validation estimate (*18*), which uses all but one sample as a calibration set to form a prediction equation, and then a prediction is made of the remaining sample. This "leave-one-out" exercise was repeated for each sample in the training set, and the standard error of prediction was determined from the predicted and actual values for samples omitted. The software for spectral reconstruction was based on algorithms obtained from Lawson and Hanson (*19*) and was programmed by using Microsoft Fortran on IBM and IBM-compatible PC-AT computers.

Solutions. Four sets of experiments were performed in which various combinations and concentrations of the reagents were dissolved in distilled water. In order to ensure accurate concentration data, the mass of reagent and the mass of water were recorded to five significant figures. Table I lists the mixtures prepared. Samples were allowed to equilibrate to room temperature before spectra were taken. These temperatures are noted in Table I.

RESULTS AND DISCUSSION

Spectroscopic Assignment. Figure 1 is the short-wave near-infrared (SW-NIR) absorption spectrum of pure water referenced to an air blank. In the spectral region 700–1150 nm, the major feature is a broad asymmetric band centered at 960 nm. This absorption band has been assigned to the combination transition $2v_1 + v_3$, where v_1 is the symmetric O–H stretch, v_3 the antisymmetric O–H stretch, and v_2 the O–H bending mode. Beginning at 1100 nm, in the pure water spectrum, is the onset of the combination band $v_1 + v_2 + v_3$. The final features that can be discerned are at 740 and 840 nm and are assigned to the combination bands $2v_1 + v_2 + v_3$ and $3v_1 + v_3$, respectively. These assignments are from refs. *6* and *20*.

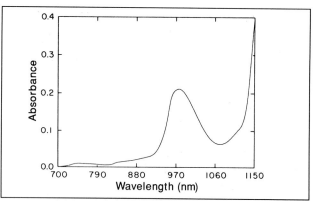

Figure 1. Short-wave near-infrared absorbance spectrum of distilled water referenced to air: path length, 1.00 cm; temperature, 22 °C. Absorption of water at 960 nm is assigned to $2v_1 + v_3$.

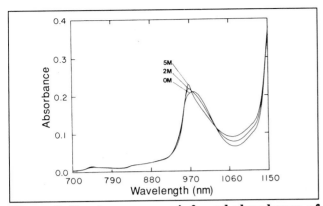

Figure 2. Short-wave near-infrared absorbance of 0.000, 2.040, and 5.027 M NaOH referenced to air: path length, 1.00 cm; temperature, 20 °C.

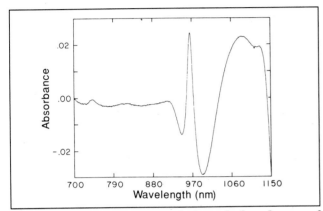

Figure 3. Short-wave near-infrared absorbance of 5.015 M NaOH solution referenced to water: path length, 1.00 cm; temperature, 25 °C. The sharp peak at 965 nm is assigned to the second overtone of the hydroxide stretching motion.

Figure 2 shows three spectra of water with successive additions of sodium hydroxide in the range 0–5 M referenced to air (from data set 1). As the hydroxide concentration increases, one observes a change in shape of the 960-nm band and an increase in the valley region between 960 nm and the feature beginning at 1100 nm. Isosbestic points are noted at 980 and 1028 nm. However, no new, distinct bands are detectable by eye. This is not unexpected, since the O–H overtone stretching transitions of hydroxide ion are expected to strongly overlap those of water, as has been observed previously in the mid-infrared and near-infrared regions. To further elucidate the hydroxide spectrum, we present in Figure 3 the spectrum of 5.015 M NaOH, now referenced against pure water. This spectrum is from data set 2, which has samples of the same concentration range as data set 1. The spectrum of Figure 3 exhibits two kinds of features: (a) positive going bands, which are due to new absorptions contributed by hydroxide ions, and (b) negative going bands, which are due either to the physical displacement of water molecules by the ions and/or to disruption of bulk water structure. The most notable feature of the hydroxide ion absorption is the very sharp band at 965 nm, which is completely contained within the bulk water envelope. This is assigned to the second overtone of the O–H

stretching motion (21). The third overtone can also be seen as a positive feature at approximately 740 nm. The broad positive feature in the region 1021–1050 nm is responsible for the "filling in" of the trough between the two major water peaks. It has previously been described as arising from the gradual disruption of the bulk water structure by the hydroxide ion (12).

Of interest are the negative absorption bands on either side of the hydroxide band at 965 nm due to the displacement of bulk water. By use of the absorption coefficient of the pure water solution, it was calculated that, for every molecule of hydroxide added to the solution, two molecules of bulk water disappeared. This could not be explained by the change in volume of the solution, since the partial molal volume of sodium hydroxide is too small and the effect is of the wrong sign. Thus, we concluded that the two water molecules were no longer participating in the bulk structure, but instead had formed a complex with the hydroxide ion. This new structure could be observed by the increase in intensity in the 1100-nm region.

Since the spectra due to hydrated hydroxide ion and the spectra due to bulk water strongly overlap, we attempted to separate the two spectra by using multivariate statistics (19). We began by assembling the various spectra into a response matrix $\mathbf{R}$ as follows:

$$\mathbf{R} \equiv \{\mathbf{r}_{ij}\} \equiv \{\vec{r}_j\} \tag{1}$$

where the $\mathbf{r}_{ij}$ are the spectral elements indexed by wavelengths i and sample number j. The columns of $\mathbf{R}$ are the spectra $\vec{r}_j$ of the jth sample. We now assume that the water and hydroxide spectra add linearly and are unaffected by each other. In this case we can represent $\mathbf{R}$ as

$$\mathbf{R} = \mathbf{SC} \tag{2}$$

where

$$\mathbf{S} \equiv \{\mathbf{s}_{ik}\} \equiv \{\vec{s}_k\},$$

and

$$\mathbf{C} \equiv \{\mathbf{c}_{kj}\} \equiv \{\vec{c}_j\} \tag{3}$$

The elements of $\mathbf{S}$ are the extinction coefficients at the ith wavelength for the kth component, while the elements of $\mathbf{C}$ are the concentration of the k components for the jth sample. The columns of $\mathbf{S}$ are the pure component spectra $\vec{s}_k$, while the rows of $\mathbf{C}$ are concentration of the components for the jth sample. From eq 2, the analytical problem can now be formulated as: Given $\mathbf{R}$ and $\mathbf{C}$, recover $\mathbf{S}$, the pure component spectra. In our case $i > j > k$, and the problem is overdetermined. Accordingly, we find an approximation to $\mathbf{S}$ from a least-squares procedure, i.e., such that

$$\|\mathbf{R} - \hat{\mathbf{S}}\mathbf{C}\|_{\min} \tag{4}$$

where $\hat{\mathbf{S}}$ is the maximum likelihood estimate for $\mathbf{S}$ (19). When eq 2 was solved for $\mathbf{S}$ under the condition of eq 4, the

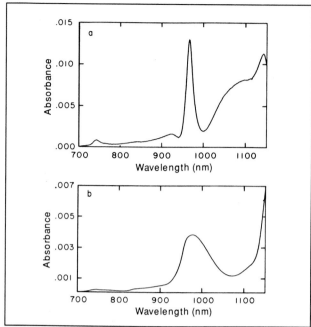

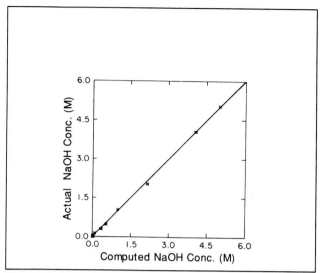

Figure 6. Actual NaOH molar concentration versus SLR-predicted NaOH molar concentration for the raw spectra of data set 1.

Figure 4. Reconstructed spectra of (a) NaOH and (b) water obtained by using the actual molar concentration of NaOH and the actual molar concentration of water minus 2 times the hydroxide concentration.

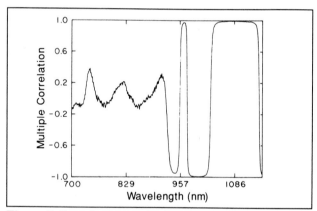

Figure 5. Correlation coefficient versus wavelength for data set 1.

two spectra of Figure 4 were recovered. The lower spectrum is clearly that of bulk water; the upper spectrum is that of the hydrated hydroxide ion. In Figure 4a, the narrow absorption band at 960 nm is assigned to the second overtone OH stretch centered on the hydroxide ion. The broad feature centered at 1100 nm arises from the combination bands of less strongly bound water molecules.

Surprisingly, there is a second very sharp feature revealed at 1140 nm. This is assigned to a stretch–bend combination band mainly centered on the hydroxide ion, but clearly involving concerted motions of the complex as a whole.

Development of Analytical Method. With a better understanding of the features of the hydroxide ion spectrum, we are now ready to develop a spectroscopic method for hydroxide ion quantitation.

Multilinear Regression. The data set represented in

Figure 2 was first analyzed by stagewise multilinear regression (SLR) because this method is rapid and provides "best wavelengths" for routine determination by fixed-wavelength instruments (filter photometers).

The wavelengths chosen also can be subject to a physical rationale for interpretation as to the basis of the analytical method. In determining how reasonable the chosen wavelengths are, we find it useful to display correlation coefficients (multiple R values) for each wavelength in graphical form. When this is done for data set 1, strong negative correlations are seen in those areas where bulk water predominates, and strong positive correlations are seen in the regions of the OH^- band (Figure 5). This data provides excellent supporting evidence for the assignment of the "continuum" region of the spectrum being due to water molecules that are lost from the bulk in binding to the hydroxide ion.

When the stagewise procedure was allowed to select wavelengths over the entire spectral range, the first wavelength selected was associated with the disappearance of bulk water (996 nm). This occurred because the spectral effect of bulk water dilution (or displacement) is larger than the spectral effect of hydroxide addition. However, real-world samples often contain more than two components. Since other solutes can also displace bulk water, this wavelength was rejected and wavelengths were searched for that were positively correlated with the hydroxide absorption band. We found that a more effective choice of wavelengths could be made by examining the reconstructed spectrum of the hydroxide ion (Figure 4). Using this information, we returned to data set 1 and constrained the SLR algorithm to choose a wavelength near the peak of the hydroxide ion (965 nm); in fact it chose 964 nm. The second wavelength selected corresponded to the displacement of water (980 nm). By this means a two-wavelength equation was developed for the data of set 1, which resulted in high correlation (multiple $R = 0.999$), a low standard error of estimation (SEE = 0.051 M), and excellent predictability by cross validation (SEP = 0.068 M) (see Figure 6). To analyze data set

data set	math	analyte	α_0	λ_1, nm	α_1	λ_2, nm	α_2	mult R	SEE, M	SEP, M
1	none	NaOH	10.90	964	88.893	980	−138.23	0.999	0.051	0.068
1	2D	NaOH	−3.10	970	−104.89			0.999	0.020	0.044
2	none	NaOH	−0.009	965	110.702	980	−130.577	0.999	0.085	0.113
2	2D	NaOH	0.029	965	−90.011			0.998	0.123	0.146
3	none	NaOH	−12.60	964	170.024	951	−165.134	0.961	0.048	0.050
3	2D	NaOH	−7.202	970	−86.038	811	−241.098	0.981	0.033	0.041
3	2D	NaCl	−5.987	1025	515.116			0.985	0.085	0.0913
4	none	NaOH	0.05	966	25.67	998	−146.22	0.991	0.042	0.048
4	2D	NaOH	0.03	966	−58.56	1135	−57.65	0.998	0.022	0.044
4	2D	NaCl	−0.18	1111	−1530.64	1093	−1947.56	0.844	0.174	0.179
4	2D	NaBr	−0.18	1020	1109.06	896	−1449.37	0.904	0.137	0.141

2, we again constrained the algorithm to the region near the absorption peak of the hydroxide ion. When the algorithm was forced to use the hydroxide overtone at 965 nm as the first wavelength and 980 nm (corresponding to the displacement of water) as the second wavelength, good correlation was again achieved (multiple $R = 0.999$; SEE = 0.085 M; SEP = 0.113 M). The results are summarized in Table II. The reader will note that the SEP for data set 2 is not as good as that for data set 1. First, referencing to water degrades the signal-to-noise ratio due to the marked temperature dependence of bulk water and problems with reproducing the cuvette position in a blank measurement. Second, the number of samples is small for data set 2. In analyzing these data sets, we noted that the second OH overtone absorption band of hydroxide was much sharper than the underlying water absorption band. Accordingly, we used a second-derivative transformation (segment 10 nm, 0 nm) on the spectra of data set 1 to see if the stagewise procedure would directly select wavelengths that corresponded to hydroxide ion absorption. Figure 7 shows the second-derivative spectra of set 1. Second-derivative smoothing parameters were chosen to yield the greatest sensitivity to sharp spectral features at the expense of broad features. Unfortunately, this procedure did not eliminate the bulk water spectral features, as shown by the 0.0 M NaOH curve. Nevertheless, the first wavelength chosen by stagewise regression corresponded more directly to hydroxide concentration. Use of the second-derivative transformation did lead to a noticeably better prediction error with only a single wavelength.

Partial Least Squares. To supplement the foregoing results, data set 1 was also analyzed by PLS (see Table III). Initially, our conception was that the hydroxide ion–water system could be approximated by two noninteracting components. Accordingly, a PLS model was formed using two latent variables. This resulted in a good fit to the data of set 1 (multiple $R = 0.994$; SEE = 0.141 M). However, it was noted that the standard error of the estimate obtained with the PLS model with two latent variables was larger than the standard error of the

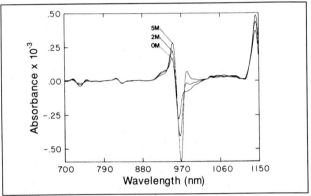

Figure 7. Second derivative of short-wave near-infrared absorbance spectra of 0.000, 2.028, and 5.015 M NaOH referenced to air: path length, 1.00 cm; temperature, 20 °C.

Table III. Regression Performance from PLS[a]

data set	analyte	no. of latent var	R	SEE, M	SEP, M
1	OH	4	1.000	0.014	0.019
2	OH	5	0.999	0.064	0.130
3	OH	4	0.997	0.014	0.024
3	Cl	5	1.00	0.007	0.014
4	OH	3	0.989	0.045	0.069
4	Cl	6	0.941	0.100	0.275
4	Br	5	0.917	0.144	0.194

[a]All data were preprocessed by mean centering.

estimate obtained with stagewise regression and two wavelengths. In order to achieve the same magnitude of standard error as that of stagewise regression, PLS required at least three components on mean-centered data, suggesting that the data contains an interaction effect or nonlinearity. Examination of the first three loading plots obtained from data set 1 is

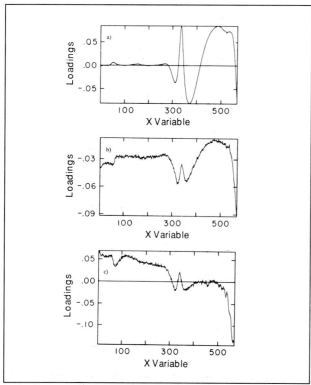

Figure 8. Loading plots of mean-centered data from data set 1 from PLS: (a) *X*-**block loading for latent variable 1; (b)** *X*-**block loading for latent variable 2; (c)** *X*-**block loading for latent variable 3.**

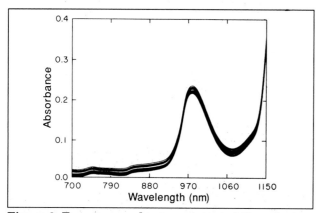

Figure 9. Twenty-one short-wave near-infrared spectra of NaOH and NaCl dissolved in water referenced to air. NaOH concentration ranged from 0 to 0.465 M, and NaCl concentration ranged from 2.33 to 3.72 M; pathlength, 1.00 cm; temperature, 20 °C.

very informative (Figure 8). The first latent variable resembles the spectrum of Figure 3, which is essentially the difference between the hydroxide ion and water spectra. This result is exactly what would be expected for a mean-centered single-component linear noninteracting model. However, this model is inadequate, as can be readily seen by examining the second and third latent variables. These appear to be accounting for changes in the number of less tightly bound water molecules as a function of concentration.

Caustic Brine. In real-world analysis, other components are

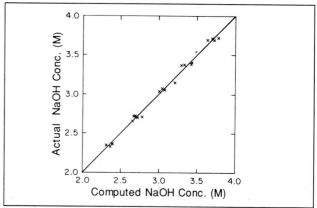

Figure 10. Actual NaOH molar concentration versus SLR-predicted NaOH molar concentration from second-derivative data of data set 3.

frequently present with caustic. One example is caustic brine. Here, a typical industrial problem is to neutralize the caustic sufficiently so that it can be disposed of safely. At first sight, the presence of variable amounts of sodium chloride might be seen to pose no problem for SW-NIR hydroxide analysis because solid NaCl has no absorption in this region. However, the sodium and chloride ions do exert considerable influence on the water absorption bands in the near-IR region due to rearrangement of the water molecules in the inner solvation sphere (8). In order to assess whether a caustic neutralization analysis could be performed in the presence of a large and varying sodium chloride background, a sample set was made up of 21 solutions where the sodium hydroxide concentration varied in the range 0–0.465 M, while that of sodium chloride varied from 2.33 to 3.72 M. Sample temperatures were allowed to equilibrate to room temperature and were run at 20 ± 1 °C. Figure 9 presents all 21 spectra referenced to the air blank (data set 3, Table I). These spectra are virtually identical. The major variance arises from the irreproducibility in the base line caused by an inability to precisely remove and replace the sample in the spectrophotometer. Nevertheless, with the use of second-derivative transformation, an excellent correlation to NaOH molar concentration was obtained by stagewise multilinear regression (see Figure 10). It should be noted that the wavelengths chosen for data set 3 were similar to those chosen for data set 1. With stagewise regression on the second-derivative spectra of data set 3, a two-wavelength equation resulted that gave a multiple $R = 0.981$, SEE = 0.033 M, and SEP = 0.041 M. PLS analysis was performed on the raw data, and a good correlation using four latent variables was attained (multiple $R = 0.997$; SEE = 0.014 M; SEP = 0.024 M).

One of the major benefits of multivariate data is that multicomponent analysis becomes possible. Accordingly, we attempted to correlate both the raw and second-derivative data with concentration of NaCl. Application of the stagewise regression algorithm on the second-derivative spectra gave a multiple $R = 0.985$ and a SEE = 0.085 M using a one-wavelength equation (see Figure 11). With five latent variables on mean-centered data, PLS was able to achieve a multiple $R = 1.00$ and a SEE = 0.007 M. We attribute the poor

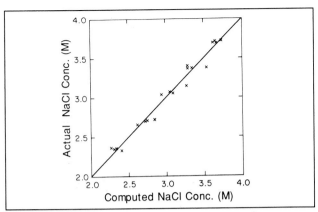

Figure 11. Actual NaCl molar concentration versus SLR-predicted NaCl molar concentration from second-derivative data of data set 3.

performance of stagewise regression to its inability to confidently choose more than two wavelengths. This point is discussed by Draper and Smith (*17*). With PLS using a cross-validation approach, we could confidently use more latent variables to describe the phenomenon. As an extension of the above analysis, NaBr was added along with NaOH and NaCl to water (data set 4). NaOH concentration ranged from 0.0 to 1 M. Good correlation to hydroxide was attainable with both stagewise regression and PLS. With stagewise regression a two-wavelength equation was obtained for NaBr with a multiple $R = 0.900$; with PLS, five latent variables were used with mean-centered data to obtain a multiple $R = 0.917$ (see Tables II, III).

CONCLUSION

The results we have obtained are significant at several levels. The first is the use of the short-wavelength near-infrared spectral region. This region allows us to use very long path lengths (path lengths of 0.5–5 cm yield absorbancies of 0.1–1.0 at 960 nm) with attendant ease of sampling. Moreover, one can employ inexpensive optics and detectors, and the measurements can be made noninvasively and nondestructively.

The second is the advantage of direct detection of hydroxide ions by spectroscopy. Our method translates absorbance measurements directly into concentration, whereas the pH electrode actually measures activity and is plagued with interferences from other ions at high pH. In addition, one must address the issue of junction potentials and the problem of sampling when using the pH electrode. On the other hand, spectroscopy is not free of environmental effects. The extinction coefficient is not guaranteed to be independent of other analytes. Yet, with multivariate calibration, an analytical method can be developed that is at least robust to known interferences as properly accounted for in the training set. In spite of the apparent nonlinear interaction terms, the SW-NIR method provides excellent measurements of caustic concentration to ±0.05 M in the concentration range 0.0–5.0 M. In the case where full spectrum methods are used, residuals

can be a powerful technique for determining whether an input spectrum is sufficiently similar to the original data set. Univariate methods do not have this feature. Multivariate methods also allowed us to reconstruct the hydroxide ion spectrum.

We wish to caution the reader that the water spectrum is temperature dependent and this source of variance must be adequately controlled. When this effect is studied and accounted for, one can not only correct for temperature variations, but measure the temperature accurately as well.

ACKNOWLEDGMENT

We thank Dr. David Burns, University of Washington, Department of Bioengineering, for his help with statistical analysis and Mrs. Louise Rose for her work on the graphic illustrations. We are also grateful to Avi Lorber for useful discussion.

LITERATURE CITED

(1) Clevitt, K. J. *Process Analyzer Technology*; John Wiley and Sons: New York, 1986; pp 506–523.
(2) Christian, G. *Analytical Chemistry*; John Wiley and Sons: New York, 1980; p 304.
(3) Callis, J. B.; Illman, D. L.; Kowalski, B. R. *Anal. Chem.* **1987**, *56*, 624A–637A.
(4) Commercial vendors include Pacific Scientific Co., Silver Spring, MD; LT Industries, Rockville, MD. See also Mayes, D. M.; Callis, J. B. *Appl. Spectrosc.* **1989**, *43*, 27–32.
(5) Beebe, K. R.; Kowalski, B. R. *Anal. Chem.* **1987**, *59*, 1007A–1017A.
(6) Buijs, K.; Choppin, G. R. *J. Chem. Phys.* **1963**, *39*, 2035–2041.
(7) Horning, D. F. *J. Chem. Phys.* **1964**, *40*, 3119–3120.
(8) Paquette, J.; Jolicoeur, C. *J. Solution Chem.* **1977**, *6*, 403–428.
(9) Georgiev, G. M.; Kalkanjiev, T. K.; Petrov, V. P.; Nickolov, Zh. *Appl. Spectrosc.* **1984**, *38*, 593–595.
(10) Kleeberg, H.; Heinje, G.; Luck, W. A. P. *J. Phys. Chem.* **1986**, *90*, 4427–4430.
(11) Struis, R. P. W. J.; de Bleijser, J.; Leyte, J. C. *J. Phys. Chem.* **1987**, *91*, 6309–6315.
(12) Giguere, D. A. *Rev. Chim. Miner.* **1983**, *20*, 588.
(13) Busing, W. R.; Horning, D. F. *J. Phys. Chem.* **1961**, *65*, 284.
(14) Moskovits, M.; Michelian, K. H. *J. Am. Chem. Soc.* **1980**, *102*, 2207.
(15) Hirschfeld, T. *Appl. Spectrosc.* **1985**, *39*, 740–741.
(16) Watson, E., Jr.; Baughman, E. H. *Spectroscopy* **1987**, *2*, 44–48.
(17) Draper, N.; Smith, H. *Applied Regression Analysis*; John Wiley and Sons: New York, 1981; p 337.
(18) Sharaf, M. A.; Illman, D. L.; Kowalski, B. R. "Chemometrics." In *Chemical Analysis, A Series of Monographs on Analytical Chemistry and Its Applications*; Elving, P. J., Winefordner, J. D., Eds.; Kolthoff, I. M., Ed. Emeritus; John Wiley and Sons: New York, 1986; Vol. 82.
(19) Lawson, C. L.; Hanson, R. J. *Solving Least Squares Problems*; Prentice-Hall, Inc.: Englewood Cliffs, NJ, 1974; p 290.
(20) Bayly, J. G.; Kartha, V. B.; Stevens, W. H. *Infrared Phys.* **1963**, *3*, 221–223.
(21) Wheeler, O. H. *Chem. Rev.* **1959**, *59*, 642–645.

Received for review September 12, 1988. Revised March 10, 1989. Accepted March 10, 1989. The authors thank the Center for Process Analytical Chemistry and the National Science Foundation for their financial support.

Reprinted from *Anal. Chem.* **1989**, *61*, 1419–24.

This article opened the door for high-performance capillary electrophoretic separation of longer oligonucleotides. Previously, most people working in CE had shied away from filling capillaries with gels, but the conventional free medium did not separate species with similar mass-to-charge ratios very well. We had been struggling to optimize separations for oligonucleotides only 15–20 bases long when this paper appeared; Karger's group proved that gels could be used effectively for much longer oligonucleotides.

Milos Novotny
Indiana University

Analytical and Micropreparative Ultrahigh Resolution of Oligonucleotides by Polyacrylamide Gel High-Performance Capillary Electrophoresis

A. Guttman, A. S. Cohen, D. N. Heiger, and B. L. Karger*
Barnett Institute and Department of Chemistry, Northeastern University, Boston, Massachusetts 02115

In continuation of previous work, the separations of polydeoxyoligonucleotides on polyacrylamide gel capillary columns are presented. The use of gel compositions with relatively low monomer content permits columns of very high plate numbers to be obtained. With a 160-mer, plate counts of 30×10^6 per meter are shown. Columns with high precision in relative migration time from run to run, day to day, and batch to batch are presented. Collection of purified fractions from high-resolution electropherograms is also shown to be feasible using field programming techniques. To accomplish this, separation is allowed to occur under high field for resolution and speed, followed by collection under low field where the velocity of the band is purposely decreased in order to widen the time-based width of the band.

INTRODUCTION

At present, there is a great deal of interest in the separation and characterization of polydeoxyoligonucleotides. The problem is a challenging one since the polydeoxyoligonucleotides may differ only slightly from one another, for example in base number or in sequence of bases. Great strides are being made in electrophoresis for various molecular weight regions, including chromosomal DNA separations (1). At the low molecular weight range, e.g., up to ca. 500 bases, there is a need for improved separation, in terms of both speed and resolving power, particularly for minute sample sizes. Rapid and significant resolving power would result in important potential advances for molecular biology including, among other areas, DNA sequencing (2) and purification of DNA probes (3).

Conventionally, polydeoxyoligonucleotide separations have been best achieved by gel electrophoresis, using flat bed or slab gels, which can be very time consuming (4). In addition, the isolation of individual fragments is inconsistent and fraught with problems.

An improved version of electrophoresis is currently being developed in the form of high-performance capillary electrophoresis (HPCE) (5–8). This instrumental approach to electrophoresis can be characterized as a rapid, high-resolution analytical technique and an effective tool in the isolation of purified materials. We have previously demonstrated the high resolution of gel-filled capillaries for the separation of low molecular weight polydeoxyoligonucleotides (9).

This paper is a continuation of the study of gel-filled capillaries for high-resolution separation of polydeoxyoligonucleotides. We first briefly present results on columns with several gel compositions. Columns of high precision in relative solute migration time have been developed. Using low monomer gel compositions and columns greater than 50 cm in effective length, we demonstrate the capability of such capillary gel columns to generate very large numbers of theoretical plates, roughly 30 million plates per meter. Finally, we illustrate a new micropreparative approach to collection of isolated fragments using electric field programming. These results indicate the significant power of capillary gel columns for the separation and analysis of polydeoxyoligonucleotides.

EXPERIMENTAL SECTION

Apparatus. Polyacrylamide gel HPCE was performed in fused-silica tubing (Polymicro Technologies, Phoenix, AZ), 75-μm i.d., 375-μm o.d., with several different column lengths. For detection purposes, a 2-mm length of polyimide coating was carefully burned off at a distance of 150–200 mm from one end of the column. A 60-kV high-voltage direct current

2855-8/94/0410$08.00/0 © American Chemical Society

power supply (Model PS/MK 60, Glassman, Whitehouse Station, NJ) was used to generate the potential across the capillary gel column. A UV detector (Model V4, ISCO, Lincoln, NE), modified as previously described (10), was employed at a wavelength of 260 nm, with a capillary slit of 0.1 mm × 0.1 mm. Cooling was achieved with a fan. Each end of the capillary was connected to a separate 3-mL vial filled with buffer (for analytical runs). For collection, the low-voltage end (cathode) was placed in a microfuge vial filled with an appropriate amount of water (~3 µL) (9, 11). Platinum wire electrodes were inserted into the two vials for connection to the electrical circuit. The electropherograms were acquired and stored on an IBM PC/AT computer via an analog to digital (A/D) interface (Model 762 SB, Nelson Analytical, Cupertino, CA).

Materials. Samples of polydeoxyadenylic acids were purchased from Pharmacia (Piscataway, NJ). The polydeoxythymidylic acids, 20–160-mers, were chemically synthesized by ligation of the 3′-3′ ends, yielding 5′-phosphate groups at the two ends of the combined polymers. Individual polydeoxythymidylic acids, differing by 20 polydeoxyoligonucleotides each (i.e., 20-mer, 40-mer, . . . 160-mer), were slab gel purified and subsequently combined to give a 20–160-mer mixture. The mixture was the kind gift of Drs. Bischoffer and Field of Genentech, South San Francisco, CA. All other reagents were of electrophoresis research grade (Schwartz/Mann Biotech, Cambridge, MA). All buffer solutions were filtered through a nylon GC filter unit of 0.2-mm pore size (Schleicher & Schuell, Keene, NH). The buffer solutions were carefully vacuum degassed before use. Samples were kept frozen at −20 °C, and the buffers were stored at 4 °C before use. Samples were heated at 65 °C for 5 min and were then injected at room temperature (9).

Procedures. Polymerization of polyacrylamide was accomplished within the capillary column by passing carefully degassed polymerizing solution slowly into the capillary (9). For stabilization, the gel was chemically bound to the silica wall by using a standard bifunctional agent, (methylacryloxypropyl)trimethoxysilane (12). Gel columns of 3–6% T and 5% C (13) were used in a buffer of 0.1 M Tris, 0.25 M boric acid, pH 7.6, and 7 M urea. The samples were electrophoretically injected into the column by dipping the cathodic end of the capillary into an aqueous solution of the sample and typically applying a field of 3 kV for 5 s for an analytical injection. In the case of the polydeoxyadenylic acids (p(dA)$_{40-60}$), the concentration of injection sample was in the range of 20–40 µg/mL.

RESULTS AND DISCUSSION

High-Resolution Separations. We have previously shown the high resolving power of capillary polyacrylamide gel electrophoresis in the separation of polydeoxyoligonucleotides using a sample mixture of polydeoxyadenylic acids from 40 to 60 bases long (p(dA)$_{40-60}$) (9). In the above-cited paper, we used a column of 7.5% T, 3.3% C gel composition.

In this work, we first briefly examined separation with several gel compositions. Figure 1A shows the separation of the p(dA)$_{40-60}$ sample at 300 V/cm, using a 6% T, 5% C gel composition with an effective column length l of 15 cm. According to slab gel reports (14), this gel composition should produce

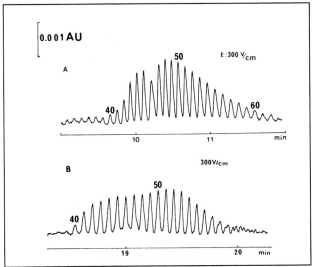

Figure 1. HPCE separation of polydeoxyadenylic acid mixture, p(dA)$_{40-60}$, as a function of gel composition and effective column length (see text): (A) capillary, 350 × 0.075 mm i.d. (effective length, 150 mm); buffer, 0.1 M Tris/0.25 M borate/7 M urea, pH 7.6; polyacrylamide gel, 6% T and 5% C; applied field, 300 V/cm, 10.8 µA. (B) capillary, 750 × 0.075 mm i.d. (effective length, 550 mm); polyacrylamide gel, 3% T and 5% C; applied field, 300 V/cm, 17.7 µA; all other conditions were the same as in A. Figure 1A is reprinted from *Nature* (Vol. 339, pp 641, 642). Copyright 1989 Macmillan Magazines Limited.

pore dimensions similar to those from 7.5% T, 3.3% C gel composition. The average plate count for the above capillary column was 2.6×10^5 for each peak, and the average resolution for any pair of solutes was 1.13.

We next turned to design gel columns of increased pore size for the separation of larger molecular weight single-strand polydeoxyoligonucleotides. This was accomplished by reducing the monomer content in the gel composition. Figure 1B shows the separation of the same p(dA)$_{40-60}$ mixture at 300 V/cm, using a 3% T, 5% C gel column. To compensate for the expected lower resolving power of the larger pore size column for the short polydeoxyoligonucleotides, the effective column length was increased from 15 to 55 cm (5, 15). Good separation was again obtained, but at the expense of longer time. The 3% T, 5% C column was an efficient column with an H value of 0.31 µm ($N = 1.8 \times 10^6$) vs an H value of 0.57 µm for the 6% T, 5% C column. The higher efficiency on the wider pore size column may be a result of a decreased restricted migration through the gel matrix. However, the resolution on the 3% T, 5% C column (1.25) was only slightly greater than on the 6% T, 5% C column (1.13), undoubtedly due to the greater size selectivity on the smaller pore, higher percentage T column.

The 3% T, 5% C column would nevertheless appear advantageous from the point of view of allowing separation of a broader molecular weight range of polydeoxyoligonucleotides. For small polydeoxyoligonucleotides, the poorer selectivity on the lower percentage T column can be compensated by larger values of N. On the other hand, on the 6% T, 5% C column, larger polydeoxyoligonucleotides, e.g., ~150-mer, yielded broad bands, presumably due to restricted migration through the gel matrix. This was not the case for the 3% T, 5% C gel, and separations of very high performance were observed for solutes in this base number range (see later).

In Figure 1B the various p(dA)'s have been identified on the basis of the use of purified single-stranded p(dA) species from 10 to 20 bases in length and the linear behavior of migration time vs base number for the p(dA)'s, $r^2 = 0.9995$. Careful studies of migration reproducibility were conducted on the 3% T, 5% C column of Figure 1B. With the 50-mer serving as the basis of study, it was found that the run-to-run reproducibility was 0.9% relative standard deviation (RSD) ($n = 5$), and the day-to-day reproducibility was 3.2% RSD ($n = 3$). We also examined the column-to-column (batch-to-batch with respect to polymerizing solution) migration time reproducibility where the columns were made by two different workers and found 4.8% RSD ($n = 11$).

The lower reproducibility from day to day and column to column was believed to arise in part from column temperature variations. Since a fan was used for cooling, ambient temperature variations could cause overall column temperature changes. Mobility, and thus migration time, are particularly sensitive to temperature changes, i.e., ~2%/°C (16), due to the sensitivity of liquid viscosity with T. The precision of absolute migration could be improved by fixing the set temperature of the capillary column by using, for example, a solid-state (Peltier) cooling system (17).

To demonstrate that temperature variation was a likely cause of the 3–5% RSD observed, we measured the migration of the 40-mer relative to that of the 50-mer. As in chromatography, we found that relative migration dramatically improved precision. Thus, in terms of relative migration with the same data, the precision from run to run was now 0.07% RSD ($n = 5$), from day to day (same column) 0.1% RSD ($n = 3$), and from column to column 0.7% RSD ($n = 11$). Given the similarity in migration behavior of the polydeoxyoligonucleotides, the use of a specific polydeoxyoligonucleotide as an internal standard can prove useful. From the results, it appears that the columns are highly reproducible. In addition, we were able to use the columns well in excess of 150 injections.

We approximated the amount of sample injected in Figure 1B (i.e., injection at 3 kV for 5 s) by first calibrating an open-tube capillary detector ($\lambda = 260$ nm) with known concentrations of the sample. A linear plot was observed with $r^2 = 0.993$. The average concentration of an individual band was next determined from the average peak height of the band in the electropherogram of Figure 1B. The volume of that given band was then measured (~8 nL) by using the time-based peak width and the migration velocity of the species. The total amount (on average) of each peak was calculated to be 600 pg or 12 ng total of p(dA)$_{40-60}$ injected (600 pg × 20 solutes = 12 ng).

As noted above, a longer column was used with the 3% T, 5% C gel column to generate a greater number of theoretical plates. The ability of such columns to maintain high resolution over a wide molecular range can be seen in Figures 2 and 3. Figure 2 shows the separation of a mixture of polydeoxythymidylic acids, p(dT)$_{20-160}$, each major species differing by 20 bases with $l = 80$ cm and $E = 200$ V/cm, 8.2 μA. Figure 3

shows the same separation on a column of $l = 50$ cm, with $E = 300$ V/cm, 12 μA. For this polydeoxythymidylic acid sample, the slab gel purified species reveal additional peaks by HPCE which migrate slightly faster than the main peak. A plot of migration time t (i.e., μ_e^{-1}) vs base number is again linear for the p(dT)$_{20-160}$ major components, $r^2 = 0.995$. On the basis of this calibration plot, each impurity peak is associated with a progressively $n - 1$ decreased number of bases, i.e., 159, 158, etc. For absolute identification it would be necessary to collect each peak and conduct further structure analysis.

The ultrahigh resolving power can be readily seen, particularly in Figures 2 and 3 for the time-expanded separation of the 160-mer. The presumed 159-mer in Figure 2 (see arrow) is calculated to contain 12×10^6 plates at 200 V/cm. (Appropriate equations were used to take account of the slight asymmetry of the band (18).) In Figure 3, the 160-mer is calculated to contain 15×10^6 plates for 50 cm or 30×10^6 plates per meter. This latter plate count with an H value of 0.03 μm is roughly 10-fold greater than previously reported for open-tube capillary electrophoresis of $(1-2) \times 10^6$, e.g., ref 19 or Figure 1B. For both Figures 2 and 3 there is a reduction in plate count as the faster migrating, shorter polydeoxyoligonucleotide species are measured. For example, on the $l = 50$ cm column, the 120-mer is calculated to contain 6×10^6 plates, whereas the 80-mer has 2×10^6 plates. Based on increased peak asymmetry with the faster migrating species, sample concentration overloading (20) may in part play a role in this decrease in plate count.

The results in Figures 1–3 demonstrate that the gel columns represent an excellent approach for resolution of complex polydeoxyoligonucleotide mixtures. They thus possess a

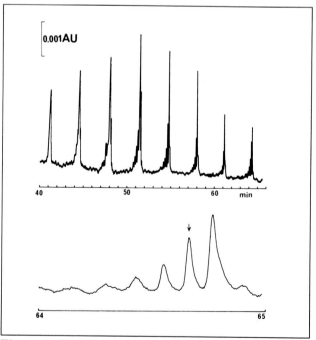

Figure 2. HPCE separation of polydeoxythymidylic acid mixture, p(dT)$_{20-160}$: capillary dimensions, 1000 × 0.075 mm i.d. (effective length, 800 mm); applied field, 200 V/cm, 8.2 μA. All other conditions were the same as in Figure 1A.

real potential for use in DNA sequencing using sensitive labeling and detection techniques (2). In addition, they can be used for rapid assessment of purity of synthesized polydeoxyoligonucleotides (9). We next turn to the use of these high-resolution columns for micropurification.

Field Programming and Micropreparative Collection. HPCE not only is an analytical tool but also can be used for micropreparative separations. Collection of substances as they migrate out of the capillary has already been demonstrated (9, 11, 21). In the case of gel columns, purified fractions can be obtained by maintaining a constant field and positioning the capillary in a microfuge vial filled with several microliters of water and/or diluted buffer. The combination of this approach with high resolution and rapid separation, where the peak widths may be only a few seconds wide, is difficult for reproducible collection. It is possible, however, to simplify collection while maintaining high-resolution separations by electric field programming.

Since the velocity of the solute can be manipulated by varying the electric field, field programming can simplify the collection process. The strategy is to separate at high field where speed and resolution (in general) are improved and then to collect at low field where the band would broaden in time units with little loss in resolution.

Figure 4 illustrates the principle of this approach. In parts A and B, separation of $p(dA)_{40-60}$ is achieved in a 6% T, 5% C gel column using 300 and 100 V/cm, respectively. The average time-based peak width in Figure 4A is 6 s and in Figure 4B, 18 s. Such narrow widths could create some difficulty in collection, particularly in manual operation.

In Figure 4C, the separation is conducted at 300 V/cm until 1 min prior to the appearance of the first peak in the detector. The field is then decreased to 30 V/cm and detection achieved. Bands of roughly 55-s peak widths are observed with close to base line separation. Note that separation in Figure 4C is comparable to that run continuously at 100 V/cm (Figure 4B).

We can apply the principles of field programming illustrated in Figure 4C to collection, as shown in Figure 5. In Figure 5A, the $p(dA)_{40-60}$ mixture is micropreparatively separated on a 3% T, 5% C gel column, effective length 55 cm, using a sample size roughly 6-fold higher than typical for the analytical runs. We attempted to collect a single species from the overloaded sample by using field programming. In this experiment, the sample traveled through the detector, as shown in Figure 5A, and the field was maintained at 300 V/cm until just before the species reached the end of the column. The field was then cut off, and the capillary column and electrode placed in a microfuge tube with 1.5 µL of water. The field was then applied at 30 V/cm and collection conducted for 60 s. The calculated peak width was 45 s, based on the time width through the detector and the applied field of 30 V/cm. The extra collection time was used in order to be certain to collect the peak, given the manual nature of the experiment.

The collected fraction was reinjected with an internal standard, $p(dA)_{20}$, and the resulting electropherogram is shown in Figure 5B. It can be seen that an individual peak has been successfully collected. The collected peak is assumed to be $p(dA)_{47}$, based on the agreement of the relative migration of the 47–mer to that of the 20-mer in Figure 5B and Figure 5C,

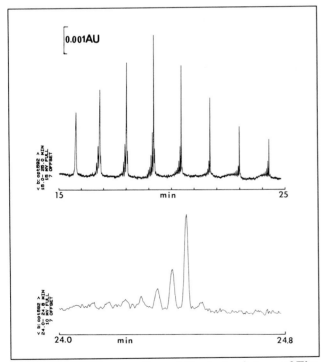

Figure 3. HPCE separation of the test mixture of Figure 2. Conditions were the same as in Figure 2 except for the following: capillary dimensions, 700 × 0.075 mm i.d. (effective length, 500 mm); applied field, 300 V/cm, 13 µA. All other conditions were the same as in Figure 1A.

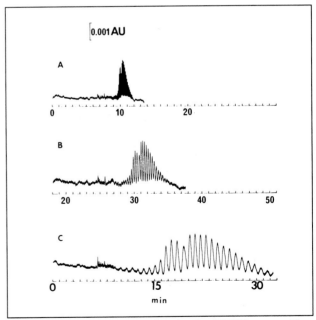

Figure 4. Field programming in the HPCE separation of polydeoxyadenylic acid mixture, $p(dA)_{40-60}$: capillary dimensions, 300 × 0.075 mm i.d. (effective length, 150 mm). Applied voltages were as follows: (A) 300 V/cm, 17.7 µA; (B) 100 V/cm, 5.6 µA; (C) 300 V/cm, 17.7 µA, 0–10 min and 30 V/cm, 1.9 µA, 10–30 min. All other conditions were the same as in Figure 1A.

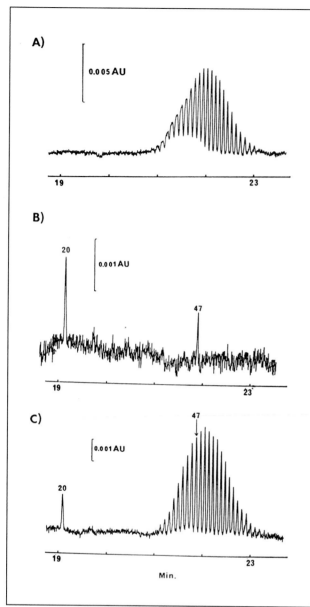

Figure 5. (A) Micropreparative HPCE separation of the polydeoxyadenylic acid test mixture, p(dA)$_{40-60}$, (B) analytical run of the isolated p(dA)$_{47}$ spiked with p(dA)$_{20}$, and (C) analytical run of p(dA)$_{40-60}$ spiked with p(dA)$_{20}$. Capillary dimensions were 800 × 0.075 mm i.d. (effective length, 600 mm). The buffer and the gel were the same as in Figure 1B. The micropreparative and the analytical injections were made electrophoretically at 9 kV for 10 s (A) and 6 kV for 1 s (B and C), respectively. In A the applied field was 300 V/cm for the run and was decreased to 30 V/cm for the collection, over 60 s. In B and C the applied field was held constant at 300 V/cm.

the latter of which is for p(dA)$_{40-60}$ and p(dA)$_{20}$ electrophoresed together. The difference in the relative migration times is less than 0.2% in B (0.874) and C (0.8725). It may also be noted that the differences in relative migration of p(dA)$_{46}$ or p(dA)$_{48}$ vs p(dA)$_{20}$ were 1.0%. The identity of p(dA)$_{47}$ in Figure 5C is again based on the linear calibration

of migration time of standard p(dA)'s vs base number.

Computer control of the collection procedure would improve this field programming approach. The mobility of the species and its peak width at the high field could be directly determined during the run as detection takes place on-column. Knowledge of the distance from the detector to the end of the column would then permit calculation of the point at which the field would be reduced. Also, the actual time-based peak width during collection could be obtained from the measured peak width at the detector and the applied field at collection. It is also to be noted that the sample migrates through the detector to the end of the column under a high field. Since, at a given field, resolution is proportional to (migration distance)$^{1/2}$, an improved resolution over that observed in the detector will result. This improvement will obviously depend on the relative distance traveled from the detection point to the end of the column. It is also important that the temperature in the postdetector section of the column be the same as in the predetector section (*17*). Since, as already noted, mobility changes by 2%/°C (*16*), a temperature difference in the postdetector section would decrease the accuracy of the time the band is predicted to reach the end of the column.

Because of the simplicity of the approach and the ease of automation, field programming should become a useful procedure for collection. It should be possible to conduct collection in a precise manner, permitting heart cutting even of high-resolution complex mixtures. Field programming can also find value in detection where the time for detection influences overall sensitivity, e.g., radioactive detection (*22*).

ACKNOWLEDGMENT

The technical assistance of A. Arai is acknowledged. The authors also thank Dr. Norbert Bischoffer and Dr. Matthew Field of Genentech for providing the chemically synthesized deoxypolythymidylic acid mixture.

LITERATURE CITED

(1) Cantor, C. R.; Smith, C. L.; Mathew, M. K. *Ann. Rev. Biophys. Chem.* **1988**, *17*, 287–304.
(2) Smith, L. M.; Sanders, J. Z.; Kaiser, R. J.; Hughes, P.; Dodd, C.; Connell, C. R.; Heiner, C.; Kent, S. B. H.; Hood, L. E. *Nature* **1986**, *321*, 674–679.
(3) *Current Protocols in Molecular Biology*; Ansabel, F. M., et al., Eds.; John Wiley and Sons: New York, 1989; 2.12.1–2.12.5.
(4) Rickwood, D.; James, B. D., Eds. *Gel Electrophoresis of Nucleic Acids: A Practical Approach*; IRC Press: Washington, DC, 1983.
(5) Karger, B. L.; Cohen, A. S.; Guttman, A. *J. Chromatogr., Biomed. Appl.* **1989**, *492*, 585–614.
(6) Jorgenson, J. W.; Lukacs, K. D. *Science* **1983**, *222*, 266–272.
(7) Gordon, M. J.; Huang, X.; Pentoney, S. L., Jr.; Zare, R. N. *Science* **1988**, *242*, 224–228.
(8) Ewing, A. G.; Wallingford, R. A.; Olefirowicz, T. M. *Anal. Chem.* **1989**, *61*, 292A–303A.
(9) Cohen, A. S.; Najarian, D. R.; Paulus, A.; Guttman, A.; Smith, J. A.; Karger, B. L. *Proc. Natl. Acad. Sci. USA* **1988**, *85*, 9660–9663.
(10) Guttman, A.; Paulus, A.; Cohen, A. S.; Grinberg, N.; Karger, B. L. *J. Chromatogr.* **1988**, *448*, 41–53.
(11) *Electrophoresis '88-Proceedings of the 6th Meeting of the International Electrophoresis Society*; Schaefer-Nielsen, C., Ed.; VCH Publishers: New York, 1988; pp 151–159.
(12) Cohen, A. S.; Karger, B. L. *J. Chromatogr.* **1987**, *397*, 409–417.
(13) Hjerten, S. *Arch. Biochem. Biophys. Suppl.* **1962**, *1*, 147–151.

(14) Stellwagen, N. *Advances in Electrophoresis*; Chrambach, A.; Dunn, M. J.; Radola, B. J., Eds.; VCH Publishers: New York, 1987; Vol. 1, pp 179–228.

(15) Terabe, S.; Yashima, T.; Tanaka, N.; Araki, M. *Anal. Chem.* **1988**, *60*, 1673–1677.

(16) Wieme, R. J. *Chromatography: A Laboratory Handbook of Chromatography and Electrophoresis Methods*, 3rd ed.; Heftmann, E., Ed.; Van Nostrand: New York, 1975; pp 228–281.

(17) Nelson, R. J.; Cohen, A. S.; Paulus, A.; Guttman, A.; Karger, B. L. *J. Chromatogr.* **1989**, *480*, 111–127.

(18) Foley, J. P.; Dorsey, J. G. *Anal. Chem.* **1983**, *55*, 730–737.

(19) Lauer, H. H.; Manigill, D. M. *Anal. Chem.* **1986**, *58*, 166–170.

(20) Mikkers, F. E. P.; Everaerts, F. M.; Verheggen, Th. P. E. M. *J. Chromatogr.* **1979**, *169*, 1–10.

(21) Rose, D. J., Jr.; Jorgenson, J. W. *J. Chromatogr.* **1988**, *327*, 23–34.

(22) Pentoney, S. L., Jr.; Zare, R. N.; Quint, J. F. *Anal. Chem.* **1989**, *61*, 1642–1647.

Received for review July 3, 1989. Accepted October 19, 1989. The authors gratefully acknowledge Beckman Instruments for support of this work. This is Barnett Institute Contribution No. 382.

Reprinted from *Anal. Chem.* **1990**, *62*, 137–41.

Subject Index

Cyclosporin A, electrospray interface for liquid chromatographs and mass spectrometers, 358f

D

Dansyl amino acids, zone electrophoresis in open-tubular glass capillaries, 326f

Data, smoothing and differentiation, 155–168

Data evaluation, introduction of students, 68,69–70f,71f,t

Degree of polarization, definition, 55

Detector, Raman spectrograph, 57–58

Detector efficiency, determination, 247f

Development of chromatogram, description, 21f

o-Dianisidine color test, procedure, 95

1,2-Dibromopropane
    addition to 1,3-dibromopropane, 16–17,18f
    identification using IR spectroscopy, 16,17f
    IR measurement in 1,3-dibromopropane, 16–17,18f

Diethylbenzene, determination in ethylbenzene using MS, 41,42t

Differentiation, use of Fourier transformations, 235,236f

Differentiation of data
    exponential function, 156,157f
    moving average, 155,156–157f
    objective, 155
    restrictions to data gathering, 155
    symmetrical exponential function, 156,157f
    symmetrical triangular function, 156,157f

Diffusion coefficient determination, thin layer electrochemical study design using controlled potential or controlled current, 185,186f,t,187t

Digital data handling of spectra using Fourier transformations
    apparatus, 232
    applications, 237
    differentiation, 235,236f
    examples, 232
    experimental procedure, 232–233
    history, 232
    information obtained, 233f,234
    resolution enhancement, 236f,237
    smoothing, 234–235f

Digitonin, analysis using laser desorption–MS, 291,292f

Digoxin, analysis using laser desorption–MS, 290,291f

Diphenylamine, determination using phosphorimetry, 110–111,112f,t

Direct-reading flame photometer, development, 81

Disubstituted aromatics, Raman spectroscopy, 62

Divalent metal ions, separation using conductimetric chromatography, 266,267f

Dropping mercury electrode, advantages, 367

Dynamic fluorescence quenching by halide ions, use in optrode system, 374

E

Edge effect, definition, 184

Efficiency, LC–MS for nonvolatile sample analysis, 310,311f

Electrical instruments, flame photometer, 51

Electrochemistry, need for stable covalently bonded surfaces, 272

Electrode determination of fluoride in water supplies, use of total ionic strength adjustment buffer, 203,204f,205

Electrokinetic separations with micellar solutions and open-tubular capillaries
    apparatus, 352
    applications, 354
    electroosmotic flow vs. electrophoretic migration, 353
    experimental description, 353
    experimental materials, 352
    experimental procedure, 352–353
    phenols, 353f
    range of elution, 354f
    retention parameters, 353–354

Electrolytic separation of copper from arsenic and antimony in ammoniacal fluoride solutions
    determination procedure for copper and antimony, 6t
    determination procedure in nitric acid solutions, 5t,6
    experimental materials, 5
    experimental procedure, 5–6
    previous studies, 5
    rapid separation procedure, 6t
    separation procedure, 6t

Electrospray interface for liquid chromatographs and mass spectrometers
    adenosine and derivatives, 357f,358
    apparatus, 356f
    cyclosporin A, 358f
    experimental conditions, 356–357
    gramicidin S, 358f
    MS signal vs. analyte concentration, 359f
    MS signal vs. sample flow rate, 359f
    operational features, 359–360
    sensitivity, 358–359

Electrospray ion
    description, 356
    development, 356

Elemental constituents in solution samples
    determination using atomic absorption and emission spectrometry, 315
    problems with MS determination, 315

Eluant, description, 21

Elution curves, analysis using telegrapher's equation, 116–117

Enzymatic radiolabeling, analysis of individual neurons, 396

Estriol-3-phosphate disodium salt, analysis using laser desorption–MS, 293,294f,t

Ethylbenzene, MS, 122,123–124f

Evaluation of data, introduction of students, 68,69–70f,71f,t

Excitation source
    atomic fluorescence spectrometry, 227
    induction-coupled plasma spectrometric, See Induction-coupled plasma spectrometric excitation source

Experiment, primary output, 155

Exponential function, smoothing of data, 156,157f

Exposure, definition, 108

F

Faradaic electrochemistry at microvoltammetric electrodes
    apparatus, 303
    application of electrodes in analysis, 305–306f
    carbon fiber electrode preparation, 302–303
    carbon fiber properties, 302t,303
    carbon fiber sources and properties, 302t
    carbon paste electrode preparation, 302
    chemical preparation, 303
    chronoamperometry using carbon fiber electrode, 304,305f
    chronoamperometry using carbon paste electrode, 304f,305
    coefficient α on fiber microelectrode tip, 302f
    voltammetry using carbon paste electrode, 303f,304

Fast atom bombardment mass spectrometer
    continuous-flow sample probe, 362–366
    continuous introduction of liquid samples, 361–362
    sample introduction using direct insertion probe, 361

Fast Fourier transform
    description, 233
    outputs, 233f

Fentanyl derivatives, structural information from tandem MS, 337–341

Fiber-optic sensors for iodide determination based on dynamic fluorescence quenching of rhodamine 6G
    detection limits, 377t
    experimental description, 374
    figures of merit, 377t,378
    fluorescence lifetime determination with sampling oscilloscope, 375–376
    interferences from other ions, 378t
    iodide optrode instrumental design, 375
    optrode design, 374,375f
    optrode preparation, 375
    precision, 377t,378
    resin porosity vs. optrode response, 376
    response times, 377t
    reversibility, 377t,378
    sensitivity, 378t
    Stern–Volmer plots of response, 376f,377f,t
    Teflon optrode preparation, 374
    time-resolved fluorescence instrument operation, 376
    XAD-4 resin optrode preparation, 375

Field-induced drift velocity, sedimentation field-flow fractionation, 251

Field ion desorption from liquids, technique to join MS to HPLC, 355–356

Flame emission, comparison of spectral interferences to those of atomic absorption, 190–191

Flame photometer
    advantages, 44
    atomizers, 50f
    burners, 50
    construction, 49,50f,51t
    electrical instruments, 51

magnitude and spatial homogeneity of primary beam, 220–221

Thin layer chronoamperometry
advantages, 187
concentration of electroactive solution species, 185
current at electrode surface, 185,187

Thin layer electrochemical study design using controlled potential or controlled current
calibration curves, 183*f*
description, 181
diffusion coefficient determination, 185,186*f,t,*187*t*
edge effect, 184
experimental procedure, 181–183
integrated current–time curves, 184*f*
intercept calculation, 183–184
lower limit to length of transition times, 185,186*f*
Lucite cover for cup of thin layer electrode, 181,182*f*
mercury-coated platinum electrode, 183
mercury effect on useful negative potential range, 184*f*
reproducibility, 184*t*
residual current effect, 184–185
schematic diagram, 181,182*f*
thin layer chronoamperometry, 185,187
thin layer micrometer electrode, 182*f*
transition time measurement, 185
transition time vs. reciprocal current, 185,186*f*

Time-of-flight mass spectrometry– gas-liquid partition chromatography
acetone MS, 122,123*f*
attachment method, 120*f,*121
benzene MS, 122,123*f*
chromatographic column recombination block, 121*f*
comparison to existing files of MS, 121,122*t*
constant current power supply for thermal conductivity cell, 121*f*
ethylbenzene MS, 122,123–124*f*
experimental apparatus for GLC, 120–121*f*
experimental apparatus for MS, 121
experimental procedure, 121–124
helium MS, 122*f*
mercury MS, 123,124*f*
perfluorinated kerosene, 123–124,125*f*
sample chromatogram of acetone, benzene, toluene, ethylbenzene, and styrene, 122*f*
styrene MS, 122,124*f*
toluene MS, 122,123*f*

Tin oxide electrode, chemically modified, *See* Chemically modified tin oxide electrode

Toluene
laser ion mobility spectra, 332–333*f*
MS, 122,123*f*

Total ionic strength adjustment buffer for electrode determination of fluoride in water supplies
advantages, 203
apparatus, 203
applications, 205
buffer preparation, 203
calibration curves, 203,204*f*
experimental procedure, 203
reproducibility, 204,205*t*
validity, 203–205

Trace element(s), determination using inductively coupled argon plasma–MS, 315–322

Trace element determination, use of semiconductor detector X-ray spectrometers, 238–249

Transient redox effects, reaction rate based determinations, 259,260–261*f*

Tridihydrogen pyrophosphatomanganiate, polarigraphic determination of manganese, 26–33

Triphenylamine, determination using phosphorimetry, 110–111,112*f,t*

Trisubstituted aromatics, Raman spectroscopy, 62

Tswett, invention of chromatographic adsorption analysis, 20

Tunable dye laser
description, 227
source of excitation for atomic fluorescence spectrometry, 227–231

U

Ultraviolet absorption spectrophotometry, advantages and disadvantages, 54

Ultraviolet laser desorption
apparatus, 385
applications, 387
experimental procedure, 385
multiply charged ions, 387
noise level, 387
sum spectra, 385,386*f,*387
width of molecular ion signal, 387

Unknowns, identification using IR spectroscopy, 14–15

Urine, human, zone electrophoresis in open-tubular glass capillaries, 326,327*f*

V

Vanadium(V), reaction rate based determination using transient redox effects, 260,261*f*

Vapor-phase chromatography
assumption of linearity, 113

capacitance of mobile phase, 114
concentrations of phases, 114
conservation equation, 114
electrical analogue of fractionating column, 113,114*f*
telegrapher's equation, 114

Velocity of electroosmotic flow, definition, 326

Vibrational pattern, phosphorescent effect, 106

Vibrationless transition frequency, phosphorescent effect, 106

Voltammetric detectors, use for high-pressure LC, 313

W

Water samples, sodium, potassium, and calcium analysis using flame photometry, 52

Water supplies, use of total ionic strength adjustment buffer for electrode determination of fluoride, 203,204*f,*205*t*

Wet gas mixture, analysis using MS, 37*t,*38

X

X-ray diffraction powder data, punched card code, 72,73*f,*74,75*t*

X-ray emission spectroscopy, advantages of semiconductor detectors, 238

Z

Zone broadening, causes, 324

Zone electrophoresis, causes of zone broadening, 324

Zone electrophoresis in narrow-bore Teflon tubes, description, 324

Zone electrophoresis in open-tubular glass capillaries
analysis time vs. applied voltage, 327*f*
apparatus, 325
applications, 328
dansyl amino acids, 326*f*
electroosmosis vs. resolution and analysis time, 328*f*
electroosmotic flow, 326–327
experimental materials, 325
experimental procedure, 325–326
fluorescamine derivatives of amines, 326*f*
fluorescamine derivatives of dipeptides, 326*f*
human urine, 326,327*f*
improvements, 328
number of theoretical plates vs. applied voltage, 327*f*
resolution, 327–328
theory, 324–325

Author Index

Picture Credits

All pictures copyright 1994 American Chemical Society except as noted.

Cover: "Fat Man" bomb, *Los Alamos Scientific Laboratory*; Moon, *NASA*

The 1930s and 1940s: Eleanor Roosevelt, *National Archives*; Boys recycling tires, *Chicago Sun-Times*; Joe DiMaggio, *Baseball Hall of Fame*; Atomic cloud, *U.S. Department of Defense*

The 1950s: Piltdown skull, *British Museum (Natural History)*; Pocket transistors, *Bell Laboratories*

The 1960s: Astronaut, *NASA*; Moon footprint, *NASA Johnson Space Flight Center*; Jackie Gleason, *New York World Telegram & Sun*; Superconductor, *Westinghouse Electric Corporation*

The 1970s: Moon rocks and astronaut, *NASA*; Frog cloning, *Volpe & McKinnel* (J. Hered, 1966, *57*, 167–74); Richard Nixon, *National Archives*

The 1980s and 1990s: Charles and Diana, *Syndication International*; Genetic Map, *John Wiley & Sons*